SYMMETRIES IN SCIENCE V

Algebraic Systems, Their Representations, Realizations, and Physical Applications

SYMMETRIES IN SCIENCE V

Algebraic Systems, Their Representations, Realizations, and Physical Applications

Edited by

Bruno Gruber
Southern Illinois University
Carbondale, Illinois

L.C. Biedenharn
Duke University
Durham, North Carolina

and

H.D. Doebner
Technical University
Clausthal, Germany

SPRINGER SCIENCE+BUSINESS MEDIA, LLC

Library of Congress Cataloging in Publication Data

Symmetries in science V: algebraic systems, their representations, realizations, and physical applications / edited by Bruno Gruber, L.C. Biedenharn, and H.D. Doebner.
p. cm.
Proceedings of a symposium held July 30–Aug. 3, 1990, Landesbildungszentrum Schloss Hofen, Lochau, Vorarlberg, Austria.
Includes bibliographical references and index.
ISBN 978-1-4613-6643-0 ISBN 978-1-4615-3696-3 (eBook)
DOI 10.1007/978-1-4615-3696-3
1. Algebra, Abstract—Congresses. I. Gruber, Bruno, 1936- . II. Biedenharn, L. C. III. Doebner, H. D. (Heinz Dieter), date. IV. Title: Symmetries in science 5. V. Title: Symmetries in science five.
QA162.S94 1991 91-9900
512′.02—dc20 CIP

Proceedings of a symposium entitled Symmetries in Science V: Algebraic Systems, Their Representations, Realizations, and Physical Applications, held July 30–August 3, 1990, at Landesbildungszentrum Schloss Hofen, Lochau, Vorarlberg, Austria

ISBN 978-1-4613-6643-0

Originally published by Plenum Press, New York in 1991
Softcover reprint of the hardcover 1st edition 1991

FOREWORD

The Symposium "Symmetries in Science V: Algebraic Systems: their Representations, Realizations and Physical Applications" was held at the Landesbildungszentrum Schloss Hofen, Vorarlberg, Austria, during the period July 30 – August 3, 1990. Leading scientists presented summaries, as well as recent research developments, concerning their own field of specialization to the assembled international group of researchers. These review-oriented presentations are included in this volume in order to make them available to the general scientific community.

As was the case for the preceding Symposia we need to thank those who made Symposium V possible through their continued and generous support. These include, on part of the State of Vorarlberg;

Dr. Martin Purtscher, Landeshauptmann
Dr. Guntram Lins, Landesrat
Ms. Elisabeth Gehrer, Landesrätin

and on part of Southern Illinois University at Carbondale;

Dr. John C. Guyon, President
Dr. Benjamin A. Shepherd, Vice President for Academic Affairs and Research
Dr. John H. Yopp, Dean of the Graduate School
Dr. Charles B. Klasek, Executive Assistant to the President for International and Economic Development
Dr. Russell R. Dutcher, Dean of the College of Science
Dr. Victoria J. Molfese, Director, Office of Research Development and Administration
Dr. James D. Quisenberry, Acting Director of International Programs and Services
Dr. Frank C. Sanders, Chairman of the Department of Physics.

Finally, we need to especially thank Dr. Hubert Regner, Vorstand der Abteilung Wissenschaft und Weiterbildung, whose support and cooperation – on a daily basis – has been the single most essential ingredient for the success of Symposium V.

Bruno Gruber
L.C. Biedenharn
H.D. Doebner

CONTENTS

DYNAMICAL SYMMETRIES OF RELATIVISTIC TWO-AND MANY BODY SYSTEMS

A. O. Barut

Department of Physics
University of Colorado
Boulder, Colorado 80309

The great success of nonrelativistic many body theory based on the Schrödinger equation in configuration space and the symmetries we have learned from it can be extended to a nonperturbative covariant many-body equation derived from field theory by a variational principle.

1. TYPES OF SYMMETRY AND SYMMETRY GROUPS

I recall the distinction between the following types of symmetry :

a. Geometric symmetry : Symmetry (usually of the Hamiltonian) under a group of coordinate transformations.

b. Dynamical symmetry : a higher symmetry of some Hamiltonians beyond the pure geometrical symmetry resulting in additional degeneracy.

c. Spectrum generating group : a symmetry which describes the complete set of states of a quantum system.

d. Dynamical group: the symmetry which includes the spectrum group and the current operators which determine the response of the system to certain external probing.

To illustrate by the example of the Coulomb potential problem the above groups are, respectively, SO(3), SO(4), SO(4,1) or E(3), and SO(4,2).

The properties of these symmetries for nonrelativistic systems have been extensively studied in the last twenty five years and they are well known.[1]

The great success of the nonrelativistic quantum physics is based on the existence of a many-body linear wave equation in configuration space

Symmetries in Science V, Edited by B. Gruber *et al.*
Plenum Press, New York, 1991

$$i\hbar\frac{\partial}{\partial t}\psi(x_1, x_2, \ldots, x_N; t) = H\psi(x_1, x_2, \ldots, x_N; t) \quad (1)$$

where the Hamiltonian is

$$H = \sum_i \frac{1}{2m_i}p_i^2 + \sum_{i<j} V_{ij}(x_i - x_j) + H_{\text{radiation}} + H_{\text{matter-Rad}}. \quad (2)$$

In studying the structure of a system, e.g. the energy levels, the last two terms may be omitted. In order to compare with relativistic theories let us emphasize that this many body equation is (i) nonperturbartive, (ii) has one time t only, (iii) is in the configuration space R^{3N} rather than R^3. With respect to the third point we remark that had we used individual wave functions for each particles in R^3 we would have obtained nonlinear coupled Hartree type equations. The distinction will be elaborated later. In the configuration space the wave equation exhibits characteristic nonlocal effects related to global symmetries of the system.

The simplest system is a two-body problem which, in nonrelativistic case, can be reduced to a one-body problem in a potential where most of the symmetry group studies have been done. Group properties of some special N body problems have also been studied.[2]

It is generally thought that no such comparable theory can be given for the relativistic N-body problem. The reasons given are that the interaction potential is a nonrelativistic concept, each particle has its own time coordinate so that the Bethe-Salpeter relations, for example, contain many times, and that center of mass and relative coordinates cannot be disentangled. Moreover for charged particles the radiation of the system and the radiation reaction must be taken into account. Hence one had to resort to approximative diagrammatical methods. Generally, one starts from a Schrödinger- or Dirac like equation with relativistic kinematics and reduced mass as an input, and then introduces radiative corrections by Feynman graphs in QED. In nuclear and hadronic phenomenology one introduces additionally effective spin- orbit type potentials . It is therefore very difficult under such circumstances to discuss precise symmetry properties of relativistic systems. What one needs is a nonperturbative relativistic equation in closed form, such as eq.(1), for relativistic systems with all the interactions included so that one can extrapolate the system to short distances and study global symmetry properties. With perturbative theories this is not possible. As a result we do not know at the present time the short distance behaviour of one of the simplest and most important system like the electron-positron two-body problem.

In the absence of an underlying dynamical microscopic theory symmetry considerations are useful and important, like crystal symmetry. In the case of elementary particles many such regularities, symmetry and symmetry groups have been discovered in the last half century, in order to classify the large number of particle states observed and their interactions. But every such global symmetry contains necessarily parameters.[3] Consequenly we must also try to understand these abstract symmetries from a dynamical point of view.

In this work we attempt to bridge field theory (dynamics) with symmetry by deriving infinite component wave equations of the type (1) which describe many of the symmetries of the relativistic composite systems very well from an underlying field theory of few basic fields.

The purpose is not just the formulation of symmetries but more importantly the possible discovery of new physics at short distances or at high energies among the old particles, like $e^+ - e^-$, from a closed-form wave equation which in principle can be extrapolated relativistically. Some surprises of relativistic dynamics at short distances may come from the increasing dominance of selfenergy-radiation reaction terms and the consequent occurence of superstrong interactions even in electrodynamics.

2. TWO-BODY RELATIVISTIC WAVE EQUATIONS

To illustrate the method I discuss the derivation of eq.(1) and the corresponding nonrelativistic Hartree equations from field theory by two different variational principles. The relativistic case will be entirely analogous.

Consider two charged scalar fields ψ_1 and ψ_2 coupled together via a scalar potential field $V(x,t)$. The action of these three coupled fields is

$$A = \int dt dx \Big\{ \psi_1^*(i\partial_t - H_1)\psi_1 + \psi_2^*(i\partial_t - H_2)\psi_2$$
$$+ (e_1\psi_1^*\psi_1 + e_2\psi_2^*\psi_2)V(x;t) \tag{3}$$
$$+ \frac{1}{2}\Delta V \cdot \Delta V - \frac{1}{2}(\partial V/\partial t)^2 \Big\}$$

It follows that V satisfies space-time Poisson's equation

$$-\frac{\partial^2 V}{\partial t^2} + \Delta V = e_1\psi_1^*\psi_1 + e_1\psi_2^*\psi_2 = \rho \tag{4}$$

the solution of which with the boundary condition that V vanishes at infinity is

$$V = \int dx' dt D(x - x')\rho(x', t') \tag{5}$$

The field V has both of the charge distributions as sources. The last two terms in the action by partial integration give

$$-\frac{1}{2}\int dx dt \partial^\mu V \partial_\mu V = +\frac{1}{2}\int dx dt V \Box V = -\frac{1}{2}\int dx dt \rho V \tag{6}$$

so that using (3) and (4) we can eliminate the field V and express the action entirely in terms of the ψ-fields

$$A - \int dt dx \Big\{ \sum_i \psi_i^*(i\partial_t - H_i)\psi_i + \frac{1}{2}\sum_{j,k} e_j\psi_j^*\psi_j \int dt' dx' D(x - x') e_k \psi_k^* \psi_k \Big\} \tag{7}$$

We see now in the interaction action a mutual interaction term with a coefficient $e_1 e_2$ and two selfinteraction (or radiation reaaction) terms proportional to e_1^2 and e_2^2.

There are two simple almost equally plausible variational principles. In one we vary the action with respect to the individual fields separately; in the other we vary the action with respect to the composite field which is the product of the two fields. These give Hartree and configuration space wave equations, respectively.

If the action is extremum under the variations of the fields ψ_1 and ψ_2 separately we obtain the coupled nonlinear equations

$$\begin{aligned}(i\partial_t - H_1)\psi_1 &= \frac{1}{2}e_1\psi_1 \int dx' dt' D(x - x')(e_1\psi_1^*\psi_1 + e_2\psi_2^*\psi_2)\\ (i\partial_t - H_2)\psi_2 &= \frac{1}{2}e_2\psi_2 \int dx' dt' D(x - x')(e_1\psi_1^*\psi_1 + e_2\psi_2^*\psi_2)\end{aligned} \tag{8}$$

Next let us introduce the composite field

$$\Phi(x_1, x_2; t) = \psi(x_1, t)\psi_2(x_2, t) \tag{9}$$

and rewrite the action (2) in terms of this composite field. In the kinetic energy and selfinteraction terms we multiply the terms suitably with the constant normalization integrals $\int \psi_2^* \psi_2 d\boldsymbol{x}'$and $\int \psi_1^* \psi_1 d\boldsymbol{x}'$ and introduce a center of mass time coordinate $i\partial_t = i\partial_{t_1} + i\partial_{t_2}$ With these operations we can write the action, neglecting the retardation since this is a nonrelativistic problem,

$$A = \int dt d\boldsymbol{x}_1 d\boldsymbol{x}_2 \Phi^*(\boldsymbol{x}_1, \boldsymbol{x}_2; t) \left(i\partial_t - H_1 - H_2 + \frac{e_1 e_2}{|\boldsymbol{x}_1 - \boldsymbol{x}_2|} + V_{\text{self}} \right) \Phi(\boldsymbol{x}_1, \boldsymbol{x}_2; t) \quad (10)$$

Now we invoke our second variational principle that the action is an extremum with respect to the variations of Φ^*. This is a weaker principle than the first one and we obtain this time a linear equation for the composite field

$$\left(i\partial_t - H_1 - H_2 + \frac{e_1 e_2}{|\boldsymbol{x}_1 - \boldsymbol{x}_2|} + V_{\text{self}} \right) \Phi(\boldsymbol{x}_1, \boldsymbol{x}_2; t) = 0 \quad (11)$$

where the nonlinear selfenergy term without retardation is

$$V_{\text{self}} = \frac{1}{2} \int dz du \Phi^*(\boldsymbol{z}, \boldsymbol{u}) \left[\frac{e_1^2}{|\boldsymbol{x}_1 - \boldsymbol{z}|} + \frac{e_2^2}{|\boldsymbol{x}_2 - \boldsymbol{u}|} \right] \Phi(\boldsymbol{z}, \boldsymbol{u})$$

This is the standard quantummechanical result (1) when the selfenergy term is dropped (its effect is presumably already taken into account in the renormalized masses). It is written down usually on the basis of a separate postulate of quantum theory that the Hilbert space of the N-body problem is an L_2-space over the tensor product of the configuration space of individual particles.

One must look at experiment to see which of the variational principles applies and when. Equation (8) should describe the separated system, and eq. (11) the nonseparated system in the terminology of quantum axiomatics.The results of the two equations may not be very large, but there are very fundamental principial differences. For two nonseparated particles a state of the total system is nonlocal in the sense that the individual particles are correlated no matter how far they are. Although we obtained eq.(11) by a variational principle from the product field (9) the solutions of (11) are no longer factorizable after the variational principle. The same applies to discrete quantum numbers : If we take the tensor product of two spins and pass to the total spin, then in a state of a definite value of the total spin the individual spins are correlated no matter how far apart they are.

The relativistic two-body problem proceeds in the same way. We shall use Dirac fields for individual particles. Because the Dirac equation is of first order, the theory is actually much simpler than the case of scalar fields, and of course more realistic for electrons and positrons interacting via the electromagnetic field. The action now is

$$A = \int dx \left\{ \left[\sum_k \bar{\psi}_k (\gamma^\mu i\partial_\mu - m_k) \psi_k - (e_k \bar{\psi}_k \gamma^\mu \psi_k) A_\mu \right] - \frac{1}{4} F_{\mu\nu} F^{\mu\nu} \right\} \quad (12)$$

Again we eliminate the field A_μ by using the field equations

$$\Box A_\mu = \sum_k e_k \bar{\psi}_k \gamma_\mu \psi_k = \sum_k j_\mu^{(k)} = j_\mu \quad (13)$$

with the solution

$$A_\mu(x) = \int dy D(x-y) j_\mu(y) \quad (14)$$

in the second and third terms of the action and obtain an action in terms of the ψ_k-fields alone

$$A = \int dx \left\{ \sum_k \bar{\psi}_k(\gamma^\mu i\partial_\mu - m_k)\psi_k - \frac{1}{2}\sum_{k,\ell} \int dy j_\mu^{(k)}(x) D(x-y) j^{\mu(\ell)}(y) \right\} \quad (15)$$

Again the variation of this action with respect to each field separately gives the Hartree-type equations

$$\begin{aligned}(\gamma^\mu i\partial_\mu - m_1)\psi_1 &= \frac{1}{2} e_1 \gamma^\mu \psi_1 \int dy D(x-y) j_\mu(y) \\ (\gamma^\mu i\partial_\mu - m_2)\psi_2 &= \frac{1}{2} e_2 \gamma^\mu \psi_1 \int dy D(x-y) j_\mu(y)\end{aligned} \quad (16)$$

But if we introduce a composite field

$$\Phi(x_1, x_2) = \psi_1(x_1)\psi_2(x_2) \quad (17)$$

we can reexpress the action in terms of this by suitable multiplying with the normalization integrals for particles 1 or 2 :

$$\int_{\sigma(s)} d\sigma \bar{\psi}\gamma \cdot n\psi = \int dx \bar{\psi}\gamma \cdot n\psi \delta\left((x \cdot n) - s\right) \quad (18)$$

where $\sigma(s)$ is the space-like surface normal to n^μ at the value of the parameter s and $d\sigma$ its volume element.

The unit vector n^μ indicates the choice of the time axis in a Lorentz covariant way. We take the same invariant time axis for both particles. We can also express the Green's function in terms of n^μ using

$$\delta\left((x_1 - x_2)^2\right) = \frac{1}{2r_\perp}\left\{\delta\left((x_1 - x_2)\cdot n - r_\perp\right) + \delta\left((x_1 - x_2)\cdot n + r_\perp\right)\right\} \quad (19)$$

where

$$r_\perp = \left[(x_1 - x_2)\cdot n - (x_1 - x_2)^2\right]^{1/2} \quad (20)$$

is the relativistic space-like distance perpendicular to n^μ. For the choice $n = (1000)$ we have $r_\perp = r$, the ordinary radial distance. Because $D(x_1 - x_2)$ therefore contains a δ-function, as in (19), we can choose the parameter s in the normalization (18) so as to match this δ-function. Then the action becomes

$$\begin{aligned}A = \int dx_1 x_2 \bar{\Phi}(x_1, x_2) \Big[&(\gamma_\mu p_1^\mu - m_1) \otimes \gamma \cdot n + \gamma \cdot n \otimes (\gamma_\mu p_2^\mu - m_2) \\ &- e_1 e_2 \gamma^\mu \otimes \gamma_\mu \frac{1}{r_\perp} + V_{\text{self}} \Big] \bar{\Phi}(x_1, x_2) \delta\left((x_1 \cdot n) - (x_2 \cdot n) - r_\perp\right)\end{aligned} \quad (21)$$

The form of V_{self} is given below. Now we can apply our second variational principle, namely varying the action A with respect to $\bar{\Phi}$ and obtain the configuration space equation

$$\left\{ (\gamma_\mu \pi_1^\mu - m_1) \otimes \gamma \cdot n + \gamma \cdot n \otimes (\gamma_\mu \pi_2^\mu - m_2) - e_1 e_2 \frac{\gamma^\mu \otimes \gamma_\mu}{r_\perp} \right\} \Phi(x_1, x_2) = 0 \quad (22)$$

where

$$\pi_k^\mu = p_k^\mu - e_k A_{k\text{self}}^\mu \quad ; \quad k = 1,2$$

with

$$\begin{aligned} A_{1,\text{self}}^\mu(x) &= e_1 \int dz du D(x-z)\bar{\Phi}(z,u)\gamma^\mu \otimes \gamma \cdot n \Phi(z,u) \\ A_{2,\text{self}}^\mu(x) &= e_2 \int dz du D(x-z)\bar{\Phi}(z,u)\gamma \cdot n \otimes \gamma^\mu \Phi(z,u) \end{aligned} \tag{23}$$

Next we introduce center of mass and relative coordinates and momenta

$$\begin{aligned} \Pi &= \pi_1 + \pi_2 \quad , \quad \pi = a\pi_1 - b\pi_2 \\ x &= x_1 - x_2 \quad , \quad X = bx_1 + ax_2 \\ a &= m_1/(m_1+m_2) \qquad b = m_2/(m_1+m_2) \end{aligned} \tag{24}$$

in terms of which our equation becomes

$$\left\{ \Gamma^\mu \Pi_\mu + k^\mu \pi_\mu - \frac{e_x e_2}{r_\perp}\gamma^\mu \otimes \gamma_\mu - m_1 I \otimes \gamma \cdot n - m_2 \gamma \cdot n \otimes I \right\} \Phi = 0 \tag{25}$$

where we have introduced

$$\begin{aligned} \Gamma^\mu &= (a\gamma^\mu \otimes \gamma \cdot n + b\gamma \cdot n \otimes \gamma^\mu) \\ k^\mu &= (\gamma^\mu \otimes \gamma \cdot n - \gamma \cdot n \otimes \gamma^\mu) \end{aligned} \tag{26}$$

All four vectors can be decomposed into components parallel and perpendicular to n^μ. For example

$$P^\mu = (P \cdot n)n^\mu + P_\perp^\mu \quad , \quad P_\perp^\mu n_\mu = 0 \tag{27}$$

The component $P_\parallel$ is the Hamiltonian, $P_\perp$ is the space-like kinetic momentum. The component of k^μ parallel to n^μ vanishes, which means that the relative momentum is just three dimensional as it should be. In other words the equation is a one-time equation, the center of mass time, the relative time drops out. The Hamiltonian is, for the choice $n^\mu = (1000)$

$$\Pi_0 \Phi = \left\{ \boldsymbol{\alpha} \cdot \boldsymbol{\Pi} + (\boldsymbol{\alpha}_1 - \boldsymbol{\alpha}_2) \cdot \boldsymbol{\pi} + \frac{e_1 e_2}{r}(1 - \boldsymbol{\alpha}_1 \cdot \boldsymbol{\alpha}_2) + m_1\beta_1 + m_2\beta_2 \right\} \Phi \tag{28}$$

where we have introduced

$$\boldsymbol{\alpha} = \gamma_{0i}\boldsymbol{\gamma} \quad , \quad \beta_i = \gamma_{0i} \quad , \quad \boldsymbol{\alpha} = a\boldsymbol{\alpha}_1 + b\boldsymbol{\alpha}_2$$

We have derived a generalized Dirac-like two body equation which is also in the form of an infinite component wave equation. Other remarkable features of this equation are : terms involving center of mass and relative momenta are separated; it reduces directly to the one-body Dirac equation for one of the masses going to infinity; it reduces to the two-body Schrödinger equation, eq(1), in the nonrelativistic limit. Furthermore angular and radial parts are exactly separable in the center of mass frame. This two-body equation has been extensively applied to positronium,muonium and hydrogen[4].

3. RELATIVISTIC MANY-BODY EQUATION WITH RADIATION REACTION

The procedure of the previous Section can be generalized to N interacting Dirac particles coupled to the electromagnetic field. Again we introduce, first for distinguishable particles, a composite field by the tensor product

$$\Phi(x_1, x_2, \ldots, x_N) = \psi_1(x_1) \cdot \psi_2(x_2) \ldots \psi_N(x_N) \tag{29}$$

and express the action in terms of it, after eliminating the electromagnetic field. The result directly generalizing (21) is[5]

$$\begin{aligned} W = \int dx_1 dx_2 \ldots dx_N \bar{\Phi}(x_1, \ldots, x_N) \Big\{ \Big[&(\gamma p_1 - m_1) \otimes \gamma \cdot n \otimes \gamma \cdot n \otimes \ldots \otimes \gamma \cdot n \\ &+ \gamma \cdot n \otimes (\gamma p_2 - m_2) \otimes \gamma \cdot n \otimes \ldots \otimes \gamma \cdot n \\ &+ \gamma \cdot n \otimes \gamma \cdot n \otimes (\gamma p_3 - m_3) \otimes \gamma \cdot n \otimes \ldots \otimes \gamma \cdot n \\ &- - - - - - - - - - - - - - - \\ &+ \gamma \cdot n \otimes \gamma \cdot n \otimes \ldots \ldots \otimes (\gamma p_N - m_N) \Big] \delta \\ - \sum_{i<j} e_i e_j \gamma \cdot n \otimes \ldots \overset{(i)}{\gamma_\nu} \ldots \overset{(j)}{\gamma^\nu} \ldots \otimes \gamma \cdot n D(x_i - x_j) \\ - \sum_i \frac{e_i^2}{2} \gamma \cdot n \otimes \ldots \overset{(i)}{\gamma_\nu} \ldots \otimes \gamma \cdot n \int dz_1 \ldots dz_N D(x_i - z_i) \bar{\Phi}(z_1 \ldots z_N) \\ \cdot \gamma \cdot n \otimes \ldots \overset{(i)}{\gamma^\nu} \ldots \otimes \gamma n \Phi(z_1 \ldots z_N) \Big\} \Phi(x_1 \ldots x_N) \end{aligned} \tag{30}$$

so that the variational principle with respect to $\bar{\Phi}(x_1 \ldots x_N)$ gives

$$\begin{aligned} \Big\{ \sum_j \gamma \cdot n \otimes \ldots \otimes (\gamma \cdot p_j - m_j) \otimes \ldots \otimes \gamma \cdot n - \sum_{i<j} \frac{e_i e_j}{r_{ij\perp}} \gamma \cdot n \ldots \overset{(i)}{\gamma_\nu} \ldots \ldots \overset{(j)}{\gamma^\nu} \ldots \gamma \cdot n \\ - \sum_i \frac{1}{2} e_i^2 \gamma \cdot n \otimes \overset{(i)}{\gamma_\nu} \ldots \otimes \gamma \cdot n \int dz_1 \ldots dz_N \bar{\Phi}(z_1 \ldots z_N) \\ \frac{\gamma \cdot n \otimes \ldots \otimes \overset{(i)}{\gamma^\nu} \otimes \ldots \otimes \gamma \cdot n}{|x_i - z_i|_\perp} \Phi(z_1 \ldots z_N) \Big\} \\ \cdot \Phi(x_1 \ldots x_N) = 0 \end{aligned} \tag{31}$$

We did not write the corresponding Hartree equations which are straightforward generalization of (8).

In eqs.(30)–(31) the spin space is 4^N-dimensional and we write the tensor product of Dirac matrices always in the ordered form with first factor being the spin matrix of the first particle, the second factor for the second particle, etc. The symmetry of the equations with respect to spin indices is evident. To elucidate these symmetries in more detail we again introduce center of mass momentum P and relative momenta $p_{ij}(i<j)$ by

$$p_\mu = \frac{1}{N}P + \frac{1}{N}\sum_{i<j} a_k^{ij} p_{ij} \tag{32}$$

where the coefficients a_k^{ij} satisfy

$$\sum_k a_k^{ij} = 0 \quad ; \quad i,j = 1,\ldots N \tag{33}$$

Inserting (32) into the wave equation and collecting terms we obtain again an infinite component wave equation

$$\left\{ \Gamma^\mu P_\mu + \Lambda^{ij}_{\mu\perp} p^\mu_{ij\perp} - \Sigma^k m_k - \sum_{i<j} e_i e_j \frac{G_{ij}}{r_{ij\perp}} - V^{\text{self}} \right\} \Phi(x_1,\ldots,.x_N) = 0 \tag{34}$$

where the following spin operators have been introduced

$$\begin{aligned}
\Gamma_\mu &= \frac{1}{N}\sum_{k=1}^{N} \gamma\cdot n \otimes \ldots \otimes \overset{(j)}{\gamma_\nu} \otimes \ldots \otimes \gamma \cdot n \\
\Sigma^k &= \sum_k \gamma \cdot n \otimes \ldots \otimes \overset{(k)}{I} \otimes \ldots \otimes \gamma \cdot m \\
G^{ij} &= \gamma \cdot n \otimes \ldots \otimes \overset{(i)}{\gamma^\nu} \otimes \ldots \otimes \overset{(j)}{\gamma_\nu} \otimes \ldots \otimes \gamma \cdot n \\
\Lambda^{ij}_\mu &= \frac{1}{N}\sum_k a_k^{ij} \gamma \cdot n \otimes \ldots \otimes \overset{(j)}{\gamma_\nu} \otimes \ldots \otimes \gamma \cdot n
\end{aligned} \tag{35}$$

The nonlinear selfenergy contribution is

$$V_{\text{self}} = \frac{1}{2}\sum_i e_i^2 \Gamma^i_\mu \int dz^N \bar{\Phi} \frac{\Gamma^{\mu i}}{(x_i - z_i)_\perp} \Phi \tag{36}$$

Only the transverse components of the relative momenta contribute because of the relation

$$\Lambda^{ij}_\mu n^\mu = 0 \tag{37}$$

which again means that we have a one-time equation, namely the center of mass time whose conjugate variable is the total Hamiltonian which, generalizing (28) is , for $n = (1000)$, given by

$$P_\| = H = \boldsymbol{\alpha} \cdot \boldsymbol{P} + H_{\text{rel}} \tag{38}$$

where the relative Hamiltonian (that in the center of mass frame) is

$$H_{\text{rel}} = \sum_{i,j} \boldsymbol{k}^{ij} \cdot \boldsymbol{p}^{ij} + \sum_k \beta_k m_k + \sum_{i<j} e_i e_j \frac{1 - \boldsymbol{\alpha}_i \cdot \boldsymbol{\alpha}_j}{r_{ij}} + \widetilde{V}_{\text{self}} \tag{39}$$

with

$$\boldsymbol{k}^{ij} = \frac{1}{N}\sum_k a_k^{ij} \boldsymbol{\alpha}^k$$

2. SPIN SYMMETRIES

Our general many-body equation has the structure of a generalized Dirac equation linear in the momenta

$$(\Gamma^\mu P_\mu - H_{\rm rel})\Phi = 0. \tag{40}$$

When $H_{\rm rel}$ is taken to be a constant such equations are known under the names of Duffin-Kemmer-Petiau[6] or Gelfand-Yaglom[7] equations and have been studied extensively. The infinite-dimensional version of (40) was introduced first by Majorana[8] and considerably extended later[1b,c].These equations have been postulated grouptheoretically in order to describe higher spin elementary particles which have no internal coordinates; the space time part of the wave function is simply an exponential $\exp(iPx)$. In contrast, in our equation (40), $H_{\rm rel}$ is an operator in the space of relative coordinates and the system is a dynamical many-body bound state. Inasmuch as all the so-called higher spin particles in nature, including some spin 1/2 particles, seem to be composite objects with internal charge or matter distribution eq.(40) is a more adequate candidate for relativistic particles of any spin except the basic Dirac particles that we have assumed to be elementary. If the bound state is extremely tight we may replace, for small energies the relative Hamiltonian by a constant.

The algebra of the spin matrices Γ^μ derives from the algebra of the Dirac matrices. For example, for $N = 2$, the matrices

$$\Gamma^\mu = \gamma^\mu \otimes I + I \otimes \gamma^\mu, \quad \text{or} \quad \Gamma^\mu_n = \gamma^\mu \otimes \gamma \cdot n + \gamma \cdot n \otimes \gamma^\mu \tag{41}$$

satisfy the Kemmer algebra

$$\beta^\nu \beta^\lambda \beta^\nu + \beta^\nu \beta^\lambda \beta^\mu = \beta^\mu \delta^{\nu\nu} + \beta^\nu \delta^{\lambda\mu} \tag{42}$$

so that our eq. (40) with constant $H_{\rm rel}$ is the socalled Kemmer equation

$$\Gamma^\mu \equiv \beta^\mu \quad , \quad (\beta^\mu P_\mu + k)\Phi = 0 \tag{43}$$

If we reduce this algebra into its irreducible parts we see that this equation describes spin zero and spin one states with 5 and 10 dimensional representations, respectively. This is immediate physically as two spin 1/2 particles can form bound states of spin 0 and 1. But they can also form bound states of all higher spins due to orbital angular momentum, and this happens when $H_{\rm rel}$ is not constant.

The question whether truly elementary particles with spin different from 1/2, which are thus not composites, can exist is an interesting one. Consistency tests imply that only spin 1/2 Dirac particles with or without mass can be considered to be elementary[9]. This is also borne out experimentally as we have noted. It turns out that, although one can formally take $H_{\rm rel}$ in (40) to be a constant, the coupling of such a system to an external field gives rise to inconsistencies simply because the external field mixes all the higher angular momentum states which have been cut off.

There is a classical model of a spinning particle which when quantized also leads to Dirac and higher spin equations precisely of the type (40). It is based on the introduction of internal spin variables to the dynamics of a relativistic point particle. These spin variables are taken ,most conveniently, to be classical four-component. c-number spinors $z(\tau), \bar{z}(\tau)$ and the action of a single particle in an external EM-field A_μ is

$$A = \int d\tau [i\lambda \bar{z}\gamma \cdot nz + p\dot{x} - \pi_\mu \bar{z}\gamma^\mu z] - \frac{1}{4}\int F_{\mu\nu}F^{\mu\nu} \tag{44}$$

where τ is an invariant time-parameter and λ a basic constant of dimension of an action. The conjugate variables are (p_μ, x_μ) and $(z, \bar{z})$. One can derive the equations of motion and the Hamiltonian from the action[10]

$$\mathcal{H} = \bar{z}\gamma^\mu z \pi_\mu \quad , \quad \pi_\mu = p_\mu - eA_\mu \tag{45}$$

The mass m is the value of the integral of motion $\mathcal{H}$. The theory can be quantized in many ways[11].In the Schrödinger quantization we take a wave function $\Phi(x,\bar{z},t)$ and represent p_μ and z as differential operators

$$p_\mu = -i\frac{\partial}{\partial x^\mu} \quad , \quad z = i\partial/\partial\bar{z}$$

If we now postulate a general Schrödinger equation

$$i\frac{\partial}{\partial t}\Phi = H\left(-i\partial/\partial_\mu, x; i\partial/\partial\bar{z}, z\right)\Phi \tag{46}$$

and expand Φ in terms of powers of $\bar{z}$ in both sides of this equation

$$\Phi(x,\bar{z},t) = \phi(x,t) + \bar{z}^\alpha\psi_\alpha(x,t) + \bar{z}^\alpha\bar{z}^\beta\psi_{\alpha\beta}(x,t) + \ldots$$

and compare coefficients we obtain the hirarchy of equations[12]

$$\begin{aligned} i\frac{\partial}{\partial t}\psi_\alpha &= \gamma^\mu(i\partial_\mu - eA_\mu)\psi_\alpha \\ i\frac{\partial}{\partial t}\psi_{\alpha\beta} &= \beta^\mu(i\partial_\mu - eA_\mu)\psi_{\alpha\beta} \end{aligned} \tag{47}$$

We see that both approaches, from field theory of quantumelectrodynamics and the classical theory, lead to the same equations. Of course the classical action (44) has very much in common with the QED action which is the quantization of it. We shall see later the classical analog of the many body equations as well.

3. DYNAMICAL SPACE-TIME SYMMETRIES

We shall now discuss the symmetries of the equations (34) with respect to the space-time canonical coordinates P's and r's and show why our equation is actually an infinite component wave equation and how in the limit we recover the well known dynamical group SO(4,2) of the Coulomb problem[1]. Consider the case $N = 2$

$$\left(\Gamma^\mu\Pi_\mu + k^\mu\pi_\mu - \frac{e_1e_2}{r_\perp}G_{12} - \Sigma^1 m_1 - \Sigma^2 m_2\right)\Phi = 0$$

multiplying it with $r_\perp$, we have

$$\left[\Gamma^\mu r_\perp\Pi_\mu + k^\mu r_\perp\pi_\perp - e_1e_2G_{12} - r_\perp(\Sigma^k m_k)\right]\Phi = 0 \tag{48}$$

In order to determine the algebra of space-time differential operators we start from the generators of the conformal algebra of space-time as differential operators in the Minkowski-space M_4[13]

$$\begin{aligned} \ell_{\mu\nu} &= x_\mu\pi_\nu - x_\nu\pi_\mu \\ \ell_{\mu 6} + \ell_{\mu 4} &= x_\mu \\ \ell_{\mu 6} - \ell_{\mu 4} &= 2(x_\mu\pi^\nu + ih)\pi_\mu - x_\mu\pi_\nu\pi^\nu \\ \ell_{64} &= -(x_\mu\pi^\mu + ih) \end{aligned} \tag{49}$$

To find the action of this algebra on a spacelike surface σ with normal n we consider the space of functions f such that $x_\mu x^\mu f = 0$, or

$$(x \cdot n)^2 f = x_\perp^2 f \quad \text{or} \quad (x \cdot n) n^\mu f = \pm x_\perp^\mu f \tag{50}$$

All four vectors can be decomposed into a component parallel to n^μ and another perpendicular to n^μ, as in (20) and (27). In this way we obtain four reduced forms of the conformal algebra. For $n = (1000)$ using (50) we have ($h = 1$)

$$\begin{aligned} \ell_{ij} &= x_i \pi_j - x_j \pi_i, \ \ell_{io} = r\pi_i \\ \ell_{i6} + \ell_{i4} &= x_i, \ \ell_{06} + \ell_{04} = r \\ \ell_{i6} - \ell_{i4} &= 2(\boldsymbol{x} \cdot \boldsymbol{\pi} - i)\pi_i - x_i \pi^2 \\ \ell_{06} - \ell_{04} &= r\pi^2 \\ \ell_{04} &= \boldsymbol{x} \cdot \boldsymbol{\pi} - i \end{aligned} \tag{51}$$

Cosequently our equation (48) can be written entirely in terms of the Lie algebra elements of three groups, two so(4,2) Dirac algebras for each spin and one so(4,2) dynamical algebra for the relative motion. In particular if we choose $n = (1000)$, the two-body Hamiltonian without the selfenergy terms is

$$H = (a\boldsymbol{\alpha}_1 + b\boldsymbol{\alpha}_2) r\boldsymbol{P} + (\boldsymbol{\alpha}_1 - \boldsymbol{\alpha}_2) \cdot r\boldsymbol{p} + e_1 e_2 (1 - \boldsymbol{\alpha}_1 \cdot \boldsymbol{\alpha}_2) + r(\beta_1 m_1 + \beta_2 m_2) \tag{52}$$

The one-body limit of this equation is obtained by setting $m_2 \to \infty$ and counting energies above m_2 and setting $\alpha_2 = p_2/m_2 \to 0$ (α is the velocity operator) and $\beta_2 = 1$. Then we find precisely the single particle Dirac Hamiltonian in a Coulomb field

$$rH_D = \boldsymbol{\alpha}_1 \cdot r\boldsymbol{p} + e_1 e_2 + r m_1 \beta_1 \tag{53}$$

On the other hand if we go to the nonrelativistic spinless limit for both particles setting $\boldsymbol{\alpha}_1 \to \boldsymbol{p}_1/m_1, \boldsymbol{\alpha}_2 \to \boldsymbol{p}_2/m_2, \beta_1 \to 1, \beta_2 \to 1$ we obtain

$$H_{NR} = \frac{1}{2M} P^2 r + \frac{1}{2\mu} \boldsymbol{p}^2 r + e_1 e_2 \left(1 - \frac{\boldsymbol{p}_1 \cdot \boldsymbol{p}_2}{m_1 m_2}\right) + M \tag{54}$$

Here we recognize the operators of an orbital SO(4,2) which is a nonrelativistic contraction of the algebra (51) as given by

$$\begin{aligned} \ell_{ij} &= x_i \pi_j - x_j \pi_i \\ \ell_{io} &= -m\pi_i, \ \ell_{i6} + \ell_{i4} = x_i; \ \ell_{06} + \ell_{04} = r \\ \ell_{i0} - \ell_{ir} &= 2(\boldsymbol{x} \cdot \boldsymbol{\pi} - ih)\pi_i - x_i \pi^2 \\ \ell_{06} - \ell_{04} &= r\pi^2, \ \ell_{04} = \boldsymbol{x} \cdot \boldsymbol{\pi} - i \end{aligned} \tag{55}$$

Finally, it is interesting to find the relativistic but spinless limit of the two-body equation (48). Using the limits we obtain the equation

$$P^2 \Phi = \left[M^2 + \frac{M^2}{\mu} \boldsymbol{p}^2 + \frac{e_1 e_2}{r} \left(\frac{P^2}{M} + \frac{\Delta m}{M} \frac{1}{\mu} \boldsymbol{P} \cdot \boldsymbol{p} + \frac{1}{\mu} \boldsymbol{p}^2 \right) \right] \Phi \tag{56}$$

here

$$P^2 = P_\mu P^\mu, M = m_1 + m_2, \Delta m = m_1 - m_2, \mu = m_1 m_2 / M.$$

6. CLASSICAL RELATIVISTIC MANY-BODY EQUATIONS WITH SPIN

We can apply similar methods to our classical Lagrangian (44) to obtain two - and many body equations for classical spinning particles. The resultant equations look almost identical because the action (44) and the action of quantumelectrodynamics are similar, in fact the latter being the quantization of the particle action. The method consists again in defining composite spinors for two particles

$$Z = z_1 \otimes z_2 \tag{57}$$

and expressing the action in terms of these composite spinors and then varying the action with respect to these composite spinors. We refer for details to the literature[14] . It is sufficient for our purpose here to give the results and to indicate that the symmetry properties of the classical system will be the same as the quantum system. For two-body problem we get the covariant Hamiltonian

$$\mathcal{H} = \bar{Z}\left[\Gamma^\mu P_\mu - k_\perp^\mu p_{\mu\perp} - \frac{e_1 e_2}{r_\perp}\gamma^\mu \otimes \gamma^\mu\right] Z \tag{58}$$

the symbols are the same as in eqs.(25)-(26).

7. REFERENCES

1. For some reviews see :

   E.B.Aronson, I.A.Malkin and V.I.Manko, *Sov.J.Particles* and *Nuclei*, 5,47 (1974)

   A.O.Barut, Dynamical Groups and Generalized Symmetries, Univ. of Canterbury Press,1972

   A.O.Barut and R.Raczka, Theory of Group Representations and Applications, Second Edition, World Scientific, 1986

   A.Bohm, A.O.Barut and Y.Neéman (edits.), Dynamical Groups and Spectrum Genarting Algebras, Vol.I and II, World Scientific, 1988

   B.G.Wybourne, Classical Groups for Physicists, J.Wiley, N.Y.,1974

   C.E.Wulfman, in Group Theory and its Applications, Vol.2, E.M.Loebl, edit., Acad. Press, N.Y. 1971

2. A.O.Barut and Y.Kitagawara, *J.Phys.* **A14**, 2581 (1981); **A15**, 117 (1982)

3. A.O.Barut, in *Symmetry in Science*, Vol. II, B.Gruber, edit. Plenum N. Y. 1987

4. A.O.Barut and N.Unal, *Physica*,**142A**, 467 and 488 (1987)

5. A.O.Barut, *J. Math. Phys.*, March 1991

6. N.Kemmer, *Helv.Phys.Acta*, **10**, 8 (1937); *Proc.Roy.Soc.* **A173**, 91 (1987)

   R.J.Duffin, *Phys.Rev.* **54**, 114 (1936)

   G.Petiau, *Acad.Roy.Soc.Belg.Cl.Sci.Mem.* **8**, 16 (1936)

7. I.M.Gelfand and A.M.Yaglom, *Zh.Eksp.Teor.Fiz.* **18**, 703, 1096, and 1105 (1948)

8. E.Majorana, *Nuovo Cimento* **9**, 335 (1932)

9. A.O.Barut and S.Malin, *Rev. Mod. Phys.* **40**, 632 (1968)

10. A.O.Barut and N.Zanghi, *Phys. Rev. Lett.* **52**, 2009 (1984)

11. A.O.Barut and M.Pavsic, *Class. Quant.Grav.* **4**, L41, L131 (1987); **5**, 707 (1988)

    A.O.Barut and I.H.Duru, *Physics Reports*, **172**, 1 (1987)

12. A.O.Barut, *Physics Letters B*, **237**, 436 (1990)

13. A.O.Barut, in Groups, Systems and Many-Body Physics, P.Kramer et al., edits., Vieweg, Braunschweig, 1980

14. A.O.Barut, C.Onem and N.Unal, *J.Phys.A*, **23**, 1113 (1990)

ALGEBRAIC TREATMENT OF MULTISTEP PROCESSES IN ELECTRON-MOLECULE SCATTERING

Roelof Bijker

Department of Physics and Astronomy
University of Utrecht, P.O. Box 80000
3508 TA Utrecht, The Netherlands

1. INTRODUCTION

The scattering of medium energy electrons from polar molecules is a complicated process involving many partial waves and the virtual excitation of many intermediate states [1]. For forward angles ($\theta < 60^0$) the scattering process is largely dominated by the long range dipole interaction between the incoming electron and the molecule. For large angles the scattering is more sensitive to short range features of the molecular dynamics, such as exchange correlations. For relatively small values of the dipole moment, $d \simeq 1$ *Debye*, the Born approximation already gives a good description of the experimental cross section. For strongly polar molecules the channel coupling between the rotational and vibrational degrees of freedom becomes increasingly important. This is illustrated in Figure 1, where the experimental cross section for electron scattering from LiF, which has a dipole moment of $d = 6.58$ *Debye*, is compared with the Born approximation. For very small angles the Born approximation is still adequate, but for larger angles multistep excitations have to be taken into account.

The standard approach to treat these multistep processes is that of a coupled channels or close-coupling calculation. Although for diatomic molecules this already presents a formidable problem, it is still feasible to do the full coupled channel calculation. For larger (tri- and polyatomic) molecules these calculations become very complicated, if not altogether impossible. Therefore it is of great interest to find a simple prescription that both exploits the simplifications that arise from the dominance of the dipole interaction at forward angles, and, at the same time, incorporates the rotational-vibrational structure of the molecule at a level sufficient to calculate the cross sections of interest at all angles.

Symmetries in Science V, Edited by B. Gruber *et al.*
Plenum Press, New York, 1991

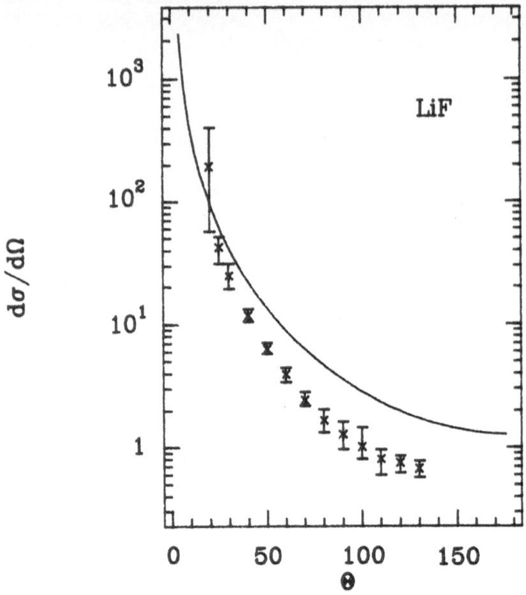

Figure 1. Differential cross section in $\mathring{A}^2/sr$ for vibrationally elastic excitation of LiF by 5.4 eV electrons. The experimental data are taken from [2]. The solid line is the Born approximation.

In this contribution I will discuss an alternative approach, in which the channel coupling can be solved in closed form to all orders in the coupling between the projectile and the target. This method, called the algebraic-eikonal approach [3], combines an algebraic description of the molecular dynamics, the vibron model [4], with eikonal scattering methods [5]. It was originally developed in nuclear physics and applied to proton-nucleus scattering [6] and electromagnetic excitation of nuclei [7]. After a review of the formalism I will discuss the application to electron scattering from LiF and HCN. Finally I suggest a possible extension to include short range correlations by matching the high partial waves of the algebraic-eikonal approach with the low partial waves of a close coupling calculation.

2. ALGEBRAIC-EIKONAL APPROACH

The algebraic-eikonal approach was proposed [3] recently to describe the scattering of high energy electrons from molecular targets. It is a combination of the eikonal scattering methods and an algebraic description of the molecular dynamics, the vibron model [4]. First it was applied to electron scattering from diatomic molecules [3, 8] and subsequently the method was extended to triatomic molecules [9] as well. Here I will discuss briefly the main features of the algebraic-eikonal approach. A more detailed account can be found in [3, 8, 9].

For medium and high energy scattering the eikonal approximation is a good approximation for both elastic and inelastic scattering. The Hamiltonian is in general given by

$$H = \frac{\hbar^2 k^2}{2\mu_e} + H_{mol}(\xi) + V(\vec{r}, \xi) \tag{1}$$

where $H_{mol}(\xi)$ is the vibron Hamiltonian describing the molecular dynamics, and $V(\vec{r}, \xi)$ represents the coupling between the incoming electron and the molecule. For small angle scattering (peripheral collisions) the long range dipole interaction is the dominant term in $V(\vec{r}, \xi)$,

$$V_d(\vec{r}, \xi) = -\frac{e}{r^2 + R_0^2}\, \hat{r} \cdot \vec{T}(\xi) \tag{2}$$

where $\vec{r}$ is the projectile coordinate and $\vec{T}(\xi)$ is the molecular dipole operator. The cut-off radius, R_0, is introduced to remove the singularity at the origin, thus crudely modeling the short range part of the electron-molecule interaction.

For scattering processes in which the projectile energy is much larger than the coupling potential and in which the projectile wavelength is small compared to the range of variation of the potential, one may use the eikonal approximation to describe the scattering. If, in addition, the molecular motion is slow (adiabatic) compared to the interaction time of the projectile electron with the molecule, one can neglect $H_{mol}(\xi)$ in Eq.(1). The molecular eigenstates are denoted by $|v, l, m>$, where v represents the vibrational and l and m the rotational quantum numbers. It is assumed that the molecule remains in its electronic ground state. Under these approximations the scattering amplitude for scattering an electron with initial momentum $\vec{k}$ from an initial molecular state $|i> = |v, l, m>$ to final momentum $\vec{k}'$ and a final state $|f> = |v', l', m'>$ can be written as [3]

$$A(i \to f|\vec{q}) = \frac{k}{2\pi i} \int d\vec{b}\, e^{i\vec{q}\cdot\vec{b}} \left[< f|e^{ig(b)\hat{b}\cdot\vec{T}(\xi)}|i> - \delta_{fi}\right] \tag{3}$$

where $\vec{q} = \vec{k} - \vec{k}'$ is the momentum transfer and $g(b)$ is the eikonal phase, that the projectile acquires as it goes by the target

$$g(b) = -\frac{\mu_e}{\hbar^2 k} \int_{-\infty}^{\infty} dz\, \frac{-e}{r^2 + R_0^2} \frac{b}{r} \tag{4}$$

In the derivation of Eq.(3) the projectile coordinate is written as $\vec{r} = \vec{b} + \vec{z}$, where the impact parameter $\vec{b}$ is perpendicular to the z-axis, which is chosen along $\hat{z} = (\vec{k} + \vec{k}')/|\vec{k} + \vec{k}'|$. The expression for the scattering amplitude can be simplified further by using the symmetry properties of the dipole operator. The transition operator can be written as a product of a rotation about an angle $(\hat{b}) = (\phi, \pi/2, 0)$ to the z axis, a simpler form of the transition operator involving only the z component of the dipole operator, and the inverse rotation about $(-\hat{b}) = (0, -\pi/2, -\phi)$

$$e^{ig(b)\hat{b}\cdot\vec{T}(\xi)} = R(\hat{b}) e^{ig(b)T_z(\xi)} R(-\hat{b}) \tag{5}$$

The angular part of the integral can now be done analytically and the scattering amplitude can be expressed in terms of a one-dimensional integral only

$$\begin{aligned} A(i \to f|\vec{q}) &= \frac{k}{i} i^{m-m'} \int_0^{\infty} b\,db\,J_{m-m'}(qb) \left[\sum_{L,M} \sqrt{\frac{4\pi}{2L+1}}\, Y^*_{L,M}(\hat{q}) \right. \\ &\times \frac{2L+1}{2l'+1} < L, M, l, m|l', m' > \sum_{m''} < L, 0, l, m''|l', m'' > \\ &\times \left. < v', l', m''|e^{ig(b)T_z(\xi)}|v, l, m'' > -\delta_{fi}\right] \end{aligned} \tag{6}$$

The transition matrix element appearing in the integrand of the scattering amplitude now has a very simple form, since it only involves the z-component of the dipole operator, $T_z(\xi)$. It contains the coupling between molecular eigenstates to *all* orders in the coupling strength $g(b)$. In the next section it will be shown how these matrix elements can be derived *exactly* in the vibron model.

3. THE VIBRON MODEL

The vibron model [4] is similar in spirit to the Interacting Boson Model [10] of nuclear physics and provides an algebraic description of the rotational and vibrational excitations in molecules. Although it cannot match the high accuracy for the energy spectrum obtained in *ab initio* calculations, the vibron model gives a good global description of the properties of rotational and vibrational excitations of diatomic molecules. It becomes particularly useful in the calculation of transition probabilities, which in general involve a complicated sum over intermediate states. Moreover, it can easily be generalized to more complex systems, such as triatomic [11] and polyatomic [12] molecules, for which *ab initio* methods become increasingly difficult. More recently the vibron model has been extended to include electronic excitations as well [13].

3.1 DIATOMIC MOLECULES

The rotational and vibrational excitations of diatomic molecules are generated in the vibron model by the spectrum generating algebra of $U(4)$. The algebra is realized in terms of four vibron operators divided into a scalar, s ($L^\pi = 0^+$), and a vector, p_μ, $\mu = \pm 1, 0$ ($L^\pi = 1^-$), boson representing the dipole degree of freedom. The vibron Hamiltonian conserves the total number of vibrons, N, and is a scalar under rotations. Keeping only terms that are at most quadratic in its generators, it can be written as

$$H_{mol} = \alpha_1\, n_p + \alpha_2\, n_p^2 + \alpha_3\, D \cdot D + \alpha_4\, L \cdot L \tag{7}$$

where n_p is the number operator for p-vibrons, L is the angular momentum operator and D is the dipole operator. In terms of the vibron operators they are given by

$$\begin{aligned} n_p &= -\sqrt{3}\,(p^\dagger \tilde{p})^{(0)} \\ D_\mu &= (s^\dagger \tilde{p} + p^\dagger s)^{(1)}_\mu \\ L_\mu &= \sqrt{2}\,(p^\dagger \tilde{p})^{(1)}_\mu \end{aligned} \tag{8}$$

with $\tilde{p}_\mu = (-1)^\mu p_{-\mu}$. When $\alpha_1 = \alpha_2 = 0$ the Hamiltonian has an $SO(4)$ dynamical symmetry. This case is of particular interest, since it corresponds [14] in the large N limit to the Morse oscillator which has been used widely in the description of diatomic molecules. The $SO(4)$ dynamical symmetry offers a natural basis to diagonalize the Hamiltonian, even if this symmetry is (slightly) broken. In this basis the molecular states are labeled by the irreducible representations of the groups in the chain

$$\left| \begin{array}{ccccccc} U(4) & \supset & SO(4) & \supset & SO(3) & \supset & SO(2) \\ \downarrow & & \downarrow & & \downarrow & & \downarrow \\ N & , & \sigma & , & l & , & m \end{array} \right\rangle \tag{9}$$

where N is the total number of vibrons, σ is related to the usual vibrational label v by $v = (N - \sigma)/2$, and l and m represent the angular momentum and its projection along a symmetry axis. The states have parity $\pi = (-)^l$.

Infrared transition rates can be calculated by expanding the $E1$ operator in terms of vibron operators [15]

$$T_\mu = d_0\, D_\mu + d_1\, \frac{1}{2}\left[e^{\lambda n_p} D_\mu + D_\mu e^{\lambda n_p}\right] \tag{10}$$

For vibrationally elastic transitions, $\Delta v = 0$, it is sufficient to keep only the first term in the transition operator

$$T_\mu = d_0\, D_\mu = d_0\, (s^\dagger \tilde{p} + p^\dagger s)^{(1)}_\mu \tag{11}$$

The matrix elements of interest in electron-molecule scattering are those of the exponentiated dipole operator, $\hat{U}(\epsilon) = \exp(i\epsilon D_z)$ with $\epsilon = d_0 g(b)$. From the symmetry properties of the dipole operator, D_z, it is easy to see that the dipole transition matrix elements are diagonal in both $\sigma = N - 2v$ and m, and furthermore do not depend on N. They can be derived in closed analytic form in the $SO(4)$ limit of the vibron model by using the isomorphism between $SO(4)$ and $SU(2) \otimes SU(2)$,

$$\begin{aligned} U_{fi}(\epsilon) \equiv &< N, \sigma', l', m'|e^{i\epsilon D_z}|N, \sigma, l, m >= \delta_{\sigma'\sigma}\delta_{m'm} \\ &\sum_\mu \left[\frac{1+(-1)^{l+l'}}{2}\cos(2\mu - m)\epsilon + \frac{1-(-1)^{l+l'}}{2} i\sin(2\mu - m)\epsilon\right] \\ &\times < \sigma/2, \mu, \sigma/2, m-\mu|l', m >< \sigma/2, \mu, \sigma/2, m-\mu|l, m > \end{aligned} \tag{12}$$

The first term only contributes to parity conserving transitions, and the second one only to parity changing transitions. These matrix elements can be interpreted as representation matrix elements of $SO(4)$ and as such are a generalization of the familiar Wigner D-matrices for $SU(2)$. A similar expression was derived in [16], using however a basis in which parity is not a good quantum number. It is interesting to note that for large values of σ, such that $\epsilon\sigma$ remains finite and $l/\sigma \ll 1$, the transition matrix element in the $SO(4)$ limit of the vibron model becomes identical to that derived in the classical rotor model [17]

$$\begin{aligned} U_{fi}(\epsilon) \rightarrow \delta_{\sigma'\sigma}\delta_{m'm}\sqrt{\frac{2l+1}{2l'+1}} &\sum_L i^L(2L+1) < l, m, L, 0|l', m > \\ &\times < l, 0, L, 0|l', 0 > j_L(\epsilon\sigma) \end{aligned} \tag{13}$$

3.2 TRIATOMIC MOLECULES

The rotational and vibrational states of triatomic molecules are described in the vibron model by the spectrum generating algebra of the direct product of two $U(4)$ groups, $G = U_1(4) \otimes U_2(4)$, one for each bond in the molecule [11]. Each of the $U(4)$ groups is realized in terms of a scalar and a vector vibron operator. To distinguish between the two sets I will use subscripts 1 and 2. The vibron Hamiltonian conserves the total number of vibrons of type 1 and 2, N_1 and N_2, separately. In addition, for rigid molecules in which $E_{rot} \ll E_{vib}$ the Hamiltonian is invariant under $SO(4)$ transformations [11]. Keeping only up to two-body terms the vibron Hamiltonian for (rigid) triatomic molecules can be written as

$$\begin{aligned} H_{mol} &= (\alpha_1+\alpha)\, D_1\cdot D_1 + (\alpha_1+\alpha+\delta)\, L_1\cdot L_1 \\ &+ (\alpha_2+\alpha)\, D_2\cdot D_2 + (\alpha_2+\alpha+\delta)\, L_2\cdot L_2 \\ &+ 2\alpha\, D_1\cdot D_2 + 2(\alpha+\delta)\, L_1\cdot L_2 \\ &+ \beta\, |(D_1+D_2)\cdot(L_1+L_2)| + \gamma\, M \end{aligned} \tag{14}$$

where L_k and D_k are the angular momentum and dipole operator for system k with $k=1,2$. The last term in H_{mol} is a Majorana type term

$$M = (p_1^\dagger s_2^\dagger - s_1^\dagger p_2^\dagger)^{(1)} \cdot (\tilde{p}_1 s_2 - s_1 \tilde{p}_2)^{(1)} - 2\,(p_1^\dagger p_2^\dagger)^{(1)} \cdot (\tilde{p}_1\tilde{p}_2)^{(1)} \tag{15}$$

If no further symmetry is present H_{mol} has to be diagonalized numerically. Since the Hamiltonian has $SO(4) \supset SO(3) \supset SO(2)$ symmetry, a convenient basis is provided by the following subgroup chain of $U_1(4)\otimes U_2(4)$,

$$\left| \begin{array}{ccccccccc} SO_1(4) & \otimes & SO_2(4) & \supset & SO(4) & \supset & SO(3) & \supset & SO(2) \\ \downarrow & & \downarrow & & \downarrow & & \downarrow & & \downarrow \\ \sigma_1 & , & \sigma_2 & , & (\sigma,\tau) & , & l & , & m \end{array} \right\rangle \tag{16}$$

The quantum numbers $\sigma_1, \sigma_2, (\sigma,\tau)$ can be related [11] to the standard spectroscopic notation v_1, v_2^K, v_3 representing the symmetric stretching (v_1), the bending (v_2) and the asymmetric stretching (v_3) vibration. K is the projection of the angular momentum along an intrisic axis. Except for $\tau=0$ the basis states of Eq.(16) are not parity eigenstates. States with good parity can be constructed by taking

$$\begin{aligned} |\sigma_1,\sigma_2,(\sigma,\tau),l(\pi),m> &= \frac{1}{\sqrt{2(1+\delta_{\tau,0})}} \Big[|\sigma_1,\sigma_2,(\sigma,\tau),l,m> \\ &+ \pi(-1)^{\sigma_1+\sigma_2+\sigma+l}\, |\sigma_1,\sigma_2,(\sigma,-\tau),l,m> \Big] \end{aligned} \tag{17}$$

For $\tau=0$ the coupling rules give $\sigma_1+\sigma_2+\sigma$ is even, and therefore the only allowed parity is $\pi=(-)^l$. For $\tau\neq 0$ one has parity doublets with $\pi=\pm$.

If $\gamma=0$ in Eq.(14) the Hamiltonian becomes diagonal in this basis and corresponds in a geometrical description to that of two coupled Morse oscillators. For $\beta=0$ the energy spectrum corresponds to that of a linear triatomic molecule, while for $\beta\neq 0$ the energy spectrum is similar to that of a non-linear triatomic molecule [11]. For the special value $\beta=2\alpha$ the Hamiltonian describes a spherically symmetric top. Summarizing, the vibron Hamiltonian of Eq.(14) presents a simple, algebraic, description of stretching and bending vibrations in both linear and non-linear triatomic molecules.

For triatomic molecules the dipole operator in the vibron model consists of two terms

$$T_\mu = d_1\, D_{1,\mu} + d_2\, D_{2,\mu} \tag{18}$$

and therefore the transition operator of interest in electron-molecule scattering is given by

$$\hat{U}(\epsilon_1,\epsilon_2) = e^{i\epsilon_1 D_{1,z} + i\epsilon_2 D_{2,z}} = e^{i\epsilon_1 D_{1,z}} e^{i\epsilon_2 D_{2,z}} \tag{19}$$

with $\epsilon_k = g(b) d_k$, $k=1,2$. If $d_1=d_2$ the dipole operator T_μ is a generator of $SO(4)$ and can only connect states which are characterized by the same values of (σ,τ) and as a consequence can only describe vibrationally elastic scattering. For $d_1 \neq d_2$ it is possible to excite bending and stretching vibrations as well. In the following I discuss two different, but equivalent, methods to calculate the matrix elements of the dipole transition operator.

Method 1
In the first method these matrix elements are evaluated in a straightforward way by expanding the coupled basis of Eq.(16) into the direct product basis by

$$|\sigma_1, \sigma_2, (\sigma, \tau), l, m> = \sum_{l_1 m_1 l_2 m_2} \left\langle \begin{matrix} \sigma_1 & \sigma_2 \\ l_1 m_1 & l_2 m_2 \end{matrix} \middle| \begin{matrix} (\sigma, \tau) \\ lm \end{matrix} \right\rangle \times |\sigma_1, l_1, m_1 > \otimes |\sigma_2, l_2, m_2 > \tag{20}$$

The expansion coefficients are the transformation brackets (or isoscalar factors) for the $SO(4) \supset SO(3) \supset SO(2)$ group reduction. Using Racah's factorization lemma [18] they can be written as the product of a $SO(4) \supset SO(3)$ (Wigner) and a $SO(3) \supset SO(2)$ (Clebsch-Gordon) coefficient

$$\left\langle \begin{matrix} \sigma_1 & \sigma_2 \\ l_1 m_1 & l_2 m_2 \end{matrix} \middle| \begin{matrix} (\sigma, \tau) \\ lm \end{matrix} \right\rangle = \sqrt{(2l_1+1)(2l_2+1)(\sigma+\tau+1)(\sigma-\tau+1)} \left\{ \begin{matrix} \sigma_1/2 & \sigma_2/2 & (\sigma+\tau)/2 \\ \sigma_1/2 & \sigma_2/2 & (\sigma-\tau)/2 \\ l_1 & l_2 & l \end{matrix} \right\} < l_1, m_1, l_2, m_2 | l, m > \tag{21}$$

Combining Eqs.(17) and (20) and using the symmetry properties of the dipole operator, the matrix element of the transition operator, $\hat{U}(\epsilon_1, \epsilon_2)$, between states of good parity can be derived as

$$\begin{aligned} U_{fi}(\epsilon_1, \epsilon_2) &\equiv < \sigma_1, \sigma_2, (\sigma', \tau'), l'(\pi'), m' | \hat{U}(\epsilon_1, \epsilon_2) | \sigma_1, \sigma_2, (\sigma, \tau), l(\pi), m > \\ &= \delta_{m'm} \frac{2}{\sqrt{(1+\delta_{\tau,0})(1+\delta_{\tau',0})}} \sum_{l_1 l_2} \delta_{\pi,(-1)^{l_1+l_2}} \sum_{l'_1 l'_2} \delta_{\pi',(-1)^{l'_1+l'_2}} \\ &\sum_{m_1 m_2} \left\langle \begin{matrix} \sigma_1 & \sigma_2 \\ l_1 m_1 & l_2 m_2 \end{matrix} \middle| \begin{matrix} (\sigma, \tau) \\ lm \end{matrix} \right\rangle \left\langle \begin{matrix} \sigma_1 & \sigma_2 \\ l'_1 m_1 & l'_2 m_2 \end{matrix} \middle| \begin{matrix} (\sigma', \tau') \\ l'm \end{matrix} \right\rangle \\ &< \sigma_1, l'_1, m_1 | e^{i\epsilon_1 D_{1,z}} | \sigma_1, l_1, m_1 >< \sigma_2, l'_2, m_2 | e^{i\epsilon_2 D_{2,z}} | \sigma_2, l_2, m_2 > \end{aligned} \tag{22}$$

The $SO(4)$ representation matrix elements appearing in the right hand side are given in Eq.(12).

The general result for the transition matrix elements still contains a five-fold sum over intermediate states which makes it very time consuming to calculate them numerically, especially since for realistic cases the number of vibrons is typically $N \approx 100$ (e.g. HCN has $N_1 = 144$ and $N_2 = 47$ [11]).

Method 2
There is however a different method that uses the symmetry properties of the dipole operator under transformations of the coupled $SO(4)$ group. It is convenient to rewrite the transition operator as

$$\hat{U}(\epsilon_1, \epsilon_2) = e^{i\epsilon D_z + i\epsilon^* D_z^*} = e^{i\epsilon D_z} e^{i\epsilon^* D_z^*} \tag{23}$$

with

$$\epsilon = (\epsilon_1 + \epsilon_2)/2, \quad D_z = D_{1,z} + D_{2,z}$$
$$\epsilon^* = (\epsilon_1 - \epsilon_2)/2, \quad D_z^* = D_{1,z} - D_{2,z} \tag{24}$$

Since D_z is a generator of the coupled $SO(4)$ group, it is diagonal in (σ, τ). Therefore in this case the sum over intermediate states in the transition matrix element, defined in Eq.(22), only involves the angular momentum and parity,

$$U_{fi}(\epsilon_1, \epsilon_2) = \delta_{m'm} \sum_{l''(\pi'')} < (\sigma', \tau'), l'(\pi'), m | e^{i\epsilon^* D_z^*} | (\sigma, \tau), l''(\pi''), m >$$
$$< (\sigma, \tau), l''(\pi''), m | e^{i\epsilon D_z} | (\sigma, \tau), l(\pi), m > \tag{25}$$

The last term is a $SO(4)$ representation matrix element, which in general can be derived as

$$< (\sigma', \tau'), l'(\pi'), m' | e^{i\epsilon D_z} | (\sigma, \tau), l(\pi), m > = \delta_{\sigma'\sigma} \delta_{\tau'\tau} \delta_{m'm}$$
$$\sum_{\mu} \left[\frac{1 + \pi\pi'}{2} \cos(2\mu - m)\epsilon + \frac{1 - \pi\pi'}{2} i \sin(2\mu - m)\epsilon \right]$$
$$\times < (\sigma + \tau)/2, \mu, (\sigma - \tau)/2, m - \mu | l', m >$$
$$\times < (\sigma + \tau)/2, \mu, (\sigma - \tau)/2, m - \mu | l, m > \tag{26}$$

For $\tau = 0$ it reduces to Eq.(12). The first term in Eq.(25) can be simplified by noting that also $D_\mu^* = D_{1,\mu} - D_{2,\mu}$ and the total angular momentum operator $L_\mu = L_{1,\mu} + L_{2,\mu}$ form a closed algebra. They provide a different realization of the $SO(4)$ algebra. The corresponding basis states can be labeled by $|(\sigma, \tau)^*, l(\pi), m>$, where the starred labels are introduced to distinguish between the two sets of $SO(4)$ basis states. Using the basis transformation

$$|(\sigma, \tau), l(\pi), m> = \sum_{(\sigma,\tau)^*} \xi_{(\sigma,\tau),l(\pi)}^{(\sigma,\tau)^*,l(\pi)} |(\sigma, \tau)^*, l(\pi), m> \tag{27}$$

and the fact that D_z^* is diagonal in $(\sigma, \tau)^*$, the transition matrix element of Eq.(25) can be expressed entirely in terms of $SO(4)$ representation matrices

$$\begin{aligned} U_{fi}(\epsilon_1, \epsilon_2) &= \delta_{m'm} \sum_{l''(\pi'')} \sum_{(\sigma,\tau)^*} \xi_{(\sigma',\tau'),l'(\pi')}^{(\sigma,\tau)^*,l'(\pi')} \xi_{(\sigma,\tau),l''(\pi'')}^{(\sigma,\tau)^*,l''(\pi'')} \\ &< (\sigma, \tau)^*, l'(\pi'), m | e^{i\epsilon^* D_z^*} | (\sigma, \tau)^*, l''(\pi''), m > \\ &< (\sigma, \tau), l''(\pi''), m | e^{i\epsilon D_z} \mid (\sigma, \tau), l(\pi), m > \end{aligned} \tag{28}$$

The transformation coefficients ξ can be obtained by expanding both sets of $SO(4)$ basis states into a direct product basis. The expansion coefficients for the $SO(4)$ basis states can be obtained by combining Eqs.(17) and (20). Those of the $SO(4)^*$ basis states pick up an extra phase factor $(-1)^{\sigma_2 + l_2}$, since these states arise from combining the $SO(4)_1$ covariant representation $(\sigma_1, 0)$ with the contravariant representation $(\bar{\sigma}_2, 0)$ of $SO(4)_2$.

The expression for the transition matrix elements in Eq.(28) is completely equivalent to Eq.(22), but by using the transformation properties of the dipole operator under rotations of the coupled $SO(4)$ group explicitly, the sum over intermediate states is reduced considerably. Moreover, since the sum over intermediate states converges rapidly, the expression of Eq.(28) becomes numerically very efficient [9].

The two parameters in the dipole operator, d_1 and d_2, can in principle be determined from the static dipole moment, $d \sim d_1 N_1 + d_2 N_2$ and a transition matrix element, for example the one to the first bending vibration, $A_d \sim d_1 - d_2$ [9]. To get a qualitative understanding of the dependence of the transition matrix elements on these two coupling strengths, it is interesting to study their behavior in the limit of $N_1, N_2 \to \infty$. This is most easily done using intrinsic states. The intrinsic matrix element of the transition operator in the ground state band [9]

$$< g.s.|e^{ig(b)T_z}|g.s. >= e^{ig(b)d} \cos[g(b)A_d] \tag{29}$$

shows that in first order the vibrationally elastic scattering is determined fully by the static dipole moment of the molecule, d, and does not depend on A_d (or equivalently $d_1 - d_2$). On the other hand vibrationally inelastic scattering depends very sensitively on the value of A_d. For $d_1 = d_2$ only vibrationally elastic transitions are allowed, while for $d_1 \neq d_2$ it is also possible to excite bending and stretching vibrations. For example the excitation of the first bending vibration is determined by the intrinsic matrix element

$$< b.v.|e^{ig(b)T_z}|g.s. >= e^{ig(b)d}\, i \sin[g(b)A_d] \tag{30}$$

which to leading order is linear in A_d.

4. APPLICATIONS

In electron-molecule scattering it is in general not possible to resolve the excitation of individual rotational states experimentally. Therefore, in order to compare with experiment one has to calculate a differential cross section which is summed over the final rotational states and averaged over the initial magnetic substates,

$$\frac{d\sigma(v \to v'|q)}{d\Omega} = \frac{2}{2l+1}\frac{k'}{k}\sum_m \sum_{l'm'} |A(i \to f|\vec{q})|^2 \tag{31}$$

Furthermore, using intrinsic states one can show [8] that this summed and averaged differential cross section does not depend on the initial angular momentum, l. This allows one to take $l = 0$, which reduces the numerical effort considerably.

As a first application I present the results for electron scattering from LiF. The parameters in the vibron Hamiltonian of Eq.(7) were determined [8] by comparing with a Dunham expansion, $\alpha_1 = 127.268$, $\alpha_2 = 0$, $\alpha_3 = -1.982$ and $\alpha_1 = -1.190\ cm^{-1}$. The number of vibrons is $N = 113$. The strength of the dipole coupling d_0, Eq.(11), is determined by the static dipole moment of LiF, $d = 6.58$ *Debye*. The cut-off parameter in the dipole interaction is $R_0 = 0.5$ Å. Figure 2a shows, that for forward angles there is good agreement with the data. A comparison with the Born approximation shows the importance of multistep exciations. The discrepancies at larger angles are a reflection of the rather crude treatment of the short range dynamics. The summed and averaged differential cross section (dashed line) shows a very smooth and regular behavior, but, unlike in the case of the Born term, this is the result of a sum over individual rotational transitions, which each separately have a much richer structure.

As a second example I discuss the application to electron scattering from HCN. The parameters in the vibron Hamiltonian, Eq.(14), are taken from a study of the vibrational spectrum of HCN [11], $\alpha_1 = -1.178$, $\alpha_2 = -10.012$, $\alpha = -1.862$, $\beta = \delta = 0$ and $\gamma = -0.033\ cm^{-1}$. The number of vibrons is $N_1 = 144$ and $N_2 = 47$. The average value

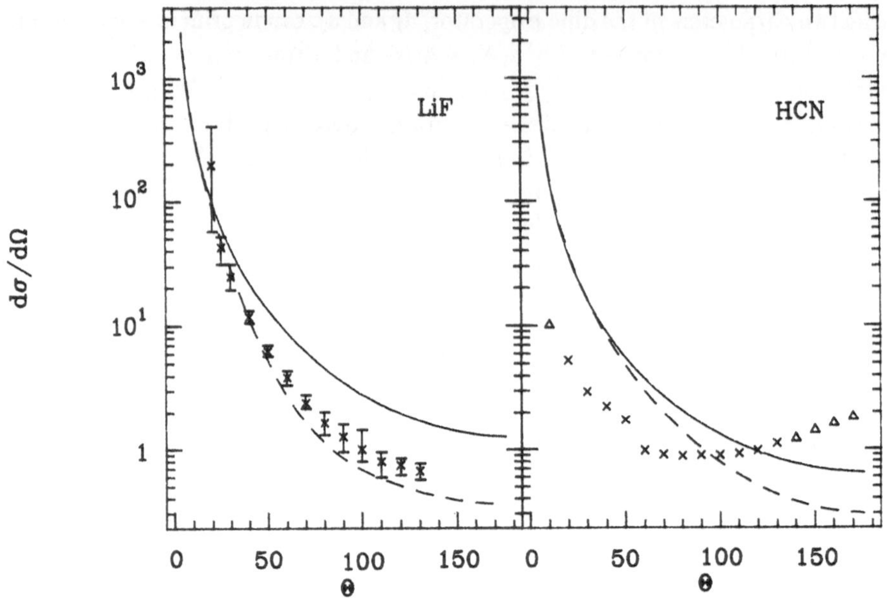

Figure 2. Differential cross section in $\AA^2/sr$ for vibrationally elastic, $v=0 \rightarrow v'=0$, excitation of (a) LiF by 5.4 eV electrons and (b) HCN by 5.0 eV electrons. The experimental data are taken from [2] and [19]. The solid line is the Born approximation and the dashed line the algebraic-eikonal approximation.

of the dipole coupling strengths, d_1 and d_2, is determined by the static dipole moment of HCN, $d = 2.98$ *Debye*. Their relative value can in principle be obtained from the transition probability of a vibrational excitation, but to the best of my knowledge no such information is available. Since the vibrationally elastic scattering only depends on the average value and not on the relative value (see Eq.(29)), the two strengths are taken equal $d_1 = d_2$. The cut-off parameter is $R_0 = 0.5$ $\AA$. Figure 2b shows a comparison between the calculated and the measured differential cross section for electron energy of 5.0 eV. The rise in the cross section at forward angles is a direct consequence of the coupling to the static dipole moment in a nearly model independent sense. For larger angles contributions from other interaction terms, such as polarization and exchange potentials become more important. In a recent *ab initio* calculation [20] a similar agreement with the data was obtained. In the present calculation only the effect of the long range dipole interaction is taken into account. In [9] it was shown how a (spherical) polarization potential can be included in eikonal distorted wave approximation. In the next section I discuss a different method to include extra interaction terms.

5. HYBRID APPROACH

The applications discussed in the previous section show that, although there is good agreement with the available experimental data for forward angles, there are still discrepancies at larger angles. This conclusion is confirmed by more extensive studies, in which the algebraic-eikonal approach was applied to electron scattering from LiF, KI [8], HCl [21] and HCN [9] for a variety of electron energies. To describe the scattering to larger angles more accurately, a more sophisticated treatment of short range correlations, such as exchange and polarization potentials, is required. One possible way to include these effects is to express the relevant short range interactions explicitly in terms of vibron operators. However, this leads to many complications and will distroy the simplicity and elegance of the method. Among other things it will give rise to non-linear terms in the eikonal phase which are very hard to calculate. Also the validity of the eikonal approximation for angles at which these effects are expected to show up becomes questionnable.

A different method [22] to incorporate short range correlations is to still use the algebraic-eikonal approach for the high partial waves, but then combined with the low partial waves of, for example, a coupled channels calculation. This proposed hybrid approach is similar to the procedure outlined in [23, 24], which uses instead the Born approximation to calculate the high partial waves.

First the differential cross section for vibrationally elastic transitions, v=0 $\rightarrow$ v'=0, summed over the final magnetic substates and averaged over the initial ones, can be expressed explicitly in terms of the T matrix elements by [25, 26]

$$\frac{d\sigma(l \rightarrow l'|q)}{d\Omega} = \frac{2}{2l+1}\frac{k'}{k}\sum_{mm'}|A(i \rightarrow f|\vec{q})|^2 = 2\sum_{\nu} A_\nu P_\nu(\cos\theta) \tag{32}$$

The coefficients A_ν contain the dependence on the T matrix elements. The differential cross section in the hybrid approach is obtained by first subtracting the low partial waves from the eikonal differential cross section and then adding back in the low partial waves from, for example, a coupled channel calculation [24]

$$\frac{d\sigma(l \rightarrow l'|q)}{d\Omega} = \left.\frac{d\sigma(l \rightarrow l'|q)}{d\Omega}\right|_{\text{eik}} + 2\sum_{\nu}\left[B_\nu(L_0) - A_\nu^{\text{eik}}\right] P_\nu(\cos\theta) \tag{33}$$

The coefficients $B_\nu(L_0)$ are obtained by replacing the T^L's in A_ν^{eik} by the close coupling T matrix elements [26] for $L < L_0$, while keeping the eikonal T matrix elements for $L \geq L_0$. The eikonal T matrix elements can be obtained by inverting the partial wave expansion of the eikonal scattering amplitude,

$$\begin{aligned} T^L(l'\lambda'|l\lambda) &= \frac{\sqrt{kk'}}{2\pi i}\sum_{m,\mu} < l,m,\lambda,\mu|L,M > \sum_{m',\mu'} < l',m',\lambda',\mu'|L,M > \\ &\times \frac{i^{\lambda'-\lambda}}{2L+1}\int d\hat{k}\, Y_{\lambda,\mu}(\hat{k})\int d\hat{k}'\, Y^*_{\lambda',\mu'}(\hat{k}')\, A(i \rightarrow f|\vec{q}) \end{aligned} \tag{34}$$

To explore the scope and applicability of this approach I discuss a set of model calculations for scattering 5 eV electrons from a rigid rotor molecule with a dipole moment of d = 6 *Debye*. Figure 3a shows the results for the $0^+ \rightarrow 1^-$ transition. The short dashed line is pure algebraic-eikonal without correcting for the low partial waves. The other curves show the results of a hybrid calculation in which the low partial waves are replaced by the exact ones from a coupled channels calculation, with the matching successively at partial wave $L_0 = 0, 1, 2$ and 10. Already for $L_0 = 1, 2$ convergence is reached. This calculation is repeated in Figure 3b, but now using the

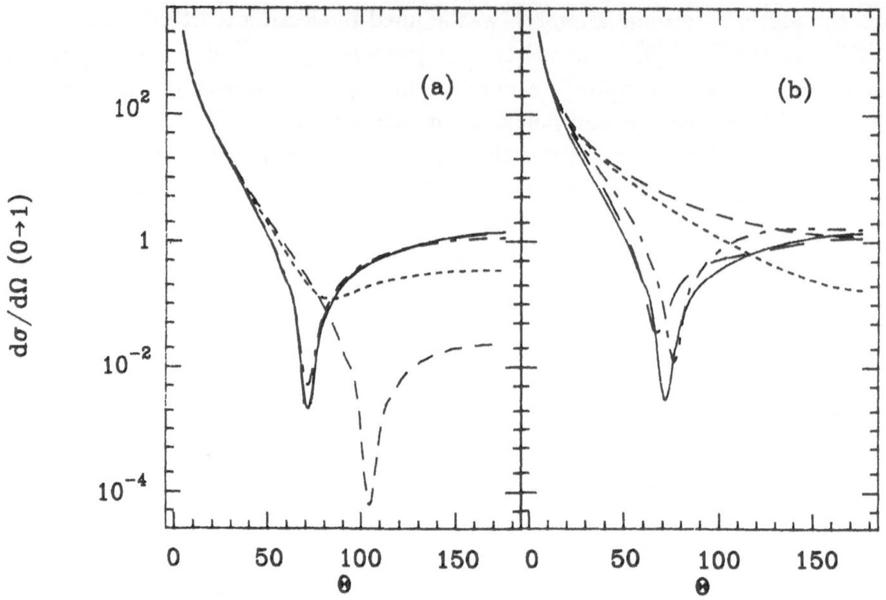

Figure 3. Differential cross section in $\AA^2/sr$ for the excitation of the first rotational state, $0^+ \to 1^-$, by scattering 5 eV electrons from a rigid rotor molecule with a dipole moment of $d = 6$ *Debye* using (a) the algebraic-eikonal approach and (b) the Born approximation (short dashed line). The other curves show the results of hybrid calculations with matching parameter $L_0 = 0$ (dotted), $L_0 = 1$ (short dash-dotted), $L_0 = 2$ (long dashed) and the "exact" result with $L_0 = 10$ (solid line).

Born approximation instead of eikonal. In this case a much higher value of L_0 is needed to reach a similar convergence ($L_0 = 7, 8$). Figure 4a and 4b show the results for the elastic transition, $0^+ \to 0^+$, in algebraic-eikonal and unitarized Born, respectively. In this case there is no contribution from the Born approximation, and therefore we compare instead with the unitarized Born [27]. In either case the result is already essentially exact for $L_0 = 2$, although the starting point of the unitarized Born is not nearly as good as that of the algebraic-eikonal.

For both transitions the hybrid approach offers a great advantage over an exact coupled channels calculation, since one only has to calculate a few T matrix elements exactly to get a good convergence. In a full coupled channels calculation of the differential cross section, according to Eq.(32), one has to sum over many partial waves to obtain a similar convergence. The example of a rigid rotor model was chosen solely to illustrate the method. It does however not contain any interesting interior dynamics. The success of the procedure suggests that the short range correlations can easily be introduced in a hybrid approach, using either the unitarized Born or the eikonal approximation for the high partial waves. However, in general the unitarized Born T matrix elements are far more difficult to calculate in a systematic way than those of the algebraic-eikonal approach, even in the simple case of the rigid rotor.

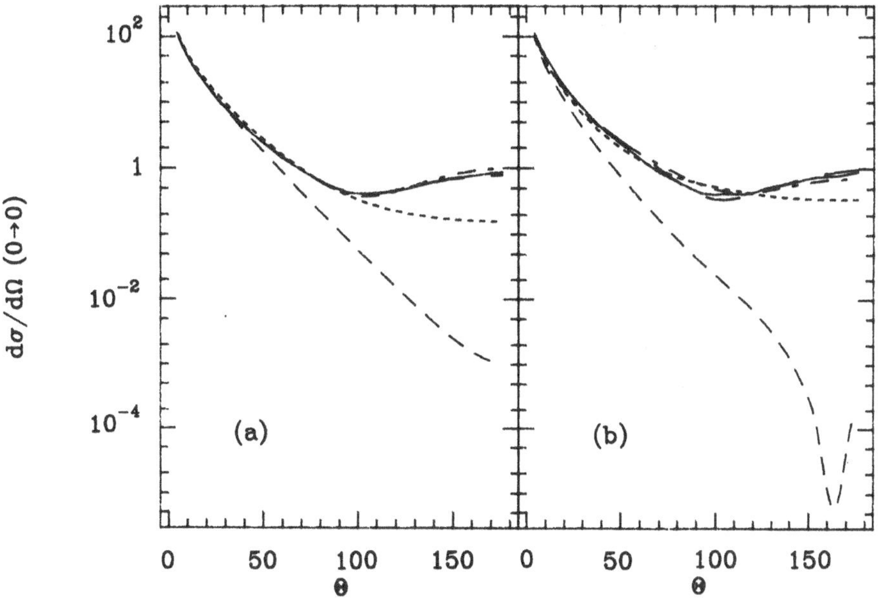

Figure 4. Same as Figure 4 but for elastic scattering, $0^+ \rightarrow 0^+$, using (a) the algebraic-eikonal approach and (b) the unitarized Born approximation.

6. SUMMARY AND CONCLUSIONS

In this contribution I discussed a new method to describe electron scattering from polar molecules for which multistep excitations play an important role. In this algebraic-eikonal approach the molecular dynamics is treated in the vibron model and the scattering equations are solved in eikonal approximation in the adiabatic limit. The scattering matrix is expressed in terms of an exponentiated dipole operator, which can be interpreted as a $SO(4)$ representation matrix. Using group theoretical methods these transition matrix elements can be obtained in closed form to all orders of the coupling constant. The method was applied to electron scattering from both diatomic (LiF) and triatomic (HCN) molecules. Especially for triatomic molecules for which the number of coupled channels becomes very large the algebraic-eikonal approach provides a useful alternative to *ab initio* calculations.

The success of the method is a consequence of the dominance of the long range dipole interaction for forward angles. The short range dynamics can be included through a hybrid approach, in which the algebraic-eikonal approach is used to sum the high partial waves, corrected for the low partial waves, which can be taken from, for example, an *ab initio* calculation. In summary, the algebraic-eikonal approach presents an alternative way to treat multistep excitations, in which the effects of channel coupling are taken into account in an elegant and numerically efficient way.

ACKNOWLEDGEMENTS

It's a pleasure to thank R.D. Amado and L.A. Collins for interesting discussions.

REFERENCES

1) L.A. Collins and D.W. Norcross, Phys. Rev. **A18**, 467 (1978);
Y. Itikawa, Phys. Rep. **46**, 117 (1978);
N.F. Lane, Rev. Mod. Phys. **52**, 29 (1980);
D.W. Norcross and L.A. Collins, Adv. A. Mol. Phys. **18**, 341 (1982);
M.A. Morrison, Aust. J. Phys. **36**, 239 (1983);
see "Electron-Molecule Interactions and their Applications", edited by L.G. Christophorou (Academic, New York, 1984), Vol 1.

2) L. Vuskovic, S.K. Srivastava and S. Trajmar, J. Phys. **B11**, 1643 (1978).

3) R. Bijker, R.D. Amado and D.A. Sparrow, Phys. Rev. **A33**, 871 (1986).

4) F. Iachello, Chem. Phys. Lett. **78**, 581 (1981);
F. Iachello and R.D. Levine, J. Chem. Phys. **77**, 3046 (1982).

5) See e.g. R.J. Glauber, in "Lectures in Theoretical Physics", edited by W.E. Brittin and L.G. Bunham (Interscience, New York, 1959).

6) R.D. Amado, J.A. McNeil and D.A. Sparrow, Phys. Rev. **C25**, 13 (1982);
G. Wenes, J.N. Ginocchio, A.E.L. Dieperink and B. van der Cammen, Nucl. Phys. **A459**, 631 (1986);
J.N. Ginocchio, G. Wenes, R.D. Amado, D.C. Cook, N.M. Hintz and M.M. Gazzaly, Phys. Rev. **C36**, 2436 (1988).

7) G. Wenes, N. Yoshinaga and A.E.L. Dieperink, Nucl. Phys. **A443**, 472 (1985).

8) R. Bijker and R.D. Amado, Phys. Rev. **A34**, 71 (1986).

9) R. Bijker and R.D. Amado, Phys. Rev. **A37**, 1425 (1988).

10) A. Arima and F. Iachello, Phys. Rev. Lett. **35**, 1069 (1975);
A. Arima and F. Iachello, Ann. Phys. (N.Y.) **99**, 253 (1976); **111**, 201 (1978); **123**, 468 (1979).

11) O.S. van Roosmalen, A.E.L. Dieperink and F. Iachello, Chem. Phys. Lett. **85**, 32 (1982);
O.S. van Roosmalen, F. Iachello, R.D. Levine and A.E.L. Dieperink, J. Chem. Phys. **79**, 2515 (1983).

12) J. Hornos and F. Iachello, J. Chem. Phys. **90**, 5284 (1989).

13) A. Frank, R. Lemus and F. Iachello, J. Chem. Phys. **91**, 29 (1989).

14) S. Levit and U. Smilansky, Nucl. Phys. **A389**, 56 (1982).

15) F. Iachello, A. Leviatan and A. Mengoni, Yale preprint YCTP-N3-90.

16) L.C. Biedenharn, J. Math. Phys. **2**, 433 (1961).

17) O. Ashihara, I. Shimamura and K. Takayanagi, J. Phys. Soc. Jpn. **38**, 1732 (1975).

18) B.G. Wybourne, "Classical Groups For Physicists" (Wiley ,New York, 1974).

19) K. Srivastava, H. Tanaka and A. Chutjian, J. Chem. Phys. **69**, 1493 (1978);
S. Trajmar, D.F. Register and A. Chutjian, Phys. Rep. **97**, 219 (1983).

20) A. Jain and D.W. Norcross, Phys. Rev. **A32**, 134 (1985).

21) A. Mengoni and T. Shirai, J. Phys. **B21**, L567 (1988).

22) R. Bijker, R.D. Amado and L.A. Collins, submitted to Phys. Rev. **A**.

23) Y. Itikawa, J. Phys. Soc. Japan **27**, 444 (1969).

24) O.H. Crawford and A. Dalgarno, J. Phys. **B4**, 494 (1971).

25) J.M. Blatt and L.C. Biedenharn, Rev. Mod. Phys. **24**, 258 (1952).

26) A.M. Arthurs and A. Dalgarno, Proc. R. Soc. **A256**, 540 (1960).

27) M.J. Seaton, Proc. Phys. Soc. **77**, 174 (1961); **89**, 469 (1966);
R.D. Levine, J. Phys. B **2**, 839 (1969).

SYMMETRY, CONSTITUTIVE LAWS OF BOUNDED SMOOTHLY DEFORMABLE MEDIA AND NEUMANN PROBLEMS

E. Binz

Lehrstuhl für Mathematik I
Universitat Mannheim
6800 Mannheim
Germany

Introduction

The purpose of these notes is to show that symmetry, in particular translational symmetry, is the base on which two notions in the theory of elasticity, associated with deformable bodies in $\mathbb{R}^n$, namely constitutive laws and a certain class of force densities, are in a one to one correspondence. This correspondence is given by a Neumann problem converting force densities into constitutive maps which in turn characterize constitutive laws.

As we will see below any constitutive law F – formulated in our fashion – yields a stress tensor, similar as in the usual treatment of elasticity (cf.[L,L]). This tensor needs not to be symmetric, however. A criterion for the symmetry is given for constitutive laws invariant under all rotations of $\mathbb{R}^n$. A relation between these two notions of stress tensors is formulated below.

To describe in short what we mean by a constitutive law, we begin by looking at a moving deformable bounded body in $\mathbb{R}^n$. The material should constitute of a deformable medium. The medium forming the boundary may differ from the one forming the inside of the body.

Let us first lay out some elements of the geometric setting. We make the geometric assumption that at any time the body is a n–dimensional, compact, connected,

---

Dedicated to Werner Greub on the occasion of his 65th *birthday.*

oriented and smooth manifold with (oriented) boundary. The boundary needs not to be connected. During the motion of the body the diffeomorphism type of the manifold with boundary is assumed to be fixed. These assumptions allow us to think of a standard body M, which from a geometrical point of view is a manifold diffeomorphic to the one moving and deforming in $\mathbb{R}^n$. Consequently a configuration is a smooth embedding from M into $\mathbb{R}^n$. The configuration space is hence the collection $E(M,\mathbb{R}^n)$ of all smooth embeddings of M into $\mathbb{R}^n$.This set equipped with Whitney's C^∞–topology is a Fréchet manifold (cf.[Bi,Sn,Fi]). A smooth motion of the body in $\mathbb{R}^n$ therefore is described by a smooth curve in $E(M,\mathbb{R}^n)$. The calculus on Fréchet manifolds used in the sequel is the one presented in [Bi,Sn,Fi], which in our setting coincides with the one developed in [Fr,Kr].

The physical qualities of the deforming medium enter certainly the work $F(J)(L)$ needed to deform (infinitesimally) the material at any configuration $J \in E(M,\mathbb{R}^n)$ in any direction L. The directions are tangent vectors to $E(M,\mathbb{R}^n)$. Since the ladder space is open in the Fréchet space $C^\infty(M,\mathbb{R}^n)$, consisting of all smooth $\mathbb{R}^n$–valued functions and being endowed with the C^∞–topology (cf.[Hi]), a tangent vector is thus nothing else but a function in $C^\infty(M,\mathbb{R}^n)$ and vice versa.

In the following we take F, which is a one–form on $E(M,\mathbb{R}^n)$, as the basic entity of our notion of a constitutive law. Throughout these notes F is assumed to be smooth. We do not discuss the question as to whether F characterizes the physical properties of the material fully or not.

To allow only internal physical properties of the material to enter F , we have to specify the notion of a constitutive law somewhat further. Basic to this specification is the fact that these sorts of constitutive properties should not be affected by the particular location of the body in $\mathbb{R}^n$. Thus F has to be invariant under the operation of the translation group $\mathbb{R}^n$ of the real vector space $\mathbb{R}^n$. Moreover we require $F(J)(L) = 0$, for any constant map L, and any $J \in E(M,\mathbb{R}^n)$ also.

The forms F, which have these two properties, can be regarded as one–forms on $\{dJ \mid J \in E(M,\mathbb{R}^n)\}$, where dJ is the differential of J. This set of differentials is equipped with the C^∞–topology as well and is denoted by $E(M,\mathbb{R}^n)/_{\mathbb{R}^n}$. The latter space is a Fréchet manifold, too. It admits a natural metric $G_{\mathbb{R}^n}$ of an L_2–type, which is closely related to the classical Dirichlet integral.

A smooth one–form on $E(M,\mathbb{R}^n)/_{\mathbb{R}^n}$ will be denoted by $F_{\mathbb{R}^n}$. Hence we deal with one–forms of the type $F = d^* F_{\mathbb{R}^n}$ which are supposed to describe the work done under any distortion. To handle such a one–form F we assume that $F_{\mathbb{R}^n}$ can be represented via an integral kernel with respect to the metric $G_{\mathbb{R}^n}$ mentioned above.

A one–form F being of the form $F=d^* F_{\mathbb{R}^n}$ and admitting an integral kernel is called a constitutive law.

It turns out that any constitutive law F is determined by some smooth map $\mathfrak{H} \in C^\infty(E(M,\mathbb{R}^n)/_{\mathbb{R}^n}, C^\infty(M,\mathbb{R}^n))$, called a constitutive map.

Hence in our setting we characterize the medium as far as the internal physical properties are encodable in the function $\mathfrak{H}$. This constitutive function $\mathfrak{H}$ determines at any $dJ \in E(M,\mathbb{R}^n)/_{\mathbb{R}^n}$ two smooth force densities $\Phi(dJ)$ and $\varphi(dJ)$ linked to $\mathfrak{H}(dJ)$ by the following system of equations:

$$\Delta(J)\mathfrak{H}(dJ) = \Phi(dJ)$$

and

$$d\mathfrak{H}(dJ)(N) = \varphi(dJ) .$$

Here $\Delta(J)$ is the Laplacian determined by the pull back under J of a fixed scalar product on $\mathbb{R}^n$. N is the J–dependent positively oriented unite normal of ∂M in M. The integrability condition, necessary to solve this Neumann problem of which the force densities are given and the function $\mathfrak{H}$ is the unknown, is equivalent with the requirement that

$$F(J)(z)=0 \qquad \forall\, J \in E(M,\mathbb{R}^n) \quad \text{and} \quad \forall\, z \in \mathbb{R}^n .$$

The constitutive function $\mathfrak{H}$ determines a stress tensor T given by

$$T(J)(X,Y) := \langle d\mathfrak{H}(J)X, dJY \rangle \qquad \forall\, J \in E(M,N) ,$$

where X,Y vary among all smooth vector fields on M . Vice versa any two–tensor yields a constitutive map via the force densities mentioned above.

Since F is also affected by the material forming the boundary, we treat in an analogous way the boundary material and exhibit in analogy to $\mathfrak{H}$ a characteristic constitutive map $\mathfrak{h}$. Thus $\Delta(j)\mathfrak{h}(dJ)$ with $j := J|\partial M$ and $J \in E(M,\mathbb{R}^n)$ describes the force density $\tilde{\varphi}(dJ)$ up to a constant force density along ∂M. However $d\mathfrak{H}(dJ)(N)$ also determines force densities which need not to be of the form $\tilde{\varphi}(dJ)$. This observation allows us to decode the influence of the whole body on the physical quality of the boundary material.

In section 8 we show that both $\mathfrak{H}$ and $\mathfrak{h}$ are structured in the following sense :
In $\mathfrak{H}$ and in $\mathfrak{h}$ is, generically and naturally encoded the work needed to deform volume, area and shape of the body and the boundary respectively. The shape is partly expressed by the unite normal vector field N(j) along the embedding j of the boundary. The procedure to decode these influences mentioned is to use an L_2–splitting of $d\mathfrak{h}(dJ)$.

Finally we discuss the symmetry of the stress tensor, which is based on the action of the rotation group of $\mathbb{R}^n$ on the configuration space..

1. Configuration and phase space, geometric preliminaries

Let us think of a deformable material body moving and deforming in the Euclidean space $\mathbb{R}^n$. We make the geometric assumption that at any time the body maintains the shape of a n–dimensional, compact, connected, oriented and smooth manifold with (oriented) boundary. The boundary shall not necessarily be connected. The physical qualities of the medium forming the boundary may differ from the ones forming the inside of the body.

Fundamental for our investigation is the assumption that during the motion of the body the diffeomorphism type of the manifold with boundary is fixed.

Hence we can think of a standard material body M of which the underlying manifold is diffeomorphic to the body in $\mathbb{R}^n$.
The standard body constitutes of a deformable medium and we use M to denote both, the manifold with boundary and the material body.

From this situation we read off what we mean by a configuration : A configuration is a smooth embedding

$$J : M \longrightarrow \mathbb{R}^n .$$

Hence the space of configurations is $E(M,\mathbb{R}^n)$, the collection of all smooth embeddings of M into $\mathbb{R}^n$. Endowed with the C^∞–topology the configuration space is a Fréchet manifold (cf.[Bi,Sn,Fi] or [Hi]).

Clearly each $J \in E(M,\mathbb{R}^n)$ induces a smooth embedding

$$J|\partial M : \partial M \to \mathbb{R}^n ,$$

a configuration of the boundary ∂M of the body. Let us denote the collection of all smooth embeddings of ∂M into $\mathbb{R}^n$ by $E(\partial M,\mathbb{R}^n)$. The latter space endowed with the C^∞–topology is also a Fréchet manifold.

Next let us determine the phase space. The set $E(M,\mathbb{R}^n)$ is obviously a subset of $C^\infty(M,\mathbb{R}^n)$, the collection of all smooth $\mathbb{R}^n$–valued maps of M. We equip it with the C^∞–topology, too. Since M is compact, $C^\infty(M,\mathbb{R}^n)$ is a complete metrizable locally convex space, a so–called Fréchet space.
The phase space is therefore

$$TE(M,\mathbb{R}^n) = E(M,\mathbb{R}^n) \times C^\infty(M,\mathbb{R}^n) .$$

Proceeding for ∂M as for M we obtain $E(\partial M,\mathbb{R}^n)$ as an open subset of the Fréchet space $C^\infty(\partial M,\mathbb{R}^n)$ (cf.[Hi]). Its tangent bundle is obviously trivial. The phase space for the boundary is hence $E(\partial M,\mathbb{R}^n) \times C^\infty(\partial M,\mathbb{R}^n)$.

In the sequel of these notes we write O_∂ instead of $\{J|\partial M \mid J \subset E(M,\mathbb{R}^n)\}$. The map assigning to any $J \in E(M,\mathbb{R}^n)$ its restriction $J|\partial M$ is called R.

On the configuration space we have two natural actions namely

$$t : E(M,\mathbb{R}^n) \times \mathbb{R}^n \to E(M,\mathbb{R}^n)$$

and

$$s : SO(n) \times E(M,\mathbb{R}^n) \to E(M,\mathbb{R}^n)$$

assigning to each $J \in E(M,\mathbb{R}^n)$ and each $z \in \mathbb{R}^n$ the embedding $J + z$ and $g \circ J$ respectively for each $g \in SO(n)$ and each J. These actions reflect the translational and the rotational symmetry of $E(M,\mathbb{R}^n)$ respectively.
t and s extend obviously to $C^\infty(M,\mathbb{R}^n)$. The groups $\mathbb{R}^n$ and SO(n) act accordingly on $E(\partial M,\mathbb{R}^n)$. These actions restrict to O_∂ and obviously both extend also to $C^\infty(\partial M,\mathbb{R}^n)$.

The orbit spaces of the respective actions of the translation group $\mathbb{R}^n$ are denoted by $C^\infty(M,\mathbb{R}^n)/_{\mathbb{R}^n}$, $C^\infty(\partial M,\mathbb{R}^n)/_{\mathbb{R}^n}$, $E(M,\mathbb{R}^n)/_{\mathbb{R}^n}$, $E(\partial M,\mathbb{R}^n)/_{\mathbb{R}^n}$ and $O_\partial/_{\mathbb{R}^n}$.

The nature of these spaces is easily understood if we introduce for any map $L \in C^\infty(M,\mathbb{R}^n)$ the differential dL which locally represented is given by the Fréchet derivative. The tangent map TL of L is (L,dL). The respective notion of $l \in C^\infty(\partial M,\mathbb{R}^n)$ is introduced accordingly.

Hence the orbit spaces mentioned above are nothing else but spaces of differentials of the elements of those spaces on which $\mathbb{R}^n$ acts.

For our later investigations we observe that M and ∂M inherit via respective embeddings into $\mathbb{R}^n$ some basic geometric structures described just below. Let us fix a scalar product $<\,,\,>$ and a normed determinant function $\underline{\Delta}$ (cf.[Gr]) on $\mathbb{R}^n$, i.e. $\mathbb{R}^n$ together with $<\,,\,>$ and $\underline{\Delta}$ is a fixed oriented Euclidean space.

Each $J \in E(M,\mathbb{R}^n)$ and each $j \in E(\partial M,\mathbb{R}^n)$ yield Riemannian metrics m(J) and m(j) on M and ∂M respectively. These metrics are defined by

(1.2) $$m(J)(X,Y) := \langle dJX, dJY \rangle\,, \quad \forall\, X,Y \in \Gamma TM$$

and

(1.3) $$m(j)(X,Y) := \langle djX, djY \rangle, \quad \forall X,Y \in \Gamma T\partial M.$$

Here ΓTQ denotes the collection of all smooth vector fields of any smooth manifold Q (with or without boundary). Both m(J) and m(j) depend smoothly on its variables J and j.

Associated with the metrics m(J) and m(j), we have the respective Levi–Civita connections $\nabla(J)$ on M and $\nabla(j)$ on ∂M. They are determined by the equations

(1.4) $$dJ\, \nabla(J)_X Y = d(dJY)(X)\,, \quad \forall\, X,Y \in \Gamma TM$$

and

(1.5) $$dj\, \nabla(j)_X Y = d(djY)(X) - m(j)(W(j)X,Y)\cdot N(j)\,, \quad \forall\, X,Y \in \Gamma T\partial M.$$

By W(j) we mean the Weingarten map given by

(1.6) $$dN(j)Z = djW(j)Z\,, \quad \forall\, Z \in \Gamma T\partial M\,,$$

where N(j) is the unite normal vector field along j , for which $j^* i_{N(j)}\underline{\Delta}$ determines the orientation class of ∂M.

For any $J \in E(M,\mathbb{R}^n)$ and any $j \in E(M,\mathbb{R}^n)$ let us denote by $\mu(J)$ and $\mu(j)$ the Riemannian volume form determined by m(J) and m(j) and the orientations of M and ∂M respectively. The positively oriented unite normal vector field on ∂M is called by N . This vector field depends on m(J) !

2. Special one–forms on $E(M,\mathbb{R}^n)$ and $E(M,\mathbb{R}^n)/_{\mathbb{R}^n}$

We will characterize the type of the material of which the body M constitutes in so far, as it affects the work caused by an infinitesimal distortion of M (cf.[He], [E,S], [Bi], [Bi,Sc,So]). This idea is formalized by giving a smooth one–form on $E(M,\mathbb{R}^n)$, i.e. a smooth map

$$(2.1) \qquad F : E(M,\mathbb{R}^n) \times C^\infty(M,\mathbb{R}^n) \longrightarrow \mathbb{R} ,$$

which varies linearly in the second argument. We interpret $F(J)(L)$ as the work done if M is distorted by $L \in C^\infty(M,\mathbb{R}^n)$ at the configuration $J \in E(M,\mathbb{R}^n)$. We call the medium described by F a smoothly deformable medium.

Next we expose F to the translational symmetry and require that

$$(2.2) \qquad F(J + z) = F(J) , \quad \forall J \in E(M,\mathbb{R}^n) , \forall z \in \mathbb{R}^n.$$

This means that the work caused by physical processes under consideration does not depend on the particular location of $J(M)$ within $\mathbb{R}^n$. Moreover a constant distortion by any $z \in \mathbb{R}^n$ is supposed to cause no work. More formally this means that we impose the following restrictions on F basic to our further development :

$$(2.3) \qquad F(J)(z) = 0 , \quad \forall J \in E(M,\mathbb{R}^n) , \forall z \in \mathbb{R}^n .$$

One–forms F on $E(M,\mathbb{R}^n)$ satisfying (2.2) and (2.3) will be the basic ingredients of our notion of constitutive laws. To implement the possibility of extracting force densities from our basic notion of work we need a little more structure associated with our forms satisfying (2.2) and (2.3). We will do this in the next section.

But first let us state the following obvious lemma :

Lemma 2.1 :

A smooth one–form $F : E(M,\mathbb{R}^n) \times C^\infty(M,\mathbb{R}^n) \longrightarrow \mathbb{R}$ satisfies (2.2) and (2.3) iff it is of the form $F = d^* F_{\mathbb{R}^n}$, that is

$$(2.4) \quad F(J)(L) = F_{\mathbb{R}^n}(dJ)(dL) , \qquad \forall J \in E(M,\mathbb{R}^n) \text{ and } \forall L \in C^\infty(M,\mathbb{R}^n) ,$$

where

(2.5)
$$F_{\mathbb{R}^n} : E(M,\mathbb{R}^n)/_{\mathbb{R}^n} \times C^\infty(M,\mathbb{R}^n)/_{\mathbb{R}^n} \longrightarrow \mathbb{R}$$

is a smooth one–form.

$$\Delta$$

In the last section of these notes we will study the effect of the rotational symmetry on our special types of one–forms, the constitutive laws.

3. The notion of a constitutive law, the Dirichlet integral

The purpose of this section is to define what is meant by an integral representation of a one–form $F_{\mathbb{R}^n}$ on $E(M,\mathbb{R}^n)/_{\mathbb{R}^n}$ and in turn to define the notion of a constitutive law.

To obtain a tool general enough to handle the underlying mathematical structures, we first will introduce a quadric structure on the trivial bundle $E(M,\mathbb{R}^n)/_{\mathbb{R}^n} \times A^1(M,\mathbb{R}^n)$. We denote the collection of all smooth $\mathbb{R}^m$–valued one–forms of any smooth manifold Q by $A^1(Q,\mathbb{R}^m)$ and equip it with the C^∞–topology in case Q is compact.

Let $\gamma \in A^1(M,\mathbb{R}^n)$ and $J \in E(M,\mathbb{R}^n)$ be given. The two–tensor $<\gamma,dJ>$ determined by γ and J shall be given by $<\gamma X,dJY>$ for all $X,Y \in \Gamma TM$. This two–tensor $<\gamma,dJ>$ yields a unique strong bundle map $A(\gamma,dJ)$ of TM defined by

(3.1)
$$<\gamma X,dJY> = m(J)(A(\gamma,dJ)X,Y) , \quad \forall X,Y \in \Gamma TM.$$

From this equation we read off :

(3.2)
$$\gamma X = dJA(\gamma,dJ)X , \quad \forall X \in \Gamma TM .$$

For any pair of one–forms γ_1, $\gamma_2 \in A^1(M,\mathbb{R}^n)$ and any embedding $J \in E(M,\mathbb{R}^n)$ we define the dot product of γ_1 and γ_2 relative to J by

(3.3)
$$\gamma_1 \cdot \gamma_2 := \operatorname{tr} A(\gamma_1,dJ) \cdot \tilde{A}(\gamma_2,dJ) ,$$

which is a smooth map on M. Here $\tilde{A}(\gamma_2,dJ)$ is the adjoint of $A(\gamma_2,dJ)$ formed fiber–wise with respect to m(J).

Associated with this product is a quadratic structure $G_{\mathbb{R}^n}(dJ)$ on $A^1(M,\mathbb{R}^n)$ defined by

$$G_{\mathbb{R}^n}(dJ)(\gamma_1,\gamma_2) := \int_M \gamma_1 \cdot \gamma_2 \, \mu(J) \,. \tag{3.4}$$

As mentioned before $\mu(J)$ denotes the Riemannian volume form determined by $m(J)$ and the given orientation of M. The real number $G_{\mathbb{R}^n}(dJ)(\gamma_1,\gamma_2)$ depends thus smoothly on all its variables dJ, γ_1 and γ_2 .

$G_{\mathbb{R}^n}$ yields obviously a metric on $E(M,\mathbb{R}^n)/_{\mathbb{R}^n}$, again denoted by $G_{\mathbb{R}^n}$.

We say that $F_{\mathbb{R}^n}$, a one–form on $E(M,\mathbb{R}^n)/_{\mathbb{R}^n}$, admits an integral representation if there exists a smooth map

$$\alpha : E(M,\mathbb{R}^n) \longrightarrow A^1(M,\mathbb{R}^n) \,,$$

called the kernel of $F_{\mathbb{R}^n}$, such that for any choices of $dJ \in E(M,\mathbb{R}^n)/_{\mathbb{R}^n}$ and $dL \in C^\infty(M,\mathbb{R}^n)/_{\mathbb{R}^n}$

$$F_{\mathbb{R}^n}(dJ)(dL) = \int_M \alpha(dJ) \cdot dL \, \mu(J) = G_{\mathbb{R}^n}(dJ)(\alpha(J),dL) \tag{3.5}$$

holds true.

We speak of a constitutive law F , if $F = d^* F_{\mathbb{R}^n}$ and if in addition $F_{\mathbb{R}^n}$ admits an integral representation with kernel α .

To discuss the uniqueness of the kernel α, associated with a constitutive law, we first prove the following :

Theorem 3.1 :

Let $\gamma \in A^1(M,\mathbb{R}^n)$ and $J \in E(M,\mathbb{R}^n)$. There exists a uniquely determined differential $d\mathfrak{H} \in C^\infty(M,\mathbb{R}^n)/_{\mathbb{R}^n}$ called the exact part of γ and a uniquely determined $\beta \in A^1(M,\mathbb{R}^n)$ such that

$$(3.6) \qquad \gamma = d\mathfrak{H} + \beta ,$$

where the exact part of β vanishes. Both $d\mathfrak{H}$ and β depend smoothly on J. If $\mathfrak{H}(p_0)$ for some $p_0 \in M$ is kept constant in J, then also $\mathfrak{H}$ varies smoothly in J.

Δ

Proof :

We will exhibit the proof to some details, in order to reveal the surrounding structure, which will allow us to handle the integral representation introduced in (3.5). First let us construct $\mathfrak{H}$ and β. To this end we fix a basis $e_1,\dots,e_n$ on $\mathbb{R}^n$, orthonormal with respect to $< , >$. With these data we may write

$$(3.7) \qquad \gamma(X) = \sum_{r=1}^{n} \gamma^r(X)\, e_r , \quad \forall X \in \Gamma TM ,$$

with $\gamma^r \in A^1(M,\mathbb{R}^n)$ for all $r=1,\dots,n$. Since for each r

$$(3.8) \qquad \gamma^r(X) = m(J)(Y^r,X) , \quad \forall X \in \Gamma TM$$

holds true for a well defined $Y^r \in \Gamma TM$, we find due to Hodge's decomposition (cf.[A,M,R]) a function $\tau^r \in C^\infty(M,\mathbb{R}^n)$ and a uniquely determined vector field $Y^r_0 \in \Gamma TM$ such that the following three equations are satisfied :

$$(3.9) \qquad Y^r = \mathrm{grad}_J \tau^r + Y^r_0$$

and

$$(3.10) \qquad \mathrm{div}_J Y^r_0 = 0$$

together with the boundary condition

$$(3.10a) \qquad m(J)(Y^r_0,N) = 0 \ \text{ along } \partial M.$$

Here the indices J in grad_J and div_J mean that the respective operations are formed with respect to m(J). The Hodge decomposition mentioned above is obtained by solving the following Neumann problem

$$(3.11) \qquad \Delta(J)\tau^r = -\,\mathrm{div}_J Y^r$$

with the boundary condition

(3.12) $$d\tau^r(N) = m(J)(Y^r,N).$$

This problem has, according to [Hö], a solution τ^r unique up to a constant.

The desired function $\mathfrak{H}$ and the form β are defined by

(3.13) $$\mathfrak{H} := \sum_r \tau^r \cdot e_r$$

and

(3.14) $$\beta(X) := \sum_r m(J)(Y_0^r,X), \quad \forall X \in \Gamma TM,$$

respectively. It is a matter of routine to show that $d\mathfrak{H}$ and β do not depend on the basis chosen. A somewhat more involved matter is to show the smoothness properties. With these notions we immediately deduce (3.6). One easily verifies that the exact part of β vanishes.

□

Some of the calculation made in the proof above allow us to look at $G_{\mathbb{R}^n}(dJ)$ from another angle. Given $\gamma \in A^1(M,\mathbb{R}^n)$ and $J \in E(M,\mathbb{R}^n)$ we have according to (3.7) and (3.8) in the proof above the following equation :

(3.15) $$\gamma(X) = dJA(\gamma,dJ)X = \sum_{r=1}^{n} m(J)(Y^r,X)\, e_r, \quad \forall X \in \Gamma TM.$$

Let us denote $(dJ)^{-1}e_r$ by E_r, for all $r=1,\dots,n$. Then we read off from equation (3.15) that

(3.16) $$Y^r = \tilde{A}(\gamma,dJ)E_r, \quad \forall r=1,\dots,n,$$

holds true. This remark yields the following observation :

Proposition 3.2 :

Given $\gamma_1,\gamma_2 \in A^1(M,\mathbb{R}^n)$, $J \in E(M,\mathbb{R}^n)$ and a fixed basis $e_1,\dots,e_n$ on $\mathbb{R}^n$ orthonormal with respect to $\langle\, ,\, \rangle$, then there exist two sets

$$Y_1^1,\dots,Y_1^n \quad \text{and} \quad Y_2^1,\dots,Y_2^n$$

of vector fields in ΓTM, such that

$$(3.17)\qquad \gamma_1\cdot\gamma_2 = \sum_{r=1}^{n} m(J)(Y_1^r, Y_2^r)$$

and hence

$$(3.18)\qquad G_{\mathbb{R}^n}(dJ)(\gamma_1,\gamma_2) = \int_M \gamma_1\cdot\gamma_2\,\mu(J) = \sum_{r=1}^{n}\int_M m(J)(Y_1^r, Y_2^r)\mu(J)\,.$$

If in addition $\gamma_1 = d\mathfrak{H}$ for some $\mathfrak{H}\in C^\infty(M,\mathbb{R}^n)$ then $G_{\mathbb{R}^n}(dJ)(d\mathfrak{H},\gamma_2) = 0$ provided that the exact part of γ_2 vanishes.

Δ

Proof:

Let $Y_i^r \in \Gamma TM$ for r=1,...,n and i=1,2 be as in (3.8). Then

$$(3.19)\qquad \gamma_1\cdot\gamma_2 = \operatorname{tr} A(\gamma_1,dJ)\cdot\tilde{A}(\gamma_2,dJ) = \sum_{r=1}^{n} m(J)(A(\gamma_1,dJ)\cdot\tilde{A}(\gamma_2,dJ)E_r,E_r)$$
$$= \sum_{r=1}^{n} m(J)(Y_1^r, Y_2^r)$$

establishing (3.17). To show the last part of the proposition we use Gauss' theorem under the three assumptions that $\mathfrak{H} = \sum_{r=1}^{n}\tau^r e_r$ and $\operatorname{div}_J Y_0^r = 0$ as well as $m(J)(Y_0^r,N) = 0$ and obtain

$$(3.20)\qquad G_{\mathbb{R}^n}(dJ)(\gamma_1,\gamma_2) = \sum_{r=1}^{n}\int_M m(J)(\operatorname{grad}_J\tau^r, Y_0^r)$$
$$= \sum_{r=1}^{n}\int_M d\tau^r(Y_0^r)\mu(J)$$
$$= \sum_{r=1}^{n}\int_M (\operatorname{div}_J(\tau^r\cdot Y_0^r) - \tau\cdot\operatorname{div}_J Y_0^r)\mu(J)$$
$$= \sum_{r=1}^{n}\int_M m(J)(\tau^r\cdot Y_0^r, N)\mu(J)$$
$$= 0\,.$$

$\square$

In case γ_1 and γ_2 in the above proposition are exact, then the respective vector fields in (3.18) are gradients. Hence the right hand side of the integral in (3.18) is the classical Dirichlet integral (cf.[J]) for $\mathbb{R}^n$–valued functions. We call the middle integral in (3.18), therefore the Dirichlet integral of any two smooth $\mathbb{R}^n$–valued forms γ_1, γ_2 relative to $J\in E(M,\mathbb{R}^n)$.

Theorem 3.1 also shows that the integral kernel of a constitutive law is not unique at all. We may add to any kernel a map which assumes as its values, one–forms of which the exact part vanishes. However the following theorem guaranties us a uniqueness of a very specific type of kernel. The proof of the following theorem is a matter of routine :

Theorem 3.3 :

Let F be a constitutive law with integral kernel. There exists a unique smooth map

(3.21) $$\alpha : E(M,\mathbb{R}^n)/_{\mathbb{R}^n} \rightarrow C^\infty(M,\mathbb{R}^n)/_{\mathbb{R}^n} \subset A^1(M,\mathbb{R}^n)\,,$$

such that for any $J \in E(M,\mathbb{R}^n)$ and any $L \in C^\infty(M,\mathbb{R}^n)$

(3.22) $$F(J)(L) = \int_M \alpha(dJ)\cdot dL\; \mu(J) = G_{\mathbb{R}^n}(dJ)(\alpha(dJ),dL)$$

holds true. In fact there is a unique smooth map

$$\mathfrak{H} : E(M,\mathbb{R}^n)/_{\mathbb{R}^n} \rightarrow C^\infty(M,\mathbb{R}^n)\,,$$

satisfying the following two equations

(3.23) $$\alpha(dJ) = d\mathfrak{H}(dJ)\,, \quad \forall\, dJ \in E(M,\mathbb{R}^n)/_{\mathbb{R}^n}$$

and

(3.24) $$\int_{\partial M} <\mathfrak{H}(dJ),z>\mu(J) = 0\,, \quad \forall\, z \in \mathbb{R}^n.$$

Δ

Equation (3.22) together with (3.23) shows that $d\mathfrak{H}$ is the generalized gradient of $F_{\mathbb{R}^n}$ with respect to $G_{\mathbb{R}^n}$. However F needs not to be exact. It is this type of gradient mentioned above which characterizes the material.

Clearly each constitutive law F forms via α and equation (3.1) a tensor on M given by

(3.25) $$\begin{aligned} T(J)(X,Y) &:= <\alpha(dJ)X, dJY> \\ &= m(J)(dJA(\alpha(dJ),dJX),Y)\,, \quad \forall\, X,Y \in \Gamma TM\,. \end{aligned}$$

We call T(J) the stress tensor at the configuration J. T(J) splits into a symmetric and a skew symmetric part (and hence is in general not identical with the one introduced in [L,L] !). We will investigate at the end of these notes under what conditions on $d\mathfrak{H}$ the skew symmetric part vanishes.

To see that the setting in [L,L] is included in the treatment presented here, we assume that the work F(J)(L) depends on the metric m(J) and the derivative Dm(J)(L) of m at J in the direction of L only rather than J and L . Moreover let us suppose that F admits an integral representation of the form

$$F(J)(L) = \int_M \mathfrak{T}(J)\cdot Dm(J)(L)\, \mu(J)\,,$$

with a smooth (symmetric) two–tensor $\mathfrak{T}(J)$ as kernel, for all variables of F. The dot–product in the integrand is defined by representing both $\mathfrak{T}(J)$ and Dm(J)(L) as strong bundle maps $K_1(J)$ and $K_2(J)(L)$ of TM via the metric m(J) and then proceeding as in (3.3), i.e.

$$\mathfrak{T}(J)\cdot Dm(J)(L)\, \mu(J) := \operatorname{tr} K_1(J)\cdot K_2(J)(L)\,.$$

Let us observe that $K_2(J)(L) = 2\cdot B(dL,dJ)$, where B(dL,dJ) is the self–adjoint part of A(dL,dJ). Using (3.6), the verification of the following is straight forward :

$$\begin{aligned}
F(J)(L) &= \int_M \operatorname{tr} K_1(J)\cdot K_2(J)(L)\, \mu(J) \\
&= 2\cdot \int_M \operatorname{tr} K_1(J)\cdot A(dL,dJ)\, \mu(J) \\
&= 2\cdot \int_M \alpha(dJ)\cdot dL\, \mu(J) \\
&= 2\cdot \int_M d\mathfrak{H}(dJ)\cdot dL\, \mu(J)\,,
\end{aligned}$$

with $\alpha(dJ) := dJK_1(J)$, holding for all the variables of F. In the setting of [L,L] the above introduced tensors $\mathfrak{T}(J)$ and $\frac{1}{2}\cdot Dm(J)(L)$ correspond respectively to the stress tensor and to the deformation tensor. The above calculation shows that to each $\mathfrak{T}(J)$ in the setting of [L,L] there is a map $\mathfrak{H}(dJ)$.

4. Force densities associated with constitutive laws

The purpose of this section is to associate with any constitutive law at any configuration some well defined force densities, one acting upon the whole body, and another one acting upon the boundary only. Vice versa any given pair of such force densities satisfying an integrability condition will be obtained via a suitable constitutive law.

Throughout this section F is a constitutive law ; its kernel is called α. By the previous theorem we may assume that $\alpha(E(M,\mathbb{R}^n)/_{\mathbb{R}^n}) \subset C^\infty(M,\mathbb{R}^n)/_{\mathbb{R}^n}$.

To construct the force densities mentioned we use F in the form

$$(4.1) \qquad F(J)(L) = \int_M \operatorname{tr} A(\alpha(dJ),dJ)\cdot\tilde{A}(dL,dJ)\,\mu(J)\,,$$

holding for all variables of F. Representing any $L \in C^\infty(M,\mathbb{R}^n)$ relative to a given $J \in E(M,\mathbb{R}^n)$ in the form

$$(4.2) \qquad L = dJ\,X(L,J)\,,$$

with a unique $X(L,J) \in \Gamma TM$, we derive

$$(4.3) \qquad dL\,Y = dJ\,\nabla(J)_Y X(L,J)\,, \qquad \forall\, Y \in \Gamma TM\,.$$

and hence obtain immediately

$$(4.4) \qquad A(dL,dJ) = \nabla(J)\,X(L,J)\,, \qquad \forall\, L \in C^\infty(M,\mathbb{R}^n)\,.$$

Thus if $e_1,\dots,e_n$ is a orthonormal basis of $\mathbb{R}^n$ for which we let $E_r \in \Gamma TM$ again be given by $dJ\,E_r = e_r$ for $r=1,\dots,n$, then (4.1) turns into

$$(4.5) \qquad F(J)(L) = \sum_{r=1}^{n} \int_M m(J)(\tilde{A}(\alpha(dJ),dJ)\cdot\nabla(J)_{E_r} X(L,J),E_r)\,\mu(J)\,.$$

Let us introduce the notion $\operatorname{div}_J T$, the divergence of a strong bundle endomorphism T of TM by

$$(4.6) \qquad \operatorname{div}_J T := \sum_{r=1}^{n} \nabla(J)_{E_r}(T)(E_r)\,.$$

This notion does not depend on the basis chosen. Equation (4.6) together with (4.5) imply

$$(4.7) \quad F(J)(L) = \int_M \operatorname{div}_J(\tilde{A}(\alpha(dJ),dJ)X(L,J))\,\mu(J) - \int_M m(J)(\operatorname{div}_J A(\alpha(dJ),dJ),X(L,J))\,\mu(J)\,.$$

To bring these formulas in a more familiar form we introduce the notions of $\Delta(J)K$ and $\Delta(J)\gamma$, the Laplacian for any $K \in C^\infty(M,\mathbb{R}^n)$ and any $\gamma \in A^1(M,\mathbb{R}^n)$. In doing so we follow [Mat]. We begin with the definition of δ by letting

$$(4.8) \qquad \delta K = 0\,.$$

If $\gamma \in A^1(M,\mathbb{R}^m)$ for some natural number m, we set

$$(4.9) \qquad \delta\gamma = -\sum_{r=1}^{n} \nabla(J)_{E_r}(\gamma)(E_r)\,.$$

Clearly

$$(4.10) \qquad \delta\gamma = -\operatorname{div}_J Y\,,$$

if

$$\gamma(X) = m(J)(Y,X)\,, \quad \forall X \in \Gamma TM.$$

$\Delta(J)$ is then defined by

$$(4.11) \qquad \Delta(J) := d\delta + \delta d\,.$$

Consequently we have

$$(4.12) \qquad \Delta(J)K = \delta dK = -\sum_{r=1}^{n} \nabla(J)_{E_r}(dK)(E_r)\,.$$

and therefore we verify the following relation :

$$(4.13) \qquad \begin{aligned} \Delta(J)K &= -\Big(\sum_{r=1}^{n} d(dJ\,A(dK,dJ\,E_r))(E_r) - dJ\,A(dK,dJ)(\nabla(J)_{E_r}E_r)\Big) \\ &= -\sum_{r=1}^{n} dJ\,\nabla(J)_{E_r}(A(dK,dJ))E_r \\ &= -dJ\,\operatorname{div}_J A(dK,dJ)\,. \end{aligned}$$

Hence equation (4.7) turns into

$$(4.14) \qquad F(J)(L) = \int_M \mathrm{div}_J(A(\alpha(dJ),dJ)X(L,J))\, \mu(J) + \int_M <\Delta(J)\mathfrak{H}(dJ),L> \mu(J)\,,$$

with $\alpha(dJ) = d\mathfrak{H}(dJ)$ for some $\mathfrak{H} \in (C^{\infty}(E(M,\mathbb{R}^n)/_{\mathbb{R}^n}, C^{\infty}(M,\mathbb{R}^n))$.

Using the theorem of Gauss, we derive with the help of theorem 3.3 the following proposition in which we adopt the convention to write L instead of $L\,|\,\partial M$ for any $L \in C^{\infty}(M,\mathbb{R}^n)$.

Proposition 4.1 :
Let F be a constitutive law. Then for each $J \in E(M,\mathbb{R}^n)$ there exists a smooth map

$$\mathfrak{H} : E(M,\mathbb{R}^n) \longrightarrow C^{\infty}(M,\mathbb{R}^n)$$

uniquely determined up to a smooth map from $E(M,\mathbb{R}^n)$ into $\mathbb{R}^n$ for which

$$(4.15) \qquad F(J)(L) = \int_M <\Delta(J)\mathfrak{H}(dJ),L> \mu(J) + \int_{\partial M} <d\mathfrak{H}(dJ)(N),L> i_N\, \mu(J)$$

and in turn Green's equation

$$(4.16) \qquad \int_M <\Delta(J)\mathfrak{H}(dJ),L> \mu(J) - \int_M <\mathfrak{H}(dJ),\Delta(J)L> \mu(J) = \int_{\partial M} <dL(N),\mathfrak{H}(dJ)> i_N\, \mu(J) - \int_{\partial M} <d\mathfrak{H}(dJ)(N),L> i_N\, \mu(J)$$

are valued for all variables of F. Here $i_N\, \mu(J)$ is the volume element on ∂M defined by $\mu(J)$ and N, the positively oriented unite normal vector field along $\partial M \subset M$.

Δ

The developments made so far show that any characteristic property of the material formulated via our notion of a constitutive law is encoded in the map $\mathfrak{H}$. We call this map $\mathfrak{H}$ therefore a constitutive map.

The above proposition motivates us to set for any $J \in E(M,\mathbb{R}^n)$

$$(4.17)\qquad \Phi(J) := \Delta(J)\, \mathfrak{H}(dJ)$$

and

$$(4.18)\qquad \varphi(dJ) := d\mathfrak{H}(dJ)(N)\,,$$

with $\mathfrak{H}(dJ)$ as in (4.15).
We call the maps Φ and φ the force densities associated with F.

To prove the following theorem is now a matter of routine again.

***Theorem 4.2*:**

Every constitutive law $F \in A^1(E(M,\mathbb{R}^n),\mathbb{R})$ admits a smooth constitutive map

$$(4.19)\qquad \mathfrak{H} : E(M,\mathbb{R}^n)/_{\mathbb{R}^n} \to C^\infty(M,\mathbb{R}^n)\,,$$

uniquely determined up to maps in $C^\infty(E(M,\mathbb{R}^n)/_{\mathbb{R}^n},\mathbb{R}^n)$, such that F can be expressed by

$$(4.20)\quad F(J)(L) = \int_M <\Delta(J)\mathfrak{H}(dJ),L>\mu(J) + \int_{\partial M} <d\mathfrak{H}(dJ)(N),L> i_N\mu(J)\,,$$

for each $J \in E(M,\mathbb{R}^n)$ and each $L \in C^\infty(M,\mathbb{R}^n)$. For all $J \in E(M,\mathbb{R}^n)$ the map $\mathfrak{H}$ defines the force densities

$$(4.21)\qquad \Phi(dJ) = \Delta(J)\, \mathfrak{H}(dJ)$$

and

$$(4.22)\qquad \varphi(dJ) = d\mathfrak{H}(dJ)(N)\,,$$

with the following property :

$$(4.23)\qquad 0 = \int_M <\Phi(dJ),z>\mu(J) + \int_{\partial M} <\varphi(dJ),z>\, i_N\mu(J)\,,\ \ \forall z \in \mathbb{R}^n\,.$$

Given vice versa two smooth maps

$$(4.24)\qquad \Phi : E(M,\mathbb{R}^n)/_{\mathbb{R}^n} \to C^\infty(M,\mathbb{R}^n)$$

and

$$(4.25)\qquad \varphi : E(M,\mathbb{R}^n)/_{\mathbb{R}^n} \to C^\infty(\partial M,\mathbb{R}^n)$$

for which the integrability condition (4.23) holds. Then there exists a smooth map

$$\mathfrak{H} : E(M,\mathbb{R}^n)/_{\mathbb{R}^n} \rightarrow C^\infty(M,\mathbb{R}^n)$$

such that $\mathfrak{H}(dJ)$ satisfies (4.21) and (4.22) and is uniquely determined up to a constant for each $J \in E(M,\mathbb{R}^n)$. Moreover $\mathfrak{H}$ is a constitutive map for the constitutive law F given by the force densities via the formula

$$(4.26) \qquad F(J)(L) = \int_M <\Phi(dJ),L>\mu(J) + \int_{\partial M} <\varphi(dJ),L>i_N\mu(J)\,,$$

holding for all $J \in E(M,\mathbb{R}^n)$ and for all $L \in C^\infty(M,\mathbb{R}^n)$.

Δ

Remark 4.3

This theorem shows that our notion of a constitutive law, based on translational invariance, is equivalent with an $\mathbb{R}^n$–valued solution of a Neumann problem formulated on M. Hence it is equivalent with a pair of force densities satisfying the integrability condition (4.23), which is obviously the analogon of (2.3).

Moreover, if for each $J \in E(M,\mathbb{R}^n)$ the one–form $\alpha(dJ)$ splits as

$$\alpha(dJ) = d\mathfrak{H}(dJ) + \beta(dJ)$$

with $d\mathfrak{H}(dJ)$ the exact part of $\alpha(dJ)$, then both $\alpha(dJ)$ and $d\mathfrak{H}(dJ)$ determine the same force densities !

5. Constitutive laws for the boundary

The task in this section is to study constitutive laws for the boundary, that is for a deformable medium forming a skin of which the underlying point set is the (n–1)–dimensional manifold ∂M. This skin, formed by a deformable material, will be studied on its own and is not regarded as boundary of some body. In doing so, we first formulate in analogy to sections two and three what is meant by a constitutive law for a skin. Again the translational invariance and an appropriate integral representation of one–forms describing the work of the material subjected to distortions are the essential tools.

Let us recall that the open set $O_\partial \subset E(\partial M,\mathbb{R}^n)$ is the collection of all $J|\partial M$ with $J \in E(M,\mathbb{R}^n)$. The constitutive laws to introduce will be formulated on any open set $O \subset E(\partial M,\mathbb{R}^n)$.

At the very first we introduce the notion corresponding to the Dirichlet integral :
Given any $l \in C^\infty(\partial M,\mathbb{R}^n)$ and any $j \in E(\partial M,\mathbb{R}^n)$, then for all $X,Y \in \Gamma T\partial M$

$$(5.1) \qquad \langle dl\, X, dj\, Y\rangle = m(j)(A(dl,dj)X,Y)$$

holds for some smooth strong bundle endomorphism A(dl,dj) of $T\partial M$. Moreover, there is a uniquely defined smooth map

$$(5.2) \qquad c(dl,dj) : \partial M \longrightarrow so(n)\,,$$

with so(n) the linear space of all skew maps of $(\mathbb{R}^n, \langle\ ,\ \rangle)$, satisfying the following two conditions

$$(5.3) \qquad c(dl,dj)dj(T_p\partial M) \subset \mathbb{R}\cdot N(j)(p)\,, \qquad \forall\, p \in \partial M$$

and

$$(5.4) \qquad c(dl,dj)N(j)(p) \subset djT_p\partial M\,, \qquad \forall\, p \in \partial M,$$

and is specified such that the equation

$$(5.5) \qquad dl\, X = c(dj,dl)dj\, X + dj\, A(dl,dj)X$$

holds true for any $X \in \Gamma TM$. We refer to [Bi,Sn,Fi] or [Bi,Sc,So] for more details. Based on (5.4) we introduce $U(dl,dj) \in \Gamma T\partial M$ by

$$(5.6) \qquad c(dl,dj)N(j) = dj\, U(dl,dj)\,.$$

This vector field U(dl,dj) is obviously uniquely determined.

Splitting A(dl,dj) into its skew–, respectively selfadjoint parts C(dl,dj) and B(dl,dj) formed pointwise with respect to m(j), we end up with

$$(5.7) \qquad dl = c(dl,dj)\cdot dj + dj(C(dl,dj) + B(dl,dj))\,.$$

This decomposition relative to any $dj \in E(\partial M,\mathbb{R}^n)$ generalizes in the obvious way to any $\gamma \in A^1(\partial M,\mathbb{R}^n)$ and in this case reads as

$$(5.7a) \qquad \gamma = c(\gamma,dj)\cdot dj + dj(C(\gamma,dj) + B(\gamma,dj))\,.$$

The quadratic structure $G^{\partial}_{\mathbb{R}^n}(dj)$ at $dj \in E(\partial M,\mathbb{R}^n)/_{\mathbb{R}^n}$ applied to any two $\gamma_1, \gamma_2 \in A^1(\partial M,\mathbb{R}^n)$ is defined by integrating the function

$$(5.8)\qquad \gamma_1\cdot\gamma_2 := -\frac{1}{2}\,\mathrm{tr}\; c(\gamma_1,dj)\cdot c(\gamma_2,dj) - \mathrm{tr}\; C(\gamma_1,dj)\cdot C(\gamma_2,dj) + \mathrm{tr}\; B(\gamma_1,dj)\cdot B(\gamma_2,dj)$$

with respect to $\mu(j)$, that is it is defined by

$$(5.9)\qquad G^{\partial}_{\mathbb{R}^n}(dj)(\gamma_1,\gamma_2) := \int_M \gamma_1\cdot\gamma_2\;\mu(j) = -\frac{1}{2}\int_M \mathrm{tr}\; c(\gamma_1,dj)\cdot c(\gamma_2,dj)\mu(j) - \int_M \mathrm{tr}\; C(\gamma_1,dj)\cdot C(\gamma_2,dj)\mu(j) + \int_M \mathrm{tr}\; B(\gamma_1,dj)\cdot B(\gamma_2,dj)\mu(j)\,.$$

Obviously $G^{\partial}_{\mathbb{R}^n}$ yields a metric on $E(M,\mathbb{R}^n)$. Let $O \subset E(\partial M,\mathbb{R}^n)$ be any open set.

We now define a constitutive law F_∂ on O in analogy to section two, that is we require

$$(5.10)\qquad F_\partial = d^* F^{\partial}_{\mathbb{R}^n}\,,$$

to hold for some one–form $F^{\partial}_{\mathbb{R}^n}$ on $O/_{\mathbb{R}^n}$ and demand furthermore that for some $\alpha \in C^\infty(O,A^1(\partial M,\mathbb{R}^n))$ the following equation is valid :

$$(5.11)\; F_\partial(dj)(dl) = \int_M \alpha(dj)\cdot dl\;\mu(j)\,,\quad \forall\, dl \in C^\infty(\partial M,\mathbb{R}^n)/_{\mathbb{R}^n}\,,\ \forall\, dj \in O/_{\mathbb{R}^n}\,.$$

We introduce for any $j \in E(\partial M,\mathbb{R}^n)$ the Laplacian $\Delta(j)$ accordingly to (4.11) but require that E_s in this case is a moving frame on ∂M.

With this notion at hand the constitutive laws on O are characterized in details by the next theorem :

Theorem 5.1 :

Let F_∂ be a constitutive law on any open set $O \subset E(\partial M,\mathbb{R}^n)$. The following are then equivalent :

(i.) F_∂ admits a kernel $\alpha \in C^\infty(O/_{\mathbb{R}^n}, A^1(\partial M, \mathbb{R}^n))$.

(ii.) There is a smooth map $\mathfrak{h} \in C^\infty(O/_{\mathbb{R}^n}, C^\infty(\partial M, \mathbb{R}^n))$ uniquely determined up to maps in $C^\infty(O/_{\mathbb{R}^n}, \mathbb{R}^n)$, such that

$$(5.12) \qquad F_\partial(j)(l) = \int_{\partial M} d\mathfrak{h}(dj) \cdot dl\, \mu(j), \qquad \forall j \in O, \text{ and } \forall l \in C^\infty(\partial M, \mathbb{R}^n).$$

(iii.) There is a smooth map $\mathfrak{h} \in C^\infty(O/_{\mathbb{R}^n}, C^\infty(\partial M, \mathbb{R}^n))$ uniquely determined up to maps in $C^\infty(E(M, \mathbb{R}^n), \mathbb{R}^n)$, such that

$$(5.13) \qquad F_\partial(j)(l) = \int_{\partial M} <\Delta(j)\mathfrak{h}(dj), l> \mu(j), \quad \forall j \in O, \forall l \in C^\infty(\partial M, \mathbb{R}^n).$$

(iv.) There is a unique smooth map $\varphi \in C^\infty(O, C^\infty(\partial M, \mathbb{R}^n))$, such that

$$(5.14) \qquad F_\partial(j)(l) = \int_{\partial M} <\varphi(dj), l> \mu(j), \quad \forall j \in O, \forall l \in C^\infty(\partial M, \mathbb{R}^n),$$

and which satisfies

$$(5.15) \qquad \int_{\partial M} <\varphi(dj), z> \mu(j) = 0, \qquad \forall j \in O, \forall z \in \mathbb{R}^n.$$

Δ

For an explicit formulation of $\Delta(j)$ in terms of the coefficients of dj we refer to [Bi,Sc,So].

Remark :
The remark at the end of section 4 translates accordingly to the situation studied in this section.

Next we will investigate a boundary material implemented to a body formed by a deformable medium.

6. The interplay between constitutive laws of boundary and body

The deformable media forming the inside of the body and the boundary respectively may differ and each separate material hence has to be described on one hand by different constitutive laws. This we have done in the previous sections. On the other hand these materials together form one body and should be describable by only one constitutive law holding for the whole body. Since constitutive laws behave additively, the comparison between the two procedures allows us to decode the influence of the whole body to the constitutive properties of the boundary material.

Let the constitutive law of the deformable medium forming the whole body be called by F again.

According to theorem 4.2, F is determined by a smooth constitutive map

$$\mathfrak{H} : E(M,\mathbb{R}^n)/_{\mathbb{R}^n} \longrightarrow C^\infty(M,\mathbb{R}^n) .$$

The following theorem exhibits its influence to the constitutive entities of the material forming the boundary of the body. We will indicate next the methods to prove it.
The map $\mathfrak{H}$ yields according to theorem 4.2 force densities

$$(6.1) \qquad \Phi : E(M,\mathbb{R}^n)/_{\mathbb{R}^n} \longrightarrow C^\infty(M,\mathbb{R}^n)$$

and

$$(6.2) \qquad \varphi : E(M,\mathbb{R}^n)/_{\mathbb{R}^n} \longrightarrow C^\infty(\partial M,\mathbb{R}^n) .$$

The force density acting on ∂M, is defined by

$$(6.3) \qquad \varphi(dJ) = d\mathfrak{H}(dJ)(N) , \quad \forall\, dJ \in E(M,\mathbb{R}^n)/_{\mathbb{R}^n} .$$

Having the integrability condition (5.15) of $\Delta(J\,|\,\partial M)$ in mind, we split this force density φ into

$$(6.4) \qquad \varphi(dJ) = \varphi_{\mathbb{R}^n}(dJ) + \psi(dJ) , \quad \forall\, dJ \in E(M,\mathbb{R}^n)/_{\mathbb{R}^n} ,$$

where $\varphi_{\mathbb{R}^n}(dJ)$ is characterized for each $dJ \in E(M,\mathbb{R}^n)/_{\mathbb{R}^n}$ by the equation

$$(6.5) \qquad \int_{\partial M} <\varphi_{\mathbb{R}^n}(dJ),z> i_N\mu(J) = 0 , \quad \forall\, z \in \mathbb{R}^n$$

and where

$$\psi : E(M,\mathbb{R}^n)/_{\mathbb{R}^n} \longrightarrow \mathbb{R}^n ,$$

is a smooth map, which lets (6.4) hold.

According to theorem 5.1 the condition (6.5) allows us to choose some map

$$\mathfrak{h} : E(M,\mathbb{R}^n)/_{\mathbb{R}^n} \longrightarrow C^\infty(\partial M,\mathbb{R}^n) , \tag{6.6}$$

such that for all $dJ \in E(M,\mathbb{R}^n)/_{\mathbb{R}^n}$ the equation

$$\varphi_{\mathbb{R}^n}(dJ) = \Delta(J\,|\,\partial M)\ \mathfrak{h}(dJ) \tag{6.7}$$

holds true. With these data we easily prove the following :

Theorem 6.1 :

Any smoothly deformable medium is characterized by a constitutive map

$$\mathfrak{H} : E(M,\mathbb{R}^n)/_{\mathbb{R}^n} \longrightarrow C^\infty(M,\mathbb{R}^n) ,$$

determining itself two smooth maps

$$\mathfrak{h} : E(M,\mathbb{R}^n)/_{\mathbb{R}^n} \longrightarrow C^\infty(\partial M,\mathbb{R}^n) \tag{6.8}$$

and

$$\psi : E(M,\mathbb{R}^n) \longrightarrow \mathbb{R}^n , \tag{6.9}$$

which are linked to $\mathfrak{H}$ by the boundary condition

$$d\mathfrak{H}(dJ)(N) = \Delta(J\,|\,\partial M)\ \mathfrak{h}(dJ) + \psi(dJ) . \tag{6.10}$$

$\mathfrak{h}$ is unique up to $\mathbb{R}^n$–valued smooth maps defined on $E(M,\mathbb{R}^n)$. The map $\psi \in C^\infty(E(M,\mathbb{R}^n)/_{\mathbb{R}^n}, \mathbb{R}^n)$ is unique. Moreover $\mathfrak{H}$ satisfies the integrability conditions

$$0 = \int_M <\Delta(J)\mathfrak{H}(dJ),z>\ \mu(J) + \int_{\partial M} <d\mathfrak{H}(dJ)(N),z> i_N\mu(J) , \tag{6.11}$$

for each $J \in E(M,\mathbb{R}^n)$ and each $z \in \mathbb{R}^n$. Equation (6.11) equivalently formulated reads as

$$(6.12) \qquad 0 = \int_M <\Delta(J)\mathfrak{H}(dJ),z>\mu(J) + \int_{\partial M} <\psi(dJ),z> i_N\mu(J) ,$$

a boundary condition holding for $\mathfrak{H}$ and ψ together. The constitutive law on $E(M,\mathbb{R}^n)$ describing the constitutive properties of the materials forming body together with its boundary is determined by $\mathfrak{H}$, $\mathfrak{h}$ and ψ and thus given via the formula

$$(6.13) \qquad F(J)(L) = \int_M <\Delta(J)\mathfrak{H}(J),L>\mu(J) + \int_{\partial M} <\Delta(J|\partial M)\mathfrak{h}(J|\partial M) + \psi(dJ)),L> i_N\mu(J) ,$$

$$\forall J \in E(M,\mathbb{R}^n), \forall L \in C^\infty(M,\mathbb{R}^n) ,$$

valid for all variables of F. The work of any distortion $l \in C^\infty(\partial M,\mathbb{R}^n)$ of the deformable material forming the boundary detached from the body is for any $J \in E(M,\mathbb{R}^n)$ given by

$$(6.14) \qquad F_{\partial M}(dJ)(l) = \int_{\partial M} <\Delta(J|\partial M)\mathfrak{h}_\partial(dJ),l> i_N\mu(J) ,$$

for some constitutive map $\mathfrak{h}_\partial \in C^\infty(O_\partial/\mathbb{R}^n, C^\infty(\partial M,\mathbb{R}^n))$.

The constitutive properties of the deformable medium of the boundary detached from the body, are given by the smooth map $\mathfrak{h}_\partial \in C^\infty(O_\partial/\mathbb{R}^n, C^\infty(\partial M,\mathbb{R}^n))$. Hence $\mathfrak{h} - \mathfrak{h}_\partial$ and ψ describe how the constitutive properties of the material forming the boundary of the body are affected by the fact that this material is incorporated into the material forming the whole body.

Δ

7. Simple examples

In this section we will study well known one–forms on $E(M,\mathbb{R}^n)$ in the light of our formalism developed above. We do so by looking at particularly simple constitutive maps determining these forms.

(i.) In our first example we specify the differential of a constitutive map $\mathfrak{H}$ by $d\mathfrak{H}(dJ) = dJ$ for all $J \in E(M,\mathbb{R}^n)$. The following calculation is easily verified :

(7.1) $\int_M dJ \cdot dL \, \mu(J)$

$$= \int_M <\Delta(J)J,L>\mu(J) + \int_{\partial M} <dJ(N),L> i_N \mu(J)$$

$$= \int_M \text{tr } A(dL,dJ)\mu(J)$$

$$= \int_M \text{tr } \nabla(J)X(L,J)\mu(J)$$

$$= \int_M \text{div}_J X(L,J)\mu(J)$$

$$= \int_{\partial M} <N(j),L>\mu(j)$$

$$= D(\int_M \mu(J))(L) ,$$

for all $j := J|\partial M$ with $J \in E(M,\mathbb{R}^n)$, for all $L \in C^\infty(M,\mathbb{R}^n)$ and with $N(j) := TJN$. Introducing the volume function

(7.2) $$\mathfrak{V} : E(M,\mathbb{R}^n) \to \mathbb{R} ,$$

assigning to any $J \in E(M,\mathbb{R}^n)$ the volume

(7.3) $$\mathfrak{V}(J) = \int_M \mu(J) ,$$

we thus read off

(7.4) $$D\mathfrak{V}(J)(L) = \int_M dJ \cdot dL \, \mu(J) ,$$

The above calculation shows

(7.5) $$\Phi(dJ) = \Delta(J)J = 0 , \quad \forall J \in E(M,\mathbb{R}^n)$$

and

(7.6) $$\varphi(dJ) = N(j) , \qquad \forall J \in E(M,\mathbb{R}^n) .$$

Clearly $l = z$ with $z \in \mathbb{R}^n$ implies

(7.7) $$\int_{\partial M} <N(j),z>\mu(j) = 0 , \quad \forall z \in \mathbb{R}^n .$$

With the notations of the previous section this shows that in this example

(7.8) $$\varphi = \varphi_{\mathbb{R}^n} .$$

There is hence a map $\mathfrak{h}_N$ determined by $\mathfrak{H}$ which in this case is given by

(7.9) $$N = \Delta(j)\mathfrak{h}_N(dJ), \qquad \forall\, J \in E(M,\mathbb{R}^n) \text{ and } j := J|\partial M .$$

(ii.) Next let us turn our attention to $\mathfrak{h}_\partial$ defined on $E(\partial M,\mathbb{R}^n)$, by the formula $d\mathfrak{h}_\partial(dj) = dj$, for all $j \in E(\partial M,\mathbb{R}^n)$. One easily verifies the following set of equations

(7.10) $$\begin{aligned}\int_{\partial M} <\Delta(j)j,l>\mu(j) &= \int_{\partial M} dj\cdot dl\ \mu(j)\\ &= \int_{\partial M} (\mathrm{div}_j X(l,j) + \theta(l,j)\cdot H(j))\mu(j)\\ &= D(\int_{\partial M} \mu(j))(l),\end{aligned}$$

with $\theta(h,j) := <N(j),l>$. Defining the area function

(7.11) $$\mathfrak{A} : E(\partial M,\mathbb{R}^n) \to \mathbb{R},$$

sending any $j \in E(\partial M,\mathbb{R}^n)$ into

(7.12) $$\mathfrak{A}(j) = \int_{\partial M} \mu(j),$$

we have

$$D\mathfrak{A}(j)(l) = \int_{\partial M} dj\cdot dl\ \mu(j)$$

for all variables of $D\mathfrak{A}$. The constitutive map $\mathfrak{h}_\partial$ determines a map $\mathfrak{H}_\partial$ given by

(7.13) $$0 = \Delta(J)\mathfrak{H}_\partial(dJ), \qquad \forall J \in E(M,\mathbb{R}^n),$$

together with

(7.14) $$d\mathfrak{H}_\partial(dJ)(N) = \Delta(j)j = H(j)\cdot N(j),$$

for each $J \in E(M,\mathbb{R}^n)$ and $j := J|\partial M$. The function $H(j)$ is the mean curvature, that is the trace of $W(j)$.

(iii.) Next let us consider quite another influence of the boundary by looking at the map $\mathfrak{h}_\partial : O_{\partial/\mathbb{R}^n} \to C^\infty(\partial M,\mathbb{R}^n)$ given by $\mathfrak{h}_\partial(dj) = N(j)$, for all $j \in O_\partial$. Then the formula

$$(7.15) \qquad \begin{aligned} \Delta(j)N(j) &= \delta dN(J) \\ &= \delta dj\, W(j) \\ &= -dj\, \mathrm{grad}_j H(j) + (\mathrm{tr}\, W(j)^2)\cdot N(j) \end{aligned}$$

holds for any $j \in O_\partial$. Let us point out that $\Delta(j)N(j) \neq 0$ even if $j(\partial M) \subset \mathbb{R}^n$ is minimal, that is to say even if $H(j) = \mathrm{const}$.

In the special case of $\dim \partial M = 2$ a topological constant, the Euler characteristic $\chi(\partial M)$, enters the constitutive law F determined by $N(j)$ for each $j \in O_\partial$. It is hidden in the formula

$$(7.16) \qquad \begin{aligned} F(j)(N(j)) &= \int_{\partial M} <\Delta(j)N(j), N(j)> \mu(j) \\ &= \int_{\partial M} \mathrm{tr}\, W(j)^2 \mu(j) . \end{aligned}$$

This is seen by applying the theorems of Cayley Hamilton (cf.[Gr]) and of Gauss Bonnet (cf.[G,H,V]) to the right hand side of equation (7.16), which yields :

$$(7.17) \qquad F(j)(N(j)) = -4\pi\, \chi(\partial M) + \int_{\partial M} H(j)^2 \mu(j) .$$

Observe that in case of $j(\partial M)$ being a two–sphere with j the inclusion map, we have $F(j)(N(j)) = 12\,\pi$.

8. A general decomposition of constitutive laws

In this section we will exhibit a decomposition of the constitutive map $\mathfrak{H}$ based on the examples of the previous section.

We will show that $D\mathfrak{V}$ and $R^* D\mathfrak{A}$ (with R as in section 1) multiplied with appropriate $\mathbb{R}$–valued functions are all part of any constitutive law F defined on $E(M,\mathbb{R}^n)$.

To get the full decomposition of a given constitutive law we broaden our scope a little and introduce first of all the Hilbert space A_j consisting of all maps

$\gamma_1, \gamma_2 : TM \to \mathbb{R}$ linear on the fibers of TM for which the right hand side of

(8.1) $$G^{\partial}_{\mathbb{R}^n}(dj)(\gamma_1,\gamma_2) := \int_{\partial M} \gamma_1 \cdot \gamma_2 \, \mu(j)$$

exists. Clearly $d\mathfrak{h}_N(dJ)$, dj and $dN(j)$ all belong to A_j and are generically linearly independent. The set O_3 of all $J \in E(M,\mathbb{R}^n)$ for which these three differentials are linearly independent form a dense open set in $E(M,\mathbb{R}^n)$. Let $j := J|\partial M$. In special case of $j(\partial M)$ being a $(n-1)$–sphere in $\mathbb{R}^n$, however, $N(j)$ is a real multiple, r say, of j and $\mathfrak{h}_N(dJ)$ is hence $\frac{r}{(n-1)} \cdot j$. In the case of linear independence the three above mentioned differentials are in general not orthogonal to each other (with respect to $G_{\mathbb{R}^n}(dj)$). However we might orthogonalize them by using the method of Schmidt. For each $J \in O_3$ we then split the differential of $\mathfrak{h}(dJ)$ into components along span of the three mentioned differentials and a component perpendicular to this span.

The next step is to extend all maps $\mathfrak{h}_N(dJ)$, j and N(j) to all of M in the following way :
Given $f \in C^\infty_j(\partial M,\mathbb{R}^n)$ we solve the following Višik problem (cf.[Hö]) :

(8.2) $$\Delta(J)f_M = 0$$
$$df_M(N) - \Delta(j)f = 0 ,$$

with $f_M \subset C^\infty(M,\mathbb{R}^n)$ and $J \in E(M,\mathbb{R}^n)$ and where $j := J|\partial M$. All the splittings and extensions done to construct j_M and $N(j)_M$ depend smoothly on $j \in E(\partial M,\mathbb{R}^n)$. The map $\mathfrak{h}_N$ satisfies therefore

(8.2a) $$d(\mathfrak{h}_N(dJ))_M = dJ , \quad \forall J \in E(M,\mathbb{R}^n) .$$

The above mentioned decomposition of $d\mathfrak{H}$ is then presented in the following theorem :

Theorem 8.1 :

Let F be a constitutive law on $E(M,\mathbb{R}^n)$ determined by $\mathfrak{H} \in C^\infty(E(M,\mathbb{R}^n)/_{\mathbb{R}^n}, C^\infty(M,\mathbb{R}^n))$. Then $\mathfrak{H}$ determines uniquely three smooth maps

$$(8.3) \qquad a_1, a_2, a_3 : O_3 \subset E(M,\mathbb{R}^n)/_{\mathbb{R}^n} \longrightarrow \mathbb{R}$$

and two smooth maps

$$\mathfrak{h}, \mathfrak{h}_2 : O_3 \subset E(M,\mathbb{R}^n)/_{\mathbb{R}^n} \longrightarrow C^\infty(\partial M,\mathbb{R}^n) ,$$

such that the following splitting holds for any $dJ \in O_3 \subset E(M,\mathbb{R}^n)$

$$(8.4) \qquad d\mathfrak{h}(dJ) = a_1(dJ)\cdot d\mathfrak{h}_N(dJ) + a_2(dJ)\cdot dj + a_3(dJ)\cdot dN(j) + d\mathfrak{h}_2(dJ)$$

with $j := J|\partial M$. The differential $d\mathfrak{h}_2(dJ)$ is orthogonal with respect to $G(dj)^{\partial}_{\mathbb{R}^n}$ to the span of $d\mathfrak{h}_N(dJ)$, dj and $dN(j)$.

The map $\mathfrak{H}(dJ)$ decomposes for each $J \in O_3$ accordingly into

$$(8.5) \qquad d\mathfrak{H}(dJ) = a_1(dJ)\cdot dJ + a_2(dJ)\cdot dj_M + a_3(dJ)\cdot dN_M(dJ) + d\mathfrak{H}_2(dJ) ,$$

with $j := J|\partial M$ and where $\mathfrak{H}_2(dJ)$ is such that (8.10) holds.

Δ

Let us illustrate the meaning of the coefficients a_1 and a_2 in (8.4) by a simple bubble model. Let $\dim M = 3$ and let us think of M as a bubble and of ∂M as the middle surface of a very thin shell bounding the bubble. Moreover we assume $a_3 = 0$ and $\mathfrak{H}_2 = 0$ as well. We suppose that there is an equilibrium configuration I in the topological closure of O_3, this is to say we suppose that $F(I) = 0$. Since ∂M represents two bounding surfaces, (8.4) turns into

$$(8.6) \qquad 0 = a_1(dI)\cdot d\mathfrak{h}_N(dI) + 2\cdot a_2(dI)\cdot di ,$$

with $i := I|\partial M$. Using (8.2a), (7.1) and (7.10) yields

$$(8.7) \qquad a_1(dI) + 2\cdot a_2(dI)\cdot H(i) = 0 .$$

Thus we observe that the mean curvature $H(i)$ is a constant map of ∂M and deduce from the classical bubble models that $a_1(dI)$ and $a_2(dI)$ can be interpreted as an internal pressure and as a capillarity respectively.

9. The rotational symmetry

Finally we investigate the effect of the SO(n), the symmetry group of the oriented Euclidean vector space $(\mathbb{R}^n, < , >, \Delta)$ to a constitutive law. In particular let us characterize those which are invariant under SO(n).

As mentioned in the first section we have the operation

$$\mathfrak{s} : SO(n) \times E(M,\mathbb{R}^n) \longrightarrow \mathbb{R}^n$$

sending any $g \in SO(n)$ and any $J \in E(M,\mathbb{R}^n)$ into

$$\mathfrak{s}(g,J) = g \circ J .$$

The induced operation $\mathfrak{s}_T$ on $TE(M,\mathbb{R}^n)$ is therefore determined by

$$\mathfrak{s}_T : SO(n) \times E(M,\mathbb{R}^n) \times C^\infty(M,\mathbb{R}^n) \longrightarrow E(M,\mathbb{R}^n) \times C^\infty(M,\mathbb{R}^n)$$

$$\mathfrak{s}_T(g,(J,L)) = (g\circ J, g\circ L) ,$$

for all $g \in SO(n)$, for all $J \in E(M,\mathbb{R}^n)$ and for all $L \in C^\infty(M,\mathbb{R}^n)$.
Invariance under SO(n) of a given constitutive law F with $\mathfrak{H}$ as a constitutive map means that the following equation

(9.1) $$F(g \circ J)(g \circ L) = F(J)(L)$$

holds for all the variables of F and for any $g \in SO(n)$. Thus the constitutive map satisfies

(9.2) $$\int_M d\mathfrak{H}(g\circ dJ)\cdot d(g\circ L)\ \mu(g\circ J) = \int_M d\mathfrak{H}(dJ)\cdot dL\ \mu(J)$$

or reformulated

(9.3) $$\int_M \operatorname{tr}\ [(\bar{g}^{-1}A(d\mathfrak{H}(g\circ dJ), g\circ dJ) - A(d\mathfrak{H}(dJ),dJ))\cdot \tilde{A}(dL,dJ)]\mu(J) = 0$$

with $\bar{g} = (dJ)^{-1} \circ g$. Thus $\bar{g}$ is an element of $SO(T_pM)$, the special orthogonal group of T_pM for each $p \in M$. The volume μ is invariant under the action $\mathfrak{s}_T$.

Hence

$$(9.4) \qquad \int_M d(g^{-1} \circ \mathfrak{H}(g \circ dJ) - \mathfrak{H}(dJ)) \cdot dL \, \mu(J) = 0 ,$$

holds for all $g \in SO(n)$, for all $J \in E(M,\mathbb{R}^n)$ and for all $L \in C^\infty(M,\mathbb{R}^n)$. From the last equation and from the general procedure of representing differentials via embeddings we read off

Proposition 9.1 :
Given a constitutive law F with constitutive map $\mathfrak{H} \in C^\infty(E(M,\mathbb{R}^n),C^\infty(M,\mathbb{R}^n))$ then F is invariant under SO(n) iff

$$(9.5) \qquad g^{-1} \circ d\mathfrak{H}(g \circ dJ) = d\mathfrak{H}(dJ) , \quad \forall G \in SO(n).$$

The validity of this equation is equivalent with the following identity

$$(9.6) \qquad A(d\mathfrak{H}(g \circ dJ), g \circ dJ) = A(d\mathfrak{H}(dJ), dJ)$$

holding for all $g \in SO(n)$ and for all $J \in E(M,\mathbb{R}^n)$. The stress tensor T(J) determined by $\mathfrak{H}$ is hence invariant under SO(n) for any $J \in E(M,\mathbb{R}^n)$.

Δ

Let F be a SO(n)–invariant constitutive law. (9.5) implies that for each $c \in so(n)$

$$(9.7) \qquad c \circ d\mathfrak{H}(dJ) = dD\mathfrak{H}(dJ)(c \circ dJ) ,$$

with D the derivative on $E(M,\mathbb{R}^n)/_{\mathbb{R}^n}$. Here so(n) is the Lie–algebra of SO(n).

Based on (9.7) it is a matter of routine to show that in case $\mathfrak{H}$ is invariant under SO(n) then $c \circ d\mathfrak{H}(dJ) = 0$ for any $c \in so(n)$ and any $J \in E(M,\mathbb{R}^n)$. In case of $\dim M = 3$, e.g. this means that the SO(3)–invariance of $\mathfrak{H}(J)$ yields $\mathfrak{H}(dJ) = \text{const.}$ and in turn that $F = 0$.
Splitting $A(d\mathfrak{H}(dJ),dJ)$ with respect to m(J) into a symmetric and a skew–symmetric part $B(d\mathfrak{H}(dJ),dJ)$ and $C(d\mathfrak{H}(dJ),dJ)$ respectively and doing the same for $A(dD\mathfrak{H}(dJ)(c \circ dJ),dJ)$, then (9.7) yields the identity

$$(9.8) \qquad \bar{c} \circ B(d\mathfrak{H}(dJ),dJ)(p) + \bar{c} \circ C(d\mathfrak{H}(dJ),dJ)(p)$$
$$= B(dD\mathfrak{H}(dJ)(c \circ dJ),dJ)(p) + C(dD\mathfrak{H}(dJ)(c \circ dJ),dJ)(p)$$

with $\bar{c} \in so(T_pM)$ for each fixed $p \in M$. If $\operatorname{tr} A(dD\mathfrak{H}(dJ)(c\circ dJ),dJ)=0$ then (9.8) yields

$$(9.9) \qquad \operatorname{tr} \bar{c}\circ C(d\mathfrak{H}(dJ),dJ)(p) = 0\,, \qquad \forall c \in so(n),\ \forall p \in M \text{ and } \forall J \in E(M,\mathbb{R}^n).$$

and vice versa. This implies the following theorem :

Theorem 9.2 :
For any SO(n)–invariant constitutive law with (smooth) constitutive map $\mathfrak{H}$ the following are equivalent:

(9.10) (i.) $C(d\mathfrak{H}(dJ),dJ) = 0\,, \quad \forall J \in E(M,\mathbb{R}^n)$

(9.11) (ii.) $\operatorname{tr} A(dD\mathfrak{H}(dJ)(c\circ dJ),dJ)=0 \quad \forall J \in E(M,\mathbb{R}^n)$

(9.12) (iii.) The stress tensor T(J) associated with $\mathfrak{H}$ is symmetric for each configuration J.

Δ

Let us illustrate (9.11) somewhat. In doing so we proceed similar as in section 8 and take the components of $d\mathfrak{H}(g\cdot dJ)$ and $d\mathfrak{H}(dJ)$ along dJ, this is to say we have the splitting

$$(9.13) \qquad d\mathfrak{H}(g\cdot dJ) = \hat{\Pi}(g\cdot dJ)\cdot dJ + d\hat{\mathfrak{H}}_1(g\cdot dJ)$$

and

$$(9.14) \qquad d\mathfrak{H}(dJ) = \Pi(dJ)\cdot dJ + d\mathfrak{H}_1(dJ)\,,$$

where both $d\hat{\mathfrak{H}}_1(dJ)$ and $d\mathfrak{H}_1(dJ)$ are orthogonal to dJ with respect to $G_{\mathbb{R}^n}(dJ)$. Both $\mathbb{R}$–valued functions Π and $\hat{\Pi}$ defined on $E(M,\mathbb{R}^n)$ can be regarded as a sort of internal pressures. All the maps in (9.13) and (9.14) are smooth. We therefore find

$$(9.15) \qquad \int_M \operatorname{tr} A(d\mathfrak{H}(g\cdot dJ),dJ)\,\mu(J) = n\cdot\hat{\Pi}(g\cdot dJ)\cdot\mathfrak{V}(J)$$
$$= n\cdot\Pi(dJ)\cdot\mathfrak{V}(J)\cdot\operatorname{tr} g + \int_M \operatorname{tr} A(g\cdot d\hat{\mathfrak{H}}_1(dJ),dJ)\,\mu(J)$$

and in turn

(9.16) $$\int_M \operatorname{tr} A(dD\mathfrak{H}(dJ)(c\cdot dJ),dJ) = n\cdot\mathfrak{V}(J)\cdot D\hat{\Pi}(dJ)(c\cdot dJ)\,.$$

Hence (9.11) requires

(9.17) $$D\hat{\Pi}(dJ)(c\cdot dJ) = 0\,, \quad \forall J \in E(M,\mathbb{R}^n) \text{ and } \forall c \in so(n)\,,$$

showing that $\hat{\Pi}$ is up to the first order invariant under SO(n).

Acknowledgements

We are indebted to T.Ackermann, A.Barut and G.Schwarz for valuable discussions and criticisms.

References

[A,M,R] R.Abraham, J.E.Marsden, T.Ratiu : Manifolds, Tensor Analysis, and Applications, Addison Wesley, Global Analysis, 1983

[Bi] E.Binz : On the Notion of the Stress Tensor Associated with $\mathbb{R}^n$–invariant Constitutive Laws Admitting Integral Representations, Reports on Mathematical Physics, Vol.87, (1989), p.p.49–57

[Bi,Sc,So] E.Binz, G.Schwarz, D.Socolescu : On a Global Differential Geometric Description of the Motions of Deformable Media, to appear in Infinite Dimensional Manifolds, Groups, and Algebras, Vol. II, Ed.H.D.Doebner, J.Hennig, 1990

[Bi,Sn,Fi] E.Binz, J.Śniatycki, H.Fischer : Geometry of Classical Fields, Mathematics Studies 154, North–Holland Verlag, Amsterdam, 1988

[Fr,Kr] A.Frölicher, A.Kriegl : Linear Spaces and Differentiation Theory, John Wiley, Chichester, England, 1988

[G,H,V] W.Greub, S.Halperin, J.Vanstone : Connections, Curvature and Cohomology, I, II, Acad. Press, New York, 1972–73

[Gr] W.Greub : Lineare Algebra I, Graduate Texts in Mathematics, Vol.23, Springer Verlag, Berlin, Heidelberg, New York, 1981

[E,S] M.Epstein, R.Segev : Differentiable Manifolds and the Principle of Virtual Work in Continuum Mechanics, Journal of Mathematical Physics, No.5, Vol.21, (1980), p.p. 1243–1245

[He] E.Hellinger : Die allgemeinen Ansätze der Mechanik der Kontinua, Enzykl. Math. Wiss. 4/4, 1914

[Hi] M.W.Hirsch : Differential Topology, Springer GTM, Berlin, 1976

[Hö] L.Hörmander : Linear Partial Differential Operations, Grundlehren der mathematischen Wissenschaften, Vol.116, Springer Verlag, Berlin, Heidelberg, New York, 1976

[J] F.John : Partial Differential Equations, Applied Mathematical Science, Vol.1, (1978)

[L,L] L.D.Landau, E.M.Lifschitz : Lehrbuch der theoretischen Physik, Vol. VII, Elastizitätstheorie, 4. Auflage, Akademie Verlag, Berlin, 1975

[Mat] Y.Matsushima : Vector Bundle Valued Canonic Forms, Osaka Journal of Mathematics, Vol.8, (1971), p.p. 309–328

Imitation of Symmetries in Local Quantum Field Theory

(On the Interplay between Locality and Spectrum Condition)

H.J. Borchers

Institut für Theoretische Physik
Universität Göttingen
Bunsenstrasse 9, D 3400 Göttingen

I. Introduction

Looking at modern foundations of physics it is a striking fact that all approaches seem to be governed by groups. It is even more remarkable that we are not dealing with spontaneous broken exact symmetries, but we have "non-symmetries", i.e., symmetries which are only present for an approximating theory.

Regarding these really broken symmetries one observes that the breaking term is present in order to make the members of a multiplet distinguishable. The electromagnetic interaction, for example, is such a breaking for the iso-spin. If we would take the approximating theory serious then all the particles of a multiplet would look alike. In such a situation there would be no way to discover the presence of the symmetry-group. (There are people who claim that in this case the group should also be observable because the Clebsch-Gordan-coefficients appear in the scattering amplitudes. I doubt whether these really can be discovered. There seems to be no careful discussion of this problem.) This consideration leads to a paradoxical statement, namely, symmetry-groups are only present because they are broken, or, with other words, real symmetries do not exist in physics. This statement is only meant for internal symmetries, but for others the situation is not much better. If we consider geometrical symmetries as space-time translations and Lorentz-transformations, then we also know that these are broken by gravitational interaction.

These discussions teach us that symmetries in physics are symmetries of approximations. In models it is allowed to switch off parts of the interaction and to violate fundamental principles of physics, which require to distinguish between states only if they can be discriminated by experiments. The last point, again, only is a problem for internal symmetries, while the approximation and the switching-off of the gravitation do not cause any problem for understanding the Poincaré-invariance in the usual theory.

So far our discussion was describing the way how symmetries appear in physics. But this does not answer the question why the groups $SU(2)$ for isospin or $SU(3)$ for the standard model of elementary particles do appear, whereas other compact groups don't. In solid state physics, for example, many of the substances condense at low temperatures in crystal structures. In this case we also have, besides the lattice-group, the discrete crystal-group as symmetry-group. These groups are sub-groups of the symmetry-group of the Euclidean space. Therefore, one can look at them as a consequence of the breaking of the symmetry. But that we obtain the particular group for a given material is a consequence of the dynamics. As we will see in a moment, the invariance under Euclidean transformations is a consequence of general principles. Therefore, special symmetries always seem to be a consequence of general principles and dynamics. Only little is known about the ingredients which lead to the gauge-field theory as used for the standard model.

By trying to imagine intelligent beings living in a crystal one realizes how difficult it might be to discover the general principles. Soon they will discover that the crystal-group they observe as a symmetry is not an exact one. But as long as they don't have enough energy to melt a large part of the crystal it needs a lot of phantasy to present the crystal-symmetry as a broken representation of an exact symmetry, namely that of the Euclidean space.

This discussion has lead us to the problem of spontaneously broken symmetries: Assume we are describing the considered theory in terms of a C*-algebra $\mathcal{A}$ and assume that a group G acts as symmetry-group, i.e., we have a representation of G by automorphisms of $\mathcal{A}$

$$\alpha : G \longrightarrow \text{Aut}\,(\mathcal{A}).$$

Describing a particular physical situation we don't deal with the abstract algebra $\mathcal{A}$ but we are looking at a particular representation of $\mathcal{A}$ on a Hilbert-space $\{\pi, \mathcal{H}\}$. We, now, talk of a spontaneously broken symmetry if there exists $g \in G$ such that the two representations $\pi(x)$ and $\pi(\alpha_g x)$ are inequivalent, i.e., if α_g cannot be implemented by unitary operators.

In most cases it has drastic macroscopie effects if a symmetry is spontaneously broken. But this is not always so, as we will see in the case of Lorentz-boosts. Lorentz-transformations have first been used by *W. Voigt* [33]. He realized that these transformations do not change the wave-equation and, therefore, can be used in order to describe the Doppler-effect, at that time known in acoustics. *A. Einstein* [19] has used the covariance under Lorentz-transformations as a principle for constructing special relativity. For a long time it was believed that Lorentz-transformations are a fundamental symmetry, which should be implemented in every reasonable representation describing particle physics. However, it has been observed by *Fröhlich, Morcio*, and *Strocci* [22] that this symmetry must be spontaneously broken in quantum-electrodynamics. A little later *Buchholz* [15] described the same phenomenon in axiomatic field theory. The astonishing fact about the breaking of the Lorentz-symmetry seems to be that all observables mostly used do not show any consequence of this symmetry breaking. In particular, the masses of charged particles fulfil the relation $p^2 = m^2$ within the measurable accuracy [21] although the existence of this relation is not implied by the spontaneously broken Lorentz-symmetry.

This representation is devoted to the understanding of the last statement. To this end we assume that we are dealing with a theory of local observables on which the space-time-translations act as a group of symmetries. Into this theory one has incorporated some principles of dynamics. First we assume that there is associated a region O in space-time to every observable. Moreover, we assume that no disturbance can spread with a velocity larger than that of light. This implies that two observations, the space-time domains of which are spacelike separated, cannot disturb each other. Accepting the principles of quantum mechanics, this implies that the operators corresponding to these observables have to commute with each other. These are the only dynamical principles we want to use in our discussion and we want to show that this is sufficient to imply the relation $p^2 = m^2$ for particles. It is not necessary to assume that our theory is Lorentz-invariant. It is this general principle of dynamics, in particular the Einstein-causality, which forces the spectrum of the space-time-translation to be located on a Lorentz-invariant set.

The only symmetry we introduced is that under space-time-translations, which is much more natural than any other symmetry. It is the basis of natural sciences that every experiment can be repeated at any other time and any other place. This implies – when describing physics in terms of observable quantities – that space-time-translations are a symmetry of our theory. Much more could be said about the special role of the space-time-translations, but we will refrain from doing so, and refer instead to the paper of *H. Ekstein* [20] and *A. Aviskai* and *H. Ekstein* [6].

Having set up the frame we must decide on the representations which describe the special physical situation. Since we want to deal with systems describing a finite number of massive particles we will use representations with a positive energy operator. As the energy is identified with the generator of the time-translations we have to deal with representations in which the time-translations are implemented, and the representation of the corresponding group has a positive generator. By Einstein's principle no particle should travel with velocity higher than that of light. Therefore, we conclude that the whole group of space-time-translations must be implemented, so that the corresponding group-representation has a spectrum in the forward light-cone.

These notes are organized as follows: In the next chapter we deal with general C^*-dynamical systems, denoted by $\{\mathcal{A}, G, \alpha\}$ where $\mathcal{A}$ is a C^*-algebra, G a topological group, and α a homomorphism of G into $\mathrm{Aut}(\mathcal{A})$, the group of automorphisms of $\mathcal{A}$. We do not require that α_g acts strongly continuous on $\mathcal{A}$. But we are interested in representations π such that we can find a continuous unitary representation of the group G implementing α_g. In the third chapter G will be the translation group of $\mathbf{R}^d$ and we will ask for the additional requirement that the spectrum of the representation is contained in a pre-given cone. The last chapter includes local quantum field theory. Here, the cone in question will be the light-cone, and we will show that under these additional conditions the support of the spectrum is a Lorentz-invariant set. As a consequence of the dynamics we find that the Lorentz-group is almost a symmetry-group, although the symmetry has not been added to our theory.

II. Dynamical Systems, Elementary Properties

II.1 Continuous Action, Crossed Products

If one applies group theory to something then one has besides the group an object upon which the group is acting. One natural object in this situation is a manifold $\mathcal{M}$ with the assumption that one has a Lie-group G, and that the group acts continuously on $\mathcal{M}$. In this situation one usually looks at the orbits and the strata, which are sets of orbits having the same structure.

In case the manifold $\mathcal{M}$ is a locally compact space then, by the Gelfand isomorphism, one can replace $\mathcal{M}$ by the C^*-algebra $\mathcal{A}(\mathcal{M})$ of continuous functions on $\mathcal{M}$. In this case one obtains an action of the group on $\mathcal{A}(\mathcal{M})$ as automorphisms

$$\alpha : G \longrightarrow \mathrm{Aut}\,\mathcal{A}(\mathcal{M}).$$

If the original action of G on $\mathcal{M}$ was a continuous action, then the action of G on $\mathcal{A}(\mathcal{M})$ is ***strongly continuous***, i.e. the function

$$G \ni g \longrightarrow \alpha_g(x)$$

is continuous on G for every $x \in \mathcal{A}(\mathcal{M})$. This definition means that to every $x \in \mathcal{A}(\mathcal{M}), \epsilon > 0$ and every $g \in G$ there exists a neighbourhood $U(g)$, such that

$$\|\alpha_g(x) - \alpha_h(x)\| < \epsilon \quad \forall \quad h \in U(g).$$

We remark that from the group properties – which are preserved since α is a group-homomorphism of G into the group Aut($\mathcal{A}(\mathcal{M})$) – it follows that the strong continuity at the identity implies the strong continuity everywhere.

By modern approaches to geometry, including also none-commutative geometry, one expresses all geometric properties by means of functions on manifolds. Therefore, I also restrict my attention to the investigation of the action of groups by automorphisms. This leads to the following

II.1.1 Notation

A. *A C^*-dynamical system is a triple*

$$\{\mathcal{A}, G, \alpha\},$$

where $\mathcal{A}$ is a C^*-algebra, $G \doteq G(\tau)$ a topological group and α a group-homomorphism

$$\alpha : G \longrightarrow \text{Aut } \mathcal{A}$$

where Aut $\mathcal{A}$ *denotes the group of automorphisms of $\mathcal{A}$.*

B. *If $\{\mathcal{A}, G, \alpha\}$ is a C^*-dynamical system, then we say that the group acts strongly continuous if the function*

$$g \longrightarrow \alpha_g(x) \qquad g \in G, \; x \in \mathcal{A}$$

is continuous on G with values in the C^*-algebra $\mathcal{A}$.

Describing physics in terms of C*-algebra one must realize that one fixed C*-algebra has to incorporate a wide range of branches of physics, ranging from particle physics to thermal physics, including gases, liquids, and solids. This large variety is possible because most C*-algebras have a large number of inequivalent representations. This implies that one has, for particular physical situations, to look for special representations, so that we understand the importance of the representation theory of C*-algebras for describing physics.

Having in addition a symmetry group one likes to know whether a given representation is a covariant representation. This means the following

II.1.2 Notation

Let $\{\mathcal{A}, G, \alpha\}$ be a C^*-dynamical system, then the triple

$$\{\pi, U, \mathcal{H}\}$$

is called a covariant representation if

(i) *π is a representation of $\mathcal{A}$ on a Hilbert-space $\mathcal{H}$, i.e. is an algebra-homomorphism of $\mathcal{A}$ into $B(\mathcal{H})$, the algebra of bounded operators on $\mathcal{H}$,*
(ii) *$U(g)$ is a continuous unitary representation of G on $\mathcal{H}$,*
(iii) *$U(g)$ implements α_g i.e., one has*

$$U(g)\pi(x)U^*(g) = \pi(\alpha_g x).$$

One has a simple situation if the group is locally compact and if it acts strongly continuous on the C*-algebra. In this case one can construct the C*-crossed product between the group G and the C*-algebra $\mathcal{A}$, denoted by $G \underset{\alpha}{\times} \mathcal{A}$. This object, sometimes also called covariance algebra, has been introduced in the case of discrete groups by *Turumaru* [32] and for continuous groups by *Doplicher, Kastler* and *Robinson* [18]. These authors have proved the following

II.1.3 Theorem

Let $\{\mathcal{A}, G, \alpha\}$ be a C-dynamical system with G a locally compact group acting strongly continuous on $\mathcal{A}$, then there is a one to one correspondence between the covariant representation $\{\pi, U, \mathcal{H}\}$ of $\mathcal{A}$ and the representations of the crossed product $G \underset{\alpha}{\times} \mathcal{A}$.*

The construction of the crossed product $G \underset{\alpha}{\times} \mathcal{A}$is similar to the construction of the convolution algebra $\mathcal{L}^1(G)$. For details see the book of *G.K.Pedersen* [29].

The crossed product is a powerful tool for studying dynamical systems, i.e. the interaction of the symmetry group with the C*-algebra and the structure of the C*-algebra implied by this interaction.

II.2 Characterization of Covariant Representations

The theorem of the last section turns out to be not very useful. In most situations one is dealing with a C*-dynamical system and a representation $\{\pi, \mathcal{H}\}$ of $\mathcal{A}$ is given. Now we are interested to know under which additional conditions on this representation we can find on $\mathcal{H}$ a continuous unitary representation $U(a)$ of the group G such that $\{\pi, U, \mathcal{H}\}$ becomes a covariant representation of $\{\mathcal{A}, G, \alpha\}$. Before answering this question one should be aware of two obstructions, one of them the problem of multiplicity: If dealing with a representation which is not a factor-representation, then the multiplicity might be different in different parts of the representation, while the unitary transformation cannot change multiplicity. A second obstruction results from the fact that the map

$$\alpha_g \longrightarrow \mathrm{ad}\, U(g) = U(g)\,.\,U^{-1}(g)$$

is not linear, which means that the representation of a group G of automorphisms by unitary operators might lead to a representation of up to a phase-factor. Both these problems can be dealt with by allowing to change the representation π of $\mathcal{A}$ to a quasi-equivalent representation $\hat{\pi}$ of $\mathcal{A}$.

Recall that two representations π and $\hat{\pi}$ are called quasi-equivalent if they have the same set of normal states.

Before giving conditions for covariant representations we want to generalize the situation. Many physicists like to use projections in their investigations. But in this case the function

$$g \longrightarrow U(g)eU^*(g)$$

is never continuous in the norm topology, except the projection commutes with $U(g)$. Therefore, one should drop the assumption of strong continuity, provided there is not too much being lost. Before stating the result we need some

II.2.1 Notations

(A) *Let $\mathcal{A}$ be a C^*-algebra then two representations $\{\pi, \mathcal{H}\}$ and $\{\hat{\pi}, \hat{\mathcal{H}}\}$ are called quasi-equivalent if π and $\hat{\pi}$ have the same kernel and if the isomorphism*

$$\pi \longleftrightarrow \hat{\pi}$$

extends to an isomorphism of the von Neumann-algebras $\pi(\mathcal{A})''$ and $\hat{\pi}(\mathcal{A})''$. This means also that the above isomorphism is continuous in the ultra-weak topology.

(B) *Let $\{\pi, \mathcal{H}\}$ be a representation then the set of normal states of π is called the folium of π. Two representations are quasi-equivalent if they have the same folium. Hence the class of quasi-equivalent representations is characterized by their folium.*

(C) *Let $\{\mathcal{A}, G, \alpha\}$ be a C^*-dynamical system and $\{\pi, \mathcal{H}\}$ be a representation of $\mathcal{A}$ then $\{\pi, \mathcal{H}\}$ is called quasi-covariant, if there exists a covariant representation $\{\pi_1, U, \mathcal{H}_1\}$ such that $\{\pi, \mathcal{H}\}$ and $\{\pi_1, \mathcal{H}_1\}$ are quasi-equivalent.*

I like to remind the reader that we do no longer require any continuity for the map $g \to \alpha_g(x)$, but we still assume that the group-representations $U(g)$ appearing in covariant representations are continuous. Assume now, we have a covariant representation $\{\pi, U, \mathcal{H}\}$ and let $\psi \in \mathcal{H}$ be a normalized vector then we can consider the vector state ω_ψ defined by

$$\omega_\psi(x) = (\psi, \pi(x)\psi).$$

In this situation one has for the transposed action

$$\begin{aligned}(\alpha'_g\omega_\psi)(x) &= \omega_\psi(\alpha_g x) = (\psi, U(g)\pi(x)U^*(g)\psi) \\ &= (U^*(g)\psi, \pi(x)U^*(g)\psi) = \omega_{U^*(g)\psi}(x).\end{aligned}$$

This equation implies the estimate

$$|(\alpha'_g\omega_\psi)(x) - \omega_\psi(x)| \leq 2\|x\|\|U^*(g)\psi - \psi\|$$

or

$$\|\alpha'_g\omega_\psi - \omega_\psi\| \leq 2\|U^*(g)\psi - \psi\|$$

from which we see that the action of α'_g on the dual space of $\pi(\mathcal{A})$ is strongly continuous. This consideration suggests to introduce the following space

II.2.2 Definition

Let $\{\mathcal{A}, G, \alpha\}$ be a C^*-dynamical system then we denote by $\mathcal{A}^*_c$ the following set:

$$\mathcal{A}^*_c = \{\phi \in \mathcal{A}^*; \alpha'_g\phi \text{ is strongly continuous at the identity of the group}\}.$$

The following properties of the set $\mathcal{A}^*_c$ are easy to prove:

II.2.3 Lemma

Let $\{\mathcal{A}, G, \alpha\}$ be a C^*-dynamical system, then the set $\mathcal{A}^*_c$ has the following properties

(a) *$\mathcal{A}^*_c$ is a vector-space.*

(b) *$\phi \in \mathcal{A}^*_c$ implies $\phi^* \in \mathcal{A}^*_c$,*

(c) $\mathcal{A}_c^*$ *is norm closed,*
(d) $\mathcal{A}_c^*$ *is invariant under the dual action of the group, i.e.*

$$\alpha_g' \mathcal{A}_c^* = \mathcal{A}_c^*, \quad \forall \quad g \in G.$$

Using the space $\mathcal{A}_c^*$ we can characterize quasi-covariant representations.

II.2.4 Theorem

Let $\{\mathcal{A}, G, \alpha\}$ *be a* C^**-dynamical system and* $\{\pi, \mathcal{H}\}$ *a representation of* $\mathcal{A}$ *then* $\{\pi, \mathcal{H}\}$ *is quasi-covariant if and only if*
(a) *The folium of* π *is invariant under the dual action* α_g', $\forall \quad g \in G$.
(b) *The folium of* π *belongs to* $\mathcal{A}_c^*$.

The necessity of these conditions is simply a consequence of the remark made before definition II.2.2. In the special case, where the group is locally compact and the group action is strongly continuous the converse direction has been proved by *Borchers* [9]. For the proof of the general case one is using standard representations of von Neumann-algebras $\mathcal{M}$ which exist as a consequence of the *Tomita-Takesaki*-Theory of modular Hilbert-algebras [31],[30]. This theory allows to construct a standard representation of the full automorphism-group having the property that this representation is continuous on those subgroups which act continuously on the normal functionals $\mathcal{M}_*$ of $\mathcal{M}$. If we now apply this result to the algebra $\pi(\mathcal{A})''$ and remember that the folium of $\pi(\mathcal{A})$ coincides with the predual of the von Neumann-algebra $\pi(\mathcal{A})''$, we see that Theorem II.2.4 is correct.

II.3 Properties of the Space $\mathcal{A}_c^*$

We saw in the last section that the properties of the group-action on the dual-space $\mathcal{A}^*$ is decisive for the existence of covariant representations. Since we want the group-representation to be a continuous one we had to introduce the subspace $\mathcal{A}_c^*$ of the dual-space $\mathcal{A}^*$. Because of its importance one has to study the properties of $\mathcal{A}_c^*$ more detailed. Including the results of Lemma II.2.3. one gets

II.3.1 Theorem

Let $\{\mathcal{A}, G, \alpha\}$ *be a* C^**-dynamical system and assume* G *is a topological group, then the space* $\mathcal{A}_c^*$ *has the following properties:*

(i) $\mathcal{A}_c^*$ *is a linear space.*
(ii) $\mathcal{A}_c^*$ *is closed in the norm topology.*
(iii) $\mathcal{A}_c^*$ *is invariant under the action of the group i.e.* $\phi \in \mathcal{A}_c^*$ *implies* $\phi \circ \alpha_g \in \mathcal{A}_c^*$ *for every* $g \in G$.
(iv) *With* $\phi \in \mathcal{A}_c^*$ *one finds also* ϕ^* *and* $|\phi|$ *belong to* $\mathcal{A}_c^*$.
(v) $\mathcal{A}_c^*$ *is generated by its positive elements.*
(vi) *Let* β *be an automorphism of* $\mathcal{A}$ *commuting with every* α_g, $g \in G$ *then* $\beta \mathcal{A}_c^* = \mathcal{A}_c^*$.
(vii) *Let* $\mathcal{A}_G^{**}$ *be the sub-von Neumann algebra of* $\mathcal{A}^{**}$ *of elements, which are* α_g *invariant for all* $g \in G$*, then* $\phi \in \mathcal{A}_c^*$ *and* $y \in \mathcal{A}_G^{**}$ *implies* $y\phi$ *and* $\phi y \in \mathcal{A}_c^*$.

(We have used the notations $(\phi y)(x) = \phi(xy)$ and $(y\phi)(x) = \phi(yx)$).

These results are obtained by *Borchers* [11]. Since $\mathcal{A}_c^*$ is generated by the positive elements belonging to $\mathcal{A}_c^*$ it is sufficient to look at those. But using the standard representations of von Neumann-algebras on some Hilbert-space $\mathcal{H}$ (in our case the von Neumann-algebra is $\mathcal{A}^{**}$) one can associate to every positive linear functional ω on $\mathcal{A}$ a vector $\psi_\omega \in \mathcal{H}$ in a unique manner. The map

$$\omega \longleftrightarrow \psi_\omega$$

is a topological homeomorphism. (For details see *H. Araki* [2],[3], *A. Connes* [16],[17], and *U. Haagerup* [25].) These vectors $\{\psi_\omega; \omega \in (\mathcal{A}_c^*)^+\}$ form a self-dual cone in a subspace of $\mathcal{H}$. If we now have a C*-dynamical system then one can look at this cone. Before stating their properties we need some notations:

II.3.2 Notation

Let $\{\mathcal{A}, G, \alpha\}$ be a C-dynamical system with G being a topological group. Let $\mathcal{H}$ be the Hilbert-space of the standard representation of $\mathcal{A}^{**}$ and let $\mathcal{H}^+$ be the natural cone associated with this representation then we denote*

(i) $\mathcal{H}_c^+ = \{\psi_\omega; \omega \in (\mathcal{A}_c^*)^+\}$.
(ii) $\mathcal{H}_c$ = *smallest sub-Hilbert-space of* $\mathcal{H}$ *containing* $\mathcal{H}_c^+$.
(iii) *Denote the canonical involution associated with the standard representation of* $\mathcal{A}^{**}$ *by* J.

With these notations we obtain

II.3.3 Proposition

With the assumptions and notations of II.3.2 one obtains

(i) $\mathcal{H}_c^+$ *is a closed cone*
(ii) *The space* $\mathcal{H}_c$ *is invariant under the canonical involution* J.
(iii) *If* $\mathcal{H}_c^J$ *denotes the vectors* $J\psi = \psi$ *then* $\mathcal{H}_c^+$ *is a self-dual cone in* $\mathcal{H}_c^J$ *and* $\mathcal{H}_c$ *is algebraically generated by* $\mathcal{H}_c^+$.
(iv) *If* E_c *denotes the projection onto* $\mathcal{H}_c$ *then for every* $\psi \in \mathcal{H}^+$ *one has* $E_c\psi \in \mathcal{H}_c^+$.

These results can be found in [11]. Knowing that the space $\mathcal{A}_c^*$ leads to a self-dual cone $\mathcal{H}_c^+$ in $\mathcal{H}_c$ one might ask whether this Banach-space also can be considered as the pre-dual of a von Neumann-algebra. In order to answer this question we recall the result of A.Connes [17] that there is a bijection between von Neumann-algebras and self-dual cones in real Hilbert-spaces, which are facially homogenious and oriented. We are able to use these techniques and can show the following result

II.3.4 Theorem

Let $\{\mathcal{A}, G, \alpha\}$ be a C-dynamical system and let $\mathcal{A}_c^*$ be the subspace of $\mathcal{A}^*$ introduced by definition II.2.2 then*

(1) $\mathcal{A}_c^*$ *is the pre-dual of a von Neumann-algebra* $\mathcal{N}_c$.
(2) $\mathcal{N}_c$ *is isomorphic to a sub-von Neumann-algebra* $\mathcal{M}_c \subset \mathcal{A}^{**}$.
(3) $\mathcal{M}_c$ *is the von Neumann-algebra generated by the support-projections of all states* $\omega \in \mathcal{A}_c^*$.
(4) *If*

$$\mathcal{M}(\mathcal{A}_c^*) = \{x \in \mathcal{A}^{**}; x\phi \in \mathcal{A}_c^*,\ \phi x \in \mathcal{A}_c^* \forall \phi \in \mathcal{A}_c^*\}$$

denotes the elements in the enveloping von Neumann-algebra which, together with their adjoints, leave $\mathcal{A}_c^*$ *invariant and if*

$$\mathrm{Ker}(\mathcal{A}_c^*) = \{x \in \mathcal{A}^{**}; x\phi = \phi x = 0 \forall \phi \in \mathcal{A}_c^*\}$$

then we have

$$\mathcal{M}_c = \mathcal{M}(\mathcal{A}_c^*)/\mathrm{Ker}(\mathcal{A}_c^*).$$

The proof of these results can be found in [13].

III Translation Group and Spectrum in a Cone

III.1 Spaces of Momentum Transfer

From now on the symmetry group will be the translation group $\mathbf{R}^d$. Since this is an Abelian group we will construct a spectral theory for these automorphisms. The spectral subspaces associated with sets in the dual group of $\mathbf{R}^d$ are known in physics by the name of spaces of momentum transfer. One difficulty in this investigation is the circumstance that the group action is not necessarily continuous on the algebra $\mathcal{A}$, and that it acts only continuous on parts of $\mathcal{A}^*$, namely on $\mathcal{A}_c^*$. One can take care of this situation by introducing the following concepts:

III.1.1 Definition

(i) *For $f \in \mathcal{L}^1(\mathbf{R}^d)$ and $x \in \mathcal{A}^{**}$ define*

$$[x(f)] = \{y \in \mathcal{A}^{**}; \phi(y) = \int \phi(\alpha_a x) f(a)\, da,\ \forall \phi \in \mathcal{A}_c^*\}.$$

This definition makes sense since for $\phi \in \mathcal{A}^*$ the function $a \to \phi(\alpha_a x)$ is a continuous one.

(ii) *By N_0 we denote the subspace of $\mathcal{A}^{**}$ which annihilates $\mathcal{A}^*$.*

(iii) *Let G be a closed set in $\mathbf{R}'^d$, then we denote by $M(G)$ the following set of operators*

$$M(G) = \{x \in \mathcal{A}^{**}; [x(f)] \subset N_0 \forall f \in \mathcal{L}^1(\mathbf{R}^d) \text{ with supp } \mathcal{F}^{-1} f \subset \mathbf{R}'^d \setminus G\}.$$

Equivalently for every $\phi \in \mathcal{A}_c^*$ one has $\mathcal{F}^{-1}\phi(\alpha_a x) \subset G$ (where the Fourier-transformation is taken in the sense of tempered distributions).

If G is the empty set then we identify $M(\emptyset)$ with N_0.

The properties of the spaces $M(G)$ are the following:

III.1.2 Proposition

The spaces $M(G)$ of momentum transfer have the following properties:

(i) *$M(G)$ is a linear space which is σ-weakly closed.*

(ii) *$M(-G) = M(G)^*$.*

(iii) *$G_1 \subset G_2$ implies $M(G_1) \subset M(G_2)$.*

(iv) *$M(\cap_\beta G_\beta) = \cap_\beta M(G_\beta)$.*

(v) *Assume $G_1 \cap G_2 = \emptyset$ and one of the G_i is compact then*

$$M(G_1 \cup G_2) = M(G_1) + M(G_2).$$

(vi) *$\alpha_a M(G) = M(G)$ for every $a \in \mathbf{R}^d$.*

(vii) *Let $\mathcal{A}_G^{**}$ be the set of elements in $\mathcal{A}^{**}$ which are pointwise invariant under the action of the group $\mathbf{R}^d$ then:*

$$\mathcal{A}_G^{**} \subset M(\{0\}).$$

(viii) *$M(\mathbf{R}'^d) = \mathcal{A}^{**}$.*

(ix) *Let β be an automorphism commuting with the translations α_g then*

$$\beta M(G) = M(G).$$

(x) *For $y \in \mathcal{A}_G^{**}$ one obtains:*

$$M(G)y \subset M(G) \quad \text{and} \quad yM(G) \subset M(G).$$

The concept of momentum transfer is common knowledge in quantum physics. In the case where the action of the group on the algebra is continuous it has been introduced into mathematics by *Arveson* [5]. The proof of the properties listed here can be found in the book of *G.K. Pedersen* [29]. The generalization to non-continuous action prsented here is streight forward.

These spaces of momentum transfer are useful, since they allow the application of harmonic analysis, which is in our case the theory of the Fourier-transfromation. An example for these methods is given in the following

III.1.3 Lemma

Denote by $G_r \subset \mathbf{R}'^d$ the ball of radius r centered at zero.

(i) *$x \in \mathcal{A}^{**}$ belongs to $M(G_r)$ if and only if for every $\phi \in \mathcal{A}_c^*$ the function*

$$a \rightarrow \phi(\alpha_a x)$$

extends to an entire analytic function $W(z)$ and this function obeys the extimate

$$|W(z)| \leq \|\phi\| \|x\| \exp\{r\|\Im z\|\}.$$

(ii) *For the above x there exixt $x_i \in M(G_r)$ such that*

$$[\alpha_a x] = \sum \frac{a^i}{i!}[x_i].$$

Here i denotes the multi-index $i = (i_1 ..., i_d)$ with $|i| = \sum i_j$, and $a^i = \prod (a_j)^{i_j}$, and $i! = \prod i_j!$.

If x is such that for every $\phi \in \mathcal{A}_c^*$ one has the estimate $|\phi(x)| \leq \|\phi\| n(x)$, then one obtains

$$|\phi(x_i)| \leq \|\phi\| n(x) r^{|i|}$$

where $n(x)$ is the semi-norm obtained by restricting x to $\mathcal{A}_c^*$.

For the proof of the last lemma one uses standard results for the theory of Fourier-transformations and the Hahn-Banach-theorem.

III.2 The One-dimensional Case

Our aim is to characterize covariant representations $\{\pi, U, \mathcal{H}\}$ of the dynamical system $\{\mathcal{A}, \mathbf{R}^d, \alpha\}$, such that the spectrum of the representation $U(a)$ is in a given convex cone. We will treat the special case $d = 1$ as an intermediate step. In this case the cone in the dual space will be the half-line $\mathbf{R}^+$. This special case has first been treated by *Borchers* [8]. But the result was not completely unexpected, since *Araki* [4] observed that a positive- energy representation of the automorphisms α_a leave the center of $\pi(\mathcal{A})''$ pointwise invariant.

The basic idea of handling this situation is extracted from the following observation: Assume we have a representation $U(a)$, $a \in \mathbf{R}$ such that it's spectrum is contained in $\mathbf{R}^+$ i.e.

$$U(a) = \int_0^\infty e^{ia\epsilon} dE(\epsilon).$$

Let Δ be a subset of $\mathbf{R}^+$ and $E(\Delta)$ the corresponding spectral projection. If now Γ is a subset of $\mathbf{R}$ and $x \in M(\Gamma)$ and if $U(a)$ is restricted to $\pi(x)E(\Delta)\mathcal{H}$ then it contains only that part of the spectrum which is contained in $(\Gamma + \Delta) \cap \mathbf{R}^+$. In particular, if $(\Gamma + \Delta) \subset (-\infty, 0)$ then $\pi(x)$ annihilates all vectors of $E(\Delta)\mathcal{H}$. This leads to the following notation:

III.2.1 Definition

(i) *For $\lambda \in \mathbf{R}$ let $E(\lambda)$ be the maximal projection in annihilating $M([-\infty, -\lambda])$ from the right, i.e. $E(\lambda)$ is the projection onto the common nullspace of all $x \in M([-\infty, -\lambda])$.*
(ii) *E^+ will denote the projection*

$$E^+ = \lim_{\lambda \to \infty} E(\lambda).$$

which we have to show exists.

These projections $E(\lambda)$ define in $E^+ \mathcal{A}^{**}$ a spectral family as shown in the following

III.2.2 Proposition

With the notation of Definition II.4.1 we obtain:

(i) *$E(\lambda)$ is monotone increasing and hence E^+ is the strong limit of the $E(\lambda)$.*
(ii) *$E(\lambda)$ is continuous from the left.*
(iii) *$E(\lambda) = 0$ for $\lambda \leq 0$.*
(iv) *$E(\lambda)$ is invariant, more precisely $E(\lambda) \in Z(\mathcal{A}_G^{**})$ - the center of $\mathcal{A}_G^{**}$.*
(v) *Let β be an automorphism of $\mathcal{A}$ which commutes with the translations α_a then one finds*

$$\beta E(\lambda) = E(\lambda).$$

The properties of the projections $E(\lambda)$ are consequences of the properties of the spaces $M(G)$ described in Proposition III.1.2. We have learned that the projections $E(\lambda)$ belong to $Z(\mathcal{A}_G^{**})$ and consequently also $E^+ = \lim_{\lambda \to \infty} E(\lambda)$ belongs to this algebra. Moreover, E^+ belongs to the center of $\mathcal{A}^{**}$. As we know from chapter II the best place for investigating this non-trivial result is the dual space. Therefore, we introduce

III.2.3 Definition

With the assumptions of this section define

(i) $\mathcal{A}^*(\mathbf{R}'^+) = \{\phi \in \mathcal{A}^*; E^+\phi = \phi E^+ = \phi\}$
(ii) $\mathcal{A}_0^*(\mathbf{R}'^+) = \{\phi \in \mathcal{A}^*; \text{there exist} 0 < \lambda < \infty \text{ with } E(\lambda)\phi = \phi E(\lambda) = \phi\}.$

$\mathcal{A}^*(\mathbf{R}'^+)$ is a subspace of $\mathcal{A}_c^*$. It's properties are given in the following

III.2.4 Proposition

With the assumptions and notations of this section one has

(i) *$\mathcal{A}^*(\mathbf{R}'^+)$ is a norm-closed linear subspace of $\mathcal{A}^*$ and $\phi \in \mathcal{A}^*(\mathbf{R}'^+)$ implies $\phi^* \in \mathcal{A}^*(\mathbf{R}'^+)$.*
(ii) *$\mathcal{A}_0^*(\mathbf{R}'^+)$ is norm-dense in $\mathcal{A}^*(\mathbf{R}'^+)$.*

(iii) $\mathcal{A}_0^*(\mathbf{R}'^+) \subset \mathcal{A}_c^*$.
(iv) *If* $\phi \in \mathcal{A}_0^*(\mathbf{R}'^+)$ *then* $a \to \alpha_a$ *extends to an entire analytic function and there exist a constant* $\lambda > 0$ *with*

$$\|\phi \circ \alpha_z\| \leq \|\phi\| e^{\lambda|\Im m\, z|}.$$

(v) $\phi \in \mathcal{A}^*(\mathbf{R}'^+)$ and $x \in \mathcal{A}^{**}$ *implies* $x\phi \in \mathcal{A}_c^*$ and $\phi x \in \mathcal{A}_c^*$.
(vi) $\phi \in \mathcal{A}_0^*(\mathbf{R}'^+)$ and $x \in M([-\mu,\mu])$ *imply* $x\phi$ *and* ϕx *both belong to* $\mathcal{A}_0^*(\mathbf{R}'^+)$.
(vii) *If* $\phi \in \mathcal{A}^*(\mathbf{R}'^+)$ and $x, y \in \mathcal{A}^{**}$ *then* $x\phi y$ *belongs to* $\mathcal{A}^*(\mathbf{R}'^+)$.
(viii) *The projection* E^+ *belongs to the center of* $\mathcal{A}^{**}$ *which is equivalent to the statement that* $\mathcal{A}^*(\mathbf{R}'^+)$ *is a folium.*

The proof of the properties (i) to (v) is more or less straight forward. To prove (vi) an essentially new feature is needed. Since one has chosen $x \in M(-\mu,\mu)$, and since $x\phi$ and ϕx is known to belong to $\mathcal{A}_c^*$, one can use the development of $\alpha_a x$ proved in Lemma III.1.3. From this we obtain

$$\phi(x\alpha_a y) = \phi(\alpha_a\{(\alpha_{-a})y\}) = \sum_{i=0}^{\infty} \phi(\alpha_a\{x_i y\}).$$

Since an estimate for $\|x_i\|$ is known, we can prove that $E(\lambda)\phi = \phi E(\lambda) = \phi$ and $x \in M(-\mu,\mu)$ implies $E(\lambda+\mu)x\phi = x\phi$. The other statements again are simple consequences hereof.

From $E^+ \in Z(\mathcal{A}^{**})$ it follows that $\mathcal{A}^*(\mathbf{R}^+)$ coincides with the set of normal states of the von Neumann-algebra $E^+\mathcal{A}^{**}$. Since E^+ also belongs to $Z(\mathcal{A}_G^{**})$, this von Neumann-algebra is invariant under the action of α_a and since $\mathcal{A}^*(\mathbf{R}'^+) \subset \mathcal{A}_c^*$ it follows that the automorphisms α_a act weakly continuous on $\mathcal{A}^{**}E^+$.

Having in $E^+\mathcal{A}^{**}$ a spectral family $\{E(\lambda)\}_0^\infty$ it is natural to ask for it's meaning. It is now our aim to show that the unitary group (in $E^+\mathcal{A}^{**}$)

$$U(a) = \int_0^\infty e^{ia\lambda}\, dE(\lambda)$$

implements the automorphisms α_a. The result is

III.2.5 Theorem

Let $\{\mathcal{A}, \mathbf{R}, \alpha\}$ *be a* C^**-dynamical system and assume the projection* E^+ *defined in II.2.1 is not zero,* $E^+ \in Z(\mathcal{A}^{**})$. *Let* $E(\lambda) \in \mathcal{A}^{**}E^+$ *be the family of projections which are defined in II.2.1. Let*

$$U(a) = \int_0^\infty e^{ia\lambda} E(\lambda)$$

which is a unitary group. Then

(i) $U(a)$ *implements the automorphisms* α_a*; namely:*

$$\alpha_a x E^+ = U(a) x U^*(a)$$

for every $x \in \mathcal{A}^{**}$.

(ii) $U(a)$ *is minimal in the following sense. Let* $\{\pi, \mathcal{H}\}$ *be a normal representation of* $\mathcal{A}E^+$. *Assume there exists on* $\mathcal{H}$ *a continuous unitary group* $V(a)$ *such that*

(α)
$$\pi(\alpha_a x) = V(a)\pi(x)V^*(a).$$

(β) *The spectrum of $V(a)$ is contained in $\mathbf{R}^+$.*

If we write $U(a) = \exp\{\imath H a\}$ and $V(a) = \exp\{\imath H' a\}$ then follows $H' \geq \pi(H)$.

Before proving this theorem one needs to know some more about the relations between the projections $E(\lambda)$ and the spaces $M(\lambda_1, \lambda_2)$ given in the following

III.2.6 Lemma

With the same assumptions and notations as before one obtains

(i) *Assume $x \in E^+ M(G_1)$ and $y \in E^+ M(G_2)$ then it follows that*
$$xy \in M(\overline{G_1 + G_2}).$$

(ii) *For every $x \in \mathcal{A}^{**}$ one has $\lambda > 0$*
$$xE(\lambda) \in M([-\lambda, \infty]).$$

(iii) *For every $x \in \mathcal{A}^{**}$ one has $(\lambda, \mu > 0)$*
$$(E^+ - E(\lambda))xE(\mu) \in M([-\mu + \lambda, \infty]).$$

(iv) *For every $x \in \mathcal{A}^{**}$, and $0 < \lambda_1 < \lambda_2$ and $0 < \mu_1 < \mu_2$ one has*
$$\big(E(\lambda_2) - E(\lambda_1)\big)x\big(E(\mu_2) - E(\mu_1)\big) \in M([-\mu_2 + \lambda_1, -\mu_1 + \lambda_2]).$$

The essential point of this last lemma is the result stated under (iv). It allows us to obtain estimates for the approximating operators. One fixes a λ_0 and approximates $U(a)E(\lambda_0)$ by a sum
$$U_I(a) \sum_{j=1}^{N} e^{\imath a \nu_j} \big(E(\mu_j) - E(\mu_{j-1})\big)$$
with $0 = \mu_0 < \mu_1 < ... < \mu_N = \lambda_0$ and $\mu_{j-1} \leq \nu_j \leq \mu_j$. If now $\epsilon(I) = \max(\mu_i - \mu_{i-1})$ then one concludes with the mentioned result of Lemma III.2.6 that
$$\operatorname{supp} \mathcal{F}^{-1} E(\lambda_0) U_I^*(a) \alpha_a x U_I(a) E(\lambda_0) \subset [-\epsilon(I), \epsilon(I)].$$

This shows us that
$$E(\lambda_0) U_I^*(a) \alpha_a x U_I(a) E(\lambda_0)$$
is an entire analytic function of exponential type $\epsilon(I)$ which is bounded on the real axis by $\|x\|$ and which takes the value x at $a = 0$. If now $\epsilon(I_n)$ converges to zero then one obtains
$$E(\lambda_0) U(a) x U^*(a) E(\lambda_0) = E(\lambda_0) \alpha_a x E(\lambda_0).$$

Next we choose a sequence $\lambda_0 \to \infty$ and obtain the desired result.

III.3 The General Problem, Spectrum in a Cone

The following situation will be investigated in this section. $\{\mathcal{A}, \mathbf{R}^d, \alpha\}$ is a C*-dynamical system and $V \subset \mathbf{R}^d$ will be cone with the properties

(i) V is a closed convex cone.
(ii) V contains interior points.
(iii) V is a proper cone i.e. $V \cap \{-V\} = \{0\}$.

The dual cone will be denoted by V' which is again a proper closed cone with interior points.

We want to characterize covariant representations $\{\mathcal{H}, \pi, U(a)\}$ such that $U(a)$ is a continuous unitary representation of the group $\mathbf{R}^d$ with spectrum $U(a)$ contained in V', and the property that it implements the automorphisms α_a. This means in particular that for every direction $t \in V$ the subgroup $\alpha_{\rho t}$, $\rho \in \mathbf{R}$ fulfills the conditions of section II.4. This implies that we have for every direction $t \in V$ a unitary group $U(\rho t)$. Since $\mathbf{R}^d$ is an Abelian group we know from proposition III.2.2 that $\alpha_a U(\rho t) = U(\rho t)$ for every $a \in \mathbf{R}^d$. Since the $U(\rho t)$ belong to $\mathcal{A}^{**}$ we conclude that the group representations for different directions commute with each other. Choosing now d linearly independent directions in the cone V we can construct a unitary representation of the group $\mathbf{R}^d$. This representation is not unique, the freedom left is a representation of the group belonging to the center of $\mathcal{A}^{**}$. It is the problem of this section to show that this freedom can be used in order to construct a representation of the group $\mathbf{R}^d$ the spectrum of which is contained in the cone V'. In order to handle this problem we introduce the following

III.3.1 Definition

Let $\{\mathcal{A}, \mathbf{R}^d, \alpha\}$ be a C-dynamical system and let $V \subset \mathbf{R}^d$ be a proper convex closed cone with interior points*

(i) *For $t \in V$, $t \neq 0$ denote by $E(t, \lambda)$ the spectral projection of the group representation $U(\rho t)$, $\rho \in \mathbf{R}$, described in section III.2. These projections are invariant under α_a by Proposition III.2.2 (v).*
(ii) *By $E^+(t)$ we denote s-$\lim_{\lambda\to\infty} E(t, \lambda)$ which belongs to $Z(\mathcal{A}^{**})$ by III.2.4 (viii).*
(iii) *Define $E(V') = \prod\{E(t)^+;\ t \in V,\ t \neq 0\}$ where the product is the limit of the decreasing net of finite products. $E(V') \in Z(\mathcal{A}^{**})$.*
(iv) *For $p \in V'$ define*

$$E(<0, p>) = \prod\{E(t, \lambda_t);\ t \in V,\ t \neq 0,\ \lambda_t = (p, t)\},$$

in which $<0, p>$ stands for the order interval $V' \cap \{p - V'\}$. These projections are again invariant under α_a.

We start the investigation with the following preparation:

III.3.2 Lemma

With the notation of Definition III.3.1 we obtain:

(i) *Let p_n be increasing relative to the order of V' such that $\bigcup_n <0, p_n>$ covers all of V'. Then one finds s-$\lim_{n\to\infty} E(<o, p_n>) = E(V')$.*
(ii) *For every $x \in \mathcal{A}^{**}$ the function*

$$a \to \alpha_a(x) E(<o, p>)$$

is weakly continuous and

$$\operatorname{supp}\mathcal{F}^{-1}\alpha_a(x)E(<o,p>) \subset -p+V'$$

For the proof of this lemma one is using standard methods of analysis and in particular of the theory of Fourier-transformations.

With help of this Lemma one is now able to construct a group representation fulfilling the spectrum condition. To this end we introduce the following

III.3.3 Definition:

With the same assumptions and notation as before:

(i) $B = \{b^1, ..., b^d\}$ *denotes a basis of* $\mathbf{R}^d$ i.e. b^i *are such that* $b^i \in V$, *and* $\{b^i\}$ *are linearly independent.*

(ii) *For every* $b^i \in B$ *let* $U(\lambda b^i) \in \mathcal{A}^{**}E(V')$ *be the minimal representation of the one parameter group fulfilling the spectrum condition and implementing the automorphisms* $\alpha_{\lambda b^i}$ *described in section* III.2. *For* $a \in \mathbf{R}^d$ *and* $a = \sum \lambda_i b^i$ *define*

$$U_B(a) = \prod_{i=1}^{d} U(\lambda_i b^i).$$

(iii) *By* V_B' *we denote the following set:*

$$V_B' = \{p; (p, b^i) \geq 0, \ i = 1, ..., d\}.$$

By construction of $U_B(a)$ *it follows that spectrum* $U_B(a) \subset V_B'$.

The representation $U_B(a)$ is certainly not the only one which implements the automorphisms α_a on $\mathcal{A}^{**}E(V')$. If $Z(a)$ is a continuous representation of translationgroup $Z(a) \in Z(\mathcal{A}^{**}E(V'))$ then $U_B(a)Z(a)$ again belongs to $\mathcal{A}^{**}E(V')$ and implements α_a. Therefore one can ask whether one can find $Z(a)$ is such a way that spectrum $U_B(a)Z(a) \subset V'$. That this is indeed the case will be the next result.

III.3.4 Theorem

Let $\{\mathcal{A}, \mathbf{R}^d, \alpha\}$ *be a* C^**-dynamical system and assume* $E(V')$ *is not zero (for the definition see* III.3.1 (iii)). *Then there exists a continuous unitary representation* $U(a)$ *of the translation group with*

(i) $U(a) \in \mathcal{A}^{**}E(V')$.

(ii) *Spectrum* $U(a) \subset V'$.

(iii) $U(a)$ *implements the automorphism* α_a *on* $\mathcal{A}^{**}E(V')$.

Since it is interesting to see how all parts obtained so far, fit together, we will present the proof:

Proof:

Let Γ be a compact set in V_B' and denote by $\Delta =< 0, p > \subset V'$ which is also compact. Denote by $F(\Gamma)$ the spectral projections of $U_B(a)$ and let $E(\Delta)$ be the projections introduced in II.6.1 (iv). All these projections belong to $\mathcal{A}_G^{**}$ and hence $F(\Gamma)E(\Delta)$ is again an invariant projection. Taking a sequence Γ_n such that $\cup_n \Gamma_n = V_B'$ and a sequence $\Delta_n =< 0, p_n >$ such that $\cup_n \Delta_n = V'$ then by construction of $F(\Gamma)$ and by Lemma II.6.2 (i) one obtains $\text{s-lim}_{n \to \infty} F(\Gamma_n)E(\Delta_n) = E(V')$. Let $X(\Gamma, \Delta)$ be the central carrier of $F(\Gamma)E(\Delta)$ then $X(\Gamma, \Delta)$ is nothing other than the common range projection of all elements of the form $xF(\Gamma)E(\Delta)$, $x \in \mathcal{A}^{**}$. Hence investigating the expression $X(\Gamma,\Delta)U_B(a)$ is the same as investigating the family of expressions

$$U_B(a)xE(\Delta)F(\Gamma) = \alpha_a(x)U_B(a)E(\Delta)F(\Gamma)$$
$$= \alpha_a(x)E(\Delta)U_B(a)F(\Gamma).$$

From the definition of $F(\Gamma)$ it follows that supp $\mathcal{F}^{-1}U_B(a)F(\Gamma) \subset \Gamma$ and from Lemma III.3.2 (ii) we know supp $\mathcal{F}^{-1}\alpha_a(x)E(\Delta) \subset -\Delta + V'$. Putting both together one has $\mathcal{F}^{-1}U_B(a)\alpha_a(x)E(\Delta)F(\Gamma) \subset \Gamma - \Delta + V'$. Since Γ and Δ are both compact we can choose $q = q(\Gamma,\Delta)$ such that $\Gamma - \Delta + q(\Gamma,\Delta) \subset V'$ with spectrum $X(\Gamma,\Delta)U_B(a)e^{i(a,q(\Gamma,\Delta))} \subset V'$. Take now a sequence Γ_n, Δ_n covering V_B' and V' then $X(\Gamma_n,\Delta_n)$ converges to $E(V')$. Define

$$W(a) = \sum_1^\infty \{X(\Gamma_{n+1},\Delta_{n+1}) - X(\Gamma_n,\Delta_n)\}e^{i(a,q(\Gamma_{n+1},\Delta_{n+1}))}$$

then $W(a)$ is a continuous unitary representation of the translations belonging to $Z(\mathcal{A}^{**}E(V'))$ therefore $U(a) = U_B(a)W(a)$ is a continuous unitary representation of the translations belonging to $\mathcal{A}^{**}E(V')$ which implements α_a. It remains to show that $U(a)$ fulfills the spectrum condition. We have

$$U(a) = U_B(a)W(a) = \sum_1^\infty \{X(\Gamma_{n+1},\Delta_{n+1}) - X(\Gamma_n,\Delta_n)\}U_B(a)e^{i(a,q(\Gamma_{n+1},\Delta_{n+1}))}.$$

Since by the above calculation and the definition of $q(\Gamma,\Delta)$ each term of the summand has its spectrum in V', it follows that spectrum $U(a) \subset V'$. □

III.4 Characterization of Positive-Energy-States

Given a C*-dynamical system $\{\mathcal{A}, \mathbf{R}^d, \alpha\}$ and a proper convex cone V in $\mathbf{R}^d$ with interior points, we have characterized those representations $\{\pi, U(a), \mathcal{H}, V'\}$ which are covariant with the property that the spectrum of $U(a)$ can be chosen to be in the cone V'. We have found that there was a projection $E(V')$ belonging to $Z(\mathcal{A}^{**})$ with the property that a representation $\{\pi, \mathcal{H}\}$ can be extended to a covariant representation fulfilling the spectrum condition iff the representation π is quasi-equivalent to a subrepresentation of $E(V')\mathcal{A}^{**}$. Usually $E(V')$ is not explicitly known. Therefore, one has to look for a better characterization of such representations.

Generally states are much easier to investigate than whole representations. Therefore, we will study the normal states of $E(V')\mathcal{A}^{**}$. To do this is also suggested by the results of section II.2 . In the following we will call the normal states of $E(V')\mathcal{A}^{**}$ positive-energy-states. In the one-dimensional case we have already introduced the space of positive-energy-states (Definition III.2.3). For the general situation we define correspondingly:

III.4.1 Definition

With the notation of this section we denote:

(i) $\mathcal{A}^*(V') = \{\phi \in \mathcal{A}^*; E(V')\phi = \phi E(V') = \phi\}$.
(ii) $\mathcal{A}_0^*(V') = \{\phi \in \mathcal{A}^*;$ there exists $p \in V'$ with $E(<0,p>)\phi = \phi E(<0,p>) = \phi\}$.

The characterization of positive-energy-states is given in the following

III.4.2 Theorem

Let $\{\mathcal{A}, \mathbf{R}^d, \alpha\}$ be a C^*-dynamical system and $V \subset \mathbf{R}^d$ a closed, convex, proper cone with interior $\overset{\circ}{V} \neq \emptyset$. Using the notation of the last definition one obtains:

(i) *$\mathcal{A}_0^*(V')$ is norm dense in $\mathcal{A}^*(V')$.*

(ii) *An element $\phi \in \mathcal{A}^*$ belongs to $\mathcal{A}_0^*(V')$ if and only if it fulfills the follwing properties:*

  (α) *$a \to \phi(x\alpha_a x)$ is a continuous function on $\mathbf{R}^d$, $\quad x, y \in \mathcal{A}$.*

  (β) *$\phi(x\alpha_a y)$ is the boundary value of an analytic function $W(z)$ holomorphic in the tube*

$$T(V) = \{z \in \mathbf{C}^d;\ \Im m\, z \in \overset{\circ}{V}\}.$$

  (γ) *There exist a constant m such that*

$$|W(z)| \leq \|\phi\|\|x\|\|y\|e^{m\|\Im m\, z\|}$$

  holds for $z \in T(V)$.

  (δ) *ϕ^* fulfills the same conditions as ϕ.*

(iii) *Let $\{\mathcal{H}, \pi\}$ be a representation of $\mathcal{A}$, then one can find a continuous unitary representation $V(a)$ acting on $\mathcal{H}$, which implements α_a with spectrum $V(a) \subset V'$ if and only if every vector state ω_ψ belongs to $\mathcal{A}^*(V')$.*

From Lemma III.3.2(i) we know that $E(< 0, p >)$ converges for $p \to \infty$ to $E(V')$. Hereof the statement (i) is easily derived. If $\phi \in \mathcal{A}_0^*(V')$ such that $\phi E(< 0, p >) = \phi$ then one has

$$\text{supp}\ \mathcal{F}^{-1}\phi(x, \alpha_a y) \subset -p + V'.$$

(See Lemma III.3.2(ii)). This implies the quoted analytic properties and the estimates. If, conversely, the conditions (α) – (δ) of (ii) are fulfilled then the expectation-value $\phi(x, \alpha_a y)$ has an analytic continuation into the tube $T(V) = \{z \in \mathbf{C}^d; \Im m\, z \in \overset{\circ}{V}\}$ and the existing estimate implies by the Palay-Wiener-theorem that there exists $p_1 \in \overset{\circ}{V}$ with

$$\text{supp}\ \mathcal{F}^{-1}\phi(x, \alpha_a y) \subset -p_1 + V'.$$

If the projections $E(\lambda)$ for the group α_{pt} used in the last section are denoted by $E(t, \lambda)$ then one concludes from the fact that ϕ annihilates $M(\{p; (p, t) \leq -(p_1, t) - \epsilon\})$ the relation $\phi E(t, (p_1, t) + \epsilon) = \phi$. Hence one finds $\phi E(< 0, (1 - \epsilon)p >) = \phi$. Since this also holds for ϕ^* one has $\phi \in \mathcal{A}_0^*(V')$. The proof of the third statement is standart again.

IV Consequences of Locality

IV.1 Introduction of Dynamics

In the last two chapters we have investigated the action of a group G on a C^*-algebra $\mathcal{A}$. We had only assumed that the group in question can be imbedded into the automorphism-group of the C^*-algebra $\mathcal{A}$. We then have studied the question of characterizing covariant representations of $\mathcal{A}$. These investigations are of cinematical nature, which means that no detailed structure of the algebra $\mathcal{A}$ has been used.

Next we want to introduce some dynamics, i.e. we will have to deal with a C^*-algebra having a structure closely related to the group. We shall make use of Einstein's principle of maximal velocity and the principle of quantum mechanics.

The assumption of quantum mechanics means that we are dealing with observables which are not always commuting. Furthermore, we assume that the observables are localized in bounded regions of space-time. The principle of relativity requires that any disturbance can spread at most with the velocity of light. This implies that two observations which are spacelike separated cannot influence each other. In the language of quantum mechanics this means that the operators representing these observables must commute with each other.

We want to deal with a theory of local observables, a so-called Araki, Haag, Kastler theory. This is a quantum field theory described in terms of bounded operators. To every bounded open region O in $\mathbf{R}^d$ is associated a C^*-algebra $\mathcal{A}(O)$. The C^* inductive limit of the increasing family $\{\mathcal{A}(O)\}$ is denoted by $\mathcal{A}$. Furthermore we have the translation group $\mathbf{R}^d$ acting on $\mathcal{A}$ as automorphism such that $\alpha_a \mathcal{A}(O) = \mathcal{A}(O+a)$ holds for $a \in \mathbf{R}^d$. If two domains O_1 and O_2 are spacelike separated then we assume, as usual, that $\mathcal{A}(O_1)$ and $\mathcal{A}(O_2)$ commute elementwise with each other. Such a system will be denoted by

$$\{\mathcal{A}(O), \mathcal{A}, \alpha, \mathbf{R}^d\}.$$

By $\mathbf{R}^d$ we always mean the Minkowski space with points $\{x_0, \vec{x}\}$ and scalar product $(x, y) = x_0 y_0 - (\vec{x}, \vec{y})$ where $(\vec{x}, \vec{y})$ is the usual Euclidean scalar product. Therefore d fulfills the restriction $d \geq 2$.

If $V \subset \mathbf{R}^d$ is an open but unbounded domain then $\mathcal{A}(V)$ will be the smallest sub C^*-algebra of $\mathcal{A}$ containing all $\mathcal{A}(O)$ provided $O \subset V$.

If V is an open set in $\mathbf{R}^d$ then we denote by V' the following set:

$$V' = \{x; x - y \quad \text{is spacelike for every} \quad y \in V\}.$$

A set D is called a double cone if it is of the form $\{a + V^+\} \cap \{b - V^-\}$. In order that this set is not empty we must have $b \in a + V^+$. If we have to distinguish between different double cones we write $D_{a,b}$. This version of quantum field theory has been introduced by ***Araki*** [1], ***Haag*** and ***Schroer*** [24], and ***Haag*** and ***Kastler*** [23].

The assumptions made so far, are of algebraic nature. What we have build into the algebra is a structure rich enough to describe physics. However, it is known that there are many branches, as for instance, particle physics, physics of condensed matter, etc. All these different branches are described by the same dynamical system. If we want to deal with a special part of physics we have to choose the right representations. In particle physics, for instance, the total energy of the system shall be an observable, which is bounded below. For this reason one identifies representations fulfilling the spectrum condition with those describing particle physics.

The open forward light-cone will be denoted by V^+. A representation $\{\pi, U, \mathcal{H}\}$ will be called a representation with spectrum condition if the spectrum of $U(a)$ is contained in the closed forward light-cone i.e.

$$\operatorname{spec} U(a) \subset \bar{V}^+$$

where the bar over a set denotes the closure of this set. Representations with spectrum condition will be denoted by

$$\{\pi, U(a), V^+, \mathcal{H}\}.$$

If $U(a)$ is a continuous unitary representation of the translation group on the Hilbert space $\mathcal{H}$, then a vector $\psi \in \mathcal{H}$ is called entire analytic for $U(a)$ if the vector valued function $U(a)\psi$ defined on $\mathbf{R}^d$ has an extension as entire analytic vector valued function defined on $\mathbf{C}^d$. If $\{\pi, U(a), V^+, \mathcal{H}\}$ is a representation fulfilling spectrum condition, and if ψ is an entire analytic vector for $U(a)$ then the vector valued function

$$U(a)\pi(A)U^{-1}(a)\psi = \pi(\alpha_a A)\psi$$

is the boundary value of a vector valued analytic function $F(z)$ which is holomorphic in the forward tube T^+:

$$T^+ = \{z \in \mathbf{C}^d; \Im m\, z \in V^+\}.$$

IV.2 Existence of a Distinguished Group Representation

Having introduced the local quantum fields in the setting of Araki, Haag, and Kastler, we want to look at the consequences of locality. Interesting results are obtained for positive-energy representations and we only will look at those. In this and the next section we only will describe the results, and in the last section we will look at the techniques one has to employ in order to prove them.

Recall: in the one-dimensional case, Section III.2, we had constructed in $\mathcal{A}^{**}E^+$ a representation $U(a)$, fulfilling the spectrum condition, which was minimal. The minimality was as follows: assume $V(a)$ is another representation of the group $\mathbf{R}$ fulfilling also the spectrum condition and which implements the automorphisms α_a. Writing $U(a) = \exp\{\imath Ha\}$ and $V(a) = \exp\{\imath H'a\}$ then H and H' are both positive operators and minimality means $H' \geq H$. This minimality criterium of the one-dimensional situation has first been introduced by *D. Olesen* and *G.K. Pedersen* [27] and *D. Olesen* [26]. In the higher dimensional case there exists in general no minimal representation except the cone has a simplex as basis. This is, for instance, automatically true in the two-dimensional situation. If the dimension is larger than 2, and if the cone is arbitary, thea we have to generalize the definition of minimal representations.

IV.2.1 Definition

Let $\{\mathcal{A}, \mathbf{R}^d, \alpha\}$ be a C^*-dynamical system and let $V \subset \mathbf{R}^d$ be a closed, proper, convex cone with interior points and denote by V' its dual cone. Let $U(a) \in \mathcal{A}^{**}E(V')$ be a continuous unitary representation of the translations with

(i) *Spectrum $U(a) \subset V'$,*
(ii) *$U(a)$ implements the automorphism α_a,*

then $U(a)$ is called minimal if for any other continuous, unitary representation $V(a) \in \mathcal{A}^{**}E(V')$ of the translations fulfilling also (i) *and* (ii) *one has spectrum $V(a)U^*(a) \subset V'$.*

Remarks

(i) If we write $U(a) = \exp\{\imath(a,P)\}$ and $V(a) = \exp\{\imath(a,Q)\}$ then $U(a)$ minimal is equivalent to the relation $(t,P) \leq (t,Q)$ for every $t \in V$.
(ii) If a minimal representation exists then it is necessarily unique.

In general one cannot expect the existence of a minimal representation, except there is a close relation between the structure of the algebra and the cone in question.

This is the case in the theory of local observables, because the two cones appearing in the axioms are dual to each other. The cone in the configuration space used for the locality condition, and the cone in the momentum space used for the spectrum condition are in both cases the forward light cone V which is self-dual with respect to the Minkowski-scalar-product. Because of these special conditions, which are a consequence of dynamics, we obtain a result first proved by *Borchers and Buchholz* [14].

IV.2.2 Theorem

Let $\{\mathcal{A}(O), \mathcal{A}, \alpha, \mathbf{R}^d\}$ be a theory of local observables. Then there exists a unique continuous unitary group representation $U(a) \in \mathcal{A}^{**}E(V^+)$ which fulfills the spectrum condition and is minimal in the sense of Definition IV.1.2.(ii).

This result is, from the point of view of physics, very satisfying. In classical physics only energy differences are measurable quantities. But with Einstein's relation $E = mc^2$ a fixed-point was introduced into the energy scale. Our result shows that in particle physics where the energy is identified with the generator of the time-translations, we have discovered the same fixing of the energy scale. As long as we are only looking at the possibility of positive-energy representations there is no way of defining a unique representation of the translation group. But at the moment we introduced the principle of Einstein causality, we obtained a distinguished representation of the translation group. By differentiation we get a unique representative for energy and momentum. Moreover, these generators belong to the von Neumann-algebra, which is generated by the representation of the local observables. If we want to measure the energy of a state then, as a consequence of the above, we can approximate this measurement with arbitrary accuracy by local measurements. This result is not correct in general. An example is given by a representation describing thermodynamics. Here we can measure only the difference of the energy with respect to a reference-state.

IV.3 Properties of the Spectrum

After we have found, as a consequence of the introduction of the main principles of dynamics, a unique representation of the group of space-time translations, we now can ask whether or not the dynamic also has some consequences for the structure of the spectrum of these translations. In particular, we might ask if the spectrum can occupy an arbitrary part of the forward light cone. The answer has been given by *Borchers* and *Buchholz* [14][10].

IV.3.1 Theorem

Let $\{\mathcal{A}(O), \mathcal{A}, \alpha, \mathbf{R}^d\}$ be a theory of local observables and assume that $U(a)$ is the unique minimal representation of the translations in $\mathcal{A}^{**}E(V^+)$ described in section IV.2. Then for every projection $G \in Z(\mathcal{A}^{**}E(V^+))$,spectrum $U(a)G$ is a set which is invariant under Lorentz transformation.

The invariant sets of the light cone are

(i) the origin $p = 0$,
(ii) the hyperboloids $p^2 = m^2$, $p_0 > 0$, and in two dimensions also the right- or lefthand side of the null-hyperboloid, i.e.
(iii) $p_1 = p_0$, $p_0 > 0$,
(iv) $p_1 = -p_0$, $p_0 > 0$.

If the spectral projection $E(0)$ associated with the point $p = 0$ is not zero, then we call this a representation with a vacuum state ω_0, which is defined by the equation $\omega_0(E(0)) = \omega_0(1)$. For vacuum representations the proof of the invariance of the spectrum's support is much easier and therefore, one can get stronger results, which are obtained in [7]

IV.3.2 Theorem

Let $\{\mathcal{A}(O), \mathcal{A}, \alpha, \mathbf{R}^d\}$ be a theory of local observables and let ω_0 be a vacuum state. Let $\{\pi, U(a), \Omega, \mathcal{H}\}$ be the G.N.S. representation given by ω_0 then the supports of the Lebesgue continuous and discontinuous part of the spectrum are both invariant under Lorentz-transformations.

Another feature to be found in a vacuum representation is the cluster property. If a is a spacelike vector and $\lambda \in \mathbf{R}$ and if ω_0 is the vacuum state then one has for $x, y \in \mathcal{A}$:

$$\lim_{\lambda \to \infty} \omega_0(x\alpha_a(y)) = \omega_0(x)\omega_0(y).$$

The cluster property implies that the spectrum is additive, as observed by Wightman

IV.3.3 Theorem:

Let $\{\mathcal{A}(O), \mathcal{A}, \alpha, \mathbf{R}^d\}$ be a theory of local observables and let $\{\pi, U(a), \Omega, \mathcal{H}\}$ be a vacuum representation then the spectrum of is $U(a)$ a semi-group, i.e.

$$\text{spec}\ U(a) + \text{spec}\ U(a) \subset \text{spec}\ U(a).$$

If we are dealing with a representation without a vacuum vector, then I conjecture that one has the relation

$$3\text{spec}\ U(a) \subset \text{spec}\ U(a).$$

The general proof of this conjecture has failed so far because of technical difficulties. But there are special situations for which it has been proved:
1.) The representation in question is connected with the vacuum representation by a charged field.
2.) The lower boundary of the spectrum is an isolated hyperboloid $p^2 = m^2$. In this situation it can be shown that the continuous part of the spectrum must start at least at $p^2 = (3m)^2$. This result can be found in [7].

IV.4 Remarks on the Techniques Applied for the Proofs

We, previously, have indicated the proofs of the results discussed in chapter II and III. There we have made use of standard methods in functional and harmonic analysis. At two places we have used results of the *Tomita-Takesaki*-theory [31][30], which is the theory of standard representations of von Neumann-algebras. For the rest the most complicated theorem used was that of *Paley* and *Wiener* [28] in II.1.

In this chapter we made use of the fact that we have properties not only in the configuration space but also in the momentum space. In the language of groups this means the functions we study are restricted not only on the group but also on the dual group. According to my knowledge, there exists no general method of characterizing such functions. In our case where the group and also it's dual group is $\mathbf{R}^d$ one can employ the theory of several complex variables, provided the properties of the functions can be reduced to support-properties. In this situation a result like the Paley-Wiener-theorem gives the analyticity. But since we are dealing with functions of several variables, we obtain holomorphic functions of several complex variables. But not every domain in the $\mathbb{C}^d$ is a natural one, which means, there are domains $G \subset \mathbb{C}^d$ such that every function holomorphic in G is automatically analytic in a larger domain $H(G)$ called the envelope of holomorphy. If, regarding our problem, this larger domain $H(G)$ contains more real points, this can be transformed into properties of the spectrum.

Let me now indicate how these ideas have been used in the theory of local observables. Assume $\{\pi, U(a), V^+, \mathcal{H}\}$ is a representation of the theory of local observables fulfilling the spectrum condition. Then $U(a)$ can be written as

$$U(a) = \int_{\bar{V}} e^{i(a,p)}\, dE(p).$$

Let now Δ be a compact set and assume $E(\Delta) \neq 0$. Let $\psi \in \mathcal{H}$ with $E(\Delta)\psi = \psi$ then we want to study the two functions

$$F(a) = (\psi, A\alpha_a(B)\psi) \quad \text{and}$$
$$G(a) = (\psi, \alpha_a(B)A\psi)$$

for A, B two operators belonging to $\pi(\mathcal{A}(O))$ for a properly choosen set O. We take, for instance, for O the set.

$$D_t := \{-t + V^+\} \cap \{t + V^+\}$$

with $t \in V^+$. If now a is spacelike with regard to $2D_t$ then the two functions coincide, i.e.

$$F(a) - G(a) = 0 \quad \text{for} \quad a \in (2D_t)'.$$

From this we learn that we can decompose this difference into two functions having restricted supports

$$F(a) - G(a) = H^+(a) - H^-(a)$$

with supp $H^+(a) \subset \{-t + V^+\}$ and supp $H^-(a) \subset \{t - V^+\}$.

Looking now at the inverse Fourier-transformed of these two functions, one observes that $\mathcal{F}^{-1}H^+(p)$ is the boundary value of a function holomorphic in $T^+ = \{z \in \mathbb{C}^d; \Im m\, z \in V^+\}$ and $\mathcal{F}^{-1}H^-(p)$ is the boundary value of a function holomorphic in $T^- = -T^+$.

If now the two functions $\mathcal{F}^{-1}H^+(p)$ and $\mathcal{F}^{-1}H^-(p)$ coincide on parts of the real space then the two analytic functions are two branches of one analytic function holomorphic in a domain which mostly can be enlarged. In order to show that $\mathcal{F}^{-1}H^+(p) = \mathcal{F}^{-1}H^-(p)$ on some set, we observe the relation $H^+ - H^- = F - G$,which also holds after Fourier-transformation. Since we have

$$\alpha_a(B) = U(a)BU^*(a)$$

and since $U(a)$ fulfills the spectrum condition one obtains by choice of the vector ψ:

$$\text{supp}\, \mathcal{F}^{-1}F(p) \subset \{-\Delta + \bar{V}^+\} \quad \text{and}$$
$$\text{supp}\, \mathcal{F}^{-1}G(p) \subset \{\Delta - \bar{V}^+\}.$$

From this we obtain

$$\mathcal{F}^{-1}(H^+ - H^-)(p) = 0 \quad \text{for} \quad p \in \complement\left\{\{-\Delta + V^+\} \cup \{\Delta - V^+\}\right\}$$

which is not empty since Δ was compact.

References

[1] H. Araki
Einführung in die Axiomatische Quantenfeldtheorie
Vorlesungen an der E.T.H. Zürich **I** W.S. 1961/62, **II** S.S. 1962

[2] H. Araki
Some Properties of Modular Conjugation Operator of a von Neumann Algebra and a Radon-Nikodym Theorem with a Chain Rule
Pac. J. Math. **50**, 309-354, (1974)

[3] H. Araki
Positive Cones for von Neumann Algebras
In: Operator Algebras and Applications
Proc. of Symposia Vol **38** part 2, A.M.S. Providence (1982)

[4] H. Araki
On the Algebra of all Local Observables
Progr. Theor. Phys. **32**, 844, (1964)

[5] W.B. Arveson
On Groups of Automorphisms of Operator Algebras
J. Functional Anal. **15**, 217-243, (1974)

[6] A. Aviskai and H. Ekstein
Presymmetry of Classical and Relativistic Fields
Phys. Rev. **D 7**, 983, (1973)

[7] H.J. Borchers
Local Rings and the Connection of Spin with Statistics
Commun. Math. Phys **1**, 281-307,(1965)

[8] H.J. Borchers
Energy and Momentum as Observables in Quantum Field Theory
Commun. Math. Phys. **2**, 49-54, (1966)

[9] H.J. Borchers
On the Implementability of Automorphism Groups
Commun. Math. Phys. **14**, 305-314, (1969)

[10] H.J. Borchers
Locality and Covariance of the Spectrum
Fizika **17**, 289-304,(1985)

[11] H.J. Borchers:
C^*-Algebras and Automorphism Groups
Commun. Math. Phys. **88**, 95-103 (1983)

[12] H.J. Borchers
A Remark on Antiparticles
Lecture notes in Physics, Vol. **257**, 268,(1986)

[13] H.J. Borchers
Covariance sub-algebras connected with symmetry groups of C^*-algebras
Preprint (1990)

[14] H.J. Borchers and D. Buchholz
The Energy-Momentum Spectrum in Local Field Theories with Broken Lorentz-Symmetries
Commun. Math. Phys. **97**, 169,(1985)

[15] D. Buchholz
Gauss'- Law and the Infraparticle Problem
Phys. Lett. **174 B**, 331-334, (1986)

[16] A. Connes
Sur le théorème de Radon Nikodym pour les poids normaux fidèles semifinis
Bull. Sci. Math 2ème Sér **97**, 253-258, (1973)

[17] A.Connes
Caractérisation des algèbres de von Neumann comme espaces vectoriels ordonés
Ann. Inst. Fourier **26**, 121-155. (1974)

[18] S. Doplicher, D. Kastler and D.W. Robinson
Covariance Algebras in Field Theory and Statistical Mechanics
Commun. Math. Phys. **3**, 1, (1966)

[19] A. Einstein
Zur Elektrodynamik bewegter Körper
Annal. Phys. **17**, 891, (1905)

[20] H. Ekstein
Presymmetry
Phys. Rev, **184**, 1315, (1969)

[21] P.S. Faragó, and L. Jánossy
Review of the Experimental Evidence for the Law of Variation of the Electron Mass with Velocity
Nuovo Cimento **5**, 1411-1436, (1957)

[22] J. Fröhlich, G. Morchio, F. Strocci
Infrared Problem and Spontaneous Breaking of the Lorentz Group
Phys. Lett. **89 B**, 61, (1979)

[23] R. Haag and D. Kastler
An Algebraic Approach to Quantum Field Theory
J. Math. Phys. **5**, 848, (1964)

[24] R. Haag and B. Schroer
Postulates of Quantum Field Theory
J. Math. Phys. **3**, 248, (1962)

[25] U. Haagerup
The standard form of von Neumann algebra
Math. Scand. **37**; 271-283, (1975)

[26] D. Olesen
On Spectral Subspaces and their Application to Automorphism Groups
Symposia Math. **20**, 253-296, (1976)

[27] D. Olesen and G.K. Pedersen
Derivations of C^*-Algebras have Semicontinuous Generators
Pacific J. Math. **53**, 563-572, (1974)

[28] R.Paley and N.Wiener
Fourier Transforms in the Complex Domain
Amer. Math. Soc., New York (1934)

[29] G.K.Pedersen
C^*-Algebras and their Automorphism Groups
Academic Press, London, New York, San Francisco, (1979)

[30] M.Takesaki
Tomita's Theory of Modular Hilbert Algebras and its Applications
Lecture Notes in Math. Vol. **128**, (1970) Springer-Verlag, Heidelberg

[31] M.Tomita
Standard Forms of von Neumann Algebras
Fifth Functional Analysis Symposium of Math. Soc. of Japan, Sendai, (1967)

[32] T. Turumaru
Crossed Products of Operator Algebras
Tôhoku Math, J. **4**, 242-251, (1952)

[33] W. Voigt
Über das Dopplers'sche Princip
Nachrichten der Göttinger Gesellschaft der Wissenschaften No **2** (1887)

[34] A.S.Wightman
Recent Achievements of Axiomatic Field Theory
Trieste Lectures (1962)

REPRESENTATIONS OF QUANTUM GROUPS *

V.K. Dobrev
Bulgarian Academy of Sciences
Institute of Nuclear Research and Nuclear Energy
72 Boul. Lenin, 1784 Sofia, Bulgaria

Introduction

The q - deformation $U_q(\mathcal{G})$ of the universal enveloping algebras $U(\mathcal{G})$ of complex simple Lie algebras $\mathcal{G}$ arose in the study of the algebraic aspects of quantum integrable systems [Fa, KR1, KS, S1, S2, S3]. (The definition of $U_q(\mathcal{G})$ is below in Section 1.) They provide a powerful tool for the solving of the quantum Yang-Baxter equations. For recent reviews we refer to [FaT, FRT1,2, J4, Ta, Vg]. The algebras $U_q(\mathcal{G})$ are called also quantum groups [D1, D2] or quantum universal enveloping algebras [Re, KiR1]. In [S3] for $\mathcal{G} = sl(2, \mathbb{C})$ and in [Dr1, J1, J2, Dr2] in general it was observed that the algebras $U_q(\mathcal{G})$ have the structure of a Hopf algebra. This brought additional mathematical interest in this new algebraic structure (see, e.g., [R1, Wo, R2, Ve, L1]). Recently, inspired by the Knizhnik-Zamolodchikov equations [KZ] Drinfeld has developed a theory of formal deformations and introduced a new notion of quasi-Hopf algebras [Dr3,4].

---

* Invited review talk, to appear in the Proceedings of the International Symposium "Symmetries in Science V: Algebraic Structures, their Representations, Realizations and Physical Applications", Schloss Hofen, Vorarlberg, Austria, 30.7.-3.8.1990

In this paper we review the representations of $U_q(\mathcal{G})$. We do not consider quantum groups in the approaches of Manin [Ma1,2] and Woronowicz [Wo]. The approach of Manin considers quantum groups which are dual objects to $U_q(\mathcal{G})$ and which act as symmetries of non-commutative, or quantum, spaces. For the connections between the different approaches we refer the reader to [R1, Mj, DHL]. Recently, the approach of Manin was pursued futher [Ma3, DMMZ] after the development of differential calculus on the quantum hyperplane with respect to the action of the quantum group $GL_q(n)$ [WeZ, We, SS]. The latter approach is actually an example of non-commutative differential geometry different from that of Connes [Co]. We should also mention that there are some other types of deformations of $U(\mathcal{G})$ introduced in the papers [S1, Mc, Wi, Fi].

The representations of $U_q(\mathcal{G})$ were studied first in in [KR1, KiR1, R2, L1, ReSt] for generic values of the deformation parameter. Actually all results from the representation theory of $\mathcal{G}$ carry over to the quantum group case. This is not so, however, if the deformation parameter q is a root of unity. Thus this case is very interesting from the mathematical point of view (see, e.g., [CKa, DJMM1,2, L2,3]). Lately, quantum groups were intensively applied (with special emphasis on the case when q is a root of unity) in rational conformal field theories [AGS1-4, MS, Wi, MR, GP, BMP, FGP, GS1,2, Td, ReS] and in two-dimensional quantum gravity [G1-3].

Many interesting developments had to be left outside this review. In particular, noncompact quantum groups [BC, Ch, VK, KuD, MNS], quantum inhomogeneous algebras [Kl1, Ch], quantum supergroups [Ku, KR2, Sa, CKu, De, BGZ, Tl, DFI, KMN, FSV, DFV, FV], q-deformation of the (super-) Witt algebra, i.e., the (super-) Virasoro algebra without central extension [CZ, CKL, Po, N].

1. Preliminaries on the quantum groups $U_q(\mathcal{G})$

Let $\mathcal{G}$ be any complex simple Lie algebra; then $U_q(\mathcal{G})$ is defined [Dr1, J1, J2, Dr2] as the associative algebra over $\mathbb{C}$ with generators $X_i^\pm$, H_i , $i = 1, \ldots, r =$ rank $\mathcal{G}$ and with relations

$$[H_i\ ,\ H_j] \ = \ 0, \quad [H_i\ ,\ X_j^\pm] \ = \ \pm a_{ij} X_j^\pm\ , \tag{1.1}$$

$$[X_i^+\ ,\ X_j^-] \ = \ \delta_{ij} \frac{q_i^{H_i/2} - q_i^{-H_i/2}}{q_i^{1/2} - q_i^{-1/2}} \ = \ \delta_{ij}\,[H_i]_{q_i}\ , \quad q_i = q^{(\alpha_i\ ,\ \alpha_i)/2}\ , \tag{1.2}$$

$$\sum_{k=o}^{n} (-1)^k \binom{n}{k}_{q_i} (X_i^{\pm})^k \; X_j^{\pm} \; (X_i^{\pm})^{n-k} = 0 \; , \; i \neq j \; , \tag{1.3}$$

where $(a_{ij}) = (2(\alpha_i \, , \, \alpha_j)/(\alpha_i \, , \alpha_i))$ is the Cartan matrix of $\mathcal{G}$, $(\cdot \, , \, \cdot)$ is the scalar product of the roots normalized so that for the short simple roots α we have $(\alpha, \alpha) = 2$, $n = 1 - a_{ij}$,

$$\binom{n}{k}_q = \frac{[n]_q!}{[k]_q! \; [n-k]_q!} \; , \; [m]_q! = [m]_q \; [m-1]_q \ldots [1]_q \; , \tag{1.4a}$$

$$[m]_q = \frac{q^{m/2} - q^{-m/2}}{q^{1/2} - q^{-1/2}} = \frac{sh(mh/2)}{sh(h/2)} = \frac{sin(\pi m \tau)}{sin(\pi \tau)} \; , \; q = e^h = e^{2\pi i \tau} \; , \; h, \tau \in \mathbb{C}, \tag{1.4b}$$

$$q_i^{a_{ij}} = q^{(\alpha_i \; , \; \alpha_j)} = q_j^{a_{ji}} . \tag{1.4c}$$

This definition is valid also for arbitrary Kac-Moody algebras [Dr1]. Further we shall omit the subscript q in $[m]_q$ if no confusion can arise. Note also that instead of q some authors use $q' = q^2$.

For $q \longrightarrow 1$, $(h \longrightarrow 0)$, we recover the standard commutation relations from (1), (2) and Serre's relations from (3) in terms of the Chevalley generators H_i , $X_i^{\pm}$. The elements H_i span the Cartan subalgebra $\mathcal{H}$ of $\mathcal{G}$, while the elements $X_i^{\pm}$ generate the subalgebras $\mathcal{G}^{\pm}$. We shall use the standard decompositions $\mathcal{G} = \mathcal{H} \oplus \underset{\beta \in \Delta}{\oplus} \mathcal{G}_\beta = \mathcal{G}^+ \oplus \mathcal{H} \oplus \mathcal{G}^-$, $\mathcal{G}^{\pm} = \underset{\beta \in \Delta^{\pm}}{\oplus} \mathcal{G}_\beta$, where $\Delta = \Delta^+ \cup \Delta^-$ is the root system of $\mathcal{G}$, Δ^+ , Δ^- , the sets of positive, negative, roots, respectively. We recall that H_i correspond to the simple roots α_i of $\mathcal{G}$, and if $\beta = \sum_i n_i \, \alpha_i$, then to β corresponds $H_\beta = \sum_i n_i \, H_i$. The elements of $\mathcal{G}$ which span $\mathcal{G}_\beta$, (recall that dim $\mathcal{G}_\beta = 1$), will be denoted by X_β . These Cartan–Weyl generators are normalized so that

$$[X_\beta \; , \; X_{-\beta}] = [H_\beta]_{q_\beta} \quad \text{for} \quad \beta \in \Delta^+ \; , \tag{1.5}$$

where $q_\beta = q_j$ for j such that α_j is the shortest simple root which enters the decomposition of $\beta = \sum n_i \alpha_i$, i.e., $n_j \neq 0$. Note that with this choice formulae (1.8), (1.9), (1.10) hold also for H_β , $X_{\pm\beta}$. Some authors prefer to use as generating elements $X_i^{\pm}$ and

$$K_i^{\pm 1} = q_i^{\pm H_i/2} , \tag{1.6}$$

thus (1.1) and (1.2) are replaced by

$$K_i K_i^{-1} = K_i^{-1} K_i = 1 , \ [K_i , K_j] = 0, \quad K_i X_j^{\pm} K_i^{-1} = q_i^{\pm a_{ij}/2} X_j^{\pm} , \tag{1.1'}$$

$$[X_i^+ , X_j^-] = \delta_{ij} \frac{K_i - K_i^{-1}}{q_i^{1/2} - q_i^{-1/2}} . \tag{1.2'}$$

One may also use instead of $X_i^{\pm}$ the generators [R3]:

$$E_i = X_i^+ q_i^{-H_i/4} = X_i^+ K_i^{-1/2} , \quad F_i = X_i^- q_i^{H_i/4} = X_i^- K_i^{1/2} . \tag{1.7}$$

In terms of these generators the Serre relations (1.3) can be rewritten as:

$$(\mathrm{ad}_q E_i)^n (E_j) = 0 = (\mathrm{ad}_q F_i)^n (F_j), \quad i \neq j. \tag{1.3'}$$

In [S3] for $\mathcal{G} = sl(2, \mathbb{C})$ and in [Dr1, J1, J2, Dr2] in general it was observed that the algebra $U_q(\mathcal{G})$ is a Hopf algebra [A] with co-multiplication δ , co-unit ε and antipode γ defined on the generators and on the unit element of $U_q(\mathcal{G})$ as follows:

$$\delta : U_q(\mathcal{G}) \longrightarrow U_q(\mathcal{G}) \otimes U_q(\mathcal{G}) , \ \delta(1) = 1 \otimes 1 , \tag{1.8a}$$

$$\delta(H_i) = H_i \otimes 1 + 1 \otimes H_i , \quad \delta(X_i^{\pm}) = X_i^{\pm} \otimes q_i^{H_i/4} + q_i^{-H_i/4} \otimes X_i^{\pm} , \tag{1.8b}$$

$$\varepsilon : U_q(\mathcal{G}) \longrightarrow \mathbb{C} , \ \varepsilon(H_i) = \varepsilon(X_i^{\pm}) = 0 , \ \varepsilon(1) = 1 , \tag{1.9}$$

$$\gamma : U_q(\mathcal{G}) \longrightarrow U_q(\mathcal{G}) , \ \gamma(H_i) = -H_i , \ \gamma(1) = 1 , \tag{1.10a}$$

$$\gamma(X_i^{\pm}) = -q_i^{\pm\hat{\rho}/2} X_i^{\pm} q_i^{-\hat{\rho}/2} = -q_i^{1/2} X_i^{\pm} , \tag{1.10b}$$

where $\hat{\rho} \in \mathcal{H}$ corresponds to $\rho = \frac{1}{2} \sum_{\alpha \in \Delta^+} \alpha$, Δ^+ is the set of positive roots, $\hat{\rho} = \frac{1}{2} \sum_{\alpha \in \Delta^+} H_\alpha$.

In terms of the generators $K_i^{\pm 1}$, E_i , F_i the above relations are rewritten as follows:

$$\delta(K_i) = K_i \otimes K_i , \quad \delta(E_i) = E_i \otimes 1 + K_i^{-1} \otimes E_i , \quad \delta(F_i) = F_i \otimes K_i + 1 \otimes F_i , \tag{1.8b'}$$

$$\varepsilon(K_i) = 1 , \ \varepsilon(E_i) = \varepsilon(F_i) = 0 , \tag{1.9'}$$

$$\gamma(K_i) = K_i^{-1} , \tag{1.10a'}$$

$$\gamma(E_i) = -K_i E_i , \ \gamma(F_i) = -F_i K_i^{-1} . \tag{1.10b'}$$

The opposite co-multiplication and antipode [J2, Dr2] $\delta' = \sigma \circ \delta$, $\gamma' = \gamma^{-1}$, where σ is the permutation in $U_q(\mathcal{G}) \otimes U_q(\mathcal{G})$, define a Hopf algebra $U_q(\mathcal{G})'$ which is connected to $U_q(\mathcal{G})$ by

$$U_q(\mathcal{G})' = U_{q^{-1}}(\mathcal{G}) . \tag{1.11}$$

These Hopf algebras are not cocommutative. This is made manifest by the so-called universal R-matrix [Dr1, Dr2] which intertwines δ and δ' :

$$R \in \bar{U}_q(\mathcal{G}) \otimes \bar{U}_q(\mathcal{G}) , \quad R\,\delta = \delta'\, R. \tag{1.12}$$

where $\bar{U}_q(\mathcal{G})$ is the completion of $U_q(\mathcal{G})$ in the h–adic topology used in [Dr1, Dr2] ($q = e^h$). The R-matrix satisfies also the relations :

$$(\delta \otimes \mathrm{id})R = R_{13}R_{23} , \quad R = R_{\cdot 3} , \tag{1.13a}$$

$$(\mathrm{id} \otimes \delta)R = R_{13}R_{12} , \quad R = R_{1\cdot} , \tag{1.13b}$$

$$(\gamma \otimes \mathrm{id})R = R^{-1} , \tag{1.13c}$$

where the indices indicate the embeddings of R into $\bar{U}_q(\mathcal{G}) \otimes \bar{U}_q(\mathcal{G}) \otimes \bar{U}_q(\mathcal{G})$. In general Hopf algebras with an element R are called *quasi-triangular Hopf algebras* [Dr2]. They are called *triangular Hopf algebras* if also the following holds

$$\sigma\, R^{-1} = R . \tag{1.14}$$

From (1.13a) or (1.13b) follows the Yang–Baxter equation for R :

$$R_{12}R_{13}R_{23} = R_{23}R_{13}R_{12} . \tag{1.15}$$

Explicit formulae for the R–matrices are reviewed below in Subsection 2.3.

For generic q the centre of $U_q(\mathcal{G})$ is generated by q–analogues of the Casimir operators [S3, J1, J2]. For $\mathcal{G} = sl(2, \mathbb{C})$ we have :

$$C_2 = [(H+1)/2]^2 + X^- X^+ . \tag{1.16}$$

For $\mathcal{G} = sl(n+1, \mathbb{C})$ we shall need more explicit expressions for the Cartan–Weyl generators. Let $\beta_{jk} \in \Delta^+$, $1 \le j \le k \le n$ be a positive root given explicitly in terms of the simple roots α_j , $j = 1, \ldots, n$ by :

$$\beta_{jk} = \alpha_j + \cdots + \alpha_k , \; j < k \tag{1.17a}$$

$$= \alpha_j , \; j = k . \tag{1.17b}$$

Let E_{ij} , $i,j = 1,\ldots,n+1$ denote the $(n+1)\times(n+1)$ matrix with only non-zero element equal to 1 at the i–th row and j–th column, i.e., $(E_{ij})_{k\ell} = \delta_{ik}\delta_{j\ell}$. Then we have explicitly:

$$X_i^+ = E_{ii+1} \ , \ X_i^- = E_{i+1i} \ , \ H_i = E_{ii} - E_{i+1i+1} \ , \ i = 1,\ldots,n \ . \qquad (1.18)$$

Then the elements $X_{jk}^\pm \equiv X_{\pm\beta_{jk}}$, $j < k$ can be defined inductively :

$$X_{jk}^\pm = \pm (q^{1/4} X_j^\pm X_{j+1k}^\pm - q^{-1/4} X_{j+1k}^\pm X_j^\pm) \ . \qquad (1.19)$$

One should check (1.5) with

$$H_{\beta_{jk}} = H_j + \cdots + H_k \ , \ j < k \qquad (1.20a)$$

$$= H_j \ , \ j = k \ . \qquad (1.20b)$$

Note that there is some inessential ambiguity in the definition (1.19), namely, $X_{jk}^{\prime\pm} = q^{\pm\nu} X_{jk}^\pm$ for generic q is also a good choice. Particularly often are used the choices $\nu = 1/4$ or $\nu = -1/4$.

Now the Casimir operator is given by [MZ] :

$$C_2 = K (\sum_{1\leq i\leq j\leq n} K_1^{-1} \ldots K_{i-1}^{-1} K_{j+1} \ldots K_n X_{ij}^- X_{ij}^+ q^{(i+j-n-1)/2} +$$

$$+ \sum_{j=0}^{n} K_1^{-1} \ldots K_j^{-1} K_{j+1} \ldots K_n q^{-j+n/2} (q^{1/2} - q^{-1/2})^{-2}) \ , \qquad (1.21)$$

where

$$K = K_1^{a_1} \ldots K_n^{a_n} \ , \ a_i = (n+1-2i)/(n+1) \ . \qquad (1.22)$$

For $n = 1$ this expression differs from (1.16) by a constant.

2. Representations of $U_q(\mathcal{G})$ for generic q

2.1. Induced highest weight modules and their irreducible subquotients

The highest weight modules V over $U_q(\mathcal{G})$ [J1] are given by their highest weight $\lambda \in \mathcal{H}^*$ and highest weight vector $v_0 \in V$ such that :

$$X_i^+ v_0 = 0 \ , \ i = 1,\ldots,r \ , \ H v_0 = \lambda(H) v_0 \ , \ H \in \mathcal{H} \ . \qquad (2.1)$$

We start with the *induced* HWM or *Verma* modules V^λ such that $V^\lambda \cong U_q(\mathcal{G}) \otimes_{U_q(\mathcal{B})} v_0 \cong U_q(\mathcal{G}^-) \otimes v_0$, where $\mathcal{B} = \mathcal{B}^+$, $\mathcal{B}^\pm = \mathcal{H} \oplus \mathcal{G}^\pm$ are Borel subalgebras of $\mathcal{G}$. (Then the algebras $U_q(\mathcal{B}^\pm)$ with generators H_i , $X_i^\pm$ are Hopf subalgebras of

$U_q(\mathcal{G})$ [Re].) Thus one should expect that the representation theory of V^λ would parallel the usual theory of Verma modules $V(\Lambda)$ over $\mathcal{G}$. ($V(\Lambda)$ is defined as the HWM over $\mathcal{G}$ induced from the one-dimensional representations of $\mathcal{B}$.) In particular, one considers the irreducible HWM L_λ over $U_q(\mathcal{G})$ as factor-modules V^λ/I^λ, where I^λ is the maximal submodule of V^λ.

We recall several facts from the case $q = 1$ (see, e.g., [Di]). The Verma module $V(\Lambda)$ over $\mathcal{G}$ is reducible iff there exists a root $\beta \in \Delta^+$ and $m \in I\!N$ such that

$$2(\Lambda + \rho, \beta) \; = \; m(\beta, \beta) \tag{2.2}$$

holds [BGG]. (In the case of affine Lie algebras (2.2) is the condition of Kac–Kazhdan [KK].) If (2.2) holds then there exists a vector $v_s \in V^\lambda$, called a *singular vector*, such that $v_s \neq v_0$, $X_i^+ v_s = 0$, $i = 1,\ldots,r$, $Hv_s \; = \; (\Lambda(H) - m\beta(H))v_s$, $H \in \mathcal{H}$. (In the case of affine Lie algebras when $(\beta,\beta) = 0$ there are $p(n)$ independent singular vectors for each $n \in I\!N$, $p(\cdot)$ being the partition function [MFF].) The space $U(\mathcal{G}^-)v_s$ is a proper submodule of $V(\Lambda)$ isomorphic to the Verma module $V(\Lambda - m\beta) = U(\mathcal{G}_-) \otimes v_0'$ where v_0' is the highest weight vector of $V(\Lambda - m\beta)$; the isomorphism being realized by $v_s \mapsto 1 \otimes v_0'$. The singular vector is given by

$$v_s^{\beta,m} \; = \; \mathcal{P}_m^\beta(X_1^-,\ldots,X_r^-) \otimes v_0 \; , \tag{2.3}$$

where $\mathcal{P}_m^\beta$ is a homogeneous polynomial in its variables of degrees mn_i, where $n_i \in Z\!\!Z_+$ come from $\beta = \sum n_i\alpha_i$, α_i - the system of simple roots [Do1, Do2]. We shall give several examples of singular vectors below in the quantized case which examples will be relevant also for the usual situation after setting $q = 1$.

The Verma module $V(\Lambda)$ contains a unique proper maximal submodule $I(\Lambda)$. Among the HWM with highest weight Λ there is a unique irreducible one, denoted by $L(\Lambda)$, i.e.,

$$L(\Lambda) \; = \; V(\Lambda)/I(\Lambda) \; . \tag{2.4}$$

If $V(\Lambda)$ is irreducible then $L(\Lambda) = V(\Lambda)$. Thus further we discuss $L(\Lambda)$ for which $V(\Lambda)$ is reducible. Consider $V(\Lambda)$ reducible with respect to (w.r.t.) to every simple root (and thus w.r.t. to all positive roots):

$$(\Lambda + \rho, \alpha_i^\vee) \; = \; m_i \; , \; i = 1,\ldots,r \; , \; \alpha^\vee = 2\alpha/(\alpha,\alpha), . \tag{2.5}$$

Then $L(\Lambda)$ is a finite-dimensional highest weight module over $\mathcal{G}$ and all such modules are obtained in this way [Di]. If we restrict $\mathcal{G}$ to its compact real form $\mathcal{G}_c$ then the set of all $L(\Lambda)$ coincides with the set of all finite-dimensional unitary irreducible representations of $\mathcal{G}_c$. (In the case of affine Lie algebras, $L(\Lambda)$ with (2.5) holding are the so-called integrable highest weight modules [Ka].) An important class of the case when (2.5) holds are the so-called *fundamental representations* $L(\Lambda_i)$, $i=1,\ldots,r$ characterized by $(\Lambda_i,\alpha_j^\vee)=\delta_{ij}$, i.e. $(\Lambda_i+\rho,\alpha_j^\vee)=1+\delta_{ij}=m_j(\Lambda_i)$.

In the case when q is not a root of unity all facts above hold also for the representations of $U_q(\mathcal{G})$ [KR1, KiR1, R2, L1]. The representations of $U_q(\mathcal{G})$ are deformations of the representations of $U(\mathcal{G})$, and the latter are obtained from the former for $q\longrightarrow 1$ [R2,L1]. This holds in particular for V^λ, L_λ and I^λ which are deformations of $V(\lambda)$, $L(\lambda)$, $I(\lambda)$, respectively.

Recently, De Concini and Kac [CKa] have given a formula for the determinant of the contravariant form on the Verma modules V^λ. For $Y=Y(q)\in \mathbb{C}(q)$, let $\bar{Y}=Y(q^{-1})$. A $\mathbb{C}$-bilinear form $\mathcal{F}$ on a vector spave V over $\mathbb{C}(q)$ with values in $\mathbb{C}(q)$ is called Hermitian if

$$\begin{gathered}\mathcal{F}(Yu,v)=\bar{Y}\mathcal{F}(u,v)\ ,\quad \mathcal{F}(u,Yv)=Y\mathcal{F}(u,v)\ ,\\ \mathcal{F}(u,v)=\bar{\mathcal{F}}(v,u)\ ,\quad Y\in\mathbb{C}(q)\ ,\quad u,v\in V\ .\end{gathered}\tag{2.6}$$

The Verma module V^λ carries a unique contravariant Hermitian form $\mathcal{F}$ such that

$$\mathcal{F}(v_0,v_0)\ =\ 1\ ,\ \ \mathcal{F}(Yu,v)\ =\ \mathcal{F}(u,\omega Y\ v)\ ,\ \ Y\in U_q(\mathcal{G})\ ,\ u,v\in V^\lambda\ ,\tag{2.7}$$

where ω is the involutive antiautomorphism such that $\omega X_i^\pm\ =\ X_i^\mp\ ,\ \omega H_i\ =\ H_i$.

Let Γ, (resp. Γ_+), be the set of all integral, (resp. integral dominant), elements of $\mathcal{H}^*$, i.e., $\lambda\in\mathcal{H}^*$ such that $(\lambda,\alpha_i^\vee)\in\mathbb{Z}$, (resp. $\mathbb{Z}_+$), for all simple roots α_i. We recall that for each invariant subspace $V\subset\mathcal{U}(\mathcal{G}^-)\otimes v_0\cong V^\lambda$ we have the following decomposition

$$V\ =\ \bigoplus_{\mu\in\Gamma_+} V_\mu\ ,\qquad V_\mu\ =\ \{u\in V\mid H_k u=(\lambda-\mu)(H_k)u,\ \forall\ H_k\}\ .\tag{2.8}$$

(Note that $V_0=\mathbb{C}v_0$.) We have $\mathcal{F}(V_\mu\ ,\ V_\nu)\ =\ 0$ if $\mu\neq\nu$. Let $\mathcal{F}_\mu$ be the restriction of $\mathcal{F}$ to V_μ , $\mu\in\Gamma^+$, and let $\det_\mu^\lambda$ denote the determinant of the matrix $\mathcal{F}_\mu$. Then we have [CKa] :

$$\det{}_\mu^\lambda\ =\ \prod_{\beta\in\Delta_+}\prod_{k=1}^{\infty}([k]_{q_\beta}[H_\beta-\rho(H_\beta)-k(\beta,\beta)/2]_{q_\beta})^{P(\mu-k\beta)}\ ,\tag{2.9}$$

where q_β is defined as above in (1.5), $P(\mu)$ is a generalized partition function, $P(\mu)=\#$ of ways μ can be presented as a sum of positive roots β_j, each root taken with its multiplicity $m_j=\dim\mathcal{G}_{\beta_j}$, (here $m_j=1$), $P(0)\equiv 1$.

This result implies in the usual way the description of irreducible subquotients of V^λ. In particular, this confirms results on the embeddings of the reducible modules V^λ [Do5]. The reducibilty conditions (2.2) are replaced by:

$$[(\lambda+\rho\,,\,\beta^\vee)-m]_{q_\beta} \;=\; [(\lambda+\rho)(H_\beta)-m]_{q_\beta} \;=\; 0\,, \qquad (2.10)$$

with q_β as in (1.5). More explicitly, as in the undeformed case, if (2.10) holds then $V^{\lambda-m\beta}$ is isomorphic to a submodules of V^λ. This situation will be denoted by $V^\lambda \longrightarrow V^{\lambda-m\beta}$, i.e., we use the usual convention that the arrows depicting the embedding maps point *to* the embedded HWM. If (2.10) holds for several pairs $(m,\beta)=(m_i,\beta_i)$, $i=1,\ldots,k$, there are other HWM modules $V^{\lambda-m_i\beta_i}$ all of which are isomorphic to submodules of V^λ.

It is convenient for the description of the embedding patterns to use the notion of a multiplet [Do1, Do2] of highest weight modules. We say that a set $\mathcal{M}$ of highest weight modules forms a *multiplet* if 1) $V\in\mathcal{M}\Rightarrow\mathcal{M}\supset\mathcal{M}_V$, where $\mathcal{M}_V$ is the set of all highest weight modules $V'\neq V$ such that either V' is isomorphic to a submodule of V or V is isomorphic to a submodule of V'; 2) $\mathcal{M}$ does not have proper subsets fulfilling 1). It is convenient to depict a multiplet by a connected oriented graph, the vertices of which correspond to the highest weight modules and the lines between the vertices correspond to the embeddings between the modules. Note that it may happen that two multiplets $\mathcal{M}^1$ and $\mathcal{M}^2$ are depicted by one and the same graph. Then we say that $\mathcal{M}^1$ and $\mathcal{M}^2$ belong to one and the same *type of multiplets* and use parametrization to distinguish multiplets belonging to a certain type. Most often only the embeddings which are not compositions of other embeddings are depicted, since these contain all the relevant information.

Let us consider some examples. We start with the case of the simple roots. Let $\beta=\alpha_j$ and we try the same expression (2.3) for the singular vector as in the case $q=1$:

$$v_s \;=\; (X_j^-)^m\otimes v_0\,. \qquad (2.11)$$

We obtain using (1.2):

$$[X_j^+,(X_j^-)^m] \;=\; \sum_{k=0}^{m-1}(X_j^-)^{m-1-k}[H_j]_{q_j}(X_j^-)^k$$

$$=\;(X_j^-)^{m-1}\sum_{k=0}^{m-1}[H_j-2k]_{q_j} \;=\; (X_j^-)^{m-1}[m]_{q_j}[H_j-m+1]_{q_j}\,. \qquad (2.12)$$

If v_s is a singular vector we should have

$$0 \;=\; X_j^+\,v_s \;=\; [X_j^+,(X_j^-)^m]\otimes v_0 \;=\; (X_j^-)^{m-1}[m]_{q_j}[\lambda(H_j)-m+1]_{q_j}\otimes v_0\,. \quad (2.13)$$

(Note that $X_k^+ v_s = 0$, for $k \neq j$.) If $q_j = q^{(\alpha_j\ ,\ \alpha_j)/2}$ is not a root of unity (2.13) gives just condition (2.2) rewritten as

$$[\lambda(H_j)+1-m]_{q_j} \;=\; [(\lambda,\alpha_j^\vee)+1-m]_{q_j} \;=\; [(\lambda+\rho,\alpha_j^\vee)-m]_{q_j} \;=\; 0, \qquad (2.14)$$

and we have used the fact that $(\rho,\alpha_j^\vee)=1$ for all α_j.

As another example we take a root β which is the sum of two simple roots of equal length : $\beta=\alpha_1+\alpha_2$, $(\beta,\beta)=(\alpha_i,\alpha_i)$, $i=1,2$, $\beta^\vee=\alpha_1^\vee+\alpha_2^\vee$. Let us have condition (2.10) fulfilled for β, but not for α_i , $i=1,2$:

$$[(\lambda+\rho,\beta^\vee)-m]_{q_\beta} \;=\; [\lambda(H_\beta)+2-m]_{q_\beta}=0\ ,\ q_\beta=q_1=q_2\ ,$$

$$[(\lambda+\rho,\alpha_j^\vee)-m]_{q_j}\neq 0\ ,\ j=1,2,\ . \qquad (2.15)$$

Then on can check that the singular vector is given by [Do5]:

$$v_s^m \;=\; \sum_{j=0}^{m} a_j\,(X_1^-)^{m-j}\,(X_2^-)^m\,(X_1^-)^j\ \otimes v_0\ , \qquad (2.16a)$$

$$a_j \;=\; (-1)^j\, a\,\frac{[\lambda(H_1)+1]}{[\lambda(H_1)+1-j]}\binom{m}{j}_q\ ,\quad j=0,\ldots,m,\quad a\neq 0, \qquad (2.16b)$$

(or by the same expression with the indices 1 and 2 interchanged). For this check we need also the following formula involving the q–hypergeometric function ${}_2F_1^q$:

$${}_2F_1^q(-k,s;\ s+1-p;\ q^{(p-k)/2}) \;=\; \delta_{p0}\frac{[k]![s]!}{[k+s]!}q^{ks/2}, \qquad (2.17a)$$

where

$${}_2F_1^q(a,b;\ c;z)\equiv\sum_{n\in\mathbb{Z}_+}\frac{[a+n]![b+n]![c]!}{[a]![b]![c+n]![n]!}z^n\ . \qquad (2.17b)$$

For $q\longrightarrow 1$ formula (2.16) goes to the correct formula in the same situation [Do1] [Do4] (cf. f-lae (8.40), (8.41)) and arbitrary m . Such q-special functions are discussed in [KiR1, MNU, Ko1, Ko2, KlG, Kl2, NM] and earlier without group-theoretic interpretations in [AA, AW].

The basis used above and in [Do1-5, Do8] for the construction of singular vectors turns out to be part of the general basis for $U_q(\mathcal{B}^-)$ introduced recently by Lusztig [L4].

Let us consider the example of $\mathcal{G}=sl(2,\mathbb{C})$; $r=1$, $X_1^\pm=X^\pm$, $H_1=H$, $\alpha_1=\alpha=\alpha^\vee=2\rho$:

$$[H\ ,\ X^\pm] \;=\; \pm 2\,X^\pm\ , \qquad (2.18)$$

$$[X^+\ ,\ X^-] \;=\; \frac{q^{H/2}-q^{-H/2}}{q^{1/2}-q^{-1/2}} \;=\; [H]_q\ . \qquad (2.19)$$

In this case the induced (Verma) module is given explicitly by $V^\lambda \cong U_q(\mathcal{G}^-) \otimes v_0$, with basis

$$(X^-)^k \otimes v_0 \ , \quad k = 0, 1, \dots \ . \tag{2.20}$$

Let us consider again the reducible case when (2.10), or more conveniently (2.14), is holding. Note that the highest weight is $\lambda = ((m-1)/2)\alpha$, $(\alpha, \alpha) = 2$. The singular vector is given precisely by (2.11) :

$$v_s \ = \ (X^-)^m \otimes v_0. \tag{2.21}$$

The submodule $U_q(\mathcal{G}^-)v_s \cong V^{\lambda - m\alpha}$ has the basis

$$(X^-)^{m+k} \otimes v_0 \ , \quad k = 0, 1, \dots \ . \tag{2.22}$$

The irreducible HWM $L_m \equiv L_\lambda$ is obtained by factorizing the submodule $U_q(\mathcal{G}^-)v_s$, i.e., by the condition

$$(X^-)^m |0> \ = \ 0 \ , \tag{2.23}$$

where $|0>$ is the highest weight vector of L_m . Explicitly, all vectors of L_m are given by :

$$v_{m,k} \equiv (X^-)^k |0> \ , \quad k = 0, 1, \dots, m-1 \ , \tag{2.24a}$$

which transform as follows :

$$\begin{aligned}
H v_{m,k} \ &= \ (m-k-1) v_{m,k} \ , \quad k = 0, 1, \dots, m-1 \ , & (2.24b)\\
X^+ v_{m,k} \ &= \ [k][m-k] v_{m,k-1} \ , \quad k = 0, 1, \dots, m-1 \ , & (2.24c)\\
X^- v_{m,k} \ &= \ v_{m,k+1} \ , \ k = 0, 1, \dots, m-2, \quad X^- v_{m,m-1} \ = \ 0. & (2.24d)
\end{aligned}$$

We note that as usual their is also a lowest weight state which is annihilated by X^-, namely, $v_{m,m-1}$; thus the lowest weight is $\lambda = 0$. (One can also introduce normalized vectors as we shall do in the next subsection.) Thus we obtain the usual result for the finite-dimensional irreducible highest weight modules over $sl(2, \mathbb{C})$, or equivalently for the finite-dimensional unitary irreducible representations of $su(2)$, namely, that they are parametrized by the positive integers $m \in \mathbb{N}$, or equivalently, by the nonnegative half integers $j = (m-1)/2$, and their dimensions are:

$$\dim L_m \ = \ m \ = \ 2j+1 \ , \quad m = 1, 2, \dots \ , \quad j = 0, 1/2, 1, \dots \ . \tag{2.25}$$

2.2. Fock type representations

In the previous subsection we considered the irreducible HWM L_λ over $U_q(\mathcal{G})$ as factor-modules V^λ / I^λ , where I^λ is the maximal submodule of V^λ. As in the undeformed case there is a dual way of directly describing at least the finite-dimensional irreducible representations by the so called Fock type representations. One particular example is the so-called bosonic realization in the Jordan–Schwinger approach [BL].

Let us recall this approach on the example of $sl(2, \mathbb{C})$, or equivalently, $su(2)$. One takes a Heisenberg algebra of a pair of independent boson operators a_i , $\bar{a}_i$, $i = 1, 2$ with commutation relations

$$[\bar{a}_i \ , \ a_j] = \delta_{ij} \tag{2.26}$$

and all other commutators vanishing. Then the approach is to map as follows :

$$X^+ \mapsto a_1\bar{a}_2 \ , \quad X^- \mapsto a_2\bar{a}_1 \ , \quad H \mapsto a_1\bar{a}_1 - a_2\bar{a}_2 \ . \tag{2.27}$$

The analogue of this construction in the deformed case was given by [Bi1,2] (see also [Mc, KuD, CE]). Relations (2.26) are replaced by

$$\bar{a}_i^q \ a_j^q - q^{1/2} a_j^q \bar{a}_i^q \ = \ \delta_{ij} q^{-\mathcal{N}_i/2} \ , \tag{2.28}$$

where $\mathcal{N}_i$ are *number operators* such that

$$[\mathcal{N}_i \ , \ a_j^q] = \delta_{ij} \ a_i^q \ , \quad [\mathcal{N}_i \ , \ \bar{a}_j^q] = -\delta_{ij} \ \bar{a}_i^q \ . \tag{2.29}$$

This algebra is the deformation of the Heisenberg algebra (2.2) which is obtained for $q = 1$. The mapping (2.27) is replaced by [Bi1]:

$$X^+ \mapsto a_1^q \bar{a}_2^q \ , \quad X^- \mapsto a_2^q \bar{a}_1^q \ , \quad H \mapsto \mathcal{N}_1 - \mathcal{N}_2 \ . \tag{2.30}$$

One uses the vacuum vector $|0>_q$ such that

$$\bar{a}_i^q \ |0>_q \ = \ 0 \ , \quad \mathcal{N}_i \ |0>_q \ = \ 0 \ . \tag{2.31}$$

Now one can introduce the eigenstates which are analogues of the undeformed angular momentum states [Bi1,2] :

$$|j, n>_q \ \equiv \ ([j+n]_q! [j-n]_q!)^{-1/2} (a_1^q)^{j+n} (a_2^q)^{j-n} |0>_q \ , \tag{2.32}$$

$$j = 0, 1/2, 1, \ldots, \quad n = -j, -j+1, \ldots, j \ .$$

One easily verifies that :

$$H \ |j, n>_q \ = \ 2n \ |j, n>_q \ , \tag{2.33a}$$

$$X^\pm \ |j, n>_q \ = \ ([j \mp n][j \pm n + 1])^{1/2} \ |j, n \pm 1>_q \ . \tag{2.33b}$$

Note that usually there is a factor of n instead of $2n$ in (2.33a). This factor is present because of factor of 2 in (2.18). Comparing with the states in (2.24) we see that $|j,n>_q$, $n=-j,-j+1,\ldots,j$ corresponds to $v_{m,k}$, $k=0,1,\ldots,m-1=2j$. One should note that the mapping (2.30) satisfies the commutation relations (2.18) only on the vectors $|j,n>_q$. For this one uses also the formula [Bi2]:

$$[H]\ |j,n>_q \ = \ ([j+n][j-n+1]-[j-n][j+n+1])|j,n>_q \ . \tag{2.34}$$

The matrix elements of the these representations can be expressed in terms of little q-Jacobi polynomials [VS, MNU, Kol].

Other Fock-type representations were constructed in [SF] for $U_q(su(n))$, in [H] for $U_q(sl(n,\mathbb{C})^{(1)})$ and in [DJM, Kw, MM]. In [DJM] the Gel'fand–Tsetlin bases become monomes in the tensor algebra of the fundamental representation of $U_q(sl(n,\mathbb{C}))$ at $q=0$, i.e., this strange choice of q provides the most simple basis. This was called crystal base [Kw] and was generalized for the integrable representations of $U_q(\mathcal{G})$ for $\mathcal{G}=A_n,B_n,C_n,D_n$ in [Kw] and for the basic representation of $U_q(sl(n,\mathbb{C})^{(1)})$ in [MM].

2.3. q–Weyl group and R–matrices

For $\mathcal{G}=sl(2,\mathbb{C})$ the R–matrix is given explicitly by [Dr2] :

$$R \ = \ R(H,X^{\pm}|q) \ = \ q^{H\otimes H/4}\sum_{n\geq 0}\frac{(1-q^{-1})^n q^{\frac{n(n-1)}{4}}}{[n]!}\,(q^{\frac{H}{4}}X^+)^n\otimes(q^{-\frac{H}{4}}X^-)^n \tag{2.35}$$

where $H=H_1$, $X^{\pm}=X_1^{\pm}$, $r=1$. For $\mathcal{G}=sl(n,\mathbb{C})$ an explicit formula for R was given in [R3] (for $n=3$ see also [Bu]).

Recently, in [KiR2, LS] was given explicit multiplicative formulas for R for any $U_q(\mathcal{G})$. For this they introduced q–version of the Weyl group for $U_q(\mathcal{G})$.

Let us recall that for $\alpha\in\Delta$,

$$s_\alpha(\lambda) \ = \ \lambda-\frac{2(\lambda,\alpha)}{(\alpha,\alpha)}\alpha, \quad \lambda\in\mathcal{H}^* \ , \tag{2.36}$$

are the standard reflections in $\mathcal{H}^*$. The Weyl group W is generated by the reflections $s_i\equiv s_{\alpha_i}$, where α_i are the simple roots. Thus every element $w\in W$ can be written as the product of simple reflections. It is said that w is written in a reduced form if it is written with the minimal possible number of simple reflections; the number of reflections of a reduced form of w is called the length of w, denoted by $\ell(w)$.

The elements of the q–Weyl group belong [KiR2] to the completion $\bar{U}_q(\mathcal{G})$ of $U_q(\mathcal{G})$. They are defined by the action of the generating elements in the irreducible representations of $U_q(\mathcal{G})$.

In the case of $sl(2,\mathbb{C})$ the nontrivial element w of W is defined to act in the representation defined, e.g., in Subsection 2.2. as [KiR2] :

$$w\ |j,n>_q = (-1)^{j-n}\ q^{(n-j(j+1))/2}\ |j,-n>_q\ . \tag{2.37}$$

It satisfies the relations [KiR2] :

$$w\ X^\pm\ w^{-1} = -q^{\pm 1/2}\ X^\mp\ ,\quad w\ H\ w^{-1} = -H\ . \tag{2.38}$$

Since $\bar{U}_q(\mathcal{G})$ is also a Hopf algebra we have [KiR2] :

$$\delta(w) = R^{-1} w \otimes w\ ,\quad \varepsilon(w) = 1\ ,\quad \gamma(w) = w\ q^{H/2}\ , \tag{2.39}$$

where R is given by (2.35). Further let us introduce the element

$$u = \sum_i \gamma(a_i) b_i\ , \tag{2.40}$$

where a_i , b_i are the coordinates of the element R :

$$R = \sum_i a_i \otimes b_i\ . \tag{2.41}$$

One may show that

$$\gamma^2(Y) = u\ Y\ u^{-1}\ , \tag{2.42}$$

and

$$v = u\ q^{-\hat{\rho}/2} \in \text{centre of } U_q(\mathcal{G})\ , \tag{2.43}$$

$\hat{\rho}$ is used in (1.10). Let ϵ be the unipotent central element, i.e., $\epsilon\ |j,n>_q = (-1)^{2j}\ |j,n>_q$, $\epsilon^2 =$id. Then [KiR2]

$$w^2 = v\epsilon = u\ q^{-\hat{\rho}/2}\epsilon\ . \tag{2.44}$$

For arbitrary $U_q(\mathcal{G})$ let V_λ be an irreducible representation of $U_q(\mathcal{G})$. Let $V_\lambda = \oplus_j (W_\lambda^j \otimes L_j)$ be the decomposition of V_λ into irreducible $(U_q(sl(2,\mathbb{C})))_j$ submodules. Define the action of w_i in V_λ as $w_i = \oplus_j(\mathrm{Id}_{W_\lambda^j} \otimes (w_i)_j)$, where $(w_i)_j$ is the action of w in L_j as in (2.37). Further one has [KiR2] :

$$w_i\ H_j\ w_i^{-1} = H_j - a_{ij} H_i\ ,\quad w_i\ X_i^\pm\ w_i^{-1} = -q^{\pm 1/2} X_i^\mp\ . \tag{2.45}$$

$$\delta\ w_i = R(i)^{-1}\ w_i \otimes w_i\ , \tag{2.46}$$

where $R(i) = R(H_i, X_i^\pm | q_i)$, cf. (2.35);

$$(w_i\ w_j)^{2-a_{ij}} = 1\ ,\quad \text{for } i \neq j\ ,\quad (w_i)^2 = 1\ , \tag{2.47a}$$

$$(\tilde{w}_i\,\tilde{w}_j)^{2-a_{ij}} = 1\,, \quad \text{for } i \neq j\,, \quad (\tilde{w}_i)^2 = 1\,, \tag{2.47b}$$

where

$$\tilde{w}_i = w_i\, q_i^{H_i^2/8}\,. \tag{2.47c}$$

Further let $s_0 = s_{i_1}\ldots s_{i_k}$ be the reduced form of the element of W with maximal length $\ell(s_0)$. It can be shown that the element

$$\tilde{w}_0 = \tilde{w}_{i_1}\ldots\tilde{w}_{i_k} \tag{2.48}$$

is well defined and does not depend on the choice of decomposition of s_0 . Finally the result of [KiR2] for the universal R-matrix is :

$$R = q^{\sum_{i,j=1}^{n}(B^{-1})_{ij}\;H_i\otimes H_j/4}\,(\tilde{w}_0\otimes\tilde{w}_0)\,\delta(\tilde{w}_0)^{-1}\,, \quad (B_{ij}) = ((\alpha_i,\alpha_j))\,, \tag{2.49a}$$

or

$$R = q^{\sum_{i,j=1}^{n}(B^{-1})_{ij}\;H_i\otimes H_j/4}\,\tilde{R}(i_k|s_{i_1}\ldots s_{i_{k-1}})\ldots\tilde{R}(i_2|s_{i_1})\,\tilde{R}(i_1)\,, \tag{2.49b}$$

where

$$\tilde{R}(i_\ell|s_{i_1}\ldots s_{i_{\ell-1}}) = (T_{i_1}^{-1}\otimes T_{i_1}^{-1})\ldots(T_{i_{\ell-1}}^{-1}\otimes T_{i_{\ell-1}}^{-1})\,\tilde{R}(i_\ell)\,, \tag{2.49c}$$

$$T_i(Y) = \tilde{w}_i^{-1}\,Y\,\tilde{w}_i\,. \tag{2.49d}$$

The same construction works for affine Lie algebras [KiR2]. Earlier work in this case includes the explicit construction for $A_1^{(1)}$ in any representation [KR1, SAA, J1, KRS], for $A_n^{(1)}$, $B_n^{(1)}$, $C_n^{(1)}$, $D_n^{(1)}$ in the vector representation [Ba, J3], for $B_n^{(1)}$, $D_n^{(1)}$ in the spinor representation [O], and for $G_2^{(1)}$ [Ku].

2.4. Vertex operators and functional realizations

Let us consider the affine quantum group $U_q(\mathcal{G}^{(1)})$ where $\mathcal{G}^{(1)}$ is the untwisted affinization of $\mathcal{G}$, rank $\mathcal{G} = r$. Let $a_i^\vee$ be the dual Kac labels, i.e., $\sum_{j=0}^{r} a_j^\vee\, a_{jk} = 0$, normalized so that $\min a_j^\vee = 1$. The element

$$\mathcal{K} = \prod_{j=0}^{r} K_j^{a_j^\vee} \tag{2.50}$$

belongs to the centre of $U_q(\mathcal{G}^{(1)})$.

Let us introduce bosonic variables y^j , $x^j(n)$, $j = 1,\ldots r = \text{rank}\,\mathcal{G}$, $n \in \mathbb{Z}$ satisfying the Heisenberg relations :

$$[x^j(m), x^k(n)] = m\delta_{jk}\,\delta_{m+n,0}\,, \tag{2.51a}$$

$$[x^j(0), y^k] = i\delta_{jk} . \tag{2.51b}$$

Further let for $q = e^h, h \in \mathbb{C}$ [FJ]

$$\Delta_{jk}(n) = (q^{n(\alpha_j,\alpha_k)/2} - q^{-n(\alpha_j,\alpha_k)/2})(\mathcal{K}^n - \mathcal{K}^{-n})/n^2h^2 . \tag{2.52}$$

Then the deformed Heisenberg algebra generators are defined by [FJ]

$$\alpha^j(n) = \sum_{j=1}^{r}(\Delta_{jk}(n))^{1/2}x^k(n) . \tag{2.53}$$

Now for each simple root α_j , $j = 1, \ldots, r$ the q–deformed vertex operators are defined by [FJ] :

$$\begin{aligned} V^j_\pm(z) &= :\ \exp\left(\pm ihQ^j_\pm(z)\right)\ : = \\ &= \exp\left(\pm ihQ^{\pm;j}_<(z)\right)\exp\left(\pm ihQ^{\pm;j}_>(z)\right)e^{\pm i(\alpha_j,y)}z^{\pm(\alpha_j,x(0))} , \end{aligned} \tag{2.54}$$

where

$$Q^j_\pm = (\alpha_j\ , y - ix(0)\ log\ z) + Q^{\pm;j}_>(z) + Q^{\pm;j}_<(z) , \tag{2.55a}$$

$$Q^{\pm;j}_>(z) = i\sum_{n>0}\frac{q^{\mp|n|/4}}{q^{n/2} - q^{-n/2}}\alpha^j(n)\ z^{-n} , \tag{2.55b}$$

$$Q^{\pm;j}_<(z) = i\sum_{n<0}\frac{q^{\mp|n|/4}}{q^{n/2} - q^{-n/2}}\alpha^j(n)\ z^{-n} . \tag{2.55b}$$

This construction is valid for the simply-laced algebras $\mathcal{G}$, for which all roots have equal length, i.e., for $\mathcal{G} = A_n , D_n , E_6 , E_7 , E_8$. Later this construction was generalized for $\mathcal{G} = B_n$ [Be]. Another construction in terms of screened vertex operators was introduced recently in [GS1, GS2].

It is natural to ask for the q-analogues of the functional realizations of classical groups representations which are best suited for writing invariants. We briefly review such a realization for $U_q(sl(2))$ following [GP] (see also [MNU, VS, Ru]).

Let $f(u)$ be a complex valued function of $u \in \mathbb{C}^\times$ and define the q - differentiation (a finite difference operator)

$$(Df)(u) = \frac{f(q^{1/2}u) - f(q^{-1/2}u)}{u(q^{1/2} - q^{-1/2})}. \tag{2.56}$$

Let at first m be an arbitrary complex number. Then the representation T_m of $U_q(sl(2))$ is defined in the space $\mathcal{C}_m$ of complex valued functions by

$$\left(q^{H/4}f\right)(u) = q^{-j/2}f(q^{1/2}u) ,\quad X^- = D , \tag{2.57a}$$

$$\left(X^{+}f\right)(u) = \frac{u}{q^{1/2}-q^{-1/2}}[q^{j/2}\left(q^{-H/4}f\right) - q^{-j/2}\left(q^{H/4}f\right)] \,, \tag{2.57b}$$

where $j = (m-1)/2$. For generic q and m the representations T_m are irreducible. For generic q and $m \in \mathbb{Z}$ the representations T_m are reducible. Moreover, for $m \in \mathbb{N}$ the representations T_m and T_{-m} are partially equivalent. This partial equivalence is realized by the operator D^m, i.e., one has

$$D^m \circ T_m \;=\; T_{-m} \circ D^m \tag{2.58}$$

The kernel $\mathcal{E}_m$ of the operator D^m is an invariant subspace of $\mathcal{C}_m$. It consists of polynomials of degree $\leq m-1$, i.e., it is a functional realization of the m-dimensional representation L_m of $U_q(sl(2))$ given in (2.24) or (2.33). An invariant bilinear form in this representation is given by

$$(f,g) \;=\; \sum_{k=0}^{2j}(-1)^k\, q^{k/2}\, D^k f(u)\big|_{u=0}\, D^{2j-k} g(u)\big|_{u=0} \,. \tag{2.59}$$

This construction is also interesting in the case when $q \neq 1$ is a root of unity which we review at the end of subsection 3.4. below. For $q = 1$ the operator D becomes differentiation with respect to u and one recovers the generators of the representations (j,0) of $SL(2,\mathbb{C})$, or for $u \in \mathbb{R}$ of the representation (m,ε) of $SL(2,\mathbb{R})$, where $\varepsilon = 0,1$ chracterizes the representations of the centre of $SL(2,\mathbb{R})$; (ε is arbitrary except when $m \in \mathbb{N}$, then $\varepsilon = (m+1) \bmod 2$). The facts about the partial equivalence and the representations $\mathcal{E}_m$ are standard in this situation. Moreover, the correspondence between the polynomials (2.3) (determining the singular vectors) and the invariant differential operators intertwining (reducible) representations of semisimple Lie groups and algebras is canonically given by replacing each X_i^- (here X^-) by a differential operator in suitable variable (here u) [Do4].

3. Representations of $U_q(\mathcal{G})$ for q a root of unity

3.1. Reducibility of Verma modules

In the case when q is a root of unity we ask what would replace the reducibility conditions (2.10). We start with the case of the simple roots. Let $\beta = \alpha_j$; we try the same expression (2.11) and we obtain (2.12) and (2.13). Now if q_j is a root of unity, $q_j^{N_j} = 1$, $N_j \in \mathbb{N}+1$ then for $k \in \mathbb{Z}$ we have :

$$[kN_j]_{q_j} \;=\; \frac{q_j^{kN_j/2} - q_j^{-kN_j/2}}{q_j^{1/2} - q_j^{-1/2}} \;=\; \frac{sin(\pi k)}{sin(\pi/N_j)} \;=\; 0 \;,\quad q_j = e^{2\pi i/N_j} \,. \tag{3.1}$$

Accordingly (2.13) gives that v_s from (2.11) is a singular vector if

$$\text{either} \quad [m]_{q_j} = 0 , \forall \lambda \in \mathcal{H}^* , \tag{3.2a}$$

$$\text{or} \quad [\lambda(H_j) + 1 - m]_{q_j} = 0 , m \neq \ell N_j . \tag{3.2b}$$

Thus we see that for $q_j^{N_j} = 1$ the HWM V^λ is always reducible. Analogously, all vectors of the form

$$v_s^{k_1,\ldots,k_r} = \prod_{j=1}^{r} (X_j^-)^{k_j N_j} \otimes v_0 , \quad k_j \in \mathbb{Z}_+ , \sum_{j=1}^{r} k_j > 0 , \tag{3.3}$$

are singular vectors. Thus the general form of the singular vector corresponding to $\beta \in \Delta^+$, $\beta = \sum n_j \alpha_j$, and $m = kN + n$, where N is such that $q_\beta = e^{2\pi i/N}$ with q_β as in (1.5), $k, n \in \mathbb{Z}_+$, $k + n \neq 0$, $n < N$, is

$$v_s^{\beta,n,k} = \prod_{j=1}^{r} (X_j^-)^{kn_j N} \mathcal{P}_n^\beta(X_1^-, \ldots, X_r^-) \otimes v_0 , \tag{3.4}$$

where $\mathcal{P}_n^\beta$ is a homogeneous polynomial as in (2.3) and if $n \neq 0$ then the condition

$$[(\lambda + \rho)(H_\beta) - n] = 0 , \quad n \in \mathbb{N}, \quad n < N , \tag{3.5}$$

is fulfilled.

3.2. Irreducible factor-modules of Verma modules

Let us say that two elements $\lambda, \lambda' \in \mathcal{H}^*$ are *equivalent* , $\lambda \cong \lambda'$, if $\lambda - \lambda' = N\beta$, where β is any element of the dual integer root lattice, i.e., $\beta = n_1\alpha_1^\vee + \cdots n_r\alpha_r^\vee$, $n_i \in \mathbb{Z}$, $\alpha_i^\vee = 2\alpha_i/(\alpha_i, \alpha_i)$, and N is such that $q_j = e^{2\pi i/N}$ for the shortest simple roots α_j whose duals enter the decomposition of β.

It is clear that if $\lambda \cong \lambda'$, then they obey or disobey (3.2b), (3.5) simultaneously. Thus the HWM V^λ , $V^{\lambda'}$ and the corresponding irreducible factor-modules $V^{\lambda'}$ have the same structure. So the irreducible HWM are described by their highest weights up to the above equivalence. Because of (15) it is also clear that the irreducible HWM of $U_q(\mathcal{G})$ are finite-dimensional. Another way to see this is to note that actually all elements $(X_j^-)^{2N_j}$, (and also $(X_j^+)^{2N_j}$), belong to the centre of $U_q(\mathcal{G})$ [Do5]. This may be seen explicitly as follows. Let α_i , α_j , $i \neq j$ be two simple roots with equal length so that $a_{ij} \neq 0$. Then using Serre relations (1.3) and $q_i = q_j$ we obtain that

$$X_i^- (X_j^-)^k = -[k-1]_{q_j} (X_j^-)^k X_i^- + [k]_{q_j} (X_j^-)^{k-1} X_i^- X_j^- . \tag{3.6a}$$

Thus if $q_j = e^{2\pi i/N_j}$ we have

$$X_i^- \, (X_j^-)^{N_j} \;=\; -(X_j^-)^{N_j} X_i^- \;, \tag{3.6b}$$

$$X_i^- \, (X_j^-)^{2N_j} \;=\; (X_j^-)^{2N_j} X_i^- \;. \tag{3.6c}$$

Recently, it was shown in [CKa] that actually all elements $(X_\beta)^{2N}$, $\beta \in \Delta$ and $(K_i^{\pm})^{2N} = q^{\pm NH_i}$ belong to the centre of $U_q(\mathcal{G})$. Their results differ by a factor of 2 since their q is the square of the one adopted here. Another difference is that their results are only in terms of q, not in terms of q_j as here. In particular, they notice some difference in the case when the two roots above are from the simple root system of the algebra B_n since then $N = 2N_j$, (these two roots are necessarily long). We should note that the same is true if the two roots above are the two long roots of F_4 .

Certainly, one may work with the finite algebra setting all elements $(X_\beta)^{2N}$, $\beta \in \Delta$, equal to zero. In some explicit realizations this is actually used [Pa, PS].

We would like to mention that recently it was proved in [CKa] that the maximal dimension of an irreducible finite-dimensional (not necessarily with highest weight) representation of $U_q(\mathcal{G})$ is equal to

$$N^{\dim \mathcal{G}^+} \;=\; N^{|\Delta^+|}.$$

This is consistent with previously obtained results on the irreducible finite-dimensional representations of $U_q(sl(2,\mathbb{C}))$ [AGS1, PS, RA] and $U_q(sl(3,\mathbb{C}))$ [Do5, Do6] which will be discussed below.

We consider the question of the irreducible representations as quotients of reducible induced HWM in the framework of embeddings between such HWM.

3.3. Embeddings of Verma modules

It is clear that that the HWM $V^{\lambda'}$ is isomorphic to a submodule of V^λ if $\lambda \cong \lambda'$ and $\lambda - \lambda' = N\beta$ for β an element of the dual nonnegative integer root lattice, i.e., $\beta = n_1\alpha_1^\vee + \cdots n_r\alpha_r^\vee$, $n_i \in \mathbb{Z}_+$. Thus to account for all other embeddings it is enough to consider the singular vectors in (3.4) with $k = 0$, $n \in \mathbb{N}$, $n < N$. As in the generic case it is clear that if (3.5) holds then $V^{\lambda - n\beta}$ is isomorphic to a submodules of V^λ. Furthermore if (3.5) holds with $\beta \in \Delta^+$ and $n \in -\mathbb{N}$ then V^λ is a submodule of $V^{\lambda + n\beta}$. Indeed, if $[(\lambda+\rho)(H_\beta)+n] = 0$ then $[(\lambda+n\beta+\rho)(H_\beta)-n] = 0$ because $\beta(H_\beta) \in 2\mathbb{N}$ for all β .

What is more interesting and in contrast to the undeformed $q = 1$ case is that if V^λ has a singular vector of type (15) with $k = 0$, $n = m \in \mathbb{N}$, $m <$

N then the embedded HWM $V^{\lambda-m\beta}$ has a singular vector of type (15) with $k=0,\ n=N-m$. The embedded HWM in $V^{\lambda-n\beta}$ is easily seen to be $V^{\lambda-N\beta}$. The latter is a submodule also of V^{λ}, however with a singular vector from (15) with $k=1,\ n=0$. The two embeddings coincide if $\beta=\alpha_i$ is a *simple* root. Before demonstrating this let us fix the usual convention that the arrows depicting the embedding maps point *to* the embedded HWM. Indeed, the first embedding is a composition of two embeddings $V^{\lambda}\longrightarrow V^{\lambda-m\beta}\longrightarrow V^{\lambda-N\beta}$; correspondingly if v_0', v_0'' are the highest weight vectors of $V^{\lambda-m\beta}$, $V^{\lambda-N\beta}$, respectively, we have $\mathcal{P}^{\beta}_{N-m}\mathcal{P}^{\beta}_{m}\otimes v_0\mapsto\mathcal{P}^{\beta}_{N-m}\otimes v_0'\mapsto 1\otimes v_0''$; the second embedding is $V^{\lambda}\longrightarrow V^{\lambda-N\beta}$; under this we have $\prod_{j=1}^{r}(X_j^-)^{n_jN}\otimes v_0\mapsto 1\otimes v_0''$, where $\beta=\sum n_j\alpha_j$. Thus if β is not a simple root we may have embedding of one and the same module in two different ways. This is similar to the affine Lie algebras case when β is an imaginary root (i.e., $(\beta,\beta)=0$).

3.4. HWM over $U_q(sl(2,\mathbb{C}))$

It turns out that all HWM V^{λ} belong to multiplets of one of the two types described below.

3.4.1. The multiplets of the first type are in 1-to-1 correspondence with those equivalent classes for which

$$\lambda(H)+n\neq 0\ ,\quad \forall n\in\mathbb{Z}\ , \tag{3.7}$$

for any representative. For a fixed class represented, say, by $\lambda\in\mathcal{H}^*$ the corresponding multiplet consists of an infinite chain of embeddings

$$\cdots\longrightarrow\tilde{V}_{-1}\longrightarrow\tilde{V}_0\longrightarrow\tilde{V}_1\longrightarrow\cdots \tag{3.8}$$

where the HWM entering the multiplet are $\tilde{V}_k=V^{\lambda-kN\alpha}$, $k\in\mathbb{Z}$, i.e., they are in 1-to-1 correspondence with the elements of the class in consideration. Each embedding in (3.8) is realized by a singular vector $v_s=(X^-)^N\otimes v_0(\tilde{V}_k)$, where $v_0(V)$ denotes the highest weight vector of the HWM V. The factor modules $\tilde{L}_k=\tilde{V}_k/\tilde{V}_{k+1}$ are isomorphic $\tilde{L}_k\cong\tilde{L}_{k'}\cong\tilde{L}$, $\forall k,k'\in\mathbb{Z}$, moreover $\dim\tilde{L}=N$ and all states of $\tilde{L}$ are given by $(X^-)^m\otimes v_0$, $m=0,\dots,N-1$. Thus the highest weight of an irreducible HWM is determined only up to the equivalence defined above.

3.4.2. The multiplets of the second type are parametrized by a positive integer, say, m such that $m\leq N/2$. Fix such an m and choose an element $\lambda\in\mathcal{H}^*$ such that

$$[\lambda(H)+1-m]\ =\ 0\ ,\quad m\in\mathbb{N}\ ,\ m\leq N/2, \tag{3.9}$$

i.e., $\lambda' = \frac{m-1}{2}\alpha$ is an element of the class of λ . If $m < N/2$ then V^λ is part of an infinite chain of embeddings

$$\cdots \longrightarrow V'^m_{-1} \longrightarrow V^m_0 \longrightarrow V'^m_0 \longrightarrow V^m_1 \longrightarrow V'^m_1 \longrightarrow \cdots \tag{3.10a}$$

where $V^m_k = V^{\lambda - kN\alpha}$, $k \in \mathbb{Z}$, $V'^m_k = V^{\lambda - m\alpha - kN\alpha}$, $k \in \mathbb{Z}$. (Thus the classes which only have elements λ for which (3.9) holds with $N/2 < m < N$ are represented by the highest weights of V'^m_k .) The embeddings $V^m_k \longrightarrow V'^m_k$ are realized by $v_s = (X^-)^m \otimes v_0(V^m_k)$, while $V'^m_k \longrightarrow V^m_{k+1}$ are realized by $v_s = (X^-)^{N-m} \otimes v_0(V'^m_k)$. The factor modules $L^m_k = V^m_k / V'^m_k$ have the same structure for all $k, \in \mathbb{Z}$; also the factor modules $L'^m_k = V'^m_k / V^m_{k+1}$ have the same structure for all $k, \in \mathbb{Z}$; moreover dim $L^m_k = m$, dim $L'^m_k = N - m$, and all states of L^m_k (respectively L'^m_k) are given by $(X^-)^n \otimes v_0$, $n = 0, \ldots, m-1$, (respectively $n = 0, \ldots, N-m-1$). If $N \in 2\mathbb{N}$ and $m = N/2$ then V^λ is part of an infinite chain of embeddings

$$\cdots \longrightarrow V^{N/2}_{-1} \longrightarrow V^{N/2}_0 \longrightarrow V'^{N/2}_1 \longrightarrow \cdots \tag{3.10b}$$

where $V^{N/2}_k = V^{\lambda - kN\alpha/2}$, $k \in \mathbb{Z}$. Everything we said above for L^m_k is valid here for $m = N/2$.

It is clear that all elements of λ and thus all HWM V^λ over $U_q(\mathcal{G})$ are accounted for in **3.4.1.** and **3.4.2.**. The results contained in Subsection 3.4. are summarized as follows [Do5]. Let $q^N = 1$, $N \in \mathbb{N} + 1$, $\mathcal{G} = sl(2, \mathbb{C})$. Then we have: (a) All HWM V^λ over $U_q(\mathcal{G})$ belong to multiplets of one of the two types described above. (b) Up to the equivalence introduced in Subsection **3.2** there are exactly N inequivalent irreducible HWM of $U_q(sl(2, \mathbb{C}))$ which have dimensions $1, 2, \ldots, N$. The last conclusion (b) was obtained by other methods in [AGS1, PS, RA]. Note that the nonuniformity in N (denoted there by m) of the results of [RA] is due to the fact that their q is a square root of ours. It is interesting to mention that (3.10a,b) have counterparts in the functional realization of [GP] we reviewed in subsection 2.4. (formulae (2.56-59)). Namely, the counterparts of V^m_k, V'^m_k are T_{m+2Nk} , T_{-m+2Nk} , respectively. Then the operators D^m intertwine T_{m+2Nk} and T_{-m+2Nk} , as in (2.58), while the operators D^{N-m} intertwine T_{-m+2Nk} and $T_{m+2N(k+1)}$. Certainly, the operators D^N intertwine T_{m+2Nk} and $T_{m+2N(k+1)}$.

Clearly the finite-dimensional HWM of $U_q(sl(2, \mathbb{C}))$ when q is an N–th root of unity have the same structure as those in the generic case discussed in Section 2. The peculiarities is that the range of dimensions m is restricted : $m \leq N$, and that for $m = N$ the highest weight λ may be arbitrary. Finally we note that finite-dimensional matrix representations of $U_q(gl(2, \mathbb{C}))$ for $q = e^{2\pi i \tau}$, $\tau \in \mathbb{R}$ were constructed in [Fl]. For q a root of unity, i.e., $\tau = 1/N$, with a N a prime integer, N–dimensional matrix representations are constructed.

3.5. HWM over $U_q(sl(3,\mathbb{C}))$

Let $\mathcal{G} = sl(3,\mathbb{C})$. Let us denote

$$X_3^{\pm} = \pm(q^{1/4} X_1^{\pm} X_2^{\pm} - q^{-1/4} X_2^{\pm} X_1^{\pm}) , \quad H_3 = H_1 + H_2 ; \qquad (3.11a)$$

where H_i correspond to the roots α_i ; $\alpha_3 = \alpha_1 + \alpha_2 = \rho$; α_1 , α_2 are the simple roots with $(\alpha_1, \alpha_2) = -1$; $(\alpha_i, \alpha_i) = 2$, $i = 1, 2, 3$. We also have (cf. (1.1),(1.2)) :

$$[H_i \, , \, X_3^{\pm}] = \pm X_3^{\pm} \ (= \pm\alpha_3(H_i)X_3^{\pm}) \ , \ i = 1, 2 \ , \qquad (3.11b)$$

$$[X_3^+ \, , X_3^-] = [H_3] \ . \qquad (3.11c)$$

The classification of the induced HWM is as follows. There are five types of multiplets of HWM.

3.5.1. The multiplets of the first type are in 1-to-1 correspondence with those equivalent classes for which

$$\lambda(H_i) + n \neq 0, \quad \forall n \in \mathbb{Z} \ , \quad \forall i = 1, 2, 3 \ . \qquad (3.12)$$

for any representative. For a fixed class represented, say, by $\lambda \in \mathcal{H}^*$ the corresponding multiplet consists of the following diagram of embeddings :

$$\begin{array}{ccccccc}
 & & \vdots & & \vdots & & \\
 & & \uparrow & & \uparrow & & \\
\cdots & \longrightarrow & V_{0,1} & \longrightarrow & V_{1,1} & \longrightarrow & \cdots \\
 & & \uparrow & & \uparrow & & \\
\cdots & \longrightarrow & V_{0,0} & \longrightarrow & V_{1,0} & \longrightarrow & \cdots \\
 & & \uparrow & & \uparrow & & \\
 & & \vdots & & \vdots & &
\end{array} \qquad (3.13)$$

where $V_{k,\ell} = V^{\lambda - kN\alpha_1 - \ell N\alpha_2}$, $k, \ell \in \mathbb{Z}$, are again parametrized by the elements equivalent to λ . These equivalence statements will be omitted below. Each embedding in (3.13) is realized by a singular vector $v_s = (X_i^-)^N \otimes v_0$, for $i = 1, 2$, respectively, when the arrow depicting the embedding is horizontal or vertical, respectively. Because of the symmetry it is clear that the factor modules $L_{k,\ell} = V_{k,\ell}/I_{k,\ell}$, where $I_{k,\ell}$ is the maximal submodule of $V_{k,\ell}$, have the same structure $\forall k,\ell, \in \mathbb{Z}$. We shall denote by L^1 any of these representations.

In (3.13) and in all diagrams below we do not depict any embeddings outside the quadrangle $(V_{0,0}\ ,\ V_{1,0}\ ,\ V_{1,1}\ ,\ V_{0,1})$ except the adjacent ones shown in (3.13).

3.5.2. The multiplets of the second type are parametrized by a positive integer, say, m_1 such that $m_1 \leq N/2$. Fix such an m_1 and choose an element $\lambda \in \mathcal{H}^*$ such that

$$[\lambda(H_1) + 1 - m_1] \;=\; 0\,, \quad \lambda(H_i) + n \;\neq\; 0\,,\ \ i = 2,3,\ \ \forall n \in \mathbb{Z}\,. \tag{3.14}$$

If $m_1 < N/2$ then V^λ is part of the following multiplet :

$$\begin{array}{ccccccccc}
 & & \vdots & & & & \vdots & & \\
 & & \uparrow & & & & \uparrow & & \\
\cdots & \longrightarrow & V_{0,1} & \longrightarrow & V^1_{0,1} & \longrightarrow & V_{1,1} & \longrightarrow & \cdots \\
 & & \uparrow & & \uparrow & & \uparrow & & \\
\cdots & \longrightarrow & V_{0,0} & \longrightarrow & V^1_{0,0} & \longrightarrow & V_{1,0} & \longrightarrow & \cdots \\
 & & \uparrow & & & & \uparrow & & \\
 & & \vdots & & & & \vdots & &
\end{array} \tag{3.15}$$

where $V_{k,\ell}$ are as above and $V^1_{k,\ell} = V^{\lambda - m_1\alpha_1 - kN\alpha_1 - \ell N\alpha_2}$, $k,\ell \in \mathbb{Z}$. The embeddings $V_{k,\ell} \longrightarrow V^1_{k,\ell}$ are realized by $v_s = (X_1^-)^m \otimes v_0(V_{k,\ell})$, while $V^1_{k,\ell} \longrightarrow V_{k+1,\ell}$ are realized by $v_s = (X_1^-)^{N-m} \otimes v_0(V^1_{k,\ell})$. The factor modules $L_{k,\ell} = V_{k,\ell}/I_{k,\ell}$ have the same structure $\forall k,\ell, \in \mathbb{Z}$; also the factor modules $L^1_{k,\ell} = V^1_{k,\ell}/I^1_{k,\ell}$ have the same structure $\forall k,\ell, \in \mathbb{Z}$. We shall denote by $L^2_{m_1}$, respectively, $L'^2_{m_1}$ any of the representations $L_{k,\ell}$, respectively, $L^1_{k,\ell}$. If $N \in 2\mathbb{N}$ and $m_1 = N/2$ then everything said above for $L^2_{m_1}$ is valid here for $m_1 = N/2$. Thus there are $N-1$ essentially different irreducible HWM with highest weights satisfying (3.14).

We do not consider separately the subcase obtained from this by exchanging the indices 1 and 2. The corresponding representations which are conjugate to $L^2_{m_1}$, $L'^2_{m_1}$, respectively, under the exchange $\alpha_1 \longleftrightarrow \alpha_2$ will be denoted by $\tilde{L}^2_{m_2}$, $\tilde{L}'^2_{m_2}$, respectively.

3.5.3. The multiplets of the third type are parametrized by a positive integer, say, m_3 such that $m_3 \leq N/2$. Fix such an m_3 and choose an element $\lambda \in \mathcal{H}^*$ such that

$$[\lambda(H_3) + 2 - m_3] = 0 \, , \quad \lambda(H_i) + n \neq 0 \, , \; i = 1, 2, \; \forall n \in \mathbb{Z} \, . \tag{3.16}$$

The HWM V^λ is part of the following multiplet :

$$\begin{array}{ccccccccc}
 & & \vdots & & & & \vdots & & \\
 & & \uparrow & & & & \uparrow & & \\
\cdots & \longrightarrow & V_{0,1} & \longrightarrow & & \longrightarrow & V_{1,1} & \longrightarrow & \cdots \\
 & & & & & \nearrow & & & \\
 & & \uparrow & & V^3_{0,0} & & \uparrow & & \\
 & & & \nearrow & & & & & \\
\cdots & \longrightarrow & V_{0,0} & \longrightarrow & & \longrightarrow & V_{1,0} & \longrightarrow & \cdots \\
 & & \uparrow & & & & \uparrow & & \\
 & & \vdots & & & & \vdots & &
\end{array} \tag{3.17}$$

where $V_{k,\ell}$ are as above and $V^3_{k,\ell} = V^{\lambda - m_3\alpha_3 - kN\alpha_1 - \ell N\alpha_2}$, $k, \ell \in \mathbb{Z}$. The embeddings $V_{k,\ell} \longrightarrow V^3_{k,\ell}$ and $V^3_{k,\ell} \longrightarrow V_{k+1,\ell+1}$ are realized by the singular vector in (2.16) for $m = m_3$, $m = N - m_3$, respectively. Formula (2.16) is valid for any $m \in \mathbb{N}$, if (3.16) holds (with m_3 replaced by m), however, if $m \geq N$, and $m = kN + t$, $k \in \mathbb{N}$, $t \in \mathbb{Z}_+$, $t < N$ it reduces to :

$$v_s^m \; = \; (X_1^-)^{kN} \, (X_2^-)^{kN} \, v_s^t \, . \tag{3.18}$$

Analogously to the previous case there are $N - 1$ essentially different irreducible HWM with highest weights satisfying (3.16), namely the factor modules $L^3_{m_3}$, $m_3 \leq N/2$, and $L'^3_{m_3}$, $m_3 < N/2$, which will denote any of the factor-modules $V_{k,\ell}/I_{k,\ell}$ $V^3_{k,\ell}/I^3_{k,\ell}$, $\forall k, \ell \in \mathbb{Z}$, respectivley.

3.5.4. The multiplets of the fourth type are parametrized by two positive integers, say, m_1 , m_2 such that $m_1 + m_2 < N$. Fix such m_1 , m_2 and choose an element $\lambda \in \mathcal{H}^*$ such that

$$[\lambda(H_i) + 1 - m_i] \; = \; 0 \, , \quad i = 1, 2, \quad \Rightarrow \; [\lambda(H_3) + 2 - m_1 - m_2] \; = \; 0 \, . \tag{3.19}$$

The HWM V^λ is part of the following multiplet :

$$
\begin{array}{ccccccccccc}
 & & \vdots & & & & & & \vdots & & \\
 & & \uparrow & & & & & & \uparrow & & \\
\cdots & \longrightarrow & V_{0,1} & \longrightarrow & V^1_{0,1} & & \longrightarrow & & V_{1,1} & \longrightarrow & \cdots \\
 & & & & \uparrow & & & \nearrow & & & \\
 & & \uparrow & & V^{12}_{0,0} & \longrightarrow & V^3_{0,0} & & \uparrow & & \\
 & & & \nearrow & & & \uparrow & & & & \\
 & & & & \uparrow & & & & & & \\
 & & V^2_{0,0} & \longrightarrow & & \longrightarrow & V^{21}_{0,0} & \longrightarrow & V^2_{1,0} & & \\
 & & & & \uparrow & & & & & & \\
 & & \uparrow & & & \nearrow & & & \uparrow & & \\
\cdots & \longrightarrow & V_{0,0} & \longrightarrow & V^1_{0,0} & & \longrightarrow & & V_{1,0} & \longrightarrow & \cdots \\
 & & \uparrow & & & & & & \uparrow & & \\
 & & \vdots & & & & & & \vdots & &
\end{array}
\tag{3.20}
$$

where $V_{k,\ell}$ are as before and

$$V^i_{k,\ell} = V^{\lambda - m_i\alpha_i - kN\alpha_1 - \ell N\alpha_2} \ , \quad i = 1,2,3, \quad m_3 = m_1 + m_2 \ , \quad k,\ell \in \mathbb{Z} \ , \tag{3.21a}$$

$$V^{ij}_{k,\ell} = V^{\lambda - m_i\alpha_i - m_3\alpha_j - kN\alpha_1 - \ell N\alpha_2} \ , \quad (ij) = (12),(21), \quad k,\ell \in \mathbb{Z} \ . \tag{3.21b}$$

We summarize the structure of the above multiplets as follows.

3.5.4.1. The embeddings $V_{k,\ell} \longrightarrow V^1_{k,\ell}$, $V^1_{k,\ell} \longrightarrow V_{k+1,\ell}$, $V^2_{k,\ell} \longrightarrow V^{21}_{k,\ell}$, $V^{21}_{k,\ell} \longrightarrow V^2_{k+1,\ell}$, $V^{12}_{k,\ell} \longrightarrow V^3_{k,\ell}$, $V^3_{k,\ell} \longrightarrow V^{12}_{k+1,\ell}$, $V_{k,\ell+1} \longrightarrow V^1_{k,\ell+1}$, $V^1_{k,\ell+1} \longrightarrow V_{k+1,\ell+1}$, respectively, are realized by singular vectors $(X_1^-)^p \otimes v_0$ with $p = m_1$, $N - m_1$, m_3 , $N - m_3$, m_2 , $N - m_2$, m_1 , $N - m_1$, respectively. The embeddings $V_{k,\ell} \longrightarrow V^2_{k,\ell}$, $V^2_{k,\ell} \longrightarrow V_{k,\ell+1}$, $V^1_{k,\ell} \longrightarrow V^{12}_{k,\ell}$, $V^{12}_{k,\ell} \longrightarrow V^1_{k,\ell+1}$, $V^{21}_{k,\ell} \longrightarrow V^3_{k,\ell}$, $V^3_{k,\ell} \longrightarrow V^{21}_{k,\ell+1}$, $V_{k+1,\ell} \longrightarrow V^2_{k+1,\ell}$, $V^2_{k+1,\ell} \longrightarrow V_{k+1,\ell+1}$, respectively, are realized by singular vectors $(X_2^-)^p \otimes v_0$ with $p = m_2$, $N - m_2$, m_3 , $N - m_3$, m_1 , $N - m_1$, m_2 , $N - m_2$, respectively. The embeddings $V^1_{k,\ell} \longrightarrow V^{21}_{k,\ell}$, $V^2_{k,\ell} \longrightarrow V^{12}_{k,\ell}$, $V^3_{k,\ell} \longrightarrow V_{k+1,\ell+1}$, respectively, are realized by singular vectors given by formula (2.10) with λ replaced by $\lambda - m_1\alpha_1$, $\lambda - m_2\alpha_2$, $\lambda - m_3\alpha_3$, respectively, and $m = m_2$, $m = m_1$, $m = N - m_3$, respectively.

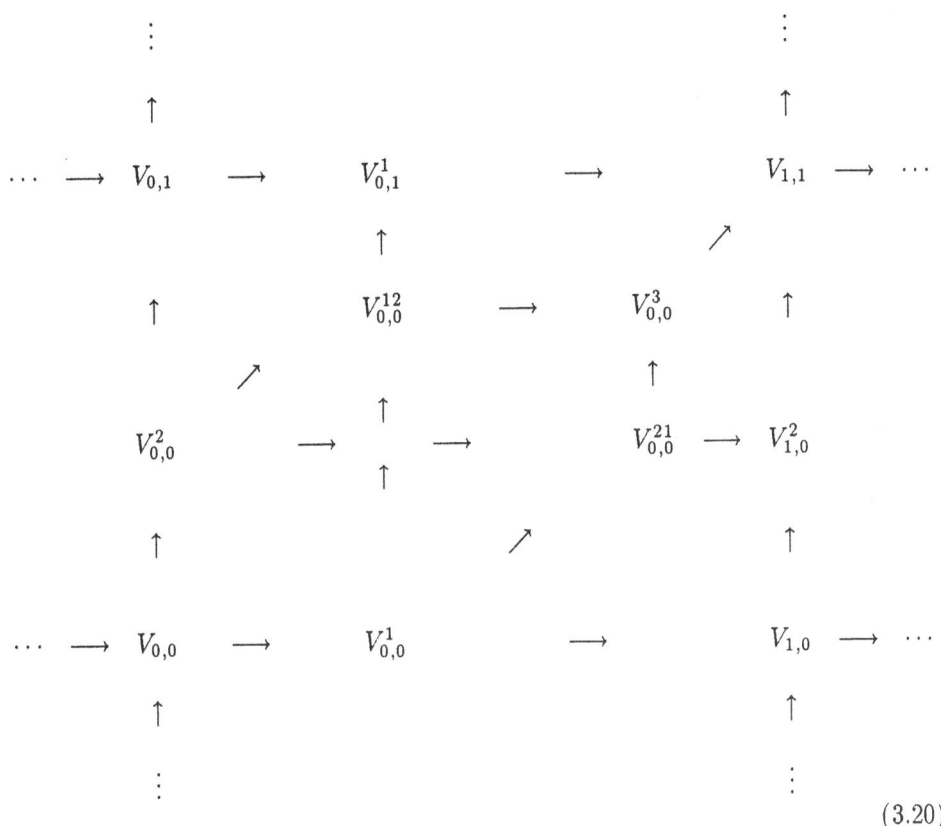

3.5.4.2. Note that the six HWM $V_{k,\ell}$, $V^i_{k,\ell}$, $V^{ij}_{k,\ell}$ for fixed k,ℓ form the basic $sl(3,\mathbb{C})$ multiplet in the case $q=1$, (cf. [Do2], formula (38)). The sextet diagram consisiting of these six HWM is commutative which one checks also here using formula (2.16). Let us say that the tip of this sextet is at $V_{k,\ell}$. This sextet shares one side with six sextets of the same type and orientation and for the same k,ℓ their tips are at $V^{21}_{k-1,\ell-1}$, $V^{12}_{k,\ell-1}$, $V^{21}_{k,\ell}$, $V^{12}_{k,\ell}$, $V^{21}_{k-1,\ell}$, $V^{12}_{k-1,\ell-1}$. The role of (m_1, m_2) in these sextets is played by $(m_2 , N-m_3)$ for $V^{12}_{*,*}$ and by $(N-m_3 , m_1)$ for $V^{21}_{*,*}$. Moreover, this structure is periodic and if we consider only such sextets then this multiplet looks like a honeycomb and resembles one of the multiplets of reducible Verma modules over the affine Lie algebra $sl(3,\mathbb{C})^{(1)}$, namely, the "maximal" multiplet in the sense that it represents the affine Weyl group W, (cf. [Do2], Proposition 2 and the Figure). However, in the affine case this honeycomb corresponding to the affine Weyl group has a distinguished point (corresponding to the unit element of W), i.e., a Verma module which contains as submodules all other Verma modules in this multiplet; (the irreducible subquotient of this distinguished Verma module is an integrable HWM and all integrable HWM over $sl(3,\mathbb{C})^{(1)}$ can be obtained in this way).

3.5.4.3. Another connection with the affine Weyl group was noticed and used in the derivation of the fusion rules in Wess–Zumino–Witten theories [FGP]. This connection appears if we look upon (3.20) as the product of the lattice in (3.13) with the basic $sl(3,\mathbb{C})$ sextet; the former is isomorphic to Γ (cf. (2.8)), the latter is isomorphic to the Weyl group W_0 of $sl(3,\mathbb{C})$, their semi-direct product (W_0 acting on Γ) is isomorphic to the Weyl group W of $sl(3,\mathbb{C})^{(1)}$.

3.5.4.4. Below we shall use also the fact that there are other sextets of HWM, namely: $V^3_{k-1,\ell-1}$, $V^{12}_{k,\ell-1}$, $V^2_{k+1,\ell}$, $V_{k+1,\ell+1}$, $V^1_{k,\ell+1}$, $V^{21}_{k-1,\ell}$, for fixed k,ℓ and containing the sextet $V_{k,\ell}$, $V^i_{k,\ell}$, $V^{ij}_{k,\ell}$. Certainly these bigger sextets are more complicated.

3.5.4.5. Thus the structure of the representations $V_{k,\ell}$, $V^{12}_{k,\ell}$, $V^{21}_{k,\ell}$, is exactly the same; moreover the range of their parameters is the same. The same holds for the representations $V^i_{k,\ell}$, $i=1,2,3$. These are situated in the sextets at the site opposite to what we called the tip. The values $(\lambda(H_1) , \lambda(H_2))$, i.e., the analogues of (m_1, m_2) , are $(N-m_1 , m_3)$, $(m_3 , N-m_2)$, $(N-m_2 , N-m_1)$, respectively, for $i=1,2,3$, respectively, and they cover the same range. Moreover, this shows that the requirement $m_1+m_2<N$ is not a restriction. Indeed, the HWM $V^i_{k,\ell}$ for one value of i exhaust all such cases.

3.5.4.6. From the above it is easy to see that there are the following essentially different irreducible HWM with highest weights satisfying (3.19), namely $L^4_{m_1m_2}$ and $L'^4_{m_1m_2}$ which will denote any of the factor modules $V_{k,\ell}/I_{k,\ell}$, $V^3_{k,\ell}/I^3_{k,\ell}$, respectively.

The multiplets of the next type can be viewed as "analytic" continuation in m_i for $m_1 + m_2 = N$.

3.5.5. The multiplets of the fifth type are parametrized by a positive integer, say, m_1 such that $m_1 \leq N/2$. Fix such m_1 and choose an element $\lambda \in \mathcal{H}^*$ such that

$$[\lambda(H_i) + 1 - m_i] \;=\; 0 \;, \quad m_2 = N - m_1 \;. \tag{3.22}$$

The HWM V^λ is part of a multiplet containing the following HWM : $V_{k,\ell}$ and $V^i_{k,\ell}$, $i = 1, 2$ given by the same formulae as in the previous case with $m_2 = N - m_1$ and $m_3 = N$. It can be depicted using (3.20) and distorting it so that $V^3_{k,\ell}$ will coincide with $V_{k+1,\ell+1}$, $V^{12}_{k,\ell}$ with $V^1_{k,\ell+1}$, $V^{21}_{k,\ell}$ with $V^2_{k+1,\ell}$. Thus the sextets with $V^{12}_{k,\ell}$, $V^{21}_{k,\ell}$ at the tips deteriorate into commutative triangles and the latter representations do not have the structure of $V_{k,\ell}$. The singular vectors depicting the embeddings are as in the previous case, however, taking into account the coincidences. It is easy to see that there are the following inequivalent irreducible HWM with highest weights satisfying (3.22), namely $L^5_{m_1}$, $L^{51}_{m_1}$, $L^{52}_{m_1}$, which will denote the factor modules $V_{k,\ell}/I_{k,\ell}$, $V^1_{k,\ell}/I^1_{k,\ell}$, $V^2_{k,\ell}/I^2_{k,\ell}$. Note that $L^{51}_{m_1}$, $L^{52}_{m_2}$ are conjugate to each other under the exchange $\alpha_1 \longleftrightarrow \alpha_2$.

The results contained in Subsection 3.5. are summarized as follows [Do5]. Let $q^N = 1$, $N \in I\!N + 2$, $\mathcal{G} = sl(3, \mathbb{C})$. Then we have: (a) All HWM V^λ over $U_q(\mathcal{G})$ belong to multiplets of one of the five types described above. (b) The list of essentially different irreducible HWM of $U_q(sl(3, \mathbb{C}))$ consists of the factor-modules L^1 from 3.5.1.; $L^2_{m_1}$ for $m_1 \leq N/2$, $L'^2_{m_1}$ for $m_1 < N/2$, $\tilde{L}^2_{m_2}$ for $m_1 \leq N/2$, $\tilde{L}'^2_{m_1}$ for $m_2 < N/2$, from 3.5.2.; $L^3_{m_3}$ for $m_3 \leq N/2$, $L'^3_{m_3}$ for $m_3 < N/2$, from 3.5.3.; $L^4_{m_1m_2}$, $L'^4_{m_1m_2}$, from 3.5.4.; $L^5_{m_1}$, $L^{51}_{m_1}$, for $m_1 \leq N/2$, $L^{52}_{m_1}$ for $m_1 < N/2$, from 3.5.5.

3.6. Cyclic representations of $U_q(\mathcal{G})$

As we saw above when q is a root of unity the $2N$-th powers of the Chevalley generators belong to the center of $U_q(\mathcal{G})$, thus they behave as Casimirs in any representation. In the Verma module approach these Casimirs are set to zero when we consider the irreducible factor-modules. However, in a more general approach, when we consider non-highest- weight representations one may give nonzero values to these Casimirs. These representations are called *cyclic* [DJMM1, DJMM2], the term referring to the fact that each root vector generates a multiplicative cyclic group.

3.6.1. Let us start with the example of $U_q(sl(2,\mathbb{C}))$ considered in [RA]. Let $q = e^{2\pi i/N}$. Their cyclic representation depends on three complex parameters a, b, μ such that $([p][\mu+1-p]+ab) \neq 0$ for $p = 1,\ldots,N-1$. It has the basis v_p , $p = 0,1,\ldots,N-1$ and transforms under the generators of $U_q(sl(2,\mathbb{C}))$ as follows:

$$H v_p \;=\; (\mu - 2p) v_p \;, \tag{3.23}$$

$$X^- v_p \;=\; ([p+1][\mu - p] + ab)^{1/2} v_{p+1} \;, p = 0,1,\ldots,N-2 \;, \tag{3.24a}$$

$$X^- v_{N-1} = a v_0 \;, \tag{3.24b}$$

$$X^+ v_p \;=\; ([p][\mu + 1 - p] + ab)^{1/2} v_{p-1} \;, p = 1,\ldots,N-1 \;, \tag{3.25a}$$

$$X^+ v_0 = b v_{N-1} \;, \tag{3.25b}$$

If $a \neq 0 \neq b$ this representations is not a highest or lowest weight module and is called cyclic because of formulae (3.24b), (3.25b). If $a = 0, b \neq 0$, $\mu - 2m + 2 \notin \mathbb{Z}_+$, ($a \neq 0, b = 0$, $\mu \notin \mathbb{Z}_+$), then it is a cyclic irreducible lowest, (highest), weight module, with lowest, (highest) weight $\lambda = (\mu - 2(m-1))\alpha/2$, ($\lambda = \mu\alpha/2$).

If $a = b$ then $X^+ = (X^-)^t$. If $a = \bar{b}$ and μ real then $X^+ = (X^-)^+$. In this last case the representations with three real parameters correspond to representations obtained in [S2].

Two such representations with parameters a, b, μ and a', b', μ' are isomorphic if and only if

$$\mu' = \mu + 2r, r \in \mathbb{Z},\; a'b = ab',\; ab - a'b' = [2r][\mu + 2r + 1] \;. \tag{3.26}$$

A class of infinite-dimensional representations (not HWM in general) of $U_q(sl(2,\mathbb{C}))$ which in some cases contain finite-dimensional subrepresentations is considered in [HS].

3.6.2. In this subsection we review the paper [DJMM2]. Let us consider $U_q = U_q(sl(n+1,\mathbb{C}))$, $n \geq 2$. In [DJMM2] is constructed (for generic q) an algebra map from U_q a $\mathbb{C}(q)$ algebra $\mathcal{W}$ determined as follows. It is generated by x_{jk}, z_{jk}, $1 \leq j \leq k \leq n$, and the inverses x_{jk}^{-1}, z_{jk}^{-1}, satisfying

$$[x_{jk}, x_{j'k'}] = [x_{jk}, z_{j'k'}] = [z_{jk}, z_{j'k'}] = 0, \text{ if } (j,k) \neq (j',k'), \tag{3.27a}$$

$$z_{jk} x_{jk} = q x_{jk} z_{jk} \;. \tag{3.27b}$$

A $\mathbb{C}(q)$ linear involution $*$ is defined by

$$(x_{jk})^* = x_{k+1-j\;k}^{-1} \;, \quad (z_{jk})^* = z_{k+1-j\;k}^{-1} \;, \tag{3.28}$$

and $\mathbb{C}$ linear involution $\hat{}$ by

$$\hat{q} = q^{-1}, \quad \hat{x}_{jk} = x_{jk} \ , \quad \hat{z} = z_{jk}^{-1} \ . \tag{3.29}$$

the analogous involutions for $U_q(sl(n+1,\mathbb{C}))$ are defined by

$$(X_i^{\pm})^* = (X_{n+1-i}^{\mp} \ , \quad (H_i)^* = -H_{n+1-i} \ , \quad (K_i)^* = K_{n+1-i}^{-1} \ , \tag{3.30}$$

$$\hat{q} = q^{-1}, \quad \hat{X}_i^{\pm} = X_i^{\pm} \ , \quad \hat{H}_i = -H_i \ , \quad \hat{K}_i = K_i^{-1} \ . \tag{3.31}$$

For $r = (r_1, \ldots, r_n) \in (\mathbb{C}^{\times})^n$ one defines $r^* = (r_n, \ldots, r_1)$, $\hat{r} = (r_1^{-1}, \ldots, r_n^{-1})$. The authors of [DJMM2] construct a family of $\mathbb{C}(q)$ homomorphisms

$$\rho_{r,s} \ : \ U_q(sl(n+1,\mathbb{C})) \ \longrightarrow \ \mathcal{W} \ , \tag{3.32}$$

depending on $r, s \in (\mathbb{C}^{\times})^n$ by the formulae :

$$\rho_{r,s}\,(X_i^+) \ = \ \sum_{k=i}^{n} \, [\zeta_{ik}] \, \xi_{ik} \ , \tag{3.33a}$$

$$\rho_{r,s}\,(X_i^-) \ = \ \rho_{s^*,r^*}\,(X_{n+1-i}^+)^* \ , \tag{3.33b}$$

$$\rho_{r,s}\,(K_i) \ = \ \frac{r_i}{s_i} \, z_{in}^2 \, z_{i+1\,n}^{-1} \ , \tag{3.33c}$$

where

$$\xi_{ik} \ = \ x_{ik} x_{ik+1} \cdots x_{in} \ , \tag{3.34a}$$

$$\zeta_{ik} \ = \ r_i z_{ik} z_{ik-1} z_{i-1k-1}^{-1} z_{i+1k}^{-1} \ . \tag{3.34b}$$

Further let $N \geq 3$ be an odd positive integer and let $q = e^{2\pi i/N}$. Let $\Phi_N(x)$ denote the N–th cyclotomic polynomial so that $\Phi_N(q) = 0$. One sets

$$\mathcal{A} \ = \ \{f \in \mathbb{C}(q) |\ f \text{ is regular at } \Phi_N(q) = 0\} \ . \tag{3.35}$$

Let $U_{\mathcal{A}}$ denote the $\mathcal{A}$–subalgebra of U_q generated by $X_i^{\pm}$, K_i , $i = 1, \ldots, n$. Let further $U = U_{\mathcal{A}} \otimes_{\mathcal{A}} \mathbb{C}$. The algebra $\mathcal{W}_{\mathcal{A}}$ is defined analogously.

Consider an N–dimensional vector space with fixed basis u_i , $i = 0, \ldots,$ $N - 1$:

$$V^1 \ = \ \oplus_{i=0}^{N-1} \, \mathbb{C} u_i \ . \tag{3.36}$$

One defines a representation σ of the Weyl algebra $\mathcal{W}_q^1$ with generators x, z by :

$$\sigma \ : \ \mathcal{W}_q^1 \longrightarrow \text{End}(V_q^1) \ , \tag{3.37a}$$

$$\sigma(x) u_i \ = \ u_{i+1} \ , \ (u_N = u_0), \quad \sigma(z) u_i \ = \ q^i u_i \ . \tag{3.37b}$$

Further let $m = n(n+1)/2 = \dim \mathcal{G}^+$ and $V = (V_q^1)^{\otimes m}$. Thus one obtains a representation $\sigma^{\otimes m} : \mathcal{W} \cong (\mathcal{W}_q^1)^{\otimes m} \longrightarrow \mathrm{End}(V)$ by letting the generators x_{jk}, z_{jk} act on the (j,k)–component of V as $\sigma(x), \sigma(z)$ and as identity on the other components. Further one defines automorphisms S_ν , T_ν of $\mathcal{W}$ for $\nu = (\nu_{jk}) \in (\mathbb{C}^\times)^m$:

$$S_\nu(x_{jk}) = \nu_{jk} x_{jk} \ , \quad S_\nu(z_{jk}) = z_{jk} \ , \tag{3.38a}$$

$$T_\nu(x_{jk}) = x_{jk} \ , \quad T_\nu(z_{jk}) = \nu_{jk} z_{jk} \ . \tag{3.38b}$$

Now the representation of U is defined by the following composition of maps :

$$\pi \ : \ U \ \overset{\rho_{r,s}}{\longrightarrow} \ \mathcal{W} \ \overset{S_g \circ T_h}{\longrightarrow} \ \mathcal{W} \ \overset{\sigma^{\otimes m}}{\longrightarrow} \ \mathrm{End}(V) \tag{3.39}$$

Besides $r, s \in (\mathbb{C}^\times)^n$, the representation π contains $n(n+1)$ arbitrary parameters $g = (g_{jk}), h = (h_{jk}) \in (\mathbb{C}^\times)^m$. Not all of these parameters are independent and one can set $s_i = 1, i+1, \ldots, n$.

Further the authors [DJMM2] show cyclicity of the representation and prove that it is irreducible for generic parameters r_i, g_{jk}, h_{jk}. For special values of the parameters they obtain invariant subspaces.

3.6.3. Let $\hat{U}_q = U_q(sl(n,\mathbb{C})^{(1)})$, $n \geq 2$. Further we consider the cyclic representations of $\hat{U}_q$ following [DJMM1]. Let $q = e^{2\pi i/N}$, $N \geq 3$. Let V be an N–dimensional vector space. Let Y, Z be two linear operators on V satifying the relation $ZY = qYZ$. Denote by Y_i (respectively, Z_i) the operator on $\mathcal{W} = V^{\otimes n}$ which acts on the i–th component as Y (repectively, Z) and as identity on the other components. Set

$$\mathcal{W}^{(0)} = \{ w \in \mathcal{W} | \left(\prod_{i=1}^{n} Z_i \right) w = w \} \ . \tag{3.40}$$

Let $a_0, \ldots, a_{n-1}$, x_0 , $\ldots, x_{n-1}$ be arbitrary nonzero numbers. An N^{n-1} - dimensional cyclic representation $\pi_{x,a}$ of $\tilde{U}_q$ on $\mathcal{W}^{(0)}_{x,a} = \mathcal{W}^{(0)}$ is defined as follows :

$$\pi_{x,a}(X_i^+) = \frac{x_i}{q^{1/2} - q^{-1/2}} (a_i Z_i^2 - a_i^{-1} Z_i^{-2}) Y_i Y_{i+1}^{-1} \ , \tag{3.41a}$$

$$\pi_{x,a}(X_i^-) = \frac{x_i^{-1}}{q^{1/2} - q^{-1/2}} (a_{i+1} Z_{i+1}^2 - a_{i+1}^{-1} Z_{i+1}^{-2}) Y_i^{-1} Y_{i+1} \ , \tag{3.41b}$$

$$\pi_{x,a}(K_i) = \frac{a_i}{a_{i+1}} Z_i Z_{i+1}^{-1} \ , \tag{3.41c}$$

where $a_0 = a_n$, $Y_0 = Y_n$, $Z_0 = Z_n$. Choose a basis v_k , $k = 0, \ldots, N-1$; $(v_N = v_0)$ of V on which Y, Z act by

$$Y v_k = q^k v_k \ , \quad Z v_k = v_{k-1} \ . \tag{3.42}$$

Then the basis vectors of $\mathcal{W}^{(0)}$ may be chosen as

$$w_{\mathbf{k}}^{(0)} = \sum_{p=0}^{N-1} v_{k_1+p} \otimes \cdots \otimes v_{k_n+p} , \tag{3.43}$$

where $\mathbf{k} = (k_1, \ldots, k_n)$, so that $\mathcal{W}_{k_1+1,\ldots,k_n+1}^{(0)} = \mathcal{W}_{k_1,\ldots,k_n}^{(0)}$.

4. Characters of highest weight modules

Let again $\mathcal{G}$ be any simple Lie algebra. Following [Ka] let $E(\mathcal{H}^*)$ be the associative abelian algebra consisting of the series $\sum_{\mu\in\mathcal{H}^*} c_\mu e(\mu)$, where $c_\mu \in \mathbb{C}$, $c_\mu = 0$ for μ outside the union of a finite number of sets of the form $D(\lambda) = \{\mu \in \mathcal{H}^* | \mu \leq \lambda\}$, using any ordering of $\mathcal{H}^*$; the formal exponents $e(\mu)$ have the properties $e(0) = 1$, $e(\mu)e(\nu) = e(\mu+\nu)$. We recall the decomposition (2.8). Then we define the character of V by :

$$ch\ V = \sum_{\mu\in\Gamma_+} (dim\ V_\mu)e(\lambda-\mu) = e(\lambda) \sum_{\mu\in\Gamma_+} (dim\ V_\mu)e(-\mu) . \tag{4.1}$$

We recall that for $V = V^\lambda$ we have $dim\ V_\mu = P(\mu)$, where the partition function was defined after (2.8). Analogously we use [Di] to obtain :

$$ch\ V^\lambda = e(\lambda) \sum_{\mu\in\Gamma_+} P(\mu)e(-\mu) = e(\lambda) \prod_{\alpha\in\Delta+} (1-e(-\alpha))^{-1} . \tag{4.2}$$

The Weyl (resp. Weyl–Kac) character formula for the finite-dimensional irreducible HWM over $\mathcal{G}$ (resp. for the integrable HWM over affine Lie algebras) has the form [Di] (resp. [Ka]) :

$$ch\ L_\lambda = ch\ V^\lambda \sum_{w\in W} (-1)^{\ell(w)} e(w\cdot\lambda - \lambda) , \tag{4.3a}$$

$$= \sum_{w\in W} (-1)^{\ell(w)}\ ch\ V^{w\cdot\lambda} , \tag{4.3b}$$

where the dot $\cdot$ action is defined by $w\cdot\lambda = w(\lambda+\rho)-\rho$. If q is not a root of unity this formula holds for the finite-dimensional irreducible HWM over $U_q(\mathcal{G})$ (this can be deduced from the results cf. [R2, L1]).

In [Do6] it was shown that the same formula holds when q is a root of unity for the irreducble HWM $L_{m_1m_2}$ and $L_{m_1}^5$ over $U_q(sl(3,\mathbb{C}))$, cf. subsection 3.5. Some other character formulae were obtained in [Do7]. These results can be summarized as follows.

Let $\mathcal{G} = sl(3,\mathbb{C})$. Denote $t_i \equiv e(-\alpha_i)$, $i = 1, 2$, then $e(-\alpha_3) = t_1t_2$. Then (4.2) can be rewritten as

$$ch\ V^\lambda = e(\lambda)/(1-t_1)(1-t_2)(1-t_1t_2) . \tag{4.4}$$

Let N be the smallest positive integer such that $q^N = 1$. Let L^1 , L^2_m for $m \le N/2$, L'^2_m for $m < N/2$, $\tilde{L}^2_m$ for $m \le N/2$, $\tilde{L}'^2_m$ for $m < N/2$, L^3_m for $m \le N/2$, L'^3_m for $m < N/2$, $L^4_{m_1 m_2}$, for $m_1 , m_2 \in I\!N$, $m_1 + m_2 \le N$, L^5_m , for $m \le N/2$, be defined as above in Subsection 3.5. Then we have :

$$
\begin{aligned}
ch\ L^1 &= ch\ V_{k,\ell}(1 - t_1^N)(1 - t_2^N)(1 - (t_1 t_2)^N) , && (4.5a) \\
ch\ L^2_m &= ch\ V_{k,\ell}(1 - t_1^m)(1 - t_2^N)(1 - (t_1 t_2)^N) , && (4.5b) \\
ch\ L'^2_m &= ch\ V^1_{k,\ell}(1 - t_1^{N-m})(1 - t_2^N)(1 - (t_1 t_2)^N) , && (4.5c) \\
ch\ \tilde{L}^2_m &= ch\ V_{k,\ell}(1 - t_1^N)(1 - t_2^m)(1 - (t_1 t_2)^N) , && (4.5d) \\
ch\ \tilde{L}'^2_m &= ch\ V^1_{k,\ell}(1 - t_1^N)(1 - t_2^{N-m})(1 - (t_1 t_2)^N) , && (4.5e) \\
ch\ L^3_m &= ch\ V_{k,\ell}(1 - t_1^N)(1 - t_2^N)(1 - (t_1 t_2)^m) , && (4.5f) \\
ch\ L'^3_m &= ch\ V^3_{k,\ell}(1 - t_1^N)(1 - t_2^N)(1 - (t_1 t_2)^{N-m}) ; && (4.5g)
\end{aligned}
$$

$$
\begin{aligned}
ch\ L^4_{m_1 m_2} &= ch\ V_{k,\ell} \sum_{w \in W} (-1)^{\ell(w)} e(w \cdot \lambda - \lambda) = && (4.6a) \\
&= ch\ V_{k,\ell} - ch\ V^1_{k,\ell} - ch\ V^2_{k,\ell} + \\
&+ ch\ V^{12}_{k,\ell} + ch\ V^{21}_{k,\ell} - ch\ V^3_{k,\ell} , && (4.6b) \\
ch\ L^5_m &= ch\ L^4_{m,N-m} . && (4.6c)
\end{aligned}
$$

Further we have that all states of the representations in (4.5) are given by :

$$
(X_1^-)^{n_1} (X_2^-)^{n_2} (X_3^-)^{n_3} \otimes v_0 , \qquad (4.7a)
$$

where

$$
\begin{array}{lll}
n_i = 0, \dots, N-1 ,\ i = 1,2,3, & \text{for } L^1 , & \\
n_1 = 0, \dots, m-1 ,\ n_i = 0, \dots, N-1 ,\ i = 2,3, & \text{for } L^2_m , & \\
n_1 = 0, \dots, N-m-1 ,\ n_i = 0, \dots, N-1 ,\ i = 2,3, & \text{for } L'^2_m , & \\
n_2 = 0, \dots, m-1 ,\ n_i = 0, \dots, N-1 ,\ i = 1,3, & \text{for } \tilde{L}^2_m , & (4.7b) \\
n_2 = 0, \dots, N-m-1 ,\ n_i = 0, \dots, N-1 ,\ i = 1,3, & \text{for } \tilde{L}'^2_m , & \\
n_3 = 0, \dots, m-1 ,\ n_i = 0, \dots, N-1 ,\ i = 1,2, & \text{for } L^3_m , & \\
n_3 = 0, \dots, N-m-1 ,\ n_i = 0, \dots, N-1 ,\ i = 1,2, & \text{for } L'^3_m . &
\end{array}
$$

Consequently we have :

$$
\begin{aligned}
&\dim L^1 = N^3 ,\ \dim L^2_m = \dim \tilde{L}^2_m = \dim L^3_m = mN^2 , \\
&\dim L'^2_m = \dim \tilde{L}'^2_m = \dim L'^3_m = (N-m)N^2 ; \qquad (4.8)
\end{aligned}
$$

$$
\begin{aligned}
\dim L^4_{m_1 m_2} &= m_1 m_2 (m_1 + m_2)/2 , && (4.9a) \\
\dim L^5_m &= m(N-m)N/2 . && (4.9b)
\end{aligned}
$$

The proof of (4.5), (4.7), (4.8) and (4.9) is given in [Do7], and the proof of (4.6) in [Do6].

The representations $L_{m_1 m_2}$ and $L^5_{m_1}$ have the same structure as the irreducible finite-dimensional HWM of $U_q(sl(3,\mathbb{C}))$ with the same highest weights for arbitrary q and also the same structure as the irreducible finite-dimensional HWM of $\mathcal{G}$ with the same highest weights. In particular, λ are integral dominant elements of $\mathcal{H}^*$. Thus it is no surprise that the character formulae (4.3) and (4.6) coincide. Note that when q is not a root of unity or when $q = 1$, formula (4.7a), however, for any m_1 , $m_2 \in \mathbb{N}$, gives the dimensions of *all* finite dimensional irreducible HWM of $U_q(sl(3,\mathbb{C}))$ or of $sl(3,\mathbb{C})$ (or equivalently of all unitary irreducible representations of $su(3)$ or the group $SU(3)$). In particular, L^4_{11} is the trivial 1-dimensional representation in all cases.

Thus we are lead to the following Conjecture. Let $\mathcal{G}$ be any simple Lie algebra, let $q^N = 1$, $N \in \mathbb{N}+1$, $\lambda \in \mathcal{H}^*$, $m_i \equiv \lambda(H_i) + 1 < N$, $i = 1,\ldots,r$, let $\tilde{\alpha}$ be the highest root of Δ . Then we conjecture that (4.3) holds if

$$m_{\tilde{\alpha}} \equiv (\lambda+\rho)(H_{\tilde{\alpha}}) \leq N . \tag{4.10}$$

The support for this conjecture is the following. If $m_{\tilde{\alpha}} = kN + n_{\tilde{\alpha}} > N$, $k, n_{\tilde{\alpha}} \in \mathbb{N}$, $n_{\tilde{\alpha}} < N$, then it is easy to see that there shall exist at least one $\beta' \in \Delta^+$ such that $m_{\beta'} \equiv (\lambda+\rho)(H'_\beta) = k'N + n_{\beta'}$ with $k' \in \mathbb{Z}_+$, $n_{\beta'} \in \mathbb{N}$, $n_{\beta'} < N$ so that $n_{\tilde{\alpha}} < n_{\beta'}$. Then the singular vector given by formula (10) with $\beta = m_{\tilde{\alpha}}$, $k = 0$ and $n = n_{\tilde{\alpha}}$ shall not factorize including as a factor the singular vector given by (3.4) with $\beta = \beta'$, $k = 0$ and $n = n_{\beta'}$. Thus the embedding pattern of the submodules of V^λ is not the same as of Verma modules $V(\lambda)$ with λ integral dominant.

In [AGS2] it is conjectured by a different motivation that (4.3) holds when $m_{\tilde{\alpha}} < N$, i.e., for the so-called regular representations. The latter can be extended (for $q = 1$) to the affine Lie algebra counterpart of $\mathcal{G}$ if $m_0 + m_{\tilde{\alpha}} \equiv k + g = N$, where k is the affine central charge, g is the dual Coxeter number, $m_0 \in \mathbb{N}$. This is natural in view of the connection (albeit in a partial case) with the affine Weyl group commented in subsections 3.5.4.2. and 3.5.4.3.

A more general Conjecture again involving the affine Weyl group was given in [L2, L3]. Further we review [L3]. Let $\in \mathbb{N}+1$. Let W be the affine Weyl group with simple reflections $s_0, \ldots, s_r$. Let E be an $\mathbb{R}$–vector space with basis $\gamma_1, \ldots, \gamma_r$. A positive definite inner product in $E \times E$ is defined by $(\gamma_i , \gamma_j) = a'_{ij}$, where

$(a'_{ij})_{1\leq i,j\leq r}$ is the matrix inverse to $(a_{ij})_{1\leq i,j\leq r}$. Further, denote :

$$\mathcal{C}_N = \{x = \sum_{i=1}^{r} c_i\gamma_i \in E \mid c_i \in \mathbb{R},\ c_i \leq -1 \text{ for } i = 1,\ldots,r,\ \sum_{j=1}^{r} m_j c_j \geq 1-N-g\}\ . \tag{4.11}$$

This is a simplex in E with $r+1$ walls given by $c_i = -1$ for $i = 1,\ldots,r$ and $m_j c_j = 1-N-g$. Denote by S_i , $i = 1,\ldots,r$ and S_0^N the orthogonal reflections in E with respect to these walls. Then $s_i \mapsto S_i$, $i = 1,\ldots,r$, and $s_0 \mapsto S_0^N$ defines an embedding $j_N : W \longrightarrow \mathrm{Aff}(E)$. Further $\mathbb{Z}^r$ is identified with a lattice in E by $(z_1,\ldots,z_r) \mapsto \sum_{i=1}^r z_i\gamma_i$. If $x \in E$, then $x \in j_N(w)(\Delta_N)$ for some $w \in W$; among such w there is a unique one, denoted by $w_{x,N}$ of maximal length. If w' is any element of W, one defines $x_{w',N} = w' w_{x,N}^{-1} \in E$.

Further, let W_0 be the finite subgroup of W generated by $s_1,\ldots,s_r$ and let $j = j_N|_{W_0}$. Let $\mathcal{C} = \{x = \sum_{i=1}^r c_i\gamma_i \in E \mid c_i \in \mathbb{R},\ c_i \leq -1 \text{ for } i = 1,\ldots,r\}$. If $x \in E$, then $x \in j(w)(\mathcal{C})$ for some $w \in W_0$; among such w there is a unique one, denoted by w_x of maximal length. If w' is any element of W_0, one defines $x'_w = w' w_x^{-1} \in E$.

Now the Conjecture of [L3] uses the Kazhdan–Lusztig polynomials $P_{y,w}$ [KL] and Bruhat order [Di] for W and W_0 . Let $\lambda \in \Gamma$, then :

$$ch\ L_\lambda = \sum_{\substack{w' \in W_0 \\ w' \leq w}} (-1)^{\ell(ww')} P_{w',w}(1)\ ch\ V^{\lambda_{w'}}\ ,\quad w = w_\lambda,\ q = 1 \text{ or } q \text{ not a root of } 1\ , \tag{4.12}$$

$$ch\ L_\lambda = \sum_{\substack{w' \in W \\ w' \leq w}} (-1)^{\ell(ww')} P_{w',w}(1)\ ch\ V^{\lambda_{w',N}}\ ,\quad w = w_{\lambda,N}\ ,\ q \text{ a primitive root of } 1\ . \tag{4.13}$$

For $q = 1$ (4.12) is the Kazhdan–Lusztig restatement of (4.3) [KL]. For q not a root of unity both follow from results of [R2, L1]. The interesting formula is (4.13). Besides the mysterious connection with affine Lie algebras there is also a mysterious connection with the representation theory of simple algebraic groups over an algebraically closed field of characteristic N [KL, L2] or with the representation theory of modular Lie algebras in characteristic N (cf. comments in [CKa]).

Let us consider (4.13) in more detail. First let us note that the simplex $\mathcal{C}_N$ correspond in our notation to the restrictions in (4.10), namely, $m_i \equiv \lambda(H_i)+1 < N$, $i = 1,\ldots,r$, $m_{\tilde{\alpha}} \equiv (\lambda+\rho)(H_{\tilde{\alpha}}) \leq N$. Thus if these restrictions hold then (4.13) coincides with (4.3) and (4.10). Let us consider again the example of $U_q(sl(3,\mathbb{C}))$ for the irregular case m_1 , $m_2 < N$, $m_{\tilde{\alpha}} = m_3 = m_1 + m_2 > N$. We recall that all Verma modules for which m_1 , $m_2 < N$, $m_3 > N$ are parametrized by $V^3_{k,\ell}$, for any $k, \ell \in \mathbb{Z}$ (cf. (3.21a)) and that the irreducible subquotients were denoted by $L'^{4}_{m_1 m_2}$, (cf. subsection 3.5.4.6.). In this case, as I was informed by V.G. Kac, (4.13) was proved in [APW]. More explicitly, they proved the following formula (which is equivalent to (4.13)):

$$ch\ L_\lambda = \sum_{w \in W_0} (-1)^{\ell(w)} \left(ch\ V^{w \cdot \lambda} - ch\ V^{w \cdot \lambda'} \right), \tag{4.14a}$$

$$\lambda' = s_3 \cdot \lambda + N\alpha_3 = \lambda - (m_3 - N)\alpha_3 ,$$

which in our notation reads

$$\begin{aligned} ch\ L'^{4}_{m_1 m_2} &= ch\ V^{3}_{k-1,\ell-1} \sum_{w \in W_0} (-1)^{\ell(w)}\ e(w \cdot \lambda - \lambda) - \\ &- ch\ V_{k,\ell} \sum_{w \in W_0} (-1)^{\ell(w)}\ e(w \cdot \lambda' - \lambda') = \qquad (4.14b) \\ &= ch\ V^{3}_{k-1,\ell-1} \sum_{w \in W_0} (-1)^{\ell(w)}\ e(w \cdot \lambda - \lambda) - ch\ L^{4}_{m'_1 m'_2}\ , \qquad (4.14c) \end{aligned}$$

where $m'_i \equiv (\lambda' + \rho,\ \alpha^\vee_i)$, and thus $m'_1 = N - m_2 < N$, $m'_2 = N - m_1 < N$. For (4.14b) we have used that if $V^\lambda = V^3_{k-1,\ell-1}$ then $V^{\lambda'} = V_{k,\ell}$. Indeed, by (3.21a) we should have $\lambda = \lambda' - m'_3 a_3 - N\alpha_1 - N\alpha_2$ with $m'_3 \equiv (\lambda' + \rho,\ \alpha^\vee_3)$ which is the case : $m'_3 = (\lambda - (m_3 - N)\alpha_3 + \rho,\ \alpha^\vee_3) = m_3 - 2(m_3 - N) = 2N - m_3 < N$. (Note that λ' is explicitly regular, i.e., $m'_i < N$, $i = 1,2,3$.)

Next we use the explicit form of the Weyl group W_0 of $sl(3, \mathbb{C})$ and the information about the bigger sextets mentioned in subsection 3.5.4.4. to rewrite (4.14b,c) in a form similar to (4.6b) :

$$\begin{aligned} ch\ L'^{4}_{m_1 m_2} &= ch\ V^{3}_{k-1,\ell-1} - ch\ V^{12}_{k,\ell-1} - ch\ V^{21}_{k-1,\ell} + \\ &+ ch\ V^{1}_{k,\ell} + ch\ V^{2}_{k,\ell} - ch\ V^{3}_{k+1,\ell+1} - \\ &- ch\ V_{k,\ell} + ch\ V^{1}_{k,\ell} + ch\ V^{2}_{k,\ell} - \\ &- ch\ V^{12}_{k,\ell} - ch\ V^{21}_{k,\ell} + ch\ V^{3}_{k,\ell}\ . \end{aligned} \tag{4.14d}$$

From (4.14b) we can derive the dimension of $L'^{4}_{m_1 m_2}$ [Do9]:

$$\begin{aligned} \dim L'^{4}_{m_1 m_2} &= m_1 m_2 (m_1 + m_2)/2 - m'_1 m'_2 (m'_1 + m'_2)/2 = \\ &= (m_1 + m_2 - N)\ (2m_1 m_2 + N(2N - m_1 - m_2))/2\ . \end{aligned} \tag{4.15}$$

Further we would like to find also the character formulae for the representations $L^{51}_{m_1}$, $L^{52}_{m_1}$, defined in 3.5.5. For this we should use the character formula for

$L'^4_{m_1 m_2}$, however, not in terms of $V^3_{k,\ell}$ but in terms of $V^1_{k,\ell}$ and $V^2_{k,\ell}$; this is possible as explained in subsection 3.5.4.5. Thus we have

$$\begin{aligned} ch\ L'^4_{m_1 m_2} &= ch\ V^1_{k,\ell} - ch\ V^{12}_{k,\ell} - ch\ V_{k+1,\ell} + \\ &+ ch\ V^3_{k+1,\ell} + ch\ V^2_{k+1,\ell+1} - ch\ V^{21}_{k+1,\ell+1} - \\ &- ch\ V^{21}_{k,\ell} + ch\ V^2_{k+1,\ell} + ch\ V^3_{k,\ell} - \\ &- ch\ V^{12}_{k+1,\ell} - ch\ V_{k+1,\ell+1} + ch\ V^1_{k+1,\ell+1} \ . \end{aligned} \tag{4.16}$$

Now to obtain the character formula for $L^{51}_{m_1}$ we should take into account the description in subsection 3.5.5., in particular the coincidences spelled out there, namely, $V^3_{k,\ell}$ coincides with $V_{k+1,\ell+1}$, $V^{12}_{k+1,\ell}$ with $V^1_{k+1,\ell+1}$, $V^{21}_{k,\ell}$ with $V^2_{k+1,\ell}$. Thus we see that the last six terms in (4.14d) or (4.16) cancell each other, i.e., the terms corresponding to the second term in (4.14a,b,c) vanish. Thus we obtain [Do9]:

$$\begin{aligned} ch\ L^{51}_{m_1} &= ch\ V^1_{k,\ell} - ch\ V^1_{k,\ell+1} - ch\ V_{k+1,\ell} + \\ &+ ch\ V_{k+2,\ell+1} + ch\ V^2_{k+1,\ell+1} - ch\ V^2_{k+2,\ell+1} \ . \end{aligned} \tag{4.17}$$

Analogous considerations give [Do9]:

$$\begin{aligned} ch\ L^{52}_{m_1} &= ch\ V^2_{k,\ell} - ch\ V^2_{k+1,\ell} - ch\ V_{k,\ell+1} + \\ &+ ch\ V_{k+1,\ell+2} + ch\ V^1_{k+1,\ell+1} - ch\ V^1_{k+1,\ell+2} \ . \end{aligned} \tag{4.18}$$

Now the formulae for the dimensions of $L^{51}_{m_1}$ and $L^{52}_{m_1}$ can be obtained from (4.15) (or formally from (4.9)) taiking into account the substitutions given in subsections 3.5.4.5. and 3.5.5., namely we have [Do9]:

$$\begin{aligned} \dim L^{51}_{m_1} &= \dim L'^4_{N-m_1,N} = \\ &= N(N-m_1)(2N-m_1)/2 \ , \quad m_1 \leq N/2 \ , \end{aligned} \tag{4.19a}$$

$$\begin{aligned} \dim L^{52}_{m_1} &= \dim L'^4_{N,m_1} = \\ &= N m_1 (N+m_1)/2 \ , \quad m_1 < N/2 \ . \end{aligned} \tag{4.19b}$$

Note that formally $\dim L^{51}_{m_1} = \dim L^4_{N-m_1,N}$, $\dim L^{52}_{m_1} = \dim L^4_{N,m_1}$.

Acknowledgments

The author would like to thank L.C. Biedenharn, C. De Concini, H.D. Doebner, D. Fairlie, J.D. Hennig, M. Jimbo, V.G. Kac, Y. Kosmann-Schwarzbach, P.P. Kulish, G. Lusztig, W. Lücke, S. Majid, V.B. Petkova, N.Yu. Reshetikhin, L. Vinet, J. Wess, B.G. Wybourne and C. Zachos for discussions. The author is thankful to Professor Dr. H.D. Doebner for hospitality at the Arnold Sommerfeld Insitiute for

Mathematical Physics of the Technical University Clausthal, where work on the final stages of preparation of this review was done. This work was partially supported also by the Ministry of Science of Bulgaria under Grants 3 and 403.

References

[A] E. Abe, *Hopf Agebras*, Cambridge Tracts in Math., N 74, (Cambridge Univ. Press, 1980).

[AA] G.E. Andrews and R. Askey, in : *Higher Combinatorics*, Ed. M. Aigner, (Reide., 1977) pp. 3-26.

[AGS1] L. Alvarez-Gaumé, C. Gómez and G. Sierra, Phys. Lett. **220B** (1989) 142-152.

[AGS2] L. Alvarez-Gaumé, C. Gómez and G. Sierra, Nucl. Phys. **B319** (1989) 155-186.

[AGS3] L. Alvarez-Gaumé, C. Gómez and G. Sierra, Nucl. Phys. **B330** (1990) 347.

[AGS4] L. Alvarez-Gaumé, C. Gómez and G. Sierra, preprint CERN-TH.5540/89 (1989), to appear in the Memorial Volume of Vadim Knizhnik, Eds. L. Brink, D. Friedan and A.M. Polyakov, World Scientific.

[APW] H.H. Andersen, P. Polo and K. Wen, Representations of quantum algebras, Aarhus University, Math. Inst. preprint series 1989/1990 No. 24 (April 1990).

[AW] R. Askey and J. Wilson, Mem. AMS **54** (1985) No. 319.

[Ba] V.V. Bazhanov, Phys. Lett. **159B** (1985) 321; Comm. Math. Phys. **113** (1987) 471-503.

[BC] D. Bernard and A. Le Clair, Phys. Lett. **227B** (1990) 417-423.

[Be] D. Bernard, Lett. Math. Phys. **17** (1989) 239-245.

[BGG] I.N. Bernstein, I.M. Gel'fand and S.I. Gel'fand, Funkts. Amal. Prilozh. **5** N1 (1971) 1-9; English translation: Funkt. Anal. Appl. **5** (1971) 1-8.

[BGZ] A.J. Bracken, M.D. Gould and R.B. Zhang, Mod. Phys. Lett. **A5** (1990) 831-840.

[Bi1] L.C. Biedenharn, J. Phys. A: Math. Gen. **22** (1989) L873-L878.

[Bi2] L.C. Biedenharn, to appear in the Proceedings of the Quantum Groups Workshop, Clausthal, (July 1989).

[BL] L.C. Biedenharn and J.D. Louck, *Angular Momentum in Quantum Physics*, Vol. 8, *Encyclopedia of Mathematics and Its Applications*, (Addison-Wesley, Reading, MA, 1981).

[BMP] P. Bouwknegt, J. McCarthy and K. Pilch, Phys. Lett. **234B** (1990) 297-303; Comm. Math. Phys. **131** (1990) 125-155.

[Bu] N. Burroughs, Comm. Math. Phys. **127** (1990) 109.

[CE] M. Chaichian and D. Ellinas, J. Phys. A: Math. Gen. **23** (1990) L291-L296.

[Ch] A. Chakrabarti, Centre de Physique Théorique de l'Ecole Polytechnique Palaiseau preprint A952.0290 (1990).

[CKa] C. De Concini and V.G. Kac, Representations of quantum groups at roots of 1, Pisa Scuola Normale Superiore math. preprint No. 75 (May 1990).

[CKL] M. Chaichian, P.P. Kulish and J. Lukierski, Phys. Lett. **237B** (1990) 401-406.

[CKu] M. Chaichian and P.P. Kulish, Phys. Lett. **234B** (1990) 72-80.

[Co] A. Connes, Publ. Math. IHES **62** (1985) 257.

[CZ] T. Curtright and C. Zachos, Phys. Lett. **243B** (1990) 237-244.

[De] C. Devchand, Freiburg University preprint THEP 89/12 (1989).

[DFI] T. Deguchi, A. Fujii and K. Ito, Phys. Lett. **238B** (1990) 242-246.

[DFV] E. D'Hoker, R. Floreanini and L. Vinet, preprint UCLA/90/TEP/28 (1990).

[DHL] H.D. Doebner, J.D. Hennig and W. Lücke, to appear in the Proceedings of the Quantum Groups Workshop, Clausthal, (July 1989).

[Di] J. Dixmier, *Enveloping Algebras*, (North Holland, New York, 1977).

[DJM] E. Date, M. Jimbo and T. Miwa, to appear in the Memorial Volume of Vadim Knizhnik, Eds. L. Brink, D. Friedan and A.M. Polyakov, World Scientific.

[DJMM1] E. Date, M. Jimbo, K. Miki and T. Miwa, R matrix for cyclic representations of $U_q(sl(3,\mathbb{C})^{(1)})$ at $q^3=1$, Kyoto preprint, RIMS - 696 (1990).

[DJMM2] E. Date, M. Jimbo, K. Miki and T. Miwa, Cyclic representations of $U_q(sl(n+1,\mathbb{C}))$ at $q^N=1$, Kyoto preprint, RIMS (1990).

[DMMZ] E.E. Demidov, Yu.I. Manin, E.E. Mukhin and D.V. Zhdanovich, Kyoto preprint, RIMS - 701 (May 1990).

[Do1] V.K. Dobrev, J.Math.Phys. **26** (1985) 235-251 and ICTP Trieste preprint IC/83/36 (1983); Lett.Math.Phys. **9** (1985) 205-211 and INRNE Sofia preprint (1983).

[Do2] V.K. Dobrev, Talk at the Conference on Algebraic Geometry and Integrable Systems, Oberwolfach (July 1984) and ICTP Trieste preprint IC/85/9 (1985).

[Do3] V.K. Dobrev, in : Proceedings of the XIII International Conference on Differential-Geometric Methods in Theoretical Physics, Shumen (1984), Eds. H.D. Doebner and T.D.Palev (World Sci., Singapore, 1986) pp.348-370; Lett.Math.Phys. **11** (1986) 225-234.

[Do4] V.K. Dobrev, Reports Math. Phys. **25** (1988) 159-181 and ICTP Trieste internal report IC/86/393 (1986).

[Do5] V.K. Dobrev, ICTP Trieste internal report IC/89/142 (1989), to appear in the Proceedings of the International Group Theory Conference at St. Andrews, 29.7.-12.8.1989.

[Do6] V.K. Dobrev, Proceedings of the 13th Johns Hopkins Workshop "Knots, Topology and Field theory", Florence, (June 1989), Ed. L. Lusanna (World Sci, Singapore, 1989) pp. 539-547.

[Do7] V.K. Dobrev, to appear in the Proceedings of the Quantum Groups Workshop, Clausthal, (July 1989).

[Do8] V.K. Dobrev, Talk at the Third Centenary Celebration of the Mathematische Gesellschaft in Hamburg, (March 1990).

[Do9] V.K. Dobrev, Talk at the International Colloquium on Group Theoretical Methods in Physics, Moscow, (June 1990).

[DP] V.K. Dobrev and V.B. Petkova, Lett. Math. Phys. **9** (1985) 287-298.

[Dr1] V.G. Drinfeld, Dokl. Akad. Nauk SSSR **283** (1985) 1060-1064 (in Russian); English translation: Soviet. Math. Dokl. **32** (1985) 254-258.

[Dr2] V.G. Drinfeld, in: Proceedings ICM, (MSRI, Berkeley, 1986) pp. 798-820.

[Dr3] V.G. Drinfeld, Quasi-Hopf algebras, Alg. Anal. **1** (1989).

[Dr4] V.G. Drinfeld, Kiev preprint ITP-89-43E (1989).

[Fa] L.D. Faddeev, Integrable models in 1+1 dimensional quantum field theory, in : Les Houches Lectures 1982, (Elsevier, Amsterdam,1984).

[FaT] L.D. Faddeev and L.A. Takhtajan, preprint LPT Paris-VII, (9185) and in: Lecture Notes in Physics, (1986) pp. 166-179.

[FRT1] L.D. Faddeev, N. Yu. Reshetikhin and L.A. Takhtajan, in: *Algebraic Analysis*, Vol. No. 1 (Academic Press, 1988) p.129.

[FRT2] L.D. Faddeev, N. Yu. Reshetikhin and L.A. Takhtajan, Alg. Anal. **1** (1989) 178-206.

[FGP] P. Furlan, A.Ch. Ganchev and V.B. Petkova, Quantum groups and fusion rules multiplicities, INFN Trieste preprint, INFN/AE-89/15 (1989).

[Fi] D. Fairlie, J. Phys. A: Math. Gen. **23** (1990) L183-L187.

[FJ] I.B. Frenkel and N. Jing, Yale preprint (1988).

[Fl] E.G. Floratos, preprint CERN-TH.5482/89 (1989).

[FSV] R. Floreanini, V.P. Spiridonov and L. Vinet, Phys. Lett. **242B** (1990) 383-386; preprint UCLA/90/TEP/21 (1990).

[FV] R. Floreanini and L. Vinet, preprint UCLA/90/TEP/30 (1990).

[G1] J.-L. Gervais, Comm. Math. Phys. **130** (1990) 257-283.

[G2] J.-L. Gervais, Phys. Lett. **243B** (1990) 85-92.

[G3] J.-L. Gervais, preprint LPTENS 90/13 (June 1990)

[GP] A.Ch. Ganchev and V.B. Petkova, Phys. Lett **B233** (1989) 374-382.

[GS1] C. Gómez and G. Sierra, Phys. Lett. **240B** (1990) 149-157.

[GS2] C. Gómez and G. Sierra, The quantum symmetry of rational conformal field theories, geneva University preprint, Dept. Theor. Physics preprint (March 1990).

[H] T. Hayashi, Comm. Math. Phys. **127** (1990) 129-144.

[HS] L.K. Hadjijivanov and D.T. Stoyanov, Sofia preprint INRNE-TH-90-2 (June 1990).

[J1] M. Jimbo, Lett. Math. Phys. **10** (1985) 63-69.

[J2] M. Jimbo, Lett. Math. Phys. **11** (1986) 247-252.

[J3] M. Jimbo, Comm. Math. Phys. **102** (1986) 537-547.

[J1] M. Jimbo, Int. J. Mod. Phys. **A4** (1989) 3759-3777.

[Ka] V.G. Kac, *Infinite-dimensional Lie algebras. An introduction*, Progr. Math. Vol. **44** (Birkhäuser, Boston, 1983).

[KK] V.G. Kac and D. Kazhdan, Adv. Math. **34** (1979) 97-108.

[KiR1] A.N. Kirillov and N.Yu. Reshetikhin, Representations of the algebra $U_q(sl(2))$, q - orthogonal polynomials and invariants of links, LOMI Leningrad preprint E-9-88 (1988).

[KiR2] A.N. Kirillov and N.Yu. Reshetikhin, q–Weyl group and multiplicative formula for universal R–matrices, MIT preprint (May 1990).

[KL] D. Kazhdan and G. Lusztig, Inv. Math. **53** (1979) 165-184.

[Kl1] A.U. Klymik, Kiev preprint ITP-90-27E (1990).

[Kl2] A.U. Klymik, Talk at the International Colloquium on Group Theoretical Methods in Physics, Moscow, (June 1990).

[KlG] A.U. Klymik and V.A. Groza, Kiev preprint ITP-89-37P (1989) (in Russian).

[KMN] S. Komata, K. Mohri and H. Nohara, Tokyo preprint UT-Komaba 90-21 (July 1990).

[Kn] A. Kuniba, Kyoto preprint RIMS-664 (1989).

[Ko1] T.H. Koornwinder, Nederrl. Akad. Wetensch. Proc. Ser. A **92** (1989) 97-117.

[Ko2] T.H. Koornwinder, in: *Orthogonal Polynomials: Theory and Practice*, NATO ASI Series C, Vol. 294, Kluwer Academic Publishers (1990) pp. 257-292.

[KR1] P.P. Kulish and N.Yu. Reshetikhin, Zap. Nauch. Semin. LOMI **101** (1981) 101-110 (in Russian); English translation: J. Soviet. Math. **23** (1983) 2435-2441.

[KR2] P.P. Kulish and N.Yu. Reshetikhin, Lett. Math. Phys. **18** (1989) 143-149.

[KRS] P.P. Kulish, N.Yu. Reshetikhin and E.K. Sklyanin, Lett. Math. Phys. **5** (1981) 393-403.

[KS] P.P. Kulish and E.K. Sklyanin, Lecture Notes in Physics, Vol. 151 (1982) pp. 61-119.

[Ku] P.P. Kulish, Kyoto preprint RIMS-615 (1988).

[KuD] P.P. Kulish and E.V. Damashinsky, J. Phys. A: Math. Gen. **23** (1990) L415-L420.

[Kw] M. Kashiwara, Kyoto University preprint RIMS-676 (1989).

[KZ] V.G. Knizhnik and A.B. Zamolodchikov, Nucl. Phys. **B247** (1984) 83-103.

[LS] S.Z. Levendorsky and Ya.S. Soibelman, Talk by Soibelman at the International Colloquium on Group Theoretical Methods in Physics, Moscow, (June 1990).

[L1] G. Lusztig, Adv. Math. **70** (1988) 237-249.

[L2] G. Lusztig, Contemporary Math. **82** (1988) 59-77.

[L3] G. Lusztig, On quantum groups, MIT preprint, to appear in J. Algebra (1990).

[L4] G. Lusztig, Lectures at the 1990 Yukawa International Seminar, Kyoto (May 1990).

[Ma1] Yu.I. Manin, Quantum groups and non-commutative geometry, Montreal University preprint, CRM-1561 (1988).

[Ma2] Yu.I. Manin, Comm. Math. Phys. **123** (1989) 163-175.

[Ma3] Yu.I. Manin, Lectures at the 1990 Yukawa International Seminar, Kyoto (May 1990); Talk at the International Colloquium on Group Theoretical Methods in Physics, Moscow, (June 1990).

[Mc] A.J. MacFarlane, J. Phys. A: Math. Gen. **22** (1989) 4581-4588.

[MFF] F.G. Malikov, B.L. Feigin and D.B. Fuchs, Funkts. Anal. Prilozh. **20** N2 (1986) 25-37; English translation: Funct. Anal. Appl. **20**, (1986) 103-113.

[Mj] S. Majid, Int. J. Mod. Phys. **A5** (1990) 1-91.

[MM] K.C. Misra and T. Miwa, Kyoto University preprint RIMS-684 (1990).

[MNS] T. Masuda, K. Mimachi, Y. Nakagami, M. Noumi, Y. Sabuti and K. Ueno, Lett. Math. Phys. **19** (1990) 187-194; Lett. Math. Phys. **19** (1990) 195-204.

[MNU] T. Masuda, K. Mimachi, Y. Nakagami, M. Noumi and K. Ueno, C.R. Acad. Sci. Paris **307** Serie I, (1988) 559-564.

[MR] G. Moore and N.Yu. Reshetikhin, Nucl. Phys. **B328** (1989) 557-574.

[MS] G. Moore and N. Seiberg, Comm. Math. Phys. **123** (1989) 177-254.

[MZ] R.L. Mkrtchyan and L.A. Zurabyan, Yerevan preprint YERPHI-1149(26)-89 (1989).

[N] F.J. Narganes-Quijano, Université Libre de Bruxelles preprint ULB-Th 90/01 (1990).

[NM] M. Nuomi and K. Mimachi, Comm. Math. Phys. **128** (1990) 521-531.

[O] M. Okado, Kyoto preprint RIMS-677 (1989).

[Pa] V. Pasquier, Nucl. Phys. **B295** [**FS21**] (1989) 491-510.

[PS] V. Pasquier and H. Saleur, Saclay preprint SPhT/89-031 (1989).

[Po] A. Polychronakos, Florida preprint, UFIFT-90-14 (1990).

[RA] P. Roche and D. Arnaudon, Lett. Math. Phys. **17** (1989) 295-300.

[Re] N.Yu. Reshetikhin, Quantized universal enveloping algebras, the Yang–Baxter equation and invariants of links I. II., LOMI Leningrad preprints E-4-87, E-17-87 (1987).

[ReS] N.Yu. Reshetikhin and F.A. Smirnov, Comm. Math. Phys. **131** (1990) 157-177.

[ReSt] N.Yu. Reshetikhin and M.A. Semenov-Tian-Shansky, Lett. Math. Phys. **19** (1990) 133-142.

[R1] M. Rosso, C.R. Acad. Sci. Paris **305** Serie I, (1987) 587-590.

[R2] M. Rosso, Comm. Math. Phys. **117** (1987) 581-593.

[R3] M. Rosso, Comm. Math. Phys. **124** (1989) 307-318.

[Ru] H. Ruegg, J. Math. Phys. **31** (1990) 1085-1087.

[Sa] H. Saleur, Nucl. Phys. **B336** (1990) 363.

[S1] E.K. Sklyanin, Funkts. Anal. Prilozh. **16** (1982) 27-34, (in Russian); English translation: Funct. Anal. Appl. **16** (1982) 263-270.

[S2] E.K. Sklyanin, Funkts. Anal. Prilozh. **17** (1983) 34-48 (in Russian); English translation: Funct. Anal. Appl. **17** (1983) 274-88.

[S3] E.K. Sklyanin, Uspekhi Mat. Nauk **40** (1985) 214 (in Russian).

[SAA] K. Sogo, Y. Akutsu and T. Abe, Progr. Theor. Phys. **70** (1983) 730.

[SF] C.-P. Sun and H.-C. Fu, J. Phys. A: Math. Gen. **22** (1989) L983-L986.

[SS] M. Schlieker and M. Scholl, Karlsruhe University preprint KA-THEP-26/89 (1989).

[Ta] L.A. Takhtajan, Adv. Stud. Pure. Math. **19** (1989) 435-457.

[Td] I.T. Todorov, to appear in the Proceedings of the Quantum Groups Workshop, Clausthal, (July 1989).

[Tl] V.N. Tolstoy, to appear in the Proceedings of the Quantum Groups Workshop, Clausthal, (July 1989).

[Ve] J.-L. Verdier, in: *Séminaire Bourbaki*, No. 685, *Astérisque*, **152-153** (1987) 305.

[Vg] H.J. de Vega, Int. J. Mod. Phys. **A4** (1989) 2371-2465.

[VK] L.L. Vaksman and L.I. Korogodsky, Harmonic analysis on quantum hyperboloids, Kiev preprint ITF-90-27P (March 1990) (in Russian).

[VS] L.L. Vaksman and Ya.S. Soibelman, Funkts. Anal. Prilozh. **22** (1988) No. 3, 1-14; English translation: Funct. Anal. Appl. **22**, (1988) 170-181.

[We] J. Wess, Talk at the Third Centenary Celebration of the Mathematische Gesellschaft in Hamburg, (March, 1990).

[WeZ] J. Wess and B. Zumino, preprint CERN-TH-5697/90, LAPP-TH-284/90 (April 1990).

[Wi] E. Witten, Nucl. Phys. **B330** (1990) 285.

[Wo] S.L. Woronowicz, Comm. Math. Phys. **111** (1987) 613-665; Publ. RIMS **23** (1987) 117-181; Comm. Math. Phys. **122** (1989) 125.

ALGEBRAS AND SYMMETRIES — QUANTUM MECHANICAL SYMMETRY BREAKING

H.D. Doebner and J. Tolar [1]

Arnold Sommerfeld Institute of Mathematical Physics
Technical University of Clausthal
D-3392 Clausthal (Fed. Rep. Germany)

1 QUANTIZATION AS ALGEBRA

The modelling of physical systems and their quantization methods are intimately related with algebras and symmetries. We discuss this relation in the following example:

Consider a (classical non-relativistic) mechanical system, e.g. a one-particle system, localized in R^3 and with a Hamiltonian h as a function on the corresponding phase space $\Gamma(R^3) = R^6$.

For a quantization of the *kinematics* of the system, e.g. momenta p_i and positions q_i, $i = 1, 2, 3$, a unitary ray representation of the Galilei group (without time translations) on a Hilbert space $\mathcal{H}$ modelled on $L^2(R^3, d^3x)$ is used; the generators of the space translations P_i are quantizations of momenta p_i, the generators of proper Galilei transformations Q_i — of positions q_i and the generators of the rotation group (if one uses a representation without spin) $L_i = \varepsilon_{ijk} Q_j P_k$ are quantizations of the angular momentum l_i. These generators span the Lie algebra of the quantum mechanical Galilei group G(3) (without time translations), which include the Heisenberg algebra $(P_i, Q_i, \mathbb{1})$:

$$[Q_i, P_j] = i\hbar\delta_{ij}\mathbb{1}, \; [Q_i, Q_j] = 0 = [P_i, P_j],$$

$$[L_i, Q_j] = i\hbar\varepsilon_{ijk}Q_k, \; [L_i, P_j] = i\hbar\varepsilon_{ijk}P_k.$$

So the kinematics can be viewed *algebraically* via a kinematical symmetry G(3).

[1] On leave from Czech Technical University, Faculty of Nuclear Science and Physical Engineering, Břehová 7, CS-115 19 Prague (Czechoslovakia).

Symmetries in Science V, Edited by B. Gruber *et al.*
Plenum Press, New York, 1991

In this context it is an old and solved problem, whether the representation of a Lie algebra g of a Lie group G will correspond to a unitary representation of the group G in the sense that it is the differential of this unitary representation. This is in general not the case. There are two obstructions. One is coming from the fact that G and all its coverings have the same Lie algebra g. The second one is due to a *domain* question in $\mathcal{H}$. In general, since the kinematical group, like the Galilei group, is a non-compact one, at least some of its generators are unbounded operators with a dense domain in $\mathcal{H}$. In a representation of g which is obtained by differentiation of a unitary representation of G, it can be proven that all the generators X_k, $k = 1, ..., n$, $n = \dim g$, are essentially self-adjoint on a common dense invariant domain D [1]. The X_k are physical observables. But not all representations of g of this type are obtained by differentiation; one can find counterexamples. One has to add a further condition to assure "integrability" to a unitary representation of G. A possible condition is to demand that the so called Nelson operator

$$\sum_{k=1}^{n} X_k^2$$

is also essentially self-adjoint on D. Non-integrable and partially integrable representations were discussed in [2], [3] and [4]. The Lie algebra of a symmetry group G is in a way an algebraization of a symmetry and is **richer** than the symmetry itself.

Concerning the *dynamics* we introduce a quantization of the Hamiltonian h formally as $H = h(Q_i, P_i)$. Because P_i and Q_i are not commutative and because of the domain questions one has difficulties to define H even if h is a polynomial in P_i, Q_i with additive terms only in P_i and only in Q_i. The lack of commutativity leads to the so called ordering problem in quantum mechanics. Different approaches to solve this problem (if the kinematics is already quantized) are known; we mention the quantization by deformation [5], [6], which gives a unique result for square-integrable $h(q_i, p_i)$, and the prime quantization [7]. If the ordering is known, or if there is, as in many examples, no ordering necessary (e.g. for $h = p^2/2m + V(q)$), one expects that the common dense invariant domain D on which P's, Q's and the Nelson operator are essentially self-adjoint is the "generic" domain also for H, because it should be possible to measure positions and momenta for the eigenstates of H; furthermore we have to demand that the Hamiltonian as a physical observable is essentially self-adjoint[2] on D. But often the Hamiltonian is only symmetric on D (or it can be symmetrized). So one has, if possible, to use a self-adjoint extension H_α of H with (invariant) domain $D(H_\alpha)$ and it could well be that the P's and the Q's, some functions of the P's and Q's, like L_i, and commutators like $[H, L_i]$, are not defined everywhere on $D(H_\alpha)$. We will comment on this in the next section.

We summarize:
The quantization of our system is modelled (including spin) on a representation of the Galilei Lie algebra. To have a unique result as well for kinematics as for the dynamics, one has to add additional physically motivated assumptions. This example resembles the general situation [12], [13], [14]. Hence, modifying Klauder's title of one of his recent articles "Quantization *is* geometry, after all" [8] we have reasons to say "Quantization *is* algebra, after all"; this puts also our article in the scope of this conference.

[2]We will not discuss the role of symmetric operators as observables.

We add a remark on domain questions. Some observables yield, because of their algebraic properties, unbounded operators after quantization; hence it is necessary to add to any quantization procedure prescriptions how to choose the domain in the abstract Hilbert space or an L^2-model space over a manifold on which the physical system is living. This prescription has a special and definite physical meaning, and has to be checked against experiment.

2 Quantization anomalies

There is a further relation between the non-relativistic system described by h and symmetry or, equivalently, algebra. We assume that the *given* Hamiltonian $h(q,p)$ in fixed coordinates q_i, p_i is *locally* [3] G-invariant, i.e. there exist functions $f_i(q,p)$, on the phase space R^6, such that

(1) The Poisson brackets $\{f_i, f_j\} = c_{ijk} f_k$ are a representation of the Lie algebra g of a Lie group G. G or g are called dynamical symmetry or dynamical algebra of $h(q,p)$; and

(2) $\{h, f_i\} = 0$. [4]

One can quantize the generators f_i similarly as h as phase space functions with known physical interpretation. This gives, with the same difficulties as for H, operators F_i on some domains D_i. The F_i are obviously observables, so they should be essentially self-adjoint on a common dense invariant domain D_g. One expects that
(a) $[F_i, F_j] = c_{ijk} F_k$ on D_g, and

(b) $[F_i, H] = 0$ on some domain $\tilde{D}$.
Concerning (a), the quantization is not necessarily a Lie-algebra homomorphism. Only for special f_i, e.g. second order polynomials in Q_i, P_i, the operators F_i will respect the Lie structure. Furthermore, even if (a) is fulfilled, the symmetry of the system expressed by (b) may be broken if e.g. $\tilde{D}$ is not dense, or if F_i, H are not self-adjoint on $\tilde{D}$. So the quantum mechanical system can have a **smaller** symmetry than the classical one. This fact is called in the literature sometimes a *quantization anomaly*. (It is related to anomalies in quantum field theory [9].) Especially, (b) is not necesarilly connected with a degeneracy of the eigenfunctions of some self-adjoint extension of H as is the case e.g. for $G = \mathrm{SU}(3)$ and the 3-dimensional harmonic oscillator in R^3. The algebraic treatment of quantization leads to an understanding also of some types of symmetry breaking.

The described effect is a special case of a general mechanism. If a group acts in a "certain way" in a classical theory and is realized "differently" after quantization, i.e. if special properties in the classical and quantum mechanical case are different,

[3] The case of global classical symmetry will not be discussed here.

[4] The symmetry of a given Hamiltonian will depend (locally) only on the dimension of the phase space if one transforms the coordinates p, q correspondingly via symplectomorphisms. However, symplectomorphisms will not commute in general with a quantization. Therefore, it is important and full of physical meaning, that one has a preferred choice of coordinates.

this is called anomaly. An example are scaling transformations of classical fields and observables (anomalous dimensions after quantization).

The question of anomalies has in the last years gained considerable attention, but almost exclusively in field theoretical problems. In order to understand the mechanism some quantum mechanical anomalies have recently been discussed [10], [11]. As it was already mentioned, quantization anomalies are intimately related to domain questions in the underlying Hilbert space. Because it is difficult to interpret a given dense domain physically, it remains after detailed discussion somewhat unclear why classical symmetries are broken after quantization.

3 QUANTIZATION OF THE HARMONIC OSCILLATOR IN A POINTED PLANE

We consider as an example for quantum anomalies the harmonic oscillator in two dimensions and we use instead of R^2 the pointed plane $\dot{R}^2 = R^2\backslash O$. Taking out one point from R^2 seems to be a harmless physical procedure but it is not. The global properties of R^2 and $\dot{R}^2$ are different, e.g. the fundamental groups are $\pi_1(R^2) = 0$ and $\pi_1(\dot{R}^2) = Z$. [5] The topological details and the result on the quantization according to [12], [13] was analyzed in detail by [14]. One meets this problem also if one uses polar coordinates, because they are naturally defined in $\dot{R}^2$, not on R^2. We discuss the quantum anomaly of the classical invariance algebra $su(2)$ of the harmonic oscillator and show that a quantum anomaly is connected with a non-trivial topology of the configuration space $\dot{R}^2$. (See in this connection [3].)

3.1 Decomposition of the Hamiltonian

Realize $\mathcal{H}$ as $L^2(\dot{R}^2, d^2x)$. The Hamiltonian is a formal differential operator

$$H = -\frac{\hbar^2}{2}\sum_{i=1}^{2}\frac{\partial^2}{\partial x_i^2} + \frac{\omega^2}{2}\sum_{i=1}^{2} x_i^2$$

We take as a domain $D = C_o^\infty(\dot{R}^2, \mathbb{C})$ because H is here symmetric and we study its possible extensions. We follow partly [14], introduce in $\dot{R}^2$ polar coordinates and have

$$L^2(\dot{R}^2, d^2x) = L^2(R^+, rdr) \otimes L^2(S^1, d\varphi) = \bigoplus_{m\in Z} L^2(R^+; rdr) \otimes \mathbb{C}\{e^{im\varphi}\}$$

The Hamiltonian decomposes as differential operator on $C_o^\infty(\dot{R}^2, \mathbb{C})$ as

$$H = -\frac{\hbar^2}{2}(\frac{\partial^2}{\partial r^2} + \frac{1}{r}\frac{\partial}{\partial r}) + \frac{1}{2}\omega^2 r^2 - \frac{\hbar^2}{2r^2}\frac{\partial^2}{\partial \varphi^2}.$$

[5] Z is the additive group of all integers.

For each $m \in Z$ (see [14]) the radial term [6]

$$H^m = \frac{\hbar^2}{2r}[-\frac{d}{dr}(r\frac{d}{dr}) + \frac{m^2}{r} + \frac{\omega^2}{\hbar^2}r^3]$$

is a symmetric operator on a domain $D(H^m) = C_o^\infty(R^+, \mathbb{C}) \subset L^2(R^+, rdr)$.

The substitution

$$k = \frac{E}{2\hbar\omega}, \; s = \frac{\omega}{\hbar}r^2, \; w(s) = ru(r)$$

transforms the eigenvalue problems

$$H^m u_{|m|,E}(r) = E u_{|m|,E}(r), \; m \in Z$$

into

$$A^m w_{|m|,k}(s) \equiv s(\frac{d^2}{ds^2} - \frac{1}{4} + \frac{1-m^2}{4s^2}) w_{|m|,k}(s) = -k w_{|m|,k}(s)$$

on $D_s^m \subset L^2(R^+, ds/s)$. This is easily identified as Whittaker's differential equation with two fundamental solutions (with $\varepsilon = \pm$; see [15]):

$$w^\varepsilon(|m|, k, s) = e^{-\varepsilon s/2}(\varepsilon s)^{\frac{|m|+1}{2}} \Psi(\frac{|m|+1}{2} - \varepsilon k, |m|+1; \varepsilon s),$$

where $\Psi(a, b, s)$ is a confluent hypergeometric function. For large $|s|$ and $|\arg s| < 3\pi/2$

$$w^+(|m|, k, s) \simeq e^{-s/2} s^k \{1 + O(|s|^{-1})\};$$

w^- grows exponentially for $|s| \to \infty$ and is not normalizable. The behaviour of w^+ for small $|s|$ is

$$w^+(|m|, k, s) \simeq \begin{cases} s^{\frac{|m|+1}{2}} \{\frac{\Gamma(|m|)}{\Gamma(a)} s^{-|m|} + R_{|m|}\} & |m| > 0; a = \frac{|m|+1}{2} - k \\ -\frac{s^{1/2}}{\Gamma(\frac{1}{2}-k)} F_k & |m| = 0; \frac{1}{2} - k \neq -n \\ s^{1/2}\{(-1)^n n! + O(s)\} & |m| = 0; \frac{1}{2} - k = -n \end{cases}$$

where

$$R_\nu = \begin{cases} O(|s|^{\nu-1}) & \nu > 1 \\ O(|\ln s|) & \nu = 1 \\ O(1) & 0 < \nu < 1 \end{cases}$$

and

$$F_k = \ln s - 2\psi(1) + \psi(\frac{1}{2} - k) + O(|s \ln s|),$$

ψ denotes the digamma function and $n = 0, 1, \ldots$

[6] H^m depends on $|m|$ only.

3.2 Extension of the radial part

To construct self-adjoint extensions of H^m we apply von Neumann's theory (see e.g. [16], [17], [18]). In a first step, we need the bases of the deficiency spaces

$$\mathcal{H}_\pm^m = \mathrm{Ker}(H^{m*} \pm i\mathbb{1}).$$

As solutions of the partial differential equations

$$H^{m*} u_\pm^{|m|}(r) = \mp i u_\pm^{|m|}(r)$$

we find (use $w^\varepsilon(|m|, \pm i/2\hbar\omega; s)$ and their behaviour for $r \to \infty$ and $r \to 0$) in $L^2(R^+, rdr)$ for $|m| > 0$ the deficiency indices $(0,0)$ and for $m = 0$ solutions ($\varepsilon = +$)

$$u_\pm^0(r) = \exp(-\frac{\omega}{2\hbar}r^2)\Psi(\frac{1}{2}(1 \pm \frac{i}{\hbar\omega}), 1; \frac{\omega}{\hbar}r^2),$$

i.e.

$$(\dim \mathcal{H}_+^m, \dim \mathcal{H}_-^m) = \begin{cases} (0,0) & \text{for} \quad |m| \geq 1 \\ (1,1) & \text{for} \quad m = 0 \end{cases}$$

In a second step we observe that, if both deficiency indices are zero, the operator H^m is already essentially self-adjoint; equality of deficiency indices $\neq 0$ yields self-adjoint extensions of the operator H^0 parametrized by unitary mappings from $\mathcal{H}_+^0$ to $\mathcal{H}_-^0$. Hence for H^m, $|m| \geq 1$ the problem is already solved. For H^0 the unitary maps U(1) are parametrized by α, $0 \leq \alpha < \pi$, and the self-adjoint extensions H_α^0 have domains

$$D(H_\alpha^0) = D(\bar{H}^0) \oplus \mathbb{C}\{u_-^0 - e^{2i\alpha} u_+^0\}.$$

The domain $D(\bar{H}^0)$ of the closure of H^0 and $D(\bar{H}^m)$ can be easily given explicitly [16].

3.3 Extensions of the Hamiltonian

The extensions of H^m, $m \in Z$ yield extensions H_α of H on a domain $D(H_\alpha)$ in $L^2(\dot{R}^2, d^2x)$. [7] The spectrum of H_α is the union of the spectra of H^m, $m \neq 0$, and of H_α^0:

$$\mathrm{Spec}\, H_\alpha = \mathrm{Spec}\, H_\alpha^0 \cup \bigcup_{m \in Z, m \neq 0} \mathrm{Spec}\, H^m \ .$$

For $|m| \geq 1$ all H^m are self-adjoint and the "canonical" spectrum for each H^m is obtained by considering only solutions w^+; we observe (behaviour of w^+ for $r \to 0$) that they are locally square-integrable at $r = 0$, if and only if $a = -n$, i.e.

$$k = n + \frac{|m|+1}{2}, \quad n = 0, 1, 2, \ldots,$$

hence (as we expect)

$$E_{nm} = (2n + |m| + 1)\hbar\omega.$$

Each energy value $N\hbar\omega$, $N = 1, 2, \ldots$ has an N-dimensional degeneracy space. The eigenfunctions are

$$\psi_{nm}(r, \varphi) = \frac{1}{r} w^+(|m|, n + \frac{|m|+1}{2}, \frac{\omega}{\hbar} r^2) e^{im\varphi} =$$

[7] This is a special class of extensions preserving rotational invariance [16].

$$= (-1)^n n! (\frac{\omega}{\hbar})^{(|m|+1)/2} r^{|m|} \exp(-\frac{\omega}{2\hbar} r^2) L_n^{|m|}(\frac{\omega}{\hbar} r^2) e^{im\varphi},$$

where $L_n^{|m|}$ are the Laguerre polynomials [15].

The rest of the H-spectrum is given by the spectra of H_α^0. For $\alpha = 0$ (the so called energy extension) the canonical spectrum is obtained:

$$E_{n0}^0 = (2n+1)\hbar\omega.$$

For $0 < \alpha < \pi$ "exotic" eigenvalues

$$E_{n0}^\alpha = 2k_n(\alpha)\hbar\omega,$$

appear, where $k_n(\alpha)$ are solutions of the transcendental equation

$$\psi(\frac{1}{2} - k) - 2\psi(1) = -\cot\alpha,$$

$-\psi(1) = 0.577...$ (Euler's constant). We can write

$$E_{n0}^\alpha = (2n+1)\hbar\omega + \epsilon_n, \; n = 1, 2, ...,$$

with

$$-2\hbar\omega < \epsilon_n < 0, \quad \lim_{n\to\infty} \epsilon_n = 0,$$

$$E_{00}^\alpha < E_{00}^0 = \hbar\omega.$$

The "exotic" eigenfunctions are

$$\psi_{n0}^\alpha(r,\varphi) = \frac{1}{r} w^+(0, k_n(\alpha); \frac{\omega}{\hbar} r^2).$$

This gives the following pattern (see [14], Fig.2), where the points • are shifted downwards if α increases.

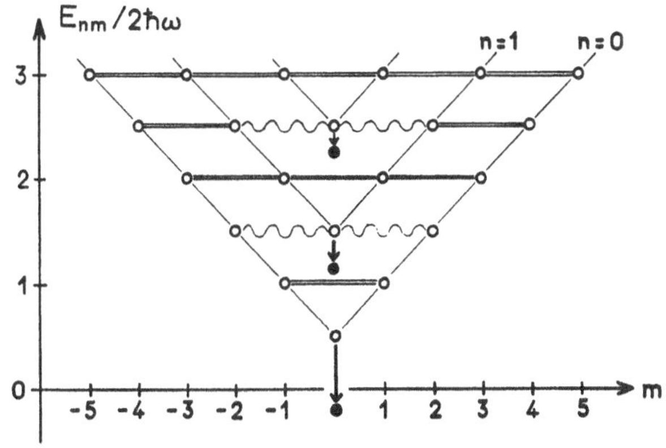

Fig. 1. Eigenvalues of self-adjoint extensions H_α (o = "canonical" spectrum, • = "exotic" spectrum, = degeneracy spaces, ~ broken degeneracy).

4 An example for quantum mechanical symmetry breaking

We will show that the above quantization of the 2-dimensional oscillator in the pointed plane has a quantization anomaly.

The classical harmonic oscillator in R^2 has an obvious geometrical SO(2) symmetry given by equal rotations in the (q_1, q_2) and the (p_1, p_2) plane. This symmetry can be enlarged to an SU(2) symmetry containing SO(2) and similar rotations in the $(q_2, p_1), (q_1, p_2)$ and in the $(q_1.p_1), (q_2, p_2)$ planes.

If one uses Cartesian coordinates (and $\omega = 1$), the Hamiltonian is $h = \frac{1}{2}(p_1^2 + q_1^2 + p_2^2 + q_2^2)$ and the generators of SU(2) are

$$j_1 = \frac{1}{2}(p_1 p_2 + q_1 q_2),$$

$$j_2 = \frac{1}{2}(q_1 p_2 - p_1 q_2),$$

$$j_3 = \frac{1}{4}(p_1^2 + q_1^2 - p_2^2 - q_2^2).$$

The SO(2) symmetry is generated by j_2. The Poisson brackets are

$$\{h, j_k\} = 0, \ \{j_k, j_l\} = \varepsilon_{klm} j_m.$$

These relations are also valid in $\Gamma(\dot{R}^2)$. The quantization of the (local) $su(2)$ symmetry is obvious. The quantized generators J_k are second order polynomials in Q_i, P_i.

If we use our decomposition of $L^2(\dot{R}^2, d^2x)$ given in section 3.1, we get the following form of J_k (for $\hbar = \omega = 1$):

$$J_1 = \frac{1}{4}\sin 2\varphi\left(r^2 - \frac{\partial^2}{\partial r^2} + \frac{1}{r}\frac{\partial}{\partial r} + \frac{1}{r^2}\frac{\partial^2}{\partial\varphi^2}\right) - \frac{1}{2r}\cos 2\varphi\left(\frac{\partial}{\partial r} - \frac{1}{r}\right)\frac{\partial}{\partial\varphi},$$

$$J_2 = -\frac{i}{2}\frac{\partial}{\partial\varphi},$$

$$J_3 = \frac{1}{4}\cos 2\varphi\left(r^2 - \frac{\partial^2}{\partial r^2} + \frac{1}{r}\frac{\partial}{\partial r} + \frac{1}{r^2}\frac{\partial^2}{\partial\varphi^2}\right) + \frac{\sin 2\varphi}{2r}\left(\frac{\partial}{\partial r} - \frac{1}{r}\right)\frac{\partial}{\partial\varphi},$$

and in the infinite-dimensional subspaces $L^2(R^+, rdr) \otimes \mathbb{C}\{e^{im\varphi}\}$

$$J_1^m = \frac{1}{4}\sin 2\varphi\left(r^2 - \frac{\partial^2}{\partial r^2} + \frac{1}{r}\frac{\partial}{\partial r} - \frac{m^2}{r^2}\right) - im\frac{\cos 2\varphi}{2r}\left(\frac{\partial}{\partial r} - \frac{1}{r}\right),$$

$$J_2^m = \frac{m}{2},$$

$$J_3^m = \frac{1}{4}\cos 2\varphi\left(r^2 - \frac{\partial^2}{\partial r^2} + \frac{1}{r}\frac{\partial}{\partial r} - \frac{m^2}{r^2}\right) + im\frac{\sin 2\varphi}{2r}\left(\frac{\partial}{\partial r} - \frac{1}{r}\right).$$

We have explained in section 2 that one expects a *symmetry* of a classical Hamiltonian to be connected with a *degeneracy* after quantization. For the harmonic oscillator in R^2 the quantization is unique. There and in $\dot{R}^2$ for $\alpha = 0$, the situation is as follows (see Fig. 1):

- The degeneracy of the spectrum is connected completely with SU(2) symmetry.

- Each degeneracy space corresponding to an energy $N\hbar\omega$, $N = 1, 2, \ldots$ spans an N-dimensional irreducible representation of SU(2); all irreducible representations appear.

- The generators J_k are unbounded. They give an integrable representation of the Lie algebra $su(2)$. Their unboundedness leads to a fully reducible infinite-dimensional representation.

- ψ^0_{nm} are eigenfunctions of J_2. Linear combinations $J_\pm = J_3 \pm iJ_1$ transform each degeneracy space as shift operators for m.

For $0 < \alpha < \pi$, i.e. for quantizations which are possible in $\dot{R}^2$, not in R^2, the situation is **different**. Because the eigenvalues E^α_{n0}, $n = 1, 2, \ldots$ are lowered, the multiplicity of the energy $N\hbar\omega$ remains N only for integer values of $E_{nm}/2\hbar\omega$, i.e. for even N. Every odd-N degeneracy space present for $\alpha = 0$ is spoiled for $\alpha > 0$ (Fig.1); it splits in two spaces with dimensions $N-1$ and 1. The eigenfunctions $\psi^\alpha_{n0}(r,\varphi)$ of H_α are eigenfunctions of J_2 with eigenvalue 0 as for $\alpha = 0$ and are in fact independent of φ. However, the action of the shift operators $J_\pm$is not the same. If $\alpha \neq 0$, for the eigenfunctions ψ^α_{nm}, $m \neq 0$ (which are independent of α), we have

$$J_\pm \psi^\alpha_{n,\mp 2} = c \cdot \psi^0_{n0} \neq \psi^\alpha_{n0}, \quad c \in \mathbb{C}.$$

Moreover, ψ^α_{n0} is singular under the action of J_1, J_3: apply the radial parts of J_1, J_3 and H_α to $\psi^\alpha_{n0}(r,\varphi)$; because of the behaviour of $w^+(0, k_n(\alpha); \frac{\omega}{\hbar} r^2)$ for $r \to 0$ and since

$$-\frac{\partial^2}{\partial r^2} \ln r = \frac{1}{r}\frac{\partial}{\partial r} \ln r = \frac{1}{r^2}$$

holds for $r > 0$, the locally at $r \to 0$ non-square-integrable contributions add up for J_1, J_3, but cancel for H. Hence ψ^α_{n0} is in the domains of H_α and J_2, but not of J_1, J_3.

We have a quantization anomaly of the type which we described in section 2. The domain of the self-adjoint extension H^0_α, the resulting domain for H_α and the domain D_g of the integrable representation of $su(2)$ (which can be easily constructed) are such that $\tilde{D}$ (see (b) in section 2) does not contain all the eigenfunctions of H_α.

The SU(2) symmetry of a quantum harmonic oscillator in R^2 can thus be broken by a topologically non-trivial change[8] of the underlying configuration space to $\dot{R}^2$. The local symmetry of classical mechanics is not changed by this procedure. There is a fine interplay between topology (i.e. $\pi_1(\dot{R}^2)$), domain questions (H and J_k are unbounded operators in $L^2(\dot{R}^2, d^2x)$) and algebra (h, j_k as a classical Lie algebra with Poisson brackets and H, J_k a quantum mechanical Lie algebra with commutators).

ACKNOWLEDGEMENTS

We would like to thank W. F. Heidenreich for valuable discussions during the preliminary stages of this work. One of the authors (J.T.) is grateful to the Arnold Sommerfeld Institute of Mathematical Physics in Clausthal for the warm hospitality extended to him.

[8]The kinematical quantization is also changed (see [14]). One gets families of unitarily inequivalent quantizations.

References

[1] A.O. Barut, R. Raczka: *Theory of Group Representations and Applications* , PWN — Polish Scientific Publishers, Warszawa (1980).

[2] H.D. Doebner, O. Melsheimer: On representations of Lie algebras with unbounded generators, Nuovo Cimento *49A*:73-98 (1967).

[3] H.D. Doebner, J.-E. Werth: Global properties of systems quantized via bundles, J. Math. Phys. *20*:1011-1014 (1979).

[4] M. Flato, D. Sternheimer: Poincaré partially integrable local representations and mass-spectrum, Commun. Math. Phys. *12*:296-303 (1969).

[5] F. Bayen, M. Flato, C. Fronsdal, A. Lichnerowicz, D. Sternheimer: Deformation theory and quantization I, II, Ann. Phys. (N.Y.) *111*:61-110, 111-151 (1978).

[6] J. Niederle, J. Tolar: Quantization as mapping and as deformation, Czech. J. Phys. *B29*:1358-1368 (1979).

[7] S. Twareque Ali, H.D. Doebner: Ordering problem in quantum mechanics: Prime quantization and a physical interpretation, Phys. Rev. *A41*:1199-1210 (1990),.

[8] J.R. Klauder: Quantization *is* geometry, after all, Ann. Phys. (N.Y.) *188*:120-141 (1988).

[9] L. Alvarez-Gaumé: Topology and anomalies. In: *Mathematics + Physics* , Vol. 2 (Ed. L. Streit) World Scientific, Singapore (1986), 50-83.

[10] N.S. Manton: The Schwinger model and its axial anomaly, Ann. Phys. (N.Y.) *159*:220-251 (1985).

[11] C. Alcalde, D. Sternheimer: Analytic vectors, anomalies and star representations, Lett. Math. Phys. *17*:117-127 (1989).

[12] B. Angermann, H.D. Doebner, J. Tolar: Quantum kinematics on smooth manifolds. In: *Non-linear Partial Differential Operators and Quantization Procedures* , Lecture Notes in Mathematics, Vol. 1037, Springer-Verlag, Berlin (1983), 171-208.

[13] H.D. Doebner, J. Tolar: Symmetry and topology of the configuration space and quantization. In: *Symmetries in Science II* (Eds. B. Gruber and R. Lenczewski), Plenum, New York (1986), 115-126.

[14] H.D. Doebner, H.J. Elmers, W.F. Heidenreich: On topological effects in quantum mechanics: The harmonic oscillator in the pointed plane, J. Math. Phys. *30*:1053-1059 (1989).

[15] *Handbook of Mathematical Functions* (Eds. M. Abramowitz and I.A. Stegun) Dover, New York (1965).

[16] C. Emmrich, H. Römer: Orbifolds as configuration spaces of systems with gauge symmetries, Commun. Math. Phys. *129*:69 (1990).

[17] M. Reed, B. Simon: *Methods of Modern Mathematical Physics*, Vols. 1, 2, Academic Press, New York (1972).

[18] J. Weidmann: *Linear Operators in Hilbert Spaces* , Springer-Verlag, Berlin (1980).

q-Analysis and Quantum Groups

D.B. Fairlie

Department of Mathematical Sciences
Durham University
Durham DH1 3LE, ENGLAND

This talk describes the manner in which the natural framework for q-analysis is a q-commutative associative field. The theory of quantum groups is developed from the observation that the q-exponential satisfies the same functional equation as does the ordinary exponential, namely $F(x)F(y) = F(x+y)$, provided the arguments x, y satisfy $xy = q^{-1}yx$.

PROPERTIES OF Q-EXPONENTIALS

I should like to start by discussing some properties of q-analysis; It seems to me that investigation of multivariable q-analysis leads naturally to the hypothesis that the base manifold is non commutative. In fact I hope to demonstrate that the whole raison d'être of quantum groups can be based upon the observation that the same functional relation for the exponential function can be extended to the q-analogue of the exponential function if their arguments do not commute. Let us begin with the Leibniz rule for ${}_qD_x$ the q-derivative with respect to x;

$$ {}_qD_x(f(x)g(x)) = ({}_qD_x f(x))g(x) + f(qx){}_qD_x g(x) \tag{1}$$

Contrary to the case with ordinary derivatives the Leibniz rule essentially dictates the form of the derivative. Since $f(x)$ and $g(x)$ are on an equal footing in (1) a second Leibniz relation also holds:

$$ {}_qD_x(f(x)g(x)) = ({}_qD_x g(x))f(x) + g(qx){}_qD_x f(x) \tag{1a}$$

Then (1) and (1a) together with the assumption ${}_qD_x\ x = 1$ gives for the choice of $g(x) = x$;

$$ {}_qD_x\ f(x) = \frac{f(qx) - f(x)}{x(q-1)} \tag{2}$$

This rule is compatible with associativity, and gives the obvious results

$$
\begin{aligned}
{}_qD_x x^n &= [n]x^{n-1}; \quad [n] = \frac{1-q^n}{1-q} \\
{}_qD_x \exp_q(\mu x)v &= \mu \exp_q(\mu x)
\end{aligned}
\tag{3}
$$

where $\exp_q(x)$ the q-analogue exponential function is given by;

$$
\exp_q(x) = \sum_0^{\infty} \frac{x^n}{[n]!} \tag{4}
$$

This function plays the role in q-analysis of the exponential in ordinary analysis. For example there is the analogue of the Taylor expansion:

$$
\exp_{\frac{1}{q}}(\lambda_q D_x)x^n = (x+\lambda)(x+\lambda q)\cdots(x+\lambda q^{n-1}) \tag{5}
$$

Although I should expect this result to be known, I have not found it in the literature. The proof runs as follows: Consider a the rth term in the expansion of $\exp_{\frac{1}{q}}(\lambda_q D_x)$ acting upon x^n Then as a consequence of Leibniz rule.

$$
\frac{q^{\frac{r(r-1)}{2}}(\lambda_q D_x)^r}{[r]!}x^n = x\frac{q^{\frac{r(r-1)}{2}}(\lambda q_q D_x)^r}{[r]!}x^{n-1} + \lambda \frac{q^{\frac{r(r-1)}{2}}(\lambda_q D_x)^{r-1}}{[r-1]!}x^{n-1} \tag{6}
$$

This equation is not immediately evident, but comes from a repeated application of (1). Hence

$$
\exp_{\frac{1}{q}}(\lambda_q D_x)x^n = (x+\lambda)\exp_{\frac{1}{q}}(\lambda q_q D_x)x^{(n-1)} \tag{7}
$$

The result (5) follows by induction. Note that the q-exponential is not defined for q a root of unity other than 1. Another characteristic property of the exponential function is its role as the unique solution to the functional equation

$$
F(a)F(b) = F(a+b) \tag{9}
$$

What is the analogue for the q-exponential? It is here that we begin to see why the extension of q-analysis to more than one variable suggests the introduction of non-commuting parameters, for remarkably equation (9) holds for the q-exponential provided a, b no longer commute but satisfy instead the quantum plane conditions

$$
ab = q^{-1}ba \tag{10}
$$

Lemma

Suppose a, b satisfy (10) and x, μ, ν are conventional variables. Then

$$
\exp_q x(a\mu + b\nu) = \exp_q(xa\mu) \times \exp_q(xb\nu) \tag{11}
$$

Proof

Applying the Leibniz rule (1) to $\exp_q(a\mu x) \times \exp_q(b\nu x)$, we find

$$\begin{aligned} {}_qD_x(\exp_q(a\mu x) \times \exp_q(b\nu x)) =& a\mu \exp_q(a\mu x) \times \exp_q(b\nu x) + \exp_q(qa\mu x) \times b\nu \exp_q(b\nu x) \\ =& (a\mu + b\nu)\exp_q(a\mu x) \times \exp_q(b\nu x) \end{aligned} \tag{12}$$

in consequence of the commutation

$$(qa\mu)^n b\nu - b\nu(a\mu)^n = 0; \quad \text{or} \quad \exp_q(qa\mu)b\nu - b\nu \exp_q(a\mu) = 0 \tag{13}$$

But equation (12) is just the defining equation for $\alpha \exp \mu(ax + by)$. The multiplicative constant α is fixed at unity by choice of $x = 0$ Hence the result follows. This result begs the question, it may be argued, because the formula (1) has been assumed to hold even when f, g do not commute. However, there is more circumstantial evidence that this is the correct way to treat q-analysis, and moreover this result may be verified, by hand for a few terms in the expansion, or further by machine. Note that the question of non commuting variables does not arise in the similar case of the finite difference operator $\Delta f(x) = \frac{f(x)-f(x-a)}{a}$, where the solution to $\Delta f(x) = f(x)$ is a true exponential, $f(x) = (1-a)^{-\frac{x}{a}}$

THE CHAINRULE

In seeking an analogue of the chain rule for differentiation of a function of a function, a restriction to powers of functions appears necessary for tasteful results. From the Leibniz rule (1)

$$ {}_qD_x\, g(x)^2 = (g(x) + g(qx))\,{}_qD_x\, g(x) \tag{14}$$

It is easy to deduce from this by induction the chain rule

$$ {}_qD_x\, g(x)^n = \left(\sum_{r=0}^{r=n-1} g(x)^{n-1-r} g(qx)^r\right) {}_qD_x\, g(x) \tag{15}$$

This result has the curious consequence that the q-exponential of x^n is $\exp_{q^n}(x^n)$ not $\exp_q(x^n)$, in the terminology of equation (4). In fact, we can take the hint from (5) and apply (1) to $(x+\lambda)(x+\lambda q)\cdots(x+\lambda q^{n-1})$ inductively to obtain

$$\begin{aligned} &{}_qD_x\ (x+\lambda)(x+\lambda q)\cdots(x+\lambda q^{n-1}) \\ &= (1+q+\cdots+q^{n-1})(x+\lambda)(x+\lambda q)\cdots(x+\lambda q^{n-2}) \end{aligned} \tag{16}$$

Thus the functions

$$(x+\lambda)_q^n = (x+\lambda)(x+\lambda q)\cdots(x+\lambda q^{n-1})$$

are the appropriate basis functions for translated polynomials; i.e. the q-analogue of $\exp(x+\lambda)$ is obtained by the replacement of $n!$ with $[n]!$ in the denominator and of $(x+\lambda)^n$ by $(x+\lambda)(x+\lambda q)\cdots(x+\lambda q^{n-1})$ in the numerator of the nth term in the power series for the exponential. In order to have the desirable property that

$$\exp_{q^{-1}}(\mu\ {}_qD_x)\exp_{q^{-1}}(\lambda\ {}_qD_x)x^n = (x+\mu+\lambda)_q^n \tag{17}$$

it is necessary to require the same commutation relation for the parameters as before;

$$\mu\lambda = q^{-1}\lambda\mu \tag{18}$$

Alternatively, these results an be formulated in the following way; Suppose λ, μ satisfy (18), and both commute with x. Then we have the identity

$$(\mu x+\mu\lambda)^n = \mu^n(x+\lambda)(x+\lambda q)\cdots(x+\lambda q^{n-1}) \tag{18}$$

Then the we can express the q-exponential of $\mu(x+\lambda)$ just like (4) as

$$\exp_q(\mu(x+\lambda)) = \sum_0^\infty \frac{(\mu x+\mu\lambda)^n}{[n]!} \tag{19}$$

Thus we see that the results of classical q-analysis may be reformulated as a natural extension of those of ordinary analysis with the introduction of non commuting variables.

THE Q-EXPONENTIAL AND THE QUANTUM GROUP

Consider

$$\begin{aligned} f(x) &= \exp_q(xa\mu + xb\lambda)\times\exp_q(xc\mu+xd\lambda) \quad \text{where} \\ M &= \begin{pmatrix} a & b \\ c & d \end{pmatrix} \end{aligned} \tag{20}$$

belongs to the quantum group $SL(2)_q$, i.e.

$$\begin{aligned} ab = q^{-1}ba;&\quad ac = q^{-1}ca \\ bd = q^{-1}db;&\quad cd = q^{-1}dc \\ ad - q^{-1}bc = 1;&\quad da - qbc = 1 \\ bc =& cb \end{aligned} \tag{21}$$

Now, in accordance with the Leibniz rule, in the notation

$${}_qD_x = \frac{q\partial}{\partial x}$$

$$\begin{aligned} {}_qD_x f(x) =& (a\mu+b\lambda)\exp_q(xa\mu+xb\lambda)\exp_q(xc\mu+xd\lambda) \\ &+\exp_q(qxa\mu+qxb\lambda)(c\mu+d\lambda)\exp_q(xc\mu+xd\lambda) \end{aligned} \tag{22}$$

Now observe, in consequence of equation (18)

$$(qxa\mu+qxb\lambda)(c\mu+d\lambda)-(c\mu+d\lambda)(xa\mu+xb\lambda)=0 \tag{23}$$

and

$$(qxa\mu+qxb\lambda)^n(c\mu+d\lambda)-(c\mu+d\lambda)(xa\mu+xb\lambda)^n=0 \tag{24}$$

that (22) may be rewritten as

$$ {}_qD_x f(x) = ((a+c)\mu + (b+d)\lambda) f(x) \tag{25} $$

Now the solution of this difference equation is simply

$$ f(x) = \alpha \exp_q((a+c)x\mu + (b+d)x\lambda) \tag{26} $$

and the boundary condition at $x = 1$ forces the choice $\alpha = 1$ upon the constant of integration. Thus, choosing $x = 1$ we obtain

$$ \exp_q(a\mu + b\lambda) \times \exp_q(c\mu + d\lambda) = \exp_q((a+c)\mu + (b+d)\lambda) \tag{28} $$

From what we have already said, we can derive this equation in another way. It will be satisfied provided that the transformed variables μ', λ' given by

$$ \begin{pmatrix} \mu' \\ \lambda' \end{pmatrix} = M \begin{pmatrix} \mu \\ \lambda \end{pmatrix} \tag{29} $$

also satisfy (18). But this is a consequence of the defining relations for M Equation (29) is in fact the starting point of Manin's approach to quantum groups. He treats them as linear transformations on what he calls a quantum plane (18). Viewed in this way, this calculation demonstrates the consistency of the transformation rule (29) with the assumptions on the properties of M. Manin's construction can be used to infer the structure of the R-matrix for the classical groups.

MANIN'S CONSTRUCTION

Manin introduces what he calls the quantum plane [1] whose elements are pairs $\boldsymbol{x} = (x, y)$, whose components x, y are assumed to satisfy the algebraic relation

$$ xy = q^{-1} yx \; , \tag{30} $$

where q is a complex number. The components neither commute nor anticommute unless $q = \pm 1$ respectively. A Grassmannian quantum plane dual to the (x, y) plane is also introduced, with elements $\boldsymbol{\xi} = (\xi, \eta)$ which are required to satisfy

$$ \xi^2 = 0 \qquad \eta^2 = 0 \qquad \xi\eta + p\eta\xi = 0 \; . \tag{31} $$

Now consider a matrix

$$ M = \begin{pmatrix} a & b \\ c & d \end{pmatrix} \tag{32} $$

which effects simultaneously linear transformations of the quantum plane and its dual,

$$ \boldsymbol{x}' = M\boldsymbol{x} \tag{33a} $$
$$ \boldsymbol{\xi}' = M\boldsymbol{\xi} \tag{33b} $$

The images $\boldsymbol{x}', \boldsymbol{\xi}'$ are supposed to lie in the appropriate planes, *i.e.* their components satisfy (30) and (31). (The elements of M are supposed to commute with x, y, ξ, η.)

This condition imposes restrictions upon M, giving the $\mathrm{GL}_q(2)$ relations

$$\begin{aligned} ab &= p^{-1}ba \,, & cd &= p^{-1}dc \,, \\ ac &= q^{-1}ca \,, & qbc &= pcb \,, \\ bd &= q^{-1}db \,, & ad - da &= (q^{-1} - p)bc \,. \end{aligned} \tag{34}$$

Using these relations it is easy to show that $\mathrm{Det}_q M = ad - q^{-1}bc$ commutes with all the elements a, b, c, d if the parameters p, q are equal and thus may be considered as a number, the "quantum determinant". The choice $\mathrm{Det}_q M = 1$ restricts the quantum 'group' to $\mathrm{SL}_q(2)$ by analogy with the classical restriction to the special linear group. Because $\mathrm{Det}_q M$ commutes with the elements of M there exists an inverse

$$M^{-1} = (\mathrm{Det}_q M)^{-1} \begin{pmatrix} d & -qb \\ -q^{-1}c & a \end{pmatrix} , \tag{35}$$

which is both a left and right inverse for M. Note that M^{-1} is a member of $\mathrm{GL}_{q^{-1}}(2)$ rather than $\mathrm{GL}_q(2)$, and thus $\mathrm{GL}_q(2)$ is not strictly speaking a group. Furthermore, it is clear that if

$$M = \begin{pmatrix} a & b \\ c & d \end{pmatrix} \quad \text{and} \quad M' = \begin{pmatrix} a' & b' \\ c' & d' \end{pmatrix} \in \mathrm{GL}_q(2),$$

and (a, b, c, d) pairwise commute with (a', b', c', d') then MM' and $M'M$ are both $\mathrm{GL}_q(2)$ matrices. Also

$$\mathrm{Det}_q(MM') = \mathrm{Det}_q(M'M) = (\mathrm{Det}_q M)(\mathrm{Det}_q M') \,, \tag{36}$$

reinforcing the identification with a determinant.

The algebra (33) is associative under multiplication and the relations may be re-expressed in a tensor product form [2]

$$R_{ijkl} M_{km} M_{ln} = M_{jl} M_{ik} R_{klmn} \,, \tag{37}$$

where R_{ijkl} is a matrix, whose explicit form is given by

$$\begin{array}{cc} & (k,l) \\ (i,j) & \begin{pmatrix} q^{-1} & 0 & 0 & 0 \\ 0 & 1 & 0 & 0 \\ 0 & q^{-1} - p & pq^{-1} & 0 \\ 0 & 0 & 0 & q^{-1} \end{pmatrix} \end{array} \tag{38}$$

where the rows are all pairs $(i, j), i, j = 1, 2$ in natural order, and similarly the columns are pairs (k, l).[3] Expression (38) is a member of a general class of R-matrices, each labelled by an additional parameter x, and each associated with one of the classical affine

Lie algebras [4]. An explicit form of the R-matrices for the classical series corresponding to $p = q$ is given by Jimbo. [5] For $\hat{A}_n$ it is

$$\begin{aligned} R(x) = (q^{-1} - xq) \sum E_{\alpha\alpha} \otimes E_{\alpha\alpha} + (1 - x) \sum_{\alpha \neq \beta} E_{\alpha\alpha} \otimes E_{\beta\beta} \\ + (q^{-1} - q) \left(\sum_{\alpha<\beta} + x \sum_{\alpha>\beta} \right) E_{\alpha\beta} \otimes E_{\beta\alpha} \ . \end{aligned} \tag{39}$$

In this expression, the indices i, j, k, l have been suppressed for the sake of clarity. The i, jth element of the matrix $E_{\alpha\beta}$ is given by

$$(E_{\alpha\beta})_{ij} = \delta_{i\alpha} \, \delta_{j\beta} \ .$$

For $\hat{A}_1$, and $x = 0$, the matrix (39) is recovered. The R-matrix [3,5] satisfies the well-known Yang-Baxter relation

$$R_{12}(x) R_{13}(xy) R_{23}(y) = R_{23}(y) R_{13}(xy) R_{12}(x) \ , \tag{40}$$

which for $x, y = 0$ is a sufficient condition for the associativity of the quantum matrices.

As Manin's construction ensures associativity, it might thus be employed as a method for the construction of Yang-Baxter R-matrices satisfying (36) by by expressing the quantum group relations in the form (36) and identifying R.

QUANTUM SUPERGROUPS

Returning to the quantum plane (x, y) and its dual (ξ, η), suppose we postulate a linear transformation $\hat{M}$ which maps the plane into its dual and vice-versa, *i.e.*

$$\begin{aligned} \boldsymbol{\xi}' &= \hat{M} \boldsymbol{x} \ , \\ \boldsymbol{x}' &= \hat{M} \boldsymbol{\xi} \ , \end{aligned} \tag{41}$$

and again impose the quantum plane conditions upon (ξ', η') and (x', y'). If the elements of $\hat{M}$ are designated by

$$\hat{M} = \begin{pmatrix} \alpha & \beta \\ \gamma & \delta \end{pmatrix} ,$$

then the constraints are ten in number:

$$\begin{aligned} \alpha\beta + q\beta\alpha &= 0 \ , & \alpha\delta + \delta\alpha &= 0 \ , \\ \alpha\gamma + p\gamma\alpha &= 0 \ , & q\beta\gamma + p\gamma\beta + (qp - 1)\delta\alpha &= 0 \ , \\ \beta\delta + p\delta\beta &= 0 \ , & \alpha^2 = \beta^2 = \gamma^2 = \delta^2 &= 0 \ . \\ \gamma\delta + q\delta\gamma &= 0 \ , \end{aligned} \tag{42}$$

These relations may be considered as a deformation of a Grassmann algebra on four elements $(\alpha, \beta, \gamma, \delta)$. In what follows we shall specialise to the case where $p = q$. As with the quantum matrix, the relations (42) may be expressed in terms of an $\hat{R}$-matrix in the form

$$\hat{R}\hat{M}\hat{M} = -\hat{M}\hat{M}\hat{R} ,$$

$$\text{where} \quad \hat{R} = \begin{pmatrix} q+q^{-1} & 0 & 0 & 0 \\ 0 & 2 & q-q^{-1} & 0 \\ 0 & -(q-q^{-1}) & 2 & 0 \\ 0 & 0 & 0 & q+q^{-1} \end{pmatrix} \tag{43}$$

Note that in the classical limit (*i.e.* $q \to 1$) $\hat{R}$ becomes twice the identity matrix. This matrix $\hat{R}$ is (39) evaluated at $x = -1$. As far as I am aware this is an observation which has not yet found any deeper significance Notice also that although the algebra (42) is an associative algebra of the matrix elements of $\hat{M}$, $\hat{R}$ does not satisfy the Yang-Baxter equation (40) thus demonstrating that the Yang-Baxter relation is not a necessary condition for associativity.

Since $\hat{M}$ is entirely Grassmannian, an inverse proper cannot exist. However, the analogue of left and right adjugate matrices can be constructed, giving:

$$\begin{pmatrix} q\delta & \beta \\ -\gamma & -q^{-1}\alpha \end{pmatrix} \begin{pmatrix} \alpha & \beta \\ \gamma & \delta \end{pmatrix} = (\beta\gamma + q\delta\alpha) \begin{pmatrix} 1 & 0 \\ 0 & 1 \end{pmatrix} , \tag{44a}$$

$$\begin{pmatrix} \alpha & \beta \\ \gamma & \delta \end{pmatrix} \begin{pmatrix} -q^{-1}\delta & \beta \\ -\gamma & q\alpha \end{pmatrix} = (\gamma\beta + q\delta\alpha) \begin{pmatrix} 1 & 0 \\ 0 & 1 \end{pmatrix} . \tag{44b}$$

The combination $\beta\gamma + q\delta\alpha$ may be thought of as a left quantum determinant and Δ_L and $\gamma\beta + q\delta\alpha$ as a right quantum determinant Δ_R. The expressions, Δ_L, Δ_R satisfy the relation

$$\Delta_L \begin{pmatrix} -q^{-1}\delta & \beta \\ -\gamma & q\alpha \end{pmatrix} = \begin{pmatrix} q\delta & \beta \\ -\gamma & -q^{-1}\alpha \end{pmatrix} \Delta_R , \tag{45}$$

which is a consequence of and and associativity.

In a similar manner one can construct the quantum analogue of GL(1|1), which we call $GL_q(1|1)$, the group of linear transformations acting upon a quantum superplane with one bosonic and one fermionic coordinate. (We use the convention of roman script for bosonic quantities, greek for fermionic.) Consider a quantum superplane and its dual;

$$\begin{pmatrix} x \\ \xi \end{pmatrix} , \quad \begin{pmatrix} \eta \\ y \end{pmatrix} ,$$

satisfying

$$\begin{matrix} x\xi - q^{-1}\xi x = 0 \\ \xi^2 = 0 \end{matrix} \quad \text{and} \quad \begin{matrix} \eta^2 = 0 \\ \eta y - qy\eta = 0 \end{matrix} . \tag{46}$$

Define a $GL_q(1|1)$ matrix

$$\mathcal{M} = \begin{pmatrix} a & \beta \\ \gamma & d \end{pmatrix} ,$$

and require

$$\mathcal{M}\begin{pmatrix} x \\ \xi \end{pmatrix} = \begin{pmatrix} x' \\ \xi' \end{pmatrix} , \qquad \mathcal{M}\begin{pmatrix} \eta \\ y \end{pmatrix} = \begin{pmatrix} \eta' \\ y' \end{pmatrix} ,$$

and impose (41) once again upon the transformed variables. We make the usual physicist's assumption that β and γ anticommute with ξ and η. Then we obtain eight relations

$$\begin{aligned} a\beta &= q^{-1}\beta a , & \beta^2 &= 0 , \\ a\gamma &= q^{-1}\gamma a , & \gamma^2 &= 0 , \\ d\beta &= q^{-1}\beta d , & \beta\gamma + \gamma\beta &= 0 , \\ d\gamma &= q^{-1}\gamma d , & ad - da + q^{-1}\beta\gamma + q\gamma\beta &= 0 . \end{aligned} \tag{47}$$

This supergroup has a curious connection with knot theory. Recently L Kauffman [6] identified an R matrix which permits an algorithmic construction of the Alexander Polynomial; the associated quantum group resembles (47) but is not identical to it. It however is the group constructed by the above procedure when the assumption that β and γ commute rather than anticommute with ξ and η is made. Again there is need for further understanding: The usual quantum group $\mathrm{GL}(2)_q$ (35) is associated with the Jones Polynomial [7], so there is scope a marriage between the two results. We continue in reference [8] by extending the results to $\mathrm{GL}(N)_q$, and give a prescription for the construction of the R matrix from its eigenvectors.

OTHER APPROACHES

As I hope to have convinced you the question of what constitutes the most general quantum group is not yet settled. At the level of groups of rank one, there are two parameter deformations of SU(2) known, one quoted above (36) [3,9] and another described in [10], which are dis-similar, but may well be equivalent. This raises the vexed question of how to determine whether or not two quantum groups are isomorphic. Recently in a collaboration with I.M. Gelfand [11],I began to study the question of what algebras could be constructed from polynomials in operators P, Q which satisfy the q-Heisenberg relations

$$PQ - qQP = i\lambda I \tag{48}$$

Note that this is a canonical form for the most general quadratic form in the operators P, Q and an identity element I which commutes with them We showed that in the case where $q = 1$ and we have the usual creation and annihilation operators the polynomials generate an infinite Lie algebra, isomorphic to the Moyal Bracket algebra [4]. Furthermore if $q^N = 1$ and $\lambda = 0$ the resulting algebra can be a finite classical Lie algebra. This construction is rather different from the 'symplecton' construction of Lie algebras from (48), which begins with infinite dimensional operators as $\lambda \neq 0$ [12] If q is not a root of unity then there exists a q-analogue of the polynomial algebra referred to above; I conclude by quoting the result.

First let me explain how to construct a generating function for the ${}_qT_{m,j}$ the quantum deformation of $T_{m,j}$, the Weyl ordered polynomial in the canonical momentum and position variables Let us suppose now that a, b are co-ordinates in a quantum plane as described by Manin [1];

$$ab = q^{-1}ba \tag{49}$$

This permits us to maintain the commutation relation

$$[aP, bQ] = i\lambda ab \tag{50}$$

even when $q \neq 1$ We also assume that a, b commute with P, Q Then (49), (50) imply

$$PQ - qQP = i\lambda,$$

and the generating function for the quantum deformation ${}_qT_{j,m}$ is

$$(aP + bQ)^s = \sum_{j=0}^{s} \frac{[s]!}{[j]![s-j]!} a^j b^{s-j} {}_qT_{j,s-j} \tag{52}$$

Here $[s]!$ means $[s][s-1]\cdots[1]$. Note that when q is a root of unity, some of the factors in (52) will vanish, so this case requires more delicate considerations. The commutator must be replaced by a deformed commutator

$$\begin{aligned} & q^{\frac{km-jn}{2}} {}_qT_{j,m} {}_qT_{k,n} - q^{\frac{jn-km}{2}} {}_qT_{k,n} {}_qT_{j,m} \\ = & \sum_{r=0}^{\min\{[\frac{j+k-1}{2}],[\frac{m+n-1}{2}]\}} \left(\frac{i}{2\lambda}\right)^{2r+1} \\ \times & \sum_{s=0}^{2r+1} (-1)^s \frac{[j]![k]![m]![n]!}{[2r+1-s]![s]![j+s-2r-1]![k-s]![m-s]![n+s-2r-1]!} \\ \times & {}_qT_{j+k-2r-1,m+n-2r-1} \end{aligned} \tag{53}$$

This is the quantum deformation of (4) in the case where q is not a root of unity. In the limit $q \to 1$ this reproduces the algebra of Bender and Dunne.[13]

REFERENCES

[1] Yu. I. Manin Annales de L'Institute Fourier, Grenoble **37** 191 (1987) Comm. Math. Phys. **123** 163 (1989)

[2] V.G. Drinfeld Soviet Math Doklady **27** 68 (1983).

[3] E.E. Deminov, Yu.I. Manin, E.E. Mukin, D. V. Zhdanovich, Non-Standard Quantum Deformations of GL(n) and Constant Solutions of the Yang-Baxter Equation RIMS-701 (1990)

[4] D.B. Fairlie, P. Fletcher and C.K. Zachos, Phys Lett. **B218** 203 (1989), J. Math. Phys.**31** 1088 (1990)
D.B. Fairlie and C.K. Zachos, Phys Lett. **B224** 101 (1989).

[5] M. Jimbo Letters in Math. Physics **10** 63 (1985)

[6] L. Kauffman Private Communication

[7] L. Kauffman *talk at Spring Workshop On Quantum Groups* Argonne (1990)

[8] E.F. Corrigan, D.B. Fairlie P. Fletcher and R. Sasaki, J. Math. Phys. **31** 776 (1990)

[9] A. Sudbery, Consistent Multiparameter Quantisation of GL(n) *talk at Spring Workshop On Quantum Groups* Argonne (1990)

[10] D.B. Fairlie, J. Phys. **A 23** L183 (1990)

[11] I.M. Gel'fand and D.B. Fairlie, The Algebra of Weyl Symmetrised Polynomials and its Quantum Extension HUTMP 90/B226, DTP 90/27 (1990)

[12] L.C. Biedenharn and D.J.Louck,*Angular Momentum in Quantum Physics*, Vol 8 *Encyclopedia of Mathematics and its applications*, Addison Wesley, Reading MA (1981)

[13] C.M. Bender and G Dunne, Phys Rev **D40** 3504 (1989).

ORTHOGONAL POLYNOMIALS AND COHERENT STATES

Philip Feinsilver

Department of Mathematics, Southern Illinois University
at Carbondale, Carbondale, IL 62901-4408

INTRODUCTION

Here we present an approach to formulating explicitly the underlying states corresponding to certain representations of (some) Lie algebras. The states are given as orthogonal polynomials in variables that arise as 'natural observables' from the point of view of quantum theory – namely, a family of commuting self-adjoint operators (elements) in a given Lie algebra, i.e. an abelian subalgebra.

The notion of coherent states is taken from Perelomov [6]. The work here is closely related to that of Hecht [4] and Zhang, Feng, and Gilmore [8]. See Klauder and Skagerstam [5] for general information on coherent states and applications. As will be seen below, Section II, we think of coherent states as generating functions for the basic states of interest. The complete story is still not available, but from the discussion in [6] as compared to the approach in [1a], it is clear that there is a strong connection with the Berezin approach to quantization on homogeneous Kähler manifolds (see [6] for discussion, Ch. 16, and references).

We present the basic theory for the orthogonal polynomials in Section I. Section II provides the main connections with coherent states. Sections III and IV give specific results — in Section III, calculation of the Ψ functions, which give the coherent state representations of the basic variables, and in Section IV we give a class of solutions to the basic equations of structure presented in Section I. We conclude with some indications concerning an infinite-dimensional theory.

We mention Tratnik [7], e.g., for related polynomials in several variables. References [1], [2], [3] provide background material, various related matter, and further references. The discussion here will, however, be self-contained, modulo proofs of some assertions which may be taken as given.

Notations. 1) We will be working on $\mathbf{R}^N$ and $\mathbf{C}^N$.

Symmetries in Science V, Edited by B. Gruber *et al.*
Plenum Press, New York, 1991

2) The summation convention is: repeated Greek indices are summed over, from 1 to N, regardless of position. E.g. $v \cdot w = v_\lambda w_\lambda = \sum_{j=1}^N v_j w_j$.

3) Contracted products, of vectors and matrices, will be abbreviated by only writing surviving indices. For clarity, underlining will be used.
E.g. $\quad v_\lambda B_{\lambda j} = \underline{vB}_j, \qquad \gamma ab = \gamma_\lambda a_\lambda b_\lambda.$

4) $B^\dagger$ denotes the transpose of the matrix B.

5) e_j denotes the multi-index with a single 1 in the jth position. So $(n_1, \ldots, n_N) = \sum n_j e_j$. $|n| = \sum n_j$ for a multi-index n. $v^n = v_1^{n_1} \cdots v_N^{n_N}$, $n! = n_1! n_2! \cdots n_N!$, and so on.

6) The vector u has components all equal to 1. Similarly, the matrix J has all entries equal to 1.

I. ORTHOGONAL POLYNOMIALS OF BERNOULLI TYPE

The orthogonal polynomials of interest here are a multivariable generalization of Meixner's classes of polynomials. Briefly, they are determined by a function $H(z)$, analytic near $0 \in \mathbf{C}^N$, with generating function

$$e^{x \cdot U(v) - tM(v)} = \sum \frac{v^n}{n!} J_n(x,t) \tag{1.1}$$

where $t \geq 0$, $x = (x_1, \ldots, x_N)$, $U = (U_1, \ldots, U_N)$, $v = (v_1, \ldots, v_N)$, $M(v) = H(U(v))$. The polynomials $J_n(x,t)$ are orthogonal with respect to a measure $p_t(dx)$ on $\mathbf{R}^N$ satisfying

$$\int_{\mathbf{R}^N} J_n(x,t) J_m(x,t) p_t(dx) = \delta_{nm} \epsilon_n \tag{1.2}$$

where the squared norms

$$\epsilon_n = n! \gamma_1^{n_1} \cdots \gamma_N^{n_N} t(t+1) \cdots (t + |n| - 1) = n! \gamma^n (t)_{|n|}, \tag{1.3}$$

for the 'generic' case, and, for the multinomial case, to be discussed in detail below, the signs switch so that

$$\epsilon_n = n! \gamma^n t(t-1) \cdots (t - |n| + 1). \tag{1.4}$$

(This will be clarified shortly.) These families arise by requiring that they give a representation of the Heisenberg-Weyl algebra

$$\begin{aligned} R^j J_n &= J_{n+e_j} \\ V_j J_n &= n_j J_{n-e_j} \end{aligned} \tag{1.5}$$

where V_j is given by

$$\frac{\partial H}{\partial z_j} = \gamma_j V_j \tag{1.6}$$

for some constants γ_j, and the function $U(v)$ is the inverse of V, $V(U(v)) = v$. On functions of x, functions such as $V(z)$ act as operators by replacing z_j by $\partial/\partial x_j$. That is, $\{z_j\}$, $\{x_j\}$ are canonical dual boson operators.

The main feature is the connection between the *recurrence formula*:

$$x_k J_n = J_{n+e_k} + a^{\mu}_{\lambda k} n_\mu J_{n+e_\lambda - e_\mu} + \gamma_k (t + |n| - 1) n_k J_{n-e_k} \tag{1.7}$$

and the *Riccati equation* (actually Riccati system):

$$\frac{\partial V_j}{\partial z_k} = \delta_{jk} + a^{\lambda}_{jk} V_\lambda + \gamma_k V_j V_k. \tag{1.8}$$

Proposition 1.1. The Riccati equation is equivalent to the recurrence relation given the basic relations

$$1) \qquad \int_{\mathbf{R}^N} e^{z\cdot x} p_t(dx) = e^{tH(z)}$$

with $H(z)$ analytic in a neighborhood of $0 \in \mathbf{C}^N$.

$$2) \qquad \frac{\partial H}{\partial z_k} = \gamma_k V_k.$$

Proof: This should clarify the whole situation. Differentiate 1) above with respect to z_j:

$$\int e^{z\cdot x} x_j p_t(dx) = t\gamma_j V_j e^{tH(z)}. \tag{1.9}$$

Successive differentiation brings down powers and forms derivatives of V. The Riccati equation allows us to replace derivatives of V by powers of V. Thus we postulate a relation of the form

$$\int e^{z\cdot x} \phi_n(x) p_t(dx) = V^n(z) e^{tH(z)}. \tag{1.10}$$

Proceeding inductively, apply $\partial/\partial z_k$:

$$\int e^{z\cdot x} x_k \phi_n p_t(dx) = \left(t\gamma_k V_k V^n + \sum n_j V^{n-e_j} \frac{\partial V_j}{\partial z_k}\right) e^{tH(z)}. \tag{1.11}$$

This gives the recurrence relation, via the Riccati equation,

$$\begin{aligned} x_k \phi_n &= t\gamma_k \phi_{n+e_k} + n_k \phi_{n-e_k} + a^{\mu}_{\lambda k} n_\lambda \phi_{n+e_\mu - e_\lambda} + \gamma_k |n| \phi_{n+e_k} \\ &= \gamma_k (t + |n|) \phi_{n+e_k} + a^{\mu}_{\lambda k} n_\lambda \phi_{n+e_\mu - e_\lambda} + n_k \phi_{n-e_k}. \end{aligned} \tag{1.12}$$

If we set $J_n = \gamma^n (t)_{|n|} \phi_n$, we get the desired recurrence relation. □

Thus, one way to find these families is to solve the Riccati equation for V. Then H may be found, and hence the measures p_t by Fourier transform. We focus on the multinomial case. Other solutions correspond to analytic continuations or limiting cases of the multinomial type. These correspond to groups with the same complex form and to group contractions respectively. In probability theory, the group contractions correspond to the classical limit theorems, e.g. contraction to the Heisenberg group corresponds to the central limit theorem.

For the multinomial case we consider a process that jumps in one of N directions with probability c_j, $1 \le j \le N$, and sits with probability $\bar{c} = 1 - \sum c_j$. The directions depend on a matrix C. This leads to the form

$$H(z) = \log\left(\sum_k c_k(\exp(c_k^{-1}C_{k\sigma}z_\sigma) - 1) + 1\right) - uCz \tag{1.13}$$

where the last term is a centering term, to normalize $H(0) = \nabla H(0) = 0$. We further assume the normalization $V(0) = \nabla V(0) = 0$. (∇ here denoting $(\partial/\partial z_1, \ldots, \partial/\partial z_N)$). The main fact we take from [2] is

Proposition 1.2. For the polynomials defined by the generating function (1.1) to be orthogonal we must have the relation

$$B^\dagger \gamma B = c^{-1} - J, \tag{1.14}$$

where γ and c are diagonal matrices with entries γ_j and c_j respectively, and $B = C^{-1}$.

The idea is that, $\langle\ \rangle$ denoting expected value, i.e. integration with respect to $p_t(dx)$,

$$\left\langle e^{xU(a)-tM(a)}e^{xU(b)-tM(b)}\right\rangle = e^{t[H(U(a)+U(b))-M(a)-M(b)]} \tag{1.15}$$

must be a function of $(a_1b_1, \ldots, a_Nb_N)$ if we are to have orthogonality:

$$\left\langle \sum \frac{a^n}{n!}J_n \sum \frac{b^m}{m!}J_m \right\rangle = \sum \frac{a^nb^n}{n!n!}\epsilon_n. \tag{1.16}$$

With $\epsilon_n = n!\gamma^n(t)_{|n|}$, this gives

$$\sum \frac{a^nb^n}{n!}\gamma^n(t)_{|n|} = (1-\gamma ab)^{-t} \tag{1.17}$$

with $\gamma ab = \gamma_\lambda a_\lambda b_\lambda$. We need some preliminary calculation. Recall that $u = (1, 1, \ldots, 1)$.

Proposition 1.3 (γ-identities). For any $a = (a_1, \ldots, a_N)$, $b = (b_1, \ldots, b_N)$

1) $\underline{\gamma aB}_j = c_j^{-1}\underline{Ca}_j - uCa$

2) $\gamma aBc = \bar{c}\, uCv$

3) $c_\mu \underline{\gamma aB}_\mu\, \underline{\gamma bB}_\mu = \gamma ab - \bar{c}(uCa)(uCb)$.

Proof: These follow by manipulating the basic relations $B^\dagger\gamma B = c^{-1} - J$, $B = C^{-1}$. □

With $H(z)$ given by (1.13), we define

$$\Delta = \sum_k c_k(\exp(c_k^{-1} C_{k\sigma} z_\sigma) - 1) + 1 \tag{1.18}$$

i.e. $H = \log \Delta$ - centering terms. One checks the following

Proposition 1.4. With $\nabla H = \gamma \nabla V$,

1) $\Delta^{-1} = 1 - \bar{c}^{-1} \gamma V B c$

2) $\Delta^{-1} = 1 - uCV$.

Proposition 1.5. The Riccati equation holds in the modified form

$$\frac{\partial V_j}{\partial z_k} = \delta_{jk} + a_{jk}^{\lambda} V_\lambda - \gamma_k V_j V_k$$

with the a-coefficients given by either of the expressions

1) $\gamma_i a_{ik}^j = c_\mu^{-1} C_{\mu i} B_{j\mu} C_{\mu k} \gamma_j - \underline{uC}_i \gamma_k \delta_{kj} - \underline{uC}_k \gamma_i \delta_{ij}$

2) $a_{ik}^j = B_{i\mu} B_{j\mu} C_{\mu k} \gamma_j - \underline{uC}_k \delta_{ij}$.

Both of these follow by direct calculation. See [2a] for the calculations where the condition of Proposition 1.2 is derived. Here we take this structure as given.

Let us calculate the condition of orthogonality

$$H(U(a) + U(b)) - H(U(a)) - H(U(b)) = \phi(ab) \tag{1.19}$$

where here ab denotes $(a_1 b_1, a_2, b_1, \ldots, a_N b_N)$. Notice that centering terms are linear and drop out. First we have

Proposition 1.6.

$$U_j(v) = \log[\prod_m \Delta^{B_{jm} c_m} (1 + \underline{\gamma v B}_m)^{B_{jm} c_m}]$$

where Δ is given in Proposition 1.4.

This gives, with $H(U(a) + U(b)) = \log \Delta_{ab} +$ linear terms,

$$\Delta_{ab} = \Delta(a)\Delta(b)[u \cdot c + \gamma a B c + \gamma b B c + c_\mu \underline{\gamma a B}_\mu \underline{\gamma b B}_\mu] + \bar{c} \tag{1.20}$$

compare with Proposition 1.4 and (1.18). By the γ-identities, Proposition 1.3, this yields

$$\begin{aligned} \frac{\Delta_{ab}}{\Delta(a)\Delta(b)} &= u \cdot c + \gamma a B c + \gamma b B c + \gamma a b - \bar{c}(uCa)(uCb) + \bar{c}(1 - uCa)(1 - uCb) \\ &= 1 + \gamma ab. \end{aligned} \tag{1.21}$$

Thus, compare with equation (1.17), with $t \to -t$, $\gamma \to -\gamma$:

Proposition 1.7. With $\phi(ab)$ defined as in (1.19),

$$\phi(ab) = 1 + \gamma ab.$$

(This change of sign, cf. (1.4), will show up later when we calculate the Ψ functions in Section III. Cf. LARRII [1b], (5.16). Here we have γ replacing $-b$ there.)

The only thing is to identify the polynomials J_n explicitly. Putting Propositions 1.4 and 1.6 together yields ([2])

Proposition 1.8. The generating function is

$$e^{xU(v)-tM(v)} = \prod_j (1 + \underline{\gamma v B}_j)^{c_j(\underline{xB}_j + t)} (1 - uCv)^{\bar{c}t - xBc}.$$

Recall the *Lauricella polynomials*

$$F_B(-r, b; t; s) = \sum \frac{(-r)_n (b)_n s^n}{(t)_{|n|} n!} \tag{1.22}$$

with N-component r, b, t a number, and $s = (s_1, \dots, s_N)$, $(-r)_n = \prod_{j=1}^N (-r_j)_{n_j}$, etc. These have the generating function [2b]

$$(1 - u \cdot v)^{u \cdot b - t} \prod_j (1 - u \cdot v + s_j v_j)^{-b_j} = \sum \frac{v^r (t)_{|r|}}{r!} F_B(-r, b; t; s). \tag{1.23}$$

We thus have

Proposition 1.9. The J_n satisfy

$$\sum \frac{v^n J_n(x,t)}{n!} = \sum \frac{(Cv)^r}{r!} (-t)_{|r|} F_B(-r, -ct - X; -t; c^{-1})$$

where $x_j = c_\lambda^{-1} X_\lambda C_{\lambda j}$.

Proof: Substitute $v \to Cv$, $b \to -ct - X$, $s \to c^{-1}$, $t \to -t$ in the generating function, equation (1.23), and compare with Proposition 1.8. □

We now have the necessary material to proceed with our study.

II. COHERENT STATES AND $sl(N+1)$

Using (1.5) we may rewrite the recurrence relations (1.7) in the form

$$x_k J_n = (R^k + a^{\mu}_{\lambda k} R^{\lambda} V_{\mu} + t\gamma_k V_k + \gamma_k R^{\lambda} V_{\lambda} V_k) J_n \tag{2.1}$$

$$(x_k - t\gamma_k V_k) J_n = (R^{\lambda} \pi_{\lambda k}(V)) J_n \tag{2.2}$$

where $\pi_{jk}(V)$ are the right-hand sides of the Riccati system (1.8). We may interpret the variables x_k as multiplication operators, a commuting family of observables,

$$X_k = t\gamma_k V_k + R^\lambda \pi_{\lambda k}(V). \tag{2.3}$$

We may also write this in the RNL form (cf. [1a]):

$$X_k = R^k + N_k + L_k \tag{2.4}$$

with

$$N_k = a^\mu_{\lambda k} R^\lambda V_\mu, \qquad L_k = \gamma_k(t+\nu)V_k \tag{2.5}$$

where we introduce the *number operator*

$$\nu = R^\lambda V_\lambda. \tag{2.6}$$

Observe that the L_j commute, i.e. $[L_i, L_j] = 0$. It is important to remark that L_j and R_j are mutually adjoint with respect to the inner product as in (1.2).

Now we can formulate conditions on the coefficients a and γ corresponding to the constraints $[X_j, X_k] = 0$. First we remark that we are dealing in general with operators from the algebra $sl(N+1)$. (This identification is thanks to Prof. Gruber). Taking R_j, L_j and $R^i V_j$ as basic elements we see that there are precisely $(N+1)^2 - 1$ of them. Notice that $R^i V_j$ can be taken as basis for a $u(N)$ subalgebra. The multinomial distribution of Section I comes specifically via $su(N+1)$.

We have ([1a]), as may be readily verified,

Proposition 2.1. We have the commutation relations

1) $[N_i, R^j] = a^j_{\lambda i} R^\lambda$

2) $[L_i, N_j] = (t+\nu)\gamma_i a^\mu_{ij} V_\mu$

3) $[L_i, R^j] = R^j \gamma_i V_i + (c+\nu)\gamma_i \delta^j_i$

4) $[N_i, N_j] = [a_i, a_j]^\mu_\lambda R^\lambda V_\mu$

where the matrix a_i has entries $(a_i)^l_k = a^l_{ki}$.

Defining the elementary matrices E^r_s by the relation

$$(E^r_s)_{\lambda\mu} R^\lambda V_\mu = R^r V_s \tag{2.7}$$

we have

Proposition 2.2. For commutativity of the X_j we require

1) $\gamma_i a^k_{ij} = \gamma_j a^k_{ji}$

2) $a^i_{kj} = a^j_{ki}$

3) $[a_i, a_j] = \gamma_j E^i_j - \gamma_i E^j_i$

Proof: $[X_i, X_j] = 0$ yields

$$0 = [R^i, N_j] + [R^i, L_j] + [N_i, N_j] + [N_i, R^j] + [N_i, L_j] + [L_i, R^j] + [L_i, N_j]. \quad (2.8)$$

The $[LN]$ terms give 1), the $[RN]$ terms give 2), and the $[RL]$ and $[NN]$ terms yield 3). These terms vanish independently since they result in terms of differing degrees acting on the basis J_n. □

One may verify that the a-coefficients given by Proposition 1.5 satisfy the above conditions, subject to the basic relation of Proposition 1.2.

Now we discuss the coherent states. As suggested in Perelomov [6], we consider

$$e^{s\cdot X}\Omega = e^{s\cdot x} J_0 \quad (2.9)$$

taking group elements generated by the X_j, with J_0 as the vacuum state Ω. Although the calculation of $e^{s\cdot X}$ may be done using group-theoretical techniques, here we utilize the generating function (1.1) and the fact that we have the X_j in spectral form as multiplication operators x_j. Thus, replacing $U(v)$ by s in (1.1) yields

$$e^{s\cdot X - tH(s)}\Omega = \sum \frac{V(s)^n}{n!} J_n = \sum \frac{V(s)^n}{n!} R^n J_0 \quad (2.10)$$

and

$$e^{s\cdot X}\Omega = e^{tH(s)} e^{V(s)\cdot R}\Omega. \quad (2.11)$$

Observe that since L is adjoint to R and $L_j\Omega = 0$, defining the vacuum expectation value $\langle e^{s\cdot X}\rangle$ by $\langle \Omega, e^{s\cdot X}\Omega\rangle$ we have

Proposition 2.3.

1) $\langle e^{s\cdot X}\rangle = e^{tH(s)}$

2) $e^{s\cdot R}\Omega = e^{X\cdot U(s) - tM(s)}\Omega = ((e^{s\cdot X}\Omega)/\langle e^{s\cdot X}\rangle)\big|_{s\to U(s)}$

Note that these are the operator formulations of 1) of Proposition 1.1, and of equation (1.1) respectively. Thus, the generating function (1.1) is indeed a normalized coherent state. Statement 1) above shows that the quantum notion of vacuum expectation value corresponds precisely to the usual notion of expected value as integration with respect to the probability measure p_t. (This way leads to the flourishing field known as "quantum probability.")

We define the *exponential vectors*

$$\psi_a = e^{a\cdot R}\Omega \quad (2.12)$$

and use these as our basic coherent states. Proposition 1.8 gives their explicit form. In the case of orthogonal polynomial states J_n, various matrix elements correspond to formulas of traditional importance in the study of special functions:

1) Matrix elements of the group elements with respect to the basis states:

$$\langle e^{s\cdot X} J_n, J_m\rangle = \sum \frac{s^l}{l!}\langle X^l J_n, J_m\rangle \tag{2.13}$$

is the generating function for relations

$$x^l J_n = \sum_m a_{nml} J_m. \tag{2.14}$$

2) Matrix elements of the basis states with respect to the coherent states, i.e. the *coherent state representation*, CSR, of the states J_n:

$$\langle J_n \psi_a, \psi_b\rangle \tag{2.15}$$

is the generating function for the linearization formula

$$J_l J_m = \sum_n c_{lm}^n J_n \tag{2.16}$$

or, equivalently, for the triple form $\int J_l J_m J_n p_t$.

Define

$$\psi_{ab} = \langle \psi_a, \psi_b\rangle \tag{2.17}$$

which corresponds to equation (1.19) and Proposition 1.7.

3) Then we have the normalized CSR's,

$$\langle J_n(X)\rangle_{ab} = \langle J_n \psi_a, \psi_b\rangle / \psi_{ab} \tag{2.18}$$

which can be interpreted as giving the (generalized) Fourier coefficients of the product $\psi_a \psi_b$ in the expansion in terms of the J_n:

$$\frac{\psi_a \psi_b}{\psi_{ab}} = \sum \frac{J_n(x,t)\langle J_n(X)\rangle_{ab}}{\epsilon_n}. \tag{2.19}$$

Finally, we remark that, as in equation (2.4), the RNL splitting — raising, lowering, neutral (and number) operators — corresponding in the Lie algebras to positive roots/Cartan subalgebra (or compact subalgebra)/negative roots is directly associated to the structure of the three-term relation characteristic of orthogonal polynomials.

III. Ψ-FUNCTIONS

The Ψ functions (cf. [1b]) arise in the CSR's of the basis states, i.e. in $\langle J_n \psi_a, \psi_b\rangle$, recalling (2.12). Proposition 2.3 and relation (2.19) suggest a way of computing these CSR's. Here we use the squared norms, equation (1.4) and Proposition 1.7, $\epsilon_n = n!\gamma^n t(t-1)\cdots(t-|n|+1)$. Thus,

$$\phi_{ab} = (1+\gamma ab)^t \tag{3.1}$$

and we make the *ansatz* $\langle J_n\rangle_{ab} = \gamma^n t\cdots(t-|n|+1)\Psi^n$:

$$\frac{\psi_a\psi_b}{\psi_{ab}} = \sum \frac{J_n\langle J_n\rangle_{ab}}{\epsilon_n} = \sum \frac{J_n\Psi^n}{n!} = e^{x\cdot U(\Psi)-tM(\Psi)}. \tag{3.2}$$

From $\psi_a = e^{x\cdot U(a)-tM(a)}$, taking logs, and recalling that, equations (1.19), (1.21),

$$t^{-1}\log\psi_{ab} = H(U(a)+U(b)) - H(U(a)) - H(U(b)) = H(U(a)+U(b)) - M(a) - M(b), \tag{3.3}$$

we see that

$$U(a)+U(b) = U(\Psi) \tag{3.4}$$

or

$$\Psi = V(U(a)+U(b)). \tag{3.5}$$

In (1.13), (1.18), set

$$E_j = \exp(c_j^{-1}C_{j\lambda}z_\lambda) \tag{3.6}$$

so that

$$H(z) = \log\Delta - uCz \tag{3.7}$$

$$\Delta = c\cdot E + \bar{c}. \tag{3.8}$$

Then differentiation of (1.13), equivalently (3.7), yields

$$\gamma_j V_j = \Delta^{-1}E_\lambda C_{\lambda j} - u_\lambda C_{\lambda j} = \Delta^{-1}\underline{EC}_j - \underline{uC}_j. \tag{3.9}$$

Proposition 3.1.

$$\Psi_j = \frac{a_j + b_j + a^\lambda_{j\rho}b_\rho a_\lambda}{1+\gamma ab}$$

Proof: From Proposition 1.6, cf. (1.20)–(1.21),

$$E_j(U(a)+U(b)) = \Delta(a)\Delta(b)(1+\underline{\gamma aB}_j)(1+\underline{\gamma bB}_j) \tag{3.10}$$

and from (1.21), $\Delta_{ab} = \Delta(a)\Delta(b)(1+\gamma ab)$. So, continuing from (3.9):

$$(1+\gamma ab)\gamma_j\Psi_j = (u_\mu + \underline{\gamma aB}_\mu)(u_\mu + \underline{\gamma bB}_\mu)C_{\mu j} - (1+\gamma ab)\underline{uC}_j. \tag{3.11}$$

Since $B = C^{-1}$, this yields

$$\gamma_j a_j + \gamma_j b_j + \gamma_\lambda a_\lambda B_{\lambda\mu}C_{\mu j}\gamma_\rho b_\rho B_{\rho\mu} - (\gamma ab)\underline{uC}_j. \tag{3.12}$$

Comparing with Proposition 1.5, 2), leads to the form

$$\gamma_j a_j + \gamma_j b_j + a^\lambda_{\rho j}\gamma_\rho b_\rho a_\lambda. \tag{3.13}$$

Finally, apply Proposition 2.2, statement 1) to arrive at

$$\gamma_j(a_j + b_j + a^\lambda_{j\rho}b_\rho a_\lambda) \tag{3.14}$$

from which the Proposition now follows. □

We can use this to find the generating function for the matrix elements $\langle e^{s\cdot X} J_n, J_m\rangle$, cf. (2.13). Consider the CSR, $\langle e^{s\cdot X}\psi_a, \psi_b\rangle$, which is a generating function for these matrix elements. We have

$$\begin{aligned}\langle e^{s\cdot X}\psi_a, \psi_b\rangle &= \langle e^{s\cdot X} e^{X\cdot(U(a)+U(b))} e^{-tM(a)-tM(b)}\rangle \\ &= e^{tH(s+U(a)+U(b))} e^{-tM(a)-tM(b)} \\ &= e^{tH(s)} e^{t\phi(V(s)\Psi)} e^{t\phi(ab)} \end{aligned} \tag{3.15}$$

as in (1.19) and Proposition 1.7, using equation (3.5). That is, from Proposition 1.7, we have

$$e^{tH(s)}(1+\gamma V\Psi)^t(1+\gamma ab)^t. \tag{3.16}$$

With Ψ given by Proposition 3.1, we have

$$(1+\gamma V\Psi)^t(1+\gamma ab)^t = (1+\gamma ab+\gamma_\lambda V_\lambda(a_\lambda+b_\lambda+a^\mu_{\lambda\rho}b_\rho a_\mu))^t. \tag{3.17}$$

Now apply $\nabla_a^n = (\partial/\partial a_1)^{n_1}\cdots(\partial/\partial a_N)^{n_N}$ at $a=0$:

$$t\cdots(t-|n|+1)(1+\gamma Vb)^{t-|n|}\prod_j(\gamma_j b_j+\gamma_j V_j+\gamma_\lambda V_\lambda a^j_{\lambda\rho}b_\rho)^{n_j} \tag{3.18}$$

By Proposition 2.2,

$$\gamma_\lambda a^j_{\lambda\rho} = \gamma_j a^\lambda_{j\rho} \tag{3.19}$$

so, factoring out $\gamma_j V_j$ from each term on the right, we can write this as

$$(\epsilon_n/n!)V^n(1+\gamma Vb)^{t-|n|}\prod_j(1+\gamma Vb+V_j^{-1}\pi_{j\sigma}(V)b_\sigma)^{n_j} \tag{3.20}$$

where $\pi_{jk}(V) = \delta_{jk}+a^\lambda_{jk}V_\lambda-\gamma_k V_k V_j$, as in equations (2.2), (1.8), with γ replaced by $-\gamma$. In the case $N=1$, cf. [1b], this may be expanded to yield the matrix elements for the group. Here, this can be done, the matrix elements being given via the application of ∇_b^n at $b=0$, resulting in types of generalized hypergeometric functions.

IV. EXAMPLES OF BERNOULLI SYSTEMS

Here we will find a class of polynomials corresponding to constant c_j — that is, the case of equiprobable directions, an isotropic random walk.

The basic relation, Proposition 1.2, is

$$B^\dagger\gamma B = c^{-1}\quad J \tag{4.1}$$

Since the matrix on the right side is symmetric, with entries all equal to -1 off the diagonal, it is diagonalizable with orthogonal B and eigenvalues γ_j. Since the spectrum of J is easily seen to consist of N with multiplicity 1 and 0 with multiplicity $N-1$, the spectrum of $c^{-1} - J$ is easy to find if $c_j = 1/\alpha$, $1 \leq j \leq N$, i.e. $c = \alpha^{-1}I$. Thus, in this case,

$$\gamma_1 = \alpha - N, \quad \gamma_2 = \alpha, \ldots, \gamma_N = \alpha \tag{4.2}$$

is the spectrum of $c^{-1} - J$. The eigenvectors of J may be taken to be:

$$\sum e_i, e_1 - e_2, e_1 - e_3, \ldots, e_1 - e_N. \tag{4.3}$$

From these one builds B in the following form:

a. First row: $(1/\sqrt{N}, 1/\sqrt{N}, \ldots, 1/\sqrt{N})$

b. Succeeding rows:

$$\begin{matrix} (1/\sqrt{2}, -1/\sqrt{2}, 0, \ldots, 0) \\ (1/\sqrt{6}, 1/\sqrt{6}, -2/\sqrt{6}, 0, \ldots, 0) \end{matrix} \tag{4.4}$$

and so on, the ith row consisting of

$$(\cdots, 1/\sqrt{i(i-1)} \ (i-1 \text{ times}), -(i-1)/\sqrt{i(i-1)}, \ \text{rest 0's}) \tag{4.5}$$

which is easily seen to yield an orthogonal matrix.

Since $C = B^{-1} = B^{\dagger}$, the a-coefficients are given according to Proposition 1.5 by

$$a_{ik}^{j} = B_{i\mu} B_{j\mu} B_{k\mu} \gamma_j - \underline{Bu_k} \delta_{ij}. \tag{4.6}$$

After some checking of cases, one can see that the matrices $(a_k)_{ij} = a_{ik}^{j}$ may be described as follows:

1. a_1 is diagonal with entries

$$(\alpha - 2N, \alpha - N, \ldots, \alpha - N)/\sqrt{N}. \tag{4.7}$$

2. a_k has this description:

   a. First row has only non-zero entry $\alpha/\sqrt{N}$ in kth column,

   b. First column has only non-zero entry $(\alpha - N)/\sqrt{N}$ in kth-row,

   c. After first row and column, until kk entry, matrix is diagonal with entry $\alpha/\sqrt{k(k-1)}$,

   d. kk entry is $\alpha(2-k)/\sqrt{k(k-1)}$,

   e. In kth column, below diagonal, entries in row $i > k$ are $\alpha/\sqrt{i(i-1)}$,

   f. In kth row, beyond diagonal, entries in column $j > k$ are $\alpha/\sqrt{j(j-1)}$.

Examples. 1. $N=2$, you have $\gamma=(\alpha-2,\alpha)$, $c_1=c_2=\alpha^{-1}$.

$$a_1=\frac{1}{\sqrt{2}}\begin{pmatrix}\alpha-4 & 0\\ 0 & \alpha-2\end{pmatrix},\quad a_2=\frac{1}{\sqrt{2}}\begin{pmatrix}0 & \alpha\\ \alpha-2 & 0\end{pmatrix} \tag{4.8}$$

$$B=\frac{1}{\sqrt{2}}\begin{pmatrix}1 & 1\\ 1 & -1\end{pmatrix} \tag{4.9}$$

2. $N=3$, $\gamma=(\alpha-3,\alpha,\alpha)$

$$a_1=\frac{1}{\sqrt{3}}\begin{pmatrix}\alpha-6 & 0 & 0\\ 0 & \alpha-3 & 0\\ 0 & 0 & \alpha-3\end{pmatrix},\quad a_2=\begin{pmatrix}0 & \alpha/\sqrt{3} & 0\\ (\alpha-3)/\sqrt{3} & 0 & \alpha/\sqrt{6}\\ 0 & \alpha/\sqrt{6} & 0\end{pmatrix},$$

$$a_3=\begin{pmatrix}0 & 0 & \alpha/\sqrt{3}\\ 0 & \alpha/\sqrt{6} & 0\\ (\alpha-3)/\sqrt{3} & 0 & -\alpha/\sqrt{6}\end{pmatrix} \tag{4.10}$$

$$B=\begin{pmatrix}1/\sqrt{3} & 1/\sqrt{3} & 1/\sqrt{3}\\ 1/\sqrt{2} & -1/\sqrt{2} & 0\\ 1/\sqrt{6} & 1/\sqrt{6} & -2/\sqrt{6}\end{pmatrix}. \tag{4.11}$$

As long as $N/\alpha<1$ you have $\bar{c}=1-\sum c_j>0$ and a veritable random walk.

V. REMARKS ON INFINITE-DIMENSIONAL PROCESSES

The form of the operator $H(z)$, (1.13), replacing $c_k^{-1}C_{k\sigma}$ by $C_{k\sigma}$, for convenience, is readily extended to the form

$$H=\log\left(\int(e^{\int C(\sigma,\tau)z(\tau)d\tau}-1)d\mu(\sigma)+1\right)-\iint C(\sigma,\tau)z(\tau)d\tau\, d\mu(\sigma)$$

corresponding to "Bernoulli fields." The function Δ, equation (1.18), takes the form

$$\Delta=1+\int(e^{Cz(\sigma)}-1)d\mu(\sigma)$$

where $d\mu$ is a given measure and C becomes an integral operator, $Cz(\sigma) = \int C(\sigma,\tau)z(\tau)d\tau$.

The generating function, Proposition 1.8, becomes

$$\exp\left(\int\log(1+\gamma VB)(xB+t)d\mu\right)\cdot\Delta^{xB-\bar{c}t}$$

where B is an inverse to C and $\bar{c}=1-\int d\mu$. The recurrence relations should yield field equations, and so on, but these topics remain open for further investigation as of now.

ACKNOWLEDGEMENT

The author would like to express his appreciation to Bruno Gruber for the fine hospitality and stimulating scientific atmosphere of the conference and related activities, and for his direct positive influence over the years at SIU.

REFERENCES

1a. Ph. Feinsilver, "Lie algebras and recurrence relations I," Acta Appl. Math., 13:291 (1988).

1b. ________, "Lie algebras and recurrence relations II," Symmetries in Science III, Plenum Press, (1989), 163–179.

2a. Ph. Feinsilver, "Bernoulli systems in several variables," Springer Lecture Notes in Math., 1064:86 (1984).

2b. ________, "Heisenberg algebras in the theory of special functions," Springer Lecture Notes in Physics, 278:423 (1987).

3. Ph. Feinsilver, "Special functions, probability semigroups, and Hamiltonian flows," Springer Lecture Notes in Math., 696 (1978).

4. K.T. Hecht, "The vector coherent state method and its application to problems of higher symmetries," Springer Lecture Notes in Physics, 290 (1987).

5. J.R. Klauder and B.S. Skagerstam, "Coherent states, applications in physics and mathematical physics," World Scientific Publishing, (1985).

6. A. Perelomov, "Generalized coherent states and applications," Springer-Verlag, (1986).

7. M.V. Tratnik, "Multivariable continuous Hahn polynomials," J. Math. Phys., 29:1529 (1988).

8. W.-M. Zhang, D.H. Feng, and R. Gilmore, "Coherent states: Theory and some applications," Rev. Mod. Phys., in press.

ALGEBRAIC MODEL FOR MOLECULAR ELECTRONIC SPECTRA

A. Frank[+], R. Lemus[+] and F. Iachello[++]

[+] Instituto de Ciencias Nucleares, UNAM
Apdo. Postal 70-543, 04510 Mexico, D.F., Mexico
[++] Center for Theoretical Physics, Sloane Laboratory, Yale University
New Haven, Connecticut 06511, USA

A unified algebraic model for rotation-vibration and electronic excitations in diatomic molecules is presented. Dynamical symmetries are constructed in the united atoms limit, which provide analytic bases for realistic calculations. The model is applied to a series of hydride molecules, determining many of the parameters from atomic calculations. A method to extract Born-Oppenheimer potentials for these molecules is introduced.

1. INTRODUCTION

The vibron model represents a powerful method to describe rotation-vibration excitations in molecules.[1)] It has been applied to diatomic,[2)] linear poli-atomic[3)] and recently, to more complicated molecules.[4)] Its simple algebraic treatment of these systems gives rise to manageable wave functions which have been successfully used in the analysis of inelastic electron scattering off molecules.[5)] This model, however, does not incorporate the molecular electronic degrees of freedom and is thus restricted to the description of rotation-vibration spectra associated to the ground electronic configuration, which in addition should be an $L = 0$ state, i.e., a Σ^+ state in the case of linear molecules.

In analogy to the case of odd-A and odd-odd nuclei in the Interacting Boson Model,[6)] we need to supplement the bosonic degrees of freedom with fermionic ones, in this case with —in principle— a large number of them, representing the $D = \Sigma_i 2(2li + 1)$ electronic single particle states. In the rest of this paper we shall be mainly concerned with the simplest possible molecular system, that of heteronuclear diatomic molecules.[7,8)]

In order to fix a basis, we choose for simplicity the united atoms limit, where we may use an $O(4)$ electronic basis. These states should be coupled with the N-boson vibron model states in order to construct a complete set of states, classified by the $[1^m] \times [N]$ irreps of $U^{rv}(4) \otimes U^e(D)$, where m is the number of active electrons and we have attached the superscripts (rv) and (e) to the vibron and electronic dynamical groups, respectively. More specifically, if we consider as a basis for the united atoms limit that of hydrogenic levels, we can take $D = 2n^2$, where n is the hydrogenic principal quantum number. As an example, we consider the case of electronic states originating from the $2s$, $2p$ levels, for which $D = 8$. In this case the basis states carry

Symmetries in Science V, Edited by B. Gruber *et al.*
Plenum Press, New York, 1991

representations of the

$$U^{rv}(4) \otimes U^{e}(8) \tag{1.1}$$

direct product group. Associated with the N interacting vibrons there is a set of creation b^{+}_{lm} and annihilation b_{lm} bosonic operators, satisfying the usual commutation relations and distributed into a scalar $s^{\dagger} \equiv b^{\dagger}_{00}$ and a vector $p^{\dagger}_{\mu} \equiv b^{\dagger}_{1\mu}$, with parity $(-)^{l}$. In addition, we define the covariant operators

$$\tilde{p}\dagger_{\mu} = (-)^{1-\mu} p_{-\mu} \ , \tag{1.2}$$

transforming under rotations as the $p^{\dagger}_{\mu}$. The bilinear products $b^{\dagger}_{l\mu} b_{l'\mu'}$ generate the unitary group $U^{rv}(4)$. Associated with each electronic level there is likewise a set of creation $a^{\dagger}_{lm\frac{1}{2}\sigma}$ and annihilation $a_{lm\frac{1}{2}\sigma}$ operators which satisfy the anticommutation relations

$$\begin{aligned} \{a_{lm\frac{1}{2}\sigma},\ a^{\dagger}_{l'm'\frac{1}{2}\sigma'}\} &= \delta_{ll'}\delta_{mm'}\delta_{\sigma\sigma'} \ , \\ \{a^{\dagger}_{lm\frac{1}{2}\sigma},\ a^{\dagger}_{l'm'\frac{1}{2}\sigma'}\} &= \{a_{lm\frac{1}{2}\sigma},\ a_{l'm'\frac{1}{2}\sigma'}\} = 0 \ , \end{aligned} \tag{1.3}$$

where l and m correspond to the orbital angular momentum and projection, and σ is the projection of the spin $s = 1/2$. For the $n = 2$ manifold $l = 0, 1$, corresponding to the $2s - 2p$ hydrogenic levels. The covariant annihilation operators are in this case given by

$$\tilde{a}_{lm\frac{1}{2}\sigma} = (-)^{l-m+\frac{1}{2}-\sigma} a_{l-m\frac{1}{2}-\sigma} \ , \tag{1.4}$$

which transform under both the orbital and spin rotations as the $a^{\dagger}_{lm\frac{1}{2}\sigma}$. The bilinear products $a^{\dagger}_{lm\frac{1}{2}\sigma}\ a_{l'm'\frac{1}{2}\sigma'}$ generate the unitary group $U^{e}(8)$. We may immediately separate the electronic degrees of freedom in their orbital and spin parts by means of the reduction

$$U^{e}(8) \supset U^{e}(4) \otimes SU_{s}(2) \ , \tag{1.5}$$

which corresponds to the $L - S$ coupling scheme. With this decomposition the group $U(4)$ appears for both the vibrons and the electrons and there are two possible chains of groups of physical interest[1)]

$$U(4) \supset 0(4) \supset O(3) \supset O(2) \ , \tag{1.6a}$$

and

$$U(4) \supset U(3) \supset O(3) \supset O(2) \ , \tag{1.6b}$$

for both systems. This fact can be exploited to construct the direct sum groups $U^{rv+e}(4)$, $O^{rv+e}(4)$, $U^{rv+e}(3)$ and $O^{rv+e}(3)$. The different decompositions of the dynamical group (1.1) followed by (1.5) is given by the different ways in which the chains (1.6) can be coupled, preserving the $O^{rv+e}(3)$ group. The latter is isomorphic to the spin group $SU_{s}(2)$ and thus can be coupled to it to yield the total angular momentum group $SU_{J}(2)$.

In the next section we introduce the general Hamiltonian of the model and discuss a complete basis for its diagonalization.

2. HAMILTONIAN AND THE $O(4)$ DYNAMICAL SYMMETRY

The full spectrum of rotation-vibration-electronic states are obtained by diagonalization of the molecular Hamiltonian

$$\hat{\mathcal{H}} = \hat{\mathcal{H}}^{rv} + \hat{\mathcal{H}}^{e} + \hat{V}^{rv-e} \ , \tag{2.1}$$

where $\hat{H}^{rv}$ is the vibron Hamiltonian,[1] $\hat{H}^{e}$ is a purely electronic contribution (formally equivalent in this case to an atomic Hamiltonian) and $\hat{V}^{rv-e}$ is the interaction term, which is the most important component of the Hamiltonian, since it determines the ordering of the electronic energy levels in the molecule. The most general forms for the latter two operators, up to two body interactions are[8]

$$\hat{H}^{e} = \sum_{l} \rho_l [a^{\dagger}_{l\frac{1}{2}} \times \tilde{a}_{l\frac{1}{2}}]^{(00)0} + \sum_{\substack{ll'l''l''' \\ LSL'S'L}} \eta^{LSL'S'\bar{L}}_{ll'l''l'''} \left[[a^{\dagger}_{l\frac{1}{2}} \times a^{\dagger}_{l'\frac{1}{2}}]^{(LS)} \times [\tilde{a}_{l''\frac{1}{2}} \times \tilde{a}_{l'''\frac{1}{2}}]^{(L'S')} \right]^{(\bar{L}\bar{L})0} \tag{2.2}$$

and

$$V^{rv-e} = \sum_{\substack{ll'l''l''' \\ LL'\bar{L}}} \xi^{LL'\bar{L}}_{ll'l''l'''} \left[[b^{\dagger}_{l} \times a^{\dagger}_{l'\frac{1}{2}}]^{(L\frac{1}{2})} \times [\tilde{b}_{l''} \times \tilde{a}_{l'''\frac{1}{2}}]^{(L'\frac{1}{2})} \right]^{(\bar{L}\bar{L})0} , \tag{2.3}$$

where the ρ, η, ξ are real parameters and the notation implies orbital and spin angular momentum coupling to (LS) and finally to $J = 0$. Besides rotational invariance, the only other restrictions imposed on (2.3) and (2.4) are hermiticity, boson and fermion number conservation and invariance under parity transformations.

The number of parameters in (2.2) and (2.3) is very large, of course, and it is necessary to invoke physical arguments to select the most important contributions to molecular spectra. There is an alternative way to write the interaction $\hat{V}^{rv-e}$ as a multiple expansion,

$$\hat{V}^{rv-e} = \sum_{k^i} v^i_k T^{(k)}_{rv,i} \cdot T^{(k)}_{e,i} , \tag{2.4}$$

where k is the multipole order and the label i accounts for the possibility of different tensors for a given k. There is a similar expression for the electronic Hamiltonian (2.2). Since in both (2.2) and (2.3) the $\bar{L} \neq 0$ terms correspond to fine structure interactions (such as spin-orbit or magnetic dipole interactions), we neglect them as a first approximation. In this case (2.4) has the simple form

$$\hat{V}^{rv-e} = V^0 \hat{n}_p \hat{n}_F + V^1_1 \hat{D}_{rv} \cdot \hat{D}_e + V^1_2 \hat{\bar{D}}_{rv} \cdot \hat{\bar{D}}_e + \hat{V}^1_3 \hat{L}_{rv} \cdot \hat{L}_e + V^2 \hat{Q}_{rv} \cdot \hat{Q}_e , \tag{2.5}$$

where

$$\begin{aligned}
\hat{n}_p &= \sqrt{3}(p^{\dagger} \times \tilde{p})^{(0)} , & \hat{n}_F &= -\sqrt{6}(a^{\dagger}_{1\frac{1}{2}} \times \tilde{a}_{1\frac{1}{2}})^{(00)} , \\
\hat{L}_{rv,\mu} &= \sqrt{2}(p^{\dagger} \times \tilde{p})^{(1)}_{\mu} , & \hat{L}_{e,\mu} &= -2(a^{\dagger}_{1\frac{1}{2}} \times \tilde{a}_{1\frac{1}{2}})^{(10)}_{\mu} , \\
\hat{Q}_{rv,\mu} &= (p^{\dagger} \times \tilde{p})^{(2)}_{\mu} , & \hat{Q}_{e,\mu} &= -\sqrt{2}(a^{\dagger}_{1\frac{1}{2}} \times \tilde{a}_{1\frac{1}{2}})^{(20)}_{\mu} , \\
D_{rv,\mu} &= (p^{\dagger} s - s^{\dagger} \tilde{p})^{(1)}_{\mu} , & D_{e,\mu} &= -\sqrt{2}(a^{\dagger}_{1\frac{1}{2}} \times \tilde{a}_{0\frac{1}{2}} - a^{\dagger}_{0\frac{1}{2}} \times \tilde{a}_{1\frac{1}{2}})^{(10)}_{\mu} , \\
\bar{D}_{rv,\mu} &= i(p^{\dagger} s + s^{\dagger} \tilde{p})^{(1)}_{\mu} , & \bar{D}_{e,\mu} &= -i\sqrt{2}(a^{\dagger}_{1\frac{1}{2}} \times \tilde{a}_{0\frac{1}{2}} + a^{\dagger}_{0\frac{1}{2}} \times \tilde{a}_{1\frac{1}{2}})^{(10)}_{\mu} .
\end{aligned} \tag{2.6}$$

Note that $\hat{n}$ and $\hat{Q}$ correspond to monopole and quadrupole operators, while the other three are $L = 1$ operators, with $\hat{D}$ being the physical dipole operator. Interactions other than these exist, but they can either be incorporated into H^{rv} and H^{e}, or they do not contribute to excitation energies.[8] In fact it turns out that three of the interactions in (2.5) dominate the spectral structure, the terms associated to V^1_1, V^1_3 and V^2. Similar considerations lead to a simple form for $H^{(e)}$, which we discuss below.

In order to construct a complete set of states for the system we select from the possible boson-fermion couplings the one dominated by the dipole-dipole interactions $\hat{D}_{rv} \cdot \hat{D}_e$, which correspond to the $O(4)$ basis. The molecular wave functions are a combination of the rotation-vibration states[1,9]

$$\begin{array}{cccc} U^{rv}(4) \supset & O^{rv}(4) \supset & O^{rv}(3) \supset & O^{rv}(2) \\ \downarrow & \downarrow & \downarrow & \downarrow \\ |\,[N]\;, & \omega\;, & l\;, & m\;>\;, \end{array} \tag{2.7}$$

and the electronic wave functions[7,8),

$$\begin{array}{cccccc} U^e(8) \supset & U^e(4) \times SU(2) \supset & & O^e(4) \times SU(2) \supset & O^e(3) \times SU(2) \supset & O^e(2) \times SU(1) \\ \downarrow & \downarrow & & \downarrow & \downarrow & \downarrow \quad \downarrow \\ |\,[1^m]\;, & S\;, & & (\tau_1\tau_2)\;, & l_e\;, & m_e\;,\quad \sigma >\;, \end{array} \tag{2.8}$$

where we indicate below each group the quantum numbers that label their representations. The full wave functions are then classified by

$$\begin{array}{l} U^{rv}(4) \times U^e(8) \supset \; U^{rv}(4) \times U^e(4) \times SU(2) \supset \; O^{rv}(4) \times O^e(4) \times SU(2) \supset \\ \quad\downarrow \qquad\quad \downarrow \qquad\qquad\qquad\qquad\qquad\qquad\qquad\quad \downarrow \qquad\quad \downarrow \\ |\,[N], \quad [1^m], \qquad\qquad\qquad\qquad\qquad\qquad\qquad \omega, \qquad (\tau_1\tau_2), \\ \qquad O(4) \times SU(2) \supset \; O(3) \times SU(2) \supset \; SU_J(2) \supset \; SU_J(1) \\ \qquad\quad \downarrow \qquad\qquad\qquad\quad \downarrow \qquad \downarrow \qquad\qquad \downarrow \qquad\quad \downarrow \\ \qquad (\sigma_1\sigma_2) \qquad\qquad\quad\;\; L, \qquad S, \qquad\quad J, \qquad M > \;. \end{array} \tag{2.9}$$

Explicitly, the coupling (2.9) is given by:

$$\begin{aligned} |[N][1^m]\omega(\tau_1\tau_2)(\sigma_1\sigma_2)SLJM >= \sum_{l,l_e,\forall m} \left\langle \begin{matrix} \omega & (\tau_1\tau_2) \\ l & l_e \end{matrix} \middle| \begin{matrix} (\sigma_1\sigma_2) \\ L \end{matrix} \right\rangle C(ll_eL; mm_em_L) \\ C(LSJ; m_L\sigma M)|[N]\omega lm > |[1^m](\tau_1\tau_2)l_em_e; S\sigma > \;, \end{aligned} \tag{2.10}$$

where $< | >$ is an isoscalar factor corresponding to the $O(4) \supset O(3)$ reduction. Taking advantage of the local isomorphism $O(4) \simeq SU(2) \times SU(2)$, the isoscalar factors can be expressed in terms of $SU(2)$ 9-j symbols. Explicit expressions for the states (2.7)-(2.9), as well as all necessary reduction rules for the corresponding quantum labels have been derived.[7,8,9) (The $U(3)$ basis has also been explicitly constructed and turns out to be relevant for the classification of homonuclear molecules[9)). The basis states (2.9) have well defined angular momentum but do not have good parity. However, simple linear combinations of these states possess good parity.[8,9) The quantum numbers in (2.9) completely characterize the molecular states, but it is necessary to correlate them with the usual spectroscopic notation. The vibrational quantum number v is given by[1)

$$v = \frac{N - \omega}{2} \tag{2.11}$$

and thus N is related to the maximum number of bound vibrational states. For the electronic part, the usual notation is based on the Born-Oppenheimer approximation, where the projection of the orbital angular momentum on the symmetry axis is conserved and denoted by $\Lambda = \Sigma, \Pi, \Delta, \ldots$. The Σ terms are denoted by Σ^+ or Σ^- depending on their behavior under reflections on a plane passing through the molecular axis. The $\Lambda \neq 0$ states are doubly degenerate and contain both signs, $\Pi^\pm$, $\Delta^\pm$, etc. The reflection parity (π_σ) is related to the total parity (p) by[9)

$$\pi_\sigma = (-)^L p \;. \tag{2.12}$$

By using the $O(4) \supset O(3)$ reduction rule $L = |\sigma_2|, |\sigma_2|+1, \ldots, |\sigma_1|$, we can identify Λ with the quantum number σ_2,

$$\Lambda = |\sigma_2| \; . \tag{2.13}$$

The quantum numbers (τ_1, τ_2) and σ_1 have no counterparts in the standard classification, which is incomplete. The identifications (2.11)-(2.13), together with the S, L, J and M quantum numbers, which are already in standard form, allow us to compare the algebraic model results with experimental data. In order to examine the kind of spectra arising from this framework, we take the $O(4)$ Hamiltonian

$$\begin{aligned}\hat{H}_{O(4)} &= \hat{H}_0 + \beta_6 \hat{C}_2(O^{rv}(4)) + f_6 \hat{C}_2(O^e(4)) + A_1 \hat{C}_2(O(4)) \\ &\quad + B_1 \hat{L}^2 + B_2 \hat{S}^2 + B_3 J^2 \; ,\end{aligned} \tag{2.14}$$

where $\hat{H}_0$ includes terms that do not contribute to the excitation energy and the $\hat{C}_2$ are Casimir operators of the different $O(4)$ groups in (2.9). We find the eigenvalue expression:

$$\begin{aligned}E_{O(4)} &= E_0 + \beta_6 \omega(\omega+2) + f_6[\tau_1(\tau_1+2) + \tau_2^2] + A_1[\sigma_1(\sigma_1+2) + \sigma_2^2] \\ &\quad + B_1 L(L+1) + B_2 S(S+1) + B_3 J(J+1) \; .\end{aligned} \tag{2.15}$$

In Figure 1, we show the corresponding electronic and rotation-vibration spectrum for $m = 1$ and 7.

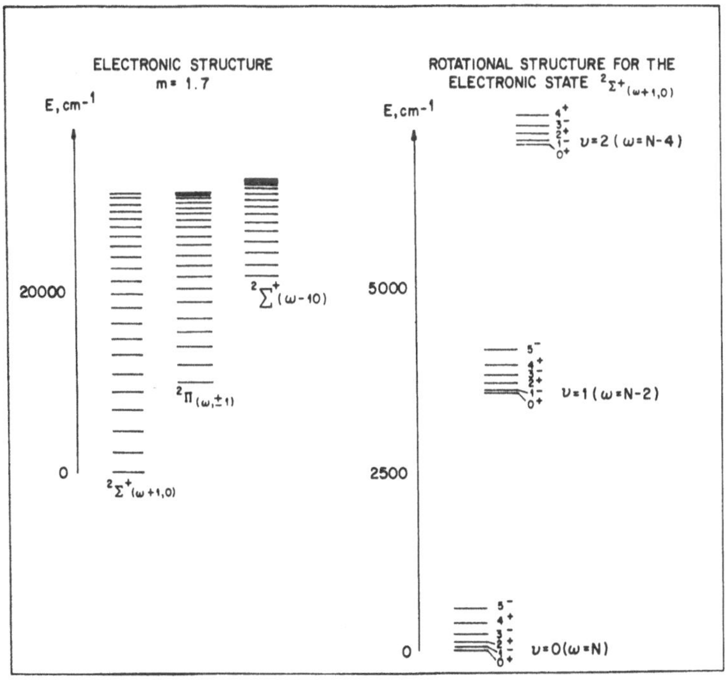

Figure 1. Schematic structure of the electronic vibration-rotation spectrum provided by an $O(4)$ dynamical symmetry for one and seven electrons. The constants β_6 and A_1 are chosen to be 157 and -174 cm^{-1}, respectively. The rotational constants are chosen to be $B_1 = 18$ cm^{-1} and $B_3 = 1$ cm^{-1}, according to Hund's case (b).

3. UNITED ATOMS CALCULATIONS AND THE HYDRIDE MOLECULES

In order to provide an accurate description of realistic spectra one needs to include in the molecular Hamiltonian terms not diagonal in $O(4)$, such as the quadrupole-quadrupole interactions in the $\hat{V}^{rv-e}$ of eq. (2.5) and analogous $U(3)$ interactions in the electronic Hamiltonian H^e. All relevant matrix elements in the $O(4)$ basis have been evaluated in terms of suitably defined tensor operators and the generalized Wigner-Eckart theorem.[1,8,9] There is, however, a large number of possible interactions and it is necessary to identify the most important terms and device ways to determine the remaining parameters. For the vibron Hamiltonian H^{rv} we may use the $O^{rv}(4)$ limit as a first approximation, since at this point we are mainly interested in the electronic structure of molecules. A fine tuning of the vibrational spectra can be achieved by introducing additional vibronic interactions.[1-3] Our main concern is thus to determine the structure of the H^e and V^{rv-e} terms in the Hamiltonian.

In the united atoms limit H^e has the structure of an atomic Hamiltonian,, whose (spin independent) terms are given by[9,10]

$$\hat{H}^e = f_6 C_2(O^e(4)) + A_2 \hat{L}_e^2 + f_3 \hat{n}_F + f_4 \hat{Q}_e \cdot \hat{Q}_e \ . \tag{3.1}$$

We have applied this Hamiltonian to the $2s-2p$ atoms, corresponding to $^4B_e(m = 2)$, $^5B(m=3)$, $^6C(m=4)$, $^7N(m=5)$ and $^8O(m=6)$ and looked for a best fit in energies and dipole transitions. The energy r.m.s. deviations turn out to be of a few percent of typical energy differences between states, as illustrated for ^4Be in Table 1. Likewise, dipole transition intensities are well described.[11]

In Figs. 2 and 3 we show the behavior of the 4 parameters in (3.1) as a function of the number of valence electrons. We note a striking uniformity and a nearly exact

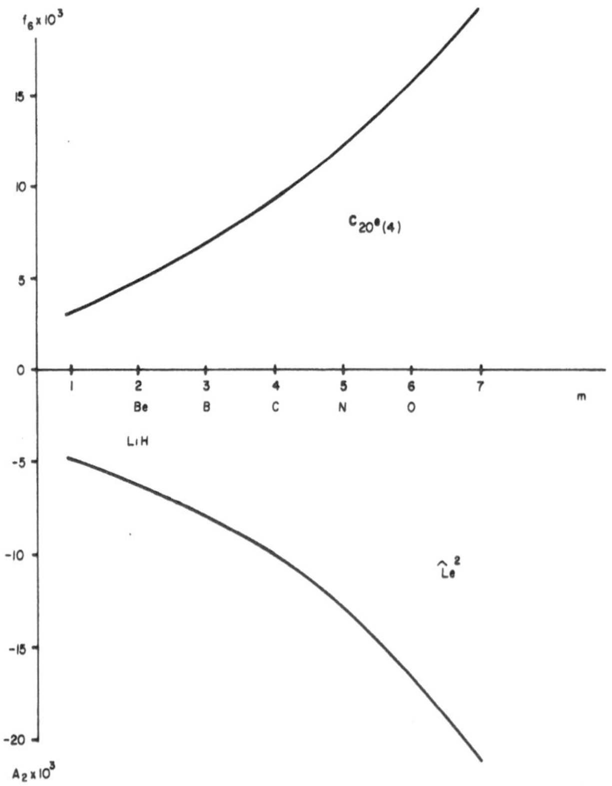

Figure 2. Parameters of $C_{2O^e(4)}$ and $\hat{L}_e^2$ as a function of the number of valence electrons m.

TABLE I.
Energies and parameters obtained in the least square fitting of the ^4Be atom. All energies are in cm^{-1}.

State	Energies exp.	Energies theor.[20]	Parameters
$^1S^+$	71498	71507	$f_6 = 5236$
$^3P^+$	59694	59825	$A_2 = -6458$
$^1D^+$	56432	56651	$f_3 = 38421$
$^1P^-$	42565	42348	$f_4 = 858$
$^3P^-$	21979	21404	$rms = 665$
$^1S^+$	0	0	

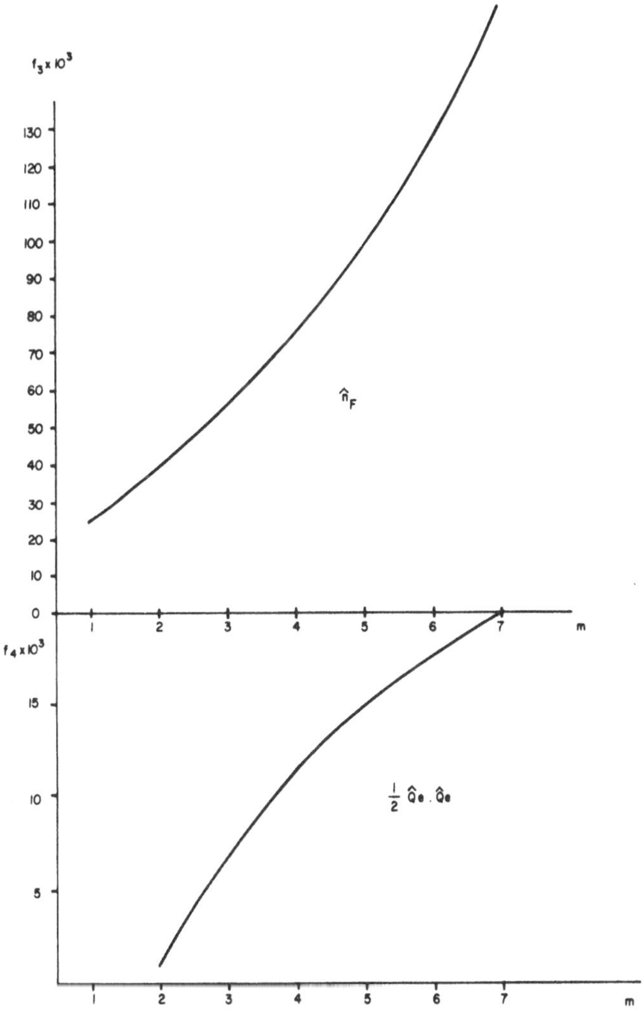

Figure 3. Parameters of $\hat{n}_F$ and $\hat{Q}_e \cdot \hat{Q}_e$ as a function of the number of valence electrons m.

cancellation of the $\hat{L}_e^2$ interaction, since $\hat{C}_2(O^e(4)) = \hat{L}_e^2 + \hat{D}^e \cdot \hat{D}^e$. This feature is expected from Coulomb interaction calculations in the $O^e(4)$ basis.[10] Since we are interested in the description of the hydride molecules: LiH, BeH, BH, CH, NH and OH, we have adopted the following prescription for the determination of the fermionic parameters in the molecular Hamiltonian: We start with the united atoms limit interactions, i.e., Be for LiH, B for BeH, etc., and move simultaneously on top of the interpolated parameter curves in the direction of the neighboring lighter atom, i.e., towards Be in BeH, for example, till we achieve a best fit in the molecule. In this way we constrict the electronic parameters and link them to the atomic calculations. For LiH and OH, however, in order to achieve a good fit we were forced to further vary the electronic parameters.

In the case of the V^{rv-e} interaction, we needed only to include the dipole and quadrupole terms. The molecular Hamiltonian of the model is thus given by

$$\hat{H} = E_O + \beta_6\hat{C}_2(O^{rv}(4)) + f_3\hat{n}_F + f_4\hat{Q}_e^2 + f_6C_2(O^e(4)) \\ + A_2\hat{L}_e^2 + A_1\hat{C}_2(O(4)) + 2k\hat{Q}_B \cdot \hat{Q}_F \,. \tag{3.2}$$

In Figs. 4-5 we show the best fits to the known electronic states of the hydride molecules, while in Table II we indicate the final parameter values employed. We have found that the quadrupole interaction parameter k is the quantity that basically determines the electronic level ordering and it is quite remarkable to note that it grows uniformly with valence electron number, as shown in Fig. 6.

The r.m.s. deviation for the states in the 6 molecules is of the order of 4000 cm^{-1} which is comparable to the accuracy of abinitio calculations.[12] We emphasize, however, that the present study includes simultaneously $\simeq$ 25 electronic levels. There are other interesting features of the calculation, like the fact that the model produces naturally all electronic states originating from the $2s - 2p$ united atomic levels, including states not known experimentally. A case in point is the BH molecule, where

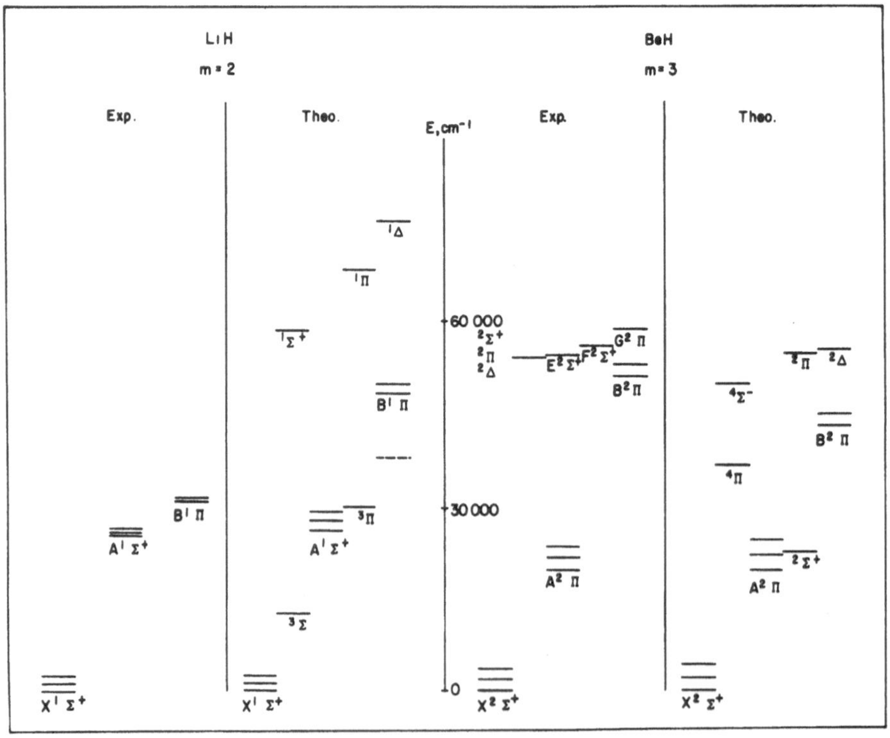

Figure 4. Comparison between calculated and experimental (Ref. 9) electronic energy levels in LiH and BeH ($m = 2$ and 3).

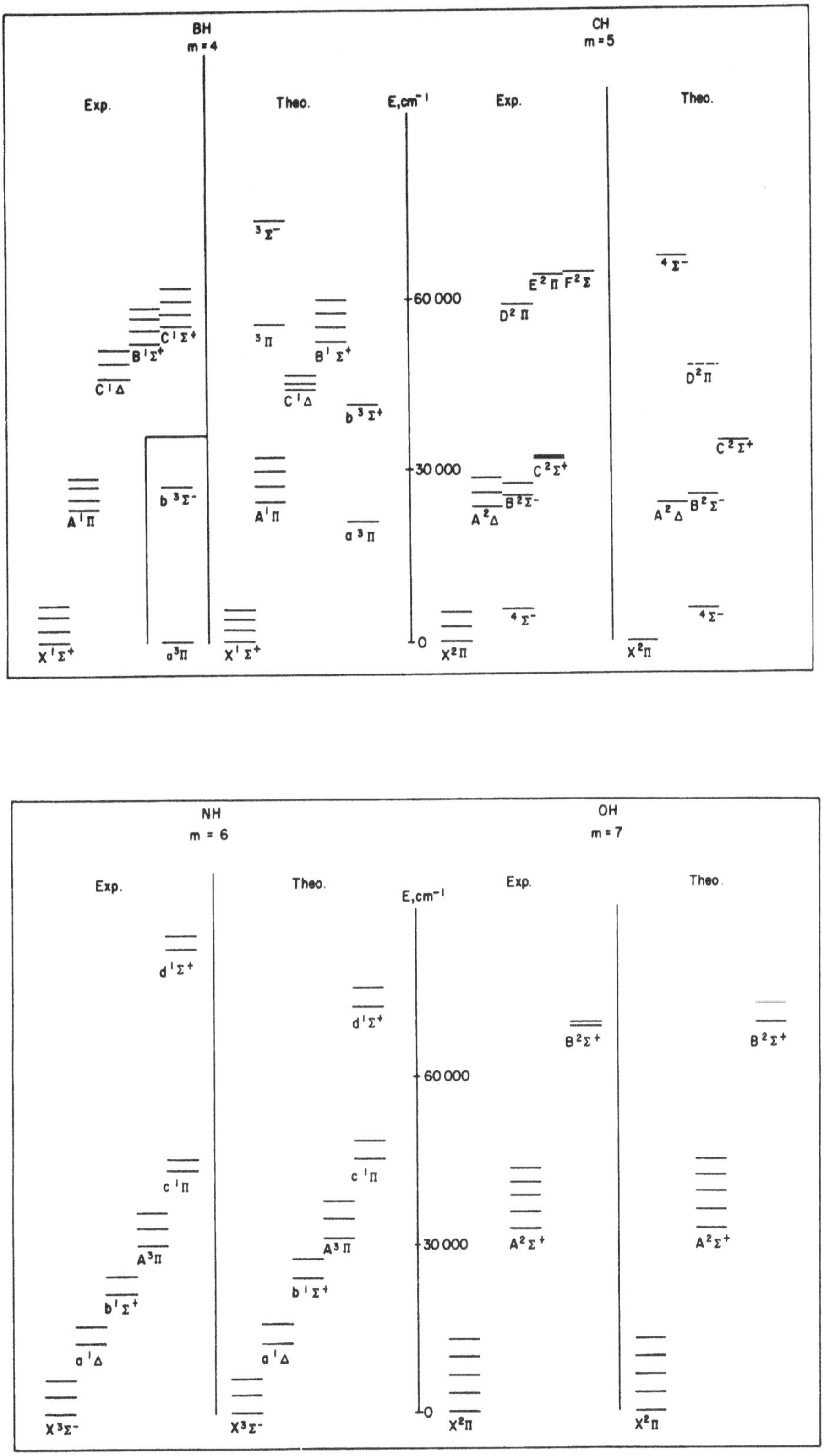

Figure 5. Comparison between calculated and experimental (Ref. 9) electronic energy levels in BH and $CH(m = 4$ and $5)$, and NH and $OH(m = 6$ and $7)$. The relative location of the triplet states $a^3\Pi$ and $b^3\Sigma^-$ to the ground state $X^1\Sigma^+$ is not known experimentally in BH. They are thus plotted separately in the figure.

TABLE II
Parameters used in the calculation of energy levels.
The vibron number is taken to be $N = 42$ for all molecules.
All parameters are in cm^{-1}.

	LiH	BeH	BH	CH	NH	OH
β_6	104	159	64	131	120	59
f_3	42000	38421	55269	86000	100000	44000
f_4	1300	858	3382	6700	7500	1600
f_6	5500	5236	7208	10800	12500	5650
A_2	−6800	−6458	−8198	−11200	−13000	−7000
A_1	−110	−170	−80	−148	−140	−80
k	−400	−450	−500	−595	−700	−820
m	2	3	4	5	6	7

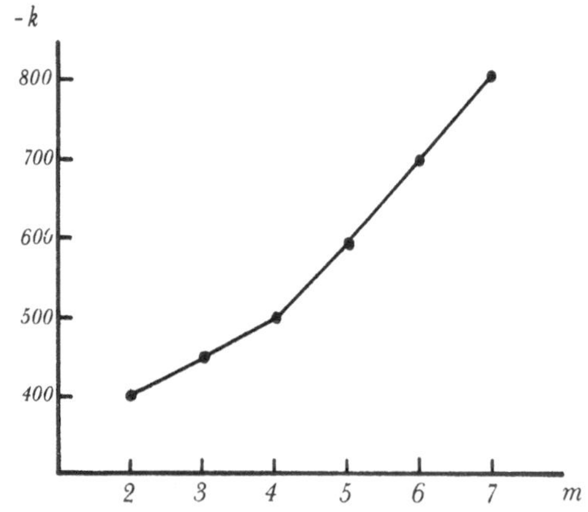

Figure 6. Parameter k as a function of the number of valence electrons m.

the position of the Spin 1 states $^3\Pi$, $^3\Sigma^-$, relative to the ground state configuration $^1\Sigma^+$ is fixed by the calculation. Another interesting result is the reproduction of the $^4\Sigma^-$ electronic state in CH, where the experimental figure has not been confirmed.[13)]

In order to further test the consistency of the model, we have also considered the evaluation of the molecular structure of the hydride ions, LiH^+, BeH^+, etc., using slight variations of the Hamiltonian parameters obtained in the fit to the neutral molecules and changing the electron number m to $m-1$. The quality of the fits turn out to be similar to that of the neutral molecules.[11)]

The evaluation of electromagnetic transition probabilities is a more stringent test for the model's wavefunctions. In particular, electric dipole transitions are associated in first approximation to the operator

$$\hat{T}^{E1}_{\mu} = \alpha \hat{D}_{e,\mu} + \beta \hat{D}_{rv,\mu} \,, \qquad (3.3)$$

where α and β are effective charges and the dipole operators are given in (2.6). The matrix elements of this operator have been explicitly evaluated in both the $O(4)$

and $U(3)$ bases and selection rules derived from its tensorial properties.[9] We are currently carrying out a systematic study of dipole transition intensities in the hydride molecules, with very encouraging preliminary results.[11]

4. POTENTIAL ENERGY CURVES

The algebraic approach appears to be useful in correlating a large amount of experimental data and in predicting states that have not been observed. It is desirable, however, to correlate this abstract procedure with the traditional approach in terms of the configuration space Schrödinger equation and, in particular, to the Born-Oppenheimer approximation which leads to the evaluation of potential energy surfaces associated to the electronic states.

To extract the potential energy surfaces associated to Hamiltonian (3.2), we proceed in analogy to the interacting Boson Model coherent state method.[14] In this case, however, we need to adapt the procedure to the vibron-electron model and, in particular, to the case of many electrons in the united atoms valence shell.

The normalized vibron model intrinsic states are defined by[14]

$$|N;\ r>= \frac{1}{\sqrt{N!\,(1+r^2)^N}}(s^+ + rp_0^+)^N|0>_v\ , \tag{4.1}$$

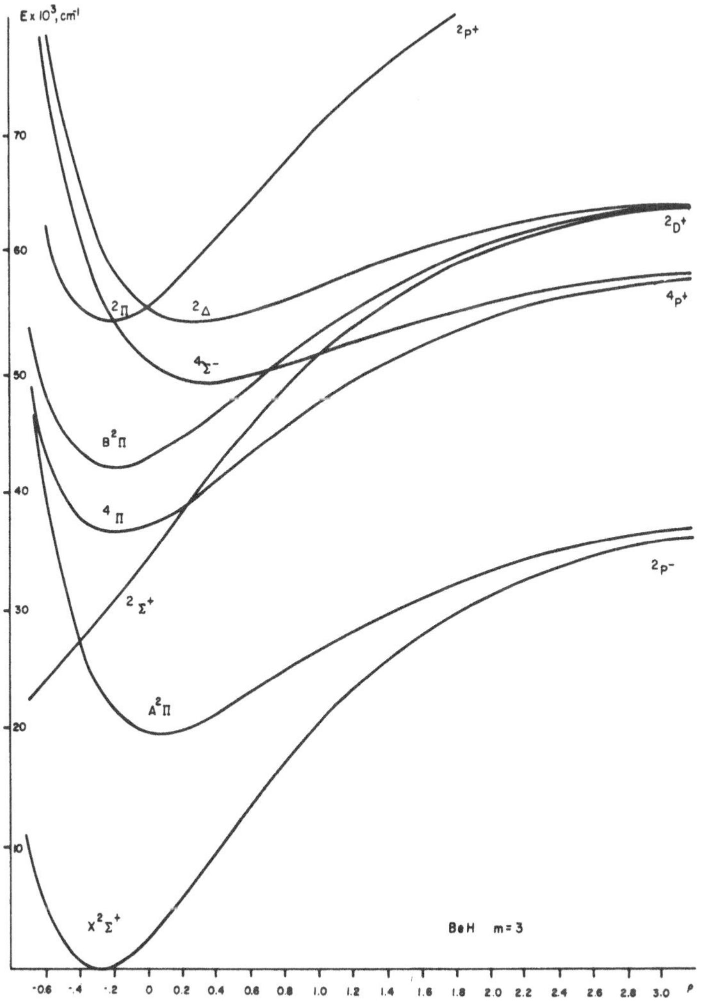

Figure 7. Theoretical potential energy curves for BeH according to (4.4).

where r is a radial coordinate and $|0>_v$ is the vibron vacuum state. It has been shown by Levit and Smilansky that the expectation value of the $O^{rv}(4)$ vibron Hamiltonian in the intrinsic states (4.1), followed by the change of variable

$$r = \sqrt{\frac{e^{-\rho}}{2 - e^{-\rho}}} , \tag{4.2}$$

leads to a Morse potential in the ρ coordinate, which should be interpreted as the physical radial variable.[15] Using the simple result

$$[b_i, f(b^+)] = \frac{\partial}{\partial b_i^+} f(b^+) , \tag{4.3}$$

for any function f of the boson creation operators, we find the following expression for the expectation value of the full molecular Hamiltonian:

$$\begin{aligned} < N; r|\hat{H}|N; r >= E_0 + (\beta_6 + A_1)[3N + \frac{4N(N-1)r^2}{(1+r^2)^2}] \\ + f_3\hat{n}_F + (f_6 + A_1)\hat{C}_2(O^e(4)) + f_4\hat{Q}_e \cdot \hat{Q}_e + A_2\hat{L}_e^2 \\ + A_1\frac{4Nr}{(1+r^2)}\hat{D}_{e,0} - 2k\sqrt{\frac{2}{3}}\frac{Nr^2}{(1+r^2)}\hat{Q}_{e,0} , \end{aligned} \tag{4.4}$$

which yields an electronic Hamiltonian with r-dependent interactions. Following the

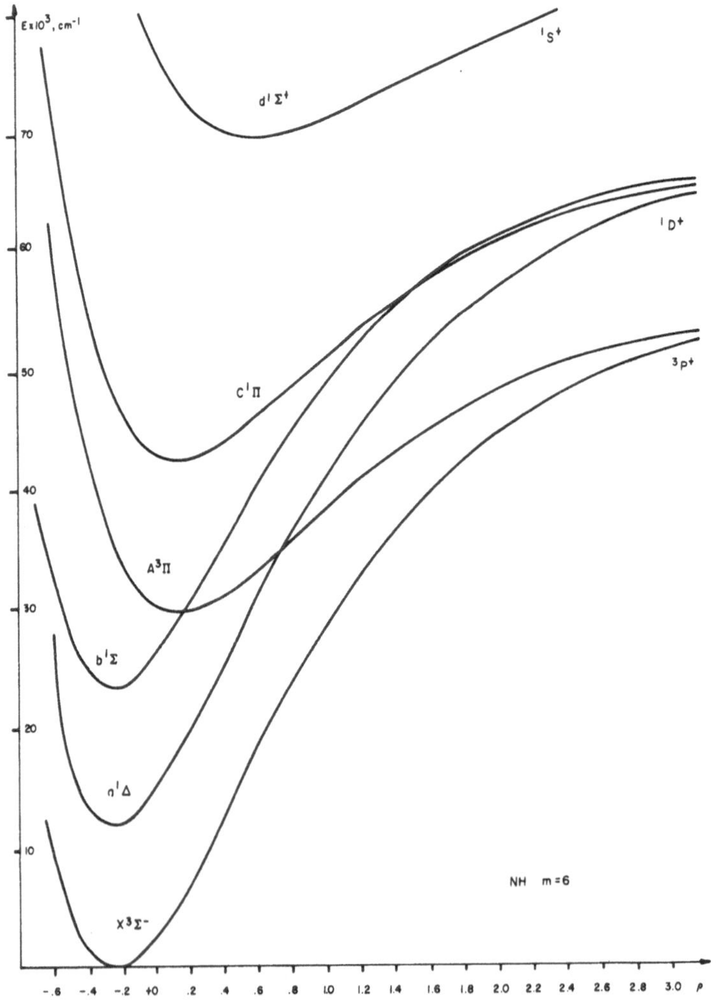

Figure 8. Theoretical potential energy curves for NH according to (4.4).

Born-Oppenheimer procedure, we may now diagonalize (4.4) in the $O^e(4)$ basis

$$|[1^m](\tau_1\tau_2)l_e m_e; S\sigma > , \tag{4.5}$$

for which we first compute the matrix elements of the electronic operators in (4.4). Again, these m.e. can be obtained in closed form.[9,15] Note that in the diagonalization $|m_e|$ may be indentified with the standard projection Λ. Also, one should carry out the change of variable (4.2) before diagonalization in order to find the physical potential energy curves. In Figs. 7 and 8 we show the results of this calculation for the BeH and NH molecules. It is remarkable that the position of the minimum in all the potentials coincides to a close approximation with the energy of the ground state configurations in the exact calculations. The Morse-like potentials are distorted by the electronic interactions in different proportions and the minima are displaced. Note also that some potentials are not bound, as the $^2\Sigma^+$ state in BeH, which is a state that has not been experimentaly observed (see Fig. 4). For comparison, in Fig. 9 we include the potential energy curves for a standard Born-Oppenheimer calculation in NH.[16]

A drawback of our calculation is that we are not able to reproduce the large ρ behavior of the potentials, since by construction we recover the united atoms limit for $\rho \to \infty (r \to 0)$. At smaller velues of the radial coordinate, though, we seem to be able to reproduce the standard potentials. A more precise correlation can only be established by means of the separated atoms limit in our model.[17]

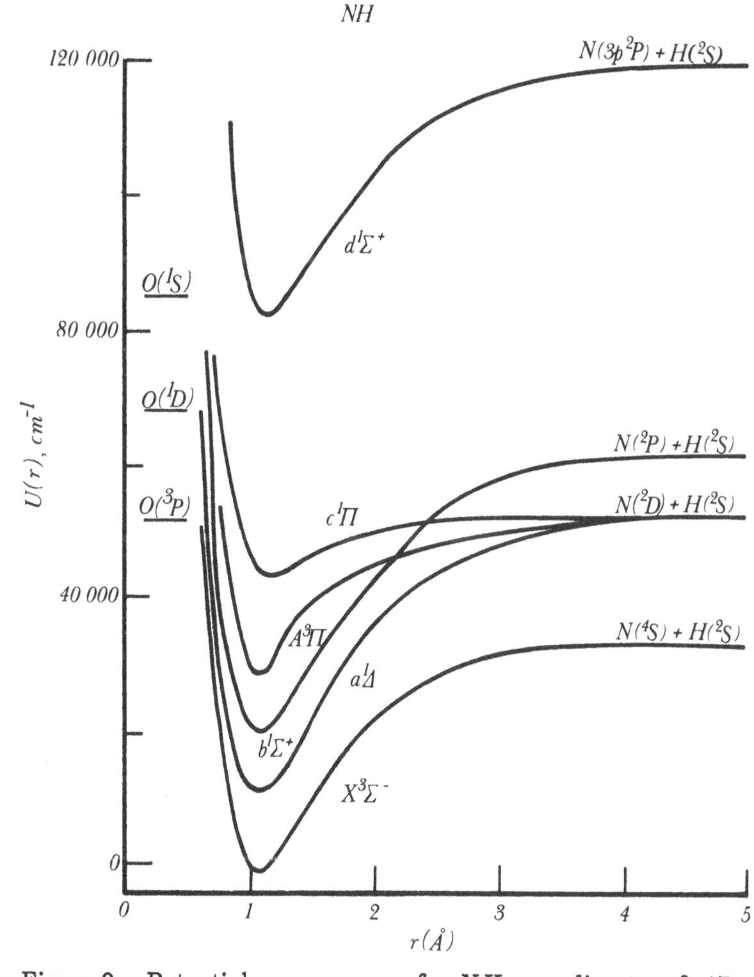

Figure 9. Potential energy curves for NH according to ref. 17.

5. CONCLUSIONS

We have presented an algebraic model capable of describing the rotation-vibration and electronic molecular spectra in a unified way. Although we have restricted our attention to heteronuclear diatomic molecules in the united atoms limit, the model can incorporate the description of homonuclear molecules, where the $U(3)$ basis gives a complete classification of states.[9]

The method correlates the spectroscopic properties of series of molecules and their ions and makes predictions for states not yet experimentally observed. Since manageable wave functions are obtained, one may compute other properties or use them in the analysis of scattering data. A coherent state approach has been devised which gives rise to potential energy curves for the electronic states.

We are presently working on the development of the separated atoms limit and the introduction of discrete symmetries in the formalism.[18] These are necessary steps for the application of the model to polyatomic molecules.

We thank A. Leviatan for his collaboration in the development of the coherent state formalism and R.D. Levine, E. Eyler and P. Van Isacker for many discussions. This work was supported in part by UNAM-DGAPA-IN10-1889 and by DOE contract number DE-A CO2-76-ER3074.

REFERENCES

1. F. Iachello, Chem. Phys. Lett. 78, 581 (1981); F. Iachello and R.D. Levine, J. Chem. Phys. 77, 576 (1982).
2. O.S. Van Roosmalen, A.E.L. Dieperink and F. Iachello, Chem. Phys. Lett. 85, 32 (1982).
3. O.S. Van Roosmalen, F. Iachello, R.D. Levine and A.E.L. Dieperink, J. Chem. Phys. 79, 2515 (1983).
4. J. Hornos and F. Iachello, J. Chem. Phys. 90, 5284 (1989); F. Iachello and S. Oss, J. Molec. Spect. (to be published).
5. R. Bijker, R.D. Amado, and D.A. Spanow, Phys. Rev. A33, 871 (1986); R. Bijker and R.D. Amado, Phys. Rev. A34, 71 (1986).
6. "The Interacting Boson-Fermion Model", Ed. F. Iachello, Plenum Press.
7. A. Frank, F. Iachello and R. Lemus, Chem. Phys. Lett. 131, 380 (1986).
8. A. Frank, R. Lemus and F. Iachello, J. Chem. Phys. 91, 29 (1989).
9. R. Lemus and A. Frank, Ann. of Phys. (to be published).
10. E. Chacón, M. Moshinsky, O. Novaro and C. Wulfman, Phys. Rev. A 3, 166 (1971).
11. R. Lemus, A. Frank and F. Iachello (to be published).
12. S.V. O'Neil and H.F. Schacfer III, J. Chem. Phys. 55, 394 (1971); A. Banerjee and F. Grein, ibid. 66, 1054 (1977); W.M. Huo, ibid. 49, 1482 (1968).
13. K.P. Huber and B. Herzberg, Molecular Spectra and Molecular Structure IV. Constants of Diatomic Molecules (Van Nostrand, NY, 1979).
14. O.S. van Roosmalen, Ph.D. Thesis, University of Groningen, The Netherlands, 1982.
15. S. Levit and U. Smilasky, Nuclear Physics A389, 56 (1982); A. Leviatan and M.W. Kirson, Annals of Physics 188, 142 (1988).
16. R. Lemus, A. Frank, F. Iachello and A. Leviatan (to be published).
17. Spectroscopic Data, Volume 1 Heteronuclear Diatomic Molecules, Edited by S.N. Suchard. Plenum.
18. R. Lemus and A. Frank, this conference.
19. R.W.B. Pearse, The Identification of Molecular Spectra (Chapman and Hall, London, 1976).
20. Atomic Energy Levels. Volume I. United States Department of Commerce. National Bureau of Standards.

HIGHEST WEIGHT UNITARY MODULES FOR NON-COMPACT GROUPS AND APPLICATIONS TO PHYSICAL PROBLEMS

Juan García-Escudero[a] and Miguel Lorente
Departamento de Física, Universidad de Oviedo
33007 Oviedo, Spain

I. INTRODUCTION

The study of unitarization of representations for non compact real forms of simple Lie Algebras has been achieved in the past decade by Jakobsen (JA81, JA83) and by Enright, Howe and Wallach (EH83) following different paths but arriving at the same final results.

In order to discuss unitarity we need to introduce a scalar product. This is done in sections II and III introducing a sesquilinear form on the universal enveloping algebra (GL90). Such a sesquilinear form was introduced by Harish-Chandra (HC55), Gel'fand and Kirillov (GK69) and Shapovalov (SH72).

The new developments given to Jakobsen method in GEL90 are contained in sections IV and V. In section VI we summarize the principal results due to Enright Howe and Wallach (EHW method). In section VII we give the possible places for unitarity including those for which the reduction level can't be higher than one and that were not considered in GEL90. We see also in this section and in a explicit way how the Jakobsen method and the EHW method give the same final results.

This type of representations have found applications in physics for a long time (see GK75, 82; GL83, 89; LO86, 89 and references contained therein)

In sections VIII and IX we consider the conformal and de Sitter algebras. In particular the construction of wave equations for conformal multispinors and the wave equations in de Sitter space are reviewed in sections X and XI.

In the Appendix we give the representations of the classical algebras in the subspace Ω - extending the results obtained in LG84 for the algebra A_l.

[a]This work contains part of the Doctoral Thesis written by one of us (J.G.E)

II. PRELIMINARIES

Let g be a semisimple Lie algebra over $\mathcal{R}$ and $g^{¢}$ its complexification. Let $B(X,Y) = \mathrm{tr}(\mathrm{ad}X\ \mathrm{ad}Y)$, $X,Y \in g^{¢}$ be the Killing form. A real form g_0 of $g^{¢}$ is called compact if $B(X,X) < 0$ for each $X \in g_0$ and an automorphism θ of $g^{¢}$ exists such that

$$\theta g_0 \subset g_0 \quad , \quad \theta g \subset g \text{ and } g = k + p \ , \ g_0 = k + ip$$

where $i = \sqrt{-1}$, k is the set of all $X \in g$ such that $\theta X = X$ and p is the set of all $Y \in g$ such that $\theta Y = -Y$.

Let $k^{¢}$ and $p^{¢}$ be the subspaces of $g^{¢}$ spanned by k, p respectively over $¢$. It holds

$$[k^{¢}, k^{¢}] \subset k^{¢} \ , \ [k^{¢}, p^{¢}] \subset p^{¢} \ , \ [p^{¢}, p^{¢}] \subset k^{¢}.$$

Let h be a Cartan subalgebra of g and $h^{¢}$ the complexification of h. Then $h^{¢}$ is a Cartan subalgebra of $g^{¢}$ and, for the cases considered here (hermitian symmetric spaces of non compact type), holds

$$[h^{¢}, k^{¢}] \subset k^{¢} \quad , \quad [h^{¢}, p^{¢}] \subset p^{¢}.$$

For given $g^{¢}$, $h^{¢}$, let Δ be the root system of $g^{¢}$ and Δ^+ the system of positive roots. We say that α is compact if $E_\alpha \in k^{¢}$ and non compact if $E_\alpha \in p^{¢}$. The set of compact and non compact roots of $g^{¢}$ with respect to $h^{¢}$ are denoted by Δ_c and Δ_n respectively. The set of compact simple roots is denoted by Σ_c, β is the only non compact simple root and γ_r is the highest root (which is a non compact positive root)

Let $k_1 = [k, k]$ and assume that k has a non empty center η of dimension one. Then $k = k_1 \oplus \eta$ and $h = (h \cap k_1) \oplus \eta$. On the other hand $h^{¢} = (h \cap k_1)^{¢} \oplus \eta^{¢}$ is an orthogonal direct sum with respect to the Killing form: for if $H_\mu \in (h \cap k_1)^{¢}$ and $H_0 \in \eta^{¢}$

$$(H_\mu, H_0) = ([E_\mu, E_{-\mu}], H_0) = (E_\mu, [E_{-\mu}, H_0]) = 0.$$

For $\gamma_1, \gamma_2 \in \Delta$ we use the notation

$$\langle \gamma_1, \gamma_2 \rangle = \frac{2(\gamma_1, \gamma_2)}{(\gamma_2, \gamma_2)} = \gamma_1(H_{\gamma_2})$$

where $(.\,,.)$ is the bilinear form on $(h^{¢})^*$ induced by the Killing form on $g^{¢}$.

Let $\mathrm{U}(g^{¢})$ be the universal enveloping algebra of $g^{¢}$, $\Lambda \in (h^{¢})^*$ and $R = \frac{1}{2} \sum_{\alpha \in \Delta^+} \alpha$.

The Verma module M_Λ of highest weight Λ (VE68) is defined to be $M_\Lambda = \mathrm{U}(g^{¢})/I_\Lambda$ where I_Λ is the left ideal generated by the elements $(H - \Lambda(H))$, $H \in h^{¢}$ and the set of generators $X\gamma$ with $\gamma \in \Delta^+$. To fix a basis on $(h^{¢})^*$ we choose the set of compact simple roots Σ_c for the space $((h \cap k_1)^{¢})^*$ and one element $\varepsilon \in (\eta^{¢})^*$ for which

$$\langle \varepsilon, \mu_i \rangle = 0 \ , \ \forall \mu \in \Sigma_c \text{ and } \langle \varepsilon, \gamma_r \rangle = 1 \ ,$$

then each $\Lambda \in (h^{¢})^*$ may be written as $\Lambda = \Lambda_0 + \lambda\varepsilon$, where Λ_0 satisfies $\langle \Lambda, \mu_i \rangle = \langle \Lambda_0, \mu_i \rangle$ $\forall \mu_i \in \Sigma_c$. If we choose a normalization for Λ_0 of the type $\langle \Lambda_0, \gamma_r \rangle = 0$, from the last decomposition of Λ we conclude that $\langle \Lambda, \gamma_r \rangle = \lambda$. The relations $\langle \Lambda, \mu_i \rangle = \langle \Lambda_0, \mu_i \rangle$ and $\langle \Lambda_0, \gamma_r \rangle = 0$ fix Λ_0 uniquely. In the following we consider Λ_0 to be k_1-dominant and integral, that is $\langle \Lambda_0, \mu_i \rangle = n_i$, where n_i are non negative integers.

Now, if M_Λ is a Verma module, $\widetilde{L_\Lambda}$ an invariant submodule, and $L_\Lambda = M_\Lambda / \widetilde{L_\Lambda}$ a quotient module and if $\rho_\Lambda = M_\Lambda$, $\widetilde{L_\Lambda}$, L_Λ is irreducible then we say that ρ_Λ is infinitesimally unitary if there exists a scalar product (,) on the carrier space V of ρ_Λ such that

$$(u, \rho_\Lambda (X) w) = - (\rho_\Lambda (X)u, w)$$

for all $X \in g$ and $u, w \in V$. The above condition is called g- invariance.

In a Verma module this scalar product is induced by a sesquilinear form.

III. SCALAR PRODUCT ON THE ENVELOPING ALGEBRA OF A SEMISIMPLE LIE ALGEBRA

Let $g^¢$ denote a complex semisimple Lie algebra of rank l with diagonal elements H_i $(i = 1, 2, \dots l)$ for its Cartan subalgebra, shift operators E_α associated to each positive root α, shift operators $E_{-\alpha}$ associated to each negative root-α. Then the following canonical commutation relations hold:

$$[H_i, E_{\pm\alpha}] = \pm\alpha_i E_{\pm\alpha}$$

$$[E_\alpha, E_{-\alpha}] = \sum_1^l \alpha_i H_i \equiv H_\alpha$$

$$[E_\alpha, E_\beta] = \begin{cases} N_{\alpha\beta} E_{\alpha+\beta}, & \text{if } \alpha + \beta \text{ is a root} \\ 0, & \text{if } \alpha + \beta \text{ is not a root } (\alpha + \beta \neq 0) \end{cases}$$

where α_i denotes the i component of the root α.

According to the Poncaré-Birkchoff-Witt theorem (DI74) a basis for the universal enveloping algebra $u(g^¢)$ of $g^¢$ can be chosen as the following set of ordered tensor products of the vectors H_α, E_α, $E_{-\alpha}$ namely,

$$\Omega = \left\{1, E^m_{-\alpha} E^n_{-\beta} \cdots E^p_{-\delta} H^q_\alpha H^r_\beta \cdots H^s_\delta E^t_\alpha E^u_\beta \cdots E^v_\delta\right\}$$

$$\equiv X(m, n, \dots, p, q, r, \dots, s, t, u, \dots, v)$$

where $m, n, p, q, r, s, t, u, v$, are non negative integers. The symbol 1 represents the identity element of the enveloping algebra, i.e. when all exponents are equal to zero.

The basis for the enveloping algebra can be written as

$$\Omega = \Omega_- \Omega_0 \Omega_+$$

where Ω_- is the enveloping algebra of the vectors $E_{-\alpha}$ associated with negative roots, Ω_+ is the enveloping algebra of the vectors E_α associated with positive roots, and Ω_0 is the enveloping algebra of the elements H_α of the Cartan subalgebra, $h^¢$.

Now for the elements X of the basis Ω we define the mapping $\gamma: \Omega \to \Omega_0$ by

$$\gamma(X) = \begin{cases} X, & \text{if } X \in \Omega_0 \\ 0, & \text{otherwise}. \end{cases}$$

Let X be a general element of the enveloping algebra $u(g^¢)$ expressed in terms of the basis Ω

$$X = \sum c_{m\,n\;p\,q\,r\,s\,t\,u\,v} X(m, n, \dots, p, q, r, \dots, s, t, u, \dots, v)$$

where the $c's$ are complex numbers; then

$$\gamma(X) = X, \quad \text{if} \quad X = \sum c_{o,\dots,\,q,r,s,o\,\dots} X(o, \dots, o, q, r, \dots, s, o, \dots)$$

$$\gamma(X) = 0, \quad \text{otherwise}$$

defines a projection of the elements of $u(g^¢)$ into the subspace expanded by the basis elements of Ω_0.

If we adscribe a weight μ to the elements of Ω by the definition

$$\mu = (t-m)\alpha + (u-m)\beta + \ldots + (v-p)\,\delta$$

it is obvious that $\gamma = 0$ if $\mu \neq 0$.

Let Λ denote a linear form in the dual space of the Cartan subalgebra such that

$$\Lambda(H_\alpha) = \frac{2(\Lambda,\alpha)}{(\alpha,\alpha)}$$

where (,) is the scalar product in the dual space $(h^{¢})^*$ induced by the Cartan-Killing form. Therefore Λ must be considered as a vector (with complex componentes) in the l - dimensional dual space. We will use the composition $\Lambda\cdot\gamma \equiv \xi_\Lambda$.

Given a real semisimple Lie algebra g and its complexification $g^{¢}$, let σ denote the conjugation of $g^{¢}$ with respect to g:

$$\sigma : g^{¢} \rightarrow g^{¢},$$

$$\sigma(X + iY) = X - iY \quad \text{for} \quad X,Y \in g.$$

Then $-\sigma$ can be extended to an antilinear antiautomorphism η of $u(g^{¢})$ as follows:

$$\eta(1) = 1,$$

$$\eta(X) = -\sigma(X),$$

$$\eta(XY\ldots Z) = \eta(Z)\ldots\eta(Y)\,\eta(X).$$

We can define a sesquilinear form S on $u(g^{¢})$ x $u(g^{¢})$ for any $\Lambda \in h^*$ as follows:

$$S : u(g^{¢}) \text{ x } u(g^{¢}) \rightarrow C,$$

$$S(X,Y) = \xi_\Lambda(\eta(X)Y), \quad X,Y \in u(g^{¢}).$$

In general, i.e., on $u(g^{¢})$ and for arbitrary Λ, S is degenerate and indefinite for a given g. However, for some Λ it may induce a scalar product on irreducible quotients of Verma modules of highest weight Λ and unitarize the finite-dimensional irreducible representations if g is compact (see HC55). On the other hand, if g is noncompact, then for some Λ it may induce a scalar product on irreducible Verma modules or on irreducible submodules of Verma modules of highest weight Λ and unitarize infinite-dimensional irreducible representations of g (KV78, HC55).

If V is the irreducible quotient of a Verma module, then π_Λ is infinitesimally unitary with respect to the scalar product induced by S if $\xi_\Lambda(\eta(z)z)$ is real and non-negative for every $z \in V$ (see HC55).

IV. JAKOBSEN METHOD

In the following we are going to reformulate the Jakobsen method to calculate the modules M_Λ that are unitarizable by using a diagramatic representation of Δ_n^+

The modules M_Λ are determinated by Λ_0 and λ where Λ_0 is k_1-dominant and integral and $\lambda \in \mathcal{R}$.

There exists a way to represent the set Δ_n^+ by means of bidimensional diagrams in the following way: one begins with β and draws an arrow originating at β for each compact simple root μ_i such that $\beta + \mu_i \in \Delta_n^+$.

Lemma 4.1 : of JA83 shows that $i \leq 2$. We suppose for simplicity that $i = 2$. Then one draws two arrows: one originating at $\beta + \mu_1$ and parallel to μ_2 and another originating at

$\beta+\mu_2$ and parallel to μ_1, both arrows point towards $\beta+\mu_1+\mu_2$ which is also a root. The next step would be to add compact simple roots to the non compact roots previously obtained by keeping those that are non compact roots. Continuing along these lines the diagram may be completed.

For the description of the possible places for unitarity, Jakobsen uses the Bernshtein, Gel'fand and Gel'fand theorem (BG71). This theorem describes the circunstances under which the irreducible quotient L_ξ of a highest weight module can occur in the Jordan-Hölder series JH(M_Λ) of another:

Definition 1: Let $\xi, \Lambda \in (h^{¢})^*$. A sequence of roots $\alpha_1,\ldots, \alpha_k \in \Delta^+$ is said to satisfy condition (A) for the pair (ξ+R, Λ+R) if

a) $\xi+R=\sigma_{\alpha_K}\ldots\sigma_{\alpha_1}(\Lambda+R)$ where σ_{α_i} is the Weyl reflexion with respect to α_i

b) Take $\xi_0\equiv\Lambda$, $\xi_i+R = \sigma_{\alpha_i}\ldots\sigma_{\alpha_1}(\Lambda+R)$

Then $\xi_{i-1}-\xi_i = n_i\,\alpha_i$, $n_i\in\mathcal{N}$.

Theorem 2: (Bernshtein, Gel'fand and Gel'fand); Let ξ , $\Lambda\in(h^{¢})^*$ and let L_ξ, M_Λ two Verma modules. Then $L_\xi\in$ JH(M_Λ) if and only if there exists a sequence $\alpha_1,\ldots,\alpha_k\in\Delta^+$ satisfying condition (A) for the pair (ξ+R, Λ+R)

On the other hand, under some conditions the α_i's may be considered as non compact ones:

Proposition 3: Let $\xi, \Lambda\in(h^{¢})^*$ and assume that the sequence $\alpha_1,\ldots, \alpha_k$ satisfies condition (A) for the pair (ξ+R, Λ+R). If ξ is k_1-dominant we may assume that $\alpha_i\in\Delta_n^+$, $i = 1\ldots k$.

Let V_{Λ_0} be an irreducible finite-dimensional $\mathfrak{u}(k_1{}^{¢})$-module with highest weight Λ_0. We first consider the $\mathfrak{u}(k_1{}^{¢})$-module $p^-\otimes V_{\Lambda_0}$. The highest weights on $p^-\otimes V_{\Lambda_0}$ are of the form $\Lambda_0-\alpha$ for certain $\alpha\in\Delta_n{}^+$ which we will describe in terms of the Jakobsen diagrams.

We now describe the method:

i) Let $\alpha\in\Delta_n^+$ and assume $\alpha-\mu_j\in\Delta_n^+$ for $\mu_j\in\Sigma_c$; $j=1,\ldots,i$ and $i\leq 2$.
Then $\Lambda_0-\alpha$ is a highest weight for the $\mathfrak{u}(k_1{}^{¢})$-module $p^-\otimes V_{\Lambda_0}$ if and only if for all $j=1,\ldots, i$ $\quad \Lambda_0(H_{\mu_j})\equiv\langle\Lambda_0,\mu_j\rangle\geq\max\{1,\langle\alpha,\mu_j\rangle\}$
Recall that Λ_0 is fixed by given integers $\langle\Lambda_0,\mu_i\rangle$, $\mu_i\in\Sigma_c$ and $\langle\Lambda_0,\gamma_r\rangle=0$.

ii) For those $\alpha\in\Delta_n^+$ of step i) let $\lambda_\alpha\in\mathcal{R}$ be determined by the equation

$$\langle\Lambda+R,\ \alpha\rangle=(\Lambda_0+\lambda_\alpha\varepsilon+R)\,(H_\alpha)=1.$$

Let λ_0 denote the samallest among those λ_α's , and let α_0 denote the corresponding element of Δ_n^+. We now define the following sets:

$$C^+_{\alpha_0}=\{\alpha\in\Delta_n^+/\alpha\geq\alpha_0\}\quad\text{and}\quad C^-_{\alpha_0}=\{\alpha\in\Delta_n^+/\alpha\leq\alpha_0\}.$$

The way in which those sets appear in the diagram of Δ_n^+ suggest that we can call $C^+_{\alpha_0}$ and $C^-_{\alpha_0}$ the forward and backward cone, respectively, at α_0.

iii) Let $\omega_q=n_1\,\alpha_1+\ldots+n_r\,\alpha_r$, $n_i\in\mathcal{N}$. If $\alpha_1,\ldots,\alpha_r\in\Delta_n^+$ satisfies condition (A) for the pair $(\Lambda-\omega_q+R\,,\Lambda+R\,)$ where $\Lambda=\Lambda_0+\lambda_q\varepsilon$ and $\Lambda_0-\omega_q$ is the weight of a highest weight vector q in the $\mathfrak{u}(k_1{}^{¢})$-module $\mathfrak{u}(p^-)\otimes V_{\Lambda_0}$ and $\lambda_q<\lambda_0$, then

$$\alpha_i\in C^+_{\alpha_0}\ \forall i=1\ldots r.$$

iv) The α_i's appearing in ω_q must satisfy certain conditions which we now describe.
Because inner products between positive noncompact roots are non negative and because

$\lambda_q < \lambda_0$, it follows that for $\alpha_i \in \Delta_n^+$

$$\langle \Lambda_0 + \lambda_0\varepsilon + R\, ,\, \alpha_i \rangle > \langle \Lambda_0 + \lambda_q\varepsilon + R\, ,\, \alpha_i \rangle > 0.$$

On the other hand, to check the k_1-dominance of $\Lambda_0 - \omega_q$ i.e.

$$\langle \Lambda_0 - \omega_q\, ,\, \mu_j \rangle \geq 0 \qquad \forall\, \mu_j \in \Sigma_c$$

is useful to have in mind that if a compact simple root μ is pointing towards a non compact positive root α in the diagram then $\langle \alpha, \mu \rangle > 0$ and if μ arises outwards α then $\langle \alpha, \mu \rangle < 0$ provided that the same μ both is not pointing towards and arises outwards α.

v) M_Λ with $\Lambda = \Lambda_0 + \lambda_0\varepsilon$ is unitarizable. The value $\lambda = \lambda_0$ is called the last possible place for unitarity, because for $\lambda > \lambda_0$ there is no unitarity. The description of the general situation follows by forming tensor products of M_Λ with the unitary module $M_{\lambda_s\varepsilon}$ corresponding to $\Lambda_0 = 0$. The restriction of $M_\Lambda \otimes M_{\lambda_s\varepsilon}$ to the diagonal is the unitarizable module $M_{\Lambda'}$ with $\Lambda' = \Lambda_0 + (\lambda_0 + \lambda_s)\varepsilon$.

This means that if we want to unitarize we must take the quotient space with respect to the invariant subspace generated by the highest weight vector corresponding to $\lambda_0 + \lambda_s$ which is a second order polynomial: this polynomial will be missing.

The modules $M_{\Lambda''}$ with $\Lambda'' = \Lambda_0 + \lambda\,\varepsilon$, $\lambda_0 + \lambda_s < \lambda < \lambda_0$ are not unitarizable.

For $\lambda_0 + 2\lambda_s$ we may have unitarity and there will be a third order missing polynomial, while there is no unitarity for $\lambda_0 + 2\lambda_s < \lambda < \lambda_0 + \lambda_s$.

Continuing along these lines we arrive at the first possible place for non unitarity which corresponds to $\lambda = \lambda_0 + u\lambda_s$ (we call $u+1$ the reduction level) and all representations with $\lambda < \lambda_0 + u\lambda_s$ are unitary.

The following diagram illustrates the possible places for unitarity (in the next section we will see that $\lambda_s < 0$).

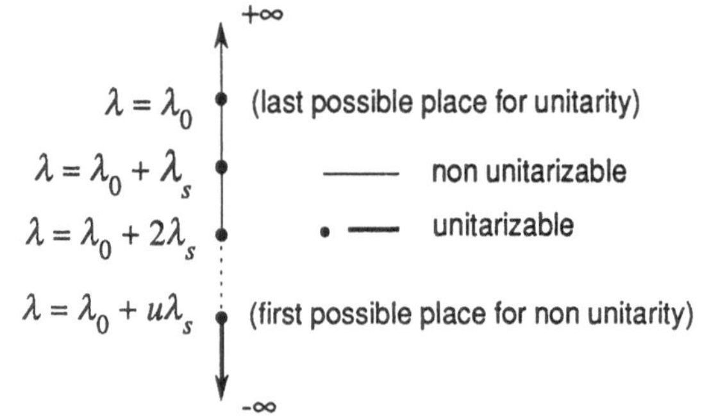

V. SOME DEFINITIONS AND NOTATIONS

From the Jakobsen diagrams (see the examples in Section V) we observe that all positive noncompact root can be expresed as

$$\alpha = \beta + \mu_{i_1} + \ldots + \mu_{i_k} \ , \quad \mu_{i_m} \in \Sigma_c \quad m = 1, \ldots k.$$

Thus, given α on this way we define its "height" $\mathcal{H}\alpha$ as $k+1$ (the roots μ_{i_m} may be repeated).

On the other hand given the decomposition $\Lambda = \Lambda_0 + \lambda\varepsilon$ we may relate the products $\langle \Lambda, \alpha \rangle$ and $\langle \Lambda_0, \alpha \rangle$, $\alpha \in \Delta_n^+$, in the following way

$$\langle \Lambda, \alpha \rangle = \langle \Lambda_0, \alpha \rangle + \lambda \frac{(\gamma_r, \gamma_r)}{(\alpha, \alpha)}$$

In fact

$$\langle \Lambda, \alpha \rangle = \langle \Lambda_0, \alpha \rangle + \lambda \langle \varepsilon, \alpha \rangle$$

and, decomposing $\alpha = \gamma_r - \sum_{\mu_i \in \Sigma_c} \mu_i$ (see Jakobsen diagrams) then

$$\langle \varepsilon,\alpha\rangle = \frac{2(\varepsilon,\gamma_r)}{(\alpha,\alpha)} = \frac{(\gamma_r,\gamma_r)}{(\alpha,\alpha)}$$

where we use the fact that $\langle \varepsilon, \gamma_r \rangle = 1$ and $\langle \varepsilon, \mu \rangle = 0 \quad \forall \mu \in \Sigma_C$.

By means of a direct calculation we obtain the following useful expresions for the products $\langle R,\alpha \rangle$ and $\langle \Lambda,\alpha \rangle$ that will be needed in next section.

a) su (p , q) , so*(2n) , so (2n-2, 2) , e_6 and e_7

$$\langle R, \alpha \rangle = \mathcal{H}_\alpha$$

$$\langle \Lambda, \alpha \rangle = (\Lambda_0 , \alpha) + \lambda .$$

b) sp (n , $\mathcal{R}$)

If α is short $\langle R, \alpha \rangle = \mathcal{H}_\alpha + 1$

$$\langle \Lambda, \alpha \rangle = (\Lambda_0 , \alpha) + 2\lambda .$$

If α is long $\langle R, \alpha \rangle = \frac{1}{2} \{\mathcal{H}_\alpha + 1\}$

$$\langle \Lambda, \alpha \rangle = \frac{1}{2} (\Lambda_0 , \alpha) + \lambda .$$

c) so (2n-1, 2)

Let α_1 be the short noncompact positive root:

$$\text{If } \alpha \text{ is long} \begin{cases} \text{i) } C^+_\alpha \subset C^+_{\alpha_1} & \begin{cases} \langle R, \alpha\rangle = \mathcal{H}_\alpha + 1 \\ \langle \Lambda, \alpha\rangle = (\Lambda_0 , \alpha) + \lambda \end{cases} \\ \text{ii) } C^+_{\alpha_1} \subset C^+_\alpha & \begin{cases} \langle R, \alpha\rangle = \mathcal{H}_\alpha \\ \langle \Lambda, \alpha\rangle = (\Lambda_0 , \alpha) + \lambda \end{cases} \end{cases}$$

$$\text{If } \alpha = \alpha_1 \begin{cases} \langle R, \alpha\rangle = 2\mathcal{H}_\alpha - 1 \\ \langle \Lambda, \alpha\rangle = 2(\Lambda_0 , \alpha) + 2\lambda . \end{cases}$$

From the Jakobsen diagrams we see that all roots of the same height are in an horizontal line.

In the figure 1 such a roots are inside a little circle and we may localize them by means of a subindex j which is equal to the height of the root and an ordenation superindex i which is equal to one, for the root placed at the right branch of the cone generated by α_0, and it increases from unit to unit when we are going toward the left branch. In this way we will write α^i_j .

In order to calculate the parameter λ_s in step **v)** we make use of the following.

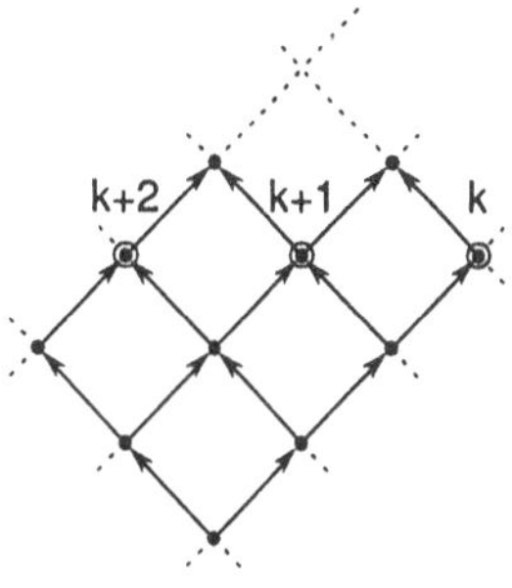

Fig. 1

Definition 4 : (Harish-Chandra) : Let γ_1 be the smallest element of Δ^+_n and, inductively, let γ_k be the smallest element of Δ^+_n that is orthogonal to $\gamma_1, \ldots ,\gamma_{k-1}$. Let $\gamma_1, \ldots ,\gamma_t$ be the maximal collection obtained. Then t is the split-rank of g .

With our notation $\gamma_1 \equiv \beta$. We use the Jakobsen diagrams to obtain the split rank.

su(p,q)

The collection $\gamma_1, \ldots ,\gamma_t$ follows by drawing a line from β as is indicated in figure 2.

The roots founded are those which are on the line:

$p \leq q: \; e_p - e_{p+1}, e_{p-1} - e_{p+2}, \ldots, e_1 - e_{2p}$

$p \geq q: \; e_p - e_{p+1}, e_{p-1} - e_{p+2}, \ldots, e_{p-(q-1)} - e_{p+q}$

So, if $p \leq q$ the split rank is p and if $p \geq q$ the split rank is q, therefore

Split rank of $\quad su(p, q) = \min\{p, q\}$.

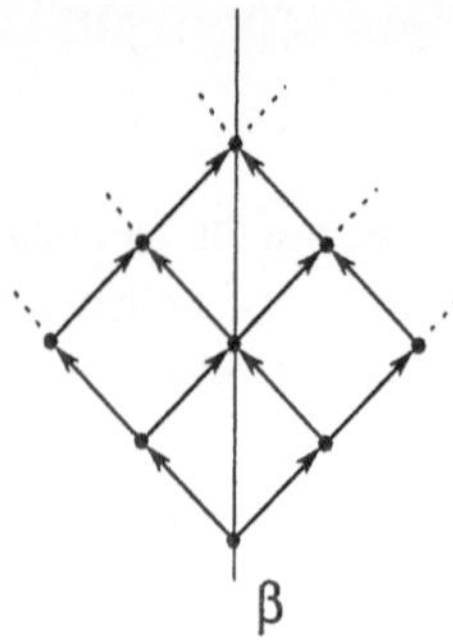

Fig. 2

sp(n,$\mathcal{R}$)

In the same way as in su (p, q) the collection obtained here is

$$2e_n, 2e_{n-1}, \ldots, 2e_1$$

Thus

$$\text{Split rank } sp(n,\mathcal{R}) = n.$$

so*(2n)

The collection is, in this case

$$e_{n-1} + e_n, \ldots, e_1 + e_2, \quad \text{if } n \text{ is even}$$
$$e_{n-1} + e_n, \ldots, e_2 + e_3, \quad \text{if } n \text{ is odd}$$

Then the split rank is $\frac{n}{2}$ if n is even and $\frac{n-1}{2}$ if n is odd:

$$\text{Split rank } so^*(2n) = \left[\frac{n}{2}\right]$$

where $[x]$ denotes the largest integer $\leq x$.

so(2n-1, 2) , so(2n-2, 2)

The split rank is, in both cases, equal to two. The collection is, in this case $e_1 - e_2, e_1 + e_2$.

$\mathfrak{e}_6, \mathfrak{e}_7$

The collection obtained is now

$$\mathfrak{e}_6 : \{\tfrac{1}{2}(e_1 - e_2 - e_3 - e_4 - e_5 - e_6 - e_7 + e_8), \tfrac{1}{2}(-e_1 + e_2 + e_3 + e_4 - e_5 - e_6 - e_7 + e_8)\}$$

then the split rank of e_6 is equal to two

$$\mathfrak{e}_7 : \{e_6 - e_5, e_6 + e_5, e_8 - e_7\}$$

thus split rank of e_7 is equal to three.

Now let $h^- = \sum_{i=1}^{t} \mathbb{C} H_{\gamma_i}$ and, for $1 \leq j \leq t$, t being the split rank, let c_j be the number of compact positive roots μ such that $\mu|_{h^-} = \frac{1}{2}(\gamma_j - \gamma_i)$, $i < j$. If we consider the most singular non trivial unitary module corresponding to $\Lambda_0 = 0$, then according to Theorem 5.10 in WA79, $\lambda_q = -\frac{1}{2} c_j$. A straightforward calculation case by case shows that

$$\lambda_q = (j - 1)\lambda_s \;, \quad 1 \leq j \leq t$$

with λ_s given in the following table:

	$su(p,q)$	$sp(n,\mathcal{R})$	$so^*(2n)$	$\mathfrak{e}_6$	$\mathfrak{e}_7$	$so(2n-2,2)$	$so(2n-1,2)$
λ_s	-1	$-\frac{1}{2}$	-2	-3	-4	$-n+2$	$-n+3/2$

VI. EHW METHOD

In this section we summarize the main results given by the EHW method (EP81, EH83). Those authors use a decomposition of the highest weight in the form $\Lambda = \Pi_0 + z\,\varepsilon$, $z \in \mathcal{R}$, $\langle \Pi_0 + R, \gamma_r \rangle = 0$, Π_0 k_1-dominant and integral, ε is given in section II (as we see the only difference between this decomposition and the one used by Jakobsen is the normalization choosed for the compact part of the highest weight).

Theorem 5: The set of real numbers z such that $M_{\Pi_0+z\varepsilon}$ is a g-module unitarizable, is given in the diagram (see Fig 3).

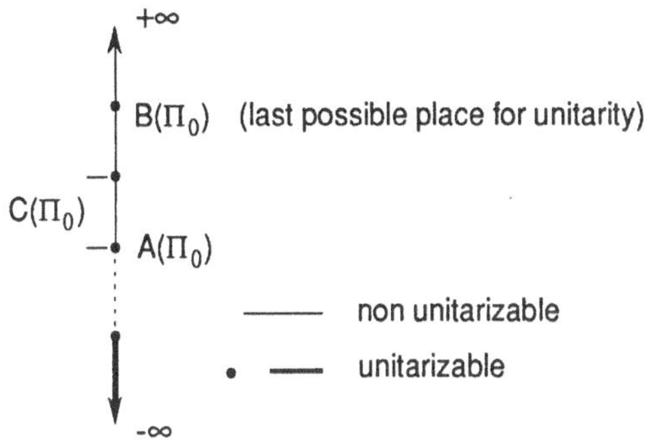

Fig. 3

a) The set includes the half line ending at $A(\Pi_0)$.
The smallest value z with $M_{\Pi_0+z\varepsilon}$ unitarizable is $z = A(\Pi_0)$, that we call the first reduction point.

b) In addition to the half line there are a number of equally spaced points in the set for which $M_{\Pi_0+z\varepsilon}$ is unitarizable. The distance between two consecutive points is $C(\Pi_0)$. The set of such a points begins at $A(\Pi_0)$ and end at $B(\Pi_0)$, and we call to its cardinal the reduction level.

In the following we will give the expressions for $A(\Pi_0)$, $B(\Pi_0)$ y $C(\Pi_0)$. To do this we associate root systems to the line $\Pi_0 + z\varepsilon$.

Let $\Delta_c(\Pi_0) = \{ \mu \in \Delta_c \;/\; \langle \Pi_0, \mu \rangle = 0 \}$ and let $\{ \pm\gamma_r , \Delta_c(\Pi_0)\}$ be the root system of Δ generated by $\pm\gamma_r$ and $\Delta_c(\Pi_0)$. Decomposing this root system into a disjoint union of simple root systems, let $Q(\Pi_0)$ be the simple root system which contains γ_r . If Δ has two lengths and if there exists short compact roots μ not orthogonal to $Q(\Pi_0)$ with $\langle \Pi_0, \mu \rangle = 1$ then let ψ be the root system generated by $\pm\gamma_r$, $\Delta_c(\Pi_0)$ and all such μ . Let $T(\Pi_0)$ be the simple component of ψ which contains γ_r . If Δ has only one length or if no such μ exist then $T(\Pi_0) = Q(\Pi_0)$.

We want to see with an example how to construct $Q(\Pi_0)$. We take the Dynkin diagram of su(p , q) and we draw a little circle around the non compact simple root β :

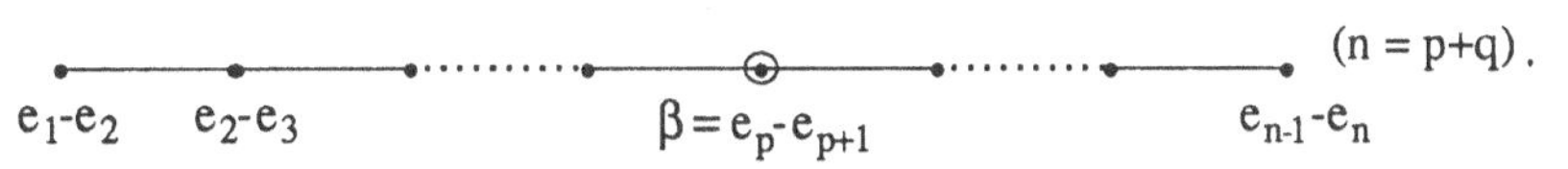

Next we omit the non compact root and we connect $-\gamma_r = e_n - e_1$ by the usual rules (Fig. 4).

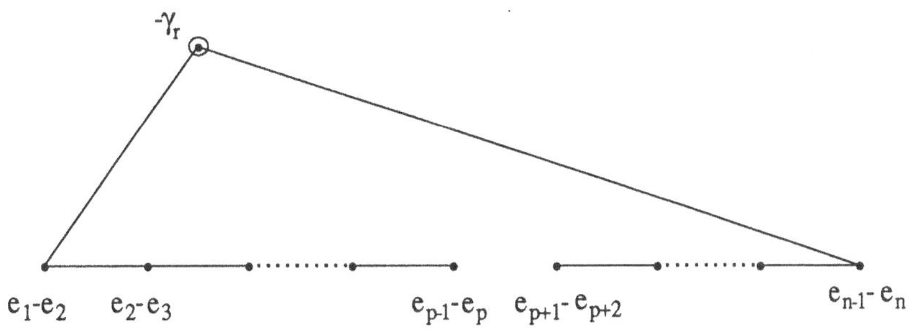

Fig. 4

Now we take the maximal connected diagram which contains $-\gamma_r$ and such that every compact simple root is orthogonal to Π_0 . This subdiagram is the diagram of $Q(\Pi_0)$.

There are only two cases where $Q(\Pi_0) \neq T(\Pi_0)$ and it corresponds to the algebras sp (n ,$\mathcal{R}$) and so (2n - 1, 2).

For the calculation of $B(\Pi_0)$ we consider $\Delta^+_{c,1} = \Delta^+_c \cap Q(\Pi_0)$ and $\Delta^+_{c,2} = \Delta^+_c \cap T(\Pi_0)$. Let $R_{c,1}$ (respectively $R_{c,2}$) be half the sum of roots in $\Delta^+_{c,1}$ (respectively $\Delta^+_{c,2}$).

Theorem 6:

a) If $g = \text{so}\ (2n - 1, 2)$ and $Q(\Pi_0) \neq T(\Pi_0)$ then $B(\Pi_0) = 1 + \langle R_{c,2}, \gamma_r \rangle$.

b) In all other cases $B(\Pi_0) = 1 + \langle R_{c,1} + R_{c,2}, \gamma_r \rangle$.

The constants $C(\Pi_0)$ only depend of g and is equal to $-\lambda_s$, where λ_s is the parameter given at the end of section V.

The first reduction point $A(\Pi_0)$ is given by:

Theorem 7:

$$A(\Pi_0) = B(\Pi_0) - \left(\text{split rank of } Q(\Pi_0) - 1\right).C(\Pi_0)$$

In the following the results for each algebra are given.

su (p , q)

We consider $\Lambda = \Pi_0 + z\varepsilon$ with $\Pi_0 = (\Pi_1, \Pi_2, \ldots, \Pi_n)$, $n = p + q$; as Π_0 is k_1-dominant and integral $\Pi_1 \geq \Pi_2 \ldots \geq \Pi_p$, $\Pi_{p+1} \geq \ldots \geq \Pi_n$ and $\Pi_i - \Pi_j \in Z^+$, $1 \leq i < j \leq p$ or $p+1 \leq i < j \leq n$. On the other hand as $\langle \Pi_0 + R, \gamma_r \rangle = 0$, $\Pi_1 - \Pi_n + n - 1 = 0$. In addition $\varepsilon = (q/n, q/n, \ldots, q/n, -p/n, \ldots, -p/n)$ with p copies of q/n and q of $-p/n$. .For su (p , q) we have:

$$2R_c = \left(p-1, p-3, \ldots, p-(2p-1), q-1, q-3, \ldots, q-(2q-1)\right).$$

We now suppose that we have the following conditions on the Π_0 components:

$$\Pi_1 = \Pi_2 = \ldots = \Pi_i \neq \Pi_{i+1}$$

$$\Pi_n = \Pi_{n-1} = \ldots = \Pi_{n-j+1} \neq \Pi_{n-j}.$$

Then the root system $Q(\Pi_0)$ has the form :

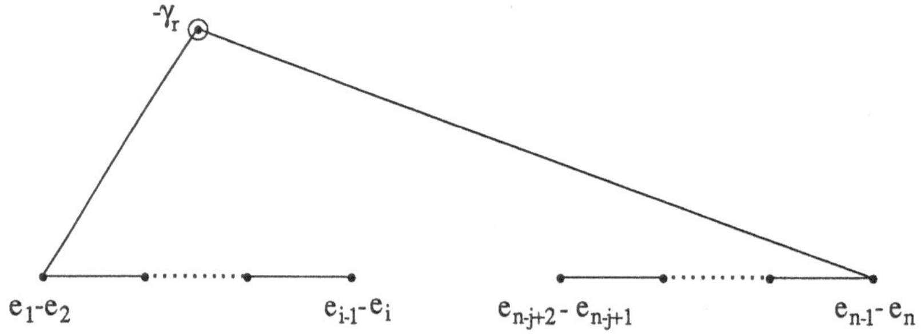

which is of type su (i , j) . Then :

$$2R_{c,1} = (i-1, \ldots, -i+1, 0, \ldots, 0, j-1, \ldots, -j+1).$$

In this case we have $T(\Pi_0) = Q(\Pi_0)$, $R_{c,1} + R_{c,2} = 2R_{c,1}$ and $\gamma_r = e_1 - e_n$; then applying Theorem 6 :

Lemma 8. The last possible place for unitarity for su (p , q) is $B(\Pi_0) = i + j - 1$

From the Theorem 7 : $A(\Pi_0) = i + j - min\ \{i , j\}$ and the reduction level is $B(\Pi_0) - A(\Pi_0) + 1$:

Lemma 9. The first possible place for non unitarity is $max\ \{i , j\}$ and the reduction level is $min\ \{i , j\}$.

Then we will have :

Theorem 10:

$M_{\Pi_0+z\varepsilon}$ is unitarizable if and only if $z \le max\,\{i\,,j\}$ or z is an integer and $z \le i + j - 1$.

sp(n , $\mathcal{R}$)

In this case $\Lambda = \Pi_0 + z\varepsilon$, $\Pi_0 = (\Pi_1 , ... , \Pi_n)$ and since it is k_1 -dominant and integral then $\Pi_i - \Pi_j \in \mathcal{N}$, $i < j$. From the condition $\langle \Pi_0 + R\,,\gamma_r\rangle = 0$ we have $\Pi_1 = -n$. On the other hand $\varepsilon = (1,1, ... ,1)$. The sum of positive compact roots is, for sp(n , $\mathcal{R}$),

$$2R_c = \big(n-1\,,n-3\,,\dots,n-(2n-1)\big).$$

We consider two cases:

Case I $\qquad \Pi_1 = \Pi_2 = ... = \Pi_i \ge \Pi_{i+1}+2$

The root sistem $Q(\Pi_0)$ is

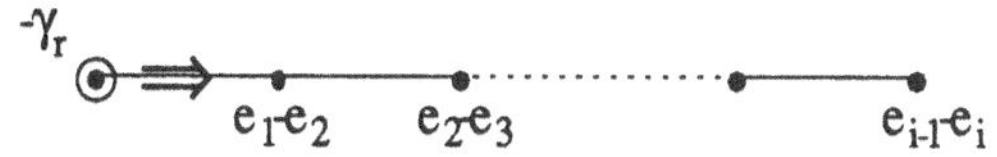

and it is of type sp(i , $\mathcal{R}$), then : $\quad 2R_{c,1} = (i-1\,,i-3\,, ...\,, -i+1\,,0,\dots,0\,)$.

In this case $T(\Pi_0) = Q(\Pi_0)$, $R_{c,1} + R_{c,2} = 2R_{c,1}$, and $\gamma_r = 2e_1$ then, from Theorem 6:

Lemma 11 . The last possible place for unitarity is $B(\Pi_0) = i$.

On the other hand $A(\Pi_0) = i - (i-1)\frac{1}{2}$ and the reduction level $2\,[B(\Pi_0) - A(\Pi_0)] + 1$ them from Theorem 7:

Lemma 12. The first possible place for non unitarity is $\frac{1}{2}(i+1)$ and at this point the reduction level is i .

Theorem 13. $M_{\Pi_0+z\varepsilon}$ is unitarizable if and only if $z \le \frac{1}{2}(i+1)$ or $2z \in Z$ with $z \le i$.

Case II $\qquad \Pi_1 = \Pi_2 = ... = \Pi_i$, $\Pi_i - \Pi_{i+1} = 1$, $\Pi_{i+1} = \Pi_{i+2} = ...= \Pi_{i+j} \ne \Pi_{i+j+1}$

The root system $Q(\Pi_0)$ is

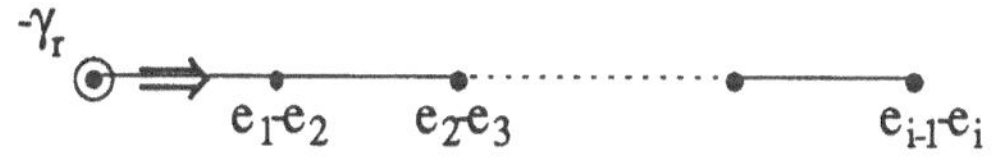

and the root system $T(\Pi_0)$

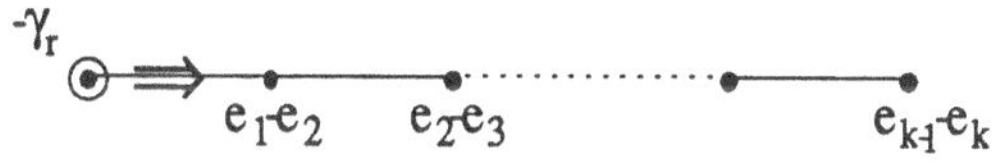

where $k = i + j$.

Then $Q(\Pi_0)$ is of type sp(i , $\mathcal{R}$) and $T(\Pi_0)$ is of type sp(k , $\mathcal{R}$), therefore

$$2R_{c,1} = (i-1\,,i-3\,, \dots, -i+1\,,0\,,\dots,0)$$
$$2R_{c,2} = (k-1\,,k-3\,, \dots, -k+1\,,0\,,\dots,0).$$

From Theorems 6 and 7:

Lemma 14. The last possible place for unitarity is $B(\Pi_0) = \frac{1}{2}(2i + j)$.

Lemma 15. The first possible place for non unitarity is $\frac{1}{2}(i + j + 1)$ and the reduction level is i .

Theorem 16: $M_{\Pi_0+z\varepsilon}$ is unitarizable if and only if $z \le \frac{1}{2}(i+j+1)$ or $2z \in Z$ with $z \le \frac{1}{2}(2i+j)$.

so*(2n)

Given $\Lambda = \Pi_0 + z\varepsilon$, $\Pi_0 = (\Pi_1, ... , \Pi_n)$ and $\varepsilon = \frac{1}{2}(\,1,\,1, \dots ,\,1)$.

As Π_0 is k_1- dominant and integral, $\Pi_i - \Pi_j \in \mathcal{N}$ for $i < j$. By the normalization $\langle \Pi_0 + R\,,\gamma_r\rangle = 0$, we have $\Pi_1 + \Pi_2 = -2n + 3$. The sum of positive compact roots is :

$$2R_c = \big(n-1\,,\,n-3\,,\,\ldots,\,n-(2n-1)\big).$$

Now, let the following conditions on the Π_0 components:

$$\Pi_1 = \Pi_2 = \ldots = \Pi_i > \Pi_{i+1} + 1 \quad , \quad i \neq 1.$$

Next we consider the root system $Q(\Pi_0) = T(\Pi_0)$ which is of type $so^*(2i)$, then

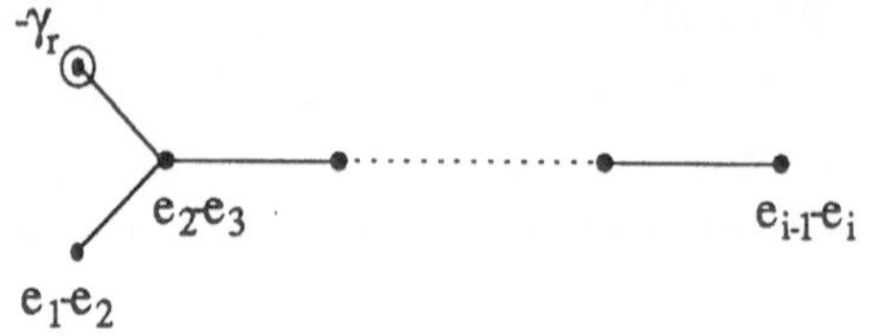

$$2R_{c,1} = (i-1\,,\,i-3\,,\,\ldots,\,-i+1\,,\,0\,,\,\ldots,\,0)$$

In this case $\gamma_r = e_1 + e_2$ therefore we will have by Theorem 6:

Lemma 17: The last possible place for unitarity is $B(\Pi_0) = 2i - 3$.

Let $[x]$ denote the biggest integer $\leq$ x. Then

$$A(\Pi_0) = 2i - 3 - 2\left(\left[\tfrac{i}{2}\right] - 1\right)$$

and the reduction level is $\dfrac{B(\Pi_0) - A(\Pi_0)}{2} + 1$, then from Theorem 7:

Lemma 18: The first possible place for non unitarity is

$$A(\Pi_0) = \begin{cases} i-1 & \text{if } i \text{ is even} \\ i & \text{if } i \text{ is odd} \end{cases} \quad \text{and the reduction level } \left[\tfrac{i}{2}\right].$$

Theorem 19: $M_{\Pi_0+z\varepsilon}$ is unitarizable if and only if $z \leq \begin{cases} i-1 & \text{if } i \text{ is even} \\ i & \text{if } i \text{ is odd} \end{cases}$ or

$z = 2i - 3 - 2m$ for some integer m with $0 \leq m \leq \left[\tfrac{i}{2}\right] - 2$.

We consider now the case $i = 1$. So, let $\Pi_1 \neq \Pi_2$ and j the biggest integer for which $\Pi_2 = \ldots = \Pi_{j+1}$.

The root system $Q(\Pi_0) = T(\Pi_0)$ is :

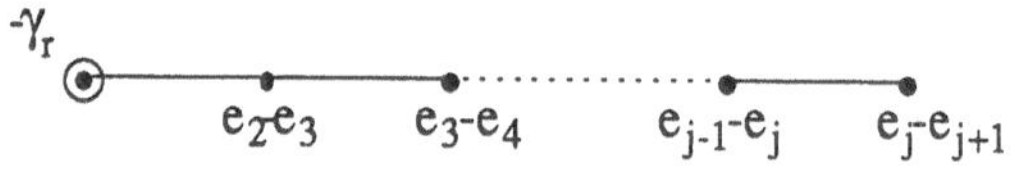

which is of type su $(1\,,j)$, then

$$2R_{c,1} = (0\,,\,j-1\,,\,j-3\,,\,\ldots,\,-j+1,\,0\,,\,\ldots,\,0)$$

so, applying theorems 6 and 7 we obtain : $A(\Pi_0) = B(\Pi_0) = j$ then the reduction level is 1 and $M_{\Pi_0+z\varepsilon}$ is unitarizable if and only if $z \leq j$.

so (2n - 1 , 2)

Let $\Lambda = \Pi_0 + z\varepsilon$, $\Pi_0 = (\Pi_1\,,\ldots,\,\Pi_n)$ and $\varepsilon = (1, 0, \ldots, 0)$; Π_0 is k_1 -dominant and integral then $\Pi_2 \geq \ldots \geq \Pi_n \geq 0$, $\Pi_i - \Pi_j \in Z$ and $2\Pi_i \in \mathcal{N}$ for all $i\,,j$ with $2 \leq i\,,j \leq n$. From the condition $\langle \Pi_0 + \mathrm{R}\,,\gamma_r \rangle = 0$ we have $\Pi_1 + \Pi_2 = -2n + 2$.

We consider two cases

Case I Let the following conditions on the Π_0 components

$$\Pi_2 = \Pi_3 = \ldots = \Pi_i \neq \Pi_{i+1} \quad i < n.$$

The root system $Q(\Pi_0) = T(\Pi_0)$ is the following

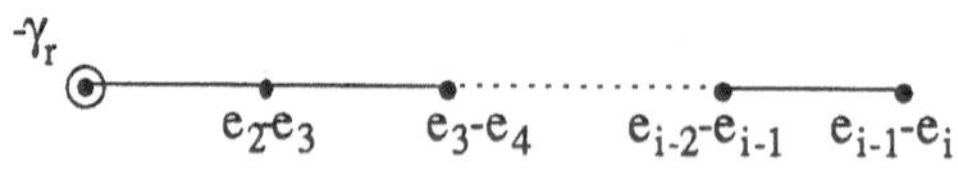

and it is of type su $(1\,,i-1)$ then

$$2R_{c,1} = (0,\,i-2\,,\,i-4\,,\,\ldots,\,-i+2\,,\,0\,,\,0\,\ldots,\,0).$$

In this case $T(\Pi_0) = Q(\Pi_0)$, $R_{c,1} + R_{c,2} = 2R_{c,1}$ and $\gamma_r = e_1 + e_2$ then we will have from Theorem 6:

Lemma 20. The last possible place for unitarity is $B(\Pi_0) = i - 1$.

From Theorem 7 $A(\Pi_0) = B(\Pi_0)$ then:

Theorem 21. $M_{\Pi_0 + z\varepsilon}$ is unitarizable if and only if $z \leq i - 1$.

Case II $\Pi_2 = \Pi_3 = \ldots = \Pi_{n-1}$ $2\Pi_n = 1$ $(\langle \Pi_0 , \mu_n \rangle = 1)$

The root system $Q(\Pi_0)$ is

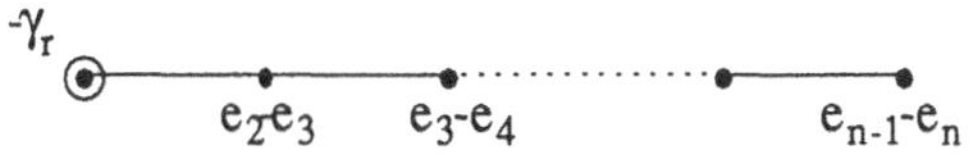

it is of type su $(1, n - 1)$
and the system $T(\Pi_0)$

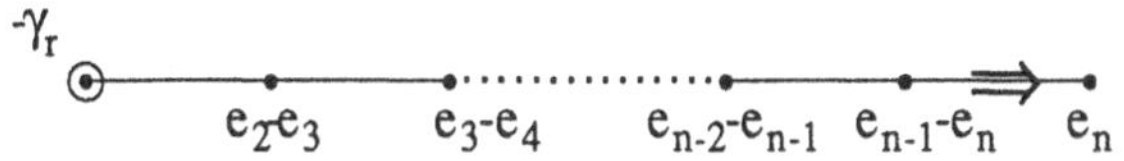

which is of type so $(2n - 1 , 2)$.

This case corresponds to the paragraph **a)** of Theorem 6, then:

Lemma 22. The last possible place for unitarity is $B(\Pi_0) = n - 1/2$.

From Theorem 7:

Theorem 23. $M_{\Pi_0 + z\varepsilon}$ is unitarizable if and only if $z \leq n - 1/2$.

so (2n - 2 , 2)

Let $\Lambda = \Pi_0 + z\varepsilon$, $\Pi_0 = (\Pi_1 , \ldots , \Pi_n)$ and $\varepsilon = (1, 0, \ldots , 0)$; Π_0 is k_1-dominant and integral, $\Pi_2 \geq \Pi_3 \geq \ldots \geq \Pi_{n-1} \geq |\Pi_n|$ and $\Pi_i - \Pi_j \in \mathcal{N}$ for $2 \leq i < j \leq n$. From the condition $\langle \Pi_0 + R , \gamma_r \rangle = 0$ it follows $\Pi_1 + \Pi_2 = -2n + 3$.

We consider two cases.

Case I Let the following conditions on the Π_0 components:

$\Pi_2 = \Pi_3 = \ldots = \Pi_i \neq \Pi_{i+1}$ $i \neq n - 1$ o $i = n - 1$ y $\Pi_{n-1} \neq -\Pi_n$

The root system $Q(\Pi_0) = T(\Pi_0)$ is

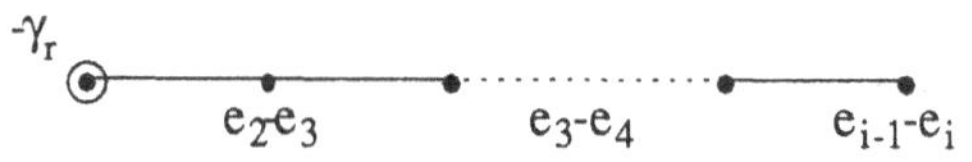

which is of type su $(1 , i - 1)$. Then by Theorem 6:

Lemma 24. The last possible place for unitarity is $B(\Pi_0) = i - 1$

From Theorem 7 $A(\Pi_0) = B(\Pi_0)$, then:

Theorem 25. $M_{\Pi_0 + z\varepsilon}$ is unitarizable if and only if $z \leq i - 1$.

Case II $\Pi_2 = \Pi_3 = \ldots = \Pi_{n-2} = \Pi_{n-1} = -\Pi_n$ $(\Pi_n < 0)$

The root system $Q(\Pi_0) = T(\Pi_0)$ is

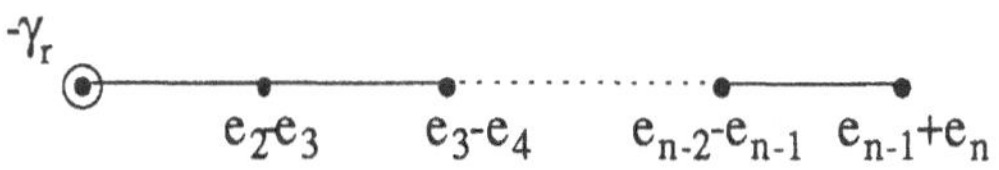

which is of type su (1, n - 1), then:

Theorem 26. $M_{\Pi_0 + z\varepsilon}$ is unitarizable if and only if $z \leq n - 1$

e_6

Let be $\Lambda = \Pi_0 + z\varepsilon$, $\Pi_0 = (\Pi_1, \ldots, \Pi_8)$ and $\varepsilon = (0, 0, 0, 0, 0, -2/3, -2/3, 2/3)$; Π_0 is k_1 -dominant and integral therefore $|\Pi_1| \leq \Pi_2 \leq \ldots \leq \Pi_5$, $\Pi_i - \Pi_j \in Z$, $2\Pi_i \in Z$, $i, j \leq 5$. From the condition $\langle \Pi_0 + R, \gamma_r \rangle = 0$ we obtain $\Pi_1 + \Pi_2 + \Pi_3 + \Pi_4 + \Pi_5 - \Pi_6 - \Pi_7 + \Pi_8 = -22$.

We consider two cases:

Case I In the following paragraphs we show, under the conditions given for Π_0 , that the root system $Q(\Pi_0) = T(\Pi_0)$.

1) $\langle \Pi_0, \mu_2 \rangle > 0$

su(1,1).

2) $\langle \Pi_0, \mu_2 \rangle = 0$, $\langle \Pi_0, \mu_4 \rangle > 0$

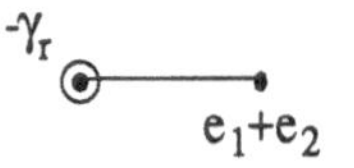

su(1,2).

3) $\langle \Pi_0, \mu_2 \rangle = \langle \Pi_0, \mu_4 \rangle = 0$, $\langle \Pi_0, \mu_5 \rangle > 0$, $\langle \Pi_0, \mu_3 \rangle > 0$

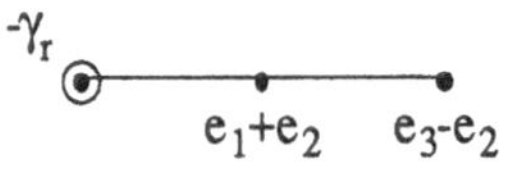

su(1,3).

4) (a) $\langle \Pi_0, \mu_2 \rangle = \langle \Pi_0, \mu_4 \rangle = \langle \Pi_0, \mu_3 \rangle = 0$, $\langle \Pi_0, \mu_5 \rangle > 0$

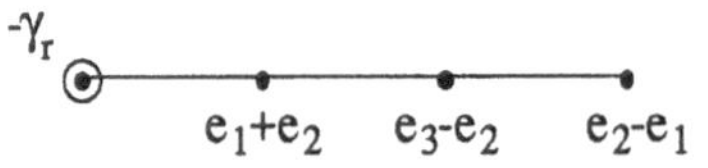

su(1,4).

(b) $\langle \Pi_0, \mu_2 \rangle = \langle \Pi_0, \mu_4 \rangle = \langle \Pi_0, \mu_5 \rangle = 0$, $\langle \Pi_0, \mu_6 \rangle > 0$ $\langle \Pi_0, \mu_3 \rangle > 0$

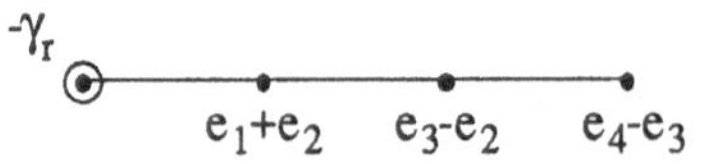

su(1,4).

5) $\langle \Pi_0, \mu_2 \rangle = \langle \Pi_0, \mu_4 \rangle = \langle \Pi_0, \mu_5 \rangle = \langle \Pi_0, \mu_6 \rangle = 0$ $\langle \Pi_0, \mu_3 \rangle > 0$

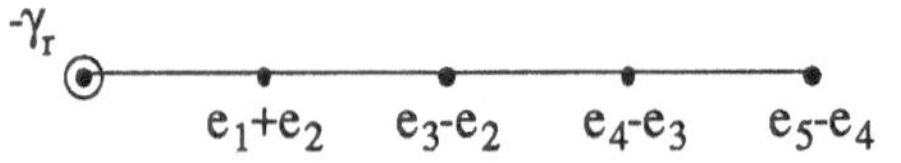

su(1,5).

Then the root system $Q(\Pi_0) = T(\Pi_0)$ is of type su $(1, i)$ with $1 \leq i \leq 5$ and by Theorem 6:

Lemma 27. The last possible place for unitarity is $B(\Pi_0) = i$.

By Theorem 7 $A(\Pi_0) = B(\Pi_0)$. Therefore

Theorem 28. $M_{\Pi_0+z\varepsilon}$ is unitarizable if and only if $z \leq i$.

Case II Let be the following conditions on the Π_0 components:

$$\langle \Pi_0, \mu_2 \rangle = \langle \Pi_0, \mu_3 \rangle = \langle \Pi_0, \mu_4 \rangle = \langle \Pi_0, \mu_5 \rangle = 0 \ , \ \langle \Pi_0, \mu_6 \rangle > 0$$

The root system $Q(\Pi_0) = T(\Pi_0)$ is

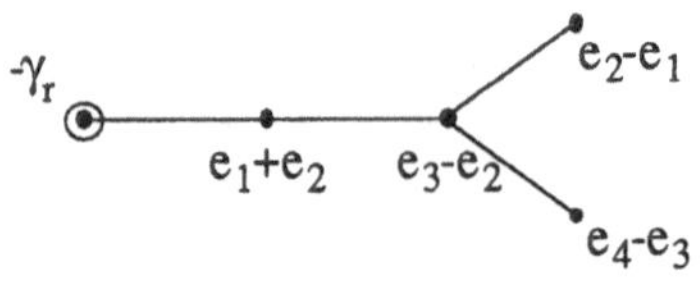

which is of type so $(8, 2)$. Then $R_{c,1} + R_{c,2} = 2R_{c,1} = 2\,(0, 1, 2, 3, 0, 0, 0, 0)$. In this algebra $\gamma_r = 1/2\,(e_1 + e_2 + e_3 + e_4 + e_5 - e_6 - e_7 + e_8)$. Therefore we will have by Theorem 6:

Lemma 29. The last possible place for unitarity is $B(\Pi_0) = 7$.
In addition $A(\Pi_0) = 7 - (2 - 1) \times 3$.

Lemma 30. The first possible place for non unitarity is $A(\Pi_0) = 4$.

Theorem 31. $M_{\Pi_0 + z\varepsilon}$ is unitarizable if and only if $z \leq 4$ or $z = 7$.

e_7

Let $\Lambda = \Pi_0 + z\varepsilon$, $\Pi_0 = (\Pi_1, \ldots, \Pi_8)$ with $\Pi_7 = -\Pi_8$ and $\varepsilon = (0, 0, 0, 0, 0, 1, -1/2, 1/2)$; Π_0 is k_1-dominant and integral therefore $|\Pi_1| \leq \Pi_2 \leq \ldots \leq \Pi_5$, $\Pi_i - \Pi_j \in Z$, $2\Pi_i \in Z$, $1 \leq i \leq j \leq 5$, and $1/2\,(\Pi_1 - \Pi_2 - \Pi_3 - \Pi_4 - \Pi_5 - \Pi_6 - \Pi_7 + \Pi_8) \in \mathcal{N}$.

The condition $\langle \Pi_0 + R, \gamma_r \rangle = 0$ fix the value $\Pi_8 = -17/2$. In all the cases

$$\mu_1 = 1/2\,(e_1 - e_2 - e_3 - e_4 - e_5 - e_6 - e_7 + e_8).$$

We consider two cases.

Case I

Given the conditions on Π_0 we can calculate the root system $Q(\Pi_0) = T(\Pi_0)$.

1) $\langle \Pi_0, \mu_1 \rangle > 0$

su(1,1).

2) $\langle \Pi_0, \mu_1 \rangle = 0$, $\langle \Pi_0, \mu_3 \rangle > 0$

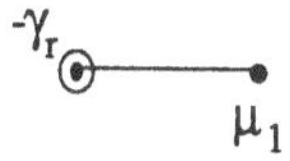

su(1,2).

3) $\langle \Pi_0, \mu_1 \rangle = \langle \Pi_0, \mu_3 \rangle = 0$, $\langle \Pi_0, \mu_4 \rangle > 0$

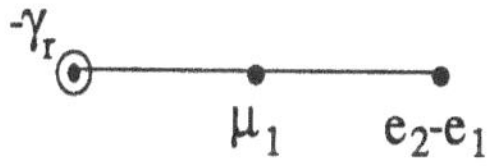

su(1,3).

4) $\langle \Pi_0, \mu_1 \rangle = \langle \Pi_0, \mu_3 \rangle = \langle \Pi_0, \mu_4 \rangle = 0$, $\langle \Pi_0, \mu_2 \rangle > 0$, $\langle \Pi_0, \mu_5 \rangle > 0$

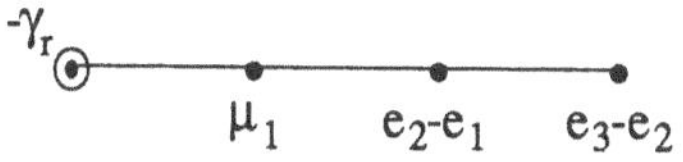

su(1,4).

5) (a) $\langle \Pi_0, \mu_1 \rangle = \langle \Pi_0, \mu_3 \rangle = \langle \Pi_0, \mu_4 \rangle = \langle \Pi_0, \mu_2 \rangle = 0$, $\langle \Pi_0, \mu_5 \rangle > 0$

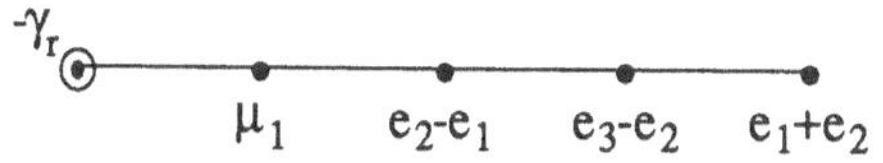

su(1,5).

(b) $\langle \Pi_0, \mu_1 \rangle = \langle \Pi_0, \mu_3 \rangle = \langle \Pi_0, \mu_4 \rangle = \langle \Pi_0, \mu_5 \rangle = 0$, $\langle \Pi_0, \mu_2 \rangle > 0$, $\langle \Pi_0, \mu_6 \rangle > 0$

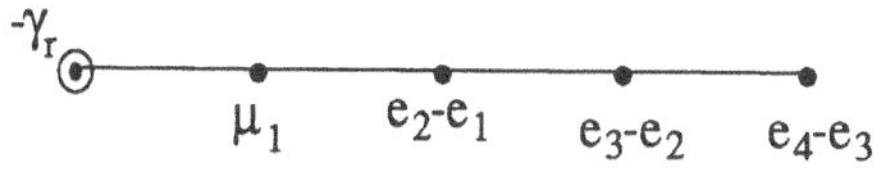

su(1,5).

6) $\langle \Pi_0, \mu_1 \rangle = \langle \Pi_0, \mu_3 \rangle = \langle \Pi_0, \mu_4 \rangle = \langle \Pi_0, \mu_5 \rangle = \langle \Pi_0, \mu_6 \rangle = 0$, $\langle \Pi_0, \mu_2 \rangle > 0$

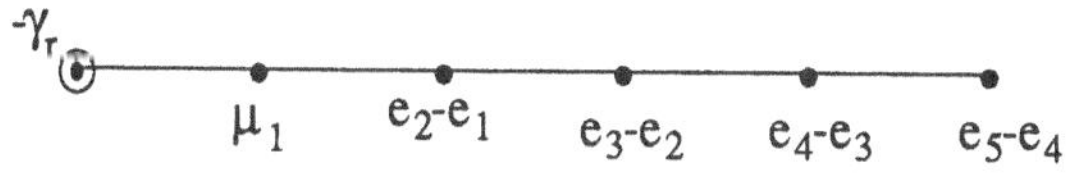

su(1,6).

Then the root system $Q(\Pi_0) = T(\Pi_0)$ is of type su (1 , i) with $1 \leq i \leq 6$ and by Theorem 6

Lemma 32. The last possible place for unitarity is $B(\Pi_0) = i$

By Theorem 7 $A(\Pi_0) = B(\Pi_0)$

Theorem 33. $M_{\Pi_0+z\varepsilon}$ is unitarizable if and only if $z \leq i$

Case II Given the following conditions on the Π_0 components:

$$\langle \Pi_0 , \mu_1 \rangle = \langle \Pi_0 , \mu_2 \rangle = \langle \Pi_0 , \mu_3 \rangle = \langle \Pi_0 , \mu_4 \rangle = \langle \Pi_0 , \mu_5 \rangle = 0 \quad , \quad \langle \Pi_0 , \mu_6 \rangle > 0$$

The root system $Q(\Pi_0) = T(\Pi_0)$ is

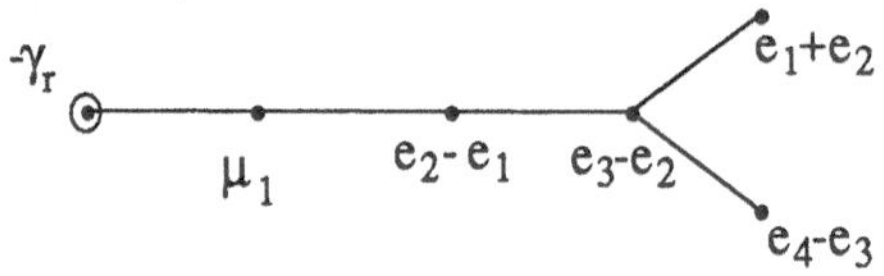

and it is of type so (10 , 2) . Then

$R_{c,1} + R_{c,2} = 2R_{c,1} = 2$ (0, 1, 2, 3, - 2, - 2, - 2, 2). For this algebra $\gamma_r = e_8 - e_7$ then we will have from Theorem 6:

Lemma 34. The last possible place for unitarity is $B(\Pi_0) = 9$.

On the other hand $A(\Pi_0) = 9 - (2 - 1) \times 4 = 5$ then:

Theorem 35. $M_{\Pi_0+z\varepsilon}$ is unitarizable if and only if $z \leq 5$ or $z = 9$.

VII. POSSIBLE PLACES FOR UNITARITY

We consider two cases:

A) Cases with more than one reduction point

su(p,q)

Dynkin diagram

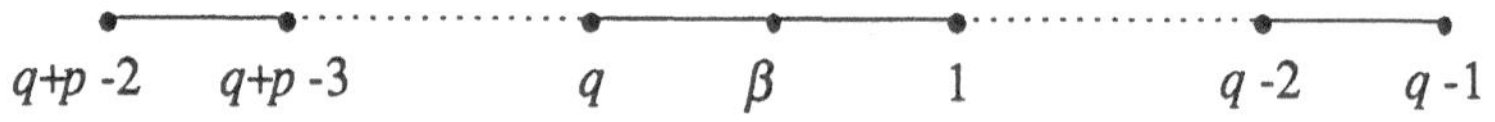

Let M_Λ be a representation for su(p, q) with $\Lambda = (\Lambda_1, \Lambda_2, ..., \Lambda_{p+q})$ with respect to the standard orthonormal basis of $\mathcal{R}^n$ $(n = p + q)$, satisfying the following conditions on its components:

$$\Lambda_1 = \Lambda_2 = ... = \Lambda_i \neq \Lambda_{i+1}$$
$$\Lambda_n = \Lambda_{n-1} = ... = \Lambda_{n-j+1} \neq \Lambda_{n-j}.$$

If we put $\Lambda = \Lambda_0 + \varepsilon\lambda$ these conditions are equivalent to the following ones:

$$\langle \Lambda_0 , \mu_{q-1} \rangle = \langle \Lambda_0 , \mu_{q-2} \rangle = ... = \langle \Lambda_0 , \mu_{t+1} \rangle = 0, \quad \langle \Lambda_0 , \mu_t \rangle \neq 0$$
$$\langle \Lambda_0 , \mu_{n-2} \rangle = \langle \Lambda_0 , \mu_{n-3} \rangle = ... = \langle \Lambda_0 , \mu_{s+1} \rangle = 0, \quad \langle \Lambda_0 , \mu_s \rangle \neq 0$$

with $t = q - j$ and $s = n - i - 1$.

Applying steps **i)** and **ii)** of Jakobsen method (section IV) we obtain (see Fig. 5).

$$\alpha_0 = \beta + \mu_1 + ... + \mu_t + \mu_q + \mu_{q+1} + ... + \mu_s$$

such that $\mathcal{H}_{\alpha_0} = t + s - q + 2$. Then $\lambda_0 = q - t - s - 1$.

Then a first order polynomial will be missing with highest weight

$$\Lambda_0 + (q - t - s - 1)\varepsilon + R - \alpha_0$$

where, in this case, $\varepsilon = (q/n , q/n , ... ,q/n , -p/n , ... ,-p/n)$ with p copies of q/n and q of $-p/n$.

Diagram su (p , q) **j > i**

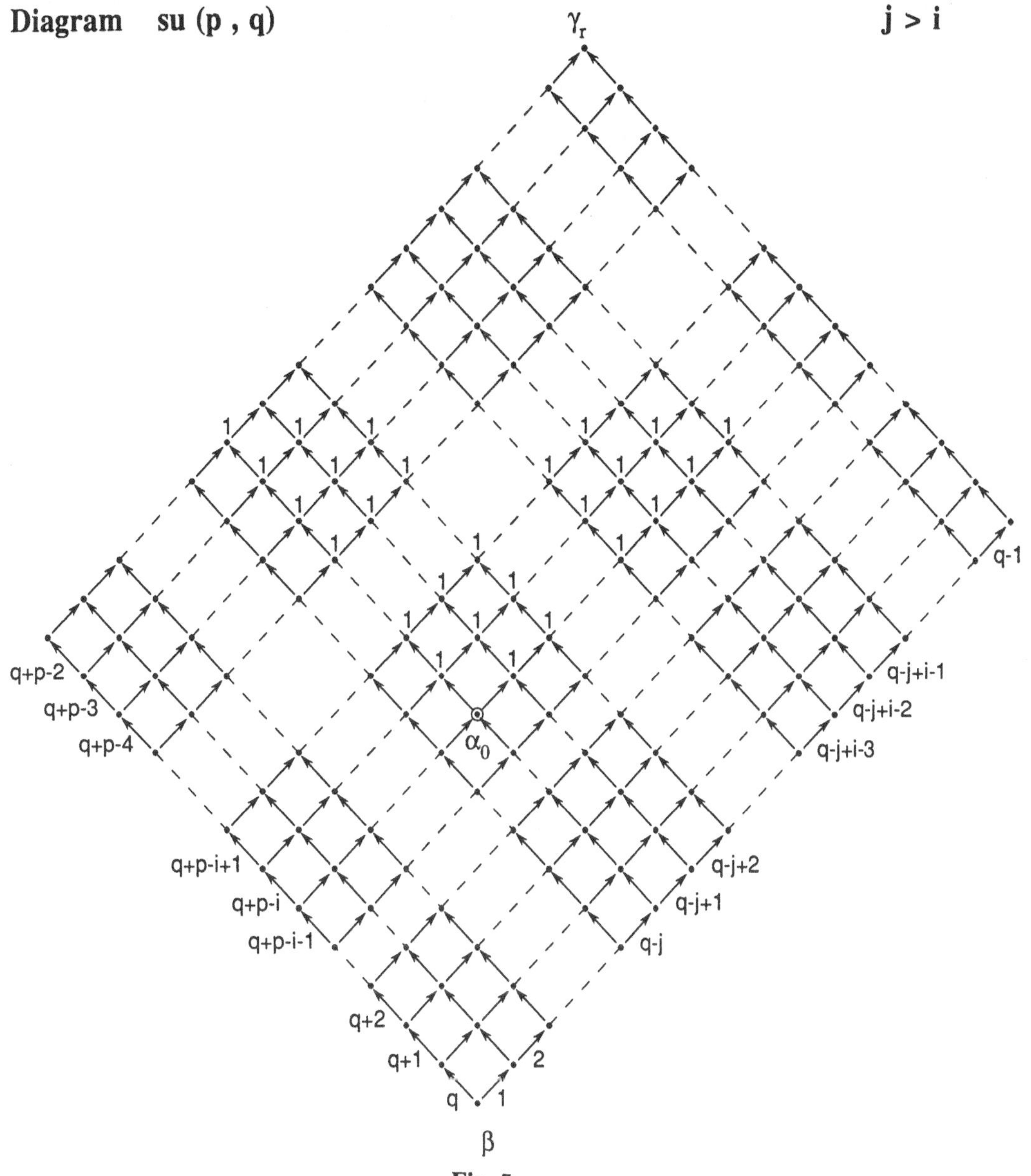

Fig. 5

For $\lambda_q = \lambda_0 + \lambda_s = \lambda_0 - 1$ we obtain from steps **iii)** and **iv)** a second order polynomial that will be missing with heighest weight

$$\Lambda_0 + (\lambda_0 - 2)\varepsilon + R - \alpha^1_{2-\lambda_0} - \alpha^2_{2-\lambda_0}.$$

Next case is $\lambda_q = \lambda_0 - 2$ where a third order polynomial with highest weight $\Lambda_0 + (\lambda_0 - 2)\varepsilon + R - \alpha_{3-\alpha_0} - \alpha_{3-\lambda_0} - \alpha_{3-\lambda_0}$ will be missing. Continuing in the same way we arrive at $\lambda_q = \lambda_0 - u$, $u = \min(q - t - 1, n - s - 2)$ where a polynomial of order $u + 1 = \min(i, j)$ will be missing. For $\lambda < \lambda_0 - u$ it is impossible to find a polynomial of order strictly higher than $\min(i, j)$ because the k_1-dominance is violated (see step **iv**).

Thus the reduction level is $\min(i, j)$. On the other hand $\lambda = \langle \Lambda, \gamma_r \rangle$ or, equivalently, $\Lambda_n = \Lambda_1 - \lambda$. Then, taking into account the possible values of λ we obtain the following diagram:

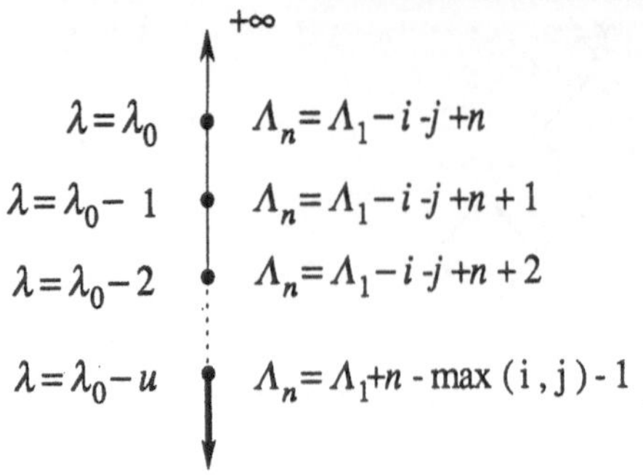

Now we apply the EHW method:

$\Lambda = \Pi_0 + z\,\varepsilon$ with $\Pi_0 = (\Pi_1, \dots, \Pi_n)$

From $\langle \Pi_0 + R, \gamma_r \rangle = 0$ we obtain

$$\Pi_1 - \Pi_n + n - 1 = 0$$

The root system $Q(\Pi_0)$ is of type su (i, j) then the last possible place for unitarity, following Lemma 8 corresponds to $z = i + j - 1$ and we will have :

$$\Lambda_1 = \Pi_1 + (i + j - 1)q/n \quad , \quad \Lambda_n = \Pi_1 + n - 1 - (i + j - 1)p/n$$

or what is the same : $\Lambda_n = \Lambda_1 - i - j + n$.

The next place for unitarity corresponds to $z = i + j - 2$ for which $\Lambda_n = \Lambda_1 - i - j + n + 1$. Continuing along this way and having in mind that the first reduction point corresponds to (Lemma 9) : $z = \max(i, j)$ we arrive at the first possible place for non unitarity for which:

$$\Lambda_1 = \Pi_1 + \max(i, j)\, q/n \quad ; \quad \Lambda_n = \Pi_1 + n - 1 - \max(i, j) p/n$$

then $\Lambda_n = \Lambda_1 - \max(i, j) - 1 + n$. We obtain, as desired, the same results as in Jakobsen method.

sp $(n, \mathcal{R})$

Dynkin diagram

Let M_Λ be a representation for sp $(n, \mathcal{R})$ with $\Lambda = (\Lambda_1, \dots, \Lambda_n)$ we put $\Lambda = \Lambda_0 + \lambda\varepsilon$ where $\varepsilon = (1, 1, \dots, 1)$. We consider two cases:

Case I

The weight Λ satisfy the following conditions on its components

$$\Lambda_1 = \Lambda_2 = \dots = \Lambda_i \geq \Lambda_{i+1} + 2$$

or, equivalently

$$\langle \Lambda_0, \mu_1 \rangle = \langle \Lambda_0, \mu_2 \rangle = \dots = \langle \Lambda_0, \mu_{i-1} \rangle = 0 \quad , \quad \langle \Lambda_0, \mu_i \rangle = n \geq 2 .$$

Applying Jakobsen method we obtain (see Fig. 6)

$$\alpha_0 = \beta + 2\mu_{n-1} + 2\mu_{n-2} + \dots + 2\mu_{n-(n-i)}$$

with $\mathcal{H}_{\alpha_0} = 2(n - i) + 1$. As α_0 is a long root, the condition $\langle \Lambda + R, \alpha_0 \rangle = 1$ implies

$$\frac{1}{2}(\Lambda_0, \alpha_0) + \lambda_0 + n - i + 1 = 1 \quad , \quad \lambda_0 = i - n$$

then a first order polynomial with highest weight $\Lambda_0 + (i - n)\varepsilon + R - \alpha_0$ will be missing when we unitarize.

For $\lambda = \lambda_0 + \lambda_s = i - n - \frac{1}{2}$ we obtain a second order polynomial which will be missing with highest weight

$$\Lambda_0 + (i - n - \tfrac{1}{2})\varepsilon + R - 2\alpha^1_{2(n-i)+2}$$

For $\lambda = \lambda_0 + 2\lambda_s$ the third order missing polynomial has highest weight

$$\Lambda_0 + (i - n - 1)\varepsilon + R - 2\alpha^1_{2(n-i)+3} - \alpha^2_{2(n-i)+3} .$$

Diagram sp (n , $\mathcal{R}$) *Case I*

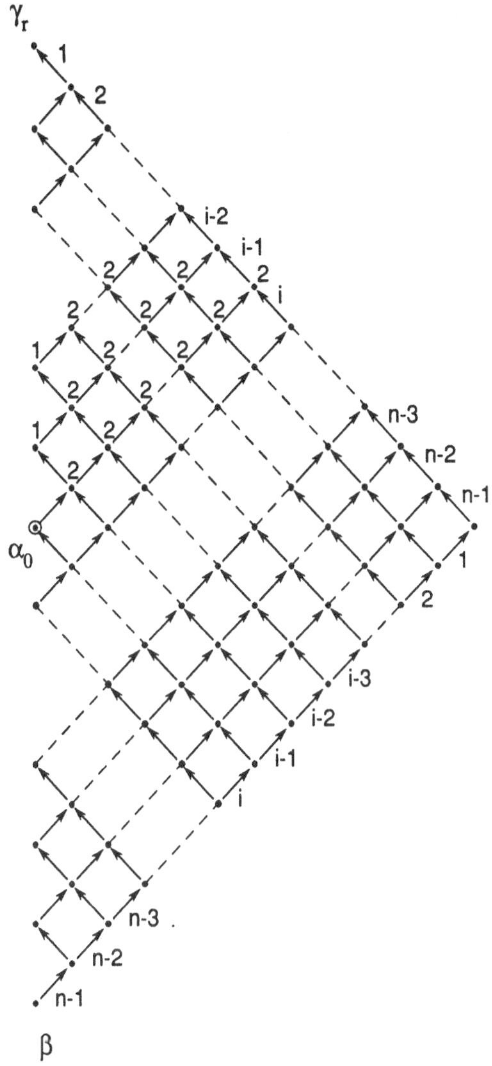

Fig. 6

Following along these lines we arrive at $\lambda = \lambda_0 + (i-1)\lambda_s = \frac{1}{2}(i+1) - n$ where a i-th order polynomial will be missing. For $\lambda < \frac{1}{2}(i+1) - n$ it is impossible to obtain polynomials of order strictly higher than i because there would not be k_1-dominance, therefore the reduction level is i . On the other hand from the condition $\lambda = \langle \Lambda, \gamma_r \rangle$ it follows that $\Lambda_1 = \lambda$ Thus, for the different values of λ we obtain the following diagram which give us the possible values of Λ_1 for unitarity:

	$+\infty$
$\lambda = \lambda_0$	$\Lambda_1 = -n + i$
$\lambda = \lambda_0 - \frac{1}{2}$	$\Lambda_1 = -n + i - \frac{1}{2}$
$\lambda = \lambda_0 - 1$	$\Lambda_1 = -n + i - 1$
$\lambda = \lambda_0 - \frac{1}{2}(i-1)$	$\Lambda_1 = -n + \frac{1}{2}(i+1)$

We now apply the EHW method

$$\Lambda = \Pi_0 + z\varepsilon \ , \ \Pi_0 = (\Pi_1, \dots, \Pi_n) \ , \ \Pi_1 = -n.$$

The root system $Q(\Pi_0)$ is of type sp $(i, \mathcal{R})$ then the last possible place for unitarity corresponds to $z = i$ following Lemma 11, then

$$\Lambda_1 = \Pi_1 + i \quad ; \quad \Lambda_1 = -n + i.$$

The next place corresponds to $z = i - \frac{1}{2}$, therefore $\Lambda_1 = -n + i - \frac{1}{2}$. Continuing along this way we obtain for the first reduction point (see Lemma 12) $z = \frac{1}{2}(i+1)$ hence the first component of the weight is $\Lambda_1 = -n + \frac{1}{2}(i+1)$, as desired.

Case II

We consider in this case the following conditions

$$\Lambda_1 = \Lambda_2 = \ldots = \Lambda_i \quad ; \quad \Lambda_i - \Lambda_{i+1} = 1 \ , \ \Lambda_{i+1} = \Lambda_{i+2} = \ldots = \Lambda_{i+j} \neq \Lambda_{i+j+1}$$

which are equivalent to the following ones:

$$\langle \Lambda_0 , \mu_1 \rangle = \langle \Lambda_0 , \mu_2 \rangle = \ldots = \langle \Lambda_0 , \mu_{i-1} \rangle = 0 \quad , \quad \langle \Lambda_0 , \mu_i \rangle = 1$$

$$\langle \Lambda_0 , \mu_{i+1} \rangle = \langle \Lambda_0 , \mu_{i+2} \rangle = \ldots = \langle \Lambda_0 , \mu_{i+j-1} \rangle = 0 \quad , \quad \langle \Lambda_0 , \mu_{i+j} \rangle = n \ \geq 1.$$

In this case (see Fig. 7 and 8) applying Jakobsen method

$$\alpha_0 = \beta + 2 \left(\mu_{n-1} + \mu_{n-2} + \ldots + \mu_{n-(n-i-j)}\right) + \mu_{n-(n-i-j)} + \ldots + \mu_{n-(n-i)}$$

Diagram sp (n , $\mathcal{R}$) ***Case II***

a) $j = i+1$

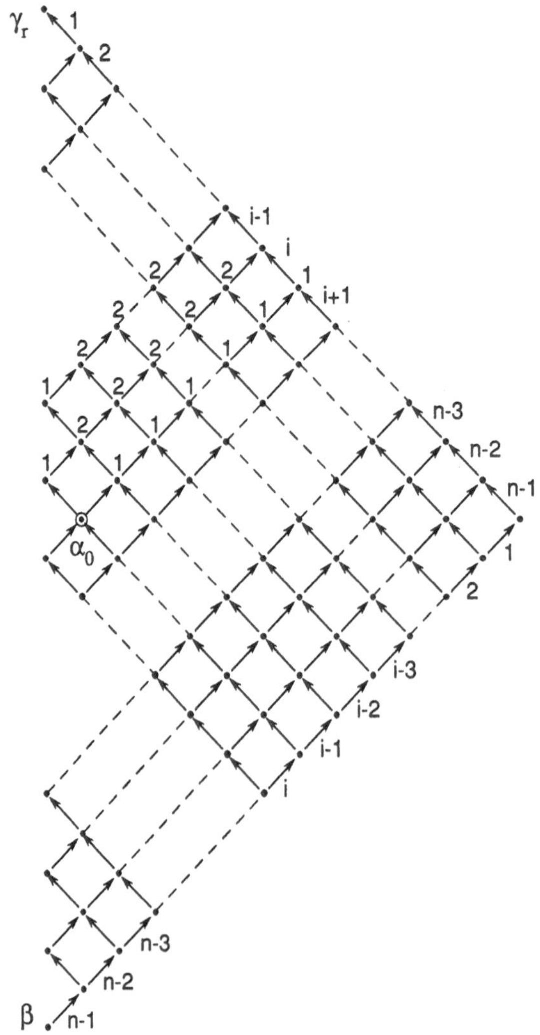

Fig. 7

b) $j > i+1$

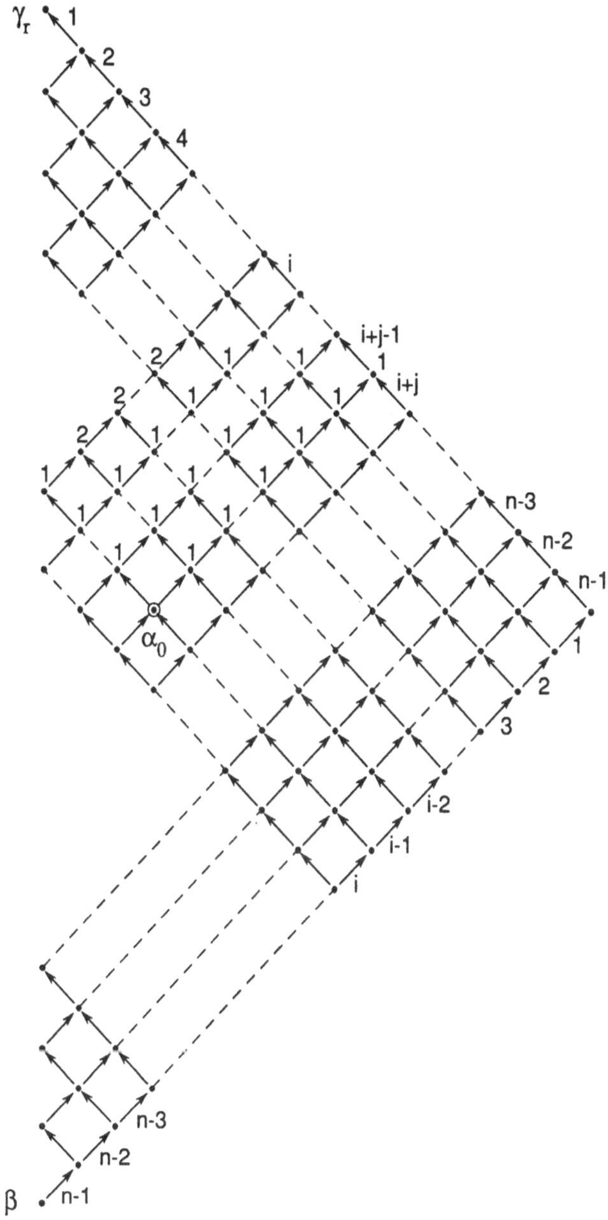

Fig. 8

the height of which is $\mathcal{H}_{\alpha_0} = 2(n-i) - j + 1$. As α_0 is a short root, the condition $\langle \Lambda + R \, , \, \alpha_0 \rangle = 1$ implies

$$(\Lambda_0, \alpha_0) + 2\lambda_0 + 2(n-i+1) - j = 1$$

and, having in mind that we can also state

$$\alpha_0 = \gamma_r - 2\,(\mu_1 + \ldots + \mu_{i-1}) - \mu_i - \mu_{i+1} - \ldots - \mu_{i+j-1}$$

we have

$$\lambda_0 = i - n + \frac{j}{2}$$

For $\lambda = \lambda_0 + \lambda_s = i - n + \frac{(j-1)}{2}$ we obtain a second order polynomial which will be missing with highest weight

$$\Lambda_0 + \left(i - n + \frac{(j-1)}{2}\right)\varepsilon + R - \alpha^1_{2(n-i+1)-j} - \alpha^2_{2(n-i+1)-j}.$$

Continuing as in case I we arrive at $\lambda = \lambda_0 + (i-1)\lambda_s = -n + \frac{1}{2}(i+j+1)$ where a i-th order polynomial will be missing. Since $\lambda < -n + \frac{1}{2}(i+j+1)$ there is no k_1-dominance therefore the reduction level is i. The diagram in this case is the following:

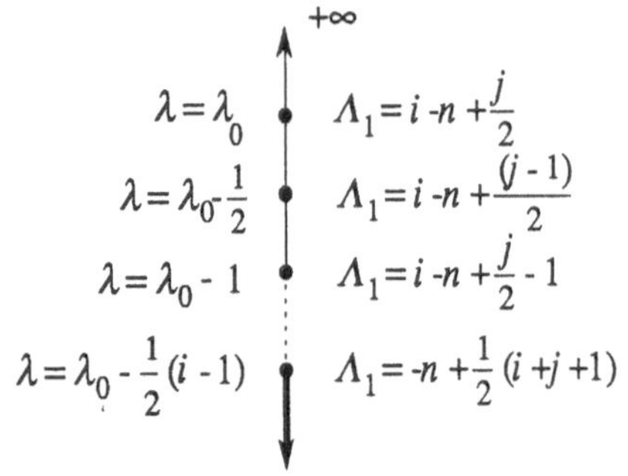

We now apply the EHW method

The root system $Q(\Pi_0)$ is of type sp $(i, \Re)$ and $T(\Pi_0)$ is of type sp $(k, \Re)$ with $k = i + j$. Then the last possible place for unitarity corresponds, by Lemma 14, to $z = \frac{1}{2}(2i + j)$ therefore in this case $\Lambda_1 = -n + i + \frac{j}{2}$.

Following along this way we obtain the first possible place for non unitarity $\Lambda_1 = -n + \frac{1}{2}(i + j + 1)$ by Lemma 15.

so* (2n)

Dynkin diagram

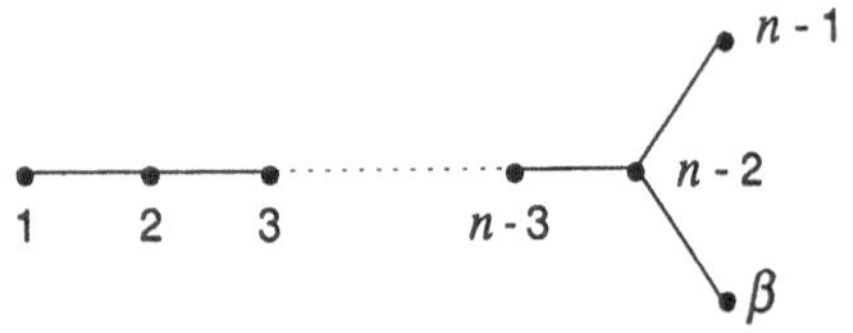

Let now M_Λ be a representation for so*(2n) with $\Lambda = (\Lambda_1, \dots, \Lambda_n)$. In this case, given $\Lambda = \Lambda_0 + \lambda\varepsilon$ we have $\varepsilon = (\frac{1}{2}, \frac{1}{2}, \dots, \frac{1}{2})$. We consider the following conditions on its components

$$\Lambda_1 = \Lambda_2 = \dots = \Lambda_i > \Lambda_{i+1} + 1 \qquad i \neq 1$$

or, equivalently,

$$\langle \Lambda_0, \mu_1 \rangle = \langle \Lambda_0, \mu_2 \rangle = \dots = \langle \Lambda_0, \mu_{i-1} \rangle = 0 \quad , \quad \langle \Lambda_0, \mu_i \rangle = n > 1.$$

From the Jakobsen method we have (see Fig. 9)

$$\alpha_0 = \beta + (\mu_{n-1} + \mu_{n-2}) + (\mu_{n-2} + \mu_{n-3}) + \dots + (\mu_{n-(n-i)} + \mu_{n-(n-i)-1})$$

with $\mathcal{H}_{\alpha_0} = 2(n-i) + 1$. The condition $\langle \Lambda + R, \alpha_0 \rangle = 1$ implies in this case

$$(\Lambda_0, \alpha_0) + \lambda_0 + 2(n-i) + 1 = 1 \quad ; \quad \lambda_0 = 2(i-n).$$

For $\lambda = \lambda_0 + \lambda_s = 2(i-n-1)$ we obtain a second order polynomial which will be missing with highest weight

$$\Lambda_0 + 2(i-n-1)\varepsilon + R - \alpha^1_{2(n-i)+3} - \alpha^2_{2(n-i)+3}.$$

Following along these lines we arrive at $\lambda = \lambda_0 + \left\{[\frac{i}{2}] - 1\right\}\lambda_s$ where a polynomial with order $[\frac{i}{2}]$ is missing. For $\lambda < 2\left(i - n - \left\{[\frac{i}{2}] - 1\right\}\right)$ there is impossible to obtain missing polynomials the order of which is strictly higher than $[\frac{i}{2}]$ because in those places the weights are not k_1-dominants therefore the reduction level is $[\frac{i}{2}]$. From the condition $\lambda = \langle \Lambda, \gamma_r \rangle$ we obtain $\Lambda_1 = \frac{\lambda}{2}$. In this way we obtain the following diagram:

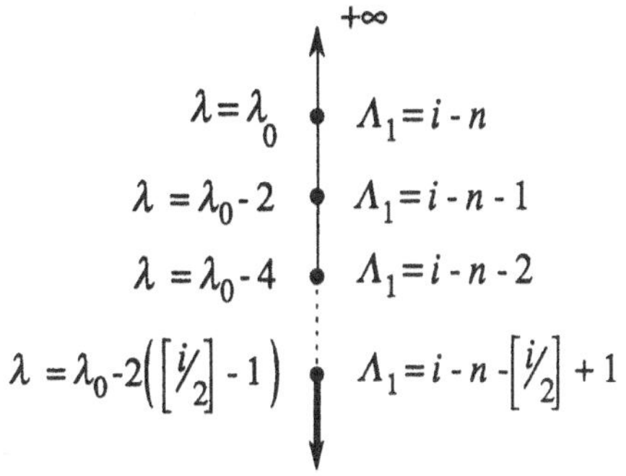

Now we apply EHW method

$\Lambda = \Pi_0 + z\varepsilon \quad , \quad \Pi_0 = (\Pi_1 , \dots , \Pi_n) \quad ,$

$2\Pi_1 = -2n + 3$

The root system is $Q(\Pi_0) = T(\Pi_0)$ which is of type so*(2i), therefore, following Lemma17, the last possible place for unitarity is $z = 2i - 3$, and at this place

$$\Lambda_1 = \Pi_1 + \frac{1}{2}(2i - 3) = i - n.$$

Diagrama so* (2n) *i par*

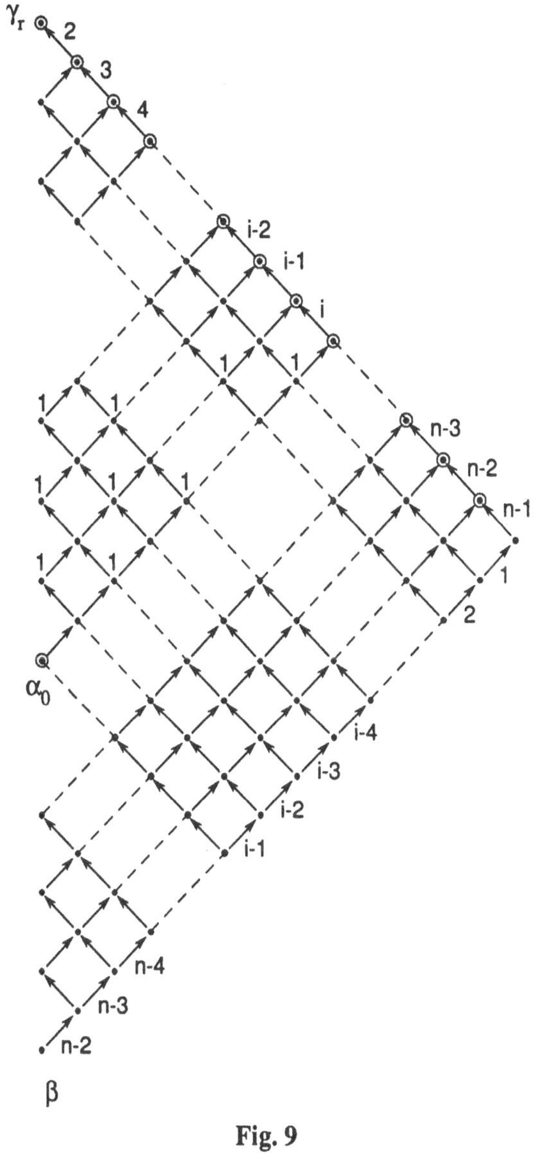

Fig. 9

The next place corresponds to $z = 2i - 5$ then in this case $\Lambda_1 = i - n - 1$. Continuing along this way we arrive at the first reduction point where by Lemma 18 $z = \begin{Bmatrix} i-1 \text{ if } i \text{ is even} \\ i \text{ if } i \text{ is odd} \end{Bmatrix}$ or, what is the same

$$z = 2i - 1 - 2[\tfrac{i}{2}].$$

At this place $\Lambda_1 = \Pi_1 + \frac{1}{2}\left(2i - 1 - 2[\frac{i}{2}]\right) = i - n - \left([\frac{i}{2}] - 1\right)$, as desired.

Let now be $i = 1$ and the following conditions:

$$\Lambda_1 \neq \Lambda_2 = \Lambda_3 = \ldots = \Lambda_j = \Lambda_{j+1} \neq \Lambda_{j+2}$$

or what is the same

$$\langle \Lambda_0, \mu_1 \rangle = m_1 > 0 \quad \langle \Lambda_0, \mu_2 \rangle = \ldots = \langle \Lambda_0, \mu_j \rangle = 0 \quad \langle \Lambda_0, \mu_{j+1} \rangle = m_{j+1} > 0$$

with $2 \leq j + 1 \leq n - 1$. In this case we have $\mathcal{H}_{\alpha_0} = 2n - j - 2$ (in Fig. 9 the possible α_0 are inside a little circle on the right branch of $C^-_{\gamma_r}$).

From condition $\langle \Lambda + R, , \alpha_0 \rangle = 1$ we obtain $\lambda_0 = 3 - 2n + j$. We will have only one reduction point for each j because $C^+_{\alpha_0}$ is unidimensional:

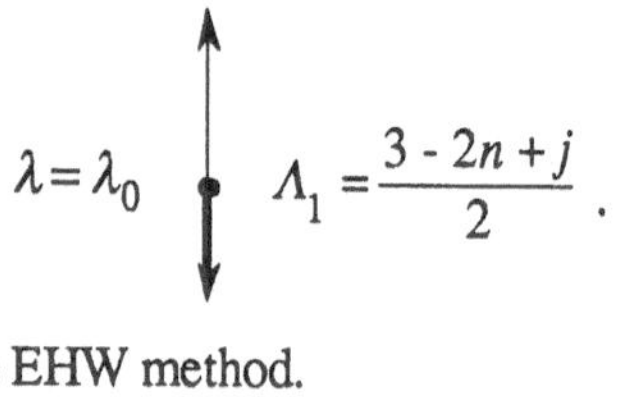

If we consider the remarks following Theorem 19 we may see that the above results are the same that those obtained by the EHW method.

e_6

Dynkin diagram

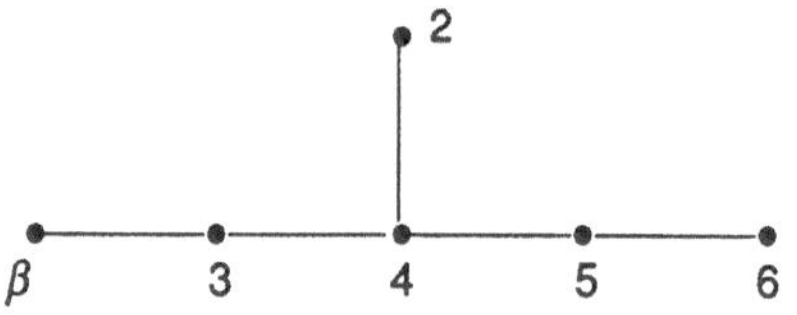

Here $\Lambda = \Lambda_0 + \lambda\varepsilon$ with $\varepsilon = \left(0, 0, 0, 0, 0, -\frac{2}{3}, -\frac{2}{3}, -\frac{2}{3}\right)$ For $\Lambda_0 \neq 0$ the only case for which the reduction level is strictly higher than one (all the rest are excluded for k_1-dominance arguments) is the following

$$\langle \Lambda_0, \mu_i \rangle = 0 \qquad 2 \leq i \leq 5$$

$$\langle \Lambda_0, \mu_6 \rangle = n > 0.$$

Applying Jakobsen method we obtain

$$\alpha_0 = \beta + \mu_3 + \mu_4 + \mu_5 + \mu_6$$

with height 5 (see Fig. 10). From the condition $\langle \Lambda + R, \alpha_0 \rangle = 1$ we obtain

$$1 = \langle \Lambda, \alpha_0 \rangle + \langle R, \alpha_0 \rangle = \langle \Lambda, \gamma_r \rangle + 5 = \lambda_0 + 5 ;$$

$$\lambda_0 = -4.$$

For $\lambda_q = \lambda_0 + \lambda_s = -7$

$$\left\langle \Lambda + R, \overset{1}{\alpha_8} \right\rangle = \left\langle \Lambda + R, \overset{2}{\alpha_8} \right\rangle = 1$$

then a second order polynomial will be missing with highest weight:

$$\Lambda_0 - 7\varepsilon + R - \overset{1}{\alpha_8} - \overset{2}{\alpha_8}.$$

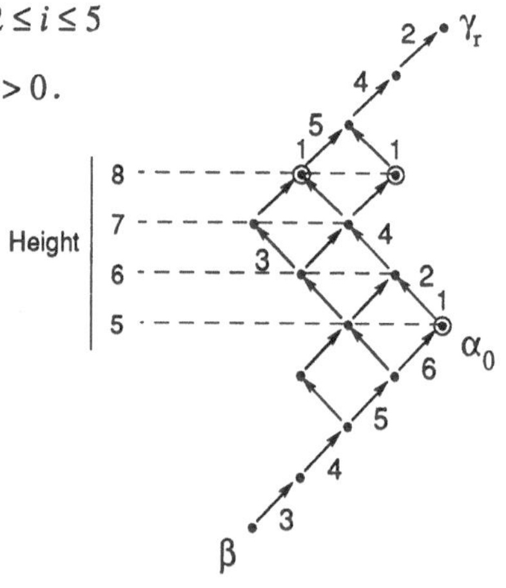

Fig. 10

The value $\lambda_q = -7$ is the first possible place for non unitarity.

The diagram is then:

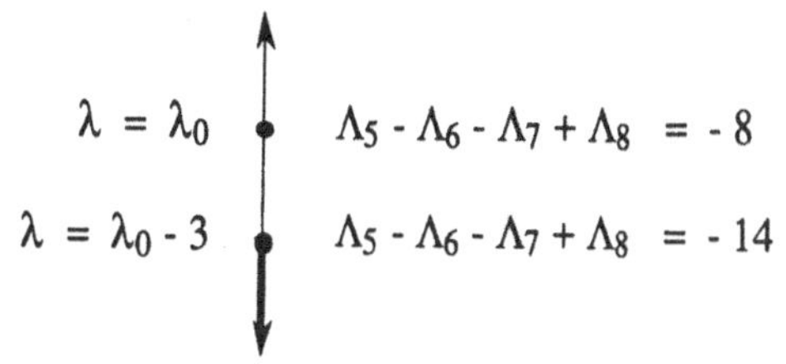

We now apply EHW method:

$$\Lambda = \Pi_0 + z\varepsilon \ , \ \Pi_0 = (\Pi_1, \ldots, \Pi_8) \ ,$$

$$\Pi_7 = -\Pi_8 = \frac{17}{2} .$$

We observe that in the last possible place for unitarity $\Lambda = \Pi_0 + 7\varepsilon$ then:

$\Lambda_5 - \Lambda_6 - \Lambda_7 + \Lambda_8 = -22 + 2 \times 7 = -8$ (we see, from the conditions on Λ_0 that $\Lambda_1 = \Lambda_2 = \Lambda_3 = \Lambda_4 = 0$) and in the first for non unitarity

$$\Lambda_5 - \Lambda_6 - \Lambda_7 + \Lambda_8 = -22 + 2 \times 4 = -14 .$$

e_7

Dynkin diagram

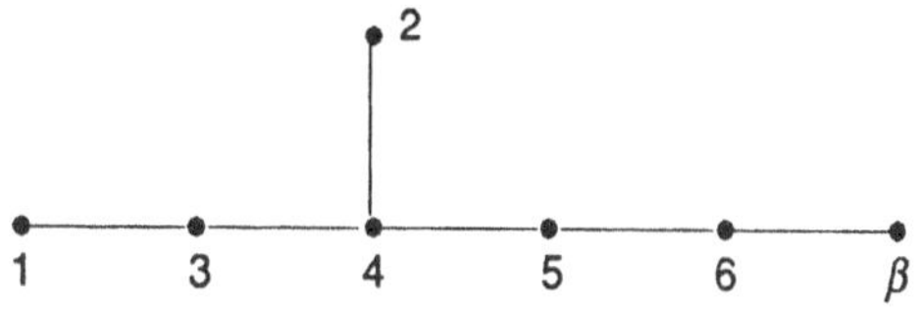

For this case $\varepsilon = \left(0, 0, 0, 0, 0, 1, -\frac{1}{2}, \frac{1}{2}\right)$. As in e_6 we must only consider one case:

$$\langle \Lambda_0, \mu_i \rangle = 0 \qquad 1 \le i \le 5$$

$$\langle \Lambda_0, \mu_6 \rangle = n > 0$$

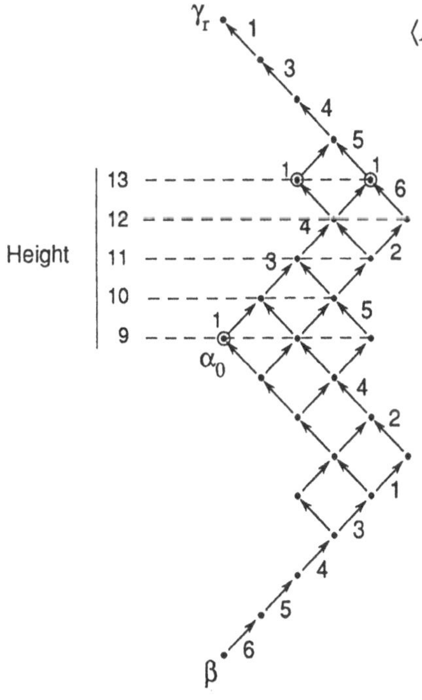

Fig. 11

with those conditions (see Fig. 11) the Jakobsen method gives

$$\alpha_0 = \beta + 2\mu_6 + 2\mu_5 + 2\mu_4 + \mu_3 + \mu_2$$

with height 9, then

$$1 = \langle \Lambda, \alpha_0 \rangle + \langle R, \alpha_0 \rangle = \langle \Lambda, \alpha_0 \rangle + 9 = \lambda_0 + 9 ;$$

$$\lambda_0 = -8 .$$

For $\lambda_q = -12$

$$\left\langle \Lambda + R, \alpha^1_{13} \right\rangle = \left\langle \Lambda + R, \alpha^2_{13} \right\rangle = 1$$

and we obtain, for $\lambda_q = -12$, a missing second order polynomial with highest weight

$$\Lambda_0 - 12\varepsilon + R - \alpha^1_{13} - \alpha^2_{13} .$$

The weight diagram is, then

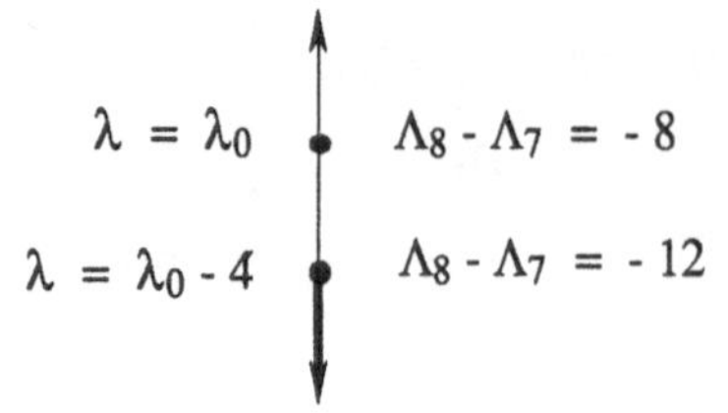

We now apply EHW method $\Lambda = \Pi_0 + z\varepsilon$, $\Pi_0 = (\Pi_1, \Pi_2, \dots, \Pi_8)$ with $\Pi_1 + \Pi_2 + \Pi_3 + \Pi_4 + \Pi_5 - \Pi_6 - \Pi_7 + \Pi_8 = -22$. We observe that in the last possible place for unitarity (Lemma 34):

$$\Lambda = \Pi_0 + 9\varepsilon \quad \Lambda_8 = \Pi_8 + \frac{9}{2} \qquad \Lambda_7 = \Pi_7 - \frac{9}{2}$$

then $\Lambda_8 - \Lambda_7 = -8$ and in the first for non unitarity by Theorem 35 , $\Lambda = \Pi_0 + 5\varepsilon$ then $\Lambda_8 - \Lambda_7 = -12$ as desired.

B) Cases with one single reduction point

In the following we want to consider the cases with one single reduction point that are not included in the general scheme studied in case **A**). The missing highest weights are all of the form $\Lambda - \alpha_0$.

e_6

With the conditions given for Λ_0, the values for α_0 (see Fig. 12) and λ_0 are the following ones (we observe that the equation $\langle \Lambda + R , \alpha_0 \rangle = 1$ implies $\lambda_0 = 1 - \mathcal{H}_{\alpha_0}$):

1) $\langle \Lambda_0 , \mu_2 \rangle > 0$

$\alpha_0 = \gamma_r$

$\lambda_0 = -10.$

2) $\langle \Lambda_0 , \mu_2 \rangle = 0 \;\; \langle \Lambda_0 , \mu_4 \rangle > 0$

$\alpha_0 = \gamma_r - \mu_2$

$\lambda_0 = -9.$

3) $\langle \Lambda_0 , \mu_2 \rangle = \langle \Lambda_0 , \mu_4 \rangle = 0$

$\langle \Lambda_0 , \mu_5 \rangle > 0 \;\; \langle \Lambda_0 , \mu_3 \rangle > 0$

$\alpha_0 = \gamma_r - \mu_2 - \mu_4$

$\lambda_0 = -8.$

4) a) $\langle \Lambda_0 , \mu_2 \rangle = \langle \Lambda_0 , \mu_4 \rangle = \langle \Lambda_0 , \mu_3 \rangle = 0$

$\langle \Lambda_0 , \mu_5 \rangle > 0$

$\alpha_0 = \gamma_r - \mu_2 - \mu_4 - \mu_3$

$\lambda_0 = -7.$

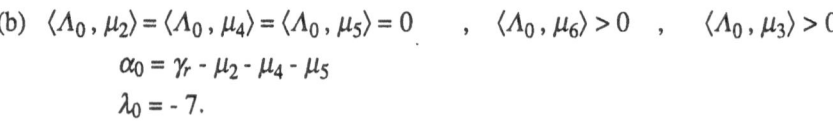

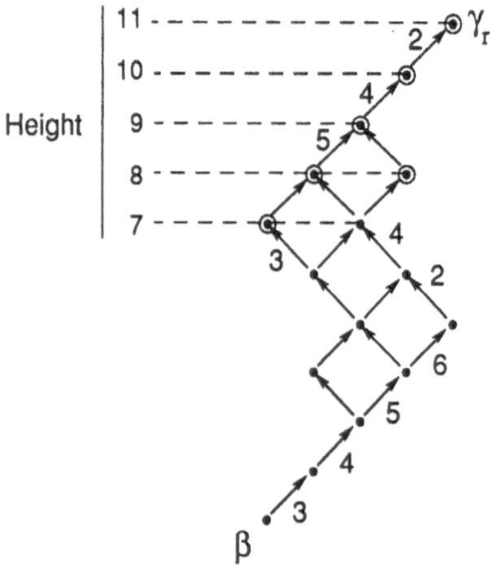

Fig. 12

(b) $\langle \Lambda_0 , \mu_2 \rangle = \langle \Lambda_0 , \mu_4 \rangle = \langle \Lambda_0 , \mu_5 \rangle = 0$, $\langle \Lambda_0 , \mu_6 \rangle > 0$, $\langle \Lambda_0 , \mu_3 \rangle > 0$

$\alpha_0 = \gamma_r - \mu_2 - \mu_4 - \mu_5$

$\lambda_0 = -7.$

5) $\langle \Lambda_0 , \mu_2 \rangle = \langle \Lambda_0 , \mu_4 \rangle = \langle \Lambda_0 , \mu_5 \rangle = \langle \Lambda_0 , \mu_6 \rangle = 0 \;\; \langle \Lambda_0 , \mu_3 \rangle > 0$

$\alpha_0 = \gamma_r - \mu_2 - \mu_4 - \mu_5 - \mu_6$

$\lambda_0 = -6.$

Given $\Lambda = (\Lambda_1 , \Lambda_2 , \dots , \Lambda_8)$ from the condition $\langle \Lambda , \gamma_r \rangle = \lambda$ we obtain

$$\Lambda_1 + \Lambda_2 + \Lambda_3 + \Lambda_4 + \Lambda_5 - \Lambda_6 - \Lambda_7 + \Lambda_8 = 2\lambda .$$

Then we will have the following diagram for the conditions i) , $1 \le i \le 5$:

$\lambda = \lambda_0$  $\Lambda_1 + \Lambda_2 + \Lambda_3 + \Lambda_4 + \Lambda_5 - \Lambda_6 - \Lambda_7 + \Lambda_8 = -22 + 2i$.

Now let $\Lambda = \Pi_0 + z\varepsilon$, as in EHW method for the algebra e_6 in A).

We have, following Theorem 28 that for the paragraph i) , $1 \le i \le 5$:

$$\Lambda_1 + \Lambda_2 + \Lambda_3 + \Lambda_4 + \Lambda_5 - \Lambda_6 - \Lambda_7 + \Lambda_8 = -22 + 2i .$$

e_7

For this algebra (Fig. 13) holds also

$$\lambda_0 = 1 - \mathcal{H}_{\alpha_0} :$$

1) $\langle \Lambda_0 , \mu_1 \rangle > 0$

$\alpha_0 = \gamma_r$

$\lambda_0 = -16$.

2) $\langle \Lambda_0 , \mu_1 \rangle = 0 \quad \langle \Lambda_0 , \mu_3 \rangle > 0$

$\alpha_0 = \gamma_r - \mu_1$

$\lambda_0 = -15$.

3) $\langle \Lambda_0 , \mu_1 \rangle = \langle \Lambda_0 , \mu_3 \rangle = 0 , \quad \langle \Lambda_0 , \mu_4 \rangle > 0$

$\alpha_0 = \gamma_r - \mu_1 - \mu_3$

$\lambda_0 = -14$.

4) $\langle \Lambda_0 , \mu_1 \rangle = \langle \Lambda_0 , \mu_3 \rangle = \langle \Lambda_0 , \mu_4 \rangle = 0$

$\langle \Lambda_0 , \mu_2 \rangle > 0 \quad \langle \Lambda_0 , \mu_5 \rangle > 0$

$\alpha_0 = \gamma_r - \mu_1 - \mu_3 - \mu_4$

$\lambda_0 = -13$.

5) (a) $\langle \Lambda_0 , \mu_1 \rangle = \langle \Lambda_0 , \mu_3 \rangle = \langle \Lambda_0 , \mu_4 \rangle =$

$= \langle \Lambda_0 , \mu_2 \rangle = 0 \quad , \quad \langle \Lambda_0 , \mu_5 \rangle > 0$

$\alpha_0 = \gamma_r - \mu_1 - \mu_3 - \mu_4 - \mu_2$

$\lambda_0 = -12$.

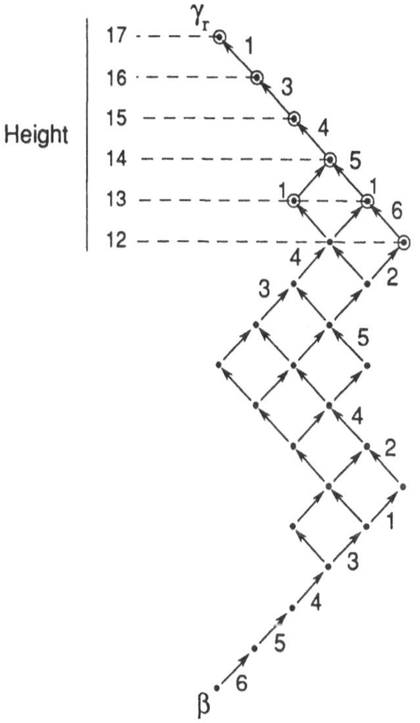

Fig. 13

(b) $\langle \Lambda_0 , \mu_1 \rangle = \langle \Lambda_0 , \mu_3 \rangle = \langle \Lambda_0 , \mu_4 \rangle = \langle \Lambda_0 , \mu_5 \rangle = 0 , \quad \langle \Lambda_0 , \mu_2 \rangle > 0 , \quad \langle \Lambda_0 , \mu_6 \rangle > 0$

$\alpha_0 = \gamma_r - \mu_1 - \mu_3 - \mu_4 - \mu_5$

$\lambda_0 = -12$.

6) $\langle \Lambda_0 , \mu_1 \rangle = \langle \Lambda_0 , \mu_3 \rangle = \langle \Lambda_0 , \mu_4 \rangle = \langle \Lambda_0 , \mu_5 \rangle = \langle \Lambda_0 , \mu_6 \rangle = 0 \quad , \quad \langle \Lambda_0 , \mu_2 \rangle > 0$

$\alpha_0 = \gamma_r - \mu_1 - \mu_3 - \mu_4 - \mu_5 - \mu_6$

$\lambda_0 = -11$.

Here also $\Lambda = (\Lambda_1 , \Lambda_2 , \dots , \Lambda_8)$ and from $\langle \Lambda , \gamma_r \rangle = \lambda$ we have $\Lambda_8 - \Lambda_7 = \lambda$.

Thus the diagram for the paragraph i) , $1 \le i \le 6$ is

$\lambda = \lambda_0$ $\quad \Lambda_8 - \Lambda_7 = -17 + i.$

Given $\Lambda = \Pi_0 + z\varepsilon$ as in the EHW method for this algebra in A) we have following Theorem 33:

$$\Lambda_8 = \Pi_8 + \frac{i}{2} \;, \quad \Lambda_7 = \Pi_7 - \frac{i}{2}$$

therefore $\Lambda_8 - \Lambda_7 = -17 + i$ for the paragraph i), $1 \le i \le 6$.

so (2n - 1, 2)

We consider two cases.

Case I

Given the following conditions:

$\langle \Lambda_0 , \mu_2 \rangle = \langle \Lambda_0 , \mu_3 \rangle = \ldots = \langle \Lambda_0 , \mu_{i-1} \rangle = 0 \qquad \langle \Lambda_0 , \mu_i \rangle > 0$, $i < n$ or $i = n$ and $\langle \Lambda_0 , \mu_n \rangle > 1$ which are equivalent to the following ones:

$$\Lambda_2 = \Lambda_3 = \ldots = \Lambda_{i-1} = \Lambda_i > \Lambda_{i+1} \qquad \Lambda_n = \frac{1}{2} \quad \text{if} \quad i = n.$$

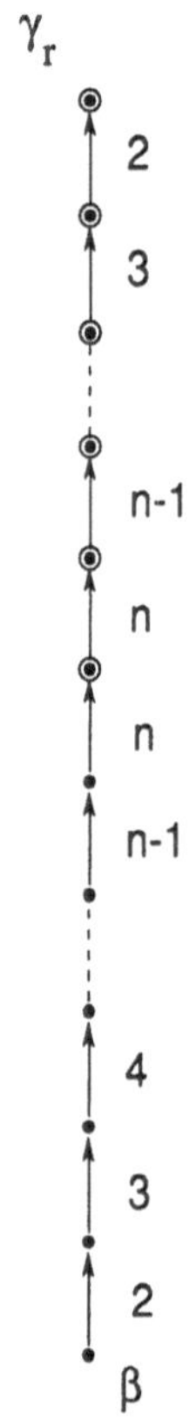

Fig. 14

In this case $\alpha_0 = \gamma_r - \mu_2 - \mu_3 - \ldots - \mu_{i-1}$ (Fig. 14) and $\lambda_0 = i - 2n + 1.$

Given

$$\Lambda = (\Lambda_1 , \Lambda_2 , \ldots , \Lambda_n)$$

from

$$\langle \Lambda , \gamma_r \rangle = \Lambda_1 + \Lambda_2 = \lambda$$

we obtain the diagram:

$\lambda = \lambda_0$ $\quad \Lambda_2 = -\Lambda_1 + i - 2n + 1.$

Using EHW method we obtain, following Theorem 21

$$\left(\Lambda_1 , \Lambda_2 , \ldots , \Lambda_n\right) = \left(\Pi_1 , \Pi_2 , \ldots , \Pi_n\right) + (i - 1)(1 , 0 , \ldots , 0)$$

and having in mind that $\Pi_1 + \Pi_2 = -2n + 2$ we have

$$\Lambda_1 + \Lambda_2 = i - 2n + 1 .$$

Case II

Let now be the following conditions on Λ_0 :

$$\langle \Lambda_0 , \mu_2 \rangle = \langle \Lambda_0 , \mu_3 \rangle = \ldots = \langle \Lambda_0 , \mu_{n-1} \rangle = 0 \quad , \quad \langle \Lambda_0 , \mu_n \rangle = 1$$

which are equivalent to the following ones:

$$\Lambda_2 = \Lambda_3 = \ldots = \Lambda_n = \frac{1}{2} .$$

In this case $\alpha_0 = \gamma_r - \mu_2 - \ldots - \mu_n$ (Fig. 14) and $\lambda_0 = -n + \frac{3}{2}$.

The weight diagram is now

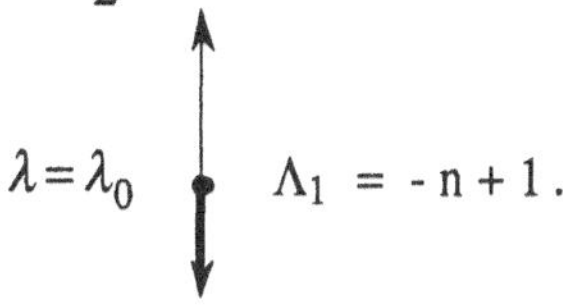

If we apply EHW method we obtain

$$\left(\Lambda_1 , \Lambda_2 , \ldots , \Lambda_n\right) = \left(\Pi_1 , \Pi_2 , \ldots , \Pi_n\right) + \left(n - \frac{1}{2}\right)(1 , 0 , \ldots , 0)$$

therefore $\Lambda_1 + \Lambda_2 = - n + \frac{3}{2}$ and having in mind that $\Lambda_2 = \Lambda_n = \frac{1}{2}$ then $\Lambda_1 = - n + 1$.

so (2n - 2, 2)

Case I

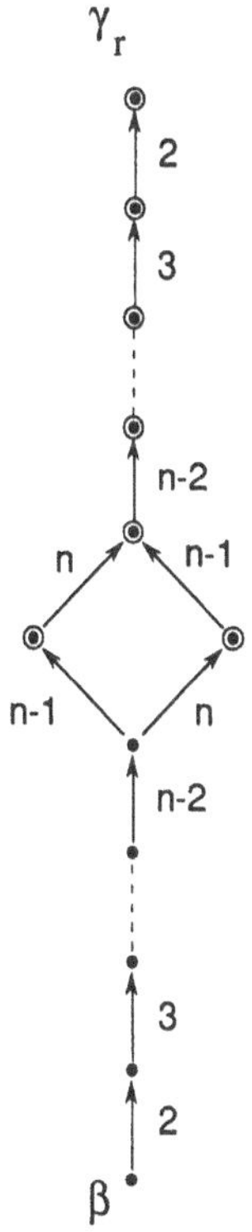

Fig. 15

Assume that

$$\langle \Lambda_0 , \mu_2 \rangle = \langle \Lambda_0 , \mu_3 \rangle = \ldots = \langle \Lambda_0 , \mu_{i-1} \rangle = 0$$
$$\langle \Lambda_0 , \mu_i \rangle > 0 \quad \text{with } i \neq n - 1 \text{ or } i = n - 1 \quad \text{and}$$
$$\langle \Lambda_0 , \mu_n \rangle > 0$$

which are equivalent to the following ones (here also $\Lambda = (\Lambda_1 , \Lambda_2 , \ldots , \Lambda_n) $)

$$\Lambda_2 = \Lambda_3 = \ldots = \Lambda_{i-1} = \Lambda_i > \Lambda_{i+1} \quad i \neq n - 1 \text{ or}$$
$$\Lambda_2 = \Lambda_3 = \ldots = \Lambda_n \geq \frac{1}{2} \quad \text{respectively.}$$

With this conditions we obtain from Jakobsen method

$$\alpha_0 = \gamma_r - \mu_2 - \mu_3 - \ldots - \mu_{i-1}$$

(Fig. 15) and $\lambda_0 = -2n + i + 2$.

The weight diagram is

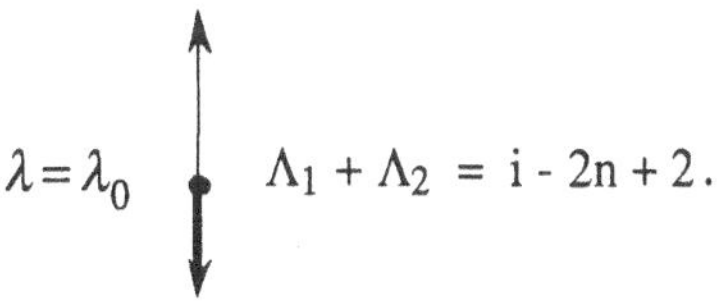

Applying the EHW method we obtain, following Theorem 25

$$\left(\Lambda_1 , \Lambda_2 , \ldots , \Lambda_n\right) = \left(\Pi_1 , \Pi_2 , \ldots , \Pi_n\right) + (i - 1)(1 , 0 , \ldots , 0)$$

and having in mind that $\Pi_1 + \Pi_2 = - 2n + 3$ we have

$$\Lambda_1 + \Lambda_2 - i - 2n + 2 \quad .$$

Case II

$$\langle \Lambda_0 , \mu_2 \rangle = \langle \Lambda_0 , \mu_3 \rangle = \ldots = \langle \Lambda_0 , \mu_{n-2} \rangle = \langle \Lambda_0 , \mu_n \rangle = 0 \quad , \qquad \langle \Lambda_0 , \mu_{n-1} \rangle > 0$$

which are equivalent to the following ones:

$$\Lambda_2 = \Lambda_3 = \ldots = \Lambda_{n\text{-}1} > \Lambda_n \quad \text{and} \quad \Lambda_{n\text{-}1} = -\Lambda_n .$$

In this case we have

$\alpha_0 = \gamma_r - \mu_2 - \mu_3 - \ldots - \mu_{n\text{-}2} - \mu_n$ and $\lambda_0 = 2 - n$.

The weight diagram is then

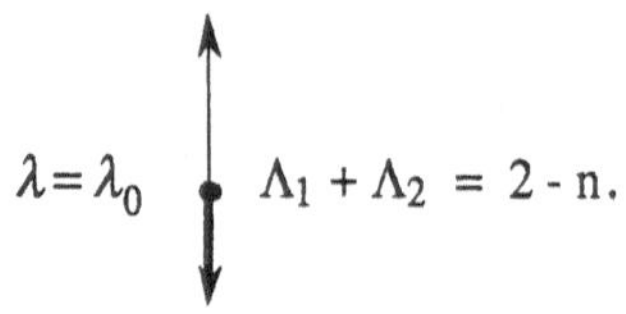

Applying the EHW method we obtain

$$\left(\Lambda_1 , \Lambda_2 , \ldots , \Lambda_n\right) = \left(\Pi_1 , \Pi_2 , \ldots , \Pi_n\right) + (n - 1)(1 , 0 , \ldots , 0)$$

by Theorem 26. Therefore $\Lambda_1 + \Lambda_2 = 2 - n$.

We complete in this way the equivalence between Jakobsen method and EHW method.

VIII. UNITARY REPRESENTATIONS OF SU(2,2)

The algebra su (2 , 2) has Dynkin diagram

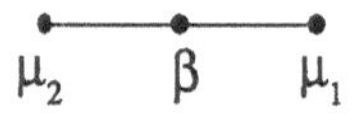

and Jakobsen diagram

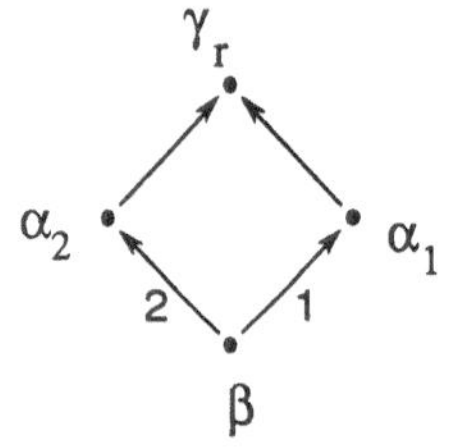

with

$\mu_1 = e_3 - e_4$, $\mu_2 = e_1 - e_2$, $\beta = e_2 - e_3$, $\gamma_r = e_1 - e_4$, $\alpha_1 = e_2 - e_4$, $\alpha_2 = e_1 - e_3$.

Given a weight Λ we may decompose it in the following way

$$\Lambda = \Lambda_0 + \lambda\varepsilon \qquad \text{with} \qquad \varepsilon = \frac{1}{2}\ (1, 1, -1, -1).$$

We consider the following cases:

Case I

Let be the following conditions on the Λ_0 components:

i) $\langle \Lambda_0 , \mu_1 \rangle = 0$

ii) $\langle \Lambda_0 , \mu_2 \rangle = n > 0$.

Having in mind that $\langle \Lambda_0 , \gamma_r \rangle = 0$ we obtain

$$\Lambda_0 = \left(\frac{n}{4} , \frac{-3n}{4} , \frac{n}{4} , \frac{n}{4}\right).$$

For this case the last possible place for unitarity is equal to the first one for non unitarity and it corresponds to $\lambda_0 = - 1$:

$$\Lambda = \left(\frac{n-2}{4}, \frac{-3n-2}{4}, \frac{n+2}{4}, \frac{n+2}{4}\right).$$

We have $\alpha_0 = e_1 - e_3$ (Fig 16) then the weight that we must exclude (missing weight) in order to obtain unitarity is:

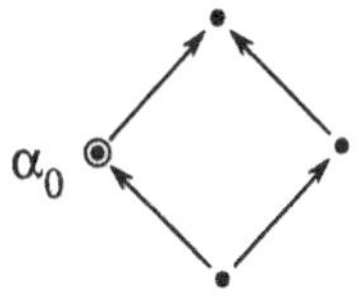

$$\Lambda - \alpha_0 = \left(\frac{n-6}{4}, \frac{-3n-2}{4}, \frac{n+6}{4}, \frac{n+2}{4}\right).$$

Fig. 16

Case II

i) $\langle \Lambda_0 , \mu_1 \rangle = n > 0$

ii) $\langle \Lambda_0 , \mu_2 \rangle = 0$

In this case

$$\Lambda_0 = \left(\frac{-n}{4}, \frac{-n}{4}, \frac{3n}{4}, \frac{-n}{4}\right).$$

Here also $\lambda_0 = -1$, then

$$\Lambda = \left(\frac{-n-2}{4}, \frac{-n-2}{4}, \frac{3n+2}{4}, \frac{-n+2}{4}\right).$$

Having in mind that $\alpha_0 = e_2 - e_4$ (Fig 17) the missing weight will be

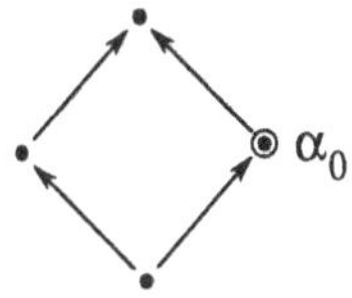

$$\Lambda - \alpha_0 = \left(\frac{-n-2}{4}, \frac{-n-6}{4}, \frac{3n+2}{4}, \frac{-n+6}{4}\right).$$

Fig. 17

Case III.

i) $\langle \Lambda_0 , \mu_1 \rangle = n > 0$

ii) $\langle \Lambda_0 , \mu_2 \rangle = m > 0$

With this conditions

$$\Lambda_0 = \left(\frac{m-n}{4}, \frac{-3m-n}{4}, \frac{m+3n}{4}, \frac{m-n}{4}\right)$$

in this case $\lambda_0 = -2$ then

$$\Lambda = \left(\frac{m-n-4}{4}, \frac{-3m-n-4}{4}, \frac{m+3n+4}{4}, \frac{m-n+4}{4}\right).$$

As we can see in Fig 18 in this case $\alpha_0 = \gamma_r$ .The missing weight will be

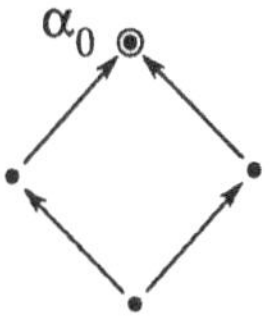

$$\Lambda - \alpha_0 = \left(\frac{m-n-8}{4}, \frac{-3m-n-4}{4}, \frac{m+3n+4}{4}, \frac{m-n+8}{4}\right).$$

Fig. 18

Having in mind the above results we can stablish the set of unitary representations (we remind that $n , m \in \mathcal{N}$ and $\lambda \in \mathfrak{R}$) for su (2 , 2):

a) $\left(\frac{n-2}{4}, \frac{-3n-2}{4}, \frac{n+2}{4}, \frac{n+2}{4}\right) \quad \forall n > 0$

Extremal vector: $Y = E_{-\beta} E_{-\mu_2} + nE_{-\alpha_2} \quad , \quad \Lambda_Y = \Lambda - \alpha_2$.

b) $\left(\frac{n+2\lambda}{4}, \frac{-3n+2\lambda}{4}, \frac{n-2\lambda}{4}, \frac{n-2\lambda}{4}\right) \quad \forall n > 0 \; , \; \forall \lambda < -1$.

c) $\left(\frac{-n-2}{4}, \frac{-n-2}{4}, \frac{3n+2}{4}, \frac{-n+2}{4}\right) \quad \forall n > 0$

Extremal vector: $Y = E_{-\beta} E_{-\mu_1} - nE_{\alpha_1} \quad , \quad \Lambda_Y = \Lambda - \alpha_1$.

d) $\left(\frac{-n+2\lambda}{4}, \frac{-n+2\lambda}{4}, \frac{3n-2\lambda}{4}, \frac{-n-2\lambda}{4}\right) \quad \forall n > 0 \; , \; \forall \lambda < -1$.

e) $\left(\frac{m-n-4}{4}, \frac{-3m-n-4}{4}, \frac{m+3n+4}{4}, \frac{m-n+4}{4}\right) \quad \forall \; m, n > 0$

Extremal vector: $Y = E_{-\beta} E_{-\mu_1} E_{-\mu_2} - nE_{-\alpha_1} E_{-\mu_2} + mE_{-\alpha_2} E_{-\mu_1} - mnE_{-\gamma_r} \quad , \quad \Lambda_Y = \Lambda - \gamma_r$.

f) $\left(\frac{m-n+2\lambda}{4}, \frac{-3m-n+2\lambda}{4}, \frac{m+3n-2\lambda}{4}, \frac{m-n-2\lambda}{4}\right) \quad \forall \; m, n > 0 \quad \forall \lambda < -2$.

The restriction to the maximal compact subgroup SU(2) x SU (2) x U (1) is given by the generators (LO86):

$$J_{\pm}^{(1)} = E_{\pm\mu_2} \quad , \quad J_3^{(1)} = \frac{1}{2}(H_1 - H_2)$$

$$J_{\pm}^{(2)} = E_{\pm\mu_1} \quad , \quad J_3^{(2)} = \frac{1}{2}(H_3 - H_4)$$

$$R_0 = \frac{1}{2}(H_1 + H_2 - H_3 - H_4) .$$

Denoting the eigenvalues of $J_3^{(1)}$, $J_3^{(2)}$ and R_0 by j_1, j_2 y d respectively we have the following correspondence with the components of the highest weights in the standard orthonormal basis:

$$j_1 = \frac{1}{2} \; (\Lambda_1 - \Lambda_2) \qquad j_2 = \frac{1}{2} \; (\Lambda_3 - \Lambda_4) \qquad d = \frac{1}{2} \; (\Lambda_1 + \Lambda_2 - \Lambda_3 - \Lambda_4) .$$

In the following we give the unitary representations of su (2 , 2) in the form $(j_1, j_2; d)$ showing its Poincaré content (MA77).

a) $\left(\frac{n}{2}, 0; \frac{-n-2}{2}\right) \qquad n = 1, 2, 3, \ldots$

Missing weight: $\left(\frac{n-1}{2}, \frac{1}{2}; \frac{-n-4}{2}\right)$

Mass = 0 , Helicity = $\frac{n}{2}$.

b) $\left(\frac{n}{2}, 0; \frac{-n+2\lambda}{2}\right) \qquad n = 1, 2, 3, \ldots, -\infty < \lambda < -1$

Mass $\neq 0$, Spin = $\frac{n}{2}$.

c) $\left(0, \frac{n}{2}; \frac{-n-2}{2}\right) \qquad n = 1, 2, 3, \ldots$

Missing weight ; $\left(\frac{1}{2}, \frac{n-1}{2}, \frac{-n-4}{2}\right)$

Mass = 0 , Helicity = $\frac{n}{2}$.

d) $\left(0, \frac{n}{2}; \frac{-n+2\lambda}{2}\right)$ $\quad n = 1, 2, 3, \ldots, \quad -\infty < \lambda < -1$

Mass $\neq 0$, Spin $= \frac{n}{2}$.

e) $\left(\frac{m}{2}, \frac{n}{2}; \frac{-m-n-4}{2}\right)$ $\quad n, m = 1, 2, 3, \ldots$

Missing weight: $\left(\frac{m-1}{2}, \frac{n-1}{2}; \frac{-m-n-6}{2}\right)$ $\quad n, m = 1, 2, 3, \ldots$

Mass $\neq 0$, Spin $= \frac{m+n}{2}$.

f) $\left(\frac{m}{2}, \frac{n}{2}; \frac{-m-n+2\lambda}{2}\right)$ $\quad n, m = 1, 2, 3, \ldots, \quad -\infty < \lambda < -2$

Mass $\neq 0$, Spin $= \left|\frac{n-m}{2}\right|, \left|\frac{n-m}{2}\right| + 1, \ldots, \left|\frac{n+m}{2}\right|$.

IX. UNITARY REPRESENTATIONS OF SO(3,2)

The Dynkin diagram of the de Sitter algebra so(3,2) is

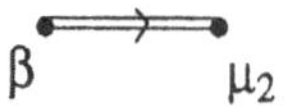

and its Jakobsen diagram

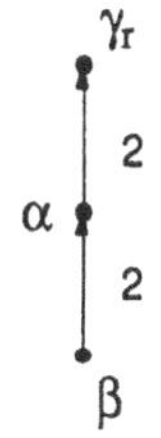

with $\mu_2 = e_2$, $\beta = e_1 - e_2$, $\gamma_r = e_1 + e_2$ and $\alpha = e_1$.

In the weight decomposition $\Lambda = \Lambda_0 + \lambda\varepsilon$ we have for this algebra $\varepsilon = (1, 0)$.

Let be now the following cases:

Case I

$\langle \Lambda_0, \mu_2 \rangle = m > 1$ or what is the same (because $\langle \Lambda_0, \gamma_r \rangle = 0$) $\Lambda_0 = \left(-\frac{m}{2}, \frac{m}{2}\right)$

In this case $\lambda_0 = -1$, then

$$\Lambda = \left(\frac{-m-2}{2}, \frac{m}{2}\right).$$

We have only one reduction point, the missing weight in this place will be (Fig. 19)

$$\Lambda - \alpha_0 = \Lambda - \gamma_r = \left(\frac{-m-4}{2}, \frac{m-2}{2}\right).$$

Fig. 19

Case II

$\langle \Lambda_0, \mu_2 \rangle = 1$ or, equivalently $\Lambda_0 = \left(-\frac{1}{2}, \frac{1}{2}\right)$

and having in mind that $\Lambda_1 = -1$

$$\Lambda = \left(-1, \frac{1}{2}\right).$$

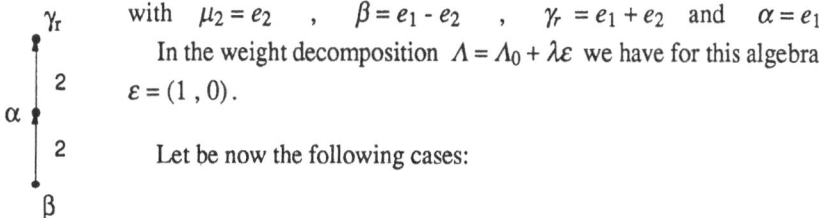

The missing weight will be (Fig. 20)

$$\Lambda - \alpha_0 = \left(-2, \frac{1}{2}\right).$$

Fig. 20

Then the set of unitary representations for the algebra so (3 , 2) ($m \in \mathcal{N}, \lambda \in \Re$) is the following:

a) $\left(\frac{-m-2}{2}, \frac{m}{2}\right)$ $\quad \forall m \geq 2$

Missing weight $\left(\frac{-m-4}{2}, \frac{m-2}{2}\right)$

Extremal vector: $E_{-\beta}E^2_{-\mu} - (m-1)E_{-\alpha}E_{-\mu} + \frac{m(m-1)}{2}E_{-\gamma_r}$.

b) $\left(\frac{-m+2\lambda}{2}, \frac{m}{2}\right)$ $\quad \forall m \geq 2 \quad \forall \lambda < -1$.

c) $\left(-1, \frac{1}{2}\right)$

Missing weight $\left(-2, \frac{1}{2}\right)$

Extremal vector: $E_{-\beta}E_{-\mu} - \frac{1}{2}E_{-\alpha}$.

d) $\left(\frac{-1+2\lambda}{2}, \frac{1}{2}\right)$ $\quad \forall \lambda < -\frac{1}{2}$.

e) $\left(\frac{-1}{2}, 0\right)$. This highest weight corresponds to the most singular non trivial unitary module $\Lambda_0 = (0, 0)$.

X. WAVE EQUATIONS FOR CONFORMAL MULTISPINORS

For the generators of the fundamental representation of the conformal group, Dirac (DI36) uses the operators:

$$\beta_a = \left(\gamma_1, \gamma_2, \gamma_3, \gamma_4, \gamma_5, i\right)$$

$$\gamma_a = \left(\gamma_1, \gamma_2, \gamma_3, \gamma_4, \gamma_5, -i\right)$$

where the β's and γ's satisfy $\beta_a\gamma_b + \beta_b\gamma_a = 2\eta_{ab}$, $\eta_{ab} = \mathrm{diag}\,(+, +, +, -, +, -,)$ being the metric tensor. On four dimensional conformal space

$$x^2 = \eta_{ab}x^a x^b = x_1^2 + x_2^2 + x_3^2 - x_4^2 + x_5^2 - x_6^2 = 0 \quad , \quad x_a = \lambda x_a$$

the conformal group acts linearly. To avoid any dynamics along the rays we fix the degree of homogeneity of all fields by imposing $x^a \partial_a \psi = n\,\psi$. The generators for the fundamental representation are then

$$J_{ab} = \frac{1}{i}\left(x_a\partial_b - x_b\partial_a\right) + \frac{1}{4i}\left(\beta_a\gamma_b - \beta_b\gamma_a\right) \equiv M_{ab} + S_{ab} .$$

The generators of the contragradient representation are obtained by interchanging the β's and γ's , namely,

$$J_{ab} = \frac{1}{i}\left(x_a\partial_b - x_b\partial_a\right) + \frac{1}{4i}\left(\gamma_a\beta_b - \gamma_b\beta_a\right) .$$

Both representations have the spin content (1/2 , 0) and (0 , 1/2) respectively. These representations act on some spinor fields of one index ψ_α and $\psi_{\dot\alpha}$ respectively.

Multispinor field transform under the direct product of the fundamental and contragradient representations (2*j* and 2*k* times, respectively). They transform under the infinitesimal generators in the following way:

$$\delta\psi_{\alpha_1\alpha_2\ldots\alpha_{2j}} = \varepsilon^{ab}\frac{1}{2i}\sum_{i=1}^{2j}\left(\beta_a\gamma_b\right)_{\alpha_i}^{\alpha'_i}\psi_{\alpha_1\ldots\alpha'_i\ldots\alpha'_{2j}}$$

$$\delta\psi_{\dot\alpha_1\dot\alpha_2\ldots\dot\alpha_{2k}} = \varepsilon^{ab}\frac{1}{2i}\sum_{i=1}^{2k}\left(\gamma_a\beta_b\right)_{\dot\alpha_i}^{\dot\alpha'_i}\psi_{\dot\alpha_1\ldots\dot\alpha_i\ldots\dot\alpha_{2k}}$$

where the multispinors are traceless and totally symmetric in the α_i and $\alpha_{\dot{i}}$. They correspond to the representations $(j, 0; j+1)$ and $(0, k; k+1)$ in the Yao (YA67) classification. All these representations are defined on the light cone $(x^2 = 0)$ and therefore, using homogeneous coordinates on a 4-dimensional manifold, when restricted to the Poincaré group, they become massless particles of helicity j and k, respectively (MA77).

The wave equation will be worked out with the help of the second order Casimir operator, i.e.

$$C_2 = \frac{1}{2}J_{ab}J^{ab} = \frac{1}{2}M_{ab}M^{ab} + \frac{1}{2}S_{ab}S^{ab} = \frac{1}{2}M^2 + MS + \frac{1}{2}S^2 .$$

Each term of the Casimir operator commutes with the others and with the operator $x^a\partial_a \equiv x\partial.$. Therefore, they must have a common set of eigenfunctions, which are at the same time, the carrier space where the representations acts.

Applying S^2 to the highest state of some irreducible representation $(j, 0; j+1)$ we get

$$\frac{1}{2}S^2 = 2j(j+1) + j(j+4)$$

and similarly for the representation $(0, k; k+1)$:

$$\frac{1}{2}S^2 = 2k(k+1) + k(k+4) .$$

The operator $(x\partial)$ has the eigenvalue n, i.e. the homogeneity degree. The operator $\frac{1}{2}M^2$ has therefore eigenvalue $n^2 + 4n$ (recall $x^2 = 0$). Collecting all the eigenvalues of the operators M^2, MS, S^2 and $(x\partial)$, we get for the Casimir operator:

$$C_2 = n^2 + 4n + m + 3j(j+2) = 3(j^2 - 1)$$

and similar expresion for the representation $(0, k; k+1)$:

$$C_2 = n^2 + 4n + m + 3k(k+2) = 3(k^2 - 1) .$$

The eigenvalue equation for the operator MS gives

$$\sum_{i=1}^{2j}(\beta x)_i(\gamma\partial)_i\,\psi = (2jn - m)\psi \quad \text{for} \quad (j, 0; j+1)$$

with two cases:

- $2jn \neq m$. Applying $\prod_i(\gamma\partial)_i$ from the left, we get

$$2j(6 + 2(n-1))\prod_i(\gamma\partial)_i\,\psi = (2jn - m)\prod_i(\gamma\partial)_i\,\psi .$$

Therefore $m = -2j(2j+1)$, $n = 2j - 3$; $m = -6j$, $n = -1$.

- $2jn = m$. From the Casimir operator, we get

$$n = -(2j+1) \;,\; m = -2j(2j+1) \;;\; n = -3 \;,\; m = -6j.$$

For $n = -(2j+1)$ we have the field ψ satisfying

$$\sum_{i=1}^{2j}(\beta x)_i(\gamma\partial)_i\,\psi \equiv Q\psi = 0 .$$

They correspond to the physical and gauge modes in the Gupta - Bleuler formalism, of electrodynamics as it will be shown.

Their defining equation corresponds to a generalized "Lorentz condition", $\partial^{\mu} A_{\mu} = 0$.

For $n = -1$, we have the field χ satisfying

$$\sum_{i=1}^{2j} (\beta x)_i (\gamma \partial)_i \chi = 4j\chi .$$

It must describe massless particle irreducibly as the field strength $F_{\mu\nu}$ does.

Between the physical and the irreducible mode we impose the condition

$$\chi = \prod_{i=1}^{2j} (\gamma x)_i \psi$$

equivalent to the relation between the field potential and the field strength

$$\left(F_{\mu\nu} = \partial_{\mu} A_{\nu} - \partial_{\nu} A_{\mu}\right).$$

Now we want an explicit realization for the solutions of the field equations. In the representation $D(j, 0; j+1)$ the lowest state corresponding to the massless physical state can be written as:

$$\psi_p^0 = \frac{1}{x_+^{2j+1}} \underbrace{\begin{pmatrix} 0 \\ 1 \\ 0 \\ 0 \end{pmatrix} \times \ldots \times \begin{pmatrix} 0 \\ 1 \\ 0 \\ 0 \end{pmatrix}}_{\text{2j terms}}$$

the weight of which is $(j, 0; j+1)$. Using the raising generators of the so(4,2) algebra we can construct all the states where the (indecomposable) representations acts. These states satisfy the same field equation as the lowest state, although they have different weights (HL88).

The lowest state of the gauge mode has the realization

$$\psi_g^0 = \frac{1}{x_+^{2j+2}} \underbrace{\begin{pmatrix} 0 \\ 0 \\ 0 \\ 1 \end{pmatrix} \times \begin{pmatrix} 0 \\ 1 \\ 0 \\ 0 \end{pmatrix} \times \ldots \times \begin{pmatrix} 0 \\ 1 \\ 0 \\ 0 \end{pmatrix}}_{\text{2j terms}} + \text{symm. terms}$$

with weight $\left(-j + \frac{1}{2}, -\frac{1}{2}; j+2\right)$. It corresponds to the extremal vector Y obtained by the method outlined before, namely

$$\psi_g^0 = \mathrm{Y} \psi_p^0 = \left(E_{\beta} E_{\mu_2} - 2j\, E_{\alpha_2}\right) \psi_p^0 .$$

Therefore it generates an invariant subspace in the representation defined by the physical state ψ_p^0. If we take the quotient space with respect to this invariant subspace, we are left with the unitary (infinite dimensional) representation corresponding to the massless particle of helicity *j*.

The field χ has the lowest state

$$\chi^0 = \frac{1}{x_+^{2j+1}} \underbrace{(\gamma x) \begin{pmatrix} 0 \\ 1 \\ 0 \\ 0 \end{pmatrix} \times \ldots \times (\gamma x) \begin{pmatrix} 0 \\ 1 \\ 0 \\ 0 \end{pmatrix}}_{\text{2j terms}}$$

with weight $(j, 0; j+1)$. It generates an irreducible representation because it does not contain extremal vectors. In fact, it can be checked:

$$Y\chi^0 = \left(E_\beta E_{\mu_2} - 2j\, E_{\alpha_2}\right)\chi^0 = 0 .$$

In order to complete the Gupta - Bleuler formalism we need also a scalar field ψ_s (corresponding to the longitudinal component of the electromagnetic field) with the following properties:

i) it has the same weight as the corresponding gauge field.

ii) it does not belong to the envelopping algebra (i.e. it can not be obtained by some raising generators applied to ψ_p^0).

iii) it is a cyclic state, i.e. the physical lowest state ψ_p^0 and the states of the (indecomposable) representation can be obtain from this scalar state. It can be realized by the state

$$\psi_s^0 = \frac{1}{x_+^{2j+2}} \underbrace{\begin{pmatrix}0\\0\\0\\1\end{pmatrix} \times \begin{pmatrix}0\\1\\0\\0\end{pmatrix} \times \dots \times \begin{pmatrix}0\\1\\0\\0\end{pmatrix}}_{2j \text{ terms}} + \text{symm. terms}$$

with weight $\left(-j+\frac{1}{2}, -\frac{1}{2}; j+2\right)$ satisfying $E_{-\alpha_2}\psi_s^0 = \psi_p^0$, as required. Finally for the scalar field we have $n = -(2j+1)$, and the following equation holds:

$$Q\,\psi_s^0 = \psi_q^0 \quad , \quad \text{hence} \quad , \quad Q^2\,\psi_s = 0 .$$

The three fields, scalar, physical and gauge have the same degree of homogeneity and satisfy the same fields equation.

Among the corresponding representation the following leakage takes place:

$$D\left(j-\frac{1}{2}, \frac{1}{2}; j+2\right) \rightarrow D\left(j, 0; j+1\right) \rightarrow D\left(j-\frac{1}{2}, \frac{1}{2}; j+2\right)$$

hence, they constitue a Gupta - Bleuler triplet.

XI. WAVE EQUATIONS IN DE SITTER SPACE

The coordinates x_a, $a = 1, 2, 3, 4, 6$ satisfy

$$x^2 = \eta_{a\beta}x^a x^b = x_1^2 + x_2^2 + x_3^2 - x_4^2 - x_6^2 < 0 \quad , \quad x = \lambda x \quad , \quad \lambda > 0 .$$

On the four dimensional representation the generators are

$$J_{ab} = -i\left(x_a \partial_b - x_b \partial_a\right) - \frac{i}{4}\left(\beta_a \gamma_b - \beta_b \gamma_a\right)$$

where $\beta_a \gamma_b + \beta_b \gamma_a = 2\eta_{ab}$.

We consider the fully symmetric multispinor fields $\psi_{\{\alpha_1 \dots \alpha_{2j}\}}(x)$ and $\psi_{\{\dot\alpha_1 \dots \dot\alpha_{2j}\}}$ which transform under de Sitter group in the same way as the conformal group. The maximal compact subgroup SO(3) x SO(2) of SO(3,2) gives us the labels of the representations $\left(E_0, j\right)$, E_0 being the energy of the system and j the angular momentum.

The wave equation is constructed with the help of the second order Casimir operator

$$C_2 = \frac{1}{2} J_{ab} J^{ab} = \frac{1}{2} M^2 + MS + \frac{1}{2} S^2 = E_0\left(E_0 - 3\right) + j\left(j+1\right)$$

with the same notation as in the conformal case; we have for the eigenvalues of the commuting operators

$$\frac{1}{2}M^2 = n^2 + 3n \quad , \quad MS = m \quad , \quad \frac{1}{2}S^2 = 2j(j+2)$$

for some common eigenstate in the representation $D(E, j)$.

Notice that we do not work in the light cone $(x^2 = 0)$ but we suppose the scalar field condition $(\partial^2 = 0)$. We have the following cases:

i) $D\left(\frac{1}{2}, 0\right) \quad m = 0 \quad , \quad n = -\frac{1}{2} \; : \; \partial^2\psi = 0.$

It corresponds to the scalar representation founded by Dirac ("RAC" in Fronsdal's notation).

ii) $D\left(1, \frac{1}{2}\right) \quad m = n = -\frac{3}{2} \; : \; (\beta x)(\gamma\partial)\,\psi = 0$

$$= -\frac{5}{2} \quad , \quad n = -\frac{1}{2} \; : \; (\beta x)(\gamma\partial)\,\chi = 2\chi\,.$$

It corresponds to the spinor representation founded by Dirac ("DI" in Fronsdal's notation).

iii) General case $D(j+1, j)$:

$$n = -1 \quad , \quad m = -4j \; : \; \sum_{i=1}^{2j} (\beta x)_i (\gamma\partial)_i\, \chi = 2j\chi$$

$$n = -(2j+1) \quad , \quad m = -2j(2j+1) \; : \; \sum_{i=1}^{2j} (\beta x)_i (\gamma\partial)_i\, \psi = 0\,.$$

It corresponds to the unitary representation of lowest weight (E_0, j) for massless particles (EV67).

An explicit realization of the solutions for these equations in the general case is given by the lowest state for the physical state (HL90)

$$\psi_p^0 = \frac{1}{x_+^{2j+1}} \underbrace{\begin{pmatrix}0\\1\\0\\0\end{pmatrix} \times \ldots \times \begin{pmatrix}0\\1\\0\\0\end{pmatrix}}_{2j \text{ terms}}$$

with weight $(j+1, j)$. This lowest state defines an indecomposable representation on the envelopping algebra of so(3 , 2).

The lowest state of the gauge field corresponding to the invariant subspace generated by the extremal vector, namely, is:

$$\psi_g^0 = \left(E_\beta E_\mu^2 + (2j-1)\, E_{\alpha_1} E_\mu + (2j-1) j\, E_{\alpha_2}\right) \psi_p^0 =$$

$$= \frac{(2j+1)\, i\left(\beta u\, \gamma_5\right)}{(2j+1)} \left[(2j-1) \underbrace{\begin{pmatrix}1\\0\\0\\0\end{pmatrix} \times \begin{pmatrix}0\\1\\0\\0\end{pmatrix} \times \ldots \times \begin{pmatrix}0\\1\\0\\0\end{pmatrix}}_{2j \text{ terms}} - \right.$$
$$\left. - \underbrace{\begin{pmatrix}0\\1\\0\\0\end{pmatrix} \times \begin{pmatrix}1\\0\\0\\0\end{pmatrix} \times \ldots \times \begin{pmatrix}0\\1\\0\\0\end{pmatrix}}_{2j \text{ terms}} - \ldots - \underbrace{\begin{pmatrix}0\\1\\0\\0\end{pmatrix} \times \ldots \times \begin{pmatrix}0\\1\\0\\0\end{pmatrix} \times \begin{pmatrix}1\\0\\0\\0\end{pmatrix}}_{2j \text{ terms}} + \text{symm.} \right]$$

with weight $(j+2, j-1)$. If we take the quotient space of the indecomposable representation with respect to this invariant subspace we are left with the unitary representation representing massless particles of spin j. The scalar state

$$\psi_s^0 = \frac{(2j+1)}{2x_+^{2j+1}} \left[\underbrace{(2j-1)\begin{pmatrix}0\\0\\1\\0\end{pmatrix}\times\begin{pmatrix}0\\1\\0\\0\end{pmatrix}\times\ldots\times\begin{pmatrix}0\\1\\0\\0\end{pmatrix}}_{2j\text{ terms}} - \underbrace{-\begin{pmatrix}0\\0\\0\\1\end{pmatrix}\times\begin{pmatrix}1\\0\\0\\0\end{pmatrix}\times\begin{pmatrix}0\\1\\0\\0\end{pmatrix}\times\ldots\times\begin{pmatrix}0\\1\\0\\0\end{pmatrix}}_{2j\text{ terms}} - \underbrace{\begin{pmatrix}0\\0\\0\\1\end{pmatrix}\times\begin{pmatrix}0\\1\\0\\0\end{pmatrix}\times\ldots\times\begin{pmatrix}0\\1\\0\\0\end{pmatrix}\times\begin{pmatrix}1\\0\\0\\0\end{pmatrix}}_{2j\text{ terms}} \right]$$

is cyclic leaking to the physical state

$$E_{-\alpha_2}\psi_s^0 = -ij\left(4j^2-1\right)\psi_p^0 .$$

The three states ψ_s^0, ψ_p^0, ψ_g^0 define a Gupta - Bleuler triplet, whose representations, are leaking in the following way:

$$D(j+2, j-1) \rightarrow D(j+1, j) \rightarrow D(j+2, j-1).$$

APPENDIX. VERMA MODULES FOR THE CLASSICAL ALGEBRAS

A_{n-1}

Generators:

$H_i = |i\rangle\langle i|$ with $1 \le i \le n$; $G_{ij} = |i\rangle\langle j|$; $\overline{G}_{ij} = |j\rangle\langle i|$ with $1 \le i < j \le n$ where $\langle i|$ denotes the file vector with a 1 at the i position and 0 at the rest of the components, $|i\rangle$ is the corresponding column vector.

Basis in Ω_-:

$$\overline{G}_{12}^{g_{12}}\,\overline{G}_{13}^{g_{13}} \ldots \overline{G}_{1n}^{g_{1n}}\,\overline{G}_{23}^{g_{23}} \ldots \overline{G}_{2n}^{g_{2n}} \ldots \overline{G}_{n-1,n}^{g_{n-1,n}} \equiv \overline{X}\,(g_{12} \ldots g_{n-1,n}).$$

Representation with highest weight $\Lambda = \left(\Lambda_1, \Lambda_2, \ldots, \Lambda_n\right)$:

$$\rho\left(H_i\right)\overline{X} = \left(\Lambda_i + \sum_{k=1}^{i-1} g_{ki} - \sum_{k=i+1}^{n} g_{ik}\right)\overline{X}$$

$$\rho\left(\overline{G}_{ij}\right)_{(j=i+1)}\overline{X} = \overline{X}\left(g_{ij}+1\right) + \sum_{k=1}^{i-1} g_{ki}\ \overline{X}\left(g_{ki}-1, g_{kj}+1\right)\left(g_{ki}-1, g_{kj}+1\right)$$

$$\begin{aligned}\rho\left(G_{ij}\right)_{(j=i+1)}\overline{X} = {} & \sum_{k=1}^{i-1} g_{kj}\,\overline{X}\left(g_{ki}+1, g_{kj}-1\right) + \\ & + g_{ij}\left(\Lambda_i - \Lambda_j + 1 + \sum_{k=j+1}^{n} g_{jk} - \sum_{k=i+1}^{n} g_{ik}\right)\overline{X}\,\left(g_{ij}-1\right) - \\ & - \sum_{k=j+1}^{n} g_{ik}\ \overline{X}\,\left(g_{ik}-1, g_{jk}+1\right)\end{aligned}$$

where, for the sake of simplicity we only show the $\overline{X}$ content when some of its coefficients change.

Extremal vectors in $\Omega\_$:

The vector Y is extremal if it satisfies $\rho(G_{ij})Y = 0$ and its weight

$M = (M_1, \ldots, M_n)$ may be obtained from the relation $\rho(H_i)Y = M_i Y$ with $\sum_{i=1}^{n} M_i = 0$.

The extremal vectors with only one generator are of the form

$$Y = \overline{G}_{ij}^{\Lambda_i - \Lambda_j + 1} \quad ; \quad (j = i + 1)$$

with weight: $M = (\Lambda_1, \Lambda_2, \ldots, \Lambda_{i-1}, \Lambda_{i+1} - 1, \Lambda_i + 1, \ldots, \Lambda_n)$.

B$_n$:

Generators :

$$H_i = |2i-1\rangle\langle 2i-1| - |2i\rangle\langle 2i|$$

$$E_i = |2i-1\rangle\langle 2n+1| - |2n+1\rangle\langle 2i| \qquad \overline{E}_i = |2n+1\rangle\langle 2i-1| - |2i\rangle\langle 2n+1|$$

$$F_{ij} = |2i-1\rangle\langle 2j| - |2j-1\rangle\langle 2i| \qquad \overline{F}_{ij} = |2j\rangle\langle 2i-1| - |2i\rangle\langle 2j-1| \qquad (i<j)$$

$$G_{ij} = |2i-1\rangle\langle 2j-1| - |2j\rangle\langle 2i| \qquad \overline{G}_{ij} = |2j-1\rangle\langle 2i-1| - |2i\rangle\langle 2j| \qquad (i<j)$$

Basis in $\Omega\_$:

$$\overline{E}_1^{e_1} \ldots \overline{E}_n^{e_n} \overline{F}_{12}^{f_{12}} \overline{F}_{13}^{f_{13}} \ldots \overline{F}_{1n}^{f_{1n}} \overline{F}_{23}^{f_{23}} \ldots \overline{F}_{n-1,n}^{f_{n-1,n}} \overline{G}_{12}^{g_{12}} \ldots \overline{G}_{n-1,n}^{g_{n-1,n}} \equiv \overline{X}(e_1, \ldots, g_{n-1,n})$$

Representation with highest weight $\Lambda = (\Lambda_1, \Lambda_2, \ldots, \Lambda_n)$:

$$\rho(H_i)\overline{X} = \left[\Lambda_i - e_i - \sum_{k=i+1}^{n} (f_{ik} + g_{ik}) - \sum_{l=1}^{i-1} (f_{li} - g_{li})\right] \overline{X}$$

$$\rho(E_n)\overline{X} = -e_n\left[-\Lambda_n + \frac{e_n - 1}{2} + \sum_{i=1}^{n-1} (e_i - g_{in})\right] \overline{X}(e_n - 1) +$$

$$+ \frac{e_n(e_n-1)}{2} \sum_{i=1}^{n-1} e_i \overline{X}(e_i - 1, e_n - 2, f_{in} + 1) +$$

$$+ e_n \sum_{i=1}^{n-2} \sum_{j=i+1}^{n-1} e_i e_j \overline{X}(e_i - 1, e_j - 1, e_n - 1, f_{ij} + 1) -$$

$$- \sum_{i=2}^{n-1} \sum_{j=1}^{i-1} e_i f_{jn} \overline{X}(e_i - 1, f_{ji} + 1, f_{jn} - 1) +$$

$$+ \sum_{i=1}^{n-2} \sum_{j=i+1}^{n-1} e_i f_{jn} \overline{X}(e_i - 1, f_{ij} + 1, f_{jn} - 1) +$$

$$+ \sum_{i=2}^{n-1} \sum_{l=1}^{i-1} e_i g_{li} \overline{X}(e_i - 1, g_{li} - 1, g_{ln} + 1) -$$

$$- \sum_{j=1}^{n-1} f_{jn} \overline{X} \left(e_j + 1, f_{jn} - 1\right) + \sum_{i=1}^{n-1} e_i \overline{X} \left(e_i - 1, g_{in} + 1\right) +$$

$$+ \sum_{j=1}^{n-2} \sum_{k=j+1}^{n-1} f_{jn} e_k \overline{X} \left(e_k - 1, f_{jk} + 1, f_{jn} - 1\right)$$

$$\underset{(m=l+1)}{\rho(G_{lm})\overline{X}} = g_{lm} \left[\Lambda_l - \Lambda_m - (g_{lm} - 1) + \sum_{j=m+1}^{n} (g_{mj} - g_{lj}) \right] \overline{X}(g_{lm} - 1) -$$

$$- e_l \overline{X} \left(e_l - 1, e_m + 1\right) + \frac{e_l(e_l - 1)}{2} \overline{X} \left(e_l - 2, f_{lm} + 1\right) -$$

$$- \sum_{j=1}^{l-1} f_{jl} \overline{X} \left(f_{jl} - 1, f_{jm} + 1\right) - \sum_{k=m+1}^{n} f_{lk} \overline{X} \left(f_{lk} - 1, f_{mk} + 1\right) -$$

$$- \sum_{j=m+1}^{n} g_{lj} \overline{X} \left(g_{lj} - 1, g_{mj} + 1\right) + \sum_{i=1}^{l-1} g_{im} \overline{X} \left(g_{il} + 1, g_{im} - 1\right)$$

$$\rho\left(\overline{E}_n\right)\overline{X} = \overline{X}\left(e_n + 1\right) - \sum_{j=1}^{n-1} e_j \overline{X} \left(e_j - 1, f_{jn} + 1\right)$$

$$\underset{(m=l+1)}{\rho\left(\overline{G}_{lm}\right)\overline{X}} = - e_m \overline{X}\left(e_l + 1, e_m - 1\right) + \frac{e_m(e_m - 1)}{2} \overline{X}\left(e_m - 2, f_{lm} + 1\right) -$$

$$- \sum_{j=1}^{l-1} f_{jm} \overline{X} \left(f_{jl} + 1, f_{jm} - 1\right) - \sum_{k=m+1}^{n} f_{mk} \overline{X} \left(f_{lk} + 1, f_{mk} - 1\right) +$$

$$+ \sum_{i=1}^{l-1} g_{il} \overline{X} \left(g_{il} - 1, g_{im} + 1\right) + \overline{X} \left(g_{lm} + 1\right)$$

Extremal vectors in Ω_- :

$$\rho(E_n)Y = \rho(G_{lm})Y = 0$$
$$\rho(H_i)Y = M_i Y$$

The solutions with only one generator :

$i)$ $Y = \overline{E}_n^{\,2\Lambda_n + 1}$; $M = \left(\Lambda_1, \ldots, \Lambda_{n-1}, -\Lambda_n - 1\right)$

$ii)$ $\underset{(j=i+1)}{Y = \overline{G}_{ij}^{\,\Lambda_i - \Lambda_j + 1}}$; $M = \left(\Lambda_1, \ldots, \Lambda_{i-1}, \Lambda_{i+1} - 1, \Lambda_i + 1, \Lambda_{i+2}, \ldots, \Lambda_n\right)$

C_n :

Generators :

$$H_i = \begin{pmatrix} |i\rangle\langle i| & 0 \\ 0 & -|i\rangle\langle i| \end{pmatrix}$$

$$E_i = \begin{pmatrix} 0 & |i\rangle\langle i| \\ 0 & 0 \end{pmatrix} \qquad \overline{E}_i = \begin{pmatrix} 0 & 0 \\ |i\rangle\langle i| & 0 \end{pmatrix} \quad 1 \le i \le n$$

$$G_{ik} = \begin{pmatrix} |i\rangle\langle k| & 0 \\ 0 & -|k\rangle\langle i| \end{pmatrix} \qquad \overline{G}_{ik} = \begin{pmatrix} |k\rangle\langle i| & 0 \\ 0 & -|i\rangle\langle k| \end{pmatrix}$$

$$F_{ik} = \begin{pmatrix} 0 & |i\rangle\langle k| + |k\rangle\langle i| \\ 0 & 0 \end{pmatrix} \qquad \overline{F}_{ik} = \begin{pmatrix} 0 & 0 \\ |i\rangle\langle k| + |k\rangle\langle i| & 0 \end{pmatrix}$$

Basis in Ω_- **:**

$$\overline{E}_1^{\,e_1} \ldots \overline{E}_n^{\,e_n}\, \overline{F}_{12}^{\,f_{12}}\, \overline{F}_{13}^{\,f_{13}} \ldots \overline{F}_{1n}^{\,f_{1n}}\, \overline{F}_{23}^{\,f_{23}} \ldots \overline{F}_{n-1,n}^{\,f_{n-1,n}}\, \overline{G}_{12}^{\,g_{12}} \ldots \overline{G}_{n-1,n}^{\,g_{n-1,n}} \equiv \overline{X}\left(e_1, \ldots, g_{n-1,n}\right)$$

Representation with highest weight $\Lambda = \left(\Lambda_1, \Lambda_2, \ldots, \Lambda_n\right)$:

$$\rho\left(H_i\right)\overline{X} = \left[\Lambda_i - 2e_i - \sum_{j=1}^{i-1}(f_{ji} - g_{ji}) - \sum_{j=i+1}^{n}(f_{ij} + g_{ij})\right]\overline{X}$$

$$\begin{aligned}
\rho\left(E_n\right)\overline{X} = &- e_n\left[-\Lambda_n + \left(e_n - 1\right) + \sum_{j=1}^{n-1}(f_{jn} - g_{jn})\right]\overline{X}\left(e_n - 1\right) - \\
&- \sum_{i=1}^{n-1} f_{in}\left(f_{in} - 1\right)\overline{X}\left(e_i + 1, f_{in} - 2\right) - \\
&- \sum_{i=1}^{n-2}\sum_{k=i+1}^{n-1} f_{in} f_{kn}\, \overline{X}\left(f_{ik} + 1, f_{in} - 1, f_{kn} - 1\right) + \\
&+ \sum_{i=2}^{n-1}\sum_{k=1}^{i-1} f_{in} g_{ki}\overline{X}\left(f_{in} - 1, g_{ki} - 1, g_{kn} + 1\right) + \\
&+ \sum_{i=1}^{n-1} f_{in}\, \overline{X}\left(f_{in} - 1, g_{in} + 1\right)
\end{aligned}$$

$$\begin{aligned}
\underset{(i=i+1)}{\rho\left(G_{ij}\right)}\overline{X} = &- e_i\, \overline{X}\left(e_i - 1, f_{ij} + 1\right) - 2f_{ij}\overline{X}\left(e_j + 1, f_{ij} - 1\right) - \\
&- \sum_{m=j+1}^{n} f_{im}\overline{X}\left(f_{im} - 1, f_{jm} + 1\right) - \sum_{l=1}^{i-1} f_{li}\overline{X}\left(f_{li} - 1, f_{lj} + 1\right) - \\
&- \sum_{m=j+1}^{n} g_{im}\overline{X}\left(g_{im} - 1, g_{jm} + 1\right) + \sum_{l=1}^{i-1} g_{lj}\overline{X}\left(g_{li} + 1, g_{lj} - 1\right) + \\
&+ g_{ij}\left[\Lambda_i - \Lambda_j - \left(g_{ij} - 1\right) + \sum_{l=j+1}^{n}(g_{jl} - g_{il})\right]\overline{X}\left(g_{ij} - 1\right)
\end{aligned}$$

$$\rho\left(\overline{E}_n\right)\overline{X} = \overline{X}\left(e_n + 1\right)$$

$$\begin{aligned}
\underset{(i=i+1)}{\rho\left(\overline{G}_{ij}\right)}\overline{X} = &- e_j\, \overline{X}\left(e_j - 1, f_{ij} + 1\right) - \sum_{l=1}^{i-1} f_{lj}\, \overline{X}\left(f_{li} + 1, f_{lj} - 1\right) - \\
&- 2f_{ij}\, \overline{X}\left(e_i + 1, f_{ij} - 1\right) - \sum_{m=j+1}^{n} f_{jm}\, \overline{X}\left(f_{im} + 1, f_{jm} - 1\right) +
\end{aligned}$$

$$+ \overline{X}\left(g_{ij}+1\right) + \sum_{k=1}^{i-1} g_{ki}\,\overline{X}\left(g_{ki}-1\,,\,g_{kj}+1\right) -$$

$$- \sum_{l=j+1}^{n} g_{jl}\,\overline{X}\left(g_{il}+1\,,\,g_{jl}-1\right)$$

Extremal vectors in Ω_- :

$$\rho\left(E_n\right)Y = \rho\left(G_{ij}\right)Y = 0$$
$$\rho\left(H_i\right)Y = M_i Y$$

The solutions with only one generator:

$$i)\ Y = \overline{E}_n^{\Lambda_n+1} \quad ;\ M = \left(\Lambda_1\,,\,\Lambda_2 \ldots\,,\,-\Lambda_n - 2\right)$$

$$ii)\ Y = \overline{G}_{i,i+1}^{\Lambda_i - \Lambda_{i+1}+1} \quad ;\ M = \left(\Lambda_1\,,\,\Lambda_3 \ldots\,,\,\Lambda_{i-1}\,,\,\Lambda_{i+1}-1\,,\,\Lambda_i+1\,,\,\Lambda_{i+2}\,,\,\ldots\,,\,\Lambda_n\right)$$

Dn :

Generators :

$$H_i = |2i-1\rangle\langle 2i-1| - |2i\rangle\langle 2i| \quad ; \quad 1 \le i \le n$$

$$F_{ij} = |2i-1\rangle\langle 2j| - |2j-1\rangle\langle 2i| \quad \overline{F}_{ij} = |2j\rangle\langle 2i-1| - |2i\rangle\langle 2j-1|$$
$$G_{ij} = |2i-1\rangle\langle 2j-1| - |2j\rangle\langle 2i| \quad \overline{G}_{ij} = |2j-1\rangle\langle 2i-1| - |2i\rangle\langle 2j| \qquad 1 \le i < j \le n$$

Basis in Ω_- :

$$\overline{F}_{12}^{f_{12}}\,\overline{F}_{13}^{f_{13}} \ldots \overline{F}_{1n}^{f_{1n}}\,\overline{F}_{23}^{f_{23}} \ldots \overline{F}_{n-1,n}^{f_{n-1,n}}\,\overline{G}_{12}^{g_{12}} \ldots \overline{G}_{n-1,n}^{g_{n-1,n}} \equiv \overline{X}\left(f_{12}\,,\,\ldots\,,\,g_{n-1,n}\right)$$

Representation with highest weight $\Lambda = \left(\Lambda_1\,,\,\Lambda_2\,,\,\ldots\,,\,\Lambda_n\right)$:

$$\rho\left(H_i\right)\overline{X} = \left[\Lambda_i - \sum_{k=i+1}^{n}\left(f_{ik}+g_{ik}\right) - \sum_{l=1}^{i-1}\left(f_{li}-g_{li}\right)\right]\overline{X}$$

$$\underset{(n=m+1)}{\rho\left(F_{m,n}\right)}\overline{X} = \sum_{i=1}^{m-1} f_{in}\,\overline{X}\left(f_{in}-1\,,\,g_{im}+1\right) +$$

$$+ \sum_{i=2}^{m-1}\sum_{k=1}^{i-1} f_{in}\,g_{ki}\left(g_{ki}-1\right)\overline{X}\left(f_{in}-1\,,\,g_{ki}-1\right) +$$

$$+ \sum_{i=2}^{m-1}\sum_{n=1}^{i-1} f_{in}\,g_{ki}\,\overline{X}\left(f_{in}-1\,,\,g_{ki}-1\,,\,g_{km}+1\right) -$$

$$- \sum_{i=1}^{m\ 2} f_{im}\,f_{m-1,n}\,\overline{X}\left(f_{i,m-1}+1\,,\,f_{im}-1\,,\,f_{m-1,n}-1\right) -$$

$$- \sum_{i=1}^{m-1} f_{im} \overline{X}\left(f_{im}-1\,,\, g_{in}+1\right) -$$

$$- \sum_{i=2}^{m-1} \sum_{k=1}^{i-1} g_{im} g_{ki} \overline{X}\left(f_{im}-1\,,\, g_{ki}-1\,,\, g_{kn}+1\right) -$$

$$- f_{mn}\left[-\Lambda_m - \Lambda_n + \left(f_{mn}-1\right) - \sum_{k=1}^{m-1}\left(g_{km}+g_{kn}\right)\right] \overline{X}\left(f_{mn}-1\right) -$$

$$- \sum_{i=1}^{m-1} f_{im} f_{mn} \overline{X}\left(f_{mn}-1\right) - \sum_{i=1}^{m-1} f_{in} f_{mn} \overline{X}\left(f_{mn}-1\right)$$

$$\underset{(m=l+1)}{\rho\left(G_{lm}\right)\overline{X}} = g_{lm}\left[\Lambda_l - \Lambda_m - \left(g_{lm}-1\right) + \sum_{j=m+1}^{n}\left(g_{mj}-g_{lj}\right)\right] \overline{X}\left(g_{lm}-1\right) -$$

$$- \sum_{j=1}^{l-1} f_{jl} \overline{X}\left(f_{jl}-1\,,\, f_{jm}+1\right) - \sum_{k=m+1}^{n} f_{lk} \overline{X}\left(f_{lk}-1\,,\, f_{mk}+1\right) -$$

$$- \sum_{j=m+1}^{n} g_{lj} \overline{X}\left(g_{lj}-1\,,\, g_{mj}+1\right) + \sum_{i=1}^{l-1} g_{im} \overline{X}\left(g_{il}+1\,,\, g_{im}-1\right)$$

$$\underset{(n=m+1)}{\rho\left(\overline{F}_{mn}\right)\overline{X}} = \overline{X}\left(f_{mn}+1\right)$$

$$\underset{(m=l+1)}{\rho\left(\overline{G}_{lm}\right)\overline{X}} = - \sum_{j=1}^{l-1} f_{jm} \overline{X}\left(f_{jl}+1\,,\, f_{jm}-1\right) -$$

$$- \sum_{k=m+1}^{n} f_{mk} \overline{X}\left(f_{lk}+1\,,\, f_{mk}-1\right) +$$

$$+ \sum_{i=1}^{l-1} g_{il} \overline{X}\left(g_{il}-1\,,\, g_{im}+1\right) + \overline{X}\left(g_{lm}+1\right)$$

Extremal vectors in Ω_- :

$$\rho\left(F_{mn}\right)Y = \rho\left(G_{lm}\right)Y = 0$$
$$\rho\left(H_i\right)Y = M_i Y$$

The solutions with only one generator :

$$i)\ Y = \overline{F}_{n-1,n}^{\,\Lambda_{n-1}+\Lambda_n+1} \quad ; \; M = \left(\Lambda_1, \Lambda_2, \ldots, -\Lambda_n - 1, -\Lambda_{n-1} - 1\right)$$

$$ii)\ Y = \overline{G}_{m-1,m}^{\,\Lambda_{m-1}-\Lambda_m+1} \quad ; \; M = \left(\Lambda_1, \Lambda_2, \ldots, \Lambda_{m-2}, \Lambda_m - 1, \Lambda_{m-1} + 1, \ldots, \Lambda_n\right)$$

A complete method to calculate all the extremal vectors in Verma modules can be found in LG84 GEL90 and DO90.

Acknowledgments

The authors want to express their gratitude to the Vicerrectorado de Investigación de la Universidad de Oviedo for financial support. This work belongs also to the common project between the Universities of Oviedo (Spain) and Clausthal (Germany) sponsored by the Volkswagen Foundation.

REFERENCES

(BG71) BERNSHTEIN, I.N.; GEL'FAND, I.M. y GEL'FAND, S.I. *"Structure of representations generated by vectors of highest weight"* Functional Anal. Appl. 5, 1-9 (1971).

(DI36) DIRAC, P.A.M. *"Wave equations in conformal space"* Ann. of Math. 37, 429-442 (1936).

(DI74) DIXMIER, J. *"Algebres enveloppantes"*. Gauthier-Villars, Paris. Colección "Cahiers scientifiques". 1974.

(DO90) DOBREV, V. *"Characteres of quantum groups representation"*. XVIII International Colloqium on Group Theoretical Methods in Physics" Moscow 1990.

(EH83) ENRIGHT, T.; HOWE,R. y WALLACH, N. *"A classification of unitary highest weight modules"* in "Proceedings of the University of Utah Conference 1982, (P.C. Trombi ed.) 40 97-143. Progress in Mathematics, Birkhäuser 1983 .

(EP81) ENRIGHT, T.; PARTHASARATHY, R. *"A proof of a conjecture of Kashiwara and Vergne"* , Lect. Notes in Mathematics. 880, 74-90. Springer (1981)

(EV67) EVANS, N.T. *"Discrete Series for the Universal Covering Group of the 3 + 2 de Sitter Group"*, J. Math. Phys. 8 170 (1967)

(GEL90) GARCIA-ESCUDERO, J.; LORENTE, M. *"Classification of unitary highest weight representations for non compact real forms"*. J. Math. Phys. 31 , 781-790 (1990)

(GK69) GEL'FAND, I.M.; KIRILLOV, A.A. *"The structure of the Lie Field Connected with a Split Lie Algebra"* Functional Anal. Appl. 3, 6-21 (1969).

(GK75) GRUBER, B.; KLIMYK, A.U. *"Properties of linear representations with a highest weight for the semisimple Lie algebras"*. J. Math. Phys. 16, 1816-1832 (1975).

(GK82) GRUBER, B.; KLIMYK, A.U. y SMIRNOV, Y.F. *"Indecomposable representations of A_2 and representations induced by them"*, Nuovo Cimento A 69, 97-127 (1982).

(GL83) GRUBER, B.; LENCZEWSKI, R. *"Indecomposable Representations of the Lorentz Algebra in an Angular Momentum Basis"* J. Phys. A 16, 3703-3722 (1983).

(GL89) GRUBER, B.; LORENTE, M. *"Wave equations invariant under Indecomposable representations. of the Lorentz group"* en Proceedings of the Symposium on Indecomposable Representations V. Cantoni, A. O. Barut ed. Instituto di Alta Matematica Universitá di Roma 1989 Vol. XXXI , 121-126.

(GL90) GRUBER, B ; LENCZEWSKI, R y LORENTE, M. *"On Induced Scalar Products and Unitarization"* . J. Math. Phys. , 31, 587-593 (1990).

(HC55) HARISH-CHANDRA. *"Representations of Semisimple Lie Groups"* , Amer. J. Math. 77, 743-777 (1955)

(HL88) HEIDENREICH, W.F.; LORENTE, M. *"Quantization of conformally invariant Bargmann-Wigner equations with gauge freedom"*. J. Math. Phys. 29, 1698-1704 (1988).

(HL90) HEIDENREICH, W.F.; LORENTE, M. *"Bargmann-Wigner equations in de Sitter space"*. J. Math. Phys. 31, 939-947 (1990).

(JA81) JAKOBSEN, H. *"The last possible place of unitarity for certain highest weight modules"*, Math. Ann. 256, 439-447 (1981)

(JA83) JAKOBSEN, H. *"Hermitian Symmetric Spaces and their Unitary Highest Weight Modules"* , J. Func. Analysis 52, 385-412 (1983)

(LG84) LORENTE, M ; GRUBER, B. *"Construction of extremal vectors for Verma submodules of Verma modules"* , J. Math. Phys. 25, 1674-1681 (1984)

(LO86) LORENTE, M. *"Wave equations for conformal multispinors"*, in Conformal Groups and Related Symmetries: Physical Results and Mathematical Background (A.O. Barut, H.D. Doebner ed.) Lect. Notes in Physics 261 185-194 Springer (1986).

(LO89) LORENTE, M. *"Extremal Vectors for Verma Modules of non compact real forms and unitarization"* in Proceedings of the Symposium on Indecomposable Representations V. Cantoni, A. O. Barut ed. Instituto di Alta Matematica Universitá di Roma 1989 Vol. XXXI, 71-83.

(MA77) MACK,G. *"All unitary ray representations of the conformal group SU(2,2) with positive energy"* Commun. Math. Phys. 55, 1-28 (1977).

(SH72) SHAPOVALOV, N. *"On a Bilinear Form on the Universal Enveloping Algebra of a Semisimple Lie Algebra"* Functional Anal. Appl. 6, 307-312 (1972).

(VE68) VERMA, D.N. *"Structure of certain induced representations of complex semisimple Lie algebras"*. Bull. Amer. Math. Soc., 74, 160-166 (1968).

(WA79) WALLACH, N.R. *"The analytic continuation of the discrete series, I,II"* , Trans.Amer.Math.Soc. 251, 1-17, 19-37 (1979)

(YA67) YAO, T. *"Unitary irreducible representations of SU(2,2)"* J. Math. Phys. 8, 1931-1954 (1967) 9, 1615-1626 (1968) 12, 315-342 (1971).

SCATTERING THEORY AND THE GROUP REPRESENTATION MATRIX

Joseph N. Ginocchio

Theoretical Division
Los Alamos National Laboratory
Los Alamos, NM 87545

1. INTRODUCTION

Group theory and symmetries have played an important role in Physics in determining allowed quantum numbers of quantum states, transition selection rules and rates between states, and energy levels for the states. Generally these properties require a knowledge of the matrix elements of the generators of the groups, but not the matrix elements of the group representation matrix. In this paper I shall discuss the application of group theory to the scattering of particles from composite systems in which the scattering function at a fixed impact parameter between different states of the system is the group representation matrix. This application is appropriate for 1) small angle scattering when the eikonal (or Glauber approximation[1]) is valid and 2) composite systems which has a dynamical symmetry. Such applications have been made for medium energy proton scattering from nuclei[2-5] and for electron scattering from molecules.[6]

2. THE EIKONAL OR GLAUBER APPROXIMATION

We now wish to consider a projectile of mass m interacting with a target. The most general Hamiltonian we can write down for this is

$$H = -\frac{\hbar^2}{2m}\nabla^2 + V(\vec{r}) + W(\vec{r}, D) + H_T, \tag{2.1}$$

where $\vec{r}$ is the projectile coordinate. $V(\vec{r})$ is the general optical potential that the projectile feels in the presence of the target. $W(\vec{r}, D)$ represents the interaction of the projectile with the target with generic degrees of freedom represented by D. W is defined to have vanishing diagonal matrix element in the ground state since that diagonal piece is already in V. Finally, H_T is the target Hamiltonian in the absence of the projectile.

The standard eikonal approximation exploits the high energy of the projectile compared with interaction energies (V and W). In addition we neglect the nuclear excitation energy (carried in H_T) which is an excellent approximation when the projectile energy is much larger than the excitation energy of the target. Neglect of H_T is equivalent to assuming that the interaction time of the projectile with the target is very short compared with the time that the constituents of the target take to transverse the size of the target. This approximation is analyzed in more detail in a related approach.[7]

If we neglect H_T, the standard eikonal treatment may be applied to the Hamiltonian to yield for the scattering amplitude from initial target state J_iM_i to final J_fM_f with the projectile going from momentum $\vec{k}$ to $\vec{k}' = \vec{k} + \vec{q}$,

$$< \vec{k}' J_f M_f | F | \vec{k} J_i M_i >= \frac{k}{2\pi i} \int d^2 b e^{i\vec{q}\cdot\vec{b}} < J_f M_f | (e^{i\psi(\vec{b})} - 1) | J_i M_i >, \tag{2.2}$$

where the eikonal phase ψ is given by

$$\psi(\vec{b}) = -\frac{m}{h^2 k} \int_{-\infty}^{\infty} dz [V(\vec{b} + \hat{k}z) + W(\vec{b} + \hat{k}z, t)], \tag{2.3}$$

and $\vec{b}$ is orthogonal to $\hat{k}$. In Figure 1 we see the situation illustrated. As the projectile transverses the nucleus it accumulates a phase along the z-direction due to the interactions with the constituents in the target

We split ψ into two parts,

$$\psi = \overline{\psi} + \hat{\psi}. \tag{2.4}$$

The first part, $\overline{\psi}$, is the average optical model phase associated with V and cannot cause transitions between excited states of the target. The second part, $\hat{\psi}$, is associated with W and is the transition phase. We can write

$$< J_f M_f | e^{i\psi(b)} - 1 | J_i M_i >= T_{fi}(\psi) - \delta_{J_f, J_i; M_f M_i}, \tag{2.5}$$

where T_{fi} is the transition matrix

$$T_{fi} = e^{i\overline{\psi}} U_{fi}(\hat{\psi}), \tag{2.6}$$

and U_{fi} is the target matrix element

$$U_{fi} =< J_f M_f | \hat{U}(\hat{\psi}) | J_i M_i > \tag{2.7}$$

of the transition operator

$$\hat{U}(\hat{\psi}) = e^{i\hat{\psi}(b)}. \tag{2.8}$$

If one expands $\hat{U}$, the first nonvanishing term $(i\hat{\psi})$ gives the distorted wave impulse approximation (DWIA) for the scattering amplitude. The phase $\overline{\psi}$ provides the distorted wave.

However, for a target nucleus described by a dynamical summetry, we can do better. We can evaluate the nuclear matrix element (2.7) completely without making a first-order approximation. We note from the definition of $\hat{\psi}(b)$, that the expectation of $\hat{\psi}(b)$ in the ground state vanishes,

$$< g.s. | \hat{\psi}(b) | g.s. >= 0 \tag{2.9}$$

and hence, there is no first-order contribution to the elastic scattering. However, for projectiles for which multiple scattering is important, there can be corrections to the DWIA even in the elastic channel.

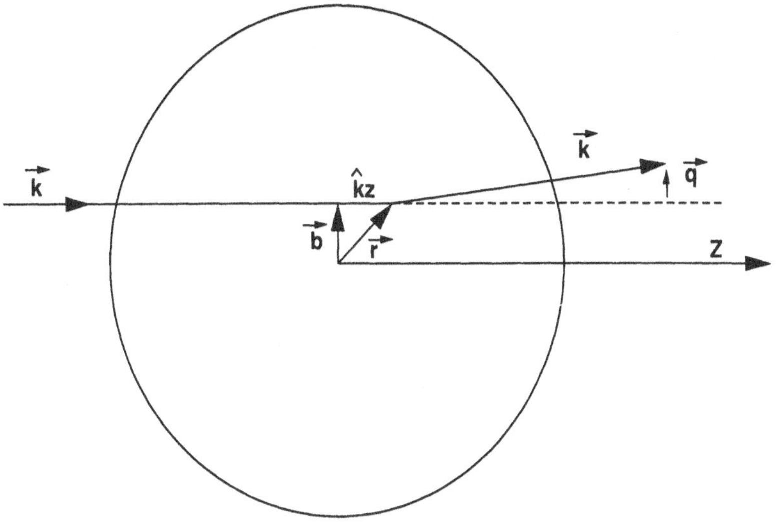

FIGURE 1. Schematic illustration of a projectile with momentum $\vec{k}$ at impact parameter $\vec{b}$ scattering from the target and leaving with momentum $\vec{k}' = \vec{k} + \vec{q}$

3. THE INTERACTING BOSON MODEL OF NUCLEI

The interacting boson model of nuclei[8] (IBM) assumes that the low-lying collective levels of nuclei are composed primarily of $J^\pi = 0^+$ and 2^+ coherent pairs of valence nucleons which are approximated as bosons. There are two versions of this model. The original version did not distinguish between neutrons and protons (IBM-1). This version may be valid for the lowest collective states of deformed nuclei since thses states are primarily symmetric in the neutron and proton degrees of freedom although deviations from total symmetry are now being measured by pion scattering.[9] The second version (IBM-2) does distinguish between neutrons and protons. This version gives a better description of the nuclear states and for some transitions is absolutely necessary. However in this paper we shall consider IBM-1 to simplify the discussion of concepts. The generalization to IBM-2 is straightforward but increases the number of indicies and hence the complexity of exposition.

The monopole ($J^\pi = 0^+$) boson creation and destruction operators are

$$s^\dagger, s, \tag{3.1a}$$

and the quadrupole ($J^\pi = 2^+$) boson operators are

$$d_m^\dagger \tilde{d}_m = (-1)^m d_{-m}; m = -2, -1, 0, 1, 2. \tag{3.1b}$$

The boson Hamiltonian which determines the nuclear wave functions and spectra is primarily composed of a single boson energy and a quadrupole interaction between bosons. This boson Hamiltonian is generally taken to be

$$H = \epsilon N_d + \kappa Q \cdot Q, \tag{3.2}$$

where ϵ is the single-boson energy and κ the interaction strength and is negative ($\kappa < 0$). The number operator

$$N_d = \sum_m d_m^\dagger d_m \tag{3.3}$$

counts the number of quadrupole bosons. Thus the first term makes the $J^\pi = 2^+$ pair higher in energy than the $J^\pi = 0^+$ and hence corresponds to the pairing energy. The quadrupole operator is

$$Q_m = d_m^\dagger s + s^\dagger \tilde{d}_m + \chi[d^\dagger \tilde{d}]_m^{(2)}. \tag{3.4}$$

The first term changes monopole bosons into quadrupole bosons and vice versa, while the second term reorients the quadrupole bosons. Hence the quadrupole interaction mixes quadrupole bosons into the ground state thereby producing deformations. Thus, this boson Hamiltonian incorporates the predominant features of the effective nuclear Hamiltonian as determined by phenomenology.

In the IBM, multipole moment operators of the nucleus are expressed in terms of boson operators:

$$T_m^{(l)} = [d^\dagger \tilde{d}]_m^{(l)},$$

$$l = 0, 1, 2, 3, 4, \quad m = -l, ..., l, \tag{3.5a}$$

and

$$P_m = s^\dagger \tilde{d}_m + d_m^\dagger s, \quad m = 2, 1, ..., -2, \tag{3.5b}$$

$$\overline{P}_m = i(s^\dagger \tilde{d}_m - d_m^\dagger s), \quad m = 2, 1, ..., -2. \tag{3.5c}$$

The multipole operators $T^{(l)}$ conserve the number of quadrupole and monopole bosons and hence are seniority conserving since senority counts the number of quadrupole bosons. The operators P_m and $\overline{P}_m$ on the other hand are senority breaking operators. These multipole operators are the 35 generators for a special untiary group in six dimensions, SU_6. The general transformation operator in SU_6 is given by

$$\hat{U}(\theta) = e^{i\theta\cdot P + i\overline{\theta}\cdot\overline{P} + i\sum_l \theta^{(l)}\cdot T^{(l)}}, \tag{3.6}$$

where θ_m, $\overline{\theta}_m$, and $\theta_m^{(l)}$ are the 35 angles of the SU_6 transformations analogous to the three Euler angles for an SU_2 transformation.

The boson Hamiltonian in (3.2) is diagonalized in the space of N bosons, where N is one-half the number of valence nucleons. This boson space forms a basis for the symmetric irreducible representation of SU_6 of rank N. Hence the martrix elements of the SU_6 transformation operator (3.6) between boson eigenstates,

$$U_{fi}(\theta) = < J_f M_f | \hat{U}(\theta) | J_i M_i >, \tag{3.7}$$

will be the representation matrix for this irreducible representation of SU_6. That is, this matrix is a generalization of the Wigner D matrix for SU_2.

The SU_6 group has three possible subgroup chains. The first group chain is

$$SU_6 \supset U_5 \supset SO_5 \supset SO_3. \tag{3.8}$$

The generators of the U_5 subgroup are the multipole operators in (3.5a). The generators of the SO_5 subgroups are those of U_5 with odd multipole operators and have $l = 1, 3$. The generators of the SO_3 subgroup are angular momentum operators and have $l = 1$. The IBM Hamiltonian (3.2) has a U_5 dynamical symmetry for $\kappa = 0$, $\kappa\chi = 0$, but $\kappa\chi^1 \neq 0$. This symmetry corresponds to the quadrupole spherical vibrator model.

The next group chain is

$$SU_6 \supset SO_6 \supset SO_5 \supset SO_3. \tag{3.9}$$

The generators of the SO_6 subgroup are the odd multipole generators $l = 1, 3$ (the SO_5 generators) plus the quadrupole operator in (3.5b). The IBM Hamiltonian has an SO_6

dynamical symmetry for $\epsilon = 0$ and $\chi = 0$. This symmetry corresponds to the γ-unstable quadrupole rotor model.

The third group chain is

$$SU_6 \supset SU_3 \supset SO_3. \tag{3.10}$$

The generators of the SU_3 subgroup are the angular momentum generators ($l = 1$) plus the quadrupole operator in (3.4) with $\chi = \pm\sqrt{7}/2$. The Hamiltonian has an SU_3 dynamical symmetry for $\epsilon = 0$ and $\chi = \pm\sqrt{7}/2$. Hence there are two possible SU_3 symmetries, one with negative χ and one with positive χ. Negative χ corresponds to a prolate deformation whereas positive χ corresponds to an oblate deformation.[10] These two choices give the same energy spectrum and transition rates but we shall show that they give different differential cross sections for cases in which multiple scattering is important.

The mapping of fermion operators, such as the density operator onto boson operators is not a completely solved problem. However, there is no doubt that the leading terms in the mapping of a one-body fermion operator, such as the density operator, will be linear in the SU_6 group generators defined in (3.5).[11] The tensor properties of these multipole operators, plus the condition that all matrix elements be real, lead to the forms of the total spin zero density,

$$\hat{\rho}^{(0)}(\vec{r}) = \rho(r) + \alpha(r)P \cdot Y^{(2)}(\hat{r}) + \sum_{l=2,4} \beta_l T^{(l)} \cdot Y^{(l)}(\hat{r}), \tag{3.11a}$$

and for the spin vector density,

$$\hat{\rho}^{(1)}(\vec{r}) = \overline{\alpha}(r)[PY^{(2)}(\hat{r})]^{(1)} + \sum_{l=2,4} \overline{\beta}_l(r)[T^{(l)}Y^{(l)}(\hat{r})]^{(1)} +$$

$$\sum_{l=1,3} \{\gamma_l(r)[T^{(l)}Y^{(l+1)}(\hat{r})]^{(1)} + \Delta_l(r)[T^{(l)}Y^{(l-1)}(\hat{r})]^{1}\}, \tag{3.11b}$$

where $Y_m^{(l)}(\hat{r})$ is the spherical tensor of rank l and projection m.

There are additional constraints on the functions if the $B(E\lambda)$ matrix elements are to be reproduced. The electric multipole moment operators are

$$Q_m^{(l)} = \int d^3r r^l Y_m^{(l)}(\hat{r})\hat{\rho}^{(0)}(r). \tag{3.12}$$

Hence the $B(E\lambda)$ are give by

$$B(E2) = \left| \int_0^\infty dr r^4 < J^\pi = 2^+||\alpha(r)P + \beta_2(r)T^{(2)}||0 > \right|^2 \tag{3.13a}$$

and

$$B(E4) = \left| \int_0^\infty dr r^6 < J^\pi = 4^+||\beta_4(r)T^{(4)}||0 > \right|^2. \tag{3.13b}$$

The microscopic theory of the IBM has made some progress in deteriming $\alpha(r)$, $\beta_l(r)$, $\gamma_l(r)$, and $\Delta_l(r)$ from the nuclear shell model.[12–14] These functions can also be determined phenomenologically from electron scattering.[15]

4. THE MARRIAGE OF THE GLAUBER APPROXIMATION AND THE IBM

To combine the IBM and the Glauber or eikonal approximation, we work in the impulse approximation and assume that the range of the projectile-nucleon interaction is short compared with the size of the nucleus. The interaction potentials V and W of (2.1) or equivalently the eikonal phase ψ of (2.3) can then be written in terms of the projectile-nucleon forward scattering amplitude $f^{(S)}$ (S is the channel spin) and the density operator $\hat{\rho}^{(S)}$ of (3.11). We then obtain for the phase

$$\psi(\vec{b}) = \frac{2\pi}{k} \sum_{\tau} \left[f_{\tau}^{(0)} \int \rho_{\tau}^{(0)}(\vec{b} + \hat{K}z)dz + f_{\tau}^{(1)} \int \rho_{\tau}^{(1)}(\vec{b} + \hat{K}z)dz \right], \tag{4.1}$$

where for completeness we have introduced an index τ that differentiates between target protons and neutrons. From (3.11) we see that $\psi(\vec{b})$ is then linear in the SU_6 generators. Hence for $\hat{\psi}$ real, the Glauber transition operator $\hat{U}$ of (2.8) will be in the form of the SU_6 transformation operator given in (3.6). Thus the Glauber transition operator will be a unitary transformation on the bosons in the nucleus , and the matrix $U_{fi}(\hat{\psi})$ given in (2.7) will be the representation matrix for the symmetric representation of SU_6. This means that, like the Wigner D matrix of SU_2, this more general matrix can be calculated in closed form, implying that the scattering in the Glauber approximation can be calculated to all orders in the projectile-nucleus coupling within the IBM space. Thus all multiple scattering in which the nucleus is virtually excited to all intermediate states of the IBM is evaluated exactly. This is equivalent to doing a distorted wave coupled channels calculation in that space of states. In general since the scattering amplitudes $f^{(S)}$ are complex, ψ will be complex and the linear transformation will then be just the analytic continuation of a unitary transformation.

If the spin degrees of freedom of the projectile are averaged over in the experiment, only the spin scalar densities and forward scattering amplitudes in (4.1) will survive. In the rest of this discussion we shall consider this case, but realize that for measurements in which projectile spin measurements are made we should use the more general formalism. For convenience then we drop the spin superscript. We note then that only the even multipole generators of SU_6 are involved in the transition operator. We also drop the distinction between neutrons and protons.

For strongly absorbed probes, the representation matrix needed can be greatly simplified. The spherical harmonic of the projectile coordinates can be expanded in powers of (z/b):

$$Y_m^{(l)}(\hat{r}) = Y_m^{(l)}(\hat{b}) + O\left(\frac{z}{b}\right). \tag{4.2}$$

For a strongly absorbing probe, the integration in (2.2) over impact parameter b will be concentrated primarily at the surface of the nucleus. However, on the surface z/b will then be small (see Fig. 1) and the first term in (4.2) will dominate. Under this assumption the transition phase becomes,

$$\hat{\psi}(b) = [g(b)P + g_2(b)T^{(2)}] \cdot Y^{(2)}(\hat{b}) + g_4(b)T^{(4)} \cdot Y^{(4)}(\hat{b}), \tag{4.3a}$$

where

$$g = \frac{2\pi}{k} f \int_{-\infty}^{\infty} dz \alpha((b^2 + z^2)^{1/2}), \tag{4.3b}$$

$$g_2 = \frac{2\pi}{k} f \int_{-\infty}^{\infty} dz \beta_2((b^2 + z^2)^{1/2}), \tag{4.3c}$$

$$g_4 = \frac{2\pi}{k} f \int_{-\infty}^{\infty} dz \beta_4((b^2 + z^2)^{1/2}). \tag{4.3d}$$

Since in this approximation the spherical tensor is completely outside the integral over z, the transition phase is just a rotation $R(\hat{b})$ through the Euler angles defined by the projectile of a reduced transition phase,

$$\hat{\psi}(\vec{b}) = R(\hat{b})\hat{\psi}^{(0)}R^{\dagger}(\hat{b}), \tag{4.4a}$$

where the reduced transition phase is,

$$\hat{\psi}^{(0)} = \tilde{g}(b)P_0 + \sum_{m=-2}^{2} \phi_m(b)d_m^\dagger d_m + \phi_0(b)d_0^\dagger d_0, \tag{4.4b}$$

with,

$$\tilde{g}(b) = \sqrt{5/4\pi}g(b), \tag{4.4c}$$

$$\phi_m(b) = \frac{(-1)^m}{\sqrt{4\pi}(1+\delta_{m,0})} \sum_{l=2,4} (2l+1) \begin{pmatrix} 2 & 2 & l \\ m & -m & 0 \end{pmatrix} g_l(b), \tag{4.4d}$$

and

$$\begin{pmatrix} 2 & 2 & l \\ m & -m & 0 \end{pmatrix}$$

is the Wigner $3-j$ symbol.[16] These phases are given by

$$\phi_0(b) = \frac{1}{\sqrt{4\pi}}[-(\frac{5}{14})^{1/2}g_2(b) + 9(\frac{1}{70})^{1/2}g_4(b)], \tag{4.5a}$$

$$\phi_2(b) = \frac{1}{\sqrt{4\pi}}[(\frac{10}{7})^{1/2}g_2(b) + 3(\frac{1}{70})^{1/2}g_4(b)], \tag{4.5b}$$

$$\phi_1(b) = -[\phi_0(b) + \phi_2(b)], \tag{4.5c}$$

$$\phi_{-m}(b) = \phi_m(b). \tag{4.5d}$$

Hence, in this approximation the transition operator (2.8) can be written as a product of three-dimensional rotation, a simpler SU_6 transformation, and finally the inverse three-dimensional rotation:

$$\hat{U}(\hat{\psi}) = R(\hat{b})\hat{U}^{(0)}R^\dagger(\hat{b}), \tag{4.6}$$

where $\hat{U}^{(0)}$ is reduced transition operator and is just

$$\hat{U}^{(0)} = e^{i\hat{\psi}(0)}. \tag{4.7}$$

The matrix $U_{fi}(\psi_l)$ then becomes

$$U_{fi}(\psi) = \sum_M D^{(J_f)}_{MM_f}(\hat{b})D^{(J_i)*}_{MM_i}(\hat{b})U^{(0)}_{fi}(\psi), \tag{4.8}$$

where the reduced transition matrix is

$$U^{(0)}_{fi}(\psi) =< J_f, M|\hat{U}^{(0)}|J_i, M >, \tag{4.9}$$

and $D^{(J)}_{MM'}(\hat{b})$ is the usual Wigner D matrix.[16]

For the target with angular momentum zero, $J_i = 0$, we can use (4.6) and the integral representation of the Bessel function, J_M, to do the azimuthal angle integration in (2.2). This leads to a one-dimensional integral:

$$< \vec{k}', J, M|F|\vec{k}, J_i = M_i = 0 >= \frac{k(-1)^{(J-1)/2}}{2^J} \frac{[(J+M)!(J-M)!]^{1/2}}{\left(\frac{J+M}{2}\right)!\left(\frac{J-M}{2}\right)!}$$

$$\int_0^\infty bdbJ_m(qb)(e^{i\overline{\psi}}U^{(0)}_{fi} - \delta_{fi}). \tag{4.10}$$

Hence our task simplifies to calculating the representation matrix for a much simpler unitary transfromation. The crucial ingredient in deriving the representation matrix is the group transformation of a single boson derived in the next section.

5. REPRESENTATION MATRIX FOR A SINGLE BOSON

The basic ingredient in deriving the representation matrix for a many-boson system is the representation matrix for a single boson. From (4.4b) we see that under the transformation by the reduced transition operator, the quadrupole bosons with $m \neq 0$ are diagonal. However, the $s^\dagger$ and $d_0^\dagger$ are transformed into each other. The best way to derive the transformation matrix is to write these bosons in terms of the eigenbosons of the transition phase:

$$[\hat{\psi}^{(0)}, B_\pm^\dagger] = e_\pm B_\pm^\dagger. \tag{5.1}$$

It is straightforward to determine that the eigenbosons are

$$B_+^\dagger = \cos u \; s^\dagger + \sin u \; d_0^\dagger, \tag{5.2a}$$

$$B_-^\dagger = -\sin u \; s^\dagger + \cos u \; d_0^\dagger, \tag{5.2b}$$

where

$$\cos u = \left(\frac{\phi - \phi_0}{2\phi}\right)^{1/2}, \tag{5.2c}$$

$$\sin u = \left(\frac{\phi + \phi_0}{2\phi}\right)^{1/2}, \tag{5.2d}$$

and

$$\phi = (\tilde{g}^2 + \phi_0^2)^{1/2}. \tag{5.3}$$

The eigenvalues are

$$e_\pm = \phi_0 \pm \phi. \tag{5.4}$$

Some algebra leads to the results:

$$\hat{U}^{(0)} s^\dagger \hat{U}^{(0)-1} = e^{i\phi_0}(X_{11} s^\dagger + X d_0^\dagger), \tag{5.5a}$$

$$\hat{U}^{(0)} d_0^\dagger \hat{U}^{(0)-1} = e^{i\phi_0}(X s^\dagger + X_{22} d_0^\dagger), \tag{5.5b}$$

$$\hat{U}^{(0)} d_m^\dagger \hat{U}^{(0)-1} = e^{i\phi_m} d_m^\dagger, \quad m \neq 0, \tag{5.5c}$$

where

$$X_{11}(b) = \cos\phi - i(\phi_0/\phi)\sin\phi, \tag{5.5d}$$

$$X_{22}(b) = \cos\phi + i(\phi_0/\phi)\sin\phi, \tag{5.5e}$$

$$X(b) = (i\tilde{g}\sin\phi)/\phi. \tag{5.5f}$$

Thus we see that the $s^\dagger$ and $d_0^\dagger$ are transformed into each other, while the other bosons only acquire a phase. It is easy to check that this transformation is unitary for ϕ, ϕ_0, and g real.

The representation matrix for many bosons can be derived from the transformation of the single bosons given in (5.5). The general result has been derived in terms of a five-dimensional integration.[17] We present in the next section some special results which can be given in closed form.

6. TRANSITION MATRIX FOR A QUADRUPOLE VIBRATOR

The dynamical symmetry group of the quadrupole spherical vibrator is the U_5 group as noted in (3.8). The ground state of the quadrupole vibrator is a boson condensate of N monopole bosons, where N is one-half the number of valence nucleons. the excited states are labeled by the number n of quadrupole bosons, which is the seniority in fermion space,[18] with the energy of the states increasing approximately linearly with n. For a given n there is a multiplet of states with the quantum number, τ, designating the number of quadrupole bosons not coupled in pairs to angular momentum zero, the quantum number, n_Δ, designating the number of quadrupole bosons coupled in triplets to angular momentum zero, and of course the angular momentum, J, and its projection, M. The allowed values to these quantum numbers are

$$\tau = n, n-2, ..., 0 \; or \; 1, \tag{6.1}$$

and for each τ the allowed angular momenta and n_Δ are determined by all possible partitions of τ,

$$\tau = 3n_\Delta + \lambda, \tag{6.2}$$

where n_Δ and λ are integers, and then the allowed J are

$$J = 2\lambda, 2\lambda - 2, 2\lambda - 3, ..., \lambda. \tag{6.3}$$

The eigenstates of the quadrupole vibrator in terms of these quantum numbers are monomials of rank $N-n$ in the monopole bosons and of rank n in quadrupole bosons:

$$|N, n, \tau, n_\Delta, J, M >= \eta_{Nn\tau}(s^\dagger)^{N-n}(d^\dagger \cdot d^\dagger)^{n-\tau/2} \times |\tau, \tau, \tau, n_\Delta, J, M > . \tag{6.4a}$$

The normalization $\eta_{Nn\tau}$ is

$$\eta_{Nn\tau} = \left[\frac{(2\tau+3)!!}{(N-n)!(n+\tau+3)!!(n-\tau)!!}\right]^{1/2}, \tag{6.4b}$$

and the state with $N = \tau$ is a complicated function of the quadrupole bosons but has the property:

$$\tilde{d} \cdot \tilde{d} |\tau, \tau, \tau, n_\Delta, J, M >= 0, \tag{6.4c}$$

which is consistent with the definition of the τ quantum number which is that it is the number of quadrupole bosons not coupled to angular mometum zero.

The ground state of the quadrupole vibrator is then

$$|\tilde{0} >\equiv |N, n = \tau = J = M = 0 >= \frac{(s^\dagger)^N}{\sqrt{N!}}|0 >, \tag{6.5}$$

where $|0>$ denotes the doubly-magic core. If we were operate on this ground state with the reduced transition operator (4.7), then

$$\hat{U}^{(0)}|\tilde{0} >= e^{iN\phi_0}\frac{(X_{11}s^\dagger + Xd_0^\dagger)^N}{\sqrt{N!}}|0 >, \tag{6.6}$$

where we have used (5.5a). If we take the matrix element with respect to a general final state with n quadrupole bosons, given in (6.4), we can use the binomial theorem to expand (6.6) in powers of quadrupole bosons. The reduced transition matrix for scattering from the ground state to an excited state is then

$$< N, n, \tau, n_\Delta, J, M = 0|\hat{U}^{(0)}|N, n = \tau = n_\Delta = J = M = 0 >=$$
$$\left[\frac{N!(2\tau+3)!!}{(N-n)!(n+\tau+3)!!(n-\tau)!!}\right]^{1/2} \times e^{iN\phi_0} X_{11}^{N-n} X^n A_{\tau n_\Delta J}, \tag{6.7a}$$

where $A_{\tau n_\Delta J}$ is a constant which is independent of the Glauber phases $\hat{\psi}$. The constant is

$$A_{\tau n_\Delta J} = \frac{< \tau, \tau, \tau, n_\Delta, J, 0|(d_0^\dagger)^\tau|0 >}{\tau!}. \tag{6.7b}$$

The coefficients $A_{\tau n_\Delta J}$ are given in Table I for the lowest values of τ.

We note that the $J^\pi = 3^+$ state cannot be excited. This result, which holds for any odd angular mometum state within the IBM space, follows from the approximation made in (4.2). Hence, the degree to which any unnatural parity state is excited is a measure of the validity of this approximation. The results in (6.7) give the amplitude for exciting the individual states in a spherical quadrupole vibrator. The average probability for exciting the states in an n-phonon multiplet has been given previously.[19]

7. GAMMA UNSTABLE ROTOR

The eigenfunctions for the SO_6 dynamical group chain do not conserve the number of monopole or qaudrupole bosons. Instead of the quantum number n used in the quadrupole vibrator, the SO_6 quantum number is σ and the allowed values are

$$\sigma = N, N-2, ..., 0 \ or \ 1. \tag{7.1}$$

Since SO_5 is also a subgroup of SO_6 as well as U_5, the remaining quantum numbers are the same as for a quadrupole vibrator with the allowed values of τ being

$$\tau = \sigma, \sigma - 1, ..., 0 \tag{7.2}$$

The eigenfunctions are given by

$$|N, \sigma, \tau, n_\Delta, J, M >= \overline{\eta}_{N\sigma}(I^\dagger)^{N-\sigma/2}|\sigma, \sigma, \tau, n_\Delta, J, M >, \tag{7.3a}$$

where

$$\overline{\eta}_{N\sigma} = \left[\frac{(2\sigma+4)!!}{(N+\sigma+4)!!(N-\sigma)!!}\right]^{1/2} \tag{7.3b}$$

and

TABLE I. Coefficients $A_{\tau n_\Delta J}$ for lowest values of τ.

τ	n_Δ	J	A
0	0	0	1
1	0	2	1
2	0	2	$-[\frac{1}{7}]^{1/2}$
2	0	4	$[\frac{9}{15}]^{1/2}$
3	0	6	$[\frac{3}{77}]^{1/2}$
3	0	4	$-3[\frac{2}{185}]^{1/2}$
3	0	3	0
3	1	0	$-[\frac{1}{105}]^{1/2}$

$$I^\dagger = (s^\dagger)^2 - d^\dagger \cdot d^\dagger \tag{7.3c}$$

is an SO_6 invariant. That is, the SO_6 generators commute with this four-boson operator. In particular, for the quadrupole operator (3.5b),

$$[P_m, I^\dagger] = 0. \tag{7.4}$$

The state for $N = \sigma$ is

$$|\sigma, \sigma, \tau, n_\Delta, J, M >= \Sigma_p D_p(\sigma\tau)(s^\dagger)^{\sigma-\tau-2p} \times (I^\dagger)^p |\tau, \tau, \tau, n_\Delta, J, M >, \tag{7.5a}$$

where

$$D_p(\sigma, \tau) = \left[\frac{2^{\sigma+1}(\sigma-\tau)!(2\tau+3)!!}{(\sigma+1)!(\sigma+\tau+3)!}\right]^{1/2} \left(-\frac{1}{4}\right)^p \times \frac{(\sigma+1-p)!}{(\sigma-\tau-2p)!p!}. \tag{7.5b}$$

Using the above eigenfunctions and the basic transformations (5.5) we can derive the reduced transition matrix. In the SO_6 limit, $\chi = 0$ and $g_4 = 0$. Hence the angles ϕ will take the simple dependence,

$$\phi_0 = 0, \quad \phi = \tilde{g}, \tag{7.6a}$$

$$X_{11} = X_{22} = \cos\phi, \quad X = i\sin\phi. \tag{7.6b}$$

Since the quadrupole operator (3.5b) is a generator of the SO_6 group, it will not connect different representations of SO_6, and consequently the transition matrix will not either. Furthermore because of (7.4) and (7.3a), the transition matrix will not depend on N. Hence we will have

$$< N, \sigma', \tau', n'_\Delta, J', M|\hat{U}^{(0)}|N, \sigma, \tau, n_\Delta, J, M >=$$

$$\delta_{\sigma',\sigma} < \sigma, \sigma, \tau', n'_\Delta, J', M|\hat{U}^{(0)}|\sigma, \sigma, \tau, n_\Delta, J, M > . \tag{7.7}$$

Thus the scattering will be entirely within the SO_6 multiplet. The ground state has $\tau = n_\Delta = J = M = 0$ and $\sigma = N$. With many steps of algebra the reduced transition matrix can be shown to be

$$< N, N, \tau, n_\Delta, J, M = 0|\hat{U}^{(0)}|N, N, \tau = n_\Delta = J = M = 0 >= \left[\frac{3 \cdot 2^{N+1} N!}{(N+3)(N+2)}\right]^{1/2}$$

$$A_{\tau n_\Delta J} X^\tau \times \sum_p D_p(N\tau)(\cos\phi)^{N-\tau-2p}. \tag{7.8}$$

This expression can be written in two additional, but equivalent ways, one in terms of Gegenbauer polynomials, $C_n^{(\alpha)}(x)$,

$$< N, N, \tau, n_\Delta, J, M = 0|\hat{U}^{(0)}|N, N, \tau = n_\Delta = J = M = 0 >=$$

$$\left[\frac{3(N-\tau)!(2\tau+3)!!}{(N+3)(N+2)(N+1)(N+\tau+3)!}\right]^{1/2} (2\tau+2)!! \times X^\tau A_{\tau n_\Delta J} C_{N-\tau}^{(\tau+2)}(\cos\phi), \tag{7.9}$$

and the hypergeometric function, $F(a, b; c; z)$,

$$< N, N, \tau, n_\Delta, J, M = 0|\hat{U}^{(0)}|N, N, \tau = n_\Delta = J = M = 0 >=$$

$$\left[\frac{3(N+\tau+3)!}{N+3)(N+2)(N+1)(N-\tau)!(2\tau+3)!!}\right]^{1/2} X^{\tau} A_{\tau n_\Delta J}$$

$$\times F\left[-\frac{(N-\tau)}{2}, \frac{N+\tau}{2}+2; \tau+\frac{5}{2}; \sin^2\phi\right]. \tag{7.10}$$

From the last expression, since $F(a,b;c;0) = 1$ and $X = i\sin\phi$, it is easy to confirm that the reduced transition matrix reduces to $\delta_{\tau,0}$ for $\phi = 0$, as it should since this is the case of no scattering.

8. AXIALLY SYMMETRIC ROTOR

In the limit of an axially symmetric rotor, the number of quadrupole bosons is not conserved just as for the γ-unstable rotor. In addition, since the SO_5 group is not in the subgroup sequence (3.10), the quantum numbers τ and n_Δ are also not conserved. The SU_3 quantum numbers (λ,μ) for the ground state band are $(\lambda,\mu) = (2N,0)$ and the allowed angular momentum are

$$J = 0, 2, 4, ..., 2N. \tag{8.1}$$

The SU_3 ground state band is generated from a boson condensate of intrinsic bosons:[10]

$$|N; i >= (B_i^\dagger)^N |0>, \tag{8.2a}$$

where for $i = o$ (oblate)

$$B_o^\dagger = s^\dagger + \frac{1}{\sqrt{2}}[d_0^\dagger + \sqrt{3}/2(d_2^\dagger + d_{-2}^\dagger)], \tag{8.2b}$$

and for $i = p$ (prolate)

$$B_p^\dagger = s^\dagger + \sqrt{2}d_0^\dagger. \tag{8.2c}$$

We use the fact that an oblate intrinsic boson can be written as a rotation about the x axis of a boson with zero z projection,

$$B_o^\dagger = R_x(s^\dagger - \sqrt{2}d_0^\dagger)R_x^\dagger. \tag{8.2d}$$

Then the SU_3 states in the ground state multiplet will be projected from this condensate of intrinsic bosons:

$$|N,(2N,0),J,M >= (\mp 1)^{J/2}\frac{2J+1}{8\pi^2\Lambda_J}\int d\Omega R(\Omega) D_{M0}^{(J)}(\Omega)(s^\dagger \mp \sqrt{2}d_0^\dagger)^N|0>, \tag{8.3a}$$

where $R(\Omega)$ is a rotation about the three Euler angles $(\theta_1,\theta_2,\theta_3)$, and $D_{M0}^{(J)}(\Omega)$ is the Winger D matrix. The normalization Λ_J is

$$\Lambda_J = \left[\frac{3^N(2J+1)N!(2N)!}{(2N-J)!!(2N+J+1)!!}\right]^{1/2}. \tag{8.3b}$$

The upper sign is for positive χ (oblate) and the lower sign is for negative χ (prolate).

In the SU_3 limit,

$$g_2 = \pm\frac{\sqrt{7}}{2}g, \quad g_4 = 0, \tag{8.4a}$$

$$\phi = \frac{3}{2\sqrt{2}}\tilde{g}, \quad \phi_0 = \mp\frac{1}{3}\phi, \quad \phi_1 = \phi_0, \quad \phi_2 = -2\phi_0, \tag{8.4b}$$

and

$$X_{11} = \cos\phi \pm \frac{i}{3}\sin\phi, \tag{8.5a}$$

$$X_{22} = \cos\phi \mp \frac{i}{3}\sin\phi, \tag{8.5b}$$

$$X = \frac{2\sqrt{2}}{3} i \sin\phi. \tag{8.5c}$$

Hence we also see the difference between the different SU_3 subgroups. While ϕ is independent of the sign of χ the ϕ_m have the same magnitude but different signs for the two choices.

Because the SU_3 eigenstates are given in terms of a boson condensate, the most convenient way to derive the representation matrix is to write the rotated intrinsic boson in terms of the eigenbosons (5.2); in fact the intrinsic boson is proportional to one of the eigenbosons. In each case, the eigenbosons transform as

$$\hat{U}^{(0)}(s^\dagger \mp \sqrt{2}d_0^\dagger)\hat{U}^{(0)\dagger} = e^{4i\phi_0}(s^\dagger \mp \sqrt{2}d_0^\dagger), \tag{8.6a}$$

$$\hat{U}^{(0)}(\sqrt{2}s^\dagger \pm d_0^\dagger)\hat{U}^{(0)\dagger} = e^{-2i\phi_0}(\sqrt{2}s^\dagger \pm d_0^\dagger). \tag{8.6b}$$

The matrix element of $\hat{U}^{(0)}$ between the states will then reduce to the four-diminsional integral:

$$< N,(2N,0),J,M|\hat{u}^{(0)}|N,(2N,0),J=M=0 >= \frac{(\mp 1)^{J/2}}{(4\pi)^2}$$

$$\left[\frac{(2J+1)(2N-J)!!(2N+J+1)!!}{(2N)!}\right]^{1/2} e^{iN\phi_0}$$

$$\times \int\int\int\int d\theta_1 d\theta_1' \sin\theta_2 d\theta_2 \sin\theta_2' d\theta_2' P_J(\cos\theta_2)$$

$$\times[\cos\theta_2\cos\theta_2' e^{(3i\phi_0/2)} + \sin\theta_2\sin\theta_2'\cos(\theta_1-\theta_1')e^{-(3i\phi_0/2)}]^{2N}. \tag{8.7}$$

Some algebra produces a reduced transition matrix in terms of a hypergeometric function:

$$< N,(2N,0),J,M=0|\hat{U}^{(0)}|N,(2n,0),J=M=0 >= (\mp 1)^{J/2}$$

$$\left[\frac{2^{J/2}(2J+1)N!(2N+J+1)!!}{(2N+1)!!\left(N-\frac{J}{2}\right)!}\right]^{1/2} \frac{(J-1)!!}{(2J+1)!!}$$

$$\times e^{-2iN\phi_0}(e^{6i\phi_0}-1)^{J/2} F\left[-\left(N-\frac{J}{2}\right), \frac{J+1}{2}; J+\frac{3}{2}; 1-e^{6i\phi_0}\right]. \tag{8.8}$$

This can also be written in terms of Jacobi polynomials, $P_n^{(\alpha,\beta)}(x)$,

$$< N,(2N,0),J,M=0|\hat{U}^(0)|N,(2n,0),J=M=0 >= (\mp 1)^{J/2}$$

$$\left[\frac{(2J+1)N!\left(N-\frac{J}{2}\right)!}{2^{J/2}(2N+1)!!(2N+J+1)!!}\right]^{1/2} (J-1)!!\, 2^N$$

$$\times e^{-2iN\phi_0}(e^{6i\phi_0}-1)^{J/2} P_{N-(J/2)}^{[J+1/2,-(N+1)]}(x), \tag{8.9a}$$

where

$$x = 2e^{6i\phi_0} - 1. \tag{8.9b}$$

The difference in oblate and prolate comes from the fact that ϕ_0 differs in sign for each case [Eq. (8.4b)].

9. THE LARGE N LIMIT

For N large the interacting boson model approaches the geometric collective model.[10] In this case the reduced transition matrix becomes a special function for each of the dynamical symmetries discussed in Sections 6-8. In taking this limit, we must pay attention to the fact that the function $\alpha(r)$ is normalized so that the $B(E2)$ from the ground state to the first excited state is reproduced [see Eq. (3.13a)]. Therefore α and hence g [Eq. (4.3b)] scale as the matrix element of the quadrupole operator. Also, for proton scattering, the forward scattering amplitude f is predominately imaginary. Hence we introduce the reduced $\bar{g}$,

$$\tilde{g} = \frac{i\bar{g}}{< 2_1^+ || Q || 0_1^+ >}. \tag{9.1}$$

The reduced matrix element $< 2_1^+||Q||0_1^+ >$ can be derived from the reduced transtion matrix by taking the $g \to 0$ limit. For U_5

$$< 2_1^+||Q||0_1^+ >_5 = \sqrt{5N}, \tag{9.2a}$$

for SU_6,

$$< 2_1^+||Q||0_1^+ >_6 = \sqrt{N(N+4)}, \tag{9.2b}$$

and for SU_3

$$< 2_1^+||Q||0_1^+ >_3 = \sqrt{2N(N+3/2)}. \tag{9.2c}$$

Therefore, since $\bar{g}$ is fixed by the $B(E2)$, for N large $\tilde{g}$ becomes small. Hence in taking the N large limit, we assume $\tilde{g}$ small, but $\bar{g}$ not necessarily small. For example in the U_5 limit,

$$\cos\tilde{g} \simeq 1 - \frac{\tilde{g}^2}{2}, \tag{9.3}$$

and then we have, assuming $\chi = 0$ and hence the relations (7.6),

$$< N, n, \tau, n_\Delta, J, M = 0|\hat{U}^{(0)}|N, n = \tau = n_\Delta = J = M = 0 > \quad \underset{N \to \infty}{\longrightarrow}$$

$$\left[\frac{(2\tau+3)!!}{(n+\tau+3)!!(n-\tau)!!}\right]^{1/2} A_{\tau n_\Delta J} \, \frac{\bar{g}^n}{\sqrt{5}} \, e^{\bar{g}^2/10}, \tag{9.4a}$$

$$\bar{g} << \sqrt{5N}, \quad n << N, \tag{9.4b}$$

which agrees with the geometrical model.[1] Hence the transition increases experimentally with the $B(E2)$ showing the importance of the coupled channels.

For SO_6 we take the limit in the hypergeometrical function using the relation that $(N)_p = \Gamma(N+p)/\Gamma(N) \to N^p$. The result is

$$< N, N, \tau, n_\Delta, J, M = 0|\hat{U}^{(0)}|N, N, \tau = n_\Delta = J = M = 0 > \quad \underset{N \to \infty}{\longrightarrow} \quad [3(2\tau+3)!!]^{1/2}$$

$$A_{\tau n_\Delta J} \, (-1)^\tau \, \frac{i_{\tau+1}(\bar{g})}{\bar{g}}, \tag{9.5a}$$

$$\bar{g} << N, \ \tau << N, \tag{9.5b}$$

where i_l is the modified spherical Bessel function of order l.

In the SU_3 limit, the hypergeometric function becomes a confluent hypergeometric function

$$< N,(2N,0),J,M=0|\hat{U}^{(0)}|N,(2N,0),J=M=0> \underset{N\to\infty}{\longrightarrow} (2J+1)^{1/2}\frac{(J-1)!!}{(2J+1)!!}$$

$$(3\bar{g})^{J/2}\, e^{\mp(\bar{g}/2)}\, M\left(\frac{J+1}{2}, J+\frac{3}{2}, \pm\frac{3\bar{g}}{2}\right), \tag{9.6a}$$

for

$$\bar{g} << \sqrt{2}N, \ J << 2N. \tag{9.6b}$$

where the different signs correspond to oblate and prolate respectively.

10. APPLICATIONS TO MEDIUM ENERGY PROTON SCATTERING

Electron scattering from nuclei can be analyzed using distorted wave Born approximation (DWBA) because the electromagnetic interaction with the nucleons inside the nucleus is small compared to the interaction between the nucleons. For hadronic scattering from nuclei the strength of the underlying interaction requires use of a distorted wave impulse approximation (DWIA), based on the two-body t matrix rather than the potential. However for 800 MeV proton scattering from collective nuclei, DWIA is not accurate at high momentum transfer.[20] These data and coupled channel calculations demonstrate that multiple scattering can be important for medium energy proton scattering from nuclei. How that importance grows with momentum transfer has also been shown theoretically.[21]

An alternative to the coupled channels approach calculates the scattering to all orders in the Glauber approximation which is a good approximation for 800 MeV protons.[1,22] Hence we shall apply the formalism discussed in Sections 2-5 to the scattering of medium energy protons from collective nuclei exciting specific states in the nucleus. These measurements were taken with the High Resolution Spectrometer at the Los Alamos Clinton P. Anderson Meson Physics Facility (LAMPF).

The following calculations were done by diagonalizing the IBM-1 Hamiltonian and using the eigenfunctions for the target ground state and excited states. The nuclei considered are in between the $U(5)$ and $SU(3)$ limits discussed above, but closer to the axially deformed $SU(3)$ limit than to the spherical quadrupole vibrator limit.

First we shall discuss the different qualitative behavior that the functions $\alpha(r)$ and $\beta(r)(=\beta_2(r))$ have with respect to the momentum transfer $\vec{q}$ (See Fig. 1). These features were derived in Ref. 4, and we summarize them here.

The contributions of $\alpha(r)$ and $\beta(r)$ to the inelastic cross section relative to the elastic cross section as a function of q are different. For a scattering amplitude with only $\alpha(r)$ contributing, the q dependence of this relative cross section for large q goes as q^2, and the next order goes as q^4. However, if only $\beta(r)$ contributes, this relative cross section would give a $q^2q^{4/3}$ dependence for large q, a dependence in between q^2 and q^4. This different behavior is illustrated in Fig. 2. We confirm the slower decrease in the slope when only $\beta(r)$ contributes as well as a slight shift in diffraction maxima. Of course, a realistic amplitude will have both form factors contributing. Hence, at some q, the slope of the differential cross section will decrease, indicating that the influence of $\beta(r)$ is increasing. The onset of this slope change depends on the ratio of the matrix elements

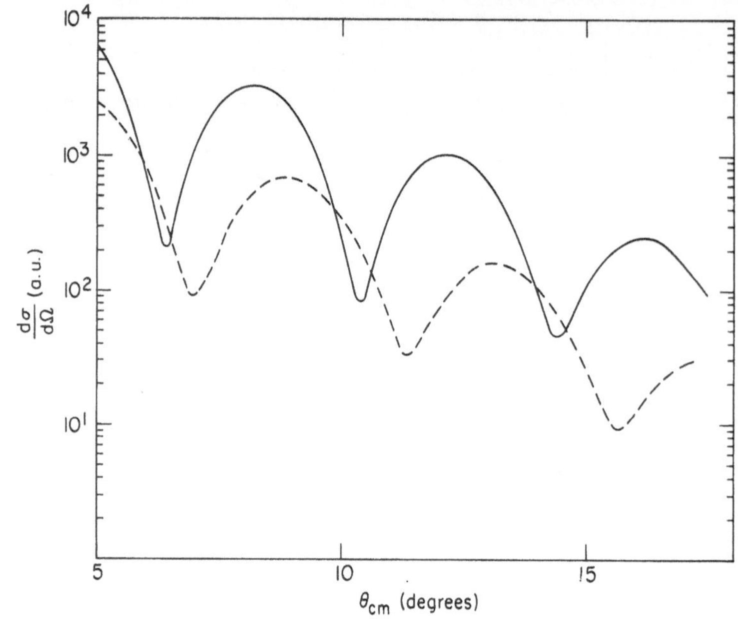

FIGURE 2. The calculated differential cross section to a 2^+ state with the $\beta(r)$ boson form factor alone (solid curve) as compared with a calculation with the $\alpha(r)$ form factor alone (dashed curve). Both curves are arbitrarily normalized.

of the seniority-conserving generators $T_m^{(l)}$ and the seniority-breaking generators P_m. This ratio will be larger for nonyrast levels than for yrast levels in general. Hence we expect to see this slope change sooner for nonyrast levels.

10.1 ^{154}Sm (p, p')

The 800 MeV (p, p') reaction on ^{154}Sm has been studied extensively experimentally and the data have been analysed in a conventional coupled channels calculation[20] and in the framework of the analytic stationary-phase method.[21] Both calculations, as well as others, have clearly demonstrated the importance of multi-step excitation processes.

The electromagnetic boson transition densities $\alpha(r)$ and $\beta_2(r)$ are extracted from inelastic electron scattering on ^{154}Sm. While the electrons scatter only from the proton bosons, the protons scatter from both the proton and neutron bosons. For the number of each kind of bosons roughly equal and large, this means that the hadronic boson transition densities $a(r)$ and $b(r)$ are twice the electromagnetic ones,[23] $a(r) = 2\alpha(r)$, $b_2(r) = 2\beta_2(r)$.

In Fig. 3 we show the results of our calculation for the $J_i^\pi = 0_1^+,\ 2_1^+$ states and compare them with the experimental data.[20] The agreement with experimental data is reasonable although we seem to underestimate the inelastic cross section for the $J_i^\pi = 2_1^+$ state by a factor of 1.5 to 3 in the region $6° < \theta_{cm} < 7.5°$. This feature is quite persistent (that is, is not easily remedied by changing the input parameters) and occurs also in the coupled channel calculation of Ref. 20. Related with this, we also note that for inelastic scattering, we calculate minima which are too deep compared with experiment. This discrepancy has been discussed in several papers.[24,25] It was found[24] that these diffractive minima in the spin-independent cross sections are <u>not</u> significantly affected by the elementary spin-orbit amplitudes although it was also pointed out that at some level one has to take into account effects of nuclear ground state deformation, transitions to magnetic substates that are not easily reached through the collective mechanism, contributions from other intermediate nuclear state and so on. Our calculation only partially addresses these suggestions: Although we take into account the nuclear ground state deformation and coupling to intermediate nuclear states we are not yet sure that (due to the small z/b approximation we have made in deriving the scattering matrix) we have not been neglecting some effects which may influence the population of the different magnetic substates.

In order to demonstrate the effect of including the coupling to excited states we also show in Figure 3 the results of calculations without them for the elastic scattering (optical model) and of including them only to first order (Born term) for the scattering to the $J_i^\pi = 2_1^+$ state. Coupled channel effects are clearly important at large q especially in the elastic channel.

In Figure 4 we show the predicted cross sections for excitation of the $J_i^\pi = 2_{2,3}^+$ states (the "β vibrational" 2_2^+ state, at 1.178 MeV and the "γ-vibrational" 2_3^+ state at 1.440 MeV). Also shown is the contribution to $d\sigma/d\Omega$ from the Born term. However, at this point, we would like to warn against taking the predicted cross sections for scattering to the $J^\pi = 2_2^+$, 2_3^+ states at face value. Indeed, IBM-1 calculations[26] do not correctly reproduce the $q \to 0$ limit or $B(E2,\ 0_1^+ \to J^\pi = 2_{2,3}^+)$, which are weak in any case, such that it may be anticipated that the curves shown in Figure 4 must be, at best, multiplied with the same factor by which the calculated $B(E2)$'s are off from the experimental values. The discrepancies for the $B(E2, 0^+ \to 2_{2,3})$ may be due to deficiencies in the IBM-1 wave functions or in the quadrupole operator (3.4). Indeed, since the $0_1^+ \to 2_{2,3}^+$ strengths originate from an appreciable cancellation between the $d^\dagger s$ and $d^\dagger d$ terms in (3.4), it is conceivable that a small change in either the IBM Hamiltonian or in the quadrupole operator could greatly improve transition strengths. The main motivation for showing these curves is to illustrate that nuclear structure effects may result in appreciable qualitative and quantitative differences in cross sections for inelastic scattering from states belonging to different bands and that consequently it is important for studying nuclear collective motion to measure these side band transitions.

Also we investigate the sensitivity of the excitation of the beta-vibrational 0_2^+ state to a monopole term ($l = 0$) in the transition phase given in (3.11). The simplest way to do this is to replace $\rho(r)$ by a density operator $\hat{\rho}(r)$,

$$\hat{\rho}(r) = \rho_{core}(r) + N\Delta\rho(r) + \alpha_0(r)\hat{n}_d, \tag{10.1}$$

where N is the total number of bosons. Here, we follow a phenomenological approach in determining ρ_{core}, $\Delta\rho$, and α_0: (i) in a first step, one relates the matrix element of $\hat{\rho}(r)$ for elastic scattering to the matter density for a nucleus with a mass number A, i.e.,

$$< A, 0_1^+|\hat{\rho}(r)|A, 0_1^+ >= \rho_{core}(r) + N\Delta\rho(r) + \alpha_0(r)n^A \equiv \rho^A(r), \tag{10.2a}$$

where

$$n^A =< A, 0_1^+|\hat{n}_d|A, 0_1^+ > \tag{10.2b}$$

which should hold for all Sm isotopes. For the vibrational ^{146}Sm, one has $n^{146} \simeq 0$ such that

$$\rho^{146}(r) = \rho_{core}(r) + 7\Delta\rho(r), \tag{10.3}$$

which leads to an extraction of $\Delta\rho(r)$, i.e.,

$$\Delta\rho(r) = \frac{1}{7}[\rho^{146}(r) - \rho_{core}(r)]. \tag{10.4}$$

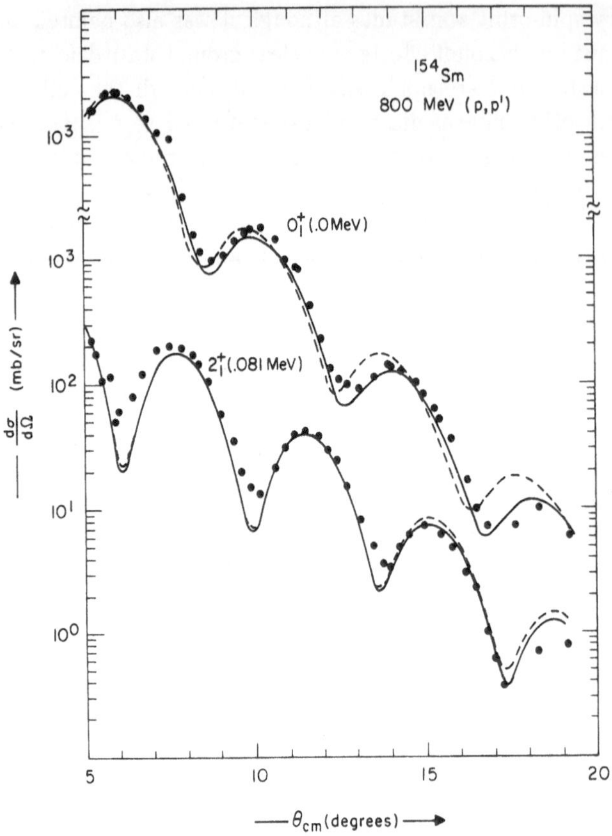

FIGURE 3. Calculation of cross sections for the $J_i^\pi = 0_1^+$, 2_1^+ state for 800 MeV (p,p') scattering on ^{154}Sm. Full curve includes channel coupling effects; dashed curve is either the optical model (in the case of $J_i^\pi = 0_1^+$) or the Born limit ($J_i^\pi = 2_1^+$).

In a second step one applies Eq. (10.2) to ^{154}Sm resulting in

$$\alpha_0(r) = \frac{1}{n^{154}}[\frac{4}{7}\rho_{core}(r) + \rho^{154}(r) - \frac{11}{7}\rho^{146}(r)] \tag{10.5}$$

which leads to the final expression for $\hat{\rho}(r)$ used in our calculations

$$\hat{\rho}(r) = \rho_{eff}(r) + \alpha_0(r)\hat{n}_d \tag{10.6a}$$

where

$$\rho_{eff}(r) = -\frac{4}{7}\rho_{core}(r) + \frac{11}{7}\rho^{146}(r) \tag{10.6b}$$

For $\rho_{core}(r)$ we take the matter distribution of ^{132}Sn which is parameterized, along with $\rho^{146}(r)$ and $\rho^{154}(r)$, in terms of a two parameter Fermi distribution.

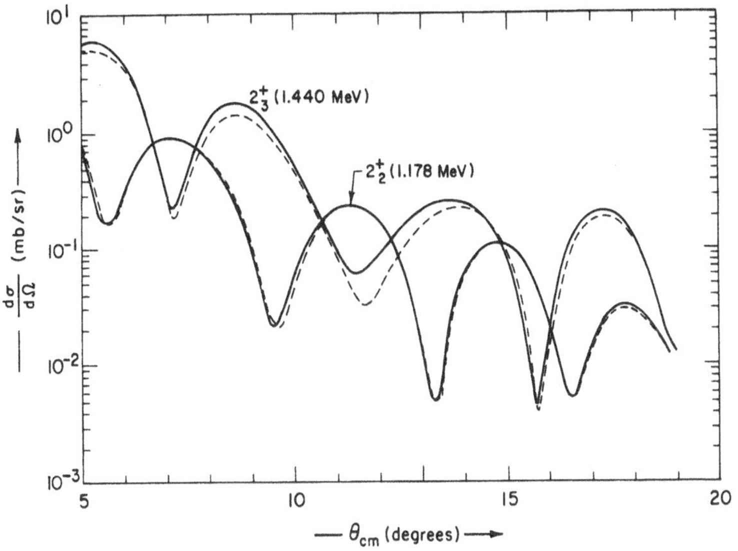

FIGURE 4. Calculated cross sections for the $J_i^\pi = 2_{2,3}^+$ states in ^{154}Sm for 800 MeV (p,p') scattering. Dashed curve is the Born limit, full curve includes channel coupling effects.

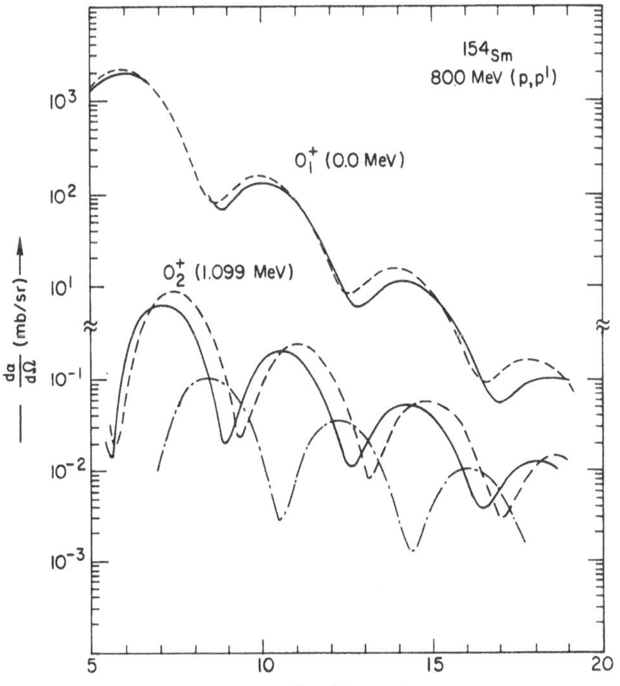

FIGURE 5. Calculated cross sections for the $J_i^\pi = 0_{1,2}^+$ states in ^{154}Sm for 800 MeV (p,p') scattering including monopole (dashed curve), quadrupole (dashed-dotted curve) only and both monopole and quadrupole (full curve) type of excitations. See text for further discussion.

In Figure 5 we show the results of our calculations of the cross sections for scattering to the $J_i^\pi = 0_1^+$ and 0_2^+ state in ^{154}Sm with a diffusity a (core) = 0.58 fm, a (^{146}Sm) = 0.62 fm, a (^{156}Sm) = 0.68 fm and $R = r_0 A^{1/3}$ where $r_0 = 1.07$ fm. The full curve includes monopole and qadrupole excitations while the dashed one includes only monopole excitations. The dotted-dashed curve is the one that is obtained using Eq. (3.11) including a quadrupole excitation term only. For the elastic scattering there is no perceptible difference between full and dotted-dashed curve which is why the latter is not shown in Figure 5. This is of course directly related with our construction of $\hat{\rho}(r)$ where we have required that in first order we obtain identical results (Eq. (2.9)). However, as could have been expected, the effect of including monopole excitations for the $J^\pi = 0_2^+$ state is more dramatic. In particular, we would like to draw attention to the fact that effects, which show up only modestly in the elastic channel, are more pronounced and often in a reversed way for the $J^\pi = 0_2^+$ state. For instance, switching the quadrupole coupling on results in shifting the maxima to smaller angles (by approximately 0.7 degrees) in contrast with the elastic scattering where cross-sections are shifted to slightly (by 0.3 degrees) larger angles when quadrupole excitations are included. Thus the coupling of the different states is clearly important for these monopole transitions.

10.2 ^{154}Gd (p, p')

We shall now use the formalism discussed in Sections 2-5 to calculate the differential cross section for 650 MeV proton scattering to the $J_i^\pi = 0_1^+, 2_{1,2,3}^+$ states in ^{154}Gd.

The IBM-1 wave functions for ^{154}Gd are taken from Ref. 27. A more detailed IBM-2 calculation for ^{154}Gd is given in Ref. 15. As can be seen from Table I, the reduced $E2$ matrix elements of the P and T transition generators with the IBM-1 calculation of Ref. 27 are very close to the IBM-2 matrix elements of the sum of the neutron and proton generators, $P_\pi + P_\nu$, $T_\pi + T_\nu$, obtained in Ref. 15.

The electromagnetic boson densities are determined from the transition densities extracted from electron scattering and shown in Fig. 6. Again the hadronic boson densities are taken to be twice the electromagnetic ones, $a(r) = 2\alpha(r)$ and $b(r) = 2\beta(r)$, where we have dropped the subscript 2. The seniority breaking boson densdity, $a(r)$, is peaking on the nuclear surface, while the senority conserving amplitude, $b(r)$, oscillates at the surface.

TABLE II. Reduced matrix elements of the P, T generator between the $J_1^\pi = 0_1^+$ and $J_f^\pi = 2_{1,2,3}^+$ states in ^{154}Gd obtained brom the IBM-1 analysis of Ref. 27 (column 1) and the IBM-2 analysis of Ref. 15 (column 3). Upper entry is for P; the lower entry is for T in columns 1 and 3. IBM-1 calculated transition rates in column 2, IBM-2 calculated transition rates in column 4, and experimental transition rates[28] in column 5.

J_f^π	IBM-1 (Ref. 27)	$B(E2)$ (e^2b^2)	IBM-2 (Ref. 15)	$B(E2)$ (e^2b^2)	$B(E2)$ (e^2b^2)
2_1^+	11.50	3.85	11.44	3.87	3.85(8)
	-1.79		-1.81		
2_2^+	-2.45	0.017	-2.49	0.023	0.015(4)
	-1.61		-1.59		0.0022(2)
2_3^+	2.76	0.11	2.74	0.13	0.14(1)
	0.55		0.59		

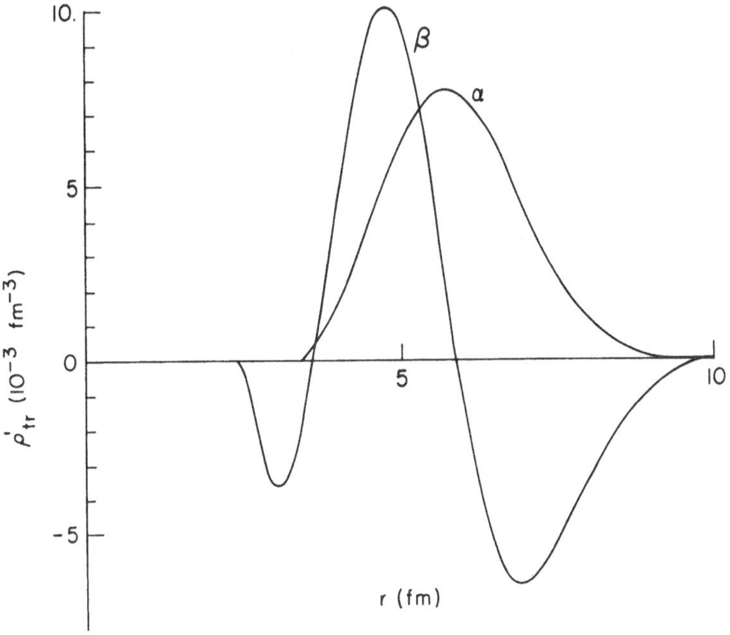

FIGURE 6. IBM-1 electromagnetic form factors $\alpha(r)$ and $\beta(r)$.

In Fig. 7 the differential cross sections for 650 MeV proton scattering on ^{154}Gd are compared with the experimental cross sections measured[4,5,29] at LAMPF. The matter density, which determines the distorted wave (2.6), is taken to be a Fermi form with half-radius 1.085 $A^{1/3}F$ and diffusivity 0.66 F. This choice was made so that the derivative of $\rho(r)$ resembles $\alpha(r)$ in shape. For elastic scattering, the calculated cross section (solid line) agrees reasonably well.

For inelastic scattering the calculated cross section (dashed line in Fig. 7) is about 20% too high for scattering to the first 2_1^+ state, but otherwise the minima and maxima are given very well. For scattering to the 2_3^+ state, the calculated cross section (dashed line) is about 20% too low and shifted forward about 0.4° in angle. However, this state depends differently than the 2_1^+ state on $b(r)$. A decrease in $a(r)$ will decrease its cross section just as for the 2_1^+, but a decrease in $b(r)$ will increase its cross section, as can be understood from an inspection of Table I. In fact, if we take $\chi \rightarrow 0.5\chi$, thereby producing a hadronic boson $b(r)$ density equal to the electromagnetic boson density rather than twice as large as suggested,[23] there is very good agreement from both states, as shown by the solid line in Fig. 7. This suggests that neutron and proton differences may be playing a role in the $b(r)$ boson form factor, since we get one-half the value expected if we assume neutrons and protons are behaving symmetrically.[23]

For the 2_1^+ state the contribution of $b(r)$ is relatively unimportant, such that the cross section for this state behaves like the ideal one as calculated with the $a(r)$ boson density alone (in slope as well as in position of the diffraction minima and maxima). For the 2_3^+ state the contribution of $b(r)$ is more important, resulting in a total transition density which peaks at slightly smaller radii and thus results in cross sections which have their diffraction maxima at slightly larger angles. The resulting slope is also slightly more gentle than for the 2_1^+ state. These features follow from our earlier observation the

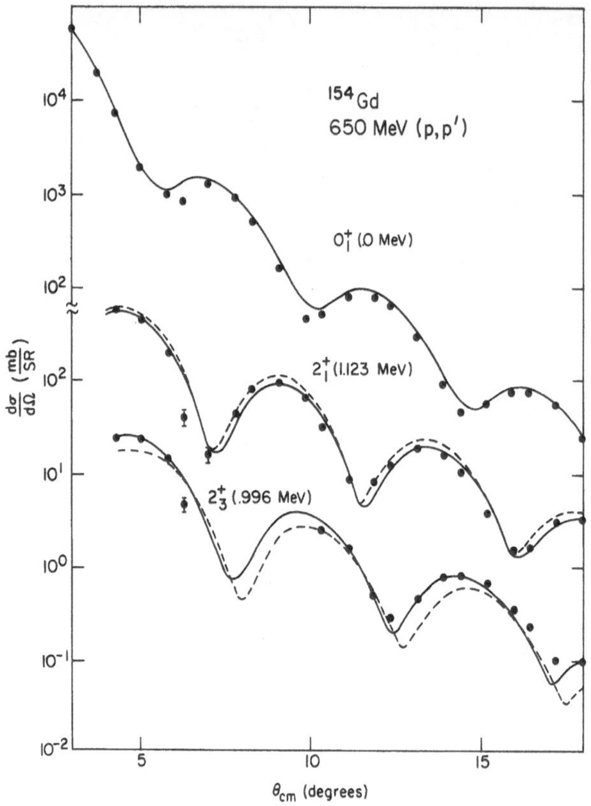

FIGURE 7. IBM-1 results for the 650MeV (p,p') reaction from the $J_i^\pi = 0_1^+, 2_{1,3}^+$ states in ^{154}Gd. Solid line is with $a(r) = 2\alpha(r)$, $b = \beta(r)$; dashed line is with $a(r) = 2\alpha(r)$, $b(r) = 2\beta(r)$. The dots are experimental data. Errors are within the experimental dots except where noted. Errors for the inelastic scattering include estimated uncertainties due to background subtraction.

ratio of the seniorty-conserving martix element to the seniority-breaking matrix element is larger for the 2_3^+ state that the 2_1^+ state.

For the 4_1^+ state we show in Figure 8 the calculated and experimental[5,29] cross sections. Use was made of a $b_4(r)$ which resemble very much the e. m. one derived by Hersman[30] but peaks at a somewhat smaller radius (6.6 F instead of 6.9 F) and has a smaller diffusivity. The quadrupole transition densities were left unaltered. In particular, note the sensitivity of the results to the interference of the two-step quadrupole process (dotted line) with the one-step hexadecupole excitation process (dashed-line). Although the two-step quadrupole excitation alone does not contribute much in absolute value to the cross section it is responsible for inducing destructive or constructive interference with the one-step hexadecupole deformation (dash-dotted versus solid line). The electron scattering does not determine the over all sign of the quadrupole boson densities and the hexadecupole boson densities. However the measured proton differential cross-section (dots) clearly agrees with the cross section in which the quadrupole and hexadecuple destructively interfere (solid line).

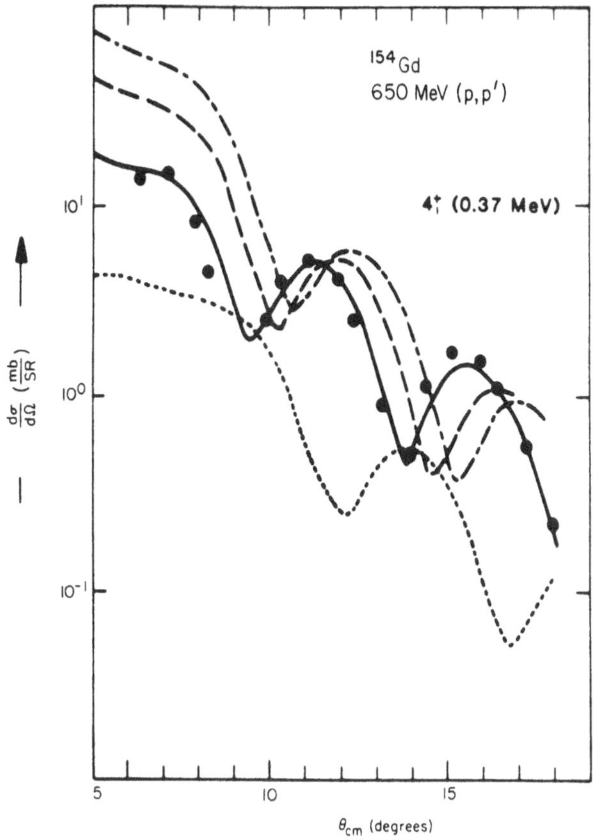

FIGURE 8. Calculated and experimental cross sections for the 650 MeV (p,p') reaction on the 4_1^+ state in ^{154}Gd. One step hexadecupole excitation (dash), two step quadrupole excitation (dotted), and total constructive (dash-dot) and destructive (solid). Dots are experimental data.

This example shows that we can successfully use electron and hadron scattering to determine the difference in neutron and proton densities. Futhermore, the scattering to the 2_2^+ state, which is the weakest state, depends on $a(r)$ and $b(r)$ state, but is even more sensitive to changes in $b(r)$. In general, this means that scattering to nonyrast states may provide more sensitive information on the $b(r)$ form factor. Clearly, in this study it would be helpful to have more systematic electron and proton scattering data to both the yrast and nonyrast 2^+ states on a series of Gd isotopes and other neighboring nuclei.

11. SUMMARY

We have shown that the scattering matrix in the Glauber approximation to scattering of a projectile at a fixed input parameter from a composite system with a dynamical symmetry is a representation matrix of the group. This scattering matrix includes all possible multistep processes within the representation space of that group. The importance of the multistep processes grow as the momentum transfer to the nucleus grows.

We have applied this formalism to medium energy proton scattering from nuclei for which the Glauber approximation is valid for small angles and for which multistep processes are important. We use the version of the Interacting Boson Model which does not distinguish between neutrons and protons and which has an $SU(6)$ dynamical symmetry. Multistep processes are found to be important for the yrast $J^\pi = 4_1^+$ level and for the nonyrast levels.

If transitions are isoscalar than the boson densities to be used for proton scattering will be twice the boson densities determined from electron scattering. From the measurements of scattering to the $J^\pi = 2_{1,3}^+$ states of ^{154}Gd we determined that this assumption is valid for the seniority breaking part of the quadrupole operator, $P_\mu = s^\dagger \tilde{d}_\mu + d_\mu^\dagger s$, but not for the seniority-conserving part, $T_\mu^{(2)} = (d^\dagger \tilde{d})_\mu^{(2)}$. In order to fit the cross-section we had to reduce the seniority-conserving boson density by <u>one-half</u>, implying both that this part of the operator has a larger isovector term, and that these states are not completely symmetric in the neutron and proton degrees of freedom, and hence IBM-2 is necessary to describe these transitions.

This non-symmetric nature of neutron and proton collective motion can be probed more deeply by single-charge exchange experiments on collective nuclei. Pion single-charge exchange has already measured the isovector part of the quadrupole transition density[31] for ^{165}Ho providing an estimate for the amount of asymmetry in the neutron-proton contributions to nuclear collective motion.[9] A more systematic determination of the isovector quadrupole transitions will be possible with the new neutral meson spectrometer being developed at LAMPF.

12. ACKNOWLEGMENTS

Geert Wenes contributed substantially to the research described in this paper. This work was supported by the U. S. Department of Energy.

13. REFERENCES

1. R. J. Glauber, in *Lectures in Theoretical Physics*, edited by W. E. Brittin and L. G. Dunham (Interscience, New York, 1959), Vol. 1, p. 315

2. J. N. Ginocchio, T. Otsuka, R. D. Amado, and D. A. Sparrow, Phys. Rev. C **33** 247 (1986).

3. G. Wenes and J. N. Ginocchio, Nucl. Phys. **A459** (1986) 631.

4. J. N. Ginocchio, G. Wenes, R. D. Amado, D. C. Cook, N. M. Hintz, and M. M. Gazzaly, Phys. Rev. C **36**, 2436 (1987).

5. G. Wenes and J. N. Ginocchio, J. Phys. G: Nucl. Phys. 14 Suppl. p. S65-S70.

6. R. Bijker and R. D. Amado, Phys. Rev. **A37**, 1425 (1988); R. Bijker, this Proceedings.

7. W. H. Bassichis, H. Feshbach, and J. Reading, Ann. Phys. (N.Y.) **68**, 462 (1971).

8. F. Iachello and A. Arima, *The Interacting Boson Model* (Cambridge University Press, Cambridge, 1987).

9. A. Leviatan, J. N. Ginocchio, and M. W. Kirson, LANL Report LA-UR-90-2560, submitted to Phys. Rev. Lett. (1990).

10. J. N. Ginocchio and M. W. Kirson, Nucl. Phys. **A350**, 31 (1980); A. Dieperink, O. Scholten, and F. Iachello, Phys. Rev. Lett. **44**, 1747 (1980).

11. A. Arima, N. Yoshida, and J. N. Ginocchio, Phys. Lett. **101B**, 209 (1981).

12. T. Otsuka, A. Arima, and F. Iachello, Nucl. Phys. **A309**, 1 (1978).

13. S. Pittel, P. D. Duval, and B. R. Barrett, Ann. Phys. (N.Y.) **144**, 168 (1982)

14. T. Otsuka, Phys. Lett. **138**, 1 (1984).

15. G. Wenes, T. Otsuka, and J. N. Ginocchio, Phys. Rev. C **37**, 1878 (1988).

16. D. A. Varshalovich, A. N. Moskalev, and V. K. Khersonskii, *Quantum Theory of Angular Momentum* (World Scientific, Singapore, 1988).

17. G. Wenes, A. E. L. Dieperink, and O. S. Van Roosmalen, Nucl. Phys. **A424**, 81 (1984).

18. J. N. Ginocchio, Ann. of Phys. **126**, 234 (1980).

19. K. Alder and A. Winther, *Electromagnetic Excitations* (North-Holland, Amsterdam, 1975).

20. M. L. Barlett, J. A. McGill, L. Ray, M. M. Barlett, G. W. Hoffmann, N. M. Hintz, G. S. Kyle, M. A. Franey, and G. Blanpied, Phys. Rev. C **22**, 408 (1980).

21. R. D. Amado, J. A. McNeil, and D. A. Sparrow, Phys. Rev. C **25**, 13 (1982).

22. S. J. Wallace, Ann. Phys. (N.Y.) **78**, 190 (1983).

23. J. N. Ginocchio and G. Wenes, Phys. Rev. C **34**, 1127 (1986).

24. G. Fäldt and P. Osland, Nucl. Phys. **A305**, 509 (1978).

25. P. Osland and R. Glauber, Nucl. Phys. **A326**, 255 (1979).

26. O. Scholten, F. Iachello and A. Arima, Ann. Phys. **115**, 325 (1978)

27. P. O. Lipas, P. Toivonen, and D. D. Warner, Phys. Lett. **115B**, 295 (1985).

28. F. W. Hersman *et al.*, Phys. Lett. 132B, 47 (1983); Phys. Rev. C **33**, 1905 (1986).

29. D. C. Cook, *et al.*, University of Minnesota Annual Progress Report, 1986, pp. 21-31.

30. F. W. Hersman, W. Bertozzi, T. N. Buti, J. M. Finn, C. E. Hyde-Wright, M. V. Hynes, J. Kelley, M. A. Kovash, S. Kowalski, J. Lichtenstadt, R. Lourie, B. Murdock, B. Pugh, F. N. Rad, C. P. Sargent, and J. B. Bellicard, Phys. Rev. C **33**, 1905 (1986).

31. J. N. Knudson, *et al*, LANL Report LA-UR-90-881, submitted to Phys. Rev. Lett. (1990).

PREDICTING "ANYONS": THE ORIGINS OF FRACTIONAL STATISTICS IN TWO-DIMENSIONAL SPACE

Gerald A. Goldin

Departments of Mathematics and Physics
Rutgers University
New Brunswick, New Jersey 08903 USA

INTRODUCTION

The possibility of particles or excitations in two-dimensional space, with statistics intermediate between those of bosons and fermions, has recently attracted considerable attention. Physical situations to which the theory of such "anyon" statistics can apply include variations of the Aharonov-Bohm effect, the fractional quantum Hall effect, the possible characterization of quantum vortices in thin superfluid films and, in general, quantum surface phenomena. Anyon statistics has been discussed in connection with the theory of monopoles, and has been proposed as a possible explanatory mechanism for high-T_c superconductivity. There is a mathematically interesting relationship between such unusual statistics and the braid group, which parallels the relationship between ordinary statistics and the symmetric group.

In this mainly qualitative survey I shall sketch the historical development of the prediction that anyons are possible, and in so doing seek a deeper physical understanding by highlighting what must be assumed for unusual statistics to occur. We shall see that some fundamental revisions are needed in the commonly held view of what particle statistics means. I shall also take the opportunity to argue for a geometric approach to quantum theory based on unitary representations of infinite-dimensional Lie groups and algebras, in particular groups of diffeomorphisms and algebras of vector fields--an approach which led my colleagues and me to an early, rigorous prediction of anyons.

APPROACHES TO QUANTUM STATISTICS

There are at least four main ways to introduce quantum statistics, which we may term loosely: (**A**) naïve, (**B**) field-theoretical, (**C**) topological, and (**D**) group-theoretical.

(**A**) The naïve approach, taken in many elementary quantum mechanics texts, introduces the many-particle wave function Ψ as a function of an ordered N-tuple of particle coordinates. When the particles are identical, one then uses the fact of exchange symmetry to remove the dependence of the probability density $\Psi^*\Psi$ on the (now superfluous) indices through an equivariance condition on Ψ. Thus, for a permutation σ in the symmetric group S_N, let us define

$$[\sigma\Psi](\mathbf{x}_1, \dots, \mathbf{x}_N) = \Psi(\mathbf{x}_{\sigma(1)}, \dots, \mathbf{x}_{\sigma(N)}). \tag{1}$$

Elements of S_N, in this approach, permute the particle *labels* or *indices*. Now the equivariance condition becomes $\sigma\Psi = V(\sigma)\Psi$, where $V(\sigma) \equiv +1$ (bosons), or $V(\sigma) = \pm 1$ according to whether σ is odd or even (fermions). These are justified as the only possibilities by the assertion that Ψ must be single-valued. Then $V(\sigma)$ defines a one-dimensional unitary representation of the group S_N of all permutations of the indices (1, ... , N); i.e., a character of the group.

Even within this naïve framework, it can be asserted that other statistics are possible. First, there is no need in principle to restrict ourselves to scalar-valued wave functions. If we took $\Psi(\mathbf{x}_1, \dots, \mathbf{x}_N)$ to have components in $\mathbb{C}^n$, then we could allow V to be an n-dimensional irreducible representation of S_N. Irreducible representations with dimension $n > 1$ exist when $N \geq 3$, and are described by means of the well-known Young tableaux--for instance when $N = 3$, we have the tableau in Fig. 1(a) for the one-dimensional totally symmetric representation (describing bosons), the tableau in Fig. 1(b) for the one-dimensional totally antisymmetric representation (fermions), and the tableau in Fig. 1(c) for the two-dimensional representation of S_3.

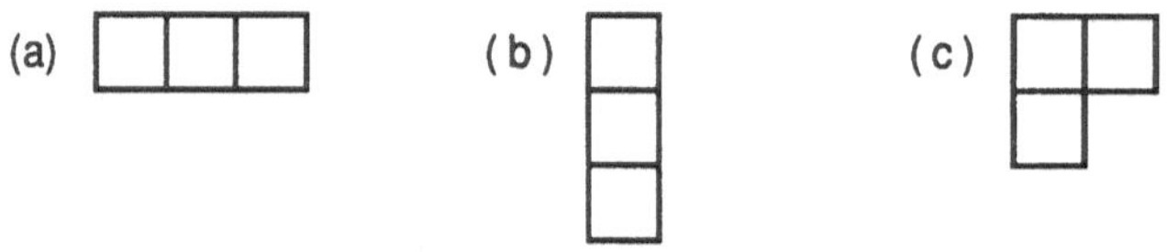

Fig. 1. Young tableaux for the symmetric group of permutations of 3 objects.

But higher-dimensional representations of S_N can enter even under the assumption of scalar-valued wave functions of $(\mathbf{x}_1, \dots, \mathbf{x}_N)$. We construct the N-particle Hilbert space $\mathcal{H}$ (naïvely) as the tensor product of one-particle Hilbert spaces, $\mathcal{H} = \otimes_{j=1,\dots,N} \mathcal{H}_j$, where $\mathcal{H}_j = \mathcal{L}^2(\mathbb{R}^3)$; then (when the particles are identical) we remove the ordering in the tensor product by reexpressing $\mathcal{H}$ as the direct sum of subspaces which transform according to irreducible representations of S_N. With $N = 3$, for instance, this procedure yields the totally symmetric and the totally antisymmetric subspaces; but it also gives two copies of the two-dimensional representation of S_3 corresponding to the Young diagrams in Fig. 2. Thus, even a naïve description of particle statistics forces a discussion of some pos-

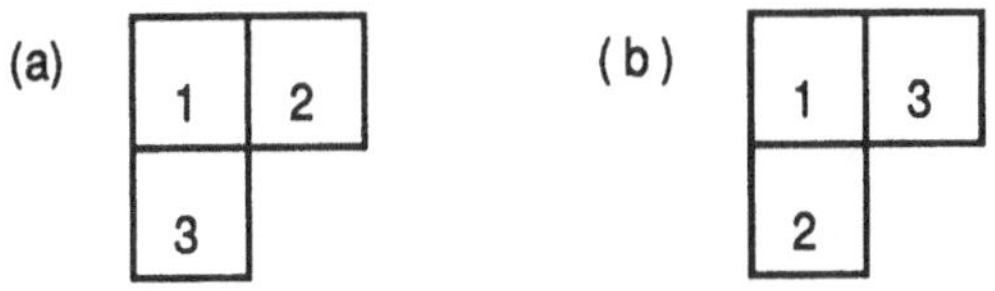

Fig. 2. Young diagrams for the two-dimensional representation of S_3.

sibilities other than Bose and Fermi statistics. The representations of S_N with dimension greater than one can be interpreted as describing paraparticles. In this picture, with Ψ reexpressed as a scalar-valued wave function of the (ordered) N-tuple of particle coordinates, the condition that $\Psi^*\Psi$ be invariant under all permutations of the coordinates appears as an extra assumption, which is sufficient to rule out parastatistics.

We shall see shortly why what I have just summarized is not only naïve, but in a certain sense incorrect. Note that what characterizes this approach is that it begins with *distinguishable* particles, and utilizes S_N to remove the distinguishability; i.e., something unphysical––the labeled coordinate description of configurations, analogous perhaps to a choice of gauge––is introduced at the start, and then some mathematics is done to remove it.

(B) In quantum field theory particles are regarded not as the fundamental objects, but as the quanta of underlying fields; and different statistics can be obtained by means of different equal-time algebras assumed for the fields in each theory. For example, (non-relativistic) bosons arise from the equal-time canonical commutation relations

$$[\psi(\mathbf{x}),\psi^*(\mathbf{y})] = \psi(\mathbf{x})\psi^*(\mathbf{y}) - \psi^*(\mathbf{y})\psi(\mathbf{x}) = \delta(\mathbf{x}-\mathbf{y}), \tag{2}$$

and fermions from the equal-time canonical anticommutation relations

$$\{\psi(\mathbf{x}),\psi^*(\mathbf{y})\} = \psi(\mathbf{x})\psi^*(\mathbf{y}) + \psi^*(\mathbf{y})\psi(\mathbf{x}) = \delta(\mathbf{x}-\mathbf{y}). \tag{3}$$

Field theories of paraparticles (parafermions and parabosons) can be obtained from various *trilinear* commutator and/or anticommutator brackets (Green, 1953; Messiah and Greenberg, 1964; Greenberg, 1966). One then constructs representations of such algebras of fields by (in general, unbounded) operators in Hilbert space; in particular, there are the usual Fock representations (e.g. Schweber, 1961) in which the N-particle subspaces for Eq. (2) contain the totally symmetric wave functions, and those for (3) the totally antisymmetric wave functions. Parafields can also be represented in Fock spaces, with some subspaces transforming according to the higher-dimensional representations of S_N.

Recently there has also been interest in more general "q-commutators" (e.g. Biedenharn, 1989), suggesting that we consider generalizations of Eqs. (2) and (3) having a form such as

$$[\psi(\mathbf{x}),\psi^*(\mathbf{y})]_q = \psi(\mathbf{x})\psi^*(\mathbf{y}) - q\,\psi^*(\mathbf{y})\psi(\mathbf{x}) = \delta(\mathbf{x}-\mathbf{y}), \tag{4}$$

for arbitrary q. Subtracting (4) from its adjoint implies that $(Im\ q)\psi^*(y)\psi(x) \equiv 0$, or for any q that is not real, $\psi^*(y)\psi(x) \equiv 0$, a too-severe constraint. For $q \in \mathbb{R}$, however, Eq. (4) is one way of "interpolating" bosons and fermions. To interpret such fields physically would require appropriate representations of the algebra by linear operators in Hilbert space.

Neither the naïve approach nor the field-theoretical approach actually led to the prediction of anyons, though Fröhlich (1976, 1980) did consider unusual commutation relations between soliton fields and meson fields in two-dimensional models exhibiting topological solitons, and observed that the Euclidean Green functions of such models are multivalued--anticipating, in a way, the braid statistics of fields. However in the late 1950's and 1960's there was considerable discussion of parastatistics in the above contexts. For indistinguishable particles, it was thought necessary that the action of permutations on many-particle wave functions be such as to *commute with all observables.* Indeed if an observable did *not* commute with some $\sigma \in S_N$ acting as in Eq. (1), thus failing to treat the arguments $(\mathbf{x}_1, \dots, \mathbf{x}_N)$ of Ψ symmetrically, the particles could be distinguished by means of that observable. Hence the result emerged that paraparticles could be regarded for physical purposes as ordinary particles to which extra quantum numbers were assigned.

A point of view implicitly challenging this conclusion follows from the work of Stolt and Taylor (1970), who argued in a field-theoretical context that we should not be using the label permutations in Eq. (1) at all; rather we should be considering another way for permutations to act (variously called *value* permutations, *wave function* permutations, or *physical* permutations). Instead of (1), consider unitary representations V of S_N given by

$$[V(\sigma)\Psi](\mathbf{x}_1, \dots, \mathbf{x}_N) = \Psi(\sigma(\mathbf{x}_1, \dots, \mathbf{x}_N)), \tag{5}$$

where on the right of Eq. (5), σ acts not on the indices but on the *values* of $(\mathbf{x}_1, \dots, \mathbf{x}_N)$. This presumes a way of ordering the N points $\{\mathbf{x}_1, \dots, \mathbf{x}_N\}$ in $\mathbb{R}^3$. While there is no natural, coordinate-independent ordering, it is easy to order the points with respect to a coordinate system--e.g. let $\mathbf{x} < \mathbf{y}$ if $x^1 < y^1$, or $x^1 = y^1$ and $x^2 < y^2$, or $x^1 = y^1$ and $x^2 = y^2$ and $x^3 < y^3$. The physics of the resulting quantum theory is nevertheless coordinate-independent, i.e., independent of the particular choice of (measurable) ordering. The action of the permutations in (5) is quite different from that in (1): for example, the permutation $\sigma = (12)$ acts in (1) by exchanging $\mathbf{x}_1$ and $\mathbf{x}_2$, regardless of their values; while in (5), it acts by exchanging the pair of coordinates having the *two lowest* values, regardless of their indices. Which coordinates are labeled $\mathbf{x}_1$ and $\mathbf{x}_2$ is arbitrary; but which have the lowest values is not (justifying the term *physical* permutations). For representations of S_N having dimension greater than one, the physical permutations do *not* commute with all observables (see Goldin, 1987); thus what would be merely a technical point for bosons or fermions (where both actions of σ give the same representation V) becomes important in the discussion of paraparticles. We shall also see that to understand the origins of intermediate statistics, it is essential to think in terms of the physical permutations.

(C) A topological approach to quantum theory was initiated by Finkelstein and Rubinstein (1968), through their theory of "kinks." In a highly original, field-theoretic context, they noted the topological possibility of exchange coefficients other than ± 1 when the spatial dimension is less than three, but did not at that time pursue the mathematical or physical implications.

Soon thereafter, much more elementary topological description of particle statistics were offered by Laidlaw and DeWitt (1971), and by Schulman (1971); and quantum mechanics on multiply connected spaces was studied by Dowker (1972). Following Laidlaw and DeWitt, we take as a starting point the configuration space for N *indistinguishable* particles, assumed to be the set $\Delta = \{\gamma \mid \gamma = \{\mathbf{x}_1, \ldots, \mathbf{x}_N\} \subset \mathbb{R}^3\}$; i.e., the set of all unordered N-tuples of distinct points (N-point subsets) in $\mathbb{R}^3$. The continuous paths in Δ can be classified by homotopy equivalence. It is easy to see that in traversing a loop based at γ, one can return to the same starting point in Δ not only by restoring to its original position each individual point $\mathbf{x}_j$ comprising γ, but also by permuting the $\mathbf{x}_j$, and that different permutations correspond to homotopically inequivalent loops. The fundamental group $\pi_1(\Delta)$, defined by the homotopy classes of such loops in Δ, is just S_N (for $\mathbb{R}^3$, and whenever the space dimension $n \geq 3$). But the elements of $\pi_1(\Delta)$ can also be placed in correspondence with homotopy classes of paths in Δ sharing the same initial and final points. In the description of quantum mechanics in which the Feynman propagator is considered as a "sum over paths," it is now possible for there to occur a relative phase (a unitary "weight") between the contributions that the different homotopy classes make to the sum--and these phases must obey a unitary representation of the group $\pi_1(\Delta)$. Therefore, assuming scalar-valued wave functions on the universal covering space of Δ, one of the two one-dimensional representations of S_N (describing bosons or fermions) characterizes any resulting quantum theory.

I shall make several brief observations about this approach:

(1) Its strength is that it succeeds mathematically in describing a topological origin for Bose and Fermi statistics, making only elementary assumptions.

(2) There are some difficulties with the decision to exclude from Δ the set $\mathcal{D}$ of so-called "diagonal" points, where two or more of the coordinates take on the same value. The fact that the fundamental group is nontrivial depends directly on this decision, and a trivial fundamental group would predict only bosons. First, there is the question (which was raised at the time) of the physical meaning of the assumption--is it a somewhat metaphysical statement asserting the impossibility *in principle* of two particles occupying the same point in spacetime, or does it depend on the existence of some kind of hard-core, repulsive potential term in the Hamiltonian? Second, there is the seeming paradox in choosing to exclude a set of Lebesgue measure zero, and obtaining important physical

consequences. What restriction prevents our arbitrarily choosing a different measure-zero set to exclude from configuration space, and introducing a more complicated fundamental group?

(3) The conclusion by Laidlaw and DeWitt that the framework ruled out parastatistics was seen as an important physical prediction. However, parastatistics are *not* eliminated if one allows vector-valued wave functions on the covering space of Δ in the Feynman path formalism. The Feynman propagator K then becomes operator-valued, and we can have matrix-valued unitary weights, transforming according to higher-dimensional representations of S_N (Goldin, Menikoff, and Sharp, 1985).

(4) The topological approach does make a tacit commitment to the Stolt-Taylor view that S_N should act *via* physical, particle permutations rather than label permutations. But this point seems to have been essentially unnoticed at the time--perhaps because the distinction between the two ways permutations can act has direct physical consequences only for paraparticles, which were believed to have been eliminated in the topological picture.

(5) The use of a similar framework to try to obtain particle spin is an interesting idea that was introduced by Hamilton and Schulman (1971) and pursued by Schulman, requiring a larger initial configuration space; it achieved limited success.

(6) Though the new possibilities that can occur in two dimensions are already implicit in the topological formalism, requiring just a few additional steps, they were not realized until several years later.

Independently of this body of earlier work, Leinaas and Myrheim (1977) considered the configuration space for a pair of indistinguishable particles in $\mathbb{R}^2$, whose relative coordinate gives the shaded half-space in Fig. 3. They advanced the case for beginning with this configuration space rather than using arbitrarily indexed coordinates, and offered in this paper what is to my present knowledge the first clear prediction of intermediate, "anyon" statistics, based on a topological argument resembling that described above for $\mathbb{R}^3$.

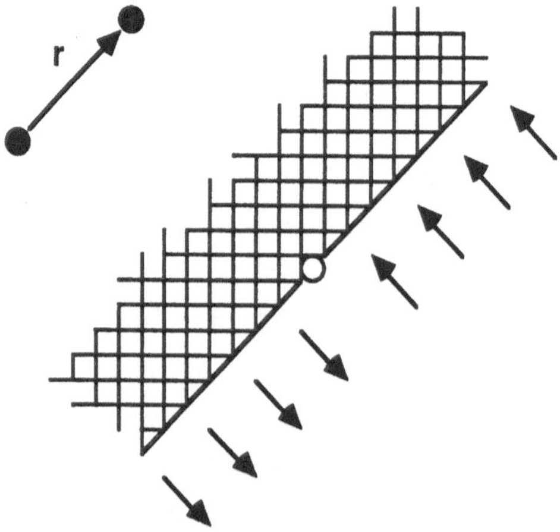

Fig. 3. The 2-particle configuration space in two dimensions (relative coordinate).

They introduced the phase change $e^{i\theta}$ as the relative coefficient of the wave function associated with a single counterclockwise exchange of the two particles, thus generating a one-dimensional representation of the fundamental group of Δ--in the case N = 2, of the group of additive integers.

Again, there arises the issue of why the "diagonal" points $\mathcal{D}$ (i.e., the origin in the plane of the relative coordinate $\mathbf{r} = \mathbf{x}_2 - \mathbf{x}_1$, excluded in Fig. 3) are ruled out of Δ. Leinaas and Myrheim argued that the identification of $-\mathbf{r}$ with $\mathbf{r}$ required that the half-plane be "wrapped around" (to match up the half-lines, in the direction of the arrows in Fig. 3), forming a cone-shaped surface. The origin then becomes a singular point, where the curvature is undefined and the surface is no longer smooth; this fact becomes the principal argument for treating it differently by excluding it. However, the reasoning on this point is not persuasive. Mathematically there is no need to introduce global curvature in order to identify the points $\mathbf{r}$ and $-\mathbf{r}$, as the geometrical properties of the space are intrinsic and do not depend on its being embedded in $\mathbb{R}^3$. Physically this would also appear unwarranted, since the particles are actually moving in a flat space. The surface resulting from the identification of points is geometrically and topologically the plane, and it remains an arbitrary decision to remove the origin.

Note once more that implicit in such topological approaches is the fact that the particles are being exchanged by *physical* permutations. Label permutations do not even make sense in the context of the enlarged fundamental group needed for intermediate statistics!

Leinaas (1978, 1980) went on to draw further implications from this formalism, exploring its consequences for charge-monopole composites ("dyons"), and relating the case of point singularities in two dimensions leading to the phase shift $e^{i\theta}$, to the case of line singularities in three dimensions which make possible the phase shift for the Aharonov-Bohm solenoid.

(D) A still different approach to quantum statistics is based on representations of infinite-dimensional Lie algebras and groups: local current algebras, including algebras of vector fields; groups of diffeomorphisms; and their semidirect products. This is a program of research which by 1980 I had been pursuing for many years (Dashen and Sharp, 1968; Goldin, 1969; Goldin and Sharp, 1970; Grodnik and Sharp, 1970a,b; Goldin, 1971; Goldin and Sharp, 1972; Goldin, Grodnik, Powers, and Sharp, 1974; Menikoff, 1974a,b; Menikoff and Sharp, 1977; for an overview, see Goldin, 1984, or Goldin and Sharp, 1989). This work, which had been applied to the quantization of hydrodynamics (e.g. Rasetti and Regge, 1975), led quite independently to the prediction of anyon statistics (Goldin, Menikoff, and Sharp, 1980; 1981). Results obtained and published by 1980-81 included: (1) a mathematically rigorous derivation of the fractional statistics of particles in two-dimensional space from the classification of unitary group representations, including the angle parameter interpolating bosons and fermions; (2) the shifted energy and angular momentum spectrum associated with anyons; (3) the relationship with the

Aharonov-Bohm effect; and (4) the N-particle configuration space as a *consequence* of the theory, with no need to assume the exclusion of diagonal points (see below). In our framework the statistics of N anyons, for arbitrary N, follows from the classification of unitary representations of a single group, Diff($\mathbb{R}^2$), or for convenience and simplicity its semidirect product with the Schwartz space $\mathcal{S}(\mathbb{R}^2)$ of C^∞ scalar functions on $\mathbb{R}^2$. There is no need to introduce different algebras of fields for different statistics. The representation theory of Diff($\mathbb{R}^2$) which led to anyon statistics, and some of its further consequences, are described briefly below.

Though the local current algebra and its associated group were obtained initially from underlying canonical fields, there are unitary group representations which do not so arise. It is therefore encouraging that such infinite-dimensional algebras and groups have been shown by very different means to provide a description of quantum systems, through fundamental considerations of quantization on manifolds (Doebner and Tolar, 1980; Angermann, Doebner, and Tolar, 1983).

SOME FURTHER HISTORY: FRACTIONAL SPIN, BRAID GROUPS, MONOPOLES, ETC.

We have seen that by 1981, a great deal was already known about the mathematical and physical foundations of intermediate statistics in two-dimensional space--both from the topological and from the group-theoretical standpoints. Much less had been asserted by then about the possibility of fractional spin. I must thank Moshe Flato for calling to my attention, some years ago, an interesting paper that advocates obtaining the (fractional) Aharonov-Bohm phase shift from a general irreducible unitary representation of the universal covering group of the two-dimensional rotation group (Martin, 1976). This covering, of course, has infinitely many sheets. Taking the infinitesimal generator L_z in the representation to be the z-component of total angular momentum, Martin in a way anticipated the interpretation of anyons as having fractional spin. There is, however, a subtle issue (which one still sees debated)--it is not really correct in this physical context to say that the *total* (canonical) angular momentum generator has its spectrum shifted from integer multiples of $\hbar$; it is the generator of the *orbital* or *kinetic* angular momentum whose spectrum takes the fractional values. A valuable discussion of angular momentum in connection with the Aharonov-Bohm effect was given not long after by Peshkin (1981), who took careful account of the return magnetic flux of the solenoid, and the angular momentum stored in the crossed **E** and **B** fields.

Wilczek (1982a,b) independently rediscovered many of the above results about fractional statistics, coining the term "anyons", in seeking to understand the origin of fractional quantum numbers. He explored the relationship with monopoles (which themselves are not anyons; see below), and introduced composites of charged particles with Aharonov-Bohm solenoids that could obey anyon statistics. Termed "cyons", such composites were further discussed by Goldhaber (1982). Wilczek also conjectured that fractional spin should in general be associated with intermediate statistics, in analogy with the well-known spin-statistics connection that associates half-integer spin with Fermi statistics and integer spin with Bose statistics. The fractional eigenvalues of L_z for cyons were attributed by Wilczek, as they had been by Martin, to the canonical rather than the kinetic angular momentum generator, but this view was shortly thereafter challenged (Lipkin and Peshkin, 1982; Jackiw and Redlich, 1983). A recent discussion by Goldhaber and MacKenzie (1988) argues explicitly the compatibility of the Jackiw-Redlich view with the generalized spin-statistics connection, by localizing part of the canonical angular momentum at infinity (as with the return flux in the Aharonov-Bohm case). Wilczek and Zee (1983) considered another example, solitons in a non-linear σ-model, where a Lagrangian term has been added that is proportional to the Hopf invariant. Here the conjectured connection between fractional spin and anyon statistics is exhibited. The papers in 1982 and 1983 by Wilczek, and by Wilczek and Zee, did a great deal to bring intermediate statistics to the attention of the wider physics community, which gradually is also becoming aware of the earlier published results described here.

Further mathematical and physical insights at that time were motivated by the study of diffeomorphism group representations. In 1983 it was observed that for N-particle configurations in $\mathbb{R}^2$, the fundamental group $\pi_1(\Delta)$ is the *braid group* B_N; and that the unitary representations of B_N both describe the particle statistics of anyons in the Leinaas-Myrheim picture, and induce unitary representations of Diff($\mathbb{R}^2$) to give the non-relativistic quantum theory of anyons *via* local current algebra (Goldin and Sharp, 1983a; Goldin, Menikoff, and Sharp, 1983b). To visualize a braid as an element of $\pi_1(\Delta)$, think of the initial configuration of the N particles as being N points on the floor of the room, the final configuration as N corresponding points on the ceiling, and the loop parameter as the height. Then an element of B_N (a homotopy class of loops in Δ) can be visualized as a braid of N rising strands, running from floor to ceiling as in Fig. 4, with each strand connecting a different pair of points. Clearly such a braid cannot be "undone" within the room as long as the strands are fastened at their ends. A pair of elements of B_N are combined in the obvious way, by taking the two braids in sequence.

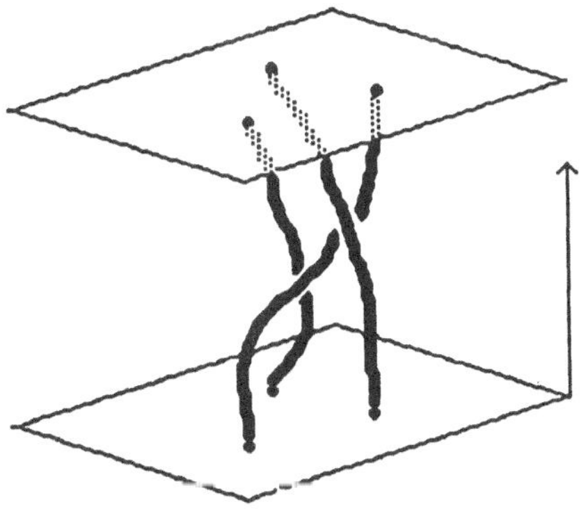

Figure 4. A braid in B_3.

The role of the braid group in the Laidlaw-DeWitt description of Feynman paths in configuration space was subsequently asserted by Wu (1984), who independently recovered the previously-obtained topological perspectives of the Leinaas-Myrheim and local current algebra formulations. Wu argued along lines similar to those Laidlaw and DeWitt had taken for S_N, that the topological framework ruled out the higher-dimensional representations of B_N, leaving the phase shift $e^{i\theta}$ (associated with a single counter-clockwise exchange of identical particles) as the only possibility. But as noted above, this conclusion depends on the assumption that the wave function is scalar-valued on the universal covering space of Δ. Without such an assumption, para-braid statistics are perfectly possible (Goldin, Menikoff, and Sharp, 1985; Goldin, 1987).

Another interesting point is that the possibility of what we are calling anyon "statistics" *does not depend on the indistinguishability of the particles* (Goldin, 1984; Goldin, Menikoff and Sharp, 1985). To see this, note that there is a natural, surjective group homomorphism $h: B_N \to S_N$, in which each braid maps to the permutation obtained by looking only at initial and final points, and disregarding the way in which the strands are intertwined. The kernel of this homomorphism is a normal subgroup of B_N, which we may call E_N; its elements correspond to homotopy classes of loops in Δ which restore *each individual point* $\mathbf{x}_j$ to its initial position. Thus E_N is the fundamental group of the *coordinate* space without the diagonal, $\{(\mathbf{x}_1, \ldots, \mathbf{x}_N) | \mathbf{x}_j \in \mathbb{R}^2\} - \mathcal{D}$, which is the distinguishable-particle configuration space. Its unitarily inequivalent representations continue to describe physically different systems.

In 1983, we proposed a construction in which towers of particle spins in $\mathbb{R}^3$, both integer and half-integer, are obtained from unitary representations of Diff($\mathbb{R}^3$), and a prescription (see below) for recovering the fixed-spin subspaces (Goldin, Menikoff and Sharp, 1983a,b; Goldin and Sharp, 1983b). It was also observed that a parallel construction permits recovery of fractional spin from unitary representations of Diff($\mathbb{R}^2$); and this essentially completed the unified framework that the diffeomorphism group provides for N-anyon systems (Goldin, 1984).

It was suggested by Ringwood and Woodward (1984), based on the topological model of Finkelstein and Rubinstein, that Polyakov-t'Hooft monopoles could admit intermediate statistics (called in their paper "parastatistics", a term much more commonly reserved for the higher- dimensional representations of the fundamental group). This paper also introduced the braid group for N particles on a sphere. The N-monopole configurations in $\mathbb{R}^3$ are associated with configurations of N points on a boundary sphere at infinity by means of strings--i.e., narrow tubes whose interiors are mapped into the unit sphere of the iso-space--and two monopoles are exchanged by shrinking along their respective strings, trading places on the two-sphere at infinity (suggesting the admissibility of braid statistics), and reexpanding the strings. However, for such a configuration of monopoles in $\mathbb{R}^3$, it is also possible to untangle the flux tubes smoothly by means of diffeomorphisms of $\mathbb{R}^3$ parameterized by t. This can be done without moving the monopole locations at all, by

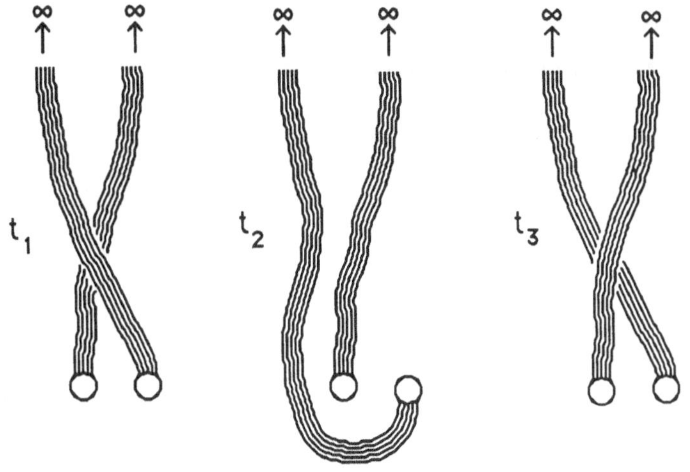

Figure 5. Looping the flux tube of one monopole around another.

looping the tube of one monopole around another (see Fig. 5, where $t_1 < t_2 < t_3$). Since the tubes are only "half-strings" there is no braid and, I would conclude, anyon statistics are not admissible after all.

Further applications of anyon statistics followed, including: the fractional quantum Hall effect (Arovas, Schrieffer, and Wilczek, 1984; Kivelson and Rocek, 1985) and the idea of a "mean-field theory of statistics" (Arovas, Schrieffer, Wilczek, and Zee, 1985; Laughlin, 1988b; Fetter, Hanna, and Laughlin, 1989; Canright, Girvin, and Brass, 1989); vortices in thin superfluid films (Chiao, Hansen, and Moulthrop, 1985; Haldane and Wu, 1985; Goldin, Menikoff, and Sharp, 1987a-e; Leinaas and Myrheim, 1988); and "anyon superconductivity" as a possible explanatory mechanism for high-temperature superconductors (Laughlin, 1988a,b; Kivelson, Rokhsar, and Sethna, 1988; Chen, Wilczek, Witten, and Halperin, 1989). The review by Canright and Girvin (1990) discusses some of these applications, and provides other useful references; but it does not include the contributions by Finkelstein and Rubinstein (1968), Martin (1976), Kivelson *et al.* (1985, 1988), the early work of Fröhlich (1976, 1980), or the group-theoretical perspective in my work with Menikoff and Sharp.

Let us now examine some of the above issues further by looking a little more closely at the representation theory of Diff($\mathbb{R}^2$). Here I shall describe just the major steps in the development.

REPRESENTATIONS OF DIFF($\mathbb{R}^2$) DESCRIBING ANYONS, FRACTIONAL SPIN, AND VORTICES

Suppose that $\psi(\mathbf{x},t)$ is a canonical field in (s+1)-dimensional space-time; i.e., an operator-valued distribution on $\mathbb{R}^{s+1}$ which, at a fixed time t, satisfies either Eq. (2) or Eq. (3). Thus ψ can be a second-quantized nonrelativistic field, or it can be a certain creation component of a canonical relativistic field. Suppressing t, define the fixed-time mass density $\rho(\mathbf{x})$ and momentum current $\mathbf{J}(\mathbf{x})$, in appropriate units ($m = \hbar = 1$), by

$$\rho(\mathbf{x}) = \psi^*(\mathbf{x})\psi(\mathbf{x}), \tag{6}$$

$$\mathbf{J}(\mathbf{x}) = (1/2i)\{\psi^*(\mathbf{x})\nabla\psi(\mathbf{x}) - [\nabla\psi^*(\mathbf{x})]\psi(\mathbf{x})\}. \tag{7}$$

Here $\rho(\mathbf{x})$ and $\mathbf{J}(\mathbf{x})$, if they exist, are themselves operator-valued distributions. To obtain actual operators, we let $f(\mathbf{x})$ be an element of the Schwartz space $\mathcal{S}(\mathbb{R}^s)$; i.e. a real-valued, C^∞ function on $\mathbb{R}^s$ which (together with all derivatives) is of rapid decrease at infinity. Then we can define the operator $\rho(f) = \int \rho(\mathbf{x})f(\mathbf{x})d\mathbf{x}$. Similarly let $\mathbf{g}$ be a vector field on $\mathbb{R}^s$ with components $g^k \in \mathcal{S}(\mathbb{R}^N)$, $k = 1, \ldots, s$ (we write $\mathbf{g} \in \mathrm{Vect}(\mathbb{R}^s)$); and set $J(\mathbf{g}) = \Sigma \int J_k(\mathbf{x})g^k(\mathbf{x})d\mathbf{x} = \int \mathbf{J}(\mathbf{x})\cdot\mathbf{g}(\mathbf{x})d\mathbf{x}$. With these definitions the commutator algebra satisfied by $\rho(f)$ and $J(\mathbf{g})$ can be calculated formally from Eq. (2) or Eq. (3); and for $f, f_1, f_2 \in \mathcal{S}(\mathbb{R}^s)$ and $\mathbf{g}, \mathbf{g}_1, \mathbf{g}_2 \in \mathrm{Vect}(\mathbb{R}^s)$, the result is the *local current algebra*

$$[\rho(f_1), \rho(f_2)] = 0, \tag{8}$$

$$[\rho(f), \mathbf{J}(\mathbf{g})] = i\rho(\mathbf{g}\cdot\nabla f), \tag{9}$$

$$[\mathbf{J}(\mathbf{g}_1), \mathbf{J}(\mathbf{g}_2)] = -i\mathbf{J}(\mathbf{g}_1\cdot\nabla\mathbf{g}_2 - \mathbf{g}_2\cdot\nabla\mathbf{g}_1). \tag{10}$$

Eqs. (8)-(10) are a self-adjoint representation of a natural *semidirect product* Lie algebra, of the (commutative) algebra of scalar functions $\mathcal{S}(\mathbb{R}^s)$, with the algebra $\mathrm{Vect}(\mathbb{R}^s)$ under the Lie bracket. It is a remarkable fact that the same Lie algebra is obtained whether we start with canonical commutation or anticommutation relations--indeed, even the q-commutators in Eq. (4) yield the same Lie algebra by formal calculation. And when Fock representations of the canonical fields $\psi(\mathbf{x})$ and $\psi^*(\mathbf{x})$ are used to calculate $\rho(f)$ and $\mathbf{J}(\mathbf{g})$, the formal calculation is borne out--both the Fermi and the Bose Fock spaces carry representations of the same Lie algebra of currents. Schwinger terms that necessarily occur in fully relativistic models do not enter here.

It is natural to exponentiate this algebra to obtain a group. For $\mathbf{g} \in \mathrm{Vect}(\mathbb{R}^s)$, there exists a unique one-parameter group of C^∞ diffeomorphisms of $\mathbb{R}^s$ generated by $\mathbf{g}$ (a *flow*) which we denote $\phi_t^{\mathbf{g}}$. That is, $\phi_t^{\mathbf{g}}: \mathbb{R}^s \to \mathbb{R}^s$ is an invertible mapping that together with its inverse $\phi_{-t}^{\mathbf{g}}$ is C^∞ (in both $\mathbf{x}$ and t), obeying the differential equation and initial condition

$$\partial_t \phi_t^{\mathbf{g}}(\mathbf{x}) = \mathbf{g}(\phi_t^{\mathbf{g}}(\mathbf{x})), \tag{11}$$

$$\phi_{t=0}^{\mathbf{g}}(\mathbf{x}) = \mathbf{x}, \tag{12}$$

as well as the group law for all $t,u \in \mathbb{R}$,

$$\phi_t^{\mathbf{g}}(\phi_u^{\mathbf{g}}(\mathbf{x})) = \phi_{t+u}^{\mathbf{g}}(\mathbf{x}). \tag{13}$$

The existence of $\phi_t^{\mathbf{g}}$ for all t is guaranteed by the good behavior of $\mathbf{g}$ at infinity. Now the unitary group obtained by exponentiating (8)-(10) can be written down. Letting $U(f) = \exp[i\rho(f)]$ and $V(\phi_t^{\mathbf{g}}) = \exp[iJ(\mathbf{g})]$, we have (with $\circ$ denoting composition of functions)

$$U(f_1)U(f_2) = U(f_1 + f_2), \tag{14}$$

$$V(\phi)U(f) = U(f\circ\phi)V(\phi), \tag{15}$$

$$V(\phi_1)V(\phi_2) = V(\phi_2\circ\phi_1), \tag{16}$$

where as before $f, f_1, f_2 \in \mathcal{S}(\mathbb{R}^s)$, and where $\phi, \phi_1, \phi_2 \in \mathrm{Diff}(\mathbb{R}^s)$ are C^∞ diffeomorphisms of $\mathbb{R}^s$ that (together with all derivatives) become rapidly trivial at infinity. We shall take $\mathrm{Diff}(\mathbb{R}^s)$ to be the group generated by the flows under composition; it becomes a topological group when endowed with a topology similar to that of Schwartz space $\mathcal{S}$. In such a topology, all the one-parameter subgroups $\phi_t^{\mathbf{g}}$ are continuous in t. Eqs. (14)-(16) can thus be interpreted as a continuous unitary representation (CUR) of the natural semidirect product of $\mathcal{S}(\mathbb{R}^s)$ with $\mathrm{Diff}(\mathbb{R}^s)$.

We have obtained an infinite-dimensional Lie group, and we have a prescription for the physical interpretation of a CUR: for each flow subgroup $\phi_t^{\mathbf{g}}$ in $\mathrm{Diff}(\mathbb{R}^s)$, one recovers $J(\mathbf{g})$ as the self-adjoint generator of $V(\phi_t^{\mathbf{g}})$, and interprets $J(\mathbf{g})$ as the momentum density averaged with the vector field $\mathbf{g}$. For example, in a bounded region of $\mathbb{R}^3$ cylindrically symmetric about the z-axis, take $g^1(\mathbf{x}) = -x^2$, $g^2(\mathbf{x}) = +x^1$, and $g^3(\mathbf{x}) = 0$ (where the superscripts denote x, y, and z components), and let $\mathbf{g}(\mathbf{x}) \to 0$ quickly and smoothly outside the region. Then the spectrum of the operator $J(\mathbf{g})$ describes measurements of L_z, the z-component of angular momentum localized within the region.

But in this model, what has become of the different particle statistics that were originally embodied in the algebra of fields from which the local currents were defined? No longer do we have a distinct algebra for each admissible type of statistics; instead, we have distinct (unitarily inequivalent) representations of a single, infinite-dimensional group! It was this observation about statistics which gradually led us to appreciate the role that Diff(M) can play as a kind of "universal group" for quantum theory--a group whose inequivalent, irreducible CUR's provide a classification of possible quantum systems. When we obtained fractional statistics in 1980, it was not because we had set out to construct exotic quantum theories, or sought fractional eigenvalues for angular momentum or particle exchange; rather the admissibility of particles in two dimensions obeying such intermediate statistics, and the fundamental physical facts about them (such as the shifted angular momentum and energy spectra), emerged as a mathematical consequence when $s = 2$ of our program of studying the unitary representations of the above group.

It should be apparent that nothing in what we have said is special to $\mathbb{R}^s$--in fact, for a quite general manifold M (compact, or σ-compact), we can now study quantum mechanics on M through the CUR's of Diff(M), or of its semidirect product with $\mathcal{S}(M)$. We thus have an extremely general, geometrically natural way of classifying quantum systems through the representations of an infinite-dimensional group, and associating self-adjoint operators with measurements in bounded regions. With M = S^2, for example, we can classify the possibilities for anyons on the 2-sphere.

I shall not try to develop the representation theory here, but will sketch just a few highlights. In a very general framework, we can realize the Hilbert space for the unitary representation as $\mathcal{H} = \mathcal{L}^2_\mu(\Delta, \mathcal{M})$, a space of functions Ψ on a configuration space Δ, taking values in a finite- or infinite-dimensional Hilbert space $\mathcal{M}$, square-integrable with respect to a countably additive measure μ on Δ. The space Δ is a G-space for Diff(M), and μ is quasiinvariant for the action of Diff(M) on Δ (i.e., the class of μ-measure zero sets is preserved by diffeomorphisms). Then the representation $V(\phi)$ is given by

$$[V(\phi)\Psi](\gamma) = \chi_\phi(\gamma)\Psi(\phi\gamma)\sqrt{\frac{d\mu_\phi}{d\mu}(\gamma)}\,, \tag{17}$$

where $\gamma \in \Delta$, μ_ϕ is the measure μ transformed by ϕ, $d\mu_\phi/d\mu$ is the Radon-Nikodym derivative (which exists because μ is quasiinvariant); and $\chi_\phi(\gamma)$ is a unitary cocycle (acting in $\mathcal{M}$) satisfying appropriate technical conditions. Cohomologous cocycles on the same configuration space describe equivalent representations, while inequivalent cocycles give rise to the statistics. The operators U(f) can act as multiplication operators in $\mathcal{H}$. The most elementary representation, associated with a single particle in $\mathbb{R}^s$ having no internal structure, corresponds to the choice $\Delta = \mathbb{R}^s$ and $\mathcal{M} = \mathbb{C}$, with μ being Lebesgue measure, $d\mu_\phi/d\mu$ the Jacobian of ϕ at $\mathbf{x}$, and $\chi_\phi(\gamma) \equiv 1$.

Two ways to construct such configuration spaces systematically deserve mention. The method of semidirect products for locally compact groups (Mackey, 1968) partially extends to the present case (Goldin, 1971; Goldin, Menikoff and Sharp, 1980). For the unitary representation U(f), f $\in \mathcal{S}$, one constructs a generating functional and a corresonding measure in the continuous dual space $\mathcal{S}'$, the space of tempered distributions (Gel'fand and Vilenkin, 1964). This space has a natural orbit structure under the action of the diffeomorphism group. Irreducible representations can now be associated with *ergodic* measures in $\mathcal{S}'$; i.e., normalized probability measures such that every invariant set has either measure zero or measure one. One type of ergodic measure is concentrated on a single orbit, which then becomes the space Δ. Another possibility (the "strictly ergodic" case) is that every orbit is of μ-measure zero; then Δ is the union of uncountably many orbits. This case typically corresponds to systems having infinitely many degrees of freedom.

A second way to construct configuration spaces systematically is through the method of coadjoint orbits (Kirillov, 1981; Goldin, Menikoff, and Sharp, 1987c,d,e). In the coadjoint representation of Diff(M), it is not the orbits themselves which become the quantum-mechanical configuration spaces, but foliations of these orbits that are obtained through geometric quantization procedures.

I have omitted mention of numerous difficulties in the representation theory, which occur because the groups in question are infinite-dimensional. They thus possess no Haar measure, and the usual techniques developed for locally compact groups cannot be applied directly. In many physical situations of practical interest, these difficulties can be overcome by identifying natural quasiinvariant measures on the configuration spaces themselves, or on fiber bundles over the configuration spaces. For important classes of configuration spaces, however (such as path and loop spaces that arise in the quantization of vorticity), the issue of how to construct measures quasiinvariant for diffeomorphisms remains to the best of my knowledge open.

To return to the case of anyons, we now consider the simplest configuration space for N particles in $\mathbb{R}^2$. We have already seen that a configuration $\gamma \in \Delta$ is an unordered set of N distinct points in $\mathbb{R}^2$, but that including the diagonal $\mathcal{D}$ in Δ eliminates the possibility of nontrivial statistics. As a subset of $\mathcal{S}'(\mathbb{R}^2)$, Δ occurs as the following orbit under the action of Diff($\mathbb{R}^2$):

$$\Delta = \left\{ \sum_{j=1}^{N} \delta_{\mathbf{x}_j} \,\middle|\, \mathbf{x}_j \in \mathbb{R}^2, \ \mathbf{x}_i \neq \mathbf{x}_j \ \text{ for } \ i \neq j \right\} \subset \mathcal{S}', \tag{18}$$

where $\delta_{\mathbf{x}}$ is the evaluation functional $< \delta_{\mathbf{x}}, f > \, = f(\mathbf{x})$. A diffeomorphism acts on this orbit by transforming the values of all the position variables.

Now we can see why the diagonal $\mathcal{D}$ *cannot* be included in Δ--it is because $\mathcal{D}$ is *outside* this orbit (it is in the weak closure of Δ in $\mathcal{S}'(\mathbb{R}^2)$. $\mathcal{D}$ is also the union of finitely many *other* orbits for Diff($\mathbb{R}^2$). If we try to include $\mathcal{D}$ in the configuration space, we obtain a multiorbit space, and the resulting unitary representation is *reducible.* It simply decomposes into a direct sum of distinct irreducible representations, some of which describe particles "stuck together" in various combinations.

To see how the statistics enters, consider next the stability subgroup K_γ of Diff($\mathbb{R}^2$) associated with a particular configuration γ. This will include diffeomorphisms that permute the points $\mathbf{x}_j$ comprising γ, and we thus obtain the natural homomorphism from K_γ to S_N quite easily. It is not difficult, moreover, to construct a homomorphism h from K_γ onto the braid group B_N, the fundamental group of Δ; e.g., by considering paths from the identity in Diff($\mathbb{R}^2$). As in the Mackey theory, CUR's of K_γ under the right conditions induce CUR's of Diff($\mathbb{R}^2$), yielding various cocycles on Δ, or equivalently, yielding various Hilbert spaces of equivariant wave functions on a fiber bundle over Δ. An interesting class of such representations are those induced by CUR's of K_γ which factor through the homomorphism h onto the braid group; the fiber bundle for these is simply the universal covering space of Δ. Thus we arrive at unitary representations of B_N inducing representations of Diff($\mathbb{R}^2$) describing anyons. It is clear that there is no *a priori* reason to restrict attention to one-dimensional representations of B_N.

The group through which we factor K_γ (S_N, B_N, or some other group) serves as a *gauge group* for the quantum theory. It is apparent that the diffeomorphisms (and hence the elements of S_N or B_N) are acting always on the *values* of the coordinates $\mathbf{x}_j$, never on the indices. The (coordinate-dependent) ordering in $\mathbb{R}^3$ that we introduced in our earlier discussion of value permutations corresponds to the choice of a measurable cross-section in a covering space of Δ. Thus the interpretation of quantum statistics by means of diffeomorphism group representations necessitates the use of physical rather than label permutations.

To derive spin from this formalism, we again consider the stability group. For simplicity, let us take just one particle, so that K_γ is the closed subgroup of Diff($\mathbb{R}^2$) whose elements leave the single point $\mathbf{x}$ fixed. Now there is a continuous homomorphism from K_γ to $GL^+(2,\mathbb{R})$, obtained by mapping each diffeomorphism ϕ such that $\phi(\mathbf{x}) = \mathbf{x}$, to its matrix of derivatives at $\mathbf{x}$. In fact, we can map K_γ to the universal covering group of $GL^+(2,\mathbb{R})$ by considering a path from infinity to $\mathbf{x}$ in $\mathbb{R}^2$, and following the matrix of derivatives of ϕ along this path; this gives a path from the identity in $GL^+(2,\mathbb{R})$ whose homotopy class is well-defined. Thus a CUR of $GL^+(2,\mathbb{R})$ or its covering will induce a CUR of Diff($\mathbb{R}^2$). But CUR's of $GL^+(2,\mathbb{R})$, or its covering, can be decomposed with respect to CUR's of the two-dimensional rotation group SO(2), or its covering! Thus we obtain towers of "fractional spin" for particles in $\mathbb{R}^2$. However, there is no necessary spin-statistics connection for anyons in this context, for the same reasons that the usual spin-statistics connection for bosons and fermions is not mandatory in nonrelativistic quantum mechanics. To associate particular spins with particular values of the θ-parameter, as conjectured by Wilczek, requires imposition of some further condition.

Finally, let me mention that the method of coadjoint orbits has proven particularly useful for the quantization of vorticity in an ideal incompressible two-dimensional fluid. Since the fluid is incompressible, we represent not Diff($\mathbb{R}^2$) in this case but its subgroup sDiff($\mathbb{R}^2$) of area-preserving diffeomorphisms. While it has been argued (based on models having finitely many degrees of freedom) that pure point vortices in $\mathbb{R}^2$ could obey intermediate statistics (Chiao, Hansen and Moulthrop, 1985; Leinaas and Myrheim, 1988), the geometric quantization based on sDiff($\mathbb{R}^2$) leads to the surprising but rigorous conclusion that a quantum theory based on pure point vortices does not exist. However, filaments or loops of vorticity in $\mathbb{R}^2$ can be quantized (Goldin, Menikoff, and Sharp, 1987c,d,e), and it is possible to describe configurations of vortex loops that obey anyon statistics. Here, however, the question of constructing appropriate quasiinvariant measures is still unresolved.

To sum up, the classification of representations of Diff(M) provides a way to unify the description of a wide variety of distinct quantum systems. This study led to one of the early predictions of the possibility of anyon statistics in two dimensions, and to many of the important physical properties of anyons. It also gives insight into the underlying mathematical and physical reasons for their existence.

B. Angermann, H. D. Doebner and J. Tolar (1983), "Quantum Kinematics on Smooth Manifolds," in S. I. Andersson & H. D. Doebner (eds.), *Nonlinear Partial Differential Operators and Quantization Procedures*, Lecture Notes in Math. **1037**, Berlin: Springer, 171-208.

D. Arovas, J. R. Schrieffer and F. Wilczek (1984), *Phys. Rev. Lett.* **53**, 722.

D. Arovas, J. R. Schrieffer, F. Wilczek and A. Zee (1985), *Nucl. Phys.* **B251**, 117.

L. C. Biedenharn (1989), *J. Phys. A: Math. Gen.* **22**, L873.

G. S. Canright and S. M. Girvin (1990), *Science* **247**, 1197-1205.

G. S. Canright, S. M. Girvin and A. Brass (1989), *Phys. Rev. Lett.* **63**, 2295.

Y.-H. Chen, F. Wilczek, E. Witten and B. I. Halperin (1989), *Int. J. Mod. Phys.* **B3**, 1001.

R. Y. Chiao, A. Hansen and A. A. Moulthrop (1985), *Phys. Rev. Lett.* **54**, 1339.

R. F. Dashen and D. H. Sharp (1968), *Phys. Rev.* **165**, 1857.

H. D. Doebner and J. Tolar (1980), "On Global Properties of Quantum Systems," in B. Gruber and R. S. Millman (eds.), *Symposium on Symmetries in Science, Carbondale, Illinois 1979*, New York: Plenum, 475-486.

J. S. Dowker (1972), *J. Phys. A* **5**, 936-943.

A. L. Fetter, C. B. Hanna and R. B. Laughlin (1989), *Phys. Rev. B* **39**, 9679.

D. Finkelstein and J. Rubinstein (1968), *J. Math. Phys.* **9**, 1762-1779.

J. Fröhlich (1976), *Commun. Math. Phys.* **47**, 269.

J. Fröhlich (1980), in G. t'Hooft *et al.* (eds.), *Recent Developments in Gauge Theories (Cargèse 1979)*, New York: Plenum.

I. M. Gel'fand and N. Ya. Vilenkin (1964), *Generalized Functions, Vol. 4*, New York: Academic.

A. S. Goldhaber (1982), *Phys. Rev. Lett.* **49**, 905.

A. S. Goldhaber and R. MacKenzie (1988), *Phys. Lett. B* **214**, 471-474.

G. A. Goldin (1969), *Current Algebras as Unitary Representations of Groups*, Ph.D. thesis, Princeton University (unpublished).

G. A. Goldin (1971), *J. Math. Phys.* **12**, 462-487.

G. A. Goldin (1984), "Diffeomorphism Groups, Semidirect Products, and Quantum Theory," in J. E. Marsden (ed.), *Fluids and Plasmas: Geometry and Dynamics*, Contemporary Mathematics **28**, Providence, Rhode Island.: American Mathematical Society, 189-207.

G. A. Goldin (1987), "Parastatistics, θ-Statistics, and Topological Quantum Mechanics from Unitary Representations of Diffeomorphism Groups," in H. D. Doebner and J. D. Hennig (eds.), *Proceedings of the XV. International Conference on Differential Geometric Methods in Theoretical Physics*, Singapore: World Scientific, 197-207.

G. A. Goldin, J. Grodnik, R. T. Powers and D. H. Sharp (1974), *J. Math. Phys.* **15**, 88-100.

G. A. Goldin, R. Menikoff and D. H. Sharp (1980), *J. Math. Phys.* **21**, 650-664.

G. A. Goldin, R. Menikoff and D. H. Sharp (1981), *J. Math. Phys.* **22**, 1664-1668.

G. A. Goldin, R. Menikoff and D. H. Sharp (1983a), *Phys. Rev. Lett.* **51**, 2246-2249.

G. A. Goldin, R. Menikoff and D. H. Sharp (1983b), *J. Phys. A: Math. Gen.* **16**, 1827-1833.

G. A. Goldin, R. Menikoff and D. H. Sharp (1985), *Phys. Rev. Lett.* **54**, 603.

G. A. Goldin, R. Menikoff and D. H. Sharp (1987a), *Phys. Rev. Lett.* **58**, 174.

G. A. Goldin, R. Menikoff and D. H. Sharp (1987b), *J. Math. Phys.* **28**, 744-746.

G. A. Goldin, R. Menikoff and D. H. Sharp (1987c), *Phys. Rev. Lett.* **58**, 2162-2164.

G. A. Goldin, R. Menikoff and D. H. Sharp (1987d), "Diffeomorphism Groups, Coadjoint Orbits, and the Quantization of Classical Fluids," in Y. S. Kim and W. W. Zachary (eds.), *Procs. of the First Int'l. Conf. on the Physics of Phase Space.* Lecture Notes in Physics **278**, Berlin: Springer, 360-362.

G. A. Goldin, R. Menikoff and D. H. Sharp (1987e), "Quantized Vortex Filaments in Incompressible Fluids," *ibid.*, 363-365.

G. A. Goldin and D. H. Sharp (1970), "Lie Algebras of Local Currents and their Representations," in V. Bargmann (ed.), *Group Representations in Mathematics and Physics: Battelle-Seattle 1969 Rencontres*, Lecture Notes in Physics **6**, Berlin: Springer, 300-310.

G. A. Goldin and D. H. Sharp (1972), "Functional Differential Equations Determining Representa- tions of Local Current Algebras," in J. R. Klauder (ed.) *Magic Without Magic: John Archibald Wheeler*, San Francisco: W.H. Freeman & Co., 171-185.

G. A. Goldin and D. H. Sharp (1983a), *Phys. Rev. D* **28** (1983), 830-832.

G. A. Goldin and D. H. Sharp (1983b), *Commun. Math. Phys.* **92**, 217-228.

G. A. Goldin and D. H. Sharp (1989), "Diffeomorphism Groups and Local Symmetries: Some Applications in Quantum Physics," in B. Gruber and F. Iachello (eds.), *Symmetries in Science III*, New York: Plenum, 181-205.

H. S. Green (1953), *Phys. Rev.* **90**, 270.

O. W. Greenberg (1966), in R. Goodman and I. Segal (eds.), *Mathematical Theory of Elementary Particles*, Cambridge, Mass.: M.I.T. Press, 29-44.

J. Grodnik and D. H. Sharp (1970a), *Phys. Rev. D* **1**, 1531.

J. Grodnik and D. H. Sharp (1970b), *Phys. Rev. D* **1**, 1546.

F. D. M. Haldane and Y. S. Wu (1985), *Phys. Rev. Lett.* **55**, 2887.

J. F. Hamilton, Jr. and L. S. Schulman (1971), *J. Math Phys.* **12**, 160-164.

R. Jackiw and A. N. Redlich (1983), *Phys. Rev. Lett.* **50**, 555.

A. A. Kirillov (1981), *Ser. Math. Sov.* **1**, 351.

S. Kivelson and M. Rocek (1985), *Phys. Lett.* **156B**, 85-88.

S. Kivelson, D. S. Rokhsar and J. P. Sethna (1988), *Europhys. Lett.* **6**, 353-358.

R. B. Laughlin (1988a), *Science* **242**, 525.

R. B. Laughlin (1988b), *Phys. Rev. Lett.* **60**, 2677.

M. G. G. Laidlaw and C. M. DeWitt (1971), *Phys. Rev. D* **3**, 1375.

J. M. Leinaas (1978), *Nuovo Cimento* **47A**, 19-34.

J. M. Leinaas (1980), *Fortschritte der Physik* **28**, 579-631.

J. M. Leinaas and J. Myrheim (1977), *Nuovo Cimento* **37B**, 1.

J. M. Leinaas and J. Myrheim (1988), *Phys. Rev. B* **37**, 9286.

H. J. Lipkin and M. Peshkin (1982), *Phys. Lett.* **118B**, 385.

G. W. Mackey (1968), *Induced Representations of Groups and Quantum Mechanics*, New York: Benjamin.

C. Martin (1976), *Lett. Math. Phys.* **1**,155.

R. Menikoff (1974a), *J. Math. Phys.* **15**, 1138.

R. Menikoff (1974b), *J. Math. Phys.* **15**, 1394.

R. Menikoff and D. H. Sharp (1977), *J. Math. Phys.* **18**, 471.

A. M. L. Messiah and O. W. Greenberg (1964), *Phys. Rev.* **136**, B248.

M. Peshkin (1981), *Phys. Rep.* **80**, 375.

M. Rasetti and T. Regge (1975), *Physica* **80A**, 217.

G. A. Ringwood and L. M. Woodward (1984), *Phys. Rev. Lett.* **53**, 1980-1983.

L. S. Schulman (1971), *J. Math. Phys.* **12**, 304.

S. Schweber (1961), *An Introduction to Relativistic Quantum Field Theory*, New York: Harper and Row.

R. H. Stolt and J. R. Taylor (1970), *Nucl. Phys. B* **19**, 1.

F. Wilczek (1982a), *Phys. Rev. Lett.* **48**, 1144.

F. Wilczek (1982b), *Phys. Rev. Lett.* **49**, 957.

F. Wilczek and A. Zee (1983), *Phys. Rev. Lett.* **51**, 2250.

Y. S. Wu (1984), *Phys. Rev. Lett.* **52**, 2103.

THE ROLE OF PARABOSE-STATISTICS IN MAKING ABSTRACT QUANTUM THEORY CONCRETE

Thomas Görnitz

Maximilianstr. 15

D-8130 Starnberg, Germany

Quantum theory is the most fruitful part of physics, no convincing experiment contradicting it has ever been found. The great success of this theory provokes the question:

Can we understand why it is so successful.

The overwhelming success of this theory in experience cannot be deduced from experience. This is so for quantum theory like any other universal law and is known in philosophy already for a long time. Perhaps some people hoped to deduce a theory from sure a priori assumptions. Then such a theory would be unfailing. But - to my knowledge - all those attempts in the past have failed.

Should this indicate not even to ask the question above? I think not. To my knowledge the only convincing attempt for a solution of this problem was made by Kant. This philosopher has clarified that all such general laws **must be necessarily** valid in experience which follow from the **preconditions** of experience alone.

Of course, it is not sure that we are able at all to formulate a priori the preconditions of experience. Then an absolute certaincy for our theories can not be concluded. But it seems to me a useful attempt to look how far it is possible to recognize such preconditions. Then it is a good working hypothesis to use them and to see which other postulates we need to deduce quantum theory from all of them together.

One attempt in this direction of understanding quantum theory and its fundamental role is made by v. Weizsäcker and his co-workers (for an overview see [1-20]) with the concept of "abstract" quantum theory. By abstract quantum theory is designated the general frame of quantum theory, without reference to a three-dimensional position space, to concepts like particle or field, or to special laws of dynamics. Even less it presupposes any set of laws of "classical" physics which would then have to be "quantized".

To understand abstract quantum theory the attempt is made to "reconstruct" it. Reconstruction does not mean a mathematically intended axioma-

Symmetries in Science V, Edited by B. Gruber *et al.*
Plenum Press, New York, 1991

tics but it is the attempt to understand as far as possible the conceptual necessities for quantum theory.

Of course, possesing the frame of abstract quantum theory does not mean that the problems of physics have been solved. The hope of our concept is that beside the principles of abstract quantum theory only one further concept is necessary. This is the idea of introducing the concept of the "ur", the quantized binary alternative.
Then we have to go the long and hard way to understand concrete quantum theory. This means to derive the properties of space-time and of the elementary particles and also of the fundamental forces between them. This is connected with some mathematical problems and also with great conceptual ones. E.g. to derive the properties of space-time from abstract quantum theory means to derive a cosmological model prior to a theory of gravitation. This is done elsewhere ([16], [17], [18]) and will not be referred in this paper. In the present paper a path will be outlined from the abstract concepts of ur theory to the construction of states for real particles.

The reconstruction of abstract quantum theory will be outlined briefly in chapter 1. In chapter 2 we explain the concept of the "ur" and its "second quantization", i.e. the theory of many urs. In the third chapter arguments are given for the use of parabose statistics for urs. The algebra of parabose operators enables us to construct massless and massive particles by the urs. This will be demonstrated on concrete examples in chapter 4.

1. RECONSTRUCTION OF "ABSTRACT QUANTUM THEORY"

"Reconstruction" means the attempt to formulate simple and plausible postulates on prediction and to derive the basic concepts of abstract quantum theory from them. The following procedure is mainly based on the paper "Reconstruction of abstract quantum theory" by Drieschner, v. Weizsäcker and myself (1987), in which the assertions are arranged in four groups: Heuristic principles, verbal definitions, basic postulates, consequences. For the sake of completeness I will give the main points in a short review:

1.1. Heuristic principles

A1. Preconditions of experience: As far as possible our postulates ought to express conditions without which we cannot expect experience to be possible at all.

A2. Simplicity: Without precisely defining simplicity, we wish for simple postulates rather than complicated ones.

A3. Innocuous generality: General rules are usually simpler than specialized ones. We shall confine ourselves to general rules as far as they give the hope of being "innocuous"; e.g. claiming the general existence of a set of states while under special conditions (like a dynamics implying a super-selection-rule) some of those might not actually come into being.

1.2. Definitions

B1. Experience: Experience means to learn from the past for the future.

B2. Facticity of the past: We speak of past events as of objective facts, independently of our actually knowing them.

B3. Possibility of the future: We are aware of future events only as possibilities.

B4. Probability: Probability is a quantification of possibility. We define it as the prediction (mathematically: the expectation value) of a relative frequency.

B5. Temporal statements: A temporal statement (briefly "statement") is a verbal proposition (or a mathematical proposition with a physical meaning) referring to a moment in time.

B6. States: States are recognizable events. A state is what is the case when some temporal statement is true. States at different times can be identical: it is meaningful to ask whether we observe now the same states at a certain time before.

B7. Conditional probability: Let x and y to be two states. Then $p(x,y)$ is the probability that, if x is a present state, y will be found as the state if searched for.

B7. Alternatives: An n-fold alternative is a set of n mutually exclusive states, exactly one of which will turn out to be present if and when an empirical test of this alternative is made.

B8. Connection: Two states x and y are called connected if there is a law of nature determining their conditional probabilities $p(x,y)$ and $p(y,x)$. If the connection is transitive, i.e. if the existence (by law of nature) of probabilities $p(x,y)$ and $p(y,z)$ implies the existence of a $p(x,z)$, then connection is an equivalence relation, defining a partition of the class of all states into subclasses of mutually connected states.

B9. Separability: Two states are called separable if they are not connected.

1.3. Postulates

C1. Separable alternatives:
There are alternatives whose states are separable from nearly all other states. "Nearly" will be defined as meaning all states not connected with the states of the alternative by postulate 2.

C2. Indeterminism:
Let x and y be two connected, mutually exclusive states $\{p(x,y) = p(y,x) = 0\}$, then there are states z, different from x and y, which are not constructed from x and y by mere logical operations and which cannot distinguished by their transformation properties from the former ones but which possess conditional probabilities $p(z,x)$ and $p(z,y)$ none of which is equal to zero or to one. Between every two of the states z there are conditional probabilities $p(z,z')$.

C3. Kinematics:
The conditional probabilities between connected states are not altered when the states change in time:

$$p[(x,t),(z,t)] = p[(x,0),(z,0)].$$

1.4. Consequences

D1. State space: We call the set of states connected with a separable alternative its state space. With innocuous generality we assume the state spaces of all separable n-fold alternatives A_n to be isomorphic: $S(n)$.
A state $z \epsilon S(n)$ defines n conditional probabilities $p(z,x_i)$ where x_i $(i = 1...n)$ are the states defining the n-fold alternative;

$$\sum_{i=1}^{n} p(z,x_i) = 1.$$

D2. Completeness: For any mathematically possible set of values $p(z,x_i)$ there is a state z in S(n). We assume this to be an example of innocuous generality.

D3. Equivalence of states: All states in S(n) are equivalent. Else their distinction would be an additional alternative connected with A_n. A_n would hence not have been separable. (This "separability" should not be confused with the mathematical concept of separability of a Hilbert space)

D4. Symmetry group: The equivalence of the elements of S(n) is expressed by a symmetry group G(n) which preserves the conditional probabilities between them. Due to D3, G(n) must be a continuous group.

D5. Alternatives in S(n): Due to D3 there exists a $p(x,y)$ between any two states x and y of S(n). The equivalence of all states in S(n) further implies that any $z \in S(n)$ is a member of a precisely n-fold alternative of mutually exclusive states of S(n).

D6. Metric in S(n): As an "assumption of simplicity" we suppose G(n) to be a simple Lie group. There are two simple Lie groups preserving a relation of mutual exclusion between precisely n normalized vectors by preserving a metric: O(n) and U(n). Thus we assume S(n) to permit a faithful irreducible representation in an n-dimensional vector space V(n), G(n) being either orthogonal or unitary. The states of S(n) will then correspond to normalized vectors in V(n), i.e. to one-dimensional subspaces.

D7. Dynamics: According to C3, the change of state in time must be a one-parameter subgroup D(t) of G(n). We call the special choice of such a subgroup the choice of a law of dynamics.

D8. Preservation of state: If a state is to be recognizable in time, there must exist a possible law of dynamics which keeps this state constant.

D9. Complexity: The generator of D(t), as defined in D7, must, according to D8, permit diagonalization. This is universally possible only if V is complex, and, due to the metric, a Hilbert space. Hence G(n) = U(n).

D10. Composition: Two alternatives A_m and A_n are simultaneously decided by deciding their Cartesian product $A_{m.n} = A_m \times A_n$. $A_{m.n}$ defines the direct product Hilbert space $V(m.n) = V(m)*V(n)$.

2. THE WAY TO CONCRETE QUANTUM THEORY: THE CONCEPT OF THE "UR"

Traditional quantum theory accepts the concepts of time, space, particle, field, hence of motion, position, momentum, energy, force from classical physics. In the reconstruction only time is used from the outset, and all concepts of objects are replaced by the logical concept of "alternative", all concepts of temporal properties of objects by the concept of "state".

Time, however, is described in a more detailed manner than in classical physics. While it is also measured by a real variable t, explicit use is made of its "modes": present, past, future, with their qualitative differences.

The "abstract" theory (which was outlined above) is general enough to serve as a frame for introducing all the above-mentioned traditional "concrete" terms. All these concepts and the theories referring to them, like relativity and particle theory, should be developed as a consequence of abstract quantum theory and of the addition of one single (and simple) idea: **the reduction of all alternatives to the successive decision of binary alternatives** (yes-no decisions, bits or "urs").

2.1. The "ur" - the quantized binary alternative

It is a trivial fact that every n-fold alternative can be decomposed in a product of binary alternatives and that every state space can be understood as a subspace of a tensor product of two-dimensional spaces.

The central dynamical postulate of the ur hypothesis is:

> For any object there is at least one decomposition into binary subobjects - called urs - such that its dynamics is invariant under the symmetry group of the urs.

An ur itself is not something like an "ultimate particle" and not even a "small object". The ur theory is to be understood as a theory of a logical atomism.

The binary (n=2) alternative defines as its state space the space of two complex dimensions C^2, possessing a SU(2) symmetry. Systems defined by a Cartesian product of n binary alternatives possess a state space which is, or is a subspace of, the tensor product of n C^2-spaces:

$$T_n = C^2 * C^2 * \cdots * C^2$$

All objects with such a type of state space will have at least SU(2) as a symmetry of their dynamics. SU(2) is locally isomorphic to SO(3), and we start from the working hypothesis that this is the reason for a three-dimensional real space offering a natural description of all objects in physics: the "position space". How relativity, and particles as irreducible representations of a relativistic group, can be derived from this hypothesis will be shown in chapter 4.

If the number of urs is not fixed then they must be described in the tensor space T_B of urs with index set {1,2} and Botzmann statistics:

$$T_B = C^2 + C^2 * C^2 + C^2 * C^2 * C^2 + \ldots = C^2 + (C^2)^{*2} + (C^2)^{*3} + \ldots + (C^2)^{*n} + \ldots = \overset{\infty}{\underset{1}{+}} T_n \quad (1)$$

thereby * is the tensor product, + the direct sum and $T_n = (C^2)^{*n}$.
In T_B there is a canonical scalar product induced from C^2. All elements from T_n are orthogonal to all of T_m for $n \neq m$.

We start with the assumption that all the urs are different. For the alternatives it must make sense to say which one of it was intended. So as yet no symmetrization is recommended.
If x_1 and x_2 constitute a basis in C^2 then the 2^n monoms

$$x_i\, x_j\, x_k \ldots x_l\, x_m\, x_n \qquad [x_i, x_j, x_k, \ldots, x_l, x_m, x_n \in \{x_1, x_2\}]$$

build up a basis for T_n.
In T_B it is possible to define mappings between T_n and T_{n+1}. This can be done by operators for rising or lowering the degree of the tensors. We define l_i^+, the operator of left-multiplication by x_i, as

$$l_i^+ x = l_i^+ x_j\, x_k \ldots x_l = x_i x_j\, x_k \ldots x_l \quad (2)$$

l_i^+ is a operator with norm 1 and therefore welldefined on T_B

$$||l_i^+|| = \sup_{||x||=1} ||l_i^+ x|| = 1 \quad (3)$$

There exists an adjoined operator l_i. It is defined for each y and all $x \in T_B$ by the condition: There is a y^*

$$y^* = l_i y \text{ such that } (y, l_i^+ x) = (y^*, x) . \quad (4)$$

l_i removes from x a left hand sided x_i or, if there is non, annihilates x. Therefore it is

$$l_i l_i^+ = id \tag{5}$$

and

$$l_i^+ l_i = P_i \tag{6}$$

$$l_k l_i^+ = 0 \qquad \text{for } i \neq k . \tag{7}$$

P_i is a projection operator which annihilates all x without a left hand sided x_i.

So l_i^+ and l_i are partial isometries. They can be extended to unitary transformations by an extension of the Hilbert space ([21] pp. 436). This imbedding of T_B into the space T can be seen analogously to the imbedding of the natural numbers into the positive and negative whole numbers. In a philosophical sense the whole numbers are to be understood as operations on the natural numbers ([22]). In the same sense the vectors of the extended space T are operations on the tensor space T_B.

The extension of the partial isometric operators to unitary operators is straight forward. Firstly we have to introduce a onedimensional subspace Ω, the "vacuum", which is defined by

$$l_i x_i = \Omega \quad \text{resp.} \qquad l_i^+\Omega = x_i \quad \text{for every } i \tag{8}$$

and further on more vectors

$$x_{(-i)} = l_i\Omega \quad \text{and so on,} \tag{9}$$

which are orthogonal to all of the vectors from T_B .

In the case of the urs we denote $x_{(-1)}$ by x_4 and $x_{(-2)}$ by x_3,

$$l_1\Omega \to l_4^+\Omega \tag{10a}$$

and

$$l_2\Omega \to l_3^+\Omega \tag{10b}$$

A basis for the space of all states in T is given by polynoms in l_i and l_i^+, $i\in\{1,2\}$ (Geyer 1973, p. 180), resp. in l_i^+ alone for $i\in\{1,2,3,4\}$ acting on Ω

$$\begin{aligned} &l_i^+ l_j^+ \ldots l_k^+ l_l^+ \Omega \\ &= x_i\, x_j \ldots x_k\, x_l \end{aligned} \qquad \text{with } i,j,\ldots,k,l \in \{1,2,3,4\} \tag{11}$$

For this Boltzmann-"ladder"-operators we have

$$l_i^+ l_i = l_i l_i^+ = id \tag{12a}$$

and therefore

$$[l_i^+, l_i] = 0 \tag{12b}$$

2.2. The introduction of symmetries between the urs

Normally, in the concept of a "second quantization" the ladder operators are used to build up generators for a symmetry group. This group acts on the objects belonging to this new step of quantization.

To construct generators for a symmetry group it is necessary to have commutation relations for the ladder operators. The same sample of vectors constitute different Boltzmann tensors which are differentiated only by the sequences of the vectors in it. Therefore to get commutation relations we have to introduce symmetry relations between such tensors. In the tensors (11) the sequence of the operators is ordered by the place of the operators in the file. To allow symmetry relations between the operators means to weaken the differences of its places in the line.

All states in T can be expressed by the formula (11). If one wants to diminish the order of the factors, states with both factors x_1 and x_4 or both factors x_2 and x_3 could possibly disappear. Giving away the identityof $l_1\Omega$ and $l_4^{+}\Omega$ and of $l_2\Omega$ and $l_3^{+}\Omega$, it becomes possible to preserve the whole manifold of vectors in the extended Boltzmann space as starting point.

We give a new definition

$$l_i \, \Omega = 0 \qquad \text{for every } i \in \{1,2,3,4\} \tag{13}$$

Every x_i in a monom (11) has a fixed position which is not marked in a extra way. To "weaken" the order we label the position by an extra upper index, i.e. we introduce "kinds of urs". Then it becomes possible to give up the demand that all of them must be different. The difference in place is moved into the difference of indices. Therefore adding a new ur, it must be put onto every possible place. The new operators (which are called "Stopf- and Rupf- Operatoren", stuff- and pull-operators [9,11]) are denoted by

$$l_i^{+} \dashrightarrow s_i \;, \quad l_i \dashrightarrow r_i \;,$$

An stuff-operator s_i^a generates an new ur of the "type a" in a state i . It is

$$s_k^b \Omega = x_k^b \tag{14a}$$

and

$$r_i^a \, x_i^a = \Omega \tag{14b}$$

and for any monom in T (not necessarily only those from tensors generated by some s_i^a out of Ω) it is defined

$$\begin{aligned} s_i^a \, x_j^b x_k^c \ldots x_l^d = N_s[\, & x_i^a x_j^b x_k^c \ldots x_l^d + \\ & +\epsilon(a,b) x_j^b x_i^a x_k^c \ldots x_l^d + \\ & +\epsilon(a,b)\epsilon(a,c) x_j^b x_k^c x_i^a \ldots x_l^d + \ldots \\ \ldots & +\epsilon(a,b)\epsilon(a,c)\ldots\epsilon(a,d) x_j^b x_k^c \ldots x_l^d x_i^a \,] \end{aligned} \tag{15}$$

The normalization factor N_s may depend on the degree of the monom. The sign factor $\epsilon(a,b)$ will be used to differentiate the upper indices:

$$\epsilon(a,b)=1 \text{ for } a=b \;, \quad \epsilon(a,b)=-1 \text{ for } a\neq b$$

For the operators r_i^a we get

$$r_i^a \, x_j^b = \delta_{ab}\delta_{ij} \, \Omega \tag{14b}$$

and for any monom in T (not necessarily generated by some s_i^a out of Ω)

$$\begin{aligned} r_i^a \, x_j^b x_k^c \ldots x_l^d x_m^e = N_r[\, & \delta_{ab}\delta_{ij} \langle x_j^b \rangle x_k^c \ldots x_l^d x_m^e + \\ & +\epsilon(a,b)\delta_{ac}\delta_{ik} x_j^b \langle x_k^c \rangle \ldots x_l^d x_m^e + \ldots \\ \ldots & +\epsilon(a,b)\epsilon(a,c)\ldots\epsilon(\ldots)\delta_{ad}\delta_{il} x_j^b x_k^c \ldots \langle x_l^d \rangle x_m^e + \\ & +\epsilon(a,b)\epsilon(a,c)\ldots\epsilon(a,d)\delta_{ae}\delta_{im} x_j^b x_k^c \ldots x_l^d \langle x_m^e \rangle \,] \end{aligned} \tag{16}$$

whereby $\langle x_j^b \rangle$ means the cancellation of term x_j^b in the monom.

We denote as usual

$$\begin{aligned} s_i^a s_j^b + s_j^b s_i^a &= \{s_i^a, s_j^b\} \\ s_i^a s_j^b - s_j^b s_i^a &= [s_i^a, s_j^b] \end{aligned} \tag{17}$$

For the operators s_i^a and r_i^a we have the following commutation relations:

$$\{s_i^a, s_j^b\} = \{s_i^a, r_j^b\} = \{r_i^a, r_j^b\} = 0 \quad \text{for } a \neq b \tag{18}$$

and

$$[s_i^a, s_j^a] = [r_i^a, r_j^a] = 0 \tag{19}$$

2.3. The connection of the ladder operators with the SU(2)-symmetry

A physically meaningful symmetry between the subspaces of T has to be connected with the symmetry inside the spaces T_n carrying a tensor representation of SU(2).

For the generators of SU(2) it is

$$[\tau_{jn}, \tau_{nl}] = \tau_{jn}\tau_{nl} - \tau_{nl}\tau_{jn} = i\tau_{jl} \qquad (j,n,l = \text{cycl. } \{1,2,3\}) \tag{20}$$

If u_j und d_k denote in this moment general up- and down- step operators, then for a connection to the SU(2) symmetry we need something like

$$d_k u_j \pm u_j d_k = l^{\pm}{}_{jk}\, \delta_{jk}\, \mathbf{id} + n^{\pm}{}_{jk} \tau_{jk} \tag{21}$$

with numbers $l^{\pm}{}_{jk}$ and $n^{\pm}{}_{jk}$ which are to specify.

There are different possibilities for such numbers. E.g. for the demand

$$N_s \equiv 1 \quad \text{and} \quad N_r \equiv 1$$

the operators $s_i{}^a$ and $r_i{}^a$ would behave as

$$[r_i{}^a, s_j{}^a] = \tau_{ij}{}^a \quad \text{for } i \neq j \ .$$

We denote by $n_i{}^a$ the number of factors $x_i{}^a$ in a monom and by

$$n^a = n_1{}^a + n_2{}^a + n_3{}^a + n_4{}^a \tag{22}$$

The total number of factors in a monom is

$$n = \Sigma_a\, n^a$$

Then it is

$$[r_i{}^a, s_i{}^a] = n_i{}^a + n + 1$$

3. PARABOSE-STATISTICS FOR URS

We will denote by p the number of the different types of urs. Till now all the urs are introduced with a clearly distinguishable type. But if the urs can belong with equal probability to every possible type we can get parabose-statistics for the urs. In this case we cannot use any monom in **T**, instead we have to restrict the theory to such tensors which are generated by the s_i out from Ω.

At first we specify the normalization factors N_s and N_r. The commutation and anticommutation relations (18) and (19) are indifferent with respect to N_s and N_r and remain valid. But $[r_i{}^a, s_i{}^a]$ will be changed.

Now we define operators $b_k{}^{a+}$ and $b_k{}^a$ by

$$b_k{}^{a+} = s_k{}^a \quad \text{with } N_s = (n^a + 1)^{-1/2} \tag{23}$$

$$b_k{}^a = r_k{}^a \quad \text{with } N_r = (n^a\ )^{-1/2} \tag{24}$$

and get by computation

$$[b_k{}^a, b_l{}^{a+}] = \delta_{kl} \qquad [b_k{}^{a+}, b_l{}^{a+}] = 0 \qquad [b_k{}^a, b_l{}^a] = 0 \tag{25}$$

and again for $b \neq a$

$$\{b_k{}^a, b_l{}^{b+}\} = 0 \qquad \{b_k{}^{a+}, b_l{}^{b+}\} = 0 \qquad \{b_k{}^a, b_l{}^b\} = 0 \tag{26}$$

Such anticommuting Bose operators are used in Green's decomposition of parabose operators $a_k{}^+$ and a_k ([11] p. 427):

$$a_k{}^+ = b_k{}^{1+} + b_k{}^{2+} + \ldots + b_k{}^{p+} \tag{27}$$

$$a_k = b_k{}^1 + b_k{}^2 + \ldots + b_k{}^p \tag{28}$$

It is well known that the $a_k{}^+$ and a_k fulfil the three-linear commutation relations

$$[\{a_r, a_s\}, a_t]=0 \qquad [\{a_r{}^+, a_s{}^+\}, a_t{}^+]=0 \qquad [\tfrac{1}{2}\{a_r, a_s{}^+\}, a_t]=-\delta_{st} a_r \tag{29}$$

The quantum theoretical interpretation of our construction is straight forward. The Green-indizes constitute a hidden classification, any ur belongs with the same probability to any class. The, strictly speaking, required normalization factor $1/p^{1/2}$ is usually neglected in the literature to make the commutation relations free of p.

The connection with the SU(2) symmetry within the subspaces T_n is given by (see formula (20))

$$\tau_{sr} = (1/2)\{a_r, a_s^+\} \qquad [\tau_{rs}, a_t^+] = \delta_{st} a_r^+ \qquad [\tau_{rs}, a_t] = -\delta_{rt} a_s \tag{30}$$

The combination of two free objects by its tensor product remains valid also for parabose urs. Because of the partial indistinguishability of the urs the tensor product of its state spaces must be constructed as a product of the creation operators over Ω and not of single tensors from T alone.

By the demand that for different objects the Green's indices should be disjoint we get a new composition rule ("Heidenreich product" [24]). This is different from the tensor product of free objects and should therefore express interaction between the objects ([26], [11] p. 432).

4. PARTICLES WITH PARABOSE-URS

Objects which can move freely in a space-time are normally called particles. Free motion means motion under a maximal group of transformations. In a fourdimensional space-time such a group can be at most 10-dimensional which happens in the case of constant curvature.

By the bilinear combinations of the parabose creation and destruction operators it is possible to construct the representations of SO(4,2) (in connection with the urs see [23, 24, 25]). So it is possible to construct all three groups of motions in a fourdimensional space-time of constant curvature, i.e. SO(4,1), SO(3,2) and the Poincare group. The combination with the linear operators also allows the representations of supersymmetries, connecting Bose- and Fermi- representations.

In ur-theory the cosmological model does not correspond to a space-time with a constant curvature, therefore there does not exist a ten-parameter transformation group. To describe particles we have to go into the tangential space, i.e. Minkowski space-time. There the irreducible representations of the Poincare group give the states of particles.

4.1. The generators of the Poincare group

We use the following abbreviations

$$\tau_{sr} = (1/2)\{a_r, a_s^+\} \tag{31a}$$
$$\alpha^+_{sr} = (1/2)\{a_r^+, a_s^+\} \tag{31b}$$
$$\alpha_{sr} = (1/2)\{a_r, a_s\} \tag{31c}$$

and define the number operator

$$\tilde{n}_r = \tau_{rr} - p/2 \tag{31d}$$

For the generators of O(4,2) (for its Bose form see e.g. [27]) we take the expression given by ([11] p. 407), where a skew-hermitean form of them is used. They are

$$M_{12} = (i/2)(+\tilde{n}_1 - \tilde{n}_2 + \tilde{n}_3 - \tilde{n}_4) \qquad N_{14} = (i/2)(+\alpha_{13} + \alpha^+_{13} - \alpha_{24} - \alpha^+_{24})$$
$$M_{13} = (1/2)(-\tau_{12} + \tau_{21} - \tau_{34} + \tau_{43}) \qquad N_{24} = (1/2)(-\alpha_{13} + \alpha^+_{13} - \alpha_{24} + \alpha^+_{24})$$
$$M_{23} - (i/2)(+\tau_{12} + \tau_{21} + \tau_{34} + \tau_{43}) \qquad N_{34} = (i/2)(-\alpha_{14} - \alpha^+_{14} - \alpha_{23} - \alpha^+_{23})$$

$$
\begin{array}{llll}
M_{15} = (i/2)(+\tau_{12} +\tau_{21} -\tau_{34} -\tau_{43}) & N_{16} = (1/2)(-\alpha_{13} +{\alpha^+}_{13} +\alpha_{24} -{\alpha^+}_{24}) & & \\
M_{25} = (1/2)(+\tau_{12} -\tau_{21} -\tau_{34} +\tau_{43}) & N_{26} = (i/2)(-\alpha_{13} -{\alpha^+}_{13} -\alpha_{24} -{\alpha^+}_{24}) & & (32) \\
M_{35} = (i/2)(+\tilde{n}_1 -\tilde{n}_2 -\tilde{n}_3 +\tilde{n}_4) & N_{36} = (1/2)(+\alpha_{14} -{\alpha^+}_{14} +\alpha_{23} -{\alpha^+}_{23}) & & \\
M_{46} = (i/2)(+\tilde{n}_1 +\tilde{n}_2 +\tilde{n}_3 +\tilde{n}_4 +2p) & N_{46} = (1/2)(+\alpha_{14} -{\alpha^+}_{14} -\alpha_{23} +{\alpha^+}_{23}) & & \\
s = (1/2)(+\tilde{n}_1 +\tilde{n}_2 -\tilde{n}_3 -\tilde{n}_4) & N_{56} = (i/2)(+\alpha_{14} +{\alpha^+}_{14} -\alpha_{23} -{\alpha^+}_{23}) & &
\end{array}
$$

For the generators of the Poincare group we use the expressions for the translations

$$P_i = M_{i5} + N_{i6} \qquad P_4 = N_{46} + M_{46} \tag{33a}$$

and the operators of SO(3,1) for rotations and Lorentz boosts:

$$M_{12},\ M_{13},\ M_{23}; \qquad N_{14},\ N_{24},\ N_{34} \tag{33b}$$

Then the momentum operators are given explicitly by

$$
\begin{array}{ll}
2P_1 & = [i(\tau_{12} +\tau_{21} -\tau_{34} -\tau_{43}) -\alpha_{13} +\alpha_{24} +{\alpha^+}_{13} -{\alpha^+}_{24}] \\
2iP_2 & = [i(\tau_{12} -\tau_{21} +\tau_{43} -\tau_{34}) +\alpha_{13} +\alpha_{24} +{\alpha^+}_{13} +{\alpha^+}_{24}] \\
2P_3 & = [i(\tilde{n}_1 -\tilde{n}_2 -\tilde{n}_3 +\tilde{n}_4) \quad +\alpha_{14} +\alpha_{23} -{\alpha^+}_{14} -{\alpha^+}_{23}] \\
2P_4 & = [i(\tilde{n}_1 +\tilde{n}_2 +\tilde{n}_3 +\tilde{n}_4 +2p) +\alpha_{14} -\alpha_{23} -{\alpha^+}_{14} +{\alpha^+}_{23}]
\end{array} \tag{34}
$$

4.2. Momentum eigenstates for particles

In (23, 24, 25) the representations are given in an abstract way. In the present paper we are interested in concrete momentum eigenstates for particles. This are states $|\Phi\rangle$ which fulfil

$$(P_1+iP_2)|\Phi\rangle = [i(\tau_{12} - \tau_{34}) + \alpha_{24} + {\alpha^+}_{13}]|\Phi\rangle = 0 \tag{35a}$$

$$(P_1-iP_2)|\Phi\rangle = [i(\tau_{21} - \tau_{43}) - \alpha_{13} - {\alpha^+}_{24}]|\Phi\rangle = 0 \tag{35b}$$

$$(P_4+P_3)|\Phi\rangle = [i(\tilde{n}_1 + \tilde{n}_4 + p) + \alpha_{14} - {\alpha^+}_{14}]|\Phi\rangle = im_+|\Phi\rangle \tag{35c}$$

$$(P_4-P_3)|\Phi\rangle = [i(\tilde{n}_2 + \tilde{n}_3 + p) - \alpha_{23} + {\alpha^+}_{23}]|\Phi\rangle = im_-|\Phi\rangle \tag{35d}$$

If m_+ and m_- are both different from zero we have a state of a massive particle and for $m_+m_-=0$ a massless one. A state with $m_+=m_-=0$ is a vacuum state.

For a state of a massless and spinless boson we make the ansatz

$$|\Phi\rangle = \Sigma_\mu\Sigma_\beta\ (-1)^{\mu+\beta} i^{\mu-\beta} (\mu!\beta!)^{-1} c(\mu,\beta) {\alpha^+}_{14}{}^{\mu} {\alpha^+}_{23}{}^{\beta}\ |\Omega\rangle \tag{36}$$

and get under the condition $m_+\neq 0$ for the coefficients $c(\mu,\beta)$ the equations

$$c(\mu,\beta)=c(\mu) \tag{37a}$$

and

$$(\mu+p)c(\mu+1) -(2\mu+p-m_+)c(\mu) -\mu c(\mu-1)= 0 \tag{37b}$$

For $p=1$, formula (37b) is the recurrence relation for the Laguerre polynomials.

A state of a massless boson with helicity σ is constructed from

$$|\phi\rangle = \Sigma_\mu\Sigma_\beta\ i^{\beta-\mu}(\beta!\mu!)^{-1} c(\mu,\sigma)\ {\alpha^+}_{11}{}^{\sigma}{\alpha^+}_{14}{}^{\mu}{\alpha^+}_{23}{}^{\beta}\ |\Omega\rangle \tag{38}$$

For the condition $m_+\neq 0$ the coefficients $c(\mu,\sigma)$ must fulfil the equations

$$(\mu+p+2\sigma)c(\mu+1,\sigma) - (2\mu+p+2\sigma-m_+)c(\mu,\sigma) + \mu c(\mu-1,\sigma) = 0 \tag{39}$$

A state of a massless fermion with helicity 1/2 is realized by

$$|\Gamma\rangle = \Sigma_\mu\Sigma_\beta\ i^{\beta-\mu}(\beta!\mu!)^{-1} c(\mu)\ {a^+}_1{\alpha^+}_{14}{}^{\mu}{\alpha^+}_{23}{}^{\beta}\ |\Omega\rangle \tag{40}$$

with $m_+\neq 0$ and

$$+(\mu+p+1)c(\mu+1) -(2\mu+p+1-m_+)c(\mu) +\mu c(\mu-1) = 0 \tag{41}$$

It is obvious that for such massless particles these states are nearly of the same form.

Now we have to look for massive particles.
Massive particles cannot made from power series in $\alpha^+_{14}{}^{\mu}$ and $\alpha^+_{23}{}^{\beta}$ alone, there must be extra factors of the type $\alpha^+_{13}{}^{\sigma}$ and $\alpha^+_{24}{}^{\sigma}$.

For a massive spinless particle we made the the ansatz

$$|\Theta\rangle = \Sigma_\sigma \Sigma_\mu \Sigma_\beta i^{\beta-\mu}(\mu!\beta!\sigma!\sigma!)^{-1} g(\mu,\beta,\sigma)\alpha^+_{13}{}^{\sigma}\alpha^+_{24}{}^{\sigma}\alpha^+_{14}{}^{\mu}\alpha^+_{23}{}^{\beta}\,|\Omega\rangle \tag{42}$$

and get from (35a-d) the conditions

$$(\sigma+1)g(\mu,\beta,\sigma) +(\sigma+1)g(\mu+1,\beta+1,\sigma) -(\sigma+1)g(\mu,\beta+1,\sigma) -(\sigma+1)g(\mu+1,\beta,\sigma)$$
$$-\mu g(\mu-1,\beta,\sigma+1) -\beta g(\mu,\beta-1,\sigma+1) +(\mu+\beta+p+\sigma)g(\mu,\beta,\sigma+1)= 0 \tag{43}$$

$$-[2\mu+2\sigma+p-m_+]g(\mu,\beta,\sigma) +(2\sigma+\mu+p)g(\mu+1,\beta,\sigma) +\mu g(\mu-1,\beta,\sigma)$$
$$= -\beta g(\mu,\beta-1,\sigma+1) \tag{44}$$

$$-[2\beta+2\sigma+p-m_-]g(\mu,\beta,\sigma) +(2\sigma+p+\beta)g(\mu,\beta+1,\sigma) +\beta g(\mu,\beta-1,\sigma)$$
$$= -\mu g(\mu-1,\beta,\sigma+1) \tag{45}$$

from this equations it follows

$$-[2\mu+2\sigma+p-m_+]g(\mu,\beta+1,\sigma) +(2\sigma+\mu+p)g(\mu+1,\beta+1,\sigma) +\mu g(\mu-1,\beta+1,\sigma)$$
$$= -(\beta+1)g(\mu,\beta,\sigma+1) \tag{46}$$

$$-[2\beta+2\sigma+p-m_-]g(\mu+1,\beta,\sigma) +(2\sigma+p+\beta)g(\mu+1,\beta+1,\sigma) +\beta g(\mu+1,\beta-1,\sigma)$$
$$= -(\mu+1)g(\mu,\beta,\sigma+1) \tag{47}$$

and we can get an equation with fixed σ:

$$(\mu+1)\{-[2\mu+2\sigma+p-m_+]g(\mu,\beta+1,\sigma)+(2\sigma+\mu+p)g(\mu+1,\beta+1,\sigma)+\mu g(\mu-1,\beta+1,\sigma)\} =$$
$$= (\beta+1)\{-[2\beta+2\sigma+p-m_-]g(\mu+1,\beta,\sigma)+(2\sigma+\beta+p)g(\mu+1,\beta+1,\sigma)+\beta g(\mu+1,\beta-1,\sigma)\} \tag{48}$$

If we want to solve it by separation of variables we get in any case mass zero equations. It seems only possible to solve it step by step. In this case we get with the ansatz

$$g(0,0,0) = 1$$

the following values for the coefficients $g(\mu,\beta,\sigma)$

$$g(1,0,0) = (p-m_+)/p$$
$$g(0,1,0) = (p-m_-)/p$$
$$g(0,0,1) = m_+m_-/p(p+1)(p-1)$$
$$g(2,0,0) = [(p-2m_+)/p +m_+^2/p(p+1)]$$
$$g(1,1,0) = [(p-m_+-m_-)/p +m_+m_-/(p+1)(p-1)]$$
$$g(0,2,0) = [(p-2m_-)/p +m_-^2/p(p+1)]$$
$$g(0,1,1) = m_+m_-(2+p-m_-)/p(p+1)(2+p)(p-1)$$
$$= m_+m_-/p(p+1)(p-1) -m_+m_-^2/p(p+1)(2+p)(p-1)$$
$$g(1,0,1) = m_+m_-(2+p-m_+)/p(p+1)(2+p)(p-1)$$
$$= m_+m_-/p(p+1)(p-1) -m_-m_+^2/p(p+1)(2+p)(p-1)$$
$$g(0,0,2) = m_+m_-/p(p^2-1)$$
$$g(1,1,1) = [(3+p-m_+-m_-)p +(2-2m_+-2m_-+m_-m_+)m_+m_-]/p(2+p)^2(p^2-1)$$
$$g(2,1,0) = [(p-2m_+)/p +m_+^2/p(1+p) -m_-(p+1-m_+)/p(1+p)$$
$$+m_+m_-(2+p-m_+)(p^2+2p-1)/p(1+p)(2+p)(p^2-1)]$$
$$g(1,2,0) = [(p-2m_-)/p +m_-^2/p(1+p) -m_+(p+1-m_-)/p(1+p)$$
$$+m_+m_-(2+p-m_-)(p^2+2p-1)/p(1+p)(2+p)(p^2-1)]$$
$$g(0,2,1) = (6+5p-6m_-+p^2-2pm_-+m_-m_-)m_+m_-/p(2+p)(3+p)(p^2-1)$$

and so on.

4.3. The introduction of the Lorentz vacuum

The last formulas became very difficult, no closed solution seems to be possible. Therefore it is reasonable to look for a simplification. This can be got by introducing the Lorentz vacuum $|\Omega_L\rangle$.

The states given above are constructed over the ur-vacuum. For this there is no ur being present, i. e. it posses not any bit of information. The meaning of the Lorentz vacuum is that there is not any real particle. The knowledge "there is no particle" is of much greater amount of information than the knowledge "there is no ur". Therefore it seems plausible and also easier if we construct the particle states over the Lorentz vacuum instead the ur vacuum.

We define the Lorentz vacuum by the equations

$$(P_1+iP_2)|\Omega_L\rangle = [i(\tau_{12} - \tau_{34}) + \alpha_{24} + \alpha^+_{13}]|\Omega_L\rangle = 0 \quad (49a)$$
$$(P_1-iP_2)|\Omega_L\rangle = [i(\tau_{21} - \tau_{43}) - \alpha_{13} - \alpha^+_{24}]|\Omega_L\rangle = 0 \quad (49b)$$
$$(P_4+P_3)|\Omega_L\rangle = [i(\tilde{n}_1 + \tilde{n}_4 + p) + \alpha_{14} - \alpha^+_{14}]|\Omega_L\rangle = 0 \quad (49c)$$
$$(P_4-P_3)|\Omega_L\rangle = [i(\tilde{n}_2 + \tilde{n}_3 + p) - \alpha_{23} + \alpha^+_{23}]|\Omega_L\rangle = 0 \quad (49d)$$

which have the solution

$$|\Omega_L\rangle = \Sigma_\mu \Sigma_\beta\ (-1)^{\mu+\beta} i^{\mu-\beta} (\mu!\beta!)^{-1} \alpha^+_{14}{}^\mu \alpha^+_{23}{}^\beta\ |\Omega\rangle \quad (50)$$
$$= exp\ i(\alpha^+_{23} - \alpha^+_{14})|\Omega\rangle$$

Now we are able to create particles out from the Lorentz vacuum instead out from the ur vacuum. The formulas for the particle states became much simpler.

4.4. Massless particles from the Lorentz vacuum

We start with a particle of mass zero and spin σ. Its state

$$\Phi(m_+,\sigma)^+|\Omega_L\rangle$$

is defined by the conditions

$$(P_1+iP_2)\Phi(m_+,\sigma)^+|\Omega_L\rangle = 0 \quad (51a)$$
$$(P_1-iP_2)\Phi(m_+,\sigma)^+|\Omega_L\rangle = 0 \quad (51b)$$
$$(P_4-P_3)\ \Phi(m_+,\sigma)^+|\Omega_L\rangle = 0 \quad (51c)$$
$$(P_4+P_3)\Phi(m_+,\sigma)^+|\Omega_L\rangle = im_+\Phi(m_+,\sigma)^+|\Omega_L\rangle \quad (51d)$$
$$M_{12}\ \Phi(m_+,\sigma)^+|\Omega_L\rangle = i\sigma\Phi(m_+,\sigma)^+|\Omega_L\rangle \quad (51e)$$

The ansatz

$$\Phi(m_+,\sigma)^+|\Omega_L\rangle = \alpha^+_{11}{}^\sigma \sum_{\mu=0}^{\infty} c(\mu,\sigma)\alpha^+_{14}{}^\mu|\Omega_L\rangle \quad (52)$$

results in

$$c(\mu,\sigma)= c(0,\sigma)(im_+)^\mu \frac{(p+2\sigma-1)!}{\mu!(p+2\sigma-1+\mu)!} \qquad (\text{for } \mu=0,1,2,\ldots) \quad (53)$$

For spin-1/2-particles we have to replace $\alpha^+_{11}{}^\sigma$ by $a^+_1\alpha^+_{11}{}^\sigma$ in formula (52) and in (53) σ by $(\sigma+1/2)$.

With the formulas (52) and (53) we have given the momentum eigenstates for massless Bosons and Fermions as expansion series in the parabose operators.

4.5. Massive spinless particles from the Lorentz vacuum

Massive Bosons without spin must fulfil the conditions

$$(P_1+iP_2)\Phi(m)^+|\Omega_L\rangle = 0 \tag{54a}$$
$$(P_1-iP_2)\Phi(m)^+|\Omega_L\rangle = 0 \tag{54b}$$
$$(P_4+P_3)\Phi(m)^+|\Omega_L\rangle = im_+\Phi(m)^+|\Omega_L\rangle \tag{54c}$$
$$(P_4-P_3)\Phi(m)^+|\Omega_L\rangle = im_-\Phi(m)^+|\Omega_L\rangle \tag{54d}$$

with $m_-m_+\neq 0$. They can be constructed from the ansatz

$$\Phi(m)^+|\Omega_L\rangle = \Sigma_\sigma \Sigma_\mu \Sigma_\beta\, h(\mu,\beta,\sigma)\alpha^+{}_{13}{}^\sigma\alpha^+{}_{24}{}^\sigma\alpha^+{}_{14}{}^\mu\alpha^+{}_{23}{}^\beta\, |\Omega_L\rangle \tag{55}$$

for which (54) give the conditions

$$-(\mu+1)(\beta+1)h(\mu+1,\beta+1,\sigma) -(\sigma+1)(p+\sigma+2)h(\mu,\beta,\sigma+1)=0$$
$$-im_+h(\mu,\beta,\sigma) +(\sigma+1)^2h(\mu,\beta-1,\sigma+1) +(\mu+1)(p+\mu+2\sigma)h(\mu+1,\beta,\sigma)=0$$
$$-im_-h(\mu,\beta,\sigma) -(\sigma+1)^2h(\mu-1,\beta,\sigma+1) -(\beta+1)(p+\beta+2\sigma)h(\mu,\beta+1,\sigma)=0$$

The coefficients of the solution have the following recurrent relations

$$h(\mu+1,\beta,\sigma)= h(\mu,\beta,\sigma)\frac{im_+}{(\mu+1)} * \frac{(p+\sigma+2)}{[(p+\mu+2\sigma)(p+\sigma+2)-\beta(\sigma+1)]} \qquad (\beta>0) \tag{56}$$

$$h(\mu,\beta+1,\sigma)= h(\mu,\beta,\sigma)\frac{-im_-}{(\beta+1)} * \frac{(p+\sigma+2)}{[(p+\beta+2\sigma)(p+\sigma+2)-\mu(\sigma+1)]} \qquad (\mu>0) \tag{57}$$

$$h(\mu,\beta,\sigma+1)= \qquad \{\text{for } \mu>0,\ \beta>0,\ \mu\neq\beta\}$$

$$= \frac{h(\mu,\beta,\sigma)\ (-m_+m_-)\ (p+\sigma+2) \qquad (p+\beta+\mu+2\sigma)}{(\sigma+1)[(p+\sigma+2)(p+\beta+2\sigma)-\mu(\sigma+1)][(p+\sigma+2)(p+\mu+2\sigma)-\beta(\sigma+1)](p+\mu+\beta+2\sigma+1)} \tag{58}$$

and $\{\text{for } \mu>0,\ \beta>0,\ \mu=\beta\}$

$$h(\mu,\mu,\sigma+1)= \frac{h(\mu,\mu,\sigma)\ (-m_+m_-)\ (p+\sigma+2)}{(\sigma+1)[(p+\mu+2\sigma)(p+\sigma+2)-\mu(\sigma+1)][(p+\mu+2\sigma)(p+\sigma+2)-(\mu+1)(\sigma+1)]} \tag{59}$$

Massive particles with spin are to describe by more complicated formulas. But in principle they are not different from the former ones.

Acknowledgement

I thank Prof. C. F. v. Weizsäcker for many discussions and for his continuous support .
I am grateful to the Stifterverband der Wissenschaft for financial support for part of this work.

REFERENCES

1. Weizsäcker, C. F. v., Komplementarität und Logik I, Naturwiss. 42, 521-529, 545-555 (1955)
2. Weizsäcker, C. F. v., Komplementarität und Logik II, Z. Naturforschung 13a, 245 (1958)
3. Scheibe, E., Süssmann, G., Weizsäcker, C. F. v., Mehrfache Quantelung, Komplementarität und Logik III, Z. Naturforschung 13a, 705 (1958)
4. Drieschner, M., Quantum Mechanics as a General Theory of Objective Prediction, Thesis, Hamburg (1970)
5. Weizsäcker, C. F. v., The unity of physics, in "Quantum Theory and Beyond", T. Bastin, ed. University Press, Cambridge (1971)
6. Weizsäcker, C. F. v., "Die Einheit der Natur", Hanser, München (1971) (engl. Ed.: "The Unity of Nature", Farrar, Straus, Giroux, New York 1980)

7. Weizsäcker, C. F. v.: A comment to Dirac's Paper, in: "The Physicist's Conception of Nature", J. Mehra, ed. Reidel, Dordrecht (1973)
8. Weizsäcker, C. F. v.: Classical and Quantum Descriptions, in: "The Physicist's Conception of Nature", J. Mehra, ed. Reidel, Dordrecht (1973)
9. Weizsäcker, C. F. v.: Technisches zum Stopfen und Rupfen, Interner Bericht, MPI Starnberg (1977)
10. Weizsäcker, C. F. v.: Programm zur Urtheorie, Interner Bericht, Starnberg (1978)
11. Weizsäcker, C. F. v., "Aufbau der Physik", Hanser, München (1985)
12. Görnitz, Th., Weizsäcker, C. F. v.: De-Sitter Representations and the Particle Concept, in an Ur-Theoretical Cosmological Model, in Barut, A. O., Doebner, H.-D. (1986)
13. Barut, A. O., Doebner, H.-D. (eds.): "Conformal Groups and Related Symmetries, Physical Results and Mathematical Background", Lect. Notes in Physics 261, Springer Verl. Berlin ect. (1986)
14. Drieschner, M., Görnitz, Th., Weizsäcker, C. F. v.,: Reconstruction of Abstract Quantum Theory, Intern. Journ. Theoret. Phys. 27, 289 - 306 (1987)
15. Görnitz, Th., Weizsäcker, C. F. v.: Quantum Interpretations; Intern. Journ. Theoret. Phys. 26, 921 (1987)
16. Görnitz, Th.,: Abstract Quantum Theory and Space-Time-Structure, Part I: Ur-Theory, Space Time Continuum and Bekenstein-Hawking-Entropy, Intern. Journ. Theoret. Phys. 27, 527 - 542 (1988)
17. Görnitz, Th.,: On Connections between Abstract Quantum Theory and Space-Time-Structure, Part II: A Model of cosmological evolution, Intern. Journ. Theoret. Phys. 27, 659 - 666 (1988)
18. Görnitz, Th., Ruhnau, E.: Connections between Abstract Quantum Theory and Space-Time-Structure, Part III: Vacuum Structure and Black Holes, Intern. Journ. Theoret. Phys. 28, 651 - 657 (1989)
19. Weizsäcker, C. F. v., Quantum Theory and space-time, in P. Lathi and P. Mittelstaedt (Eds.): "Symp. on the foundations of modern physics", Joensuu ; World Scientific (1985)
20. Weizsäcker, C. F. v., Reconstruction of Quantum Theory, in L. Castell and C. F. v. Weizsäcker (eds.): "Quantum Theory and the Structures of Space and Time VI", München, Hanser (1986)
21. Riesz,F., Sz.-Nagy, B.: "Vorlesungen über Funktionalanalysis", Deutscher Verlag der Wissenschaften, Berlin (1981),
22. Görnitz, Th., Rhunau, E., v. Weizsäcker, C. F. , in preparation
23. Jacob, P.: Thesis, MPI Starnberg (1977)
24. Heidenreich, W.: Thesis, TU München (1981)
25. Künemund, Th.: Thesis, TU München (1982)
26. Görnitz, Th., Graudenz, D., v. Weizsäcker, C.F.: in preparation
27. Castell, L. in "Quantum Theory and the Structures of Space and Time II", München, Hanser (1975)

ON QUANTIZED VERMA MODULES

B. Gruber[1] and Yu.F. Smirnov[2]

[1] Department of Physics and Molecular Science Program
Southern Illinois University
Carbondale, Illinois, USA 62901–4401
and
Arnold Sommerfeld Institute for Mathematical Physics
Technical University Clausthal
D-3392 Clausthal, Germany

[2] Nuclear Physics Institute
Moscow State University, Moscow, USSR

1. INTRODUCTION

In this article we present the q-analogue of results which were obtained in two earlier articles [1,2] for representations of the algebras $su(2)$ and $su(1,1)$, including the unitarization of these representations. The carrier spaces for the representations of $su(2)$ and $su(1,1)$,which were discussed in [1,2], are Verma modules and certain subspaces and quotient spaces of Verma modules. The results obtained in [1,2] for $su(2)$ and $su(1,1)$ are re-evaluated in the language of the q-calculus for the corresponding quantum algebras $su_q(2)$ and $su_q(1,1)$ on "quantized" Verma modules (sections 2,3).

In section 4 we consider tensor products of unitary irreducible representations of the (positive) discrete series of representations of the quantum algebras $su_q(2)$ and $su_q(1,1)$ obtained in sections 2,3. The properties of the q-Clebsch-Gordan coefficients and q-Racah coefficients are discussed.

In section 5 we discuss the q-analogue of indecomposable pairs of representations of $su(2)$ and $su(1,1)$ which were obtained in references [3,4].

Symmetries in Science V, Edited by B. Gruber *et al.*
Plenum Press, New York, 1991

2. QUANTIZED VERMA MODULES FOR $su_q(2)$

In order to keep the discussion of the q-analogue of the algebra $su(2)$ as simple as possible we choose for $su(2)$ a basis

$$su(2) : \{J_+, J_-, J_0\} \tag{2.1}$$

with the following Lie products

$$[J_0, J_+] = J_+, \quad [J_0, J_-] = -J_-, \quad [J_+, J_-] = 2J_0. \tag{2.2}$$

In the Lie products we have introduced — in distinction to ref. [1,2] — a factor 2. This factor is of advantage for the q-calculus, while it is insignificant for $su(2)$. The quantum algebra $su_q(2)$ is then defined by the Lie products

$$[J_0, J_+] = J_+, \quad [J_0, J_-] = -J_-, \quad [J_+, J_-] = [2J_0], \tag{2.3}$$

where

$$\begin{aligned} [x] &\equiv \frac{q^x - q^{-x}}{q - q^{-1}}, \quad q \in \mathbf{C} \\ [0] = 0, &\quad [1] = 1, \quad [-x] = -[x]. \end{aligned} \tag{2.4}$$

In this article $q \neq \exp(i\frac{2\pi}{N})$. For a detailed description of q-analysis we refer to ref. [5], while ref. [6] contains a brief summary. Using eq. (2.3) and eq. (2.4) we obtain

$$[J_+, J_-^n] = J_-^{n-1}[n][2J_0 - n + 1] \tag{2.5}$$

and for the Casimir operator C,

$$C = J_-J_+ + [J_0][J_0 + 1]. \tag{2.6}$$

We consider the space given by the linear span of

$$\{X(m,n) = J_-^m J_+^n, \quad m,n \in \mathbf{N}\}, \quad X(0,0) = \mathbb{1}. \tag{2.7}$$

On this space one obtains a representation ρ of $su_q(2)$, (analogous to eq. (4.5) of ref. [2])

$$\begin{aligned} &\rho(J_0)\mathbb{1} = j\mathbb{1}, \quad j \in \mathbf{C}, \\ &\rho(J_0)X(m,n) = (j + n - m)X(m,n), \\ &\rho(J_-)X(m,n) = X(m+1,n), \\ &\rho(J_+)X(m,n) = X(m,n+1) + [m][2j + 2n - m + 1]X(m-1,n), \\ &\rho(C)X(m,n) = X(m+1,n+1) + [j+m][j+m+1]X(m,n). \end{aligned} \tag{2.8}$$

A basis for the quotient space with respect to the left ideal generated by the relation $J_+\mathbb{1} = 0$ is given by

$$\{X(n) = J_-^n, \quad n \in \mathbf{N}\}, \quad X(0) = \mathbb{1}. \tag{2.9}$$

On this space one obtains a representation (Verma module) d_j, (analogous to section III of [1])

$$\begin{aligned}
\rho(J_0)\mathbb{1} &= j\mathbb{1},\\
\rho(J_0)X(n) &= (j-n)X(n),\\
\rho(J_+)X(n) &= [n][2j-n+1]X(n-1),\\
\rho(J_-)X(n) &= X(n+1)
\end{aligned} \tag{2.10}$$

where $j \in \mathbf{C}$ is the highest weight of d_j. For values

$$j \neq 0, \frac{1}{2}, 1, \frac{3}{2}, \ldots$$

the Verma module d_j is irreducible. For values

$$j = 0, \frac{1}{2}, 1, \frac{3}{2}, \ldots$$

the Verma module is indecomposable. For this case there exists in d_j, apart from $\mathbb{1}$, a second extremal vector $y_0 = J_-^{2j+1}$ which generates an irreducible submodule $\widetilde{\pi}_j$ with basis

$$\{y_k = J_-^k y_0, \quad y_0 = J_-^{2j+1}; \quad k \in \mathbf{N}\} \tag{2.11}$$

On the submodule $\widetilde{\pi}_j$ one obtains the irreducible representation

$$\begin{aligned}
\rho(J_0)y_k &= (-j-1-k)y_k,\\
\rho(J_-)y_k &= y_{k+1},\\
\rho(J_+)y_k &= -[k][2j+1+k]y_{k-1},\\
\rho(C)y_k &= [j][j+1]y_k.
\end{aligned} \tag{2.12}$$

The sesquilinear form S_0, introduced in [1] for $su(2)$, can also be extended via the q-calculus to the quantum algebra $su_q(2)$,

$$\begin{aligned}
S_0(J_-^k, J_-^n) &= \xi_\Lambda(\eta_0(J_-^k), J_-^n) = \delta_{kn}\Lambda \cdot \phi(J_+^n J_-^n)\\
&= \delta_{kn}\Lambda \cdot \phi\{J_+^{n-1}(J_-^n J_+ + [n]J_-^{n-1}[2J_0 - n + 1]\mathbb{1}\}\\
&= \delta_{kn}\Lambda \cdot \prod_{t=0}^{n-1}\{[n-t][2J_0 - n + t + 1]\}\\
&= \delta_{kn}[n]! \prod_{s=1}^{n}[2j - s + 1]
\end{aligned} \tag{2.13a}$$

with the conjugation law

$$\sigma_0(J_+) = -J_-, \quad \sigma_0(J_-) = -J_+, \quad \sigma_0(J_0) = -J_0, \quad \eta_0 = -\sigma_0. \tag{2.13b}$$

We consider here the case for q, j real. Then the complex valued sesquilinear form S_0 becomes real valued. If j is real, but not a non-negative integer or halfinteger, then S_0 is indefinite. If j is a non-negative integer or halfinteger, then S_0 becomes degenerate and vanishes on the submodule π_j generated by the extremal vector

$y_0 = J_-^{2j+1}$ and spanned by $U(J_-)y_0$. The irreducible quotient module $\pi_j = d_j/\tilde{\pi}_j$, is spanned by

$$\{J_-^k, \quad 0 \leq k \leq 2j\}.$$

We introduce for π_j a new basis, orthonormal with respect to S_0,

$$\begin{aligned} |jm\rangle &= \|X(j-m)\|^{-1} X(j-m), \\ n = j-m, &\quad m = -j, -j+1, \ldots, j \\ \|X(j-m)\| &= \left(\frac{[j-m]![2j]!}{[j+m]!} \right)^{1/2} \end{aligned} \tag{2.14}$$

With respect to this new basis we obtain on π_j the representation

$$\begin{aligned} J_0 \; |jm\rangle &= m \; |jm\rangle \\ J_+ \; |jm\rangle &= \sqrt{[j][j+1]-[m][m+1]} \; |j, m+1\rangle \\ &= \sqrt{[j-m][j+m+1]} \; |j, m+1\rangle \\ J_- \; |jm\rangle &= \sqrt{[j][j+1]-[m][m-1]} \; |j, m-1\rangle \\ &= \sqrt{[j+m][j-m+1]} \; |j, m-1\rangle \; , \end{aligned}$$

i.e., the standard form for the matrix elements of the generators of the quantum algebra $su_q(2)$.

3. QUANTIZED VERMA MODULES FOR $su_q(1,1)$

In this section we discuss the case of $su_q(1,1)$, along the lines of discussion of $su_q(2)$ in section 2. However, in distinction to ref. [1], we use for $su(1,1)$ the same commutation rules as for $su(2)$, eq. (2.3), and change the conjugation law instead. The modified conjugation law takes on the form

$$\sigma_1(J_+) = J_-, \quad \sigma_1(J_-) = J_+, \quad \sigma_1(J_0) = -J_0, \quad \eta_1 = -\sigma_1. \tag{3.1}$$

The sesquilinear form $S_1(x,y) = \xi_\Lambda(\eta_1(x), y)$ is in the $su_q(1,1)$ basis obtained as

$$\begin{aligned} &S_1(\mathbb{1}, \mathbb{1}) = 1, \quad n = 0, \\ &S_1(J_-^k, J_-^n) = \delta_{kn}(-1)^n [n]! \prod_{s=1}^{n} [2j-s+1], \quad k+n > 0. \end{aligned} \tag{3.2}$$

We consider again the case j, q real. Then the sesquilinear form S_1 becomes real valued. We discuss three cases.

<u>Case A</u>: $j > 0, \quad j \neq k/2, \quad k \in \mathbb{N}_+$.

For these values S_1 is non-degenerate, but indefinite.

Case B: $j \leq -l-1 < 0$.

For these values S_1 becomes a scalar product. It holds

$$\begin{aligned} S_1(\mathbb{1},\mathbb{1}) &= 1, \quad n=0, \\ S_1(J_-^k, J_-^n) &= \delta_{kn}[n]! \prod_{s=1}^{n}[2l+s+1] \\ &= \delta_{kn}[n]! \frac{\Gamma_q(2l+n+2)}{\Gamma_q(2l+2)} \\ &= \delta_{kn} \|J_-^n\|^q, \quad k+n>0, \end{aligned} \tag{3.3}$$

where we used $\Gamma_q(x+1) = x\Gamma_q(x)$. The basis for the infinitesimally unitary Verma module $d_j \equiv d_{-l-1}$ is given by

$$\begin{aligned} |-l-1,-l-1\rangle &= \mathbb{1}, \quad m=-l-1, \\ |-l-1,m\rangle &= \|J_-^{-l-1-m}\|^{-1} J_-^{-l-1-m} \\ &= \Big\{[-l-1-m]! \prod_{s=1}^{-l-1-m}[2l-s+1]\Big\}^{-1/2} J_-^{-l-1-m}, \\ l>0, &\quad m=-l-1,-l-2,\ldots\,. \end{aligned} \tag{3.4}$$

Substituting the new basis states $|lm\rangle$ into the $su_q(1,1)$ relations

$$\begin{aligned} \rho(J_0)X(n) &= (-l-1-n)X(n), \\ \rho(J_-)X(n) &= X(n+1), \\ \rho(J_+)X(n) &= -[n][2l+n+1]X(n-1), \\ X(n) &= J_-^n, \quad n=-l-1-m, \end{aligned} \tag{3.5}$$

one obtains the $su_q(1,1)$ representation as

$$\begin{aligned} \rho(J_0)\,|-l-1,m\rangle &= m\,|-l-1,m\rangle\,, \\ \rho(J_-)\,|-l-1,m\rangle &= \sqrt{[-l-m][l-m+1]}\,|-l-1,m-1\rangle \\ &= \sqrt{-[l][l+1]+[m][m-1]}\,|-l-1,m-1\rangle\,, \\ \rho(J_+)\,|-l-1,m\rangle &= \sqrt{[-l-m-1][l-m]}\,|-l-1,m+1\rangle \\ &= -\sqrt{-[l][l+1]+[m][m+1]}\,|-l-1,m+1\rangle \\ m &= -l-1,-l-2,-l-3,\ldots\,. \end{aligned} \tag{3.6}$$

Case C: $j>0, \quad j=\frac{1}{2}k, \quad k \in \mathbf{N}$.

For this case the sesquilinear form S_1 is degenerate and vanishes on the irreducible submodule π_j of d_j generated by the (second) extremal vector $y_0 = J_-^{2j+1}$

and spanned by $U(J_-)y_0$. However, S_1 induces a scalar product on the submodule $\widetilde{\pi}_j$ as will be shown.

Consider the Verma module $d_{j+i\epsilon}$, where $i = \sqrt{-1}$, j a non-negative integer or halfinteger and $\epsilon > 0$. The sesquilinear form can be factorized

$$S_1(J_-^k y_0, J_-^n y_0) = S_1(y_0, y_0) S_1^*(J_-^k y_0, J_-^n y_0) \tag{3.7}$$

with

$$S_1(y_0, y_0) = (-1)^{2l+1}[2l+1]! \prod_{s=1}^{2l+1} [2j - s + 1 + 2i\epsilon] \neq 0.$$

It follows that

$$\begin{aligned} S_1^*(y_0, y_0) &= 1, \quad n = 0, \\ S_1^*(J_-^k y_0, J_-^n y_0) &= \delta_{kn}(-1)^n \frac{[2j+n+1]!}{[2j+1]!} \prod_{s=2j+2}^{2j+n+1} [2j - s + 1 + 2i\epsilon] \\ &= \delta_{kn} \frac{[2j+n+1]!}{[2j+1]!} \prod_{t=1}^{n} [t - 2i\epsilon], \quad k + n > 0. \end{aligned} \tag{3.8}$$

Thus, for $\epsilon \to 0$, the sesquilinear form S_1^* becomes a scalar product on $\widetilde{\pi}_j$,

$$\begin{aligned} S_1^*(y_0, y_0) &= 1, \\ S_1^*(J_-^k y_0, J_-^n y_0) &= \delta_{kn} \frac{[2j+n+1]!}{[2j+1]!}[n]! = \|J_-^n y_0\|^2, \quad k + n > 0. \end{aligned} \tag{3.9}$$

If a new basis is defined on $\widetilde{\pi}_j$,

$$\begin{aligned} |j, -j-1\rangle &= y_0, \\ |j, m\rangle &= \left(\frac{[2j+1]!}{[-j-1+m]![j-m]!} \right)^{1/2} J_-^{-j-1-m} y_0, \\ n &= -j-1-m, \quad m = -j-1, -j-2, -j-3, \ldots \end{aligned} \tag{3.10}$$

then the infintesimally unitary representation of $su_q(1,1)$ on $\widetilde{\pi}_j$ takes on the form

$$\begin{aligned} J_0 \, |jm\rangle &= m \, |jm\rangle \, , \\ J_- \, |jm\rangle &= \sqrt{[-j-m][j-m+1]} \, |j, m-1\rangle \, , \\ J_+ \, |jm\rangle &= -\sqrt{[-j-m-1][j-m]} \, |j, m+1\rangle \, , \\ j = k/2, \quad & k \in \mathbf{N}, \quad m = -j-1, -j-2, -j-3, \ldots . \end{aligned} \tag{3.11}$$

Eq. (3.11) are however identical to Case B, eq. (3.6) with $j \to l$.

So far Verma modules with a highest weight were considered for the algebra $su_q(1,1)$ and an orthonormal basis $|jm\rangle$, $m = -j-1, -j-2, -j-3, \ldots$ was constructed for the irreducible representations D^{j-} of the negative discrete series

$(j = k/2,\ k \in \mathbf{N})$. Similarly one can construct Verma modules with lowest weight. The unitary irreducible representations of the positive discrete series representations D^{j+} $(j = k/2,\ k \in \mathbf{N})$ are given as follows. For a lowest weight vector $y_0 = |j, j+1\rangle$,

$$\begin{aligned} J_0\ |j,j+1\rangle &= (j+1)\ |j,j+1\rangle\ , \\ J_+\ |j,j\rangle &= 0, \end{aligned} \tag{3.12}$$

one obtains the orthonormal basis

$$\begin{aligned} |j,m\rangle &= \left(\frac{[2j+1]!}{[j+m]![j-m]!}\right)^{1/2} J_+^{m-j-1}\ |j,j+1\rangle\ , \\ m &= j+1, j+2, j+3, \dots\ . \end{aligned} \tag{3.13}$$

Then D^{j+} is obtained as

$$\begin{aligned} J_0\ |jm\rangle &= m(\ |jm\rangle \\ J_+\ |jm\rangle &= \sqrt{[m-j][j+m+1]}\ |j,m+1\rangle \\ J_-\ |jm\rangle &= -\sqrt{[j+m][m-j-1]}\ |j,m-1\rangle\ . \end{aligned} \tag{3.14}$$

4. TENSOR PRODUCTS AND CGC'S

The tensor product of two unitary irreducible representations belonging to the positive discrete series can be expanded into irreducible components in the same manner as for the classical $su(1,1)$ algebra. Namely

$$D^{j_1+} \otimes D^{j_2+} = \sum_{j=j_1+j_2+1}^{\infty} \oplus\, D^{j+}. \tag{4.1}$$

The CGC's involved in the expansion of the tensor product

$$|j_1 j_2; jm\rangle_q = \sum_{m_1,m_2} \langle j_1 m_1 j_2 m_2 | jm\rangle_q\ |j_1 m_1\rangle\ |j_2 m_2\rangle \tag{4.2}$$

are obtained by the projection operator technique, analogous to the case of the CGC's for the $su_q(2)$ quantum algebra [7]. Following the procedure given in ref. [7] one obtains

$$\begin{aligned} \langle j_1 m_1 j_2 m_2 | jm\rangle_q &= q^{-\frac{1}{2}(j-j_1-j_2-1)(j+j_1+j_2+2)-m_2(j_1+1)+m_1(j_2+1)} \\ &\times \left(\frac{[2j+1][j_1+m_1]![j_2+m_2]![j-j_1-j_2-1]![j+j_2-j_1]![m_2-j_2-1]!}{[j+m]![m-j-1]![m_1-j_1-1]![j+j_1-j_2]![j_1+j_2+j+1]!}\right)^{1/2} \times (-1)^{j-j_1-j_2-1} \\ &\times \sum_z (-1)^z \frac{q^{z(m_1+j_1+1)}[j+j_1+j_2+1+z]![m-j_1-j_2-2-z]!}{[z]![m_2-j_2-1-z]![j-j_1-j_2-1-z]![2j_2+1+z]!}\ , \end{aligned} \tag{4.3a}$$

with

$$\langle j_1 m_1 j_2 m_2 | jm\rangle_q = (-1)^{j-j_1-j_2-1}\ \langle j_2 m_2 j_1 m_1 | jm\rangle_{q^{-1}}. \tag{4.3b}$$

The familiar relationship between the $su(2)$ and $su(1,1)$ CGC's found by Chacon, Moshinsky and Rasmussen also holds for the corresponding quantum algebras,

$$| \langle j_1 m_1 j_2 m_2 | jm \rangle_q | = |(J_1 M_1 J_2 M_2 | JM)_q|$$

where the quantum numbers of the $su_q(2)$ CGC's are expressed in terms of the quantum numbers of the $su_q(1,1)$ CGC's as follows

$$\begin{aligned} J_1 &= \frac{1}{2}(m_1 + m_2 + j_1 - j_2 - 1), \quad M_1 = \frac{1}{2}(m_1 - m_2 + j_1 + j_2 + 1), \\ J_2 &= \frac{1}{2}(m_1 + m_2 - j_1 + j_2 - 1), \quad M_2 = \frac{1}{2}(-m_1 + m_2 + j_1 + j_2 + 1), \\ J &= j, \quad M = j_1 + j_2 + 1. \end{aligned}$$

The Racah coefficients are also obtained in the standard manner

$$|j_1 j_2 (j_{12}), j_3; jm\rangle_{qq} = \sum_{j_{23}} U(j_1 j_2 j j_3; j_{12} j_{23})_q \; |j_1, j_2 j_3 (j_{23}); jm\rangle_{qq}$$

with

$$\begin{aligned} &U(j_1 j_2 j j_3; j_{12} j_{23})_q \\ &= \left(\frac{[2j_{12}+1][2j_{23}+1][j_{12}-j_1-j_2-1]![-j_1+j_2+j_{12}]![j_1-j_2+j_{12}]![j_1+j_2+j_{12}+1]!}{[-j_1+j_{23}+j]![j_1+j_{23}+j+1]![j_{12}-j_3+j]![j_{12}+j_3+j+1]!} \right)^{1/2} \\ &\times \left([j_{23} - j_2 - j_3 - 1]![-j_2 + j_3 + j_{23}]![j_2 - j_3 + j_{23}]![j_2 + j_3 + j_{23} + 1]! \right)^{1/2} \\ &\times \left([j - j_{23} - j_1 - 1]![j_1 + j - j_{23}]![j - j_{23} - j_3 - 1]![j - j_{12} + j_3]! \right)^{1/2} \\ &\times \sum_r \frac{(-1)^{j_{12}-j_2-j+j_{23}+r}[2j - r]!}{[r]![j_1 - j_{23} + j - r]![j - j_{23} - j_1 - r - 1]![j_3 - j_{12} + j - r]!} \\ &\times \frac{1}{[j - j_{12} - j_3 - r - 1]![j_{12} + j_{23} - j_2 - j + r]![j_{12} + j_{23} + j_2 - j + r + 1]!} . \end{aligned}$$

These coefficients have the obvious symmetry property

$$U(j_1 j_2 j j_3; j_{12} j_{23})_q = U(j_3 j_2 j j_1; j_{23} j_{12})_q.$$

Moreover they are invariant with respect to the substitution $q \to q^{-1}$.

5. INDECOMPOSABLE PAIRS

In ref. [3,4] indecomposable pairs of representations $D^{(j,-j-1)}$, $j = \frac{k}{2}$, $k \in \mathbf{N}$, of the classical algebra $su(1,1)$ were discussed. In this section we will discuss the corresponding indecomposable pairs for the quantum algebra $su_q(1,1)$.

We consider an $su(1,1)$ indecomposable pair which is bounded above. The two extremal vectors A and D correspond to the weights j and $-j - 1$ respectively,

$j = k/2,\ k \in \mathbf{N}$. These two vectors belong to the "left chain", while the vector F belongs to the highest weight of the "right chain". See Fig. 1. Making use of the known properties of the $su(1,1)$ indecomposable pair we proceed to construct the indecomposable pair for the quantum algebra $su_q(1,1)$.

We assume

$$\rho(J_0)\mathbb{1} = m'\mathbb{1}, \tag{5.1}$$

that is, the representation space is Ω/I_3, where Ω denotes the universal enveloping algebra of $su(1,1)$ and I_3 the left ideal generated by the relation (5.1). Then

$$A = \sum_{r=0}^{\infty} c_r J_-^r J_+^{r+j} \tag{5.2}$$

with

$$\rho(J_0)A = jA, \tag{5.3a}$$

$$\rho(J_+)A = 0, \tag{5.3b}$$

i.e., A is to be an extremal vector. From eq. (5.3b) follows

$$\begin{aligned} c_{r-1} + [r][2j+r+1]c_r &= 0, \\ c_r = (-1)^r \frac{[2j+1]!}{[r]![2j+r+1]!} \end{aligned} \tag{5.4}$$

It is clear that the general vector of the left chain has the form

$$|j,m\rangle = J_-^n A, \quad n = j - m. \tag{5.5}$$

It follows

$$\begin{aligned} \rho(J_0)\ |jm\rangle &= m\ |jm\rangle \\ \rho(J_-)\ |jm\rangle &= |j, m-1\rangle \\ \rho(J_+)\ |jm\rangle &= [n][2j-n+1]\ |j, m+1\rangle \\ &= [j-m][j+m+1]\ |j, m+1\rangle\ . \end{aligned} \tag{5.6}$$

In particular

$$B = J_-^{2j} A = \sum_{r=0}^{\infty} c_r J_-^{r+2j} J_+^{r+j} \tag{5.7}$$

and

$$D = J_-^{2j+1}A, \quad \rho(J_0)D = (-j-1)D, \quad \rho(J_+)D = 0, \tag{5.8}$$

i.e., D is a second extremal vector. The vector F needs to be of the form

$$F = \sum_{k=0}^{\infty} \tau_k J_-^k J_+^{k+j}, \tag{5.9}$$

with the conditions

$$\begin{aligned} \rho(J_0)F &= (-j-1)F, \\ \rho(J_+)F &= gB. \end{aligned} \tag{5.10}$$

From these conditions one obtains

$$\tau_{k-1} - [k][2j-k+1]\tau_k = 0, \quad \text{if} \quad k \leq 2j, \tag{5.11a}$$

$$\tau_{k-1} - [k][2j-k+1]\tau_k = gc_r, \quad r = k-2j-1, \quad r > 0, \tag{5.11b}$$

while for $k = 2j+1$ the coefficient is indefinite and we set

$$\tau_{2j+1} = 0. \tag{5.11c}$$

Solving eq. (5.11a), with the initial condition $\tau_0 = 1$, one obtains

$$\tau_k = \frac{[2j-k]!}{[k]![2j]!}, \quad k = 0,1,2,\ldots,2j. \tag{5.12}$$

For $k = 2j+1$ eq. (5.11b) yields

$$\tau_{2j} = gc_0,$$

i.e.

$$g = \frac{1}{[2j]![2j]!}. \tag{5.13}$$

For $k = 2j+2$ one obtains

$$[2j+2]\tau_{2j+2} = -g\frac{1}{[2j+2]}$$

i.e.

$$\tau_{2j+2} = -g\frac{1}{[2j+2][2j+2]}.$$

For $k = 2j+3$ one obtains from eq. (5.11b)

$$\tau_{2j+3} = \frac{g}{[2][2j+2][2j+3]}\left\{\frac{1}{[2j+2]} + \frac{1}{[2][2j+3]}\right\}.$$

Continuing in this manner one obtains

$$\tau_{2j+1+r} = gc_r\left(\sum_{s=1}^{r}\frac{1}{[s][2j+1+s]}\right), \quad r \geq 1. \tag{5.14}$$

In the classical limit $q \to 1$ the equations obtained above for the quantum algebra take on the form

$$\begin{aligned} \tau_k &= \frac{(2j-k)!}{k!(2j)!}, \quad k = 0,1,\ldots,2j \\ \tau_{2j+1} &= 0 \\ \tau_{2j+r} &= gc_r\left(\sum_{s=1}^{r}\frac{1}{s(2j+s+1)}\right) \end{aligned} \tag{5.15}$$

with

$$g = \frac{1}{(2j)!(2j)!}, \quad c_r = (-1)^r\frac{(2j+1)!}{r!(2j+r+1)!}.$$

In terms of ψ-functions one has in the classical case

$$\sum_{s=1}^{r} \frac{1}{s(2j+s+1)} = \frac{1}{2j+1}(\psi(r) - \psi(1) - \psi(2j+r+1) + \psi(2j+2)). \tag{5.16}$$

It would be of interest to connect the corresponding sum for the quantum case with the q-analogue of the ψ-function.

The vectors of the right chain of Fig. 1 are of the form

$$|-j-1, m\rangle = J_-^{-j-1-m} F, \quad F \equiv |-j-1, -j-1\rangle \ .$$

One obtains

$$\begin{aligned}
\rho(J_0)\,|-j-1,m\rangle &= m\,|-j-1,m\rangle \ , \\
\rho(J_-)\,|-j-1,m\rangle &= |-j-1,m-1\rangle \ , \\
\rho(J_+)\,|-j-1,m\rangle &= g\,|j,m+1\rangle + [-j-1-m][-j+m]\,|-j-1,m+1\rangle \ .
\end{aligned} \tag{5.17}$$

Therefore the structure of the indecomposable pair $D^{(j,-j-1)}$, and the explicit form for this representation, have been obtained for the general case of arbitrary $j = k/2$, $k \in \mathbf{N}$.

In concluding we want to state the result for the tensor product of two indecomposable pairs, $j_1 \geq j_2$,

$$\begin{aligned}
D^{j_1,-j_1-1} &\otimes D^{j_2,-j_2-1} \\
&= \sum_{j=j_1-j_2}^{j_1+j_2} D^{j,-j-1} + 2 \sum_{j=\frac{1}{2} \text{ or } 0}^{j_1-j_2-1} D^{j,-j-1} \\
&+ 4 \sum_{-j-1=-j_1-j_2-2,-j_1-j_2-3,\ldots} D^{-j-1} \\
&+ 2 \sum_{-j-1=-j_1-j_2-1}^{-j_1+j_2-1} D^{-j-1},
\end{aligned}$$

where

$D^{j,-j-1}$, $j = k/2$, $k \in \mathbf{N}$, is an indecomposable pair and
D^{-j-1}, $j = k/2$, $k \in \mathbf{N}$, is an irreducible representation with highest weight $(-j-1)$.

The (non-orthonormalized) CGC's can be calculated for these tensor products in the same manner as it is done for the UIR's of $su_q(2)$ in ref. [8].

REFERENCES

1. B. Gruber, R. Lenczewski and M. Lorente, On induced scalar products and unitarization, J. Math. Phys. no. 3 31:587 (1990).

2. H.D. Doebner, B. Gruber and M. Lorente, Boson operator realizations of $su(2)$ and $su(1,1)$ and unitarization, J. Math. Phys. no. 3 30:594 (1989).
3. B. Gruber and A.U. Klimyk, Multiplicity free and finite multiplicity indecomposable representations of the algebra $su(1,1)$, J. Math. Phys. no. 10 19:2009 (1978).
4. B. Gruber and A.U. Klimyk, Matrix elements for indecomposable representations of complex $su(2)$, J. Math. Phys. no. 4 25:755 (1984).
5. D.B. Fairlie, q-Analysis and quantum groups, in "Symmetries in Science V", this volume.
6. A.U. Klimyk, Yu.F. Smirnov and B. Gruber, Representations of the quantum algebras $U_q(su(2))$ and $U_q(su(1,1))$, in "Symmetries in Science V", this volume.
7. Yu. F. Smirnov, V.N. Tolstoy and Yu.I. Kharitonov, Projection Operator Method and Q-Analog of Angular Momentum Theory, this volume.

THE ROLE OF SPECTRUM GENERATING ALGEBRAS AND DYNAMIC SYMMETRIES IN MOLECULAR PHYSICS

F. Iachello

Center for Theoretical Physics, Yale University
New Haven, Connecticut 06511

ABSTRACT

The use of spectrum generating algebras and dynamic symmetries in molecular physics is reviewed. The special role played by the algebra of SO(4) both in rotation-vibration and electronic spectra is emphasized. Examples are presented.

1. INTRODUCTION

Algebraic theory is a general framework that can be used to attack any problem in physics and chemistry. It makes use of an algebraic structure. Most applications so far have made use of Lie algebras, but one can, in principle use (and in some cases has used) other structures such as graded Lie algebras, infinite dimensional algebras, q-algebras, etc.

The key ingredients of algebraic theory are the notions of:

(a) Spectrum generating algebra (SGA).

This is the algebra, $\mathcal{G}$, onto which all operators of interests (in particular the Hamiltonian, H) are expanded,

$$H = f(G_\alpha) \quad , \quad G_\alpha \in \mathcal{G} \quad , \tag{1.1}$$

i.e. all operators are in the enveloping algebra of $\mathcal{G}$. Here G_α denotes generically an element of $\mathcal{G}$.

(b) Dynamic Symmetry (DS).

This is a situation in which H contains only invariant (Casimir) operators of $\mathcal{G}$ and of its subalgebras in the chain $\mathcal{G} \supset \mathcal{G}' \supset \mathcal{G}'' \supset \ldots$

Symmetries in Science V, Edited by B. Gruber *et al.*
Plenum Press, New York, 1991

$$H = f(C_i) \quad , \qquad (1.2)$$

where C_i denotes generically a Casimir operator. Since all Casimir operators are in the enveloping algebra of $\mathcal{G}$, (b) is a special case of (a).

Algebraic theory has been used in the solution of a variety of problems in physics and chemistry, at all scales: (i) Molecules, (ii) Atoms, (iii) Nuclei, (iv) Hadrons, (v) Quarks, ... In this article, I will briefly review applications of algebraic theory to molecular physics. Molecules have three types of spectra: (i) Rotational, (ii) Vibrational and (iii) Electronic. In view of the fact that rotational and vibrational spectra are intimately connected, they can be treated by making use of a single spectrum generating algebra. This will be discussed in Sects. 2 and 3. The spectrum generating algebra of electronic spectra will instead be discussed in Sect. 4.

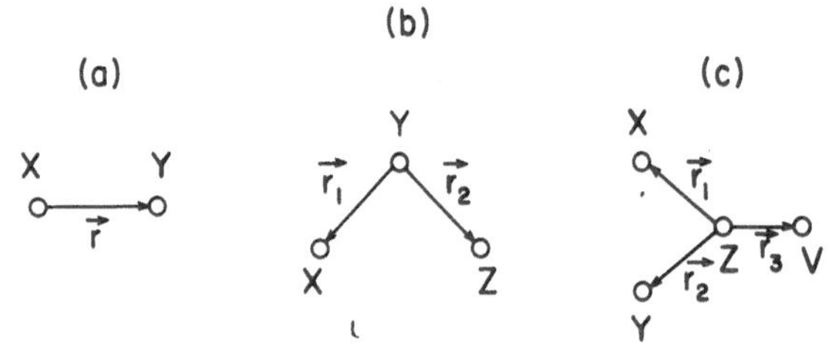

Fig.1. Schematic representation of: (a) diatomic molecules; (b) triatomic molecules; (c) tetratomic molecules.

2. THE SPECTRUM GENERATING ALGEBRA, $\mathcal{G}$, OF MOLECULAR ROTATION-VIBRATION SPECTRA

2.1. Diatomic molecules

We begin by considering the case of a diatomic molecule composed of two atoms X and Y separated by a distance $\vec{r}$, Fig. 1a. The degrees of freedom here are the three components of the vector $\vec{r} = (r,\theta,\phi)$. It has been suggested that the spectrum generating algebra appropriate to non-relativistic problems with n degrees of freedom is U(n+1). In this case, since n=3, one has[1,2] $\mathcal{G} = U(4)$. In addition to the algebra $\mathcal{G}$ one must specify the representation within which it acts. This is the totally symmetric representation[1,2] $[N,\dot{0}] = [N,0,0,0]$. The physical meaning of the quantum number N will become apparent at the end of the section.

An explicit realization of the algebra $\mathcal{G}$ can be obtained in terms of boson operators (called vibrons)

$$b_\alpha^\dagger \; ; \; b_\alpha \quad ; \qquad \alpha = 1,\ldots,4 \quad . \tag{2.1}$$

The generators of U(4) are

$$\mathcal{G}: \; G_{\alpha\beta} = b_\alpha^\dagger b_\beta \quad , \qquad \alpha,\beta = 1,\ldots,4 \quad , \tag{2.2}$$

with basis

$$\mathcal{B}: \; b_\alpha^\dagger b_\beta^\dagger \ldots |0\rangle \quad , \tag{2.3}$$

where $|0\rangle$ denotes the boson vacuum.

Algebraic theory is an expansion of all molecular rotation-vibration operators in terms of elements of $\mathcal{G} \equiv U(4)$. For example, the Hamiltonian operator has the form

$$H = E_0 + \sum_{\alpha\beta} \epsilon_{\alpha\beta} G_{\alpha\beta} + \frac{1}{2} \sum_{\alpha\beta\gamma\delta} u_{\alpha\beta\gamma\delta} G_{\alpha\beta} G_{\gamma\delta} + \ldots \quad . \tag{2.4}$$

Since H is built in terms of vibron operators, algebraic theory for molecules is often referred to as "the vibron model". Having specified $\mathcal{G}$ one can now study all the possible dynamic symmetries of the model by studying all the possible branchings of $\mathcal{G}$ into subalgebras. One must remember, however, that one wants to consider states with good angular momentum and thus that the physical rotation algebra, SO(3), must be contained in the chain of subalgebras of $\mathcal{G}$. A convenient way to study this problem is to divide the boson operators into operators with definite transformation properties under rotations[1,2], i.e. into a scalar, σ, and a vector π_μ (μ=0,±1),

$$b_\alpha^\dagger \equiv (\sigma^\dagger, \pi_\mu^\dagger) \quad . \tag{2.5}$$

It then becomes clear that there are two possible branchings

$$U(4) \begin{cases} U(3) \supset SO(3) \supset SO(2) & , \quad (I) \\ SO(4) \supset SO(3) \supset SO(2) & . \quad (II) \end{cases} \qquad . \quad (2.6)$$

For each of the two branchings one can write down a classification scheme. For the first branch it is

$$(I): \left| \begin{array}{ccccccc} U(4) & \supset & U(3) & \supset & SO(3) & \supset & SO(2 \\ \downarrow & & \downarrow & & \downarrow & & \downarrow \\ [N,0,0,0] & & [n_\pi,0,0] & & L & & M_L \end{array} \right\rangle \quad , \quad (2.7)$$

where n_π=N, N-1, ..., 0, L=n_π, n_π-2, ..., 1 or 0 (n_π=odd or even) and $-L \leq M_L \leq +L$. For the second branch it is

$$(II): \left| \begin{array}{ccccccc} U(4) & \supset & SO(4) & \supset & SO(3) & \supset & SO(2) \\ \downarrow & & \downarrow & & \downarrow & & \downarrow \\ [N,0,0,0] & & (\omega,0) & & L & & M_L \end{array} \right\rangle \quad , \quad (2.8)$$

where ω=N, N-2, 1 or 0 (N=odd or even), L=ω, ω-1, ..., 0 and $-L \leq M_L \leq +L$. The quantum numbers L and M_L represent the angular momentum and its z-component.

Dynamic symmetries are obtained by writing the Hamiltonian in terms of invariant operators of the algebras in (2.6). If H contains only terms up to quadratic in $G_{\alpha\beta}$, one has

$$\begin{cases} H^{(I)} = E_0 + \alpha\, C_1(U3) + \beta\, C_2(U3) + \gamma\, C_2(SO3) & , \quad (I) \\ H^{(II)} = E_0 + A\, C_2(SO4) + B\, C_2(SO3) & , \quad (II) \end{cases} \qquad (2.9)$$

where $C_i(\mathcal{G})$ denotes a Casimir operator of order i of the algebra $\mathcal{G}$. Casimir operators of SO(2) have not been introduced in (2.9) since these will appear only if the molecule is placed in an external field. The eigenvalues of (2.9) in the representations (2.7) and (2.8) are given by

$$E^{(I)}(N,n_\pi,L,M_L) = E_0 + \alpha\, n_\pi + \beta\, n_\pi(n_\pi+1) + \gamma\, L(L+1) \quad , \quad (I)$$

$$(2.10)$$

$$E^{(II)}(N,\omega,L,M_L) = E_0 + A\,\omega(\omega+2) + B\,L(L+1) \quad . \quad (II)$$

The occurrence of explicit formulas for the eigenvalues of H in terms of the quantum numbers characterizing the states are the important feature of dynamic symmetries. They allow a simple analysis of the experimental situation. It turns out that the symmetry (I) describes soft molecules, while symmetry (II) describes rigid molecules[2]. Since most molecules observed experimentally are of the rigid type, the dynamic symmetry (II), going through the SO(4) chain, plays a very important role in molecular physics.

The reason why SO(4) is the appropriate dynamic symmetry of diatomic molecules can also be understood in a different way by returning to the treatment of rotation-vibration spectra of molecules in terms of the Schrödinger equation with a potential V(r). The energy eigenvalues are obtained by solving the equation

$$\left[-\frac{\hbar^2}{2\mu}\nabla^2 + V(r)\right]\psi(\vec{r}) = E\psi(\vec{r}) \quad , \qquad (2.11)$$

with appropriate boundary conditions. The interatomic potential V(r) has the shape of Fig. 2, and is often written as

$$V(r) = V_0\left[e^{-2\alpha(r-r_0)} - 2e^{-\alpha(r-r_0)}\right] \quad , \qquad (2.12)$$

called the Morse potential[3]. The solution of (2.11) is of the form

$$E(v,L) = E_0 - 4A\left[(N+1)v - v^2\right] + B\,L(L+1) + $$
$$+ \text{ small corrections} \quad , \qquad (2.13)$$

with $v = 0, 1, \ldots, \frac{N}{2}$ or $\frac{N-1}{2}$ (N=even or odd). One can see that by replacing the quantum number ω in (2.10),II by

$$\omega = N-2v \quad , \qquad (2.14)$$

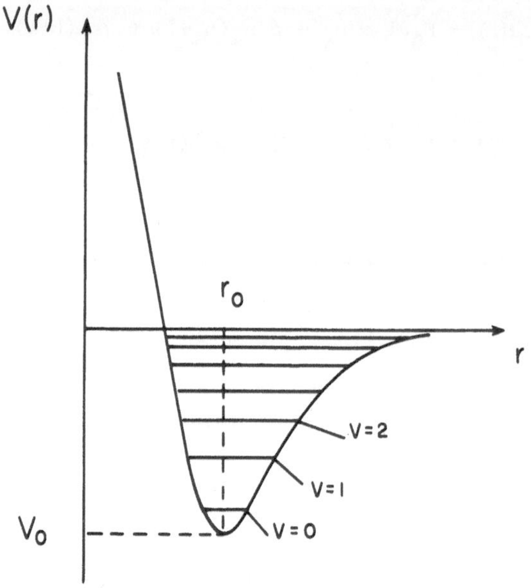

Fig.2. Typical behavior of the interatomic potential V(r) as a function of r.

one obtains precisely Eq. (2.13), apart from small corrections. Thus, the 3-dimensional Morse potential has an (almost) exact dynamic symmetry

$$U(4) \supset SO(4) \supset SO(3) \supset SO(2) \quad , \qquad (2.15)$$

a result of crucial importance for molecular physics. The number N (called vibron number) characterizing the representations of U(4) is related to the number of bound vibrational states.

Experimental data appear to indicate that the Morse potential (and hence SO(4) symmetry) is a good approximation to the observed situation, as shown in Fig. 3 for the electronic ground state of the H_2 molecule. A more accurate description can be obtained by slightly breaking the SO(4) symmetry and/or adding higher order terms in the $G_{\alpha\beta}$'s.

2.2 Triatomic molecules

Algebraic theory can be generalized to describe molecules with more than two atoms. Consider, for example, triatomic molecules composed of three atoms X, Y and Z, Fig. 1b. The degrees of freedom here are the components of the vectors $\vec{r}_1 \equiv (r_1, \theta_1, \phi_1)$ and $\vec{r}_2 \equiv (r_2, \theta_2, \phi_2)$. To each of

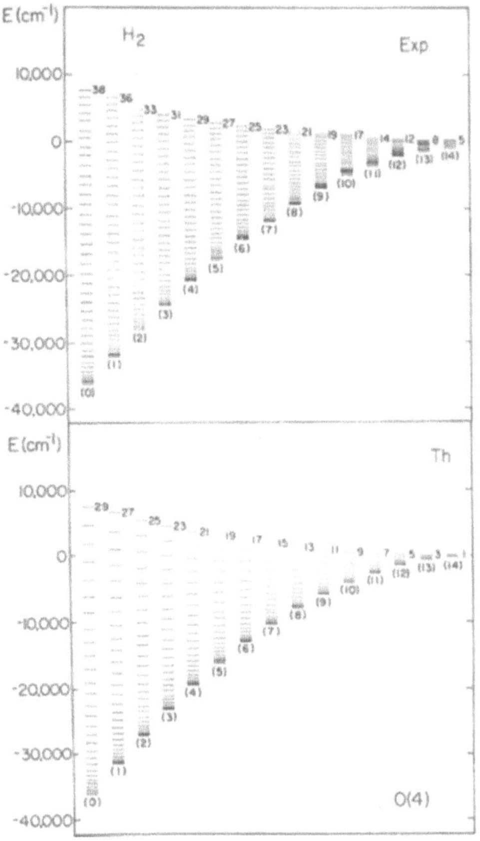

Fig.3. An example of dynamic symmetry in molecules: the rotation-vibration spectrum of the H_2 molecule in its ground electronic state.

these one can associate a SGA. The total spectrum generating algebra is thus[4,5] $\mathcal{G} \equiv U_1(4) \oplus U_2(4)$, where the indices refer to the bonds in Fig. 1b. The algebra acts on the direct product of two totally symmetric representations $[N_1,\dot{0}] \otimes [N_1,\dot{0}]$, with vibron number N_1 and N_2 respectively.

An explicit realization of $\mathcal{G}$ can be obtained again in terms of boson operators

$$b^{\dagger}_{\alpha,i} \;;\; b_{\alpha,i} \quad ; \quad \alpha=1,\ldots,4 \quad ; \quad i=1,2 \tag{2.16}$$

where α is the vibron index and i denotes the bond. Boson operators with different i commute. The generators of $U_1(4)$ and $U_2(4)$ are

$$G_{1\alpha\beta} = b^{\dagger}_{\alpha 1}\, b_{\beta 1} \quad , \quad G_{2\alpha\beta} = b^{\dagger}_{\alpha 2}\, b_{\beta 2} \quad . \tag{2.17}$$

The Hamiltonian operator is now expanded into elements of $\mathcal{G}$. It is of the form

$$H = H_1 + H_2 + V_{12} \quad , \tag{2.18}$$

where H_1 and H_2 describe bond 1 and 2 and are of the type (2.4), while V_{12} represents the bond-bond interaction and can be expanded into number conserving products of operators of 1 and 2,

$$V_{12} = \sum_{\alpha\beta\gamma\delta} w_{\alpha\beta\gamma\delta}\, G_{1\alpha\beta}\, G_{2\alpha\beta} \tag{2.19}$$

There are several possible dynamic symmetries here, two of which are of particular interest, corresponding to the chains

$$U_1(4) \otimes U_2(4) \begin{cases} U_1(3) \otimes U_2(3) \supset U_{12}(3) \supset SO_{12}(3) \supset SO_{12}(2) \quad , & \text{(I)} \\ SO_1(4) \otimes SO_2(4) \supset SO_{12}(4) \supset SO_{12}(3) \supset SO_{12}(2) \quad . & \text{(II)} \end{cases} \tag{2.20}$$

Here $SO_{12}(4)$ denotes the algebra obtained by summing the corresponding operators of $SO_1(4)$ and $SO_2(4)$. It has become customary to use the multiplication sign,$\otimes$, rather than the summation sign, $\oplus$, since the chain is usually written when discussing states (representations) rather than algebras. For rigid triatomic molecules the important chain is II, with basis states

$$\left| \begin{matrix} U_1(4) & \otimes & U_2(4) & \supset & SO_1(4) & \otimes & SO_2(4) & \supset & SO_{12}(4) & \supset & SO_{12}(3) & \supset & SO_{12}(2) \\ \downarrow & & \downarrow & & \downarrow & & \downarrow & & \downarrow & & \downarrow & & \downarrow \\ [N_1,\dot{0}] & & [N_2,\dot{0}] & & (\omega_1,0) & & (\omega_2,0) & & (\tau_1,\tau_2) & & L & & M_L \end{matrix} \right\rangle . \tag{2.21}$$

By writing the Hamiltonian in terms of Casimir operators of the algebras in (2.21) one obtains the eigenvalue expression[4,5]

$$E(N_1,N_2,\omega_1,\omega_2,\tau_1,\tau_2,L,M_L) = E_0 + A_1\omega_1(\omega_1+2) + A_2\omega_2(\omega_2+2) +$$

$$+ A_{12}\left[\tau_1(\tau_1+2)+\tau_2^2\right] + B\ L(L+1) \quad . \tag{2.22}$$

The dynamic symmetry chain (2.21) provides a basis in which more elaborate calculations can be performed. In order to do these calculations one needs the Wigner-Racah calculus of SO(4). This is aided by the isomorphism SO(4) ≈ SO(3) ⊗ SO(3). For example, the Kronecker products

$$\mathcal{D}_1 \otimes \mathcal{D}_2 = \sum_3 \oplus \mathcal{D}_3 \quad , \quad \text{i.e.}$$

$$(\eta_1,\eta_2) \otimes (\tau_1,\tau_2) = \sum_{\zeta_1\zeta_2} \oplus (\zeta_1,\zeta_2) \quad , \tag{2.23}$$

can be easily evaluated.

The Wigner-Racah calculus allows one to compute matrix elements of operators that are not diagonal in the basis (2.21). A diagonalization of the corresponding secular matrix produces the eigenvalues that can be compared with experiment. An example[7] is shown in Table I, where it is seen that algebraic theory can provided a description of the observed energies with occurances of the order 10^{-4}.

2.3. Polyatomic molecules

The vibron model can be further generalized to molecules with any number of atoms, Fig. 1c. The SGA is here the direct sum of U(4) bond algebras

$$\text{SGA:} \quad U_1(4) \oplus U_2(4) \oplus \ldots \oplus U_\nu(4) \quad , \tag{2.24}$$

and acts upon the product representation

$$[N_1,\dot{0}] \otimes [N_2,\dot{0}] \otimes \ldots \otimes [N_\nu,0] \quad . \tag{2.25}$$

Table I. Comparison[7] between experimental and calculated vibrational levels (v_1,v_2,v_3) of H_2S. All energies in cm^{-1}.

$v_1v_2v_3$	Exp	Calc	Calc-Exp
010	1182.6	1184.1	1.5
100	2614.4	2615.4	1.0
001	2628.5	2626.4	-2.1
020	2354.0	2353.6	-0.4
110	3779.2	3778.0	-1.2
011	3789.3	3790.9	1.6
200	5145.1	5145.3	0.2
101	5147.4	5146.9	-0.5
021	4939.2	4939.3	0.1
210	6288.2	6287.4	-0.8
111	6289.2	6289.5	0.3
012	6388.7	6386.2	-2.5
300	7576.3	7576.3	0.0
201	7576.3	7576.4	0.1
102	7751.9	7753.0	1.1
003	7779.2	7780.8	1.6
211	8697.3	8697.6	0.3
301	9911.1	9909.9	-1.2
103	10194.5	10193.8	-0.7
311	11008.8	11009.5	0.7

The basis states for rigid molecules are constructed by successive couplings of SO(4) representation. For example for the tetratomic molecule of Fig. 1c, one has

$$\left| \begin{array}{l} \begin{array}{cccccccccccc} U_1(4) & \otimes & U_2(4) & \otimes & U_3(4) & \supset & SO_1(4) & \otimes & SO_2(4) & \otimes & SO_3(4) & \supset \\ \downarrow & & \downarrow & & \downarrow & & \downarrow & & \downarrow & & \downarrow & \\ N_1 & & N_2 & & N_3 & & (\omega_1,0) & & (\omega_2,0) & & (\omega_3,0) & \end{array} \\ \\ \begin{array}{cccccccccc} \supset SO_{12}(4) & \otimes & SO_3(4) & \supset & SO_{123}(4) & \supset & SO_{123}(3) & \supset & SO_{123}(2) \\ \downarrow & & & & \downarrow & & \downarrow & & \downarrow \\ (\tau_1,\tau_2) & & & & (\eta_1,\eta_2) & & L & & M_L \end{array} \end{array} \right\rangle , \quad (2.26)$$

where bonds 1 and 2 are first coupled to 12 and bond 3 is successively coupled to 123. The Hamiltonian operator is now of the form

$$H = \sum_{i=1}^{\nu} H_i + \sum_{i<j}^{\nu} V_{ij} \quad , \tag{2.27}$$

with

$$V_{ij} = \sum_{\alpha\beta\gamma\delta} w^{ij}_{\alpha\beta\gamma\delta} G_{i\alpha\beta} G_{j\gamma\delta} \quad . \tag{2.28}$$

The summations in (2.27) go up to ν, the total number of bonds. This number is the number of atoms in the molecule minus one.

The method can in principle be applied to molecules with any number of atoms, although it has been applied, up to now, to molecules with at most four atoms[8]. The structure of complex molecules, such as the benzene molecule of Fig. 4, is one of the most interesting problems in present day molecular spectroscopy and one for which algebraic methods can be particularly useful.

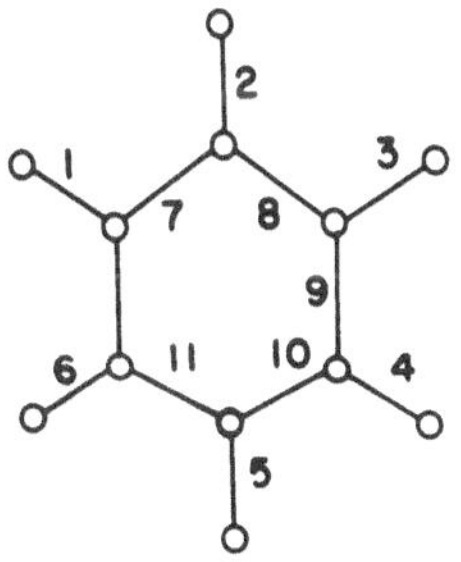

Fig.4. The benzene molecule.

2.4. Infinite chains: polymers

Polymers can be considered as the continuum limit of long linear molecules, Fig. 5. They can thus be described by a spectrum generating algebra which is the infinite sum of (U(4) algebras

$$\text{SGA:} \quad \sum_{i=1}^{\infty} \oplus U_i(4) \supset \sum_{i=1}^{\infty} \oplus SO_i(4) \quad , \tag{2.29}$$

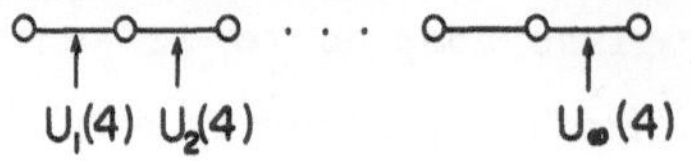

Fig.5. Schematic representation of a polymer.

with representations which are the infinite product

$$\prod_{i=1}^{\infty} \otimes \left[N_i, \dot{0}\right] \qquad . \qquad (2.30)$$

Implications of (2.29) on specific properties of polymers have not been worked out yet, nor has the relationship of this scheme with infinite dimensional Kac-Moody algebras been elucidated. It remains one of the interesting areas for further development of algebraic theory.

3. MATRIX ELEMENTS OF OPERATORS

In addition to the energy spectrum, one is often interested in the evaluation of matrix elements of operators between an initial and a final state. Algebraic theory provides a way to compute these matrix elements. There are two types of matrix elements of particular interest in molecules, those of electromagnetic transition operators and those of scattering operators.

(i) Electromagnetic transition operators.

In algebraic theory, one expands these operators also into the elements of the algebra $\mathcal{G}$

$$\hat{T} = t(G_\alpha) \qquad , \qquad G_\alpha \in \mathcal{G} \qquad . \qquad (3.1)$$

The operators $\hat{T}$ are thus in the enveloping algebra of $\mathcal{G}$, as the Hamiltonian H. Here G_α denotes generically an element of $\mathcal{G}$. Since the states are representations of $\mathcal{G}$ one then needs to compute matrix elements of the type

$$< \text{Irrep of } \mathcal{G} \mid \hat{T} \mid \text{Irrep of } \mathcal{G} > \qquad . \qquad (3.2)$$

To be more specific, in the case of diatomic molecules, $\mathcal{G} \equiv U(4)$ with generators given by (2.2). The electromagnetic transition operators are taken to be of the form

$$\hat{T} = t_0 + \sum_{\alpha\beta} t_{\alpha\beta} G_{\alpha\beta} + \ldots \quad . \tag{3.3}$$

When a dynamic symmetry (DS) exists, the states are not only representations of $\mathcal{G}$ but can be classified according to one of the subalgebra chains, for example (2.8). In this case it is convenient to write $\hat{T}$ as a sum of irreducible tensors with respect to the subalgebra chain (2.6) II. The matrix element

$$< N, \omega', L', M'_L \mid \hat{T} \mid N, \omega, L, M_L > \quad , \tag{3.4}$$

is then given in terms of isocalar factors for the chain[3]. Explicit expressions for some of the matrix elements of interest have been worked out[2,4]. When the dynamic symmetry is broken, states can be written as linear combinatons of those in (2.8),

$$\mid N, i, L, M_L > = \sum_{\omega} c^i_{\omega} \mid N, \omega, L, M_L > \quad , \tag{3.5}$$

where the coefficients c^i_{ω} are obtained from a diagonalization of the Hamiltonian, H. The final expression for the matrix elements is then of the form

$$T_{i\to i'} = < N, i', L', M'_L \mid \hat{T} \mid N, i, L, M_L > =$$

$$= \sum_{\omega,\omega'} c^i_{\omega} c^{i'}_{\omega'} < N, \omega', L', M'_L \mid \hat{T} \mid N, \omega, L, M_L > \quad . \tag{3.6}$$

(ii) Scattering operators

These operators appear in the calculation of scattering cross sections. These operators usually involve the exponentiation of the operators $\hat{T}$,

$$\hat{S} = e^{iq\hat{T}} \quad , \tag{3.7}$$

where q is a parameter determined from the scattering conditions. The evaluations of the matrix elements of $\hat{S}$ is obviously more difficult than that of the matrix elements of $\hat{T}$. If $\hat{T}$ is of the form (3.3) i.e. linear in the elements of $\mathcal{G}$, the matrix elements of $\hat{S}$ are nothing but the group elements of $\mathcal{G}$. For example, the matrix elements

$$< N, \omega', L', M'_L \mid \hat{S} \mid N, \omega, L, M_L > \quad , \tag{3.8}$$

are the group elements of U(4) when $\hat{T}$ is

$$\hat{T} = \sum_{\alpha\beta} t_{\alpha\beta} G_{\alpha\beta} \quad , \tag{3.9}$$

i.e. they are the Wigner matrices

$$\mathcal{D}^{[N]}_{\omega',L',M'_L;\ \omega,L,M_L}(t_{\alpha\beta}) = \tag{3.10}$$

$$= < N, \omega', L', M'_L \mid e^{i \sum_{\alpha\beta} t_{\alpha\beta} G_{\alpha\beta}} \mid N, \omega, L, M_L > \quad .$$

These matrices have been evaluated for some cases of interest[10].
If $\hat{T}$ itself is not in $\mathcal{G}$, as in (3.9), but in its enveloping algebra, the evaluation of the matrix elements of $\hat{S}$ is even more difficult and only very few results have been obtained so far.

3.1. Polyatomic molecules

The discussion following Eq. (3.2) has been confined to the case of diatomic molecules, for which $\mathcal{G} \equiv U(4)$ and $\hat{T}$ is of the form (3.3). In general, for polyatomic molecules with ν bonds, $\mathcal{G}$ is as in Eq. (2.25) and the electromagnetic transition operators are of the form

$$\hat{T} = \sum_{i=1}^{\nu} \hat{T}_i + \sum_{i<j}^{\nu} \hat{T}_{ij} \quad , \tag{3.11}$$

where the $\hat{T}_i$ are as in (3.3), while the $\hat{T}_{ij}$ are of the type

$$\hat{T}_{ij} = \sum_{\alpha\beta\gamma\delta} t^{ij}_{\alpha\beta\gamma\delta} \, G_{i\alpha\beta} \, G_{j\gamma\delta} \quad . \tag{3.12}$$

Use of algebraic theory allows one to compute the matrix elements of these rather complex operators.

4. SPECTRUM GENERATING ALGEBRAS OF MOLECULAR ELECTRONIC SPECTRA

In addition to vibrational and rotational spectra, molecules display also electronic excitations. These appear at much larger energies but, nonetheless, can and in some cases have been measured. Algebraic methods can be used to describe also these excitations. Here one has, in addition to the bond degrees of freedom, also the degrees of freedom of electrons, Fig. 6. A possible treatment of these degrees of freedom is in terms of

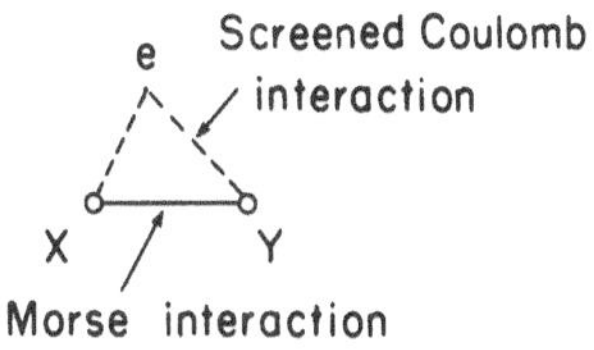

Fig. 6. Schematic representation of a diatomic molecule including electronic degrees of freedom.

the unitary algebra approach[13,14], developed for atoms, and adapted to the present case. This approach, which at the moment has been developed only for diatomic molecules in the united atom limit, starts from the spectrum of the screened Coulomb interactions, shown in Fig. 7. Electrons are placed in the single particle orbitals of Fig. 7 satisfying the Pauli principle. Since in the single particle spectrum there are gaps, one can approximate the treatment of the electronic degrees of freedom by considering only the valence shells. Shells with a given principal quantum number, n span a $2n^2$ dimensional space, the factor of 2 coming from the spin of the electron. This space forms the basis for a representation of $U(2n^2)$. the representations of $U(2n^2)$ are classified according to the chain

$$U(2n^2) \supset U(n^2) \otimes SU_S(2) \supset \ldots \supset SO(4) \otimes SU_S(2) \supset$$

$$\supset SO_L(3) \otimes SU_S(2) \supset Spin_J(3) \supset Spin_J(2) \quad , \qquad (4.1)$$

where $SU_S(2)$ denotes the spin algebra, $SO_L(3)$ the orbital angular momentum algebra and $Spin_J(3)$ the algebra obtained by combining the orbital, $\vec{L}$, and spin, $\vec{S}$, angular momenta into $\vec{J} = \vec{L} + \vec{S}$. The dots in (4.1) represent further algebras needed to classify the states for $n > 2$.

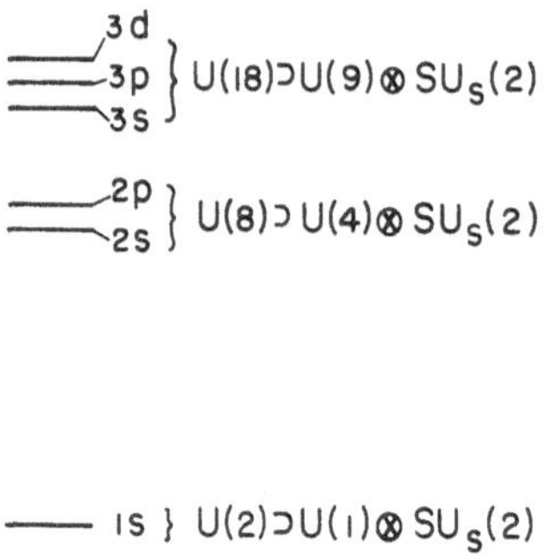

Fig.7. The single particle spectrum of a screened Coulomb interaction.

All operators describing electrons can be constructed in terms of creation and annihilation operators for fermions

$$a_i^\dagger \, , \; a_i \quad ; \qquad i = 1,\ldots,2n^2 \quad . \qquad (4.2)$$

These operators satisfy anticommutation relations. The elements of the Lie algebra $U(2n^2)$ are now

$$G_{ij} = a_i^\dagger a_j \quad ; \qquad i, j = 1,\ldots,2n^2 \quad . \qquad (4.3)$$

The representations of $U(2n^2)$ that one is interested in are the totally antisymmetric representations, characterized by the Young tableau

$$\left.\begin{array}{c}\square\\ \square\\ \cdot\\ \square\end{array}\right\} N_F \equiv \underbrace{[1,1,\ldots,1]}_{N_F} \quad . \tag{4.4}$$

When considering simultaneously rotation-vibration and electronic excitations one is led to introduce a combined Hamiltonian

$$H = H^{(B)} + H^{(F)} + V^{(BF)} \quad , \tag{4.5}$$

where $H^{(B)}$ describes the rotation-vibration part in terms of boson (B) operators, $H^{(F)}$ describes the electronic part in terms of fermion (F) operators and $V^{(BF)}$ their interaction. Use of the Hamiltonian (4.5) is often referred to as "vibron-electron" model. Its spectrum generating algebra is

$$\text{SGA:} \quad U^{(B)}(4) \oplus U^{(F)}(2n^2) \quad , \tag{4.6}$$

where the superscript (B) and (F) refers to bosons and fermions respectively. It is interesting to note that, since the Coulomb algebra SO(4) is contained in (4.1) one can combine it with the SO(4) algebra contained in (2.6). This combination provides a considerable simplification in the treatment of electronic excitations, as discussed in the accompanying article[15] which describes in detail the situation when n=2.

5. CONCLUSIONS

Algebraic theory appears to be a very powerful tool to study complex situations. In applications to the study of molecular spectra a dominant role is played by the algebra of SO(4) since this appears as an approximate dynamic symmetry of both rotation-vibration and electronic spectra.

A suitable algebraic framework for all of chemistry is

$$\sum_{i=1}^{\nu} \oplus U_i(4) \oplus \sum_{k=1}^{\mu} \oplus U\left(2n_k^2\right) \quad , \tag{5.1}$$

where the first sum describes the algebra of bonds and the second sum the algebra of the electronic space including several shells.

For practical applications the first part is the most interesting one, since molecular spectroscopy deals, to a large extent, with rotation-vibration spectra. Indeed there are several aspects of molecular spectroscopy which are particularly relevant at the present time in view of its impact on environmental (especially the structure of ozone, O_3, water, H_2O, and carbon dioxide, CO_2), planetary (especially the structure of acetylene, C_2H_2) and galactic science. It is for the solution of these problems that algebraic methods are particularly useful.

REFERENCES

1. F. Iachello, Chem. Phys. Lett. 78, 581 (1981).
2. F. Iachello and R.D. Levine, J. Chem. Phys. 77, 3046 (1982).
3. P.M. Morse, Phys. Rev. 34, 57 (1929).
4. O.S. van Roosmalen, A.E.L. Dieperink and F. Iachello, Chem. Phys. Lett. 85, 32 (1982).
5. O.S. van Roosmalen, F. Iachello, R.D. Levine and A.E.L. Dieperink, J. Chem. Phys. 79, 2515 (1983).
6. G. Herzberg, "Infrared and Raman Spectra of Polyatomic Molecules" (Van Nostrand, New York, 1950).
7. F. Iachello and S. Oss, J. Molec. Spectrosc. 142, 85 (1990).
8. J. Hornos and F. Iachello, J. Chem. Phys. 90, 5284 (1989).
9. B.G. Wybourne, "Classical Group for Physicists", (J. Wiley and Son, New York, 1974).
10. R. Bijker, These Proceedings.
11. A.Frank, F.Iachello and R.Lemus, Chem. Phys. Lett. 131, 380 (1986).
12. A. Frank, R. Lemus and F. Iachello, J. Chem. Phys. 91, 29 (1989).
13. J. Paldus "Group Theoretical Methods in Physics, Lecture Notes in Physics", edited by W. Berglböck, A. Böhm and E. Takasayi, (Springer, Berlin, 1978), Vol. 94, p.51.
14. B.R. Judd; in "Group Theory and Its Applications", edited by E.M. Loebl (Academic, New York, 1968), p. 183.
15. A. Frank, These Proceedings.

O(4) SYMMETRY AND ANGULAR MOMENTUM THEORY

IN FOUR DIMENSIONS

P. Van Isacker

SERC Daresbury Laboratory

Warrington WA4 4AD, England

The theory of angular momentum in four-dimensional space is developed by studying in detail the properties of O(4), the group of rotations in four dimensions. Properties of O(4) Clebsch–Gordan coefficients are reviewed. Making use of the isomorphism $O(4) \approx O(3) \otimes O(3)$, general expressions are derived for O(4) recoupling coefficients in terms of the corresponding ones in O(3) ($6j$-, $9j$-,... symbols). The Wigner–Eckart theorem is discussed for O(4) and explicit formulas are given for the associated tensor calculus. The Pauli principle is generalized to four-dimensional space by establishing a connection between O(4) tensor character and statistics (Bose or Fermi). The equations for computing coefficients of fractional parentage in O(4) are derived. The applicability of this formalism to atomic and molecular physics problems is discussed.

1. INTRODUCTION

The development of angular momentum theory has played an important role in the application of quantum mechanics to physical problems. In many problems in virtually all branches of physics where the quantum mechanical regime applies, the coupling of angular momenta—or the recoupling, if more than two are involved—is required, and the fact that closed expressions have become available for the associated (re)coupling coefficients has significantly increased the predictive power of quantum mechanical theories and models. Another important result of angular momentum theory is expressed by the Wigner–Eckart theorem which states that, for systems with rotational symmetry, the matrix element of a tensor operator can be written as the product of a geometrical factor (essentially a vector-addition or Clebsch–Gordan coefficient) and a so-called reduced matrix element, which only depends on the angular momenta involved and not on the magnetic quantum numbers (projections of the angular momenta on the z-axis). On the basis of this theorem a tensor calculus of rotations in three dimensions can be formulated which leads to a drastic simplification in the evaluation—both formal and numerical—of matrix elements of any operator associated to physically observable quantities.

Over the years numerous physicists have made a vast number of contributions related to angular momentum theory. Among them the work of Wigner [1] and of Racah [2] deserves a special mention. The subject also has been treated comprehensively in several textbooks including the ones by Rose [3], Edmonds [4], Brink and Satchler [5], Yutsis, Levinson and Vanagas [6] and de-Shalit and Talmi [7].

From the beginning it was realized [1] that many of the properties of angular momentum can be studied via the theory of Lie algebras and groups and, more specifically, that they are related to the group of rotations in three dimensions, denoted as O(3), and its subgroup O(2) of rotations around a single axis. The results of angular momentum theory are thus specific to $O(3) \supset O(2)$ and are particular examples of results valid for the general reduction $G \supset G'$, where G and G' are two classical Lie groups.

Although many of the results of angular momentum theory can be generalized to an arbitrary reduction $G \supset G'$, it should be stressed that this generalization is only a formal one. There still remains the task of the explicit evaluation of the (re)coupling coefficients which are different in each case (i.e., depend on G and G') and without which much of the computational power of group theory is lost. The choice of the groups G and G' depends on the nature of the physical problem under consideration and, in particular, on the symmetry character of the Hamiltonian.

In this paper the angular momentum theory associated with the group O(4) and its reduction to O(3) and O(2) is developed. Since O(4) is the group of rotations in four dimensions, the formalism can be considered as a theory of angular momentum in four-dimensional space. The group O(4) is relevant in the description of the hydrogen atom [8] where it appears as the symmetry group of the Hamiltonian which is responsible for the degeneracy of levels into groups $(1s)$, $(2s, 2p)$, $(3s, 3p, 3d)$, etc. Hence the quantum mechanical solution of the problem of a single particle moving in a Coulomb potential exhibits two types of degeneracy: the first, present for any central potential, is the usual one of magnetic substates $M = -L, -L+1, \ldots, L$, while the second, the degeneracy of states with the same principal quantum number n, only occurs for $1/r$ potentials. The associated conserved quantities in the corresponding classical problem are the angular momentum vector $\vec{L}$ and the Runge–Lenz vector $\vec{A} = \vec{p} \times \vec{L} - Z\vec{r}/r$ (which characterizes the orientation of the major axis in the orbital plane), respectively. Because O(4) symmetry is of central importance in the description of the hydrogen atom, it was initially suggested [9] that its relevance would extend to the many-electron problem. Analyzing the case of two electrons in the $(2s, 2p)$-shell, Butler and Wybourne [10] showed, however, that the O(4) symmetry is strongly broken for this system, a conclusion that can be generalized to the many-electron problem. Thus applications to atomic many-electron physics based on the group O(4) appear to be limited in scope.

More recently Iachello and collaborators [11,12] suggested an algebraic approach to problems in molecular structure physics in which the group O(4) plays a central role. Their model, referred to as the vibron model, describes molecular rotation–vibration states in terms of a system of interacting bosons that can be either in an $L = 0$ (s-boson) or an $L = 1$ state (p-boson). The resulting algebraic structure is the one of U(4) and has two canonical group reductions [11]: the first containing U(3) and the second containing O(4). Furthermore, analysis of triatomic molecular spectra [12] reveals that many of their properties can be explained assuming an approximate O(4) symmetry. This result indicates the possible relevance of O(4) in the description of complex polyatomic molecules, that are outside the scope of *ab initio* calculations. The vibron model recently has also been extended in a different direction [13] to include electronic molecular excitations. It appears that in all these applications—present or future—the group O(4) plays or will play a central role. This observation constitutes the main incentive for developing the theory of angular momentum in four-dimensional space.

The material discussed in this paper is presented as follows. In §§2 and 3 the O(4) group is defined and its representations are discussed. The O(4) coupling (Clebsch–Gordan) coefficients are introduced in §4. Most of the material presented in §§2–4 is known (see, e.g. Wybourne [8]), but is given here to introduce notation and conventions. The recoupling of several (3, 4 and n) angular momenta in four-dimensional space is discussed in §§5–7. Some of these results have been derived elsewhere [14,15] making use of a sophisticated tensor formalism. The formalism and notation employed in this paper is less general and more specific to O(4), but, because of its similarity to the notation in conventional angular momentum theory, it is easier to comprehend and better suited for practical computations. In §§8–10 tensors in O(4) are introduced, the Wigner–Eckart theorem for O(4) is discussed and the O(4) tensor calculus is developed. The results given in these sections should enable one to achieve a considerable reduction in complexity in problems with O(4) symmetry. The relation between statistics (Bose or Fermi) and the O(4) tensor character of a particle is discussed in §11. This relation is used in §12 to construct O(4)-coupled symmetric or antisymmetric many-particle wavefunctions and to derive expressions for the coefficients of fractional parentage in O(4). Finally, the conclusions of this work are presented in §13.

2. THE LIE ALGEBRA O(4)

Rotations in four-dimensional space are generated by the infinitesimal rotation operators

$$J_{\lambda\mu} = -i\left(x_\lambda \frac{\partial}{\partial x_\mu} - x_\mu \frac{\partial}{\partial x_\lambda}\right) \qquad (\lambda \neq \mu = 1,2,3,4). \tag{2.1}$$

These are antisymmetric in the indices λ and μ, $J_{\lambda\mu} = -J_{\mu\lambda}$, and hence (2.1) defines six independent operators which together form o(4), the rotation algebra in four dimensions. Introducing the notation

$$\begin{aligned} L_1 &= J_{23} & L_2 &= J_{31} & L_3 &= J_{12} \\ L'_1 &= J_{41} & L'_2 &= J_{42} & L'_3 &= J_{43} \end{aligned} \tag{2.2}$$

these generators can be shown to satisfy the commutation relations

$$[L_i, L_j] = i\epsilon_{ijk} L_k \qquad [L_i, L'_j] = i\epsilon_{ijk} L'_k \qquad [L'_i, L'_j] = i\epsilon_{ijk} L_k \tag{2.3}$$

where $\epsilon_{ijk} = +1$ if $\{ijk\}$ is an even permutation of $\{123\}$, $\epsilon_{ijk} = -1$ if $\{ijk\}$ is an odd permutation of $\{123\}$ and $\epsilon_{ijk} = 0$ otherwise.

The set of operators $\{L_i, L'_j\}$ represents one possible basis of the Lie algebra o(4). Another basis consists of the operators

$$J_i = \tfrac{1}{2}(L_i + L'_i) \qquad J'_i = \tfrac{1}{2}(L_i - L'_i) \tag{2.4}$$

which satisfy the commutation relations

$$[J_i, J_j] = i\epsilon_{ijk} J_k \qquad [J'_i, J'_j] = i\epsilon_{ijk} J'_k \qquad [J_i, J'_j] = 0. \tag{2.5}$$

The operators $\{J_i, J'_j\}$ form again the Lie algebra o(4), but at the same time (2.5) shows that $\{J_i\}$ and $\{J'_i\}$ separately form two commuting o(3) algebras. This shows that o(4) is the direct sum of two o(3) algebras, $o(4) = o_J(3) \oplus o_{J'}(3)$. The corresponding group property is that the Lie group O(4) is (locally) isomorphic [16] to the direct product of two O(3) groups, $O(4) \approx O_J(3) \otimes O_{J'}(3)$.

3. REPRESENTATIONS OF O(4) AND ITS SUBGROUPS

To label the representations of O(4) either of the two bases described in the previous section can be used. The first basis, $\{L_i, L'_j\}$, leads to the usual Cartan labels [8]

$$\boldsymbol{\omega} \equiv (\omega, \omega') \qquad \omega \geq |\omega'| \tag{3.1}$$

where ω and ω' are both integer or both half integer. The second basis, $\{J_i, J'_j\}$, gives two labels, associated with $O_J(3)$ and $O_{J'}(3)$,

$$\{j, j'\} \qquad j, j' \geq 0 \tag{3.2}$$

with again j and j' integer or half integer. As a notational convention the Cartan labels (3.1) are enclosed by round brackets while the labels (3.2) are put between curly brackets. Furthermore, to obtain a more compact notation the two Cartan labels associated with an O(4) representation often will be represented as a single boldface symbol. The two classifications are related by

$$\omega = j + j' \qquad \omega' = j - j' \tag{3.3}$$

or

$$j = \tfrac{1}{2}(\omega + \omega') \qquad j' = \tfrac{1}{2}(\omega - \omega') \tag{3.4}$$

as can be seen from (2.4).

The group O(4) has several O(3) subgroups. The most important one is the O(3) group of rotations in three-dimensional, physical space. We assume it to be associated with the indices $\lambda, \mu = 1, 2, 3$ and, consequently, this rotation group in three dimensions is generated by the operators $\{L_i\} = \{J_i + J'_i\}$ and can be denoted as $O_L(3)$ or $O_{J+J'}(3)$. Similarly, the operator $L_3 = J_3 + J'_3$ generates rotations in two dimensions which form the group $O_L(2) = O_{J+J'}(2)$. These subgroups of O(4) can be used to establish the classification scheme

$$\begin{array}{ccccccccc} O(4) & \approx & O_J(3) & \otimes & O_{J'}(3) & \supset & O_{J+J'}(3) & \supset & O_{J+J'}(2) \\ \downarrow & & \downarrow & & \downarrow & & \downarrow & & \downarrow \\ (\omega, \omega') & & j & & j' & & l & & m_l \end{array} . \tag{3.5}$$

The values of l contained in the representation (ω, ω') of O(4) are obtained from the multiplication $j \times j'$:

$$\begin{aligned} (\omega, \omega') \approx \{j, j'\} \mapsto l &= j + j', j + j' - 1, \ldots, |j - j'| \\ &= \omega, \omega - 1, \ldots, |\omega'| \end{aligned} \tag{3.6}$$

and the usual $O(3) \supset O(2)$ branching rule is valid:

$$l \mapsto m_l = l, l - 1, \ldots, -l. \tag{3.7}$$

Instead of (3.5) the following classification scheme can be used:

$$\begin{array}{ccccccccccc} O(4) & \approx & O_J(3) & \otimes & O_{J'}(3) & \supset & O_J(2) & \otimes & O_{J'}(2) & \supset & O_{J+J'}(2) \\ \downarrow & & \downarrow & & \downarrow & & \downarrow & & \downarrow & & \downarrow \\ (\omega, \omega') & & j & & j' & & m_j & & m'_j & & m_l \end{array} . \tag{3.8}$$

Clearly, the states (3.5) and (3.8) are related through

$$|(\omega, \omega') l m_l\rangle = \sum_{m_j m'_j} (j m_j \, j' m'_j | l m_l) |(\omega, \omega') m_j m'_j m_l\rangle \tag{3.9}$$

where $(jm_j\, j'm'_j|lm_l)$ is a Clebsch–Gordan coefficient for $O(3) \supset O(2)$ [4].

In the following the two- and three-dimensional rotation groups are, for notational simplicity, denoted as $O(2)$ and $O(3)$, omitting the subscript L or $J + J'$.

4. O(4) COUPLING COEFFICIENTS

Coupling coefficients for $O(4)$ arise in the reduction of the Kronecker product $\boldsymbol{\omega}_1 \times \boldsymbol{\omega}_2 \equiv (\omega_1, \omega_1') \times (\omega_2, \omega_2')$. The Clebsch–Gordan series is found using the isomorphism $O(4) \approx O_J(3) \otimes O_{J'}(3)$ for each of the $O(4)$ groups involved. One has

$$\{j_1, j_1'\} \times \{j_2, j_2'\} = \sum_{\alpha=|j_1-j_2|}^{j_1+j_2} \sum_{\beta=|j_1'-j_2'|}^{j_1'+j_2'} \{\alpha, \beta\} \tag{4.1}$$

and, using the relations (3.3) and (3.4), this can be changed to [8]

$$(\omega_1, \omega_1') \times (\omega_2, \omega_2') = \sum_{\alpha=0}^{\omega_+} \sum_{\beta=0}^{\omega_-} (\omega_1 + \omega_2 - \alpha - \beta, \omega_1' + \omega_2' - \alpha + \beta) \tag{4.2}$$

where $\omega_\pm = \min(\omega_1 \pm \omega_1', \omega_2 \pm \omega_2')$. From this Clebsch–Gordan series it is seen that the product $O(4) \otimes O(4)$ is simply reducible, that is, no $O(4)$ representation occurs twice in the product, a fact which greatly simplifies four-dimensional angular momentum theory.

The coupling of representations can be carried out not only at the level of $O(4)$, but also at the level of the $O(3)$ or $O(2)$ subgroups. When two $O(4)$ groups are involved, $O_1(4)$ and $O_2(4)$, three possible classification schemes exist:

$$\begin{array}{ll}
(a) & O_1(4) \otimes O_2(4) \supset O(4) \supset O(3) \supset O(2) \\
(b) & O_1(4) \otimes O_2(4) \supset O_1(3) \otimes O_2(3) \supset O(3) \supset O(2) \\
(c) & O_1(4) \otimes O_2(4) \supset O_1(3) \otimes O_2(3) \supset O_1(2) \otimes O_2(2) \supset O(2)
\end{array} \tag{4.3}$$

where the $O(n)$ groups ($n = 4, 3, 2$) are obtained by adding the generators of $O_1(n)$ and $O_2(n)$. The above classification schemes conveniently can be summarized in the following *lattice*:

$$\begin{array}{ccccccc}
O_1(4) & & & \otimes & & & O_2(4) \\
\downarrow & & \overset{a}{\searrow} & & \overset{a}{\swarrow} & & \downarrow \\
O_1(3) & & & O(4) & & & O_2(3) \\
\downarrow & & \overset{b}{\searrow} & \downarrow & \overset{b}{\swarrow} & & \downarrow \\
O_1(2) & & & O(3) & & & O_2(2) \\
 & & \overset{c}{\searrow} & \downarrow & \overset{c}{\swarrow} & & \\
 & & & O(2) & & &
\end{array} \tag{4.4}$$

where the different routes a, b and c, indicated in the lattice, correspond to the reduction schemes given in (4.3). Labels can be introduced for each of the three schemes. In the notation of the previous section one has

$$\begin{array}{ccccccccc} O_1(4) & \otimes & O_2(4) & \supset & O(4) & \supset & O(3) & \supset & O(2) \\ \downarrow & & \downarrow & & \downarrow & & \downarrow & & \downarrow \\ \omega_1 & & \omega_2 & & \Omega & & L & & M_L \end{array}$$

$$\begin{array}{ccccccccccc} O_1(4) & \otimes & O_2(4) & \supset & O_1(3) & \otimes & O_2(3) & \supset & O(3) & \supset & O(2) \\ \downarrow & & \downarrow & & \downarrow & & \downarrow & & \downarrow & & \downarrow \\ \omega_1 & & \omega_2 & & l_1 & & l_2 & & L & & M_L \end{array} \tag{4.5}$$

$$\begin{array}{ccccccccccccc} O_1(4) & \otimes & O_2(4) & \supset & O_1(3) & \otimes & O_2(3) & \supset & O_1(2) & \otimes & O_2(2) & \supset & O(2) \\ \downarrow & & \downarrow & & \downarrow & & \downarrow & & \downarrow & & \downarrow & & \downarrow \\ \omega_1 & & \omega_2 & & l_1 & & l_2 & & m_1 & & m_2 & & M_L \end{array}$$

and the corresponding basis states are thus denoted as

$$\begin{aligned} &(a) \quad |\omega_1, \omega_2; \Omega L M_L\rangle \\ &(b) \quad |\omega_1 l_1, \omega_2 l_2; L M_L\rangle \\ &(c) \quad |\omega_1 l_1 m_1, \omega_2 l_2 m_2; M_L\rangle \equiv |\omega_1 l_1 m_1\rangle |\omega_2 l_2 m_2\rangle. \end{aligned} \tag{4.6}$$

The transformation matrix from (b) to (c) is readily established since it only involves an $O(3) \supset O(2)$ coupling (Clebsch–Gordan) coefficient:

$$|\omega_1 l_1, \omega_2 l_2; L M_L\rangle = \sum_{m_1 m_2} (l_1 m_1\, l_2 m_2 | L M_L) |\omega_1 l_1 m_1\rangle |\omega_2 l_2 m_2\rangle. \tag{4.7}$$

The transformation from (a) to (b) is more easily found by first changing to the $O_J(3) \otimes O_{J'}(3)$ notation:

$$\begin{aligned} &(a) \quad |\{j_1, j_1'\}, \{j_2, j_2'\}; \{J, J'\} L M_L\rangle \equiv |(j_1, j_2)J, (j_1', j_2')J'; L M_L\rangle \\ &(b) \quad |\{j_1, j_1'\} l_1, \{j_2, j_2'\} l_2; L M_L\rangle \equiv |(j_1, j_1') l_1, (j_2, j_2') l_2; L M_L\rangle \end{aligned} \tag{4.8}$$

where, on the right-hand side, the notation $(j, j')l$ is used to indicate that j and j' are coupled (in O(3)) to l. This shows [17] that the transformation requires the recoupling of four angular momenta in O(3), known from standard angular momentum theory [4]:

$$\begin{aligned} &|\{j_1, j_1'\}, \{j_2, j_2'\}; \{J, J'\} L M_L\rangle \\ &\qquad = \sum_{l_1 l_2} X \begin{bmatrix} j_1 & j_2 & J \\ j_1' & j_2' & J' \\ l_1 & l_2 & L \end{bmatrix} |\{j_1, j_1'\} l_1, \{j_2, j_2'\} l_2; L M_L\rangle. \end{aligned} \tag{4.9}$$

The X-coefficient in (4.9) is identical to the overlap matrix encountered in the recoupling of four angular momenta and is related to the $9j$-symbol in the following way:

$$X \begin{bmatrix} j_1 & j_2 & J \\ j_1' & j_2' & J' \\ l_1 & l_2 & L \end{bmatrix} = \sqrt{(2J+1)(2J'+1)(2l_1+1)(2l_2+1)} \begin{Bmatrix} j_1 & j_2 & J \\ j_1' & j_2' & J' \\ l_1 & l_2 & L \end{Bmatrix}. \tag{4.10}$$

It should be emphasized that the transformation (4.9), describing the recoupling of *four* angular momenta in O(3), at the same time represents a coupling coefficient for *two* representations in O(4). To bring out more clearly the coupling in O(4) the relation (4.9) can be rewritten as

$$|\omega_1, \omega_2; \Omega L M_L\rangle = \sum_{l_1 l_2} (\omega_1 l_1\, \omega_2 l_2 | \Omega L) |\omega_1 l_1, \omega_2 l_2; L M_L\rangle \tag{4.11}$$

with

$$(\omega_1 l_1\, \omega_2 l_2|\Omega L) = X \begin{bmatrix} j_1 & j_2 & J \\ j_1' & j_2' & J' \\ l_1 & l_2 & L \end{bmatrix} \tag{4.12}$$

where the relations $\Omega = (\Omega, \Omega') \approx \{J, J'\}$ and $\omega_i = (\omega_i, \omega_i') \approx \{j_i, j_i'\}, i = 1,2$ are assumed implicitly. The coefficient $(\omega_1 l_1\, \omega_2 l_2|\Omega L)$ has the same structure as the familiar Clebsch–Gordan coefficient $(l_1 m_1\, l_2 m_2|L M_L)$, but the latter pertains to the reduction O(3) ⊃ O(2) whereas (4.12) is related to O(4) ⊃ O(3). Its properties can be deduced from those of the X-coefficient. A particularly important one concerns the orthonormality of the O(4) ⊃ O(3) coupling coefficients:

$$\sum_{l_1 l_2} (\omega_1 l_1\, \omega_2 l_2|\Omega_a L)(\omega_1 l_1\, \omega_2 l_2|\Omega_b L) = \delta_{\Omega_a \Omega_b}. \tag{4.13}$$

Alternatively, a relation expressing the orthonormality in l_1 and l_2 can be derived:

$$\sum_{\Omega} (\omega_1 l_{1a}\, \omega_2 l_{2a}|\Omega L)(\omega_1 l_{1b}\, \omega_2 l_{2b}|\Omega L) = \delta_{l_{1a} l_{1b}} \delta_{l_{2a} l_{2b}}. \tag{4.14}$$

Other useful properties are concerned with the symmetries of the O(4) ⊃ O(3) coupling coefficients under the exchange of its arguments:

$$(\omega_2 l_2\, \omega_1 l_1|\Omega L) = (-)^{\omega_1 + \omega_2 + \Omega + l_1 + l_2 + L} (\omega_1 l_1\, \omega_2 l_2|\Omega L) \tag{4.15}$$

and

$$\begin{aligned} (\omega_2 l_2\, \Omega L|\omega_1 l_1) &= \sqrt{\frac{2L+1}{2l_1+1}} \sqrt{\frac{(2j_1+1)(2j_1'+1)}{(2J+1)(2J'+1)}} (\omega_1 l_1\, \omega_2 l_2|\Omega L) \\ &= \sqrt{\frac{2L+1}{2l_1+1}} \sqrt{\frac{2\omega_1+1}{2\Omega+1}} (\omega_1 l_1\, \omega_2 l_2|\Omega L) \end{aligned} \tag{4.16}$$

where the notation $2\omega_x + 1 \equiv (2j_x+1)(2j_x'+1)$ is introduced to denote the dimensionality of a representation ω_x of O(4).

Finally, the transformation from (a) to (c) can be found by combining (4.7) and (4.11):

$$|\omega_1, \omega_2; \Omega L M_L\rangle = \sum_{\substack{l_1 l_2 \\ m_1 m_2}} (\omega_1 l_1 m_1\, \omega_2 l_2 m_2|\Omega L M_L)|\omega_1 l_1 m_1\rangle|\omega_2 l_2 m_2\rangle \tag{4.17}$$

where

$$(\omega_1 l_1 m_1\, \omega_2 l_2 m_2|\Omega L M_L) = (\omega_1 l_1\, \omega_2 l_2|\Omega L)(l_1 m_1\, l_2 m_2|L M_L). \tag{4.18}$$

The coupling coefficient in (4.17) is associated with the full reduction O(4) ⊃ O(3) ⊃ O(2). According to (4.18) this coefficient can be written as the product of two factors, the first one associated with O(4) ⊃ O(3) (sometimes referred to as an isoscalar factor) and the second associated with O(3) ⊃ O(2) (the usual Clebsch–Gordan coefficient). As such relation (4.18) provides an example of Racah's celebrated factorization lemma [2].

5. RECOUPLING OF THREE ω'S

In analogy with the three-dimensional case a theory of angular momentum can now be developed in four dimensions. Of basic importance are the transformation coefficients arising from the recoupling of several O(4) representations and we begin our discussion with the simplest case, the recoupling of three O(4) representations. Here and in the following sections the notation $\Omega_x = (\Omega_x, \Omega'_x) \approx \{J_x, J'_x\}$ or $\omega_x = (\omega_x, \omega'_x) \approx \{j_x, j'_x\}$ is used to denote O(4) representations, where x represents any index.

Three O(4) representations ω_1, ω_2 and ω_3 can be combined to a total O(4) representation Ω in two different coupling orders. In the first ω_1 and ω_2 are coupled to Ω_{12}, which in turn is coupled to ω_3 to give the total Ω. The corresponding state thus reads

$$
\begin{aligned}
&|(\omega_1, \omega_2)\Omega_{12}, \omega_3; \Omega L M_L\rangle \\
&\quad = \sum_{\substack{l_1 l_2 l_3 L_{12} \\ m_1 m_2 m_3 M_{12}}} (\omega_1 l_1 m_1\, \omega_2 l_2 m_2 | \Omega_{12} L_{12} M_{12})(\Omega_{12} L_{12} M_{12}\, \omega_3 l_3 m_3 | \Omega L M_L) \\
&\qquad\qquad \times |\omega_1 l_1 m_1\rangle |\omega_2 l_2 m_2\rangle |\omega_3 l_3 m_3\rangle \\
&\quad = \sum_{l_1 l_2 L_{12} l_3} (\omega_1 l_1\, \omega_2 l_2 | \Omega_{12} L_{12})(\Omega_{12} L_{12}\, \omega_3 l_3 | \Omega L) \\
&\qquad\qquad \times |(\omega_1 l_1, \omega_2 l_2) L_{12}, \omega_3 l_3; L M_L\rangle.
\end{aligned}
\tag{5.1}
$$

In the second coupling order ω_1 is coupled to Ω_{23}, obtained from ω_2 and ω_3,

$$
\begin{aligned}
&|\omega_1, (\omega_2, \omega_3)\Omega_{23}; \Omega L M_L\rangle \\
&\quad = \sum_{\substack{l_1 l_2 l_3 L_{23} \\ m_1 m_2 m_3 M_{23}}} (\omega_1 l_1 m_1\, \Omega_{23} L_{23} M_{23} | \Omega L M_L)(\omega_2 l_2 m_2\, \omega_3 l_3 m_3 | \Omega_{23} L_{23} M_{23}) \\
&\qquad\qquad \times |\omega_1 l_1 m_1\rangle |\omega_2 l_2 m_2\rangle |\omega_3 l_3 m_3\rangle \\
&\quad = \sum_{l_1 l_2 l_3 L_{23}} (\omega_1 l_1\, \Omega_{23} L_{23} | \Omega L)(\omega_2 l_2\, \omega_3 l_3 | \Omega_{23} L_{23}) \\
&\qquad\qquad \times |\omega_1 l_1, (\omega_2 l_2, \omega_3 l_3) L_{23}; L M_L\rangle.
\end{aligned}
\tag{5.2}
$$

The transformation between the two bases formally can be written as

$$
\begin{aligned}
&|(\omega_1, \omega_2)\Omega_{12}, \omega_3; \Omega L M_L\rangle \\
&\quad = \sum_{\Omega_{23}} \langle(\omega_1, \omega_2)\Omega_{12}, \omega_3; \Omega | \omega_1, (\omega_2, \omega_3)\Omega_{23}; \Omega\rangle |\omega_1, (\omega_2, \omega_3)\Omega_{23}; \Omega L M_L\rangle
\end{aligned}
\tag{5.3}
$$

where the transformation matrix is a U-coefficient in O(4):

$$
\langle(\omega_1, \omega_2)\Omega_{12}, \omega_3; \Omega | \omega_1, (\omega_2, \omega_3)\Omega_{23}; \Omega\rangle = U(\omega_1 \omega_2 \Omega \omega_3; \Omega_{12} \Omega_{23}).
\tag{5.4}
$$

Note that this transformation matrix is independent of L and M_L, as will be shown below. From (5.1) and (5.2) an expression for the O(4) U-coefficient in terms of O(4) $\supset$ O(3) recoupling coefficients is obtained:

$$
\begin{aligned}
&U(\omega_1 \omega_2 \Omega \omega_3; \Omega_{12} \Omega_{23}) \\
&\qquad = \sum_{\substack{l_1 l_2 l_3 \\ L_{12} L_{23}}} U(l_1 l_2 L l_3; L_{12} L_{23}) \\
&\qquad\qquad \times (\omega_1 l_1\, \omega_2 l_2 | \Omega_{12} L_{12})(\Omega_{12} L_{12}\, \omega_3 l_3 | \Omega L) \\
&\qquad\qquad \times (\omega_1 l_1\, \Omega_{23} L_{23} | \Omega L)(\omega_2 l_2\, \omega_3 l_3 | \Omega_{23} L_{23}).
\end{aligned}
\tag{5.5}
$$

Using the identity (4.12) the O(4) U-coefficient can thus be written as a sum of products of four $9j$-symbols and a U-coefficient in O(3). A substantially simplified expression is found by changing to the $O_J(3) \otimes O_{J'}(3)$ notation:

$$\begin{aligned}
&\langle(\omega_1,\omega_2)\Omega_{12},\omega_3;\Omega|\omega_1,(\omega_2,\omega_3)\Omega_{23};\Omega\rangle \\
&\quad\equiv \langle(\{j_1,j_1'\},\{j_2,j_2'\})\{J_{12},J_{12}'\},\{j_3,j_3'\};\{J,J'\}LM_L| \\
&\qquad |\{j_1,j_1'\},(\{j_2,j_2'\},\{j_3,j_3'\})\{J_{23},J_{23}'\};\{J,J'\}LM_L\rangle \\
&\quad= \langle((j_1,j_2)J_{12},j_3)J,((j_1',j_2')J_{12}',j_3')J';LM_L| \\
&\qquad |(j_1,(j_2,j_3)J_{23})J,(j_1',(j_2',j_3')J_{23}')J';LM_L\rangle \qquad (5.6)
\end{aligned}$$

where the angular momentum L and its z-projection M_L are indicated. According to standard O(3) angular momentum theory [6] the last matrix element in (5.6) is independent of L and M_L and can be written as the product of two overlap matrices involving the unprimed and the primed angular momenta, respectively. Thus one obtains the relation

$$U(\omega_1\omega_2\Omega\omega_3;\Omega_{12}\Omega_{23}) = U(j_1j_2Jj_3;J_{12}J_{23})U(j_1'j_2'J'j_3';J_{12}'J_{23}'). \qquad (5.7)$$

In analogy with the O(3) case it is of interest to define the 6ω-symbol, which can be written as the product of two $6j$-symbols:

$$\left\{\begin{matrix}\omega_1 & \omega_2 & \Omega_{12}\\ \omega_3 & \Omega & \Omega_{23}\end{matrix}\right\} = \left\{\begin{matrix}j_1 & j_2 & J_{12}\\ j_3 & J & J_{23}\end{matrix}\right\}\left\{\begin{matrix}j_1' & j_2' & J_{12}'\\ j_3' & J' & J_{23}'\end{matrix}\right\}. \qquad (5.8)$$

The relation between the 6ω-symbol and the O(4) U-coefficient is defined according to the corresponding one in O(3) [4]:

$$\begin{aligned}
&U(\omega_1\omega_2\Omega\omega_3;\Omega_{12}\Omega_{23}) \\
&\quad= (-)^{\omega_1+\omega_2+\omega_3+\Omega}\sqrt{(2\Omega_{12}+1)(2\Omega_{23}+1)}\left\{\begin{matrix}\omega_1 & \omega_2 & \Omega_{12}\\ \omega_3 & \Omega & \Omega_{23}\end{matrix}\right\}. \qquad (5.9)
\end{aligned}$$

Another recoupling coefficient of three ω's frequently encountered transforms the basis states

$$|(\omega_1,\omega_2)\Omega_{12},\omega_3;\Omega LM_L\rangle \qquad (5.10)$$

into

$$|(\omega_1,\omega_3)\Omega_{13},\omega_2;\Omega LM_L\rangle. \qquad (5.11)$$

The associated transformation matrix equals

$$\begin{aligned}
&\langle(\omega_1,\omega_2)\Omega_{12},\omega_3;\Omega|(\omega_1,\omega_3)\Omega_{13},\omega_2;\Omega\rangle \\
&\quad= (-)^{\omega_2+\omega_3+\Omega_{12}+\Omega_{13}}\sqrt{(2\Omega_{12}+1)(2\Omega_{13}+1)}\left\{\begin{matrix}\omega_1 & \omega_2 & \Omega_{12}\\ \Omega & \omega_3 & \Omega_{13}\end{matrix}\right\}. \qquad (5.12)
\end{aligned}$$

The (symmetry) properties of the 6ω-symbol can be derived from those of the $6j$-symbol.

6. RECOUPLING OF FOUR ω'S

The recoupling of four O(4) representations can be dealt with in much the same way as the previous case. Only the most important formulas are given here.

The basic quantity to calculate in this case is the overlap matrix between the states

$$\begin{aligned}
&|(\omega_1,\omega_2)\Omega_{12},(\omega_3,\omega_4)\Omega_{34};\Omega LM_L\rangle \\
&\quad= \sum_{\substack{l_1l_2l_3l_4\\ L_{12}L_{34}}} (\omega_1l_1\,\omega_2l_2|\Omega_{12}L_{12})(\omega_3l_3\,\omega_4l_4|\Omega_{34}L_{34})(\Omega_{12}L_{12}\,\Omega_{34}L_{34}|\Omega L) \\
&\qquad \times|(\omega_1l_1,\omega_2l_2)L_{12},(\omega_3l_3,\omega_4l_4)L_{34};LM_L\rangle \qquad (6.1)
\end{aligned}$$

and

$$|(\omega_1,\omega_3)\Omega_{13},(\omega_2,\omega_4)\Omega_{24};\Omega LM_L\rangle$$

$$= \sum_{\substack{l_1l_2l_3l_4 \\ L_{13}L_{24}}} (\omega_1 l_1\ \omega_3 l_3|\Omega_{13}L_{13})(\omega_2 l_2\ \omega_4 l_4|\Omega_{24}L_{24})(\Omega_{13}L_{13}\ \Omega_{24}L_{24}|\Omega L)$$

$$\times|(\omega_1 l_1,\omega_3 l_3)L_{13},(\omega_2 l_2,\omega_4 l_4)L_{24};LM_L\rangle \tag{6.2}$$

which is, in analogy with the notation introduced in §4, denoted as an O(4) X-coefficient:

$$|(\omega_1,\omega_2)\Omega_{12},(\omega_3,\omega_4)\Omega_{34};\Omega LM_L\rangle$$

$$= \sum_{\Omega_{13}\Omega_{24}} X\begin{bmatrix} \omega_1 & \omega_2 & \Omega_{12} \\ \omega_3 & \omega_4 & \Omega_{34} \\ \Omega_{13} & \Omega_{24} & \Omega \end{bmatrix} |(\omega_1,\omega_3)\Omega_{13},(\omega_2,\omega_4)\Omega_{24};\Omega LM_L\rangle. \tag{6.3}$$

This overlap matrix can be expressed in terms of O(4) ⊃ O(3) coupling coefficients:

$$X\begin{bmatrix} \omega_1 & \omega_2 & \Omega_{12} \\ \omega_3 & \omega_4 & \Omega_{34} \\ \Omega_{13} & \Omega_{24} & \Omega \end{bmatrix}$$

$$= \sum_{\substack{l_1l_2l_3l_4 \\ L_{12}L_{34}L_{13}L_{24}}} X\begin{bmatrix} l_1 & l_2 & L_{12} \\ l_3 & l_4 & L_{34} \\ L_{13} & L_{24} & L \end{bmatrix}$$

$$\times(\omega_1 l_1\ \omega_2 l_2|\Omega_{12}L_{12})(\omega_3 l_3\ \omega_4 l_4|\Omega_{34}L_{34})(\Omega_{12}L_{12}\ \Omega_{34}L_{34}|\Omega L)$$

$$\times(\omega_1 l_1\ \omega_3 l_3|\Omega_{13}L_{13})(\omega_2 l_2\ \omega_4 l_4|\Omega_{24}L_{24})(\Omega_{13}L_{13}\ \Omega_{24}L_{24}|\Omega L) \tag{6.4}$$

which can be summed to give

$$X\begin{bmatrix} \omega_1 & \omega_2 & \Omega_{12} \\ \omega_3 & \omega_4 & \Omega_{34} \\ \Omega_{13} & \Omega_{24} & \Omega \end{bmatrix} = X\begin{bmatrix} j_1 & j_2 & J_{12} \\ j_3 & j_4 & J_{34} \\ J_{13} & J_{24} & J \end{bmatrix} X\begin{bmatrix} j_1' & j_2' & J_{12}' \\ j_3' & j_4' & J_{34}' \\ J_{13}' & J_{24}' & J' \end{bmatrix}. \tag{6.5}$$

In analogy with the O(3) case it is of interest to define the 9ω-symbol, which can be written as the product of two $9j$-symbols:

$$\begin{Bmatrix} \omega_1 & \omega_2 & \Omega_{12} \\ \omega_3 & \omega_4 & \Omega_{34} \\ \Omega_{13} & \Omega_{24} & \Omega \end{Bmatrix} = \begin{Bmatrix} j_1 & j_2 & J_{12} \\ j_3 & j_4 & J_{34} \\ J_{13} & J_{24} & J \end{Bmatrix} \begin{Bmatrix} j_1' & j_2' & J_{12}' \\ j_3' & j_4' & J_{34}' \\ J_{13}' & J_{24}' & J' \end{Bmatrix} \tag{6.6}$$

and its (symmetry) properties can thus be derived from those of the $9j$-symbol.

7. RECOUPLING OF n ω'S

The previous results can be generalized to the recoupling of an arbitrary number of O(4) representations, leading to the following.

Theorem. All recoupling coefficients in $O(4) \approx O_J(3) \otimes O_{J'}(3)$ can be expressed as the product of two corresponding recoupling coefficients in $O_J(3)$ and $O_{J'}(3)$:

$$\langle((\omega_1,\omega_2)\Omega_{12},\omega_3)\Omega_{123}\ldots|((\omega_k,\omega_l)\Omega_{kl},\omega_m)\Omega_{klm}\ldots\rangle$$

$$= \langle((j_1,j_2)J_{12},j_3)J_{123}\ldots|((j_k,j_l)J_{kl},j_m)J_{klm}\ldots\rangle$$

$$\times\langle((j_1',j_2')J_{12}',j_3')J_{123}'\ldots|((j_k',j_l')J_{kl}',j_m')J_{klm}'\ldots\rangle \tag{7.1}$$

where $\Omega_x = (\Omega_x, \Omega'_x) \approx \{J_x, J'_x\}$ and $\omega_x = (\omega_x, \omega'_x) \approx \{j_x, j'_x\}$.

With this theorem the recoupling of angular momenta in O(4) is thus reduced to (twice) the corresponding recoupling in O(3).

8. O(4) TENSORS

Starting from some "elementary" operators that transform as tensors under O(4) ⊃ O(3) ⊃ O(2) and using the O(4) coupling coefficients introduced in §4, more complex tensors can be built. One has, in general,

$$\begin{aligned} [T^{\Omega_1} \times T^{\Omega_2}]^{\Omega}_{LM_L} &= \sum_{\substack{L_1 L_2 \\ M_1 M_2}} (\Omega_1 L_1\, \Omega_2 L_2 | \Omega L)(L_1 M_1\, L_2 M_2 | L M_L) T^{\Omega_1}_{L_1 M_1} T^{\Omega_2}_{L_2 M_2} \\ &= \sum_{L_1 L_2} (\Omega_1 L_1\, \Omega_2 L_2 | \Omega L)[T^{\Omega_1}_{L_1} \times T^{\Omega_2}_{L_2}]_{LM_L} \end{aligned} \tag{8.1}$$

where the $T^{\Omega_i}_{L_i M_i}$ are the elementary tensors carrying O(4), O(3) and O(2) labels, or are combinations of them.

9. WIGNER–ECKART THEOREM

For rotations in three-dimensional space the Wigner–Eckart theorem allows one to write the matrix element of a tensor operator between O(3) ⊃ O(2) states as a geometrical factor (essentially a Clebsch–Gordan coefficient) times a reduced matrix element:

$$\begin{aligned} \langle L_a M_a | T_{LM_L} | L_b M_b \rangle &= (-)^{L_a - M_a} \begin{pmatrix} L_a & L & L_b \\ -M_a & M_L & M_b \end{pmatrix} \langle L_a \parallel T_L \parallel L_b \rangle_{O(3)} \\ &= (-)^{2L} (L_b M_b\, L M_L | L_a M_a) \frac{\langle L_a \parallel T_L \parallel L_b \rangle_{O(3)}}{\sqrt{2L_a + 1}}. \end{aligned} \tag{9.1}$$

The essential result is that the entire dependence on the magnetic quantum numbers M_a, M_b and M_L is contained in the Clebsch–Gordan coefficient and thus is well known.

In a similar way a Wigner–Eckart theorem can be defined for O(4) ⊃ O(3) ⊃ O(2):

$$\begin{aligned} &\langle \omega_a L_a M_a | T^{\Omega}_{LM_L} | \omega_b L_b M_b \rangle \\ &= (-)^{2L} (L_b M_b\, L M_L | L_a M_a) \frac{\langle \omega_a L_a \parallel T^{\Omega}_L \parallel \omega_b L_b \rangle_{O(3)}}{\sqrt{2L_a + 1}} \\ &= (-)^{2L} (L_b M_b\, L M_L | L_a M_a)(\omega_b L_b\, \Omega L | \omega_a L_a) \frac{\langle \omega_a \parallel T^{\Omega} \parallel \omega_b \rangle_{O(4)}}{\sqrt{2\omega_a + 1}} \end{aligned} \tag{9.2}$$

where the square root factors take account of the dimensionality of O(4) and O(3) representations and are needed to make the definition of the reduced matrix element more symmetric. Note that the reduced matrix elements in (9.1) and (9.2) are written with a subscript to indicate with respect to which group, O(4) or O(3), they are reduced. They are related through

$$\begin{aligned} &\langle \omega_a L_a \parallel T^{\Omega}_L \parallel \omega_b L_b \rangle_{O(3)} \\ &= \sqrt{\frac{2L_a + 1}{2\omega_a + 1}} (\omega_b L_b\, \Omega L | \omega_a L_a) \langle \omega_a \parallel T^{\Omega} \parallel \omega_b \rangle_{O(4)} \\ &= \sqrt{\frac{2L + 1}{2\Omega + 1}} (\omega_a L_a\, \omega_b L_b | \Omega L) \langle \omega_a \parallel T^{\Omega} \parallel \omega_b \rangle_{O(4)} \end{aligned} \tag{9.3}$$

or, inversely,

$$\langle \omega_a \| T^{\Omega} \| \omega_b \rangle_{O(4)} = \sqrt{\frac{2\Omega+1}{2L+1}} \sum_{L_a L_b} (\omega_a L_a\, \omega_b L_b | \Omega L) \langle \omega_a L_a \| T^{\Omega}_L \| \omega_b L_b \rangle_{O(3)} \tag{9.4}$$

where use has been made of the symmetry properties and orthonormality of the O(4) $\supset$ O(3) coupling coefficients.

The usefulness of (9.2) is clear: the entire dependence of the matrix elements on the angular momenta L_a, L_b and L and their z-projections M_a, M_b and M_L, is contained in the O(4) $\supset$ O(3) and O(3) $\supset$ O(2) coupling coefficients, for which general expressions are available.

10. TENSOR CALCULUS IN O(4)

By virtue of the Wigner–Eckart theorem the calculation of matrix elements of O(4) tensors between states in an O(4) $\supset$ O(3) $\supset$ O(2) basis is reduced to the calculation of O(4)-reduced matrix elements. In order to do so, the same powerful techniques can be applied as those used in three dimensions, as will be illustrated in this section.

One of the cases most frequently encountered concerns the matrix elements of the tensor product $[T_1^{\Omega_1} \times T_2^{\Omega_2}]^{\Omega}_{LM_L}$ where the tensors T_1 and T_2 act on different coordinate systems 1 and 2. An expression for the O(4)-reduced matrix element of this tensor product is obtained by converting it to the O(3)-reduced matrix element, (9.4), using the decoupling formula in O(3), converting back to O(4)-reduced matrix elements, (9.3), and using the expression (6.4) for the X-coefficient in O(4). The final result is

$$\begin{aligned} &\langle \omega_{1a}, \omega_{2a}; \Omega_a \| [T_1^{\Omega_1} \times T_2^{\Omega_2}]^{\Omega} \| \omega_{1b}, \omega_{2b}; \Omega_b \rangle_{O(4)} \\ &= \sqrt{\frac{2\Omega+1}{(2\Omega_1+1)(2\Omega_2+1)}} X \begin{bmatrix} \omega_{1a} & \omega_{2a} & \Omega_a \\ \omega_{1b} & \omega_{2b} & \Omega_b \\ \Omega_1 & \Omega_2 & \Omega \end{bmatrix} \\ &\quad \times \langle \omega_{1a} \| T_1^{\Omega_1} \| \omega_{1b} \rangle_{O(4)} \langle \omega_{2a} \| T_2^{\Omega_2} \| \omega_{2b} \rangle_{O(4)} \\ &= \sqrt{(2\Omega_a+1)(2\Omega_b+1)(2\Omega+1)} \begin{Bmatrix} \omega_{1a} & \omega_{2a} & \Omega_a \\ \omega_{1b} & \omega_{2b} & \Omega_b \\ \Omega_1 & \Omega_2 & \Omega \end{Bmatrix} \\ &\quad \times \langle \omega_{1a} \| T_1^{\Omega_1} \| \omega_{1b} \rangle_{O(4)} \langle \omega_{2a} \| T_2^{\Omega_2} \| \omega_{2b} \rangle_{O(4)}. \end{aligned} \tag{10.1}$$

An important special case occurs for $\Omega = (0,0)$, the scalar O(4) representation, for which the expression (10.1) reduces to

$$\begin{aligned} &\langle \omega_{1a}, \omega_{2a}; \Omega_a \| [T_1^{\Omega} \times T_2^{\Omega}]^{(0,0)} \| \omega_{1b}, \omega_{2b}; \Omega_b \rangle_{O(4)} \\ &= (-)^{\omega_{1b}+\omega_{2a}+\Omega_a+\Omega} \sqrt{\frac{2\Omega_a+1}{2\Omega+1}} \begin{Bmatrix} \omega_{1a} & \omega_{2a} & \Omega_a \\ \omega_{2b} & \omega_{1b} & \Omega \end{Bmatrix} \\ &\quad \times \langle \omega_{1a} \| T_1^{\Omega} \| \omega_{1b} \rangle_{O(4)} \langle \omega_{2a} \| T_2^{\Omega} \| \omega_{2b} \rangle_{O(4)} \delta_{\Omega_a \Omega_b}. \end{aligned} \tag{10.2}$$

Two other special cases occur if one of the tensors, T_1 or T_2, is the scalar unit tensor. The general formula (10.1) then becomes

$$\langle \alpha_{1a}\omega_{1a}, \alpha_{2a}\omega_{2a}; \Omega_a \| T_1^{\Omega} \| \alpha_{1b}\omega_{1b}, \alpha_{2b}\omega_{2b}; \Omega_b \rangle_{O(4)}$$

$$= (-)^{\omega_{1a}+\omega_{2a}+\Omega_b+\Omega}\sqrt{(2\Omega_a+1)(2\Omega_b+1)}\left\{\begin{array}{ccc}\omega_{1a} & \Omega_a & \omega_{2a}\\ \Omega_b & \omega_{1b} & \Omega\end{array}\right\}$$
$$\times\langle\alpha_{1a}\omega_{1a} \parallel T_1^{\Omega} \parallel \alpha_{1b}\omega_{1b}\rangle_{O(4)}\delta_{\alpha_{2a}\alpha_{2b}}\delta_{\omega_{2a}\omega_{2b}} \tag{10.3}$$

and

$$\langle\alpha_{1a}\omega_{1a}, \alpha_{2a}\omega_{2a}; \Omega_a \parallel T_2^{\Omega} \parallel \alpha_{1b}\omega_{1b}, \alpha_{2b}\omega_{2b}; \Omega_b\rangle_{O(4)}$$
$$= (-)^{\omega_{1a}+\omega_{2b}+\Omega_a+\Omega}\sqrt{(2\Omega_a+1)(2\Omega_b+1)}\left\{\begin{array}{ccc}\omega_{2a} & \Omega_a & \omega_{1a}\\ \Omega_b & \omega_{2b} & \Omega\end{array}\right\}$$
$$\times\langle\alpha_{2a}\omega_{2a} \parallel T_2^{\Omega} \parallel \alpha_{2b}\omega_{2b}\rangle_{O(4)}\delta_{\alpha_{1a}\alpha_{1b}}\delta_{\omega_{1a}\omega_{1b}} \tag{10.4}$$

where α_x denotes all additional quantum numbers (besides ω, l and m_l) to completely characterize the states.

11. BOSE AND FERMI STATISTICS

One of the most fundamental properties of a particle is whether it obeys Bose or Fermi statistics. In three-dimensional space statistics is governed by the Pauli principle which states that particles with integer spin satisfy Bose commutation relations and particles with half-integer spin satisfy Fermi anticommutation relations. Thus the Pauli principle establishes a relation between the character of the O(3) representation (integer or half-integer) according to which a particle transforms, and the statistics it obeys.

In this section the relation between spin and statistics is extended to four-dimensional space. The principal aim is to find the connection between the O(4) representation ω associated with a particle and the statistics this particle obeys. This connection is immediately established by requiring the Pauli principle in four dimensions to be consistent with that in three dimensions: because of the $O(4) \supset O(3)$ branching rule (3.6), integer representations of O(4) (ω and ω' integer) contain integer O(3) angular momenta and half-integer representations of O(4) (ω and ω' half-integer) contain half-integer angular momenta. Hence one arrives to the following.

Conjecture. Operators transforming as integer representations of O(4) (ω and ω' integer) satisfy Bose commutation relations; operators transforming as half-integer (or spinor) representations of O(4) (ω and ω' half-integer) satisfy Fermi anticommutation relations.

To illustrate the action of this "generalized" Pauli principle consider a two-particle state

$$|\omega^2; \Omega L M_L\rangle_{\text{nn}} = \sum_{\substack{l_1 l_2\\ m_1 m_2}} (\omega l_1\, \omega l_2|\Omega L)(l_1 m_1\, l_2 m_2|L M_L)|\omega l_1 m_1\rangle|\omega l_2 m_2\rangle \tag{11.1}$$

where the subscript "nn" indicates that the state is non-normalized. Introducing the label σ which equals +1 in the case of Bose statistics and −1 in the case of Fermi statistics, one finds the overlap

$$_{\text{nn}}\langle\omega^2; \Omega L M_L|\omega^2; \Omega L M_L\rangle_{\text{nn}}$$
$$= \sum_{\substack{l_1 l_2\\ m_1 m_2}} (\omega l_1\, \omega l_2|\Omega L)^2(l_1 m_1\, l_2 m_2|L M_L)^2$$
$$\times\left[1+\sigma(-)^{\omega+\omega+\Omega+l_1+l_2+L}(-)^{l_1+l_2-L}\right]$$
$$= 1+\sigma(-)^{2\omega+\Omega}. \tag{11.2}$$

In the derivation of (11.2) use has been made of the orthogonality and symmetry properties of the O(4) coupling coefficients and also of the fact that $(-)^{2l_1+2l_2}=+1$, because l_1 and l_2 are both integer or both half-integer. Since, on the basis of the generalized Pauli principle,

$$\sigma = (-)^{2\omega} \tag{11.3}$$

one finds that the state (11.1) vanishes unless $\Omega = J + J'$ is even. Furthermore, for these allowed O(4) representations the normalized state equals

$$|\omega^2; \Omega L M_L\rangle = \tfrac{1}{\sqrt{2}}|\omega^2; \Omega L M_L\rangle_{\rm nn}. \tag{11.4}$$

12. COEFFICIENTS OF FRACTIONAL PARENTAGE IN O(4)

In the previous section the (anti)symmetric coupling of two identical O(4) representations was investigated and it was shown that the requirement of (anti)symmetry restricts the allowed values for the coupled representation. In this section a system consisting of n identical particles is considered and the construction of (anti)symmetric wavefunctions coupled to a well-defined, total O(4) representation is discussed. All results are derived under the assumption of the generalized Pauli principle.

As an example the case of three particles is considered first and afterwards the extension to a general n-particle system is given. A(n) (anti)symmetric, non-normalized three-particle state is of the form

$$|\omega^3; \Omega L M_L\rangle_{\rm nn} = \mathcal{P}_\sigma |(\omega_1, \omega_2)\Omega_0, \omega_3; \Omega L M_L\rangle \tag{12.1}$$

where $\mathcal{P}_\sigma$ is a symmetrizer for $\sigma = +1$ and an antisymmetrizer for $\sigma = -1$. Although all ω's have the same numerical value, $\omega_1 = \omega_2 = \omega_3 = \omega$, a subscript is attached to them to follow the action of the (anti)symmetrizer. Written explicitly, (12.1) becomes

$$\begin{aligned}|\omega^3; \Omega L M_L\rangle_{\rm nn} &= |(\omega_{12}^2)\Omega_0, \omega_3; \Omega L M_L\rangle + \sigma|(\omega_{13}^2)\Omega_0, \omega_2; \Omega L M_L\rangle \\ &\quad + |(\omega_{23}^2)\Omega_0, \omega_1; \Omega L M_L\rangle. \end{aligned} \tag{12.2}$$

According to the discussion in the previous section a non-vanishing state can only occur if the O(4) representation Ω_0 has even Ω_0. The Ω_0 is left unspecified otherwise and can be used as a classification label for the ω^3 state. The expression (12.2) can be simplified by a recoupling in O(4):

$$\begin{aligned}&|\omega^3; \Omega L M_L\rangle_{\rm nn} \\ &= |(\omega_{12}^2)\Omega_0, \omega_3; \Omega L M_L\rangle \\ &\quad + \sigma \sum_{\Omega_1} \langle(\omega,\omega)\Omega_0, \omega; \Omega|(\omega,\omega)\Omega_1, \omega; \Omega\rangle |(\omega_{12}^2)\Omega_1, \omega_3; \Omega L M_L\rangle \\ &\quad + \sum_{\Omega_1} \langle(\omega,\omega)\Omega_0, \omega; \Omega|(\omega,\omega)\Omega_1, \omega; \Omega\rangle |(\omega_{21}^2)\Omega_1, \omega_3; \Omega L M_L\rangle. \end{aligned} \tag{12.3}$$

Introducing the expression (5.12) for the recoupling matrix each term in the summation (12.3) becomes

$$\delta_{\Omega_0\Omega_1} + \left(\sigma + (-)^{2\omega+\Omega_1}\right)(-)^{2\omega+\Omega_0+\Omega_1}\sqrt{(2\Omega_0+1)(2\Omega_1+1)}\left\{\begin{matrix}\omega & \omega & \Omega_0 \\ \Omega & \omega & \Omega_1\end{matrix}\right\}. \tag{12.4}$$

Since $\sigma = (-)^{2\omega}$ and Ω_0 is even, only terms with even Ω_1 contribute and the ω^3 state becomes

$$
\begin{aligned}
&|\omega^3; \Omega L M_L\rangle_{\rm nn} \\
&\quad = \sum_{\substack{\Omega_1 \\ \Omega_1 \text{even}}} \left[\delta_{\Omega_0\Omega_1} + 2\sqrt{(2\Omega_0+1)(2\Omega_1+1)} \left\{ \begin{array}{ccc} \omega & \omega & \Omega_0 \\ \Omega & \omega & \Omega_1 \end{array} \right\}\right] \\
&\qquad \times |(\omega_{12}^2)\Omega_1, \omega_3; \Omega L M_L\rangle. \qquad (12.5)
\end{aligned}
$$

The corresponding normalized state can be found by making use of the result

$$
\begin{aligned}
&\sum_{\substack{\Omega_1 \\ \Omega_1 \text{even}}} \left[\delta_{\Omega_0\Omega_1} + 2\sqrt{(2\Omega_0+1)(2\Omega_1+1)} \left\{ \begin{array}{ccc} \omega & \omega & \Omega_0 \\ \Omega & \omega & \Omega_1 \end{array} \right\}\right]^2 \\
&\qquad = 3 + 6(2\Omega_0+1) \left\{ \begin{array}{ccc} \omega & \omega & \Omega_0 \\ \Omega & \omega & \Omega_0 \end{array} \right\}. \qquad (12.6)
\end{aligned}
$$

In general, a normalized, (anti)symmetric ω^3 state can be written in terms of *coefficients of fractional parentage* or cfp's associated with O(4):

$$
|\omega^3[\Omega_0]; \Omega L M_L\rangle = \sum_{\Omega_1} [(\omega^2)\Omega_1, \omega; \Omega|\}\omega^3[\Omega_0]; \Omega] |(\omega_{12}^2)\Omega_1, \omega_3; \Omega L M_L\rangle \qquad (12.7)
$$

where the notation $[\Omega_0]$ indicates that the state is obtained after (anti)symmetrization of $|(\omega_{12}^2)\Omega_0, \omega_3; \Omega L M_L\rangle$. The previous derivation can thus be summarized by giving the following closed expression for the $\omega^3 \to \omega^2$ cfp:

$$
\begin{aligned}
&[(\omega^2)\Omega_1, \omega; \Omega|\}\omega^3[\Omega_0]; \Omega] \\
&\quad = \frac{\delta_{\Omega_0\Omega_1} + 2\sqrt{(2\Omega_0+1)(2\Omega_1+1)} \left\{ \begin{array}{ccc} \omega & \omega & \Omega_0 \\ \Omega & \omega & \Omega_1 \end{array} \right\}}{\left[3 + 6(2\Omega_0+1) \left\{ \begin{array}{ccc} \omega & \omega & \Omega_0 \\ \Omega & \omega & \Omega_0 \end{array} \right\}\right]^{1/2}}. \qquad (12.8)
\end{aligned}
$$

The $\omega^3 \to \omega^2$ cfp in O(4) has exactly the same structure as the $j^3 \to j^2$ cfp in O(3) (see, e.g. de-Shalit and Talmi [7]). Note, however, that the theorem which was derived for O(4) recoupling coefficients is *not* valid for O(4) cfp's, that is, an O(4) cfp *cannot* be written as the product of two O(3) cfp's. The expression (12.8) provides us with an example of this result since, together with the corresponding expression in O(3), it shows that

$$
[(\omega^2)\Omega_1, \omega; \Omega|\}\omega^3[\Omega_0]; \Omega] \neq [(j^2)J_1, j; J|\}j^3[J_0]; J][(j'^2)J'_1, j'; J'|\}j'^3[J'_0]; J']. \qquad (12.9)
$$

The above discussion can be generalized to $\omega^n \to \omega^{n-1}$ cfp's. The derivations are rather lengthy but almost identical to the corresponding ones in O(3) (see chapter 26 of the book by de-Shalit and Talmi [7]). Only the most important results are summarized here.

The general expansion in terms of $\omega^n \to \omega^{n-1}$ cfp's reads

$$
|\omega^n\alpha; \Omega L M_L\rangle = \sum_{\alpha_1\Omega_1} [(\omega^{n-1})\alpha_1\Omega_1, \omega; \Omega|\}\omega^n\alpha; \Omega] |(\omega_{1\ldots n-1}^{n-1})\alpha_1\Omega_1, \omega_n; \Omega L M_L\rangle. \qquad (12.10)
$$

The $\omega^n \to \omega^{n-1}$ cfp's satisfy the orthogonality relation

$$
\sum_{\alpha_1\Omega_1} [(\omega^{n-1})\alpha_1\Omega_1, \omega; \Omega|\}\omega^n\alpha; \Omega][(\omega^{n-1})\alpha_1\Omega_1, \omega; \Omega|\}\omega^n\alpha'; \Omega] = \delta_{\alpha\alpha'} \qquad (12.11)
$$

and can be computed via a recurrence relation of the form

$$n[(\omega^{n-1})\alpha_0\Omega_0,\omega;\Omega|\}\omega^n[\alpha_0,\Omega_0];\Omega][(\omega^{n-1})\alpha_1\Omega_1,\omega;\Omega|\}\omega^n[\alpha_0,\Omega_0];\Omega]$$

$$= \delta_{\alpha_0\alpha_1}\delta_{\Omega_0\Omega_1} + (n-1)\sum_{\alpha_2\Omega_2}(-)^{\Omega_0+\Omega_1}\sqrt{(2\Omega_0+1)(2\Omega_1+1)}\left\{\begin{array}{ccc}\Omega_2 & \omega & \Omega_0\\ \Omega & \omega & \Omega_1\end{array}\right\}$$
$$\times[(\omega^{n-2})\alpha_2\Omega_2,\omega;\Omega_0|\}\omega^{n-1}\alpha_0;\Omega_0]$$
$$\times[(\omega^{n-2})\alpha_2\Omega_2,\omega;\Omega_1|\}\omega^{n-1}\alpha_1;\Omega_1]. \qquad (12.12)$$

Similarly, an expansion in terms of $\omega^n \to \omega^{n-2}$ cfp's can be given:

$$|\omega^n\alpha;\Omega LM_L\rangle = \sum_{\Omega_0\alpha_1\Omega_1}[(\omega^{n-2})\alpha_1\Omega_1,(\omega^2)\Omega_0;\Omega|\}\omega^n\alpha;\Omega]$$
$$\times|(\omega^{n-2}_{1\ldots n-2})\alpha_1\Omega_1,(\omega^2_{n-1,n})\Omega_0;\Omega LM_L\rangle. \qquad (12.13)$$

The $\omega^n \to \omega^{n-2}$ cfp's can be computed from $\omega^n \to \omega^{n-1}$ cfp's using the relation

$$[(\omega^{n-2})\alpha_2\Omega_2,(\omega^2)\Omega_0;\Omega|\}\omega^n\alpha;\Omega]$$
$$= \sum_{\alpha_1\Omega_1}(-)^{\Omega_2+\Omega+2\omega}\sqrt{(2\Omega_0+1)(2\Omega_1+1)}\left\{\begin{array}{ccc}\Omega_2 & \Omega & \Omega_0\\ \omega & \omega & \Omega_1\end{array}\right\}$$
$$\times[(\omega^{n-2})\alpha_2\Omega_2,\omega;\Omega_1|\}\omega^{n-1}\alpha_1;\Omega_1]$$
$$\times[(\omega^{n-1})\alpha_1\Omega_1,\omega;\Omega|\}\omega^n\alpha;\Omega]. \qquad (12.14)$$

The matrix element of a one-body operator between (anti)symmetric ω^n states is obtained in terms of $\omega^n \to \omega^{n-1}$ cfp's:

$$\langle\omega^n\alpha_a;\Omega_a\,\|\sum_{i=1}^{n}f_i^{\Omega}\,\|\,\omega^n\alpha_b;\Omega_b\rangle_{O(4)}$$
$$= n\langle\omega\,\|\,f_1^{\Omega}\,\|\,\omega\rangle_{O(4)}$$
$$\times\sum_{\alpha_1\Omega_1}(-)^{\Omega_1+\omega+\Omega_a+\Omega}\sqrt{(2\Omega_a+1)(2\Omega_b+1)}\left\{\begin{array}{ccc}\omega & \omega & \Omega\\ \Omega_a & \Omega_b & \Omega_1\end{array}\right\}$$
$$\times[(\omega^{n-1})\alpha_1\Omega_1,\omega;\Omega_a|\}\omega^n\alpha_a;\Omega_a]$$
$$\times[(\omega^{n-1})\alpha_1\Omega_1,\omega;\Omega_b|\}\omega^n\alpha_b;\Omega_b]. \qquad (12.15)$$

Similarly, the matrix element of a two-body operator involves $\omega^n \to \omega^{n-2}$ cfp's. For an O(4) scalar operator one finds

$$\langle\omega^n\alpha_a;\Omega LM_L|\sum_{i\leq j}^{n}g_{ij}^{(0,0)}|\omega^n\alpha_b;\Omega LM_L\rangle$$
$$= \frac{n(n-1)}{2}\sum_{\Omega_0\alpha_1\Omega_1}\langle\omega^2;\Omega_0L_0M_0|g_{12}^{(0,0)}|\omega^2;\Omega_0L_0M_0\rangle$$
$$\times[(\omega^{n-2})\alpha_1\Omega_1,(\omega^2)\Omega_0;\Omega|\}\omega^n\alpha_a;\Omega]$$
$$\times[(\omega^{n-2})\alpha_1\Omega_1,(\omega^2)\Omega_0;\Omega|\}\omega^n\alpha_b;\Omega]. \qquad (12.16)$$

The relations (12.15) and (12.16) express the matrix elements of one-body and two-body operators between (anti)symmetric n-particle states in terms of matrix elements between one- and two-particle states and $\omega^n \to \omega^{n-1}$ and $\omega^n \to \omega^{n-2}$ cfp's. For $n = 3$

closed expressions are available for the latter, relations (12.8) and (12.14), and for $n > 3$ they can be computed recursively with the help of (12.12) and (12.14).

13. CONCLUSION

The main objective of this paper was to develop the Wigner–Racah algebra associated with O(4), the group of rotations in four dimensions. Explicit expressions were given for O(4) coupling and recoupling coefficients of interest and the O(4) tensor calculus was developed in detail, giving many of the relations that could be of use in practical applications. As mentioned in the introduction, it appears that O(4), although of fundamental importance in the description of the hydrogen atom, has little relevance in many-electron theory. On the other hand, calculations of rotation–vibration properties and of electronic properties of diatomic and triatomic molecules in the framework of the vibron model show that O(4) can be considered as an approximate (dynamical) symmetry for such systems. The formalism presented here aims primarily at applications in molecular structure and can, for instance, already be used to simplify existing calculations for diatomic and triatomic molecules. It conceivably could become an essential element in the algebraic description of complex polyatomic molecules: each of the bonds in the molecule can be associated with an O(4) group and, in the case several bonds, a proper treatment of the (re)coupling of O(4) representations is indispensable. In addition, for homonuclear molecules, considerations related to the permutation symmetry of the molecular wavefunction become important. The connection between O(4) and (anti)symmetry was established in §11 and in §12 the formalism of coefficients of fractional parentage in O(4) was developed to construct states with definite O(4) symmetry and definite permutation symmetry. All these results should facilitate future developments in an algebraic theory of polyatomic molecules.

ACKNOWLEDGEMENTS

I wish to thank F. Iachello for suggesting the problem discussed in this paper and A. Frank for encouraging me to write the paper, and both for helpful discussions.

REFERENCES

1. E.P. Wigner, *Gruppentheorie und ihre Anwendung auf die Quantenmechanik der Atomspektren,* Vieweg, (1931).
2. G. Racah, *Phys. Rev.* **61**, 186 (1942); *ibid.* **62**, 438 (1942); *ibid.* **63**, 367 (1943); *ibid.* **76**, 1352 (1949).
3. M.E. Rose, *Elementary Theory of Angular Momentum,* Wiley, (1957).
4. A.R. Edmonds, *Angular Momentum in Quantum Mechanics,* Princeton University Press, (1957).
5. D.M. Brink and G.R. Satchler, *Angular Momentum,* Oxford University Press, (1962).
6. A.P. Yutsis, I.B. Levinson and V.V. Vanagas, *Mathematical Apparatus of the Theory of Angular Momentum,* Israel Program for Scientific Translations, (1962).
7. A. de-Shalit and I. Talmi, *Nuclear Shell Theory,* Academic, (1963).
8. B.G. Wybourne, *Classical Groups for Physicists,* Wiley–Interscience, (1974).
9. J.S. Alper, *Phys. Rev.* **177**, 86 (1969).

10. P.H. Butler and B.G. Wybourne, *J. Math. Phys.* **11**, 2519 (1970).

11. F. Iachello and R.D. Levine, *J. Chem. Phys.* **77**, 8046 (1982).

12. O.S. van Roosmalen, F. Iachello, R.D. Levine, and A.E.L. Dieperink, *J. Chem. Phys.* **79**, 2515 (1983).

13. A. Frank, F. Iachello, and R. Lemus, *Chem. Phys. Lett.* **131**, 380 (1986).

14. L.C. Biedenharn and J.D. Louck, *Angular Momentum in Quantum Physics*, Addison-Wesley, (1981).

15. L.C. Biedenharn and J.D. Louck, *The Racah–Wigner Algebra in Quantum Theory*, Addison-Wesley, (1981).

16. R. Gilmore, *Lie Groups, Lie Algebras, and Some of Their Applications*, Wiley-Interscience, (1974).

17. L.C. Biedenharn, *J. Math. Phys.* **2**, 433 (1961).

REPRESENTATIONS OF THE QUANTUM ALGEBRAS $U_q(su(2))$ AND $U_q(su(1,1))$

A.U. Klimyk[1] , Yu.F. Smirnov[2] , B. Gruber[3]

[1] Institute for Theoretical Physics
Ukrainian Academy of Sciences, Kiev 130, USSR

[2] Nuclear Physics Institute
Moscow State University, Moscow, USSR

[3] Department of Physics and Molecular Science Program
Southern Illinois University
Carbondale, Illinois, USA 62901-4401
and
Arnold Sommerfeld Institute for Mathematical Physics
Technical University Clausthal
D-3392 Clausthal, Germany

1. INTRODUCTION

Quantum Groups and Algebras made their appearance in the quantum methodical treatment of the inverse scattering problem [1–3]. Presently the representation theory for the quantum groups and algebras is being actively and vigorously pursued. The results to be obtained are of considerable interest to quantum field theory as well as to statistical physics [4]. Recent papers [5-9] also show that representations of quantum groups and algebras are closely related to the so-called basic hypergeometric functions.

Irreducible finite dimensional representations for the quantum algebra $U_q(su(2))$ were constructed in [3]. Representations of the quantum group $SU_q(2)$ are discussed in [5,10]. Kirillov and Reshetickin [11] have studied the Clebsch-Gordon coefficients (CGC's) and Racah coefficients of the quantum algebra $U_q(su(2))$. In references [12–15] the representation theory is further developed.

Biedenharn [16] and Macfarlane [17] have considered a q-analogue of the quantum harmonic oscillator, which is related to the quantum algebra $U_q(su(2))$. Biedenharn

[18] has also initiated a study of the q-analogue of tensor operators which are related to the CGC's and Racah coefficients of $U_q(su(2))$.

The quantum algebra $U_q(su(2))$ is closely related to the "non-compact" quantum algebra $U_q(su(1,1))$. Representations of this quantum algebra were studied in [19–21]. The classification of the irreducible and unitary representations of $U_q(su(1,1))$, for which the diagonal operator has integral or half-integral eigenvalues, is obtained in these articles. Apart from these unitary representations, the quantum algebra $U_q(su(1,1))$ has a series of irreducible representations with highest (lowest) weights. These are the analogues of the representations with highest (lowest) weight of the classical algebra $su(1,1)$ [22,23]. In this article we construct these representations for the quantum algebra $U_q(su(1,1))$ and moreover we review the results which have been mentioned above.

The representation theory of $U_q(su(2))$ and $U_q(su(1,1))$ is closely connected with the so-called q-calculus (the q-analogue of the usual differential and integral calculus). Hence, we give a brief review of the q-calculus in the following section 2.

2. THE q-CALCULUS AND BASIC HYPERGEOMETRIC FUNCTIONS

The expression

$$(a;q)_n = \prod_{j=0}^{n-1}(1-aq^j),\ (a;q)_0 = 1,\ a \in \mathbf{C},\ n \in Z_+, \tag{1}$$

where q is a fixed complex number and Z_+ is the set of non-negative integers, is important for the q-calculus. We have

$$\lim_{q\to 1}\frac{(q^\alpha;q)_n}{(q^\beta;q)_n} = \frac{(\alpha)_n}{(\beta)_n}, \quad \text{where} \quad (\alpha)_n = \frac{\Gamma(\alpha+n)}{\Gamma(\alpha)}.$$

In the representation theory of quantum groups and algebras it is convenient to use the expression

$$[m] = \frac{q^{m/2}-q^{-m/2}}{q^{1/2}-q^{-1/2}} = q^{-(m-1)/2}\frac{1-q^m}{1-q} \tag{2}$$

instead of $(1-q^m)$. We have

$$[m+n] = q^{n/2}[m] + q^{-m/2}[n] = q^{-n/2}[m] + q^{m/2}[n],$$
$$[m-n] = q^{n/2}[m] - q^{m/2}[n] = q^{-n/2}[m] - q^{-m/2}[n].$$

Instead of the usual factorial $m!$ we have in the theory of quantum groups and algebras the q-factorial

$$[m]! = [1][2][3]\dots[m] = \frac{q^{-m(m-1)/4}}{(1-q)^m}(q;q)_m, \tag{3}$$

where $m \in Z_+$, with $[0]! = 1$. The expressions $[m]!$ and $(q^\alpha; q)_m$ are connected by the formulas

$$(q^{N+1}; q)_r = \frac{[N+r]!}{[N]!}(1-q)^r q^{r(r+2N-1)/4}, \tag{4}$$

$$(q^{-N}; q)_r = \frac{[N]!}{[N-r]!}(-1)^r (1-q)^r q^{r(r-2N-3)/4}, \tag{5}$$

$$(q^{-1}; q^{-1})^r = (-1)^r q^{-r(r+1)/2}(q; q)_r, \tag{6}$$

$$\lim_{r\to\infty} \frac{[r+\alpha]}{[r+\beta]} = q^{-(\alpha-\beta)/2}, \quad |q| < 1, \tag{7}$$

$$\lim_{r\to\infty} \frac{[r+\alpha]}{[r+\beta]} = q^{(\alpha-\beta)/2}, \quad |q| > 1. \tag{8}$$

The q-analogue of the binomial formula has the form

$$(1-x)^{(n)} = (1-x')(1-qx')\ldots(1-q^{n-1}x') = (x', q)_n =$$
$$= \sum_{r=0}^{n} \begin{bmatrix} n \\ r \end{bmatrix}_q (-x)^r q^{r(r-2n-1)/2}, \quad x' = q^{-n}x, \tag{9}$$

where $\begin{bmatrix} n \\ r \end{bmatrix}_q$ is the q-analogue of the binomial coefficients:

$$\begin{bmatrix} n \\ r \end{bmatrix}_q = \frac{(q;q)_n}{(q;q)_r (q;q)_{n-r}} = \frac{[n]! q^{r(n-r)/2}}{[r]![n-r]!}. \tag{10}$$

For $(b-x)^{(n)}$ we have

$$(b-x)^{(n)} = b^n \left(1 - \frac{x}{b}\right)^{(n)} = \sum_{r=0}^{n} \frac{(q^{-n}; q)_r}{(q;q)_r} x^r b^{n-r}. \tag{11}$$

In the q-calculus the differentiation operator is replaced by the difference operator $\widehat{B}_x$, where

$$\widehat{B}_x f(x) = \frac{f(qx) - f(x)}{x(q-1)}. \tag{12}$$

If $q \to 1$, then $\widehat{B}_x \to d/dx$. For $\widehat{B}_x^n$ we have

$$q^{n(n-1)/2} x^n (q-1)^n \widehat{B}_x^n f(x) =$$
$$= \sum_{r=0}^{\infty} \begin{bmatrix} r \\ n \end{bmatrix}_q (-1)^r q^{r(r-1)/2} f(q^{n-r}x). \tag{13}$$

The inverse of the operator $\widehat{B}_x$ is the q-integration operator. For $0 < q < 1$ the definite q-integral is defined as

$$\int_0^c f(x)d_qx = c(1-q)\sum_{r=0}^{\infty} q^r f(q^r c) = \sum_{r=0}^{\infty}(x_r - x_{r-1})f(x_r), \tag{14}$$

where $x_r = cq^r$ and $0 < c < \infty$. For $q > 1$ we have

$$\int_c^{\infty} f(x)d_qx = c(q-1)\sum_{r=0}^{\infty} q^r f(q^r c) = \sum_{r=0}^{\infty}(x_{r+1} - x_r)f(x_r). \tag{15}$$

In particular

$$\widehat{B}_x x^{\mu} = [[\mu]]x^{\mu-1}, \quad \widehat{B}_x(1+ax)^{(\nu)} = a[[\nu]](1+aqx)^{(\nu-1)},$$

where $[[\mu]] = (q^{\mu}-1)/(q-1) = q^{(\mu-1)/2}[\mu]$.

The q-analogue of the exponential function is

$$E_q(x) = \sum_{r=0}^{\infty} \frac{x^r}{[[r]]!}, \tag{16}$$

where $[[r]]! = [[1]][[2]]\ldots[[r]] = (q;q)_r/(1-q)^r$. We have

$$\widehat{B}_x E_q(ax) = aE_q(ax).$$

The q-analogue of gamma-function is defined in the same way as in the classical case with the help of q-integral:

$$\int_0^{\infty} x^{\nu-1}E_q(-x)d_qx = q^{\nu(\nu+1)/2}\Gamma_q(\nu),$$

where

$$\int_0^{\infty} f(x)d_qx = (1-q)\sum_{n=-\infty}^{\infty} f(q^n)q^n.$$

This function has the properties

$$\begin{aligned} \Gamma_q(n) &= [[n-1]]! = (q;q)_{n-1}/(1-q)^{n-1}, \quad n \in Z_+, \\ \Gamma_q(\nu+1) &= [[\nu]]\Gamma_q(\nu), \quad \nu \in \mathbf{C}. \end{aligned} \tag{17}$$

The q-analogue of the Taylor formula has the form

$$f(x) = f(a) + \frac{(x-a)^{(1)}}{[[1]]!}\widehat{B}_x f(a) + \frac{(x-a)^{(2)}}{[[2]]!}\widehat{B}_x^2 f(a) + \ldots, \tag{18}$$

where $(x-a)^{(n)}$ is of the form (11) and

$$\widehat{B}_x^n f(a) = \{\widehat{B}_x^n f(x)\}\Big|_{x=a}.$$

The usual hypergeometric functions are replaced in the q-calculus by the so-called basic hypergeometric functions. These are defined by means of $(a,q)_n$:

$$\begin{aligned} &{}_{n+1}\varphi_n(a_1,\dots,a_{n+1};b_1,\dots,b_n;q;z) \\ &\equiv {}_{n+1}\varphi_n\left(\begin{matrix} a_1,\dots,a_{n+1} \\ b_1,\dots,b_n \end{matrix}\Bigg| q,z\right) \\ &= \sum_{r=0}^{\infty} \frac{(a_1;q)_r\cdots(a_{n+1};q)_r}{(b_1;q)_r\cdots(b_n;q)_r} \, \frac{z^r}{(q;q)_r}. \end{aligned} \tag{19}$$

Sometimes it is convenient to use the notation

$$\begin{aligned} &{}_{n+1}\Phi_n(\alpha_1,\dots,\alpha_{n+1};\beta_1,\dots,\beta_n;q,z) \\ &= {}_{n+1}\varphi_n(q^{\alpha_1},\dots,q^{\alpha_{n+1}};q^{\beta_1},\dots,q^{\beta_n};q,z). \end{aligned} \tag{20}$$

For $-\alpha_i \in Z_+$ this series terminates and we have a polynomial. The following formula holds,

$$(1-x)^{(n)} = {}_1\varphi_0(q^{-n};q,x).$$

Therefore, for $a \in \mathbf{C}$ we have

$$(1-x)^{(a)} = {}_1\varphi_0(q^{-a};q,x).$$

The formula

$$\begin{aligned} &\int_0^1 u^{a-1}(1-qu)^{(c-a-1)}{}_n\Phi_{n-1}(d_1,\dots,d_n;e_1,\dots,e_{n-1};q,ux)d_q u \\ &= \frac{\Gamma_q(a)\Gamma_q(c-a)}{\Gamma_q(c)} {}_{n+1}\Phi_n(d_1,\dots,d_n,a;e_1,\dots,e_{n-1},c;q,x), \end{aligned} \tag{21}$$

where Re $a>0$, Re$(c-a)>0$, is the q-analogue of a well-known formula for the usual hypergeometric functions. In particular,

$$\begin{aligned} {}_2\Phi_1(a,b;c;q,x) &= \big(\Gamma_q(c)/\Gamma_q(a)\Gamma_q(c-a)\big) \\ &\times \int_0^1 u^{a-1}(1-qu)^{(c-a-1)}(1-q^b xu)^{(-b)} d_q u \end{aligned} \tag{22}$$

This is the q-analogue of the well-known integral representation of the hypergeometric function ${}_2F_1(a,b;c;x)$.

The relations

$$\begin{aligned} {}_3\varphi_2(q^{-n},a,b;c,d;q,q) &= a^n(c/a;q)_n/(c;q)_n \\ &\times {}_3\varphi_2(q^{-n},a,d/b;aq^{-n+1}/c,d;q,bq/c), \end{aligned} \tag{23}$$

$$ {}_4\varphi_3\left(\begin{matrix} q^{-n}, a, b, c \\ d, e, f \end{matrix}\middle| q, q\right) = \left(\frac{bc}{d}\right)^n \frac{(aq^{1-n}/e;q)_n (aq^{1-n}/f;q)_n}{(e;q)_n (f;q)_n} $$

$$ \times {}_4\varphi_3\left(\begin{matrix} q^{-n}, a, d/b, d/c \\ d, aq^{1-n}/e, aq^{1-n}/f \end{matrix}\middle| q, q\right), \qquad q^{1-n}abc = def, \tag{24} $$

are of great importance for the theory of Clebsch-Gordan and Racah coefficients of $U_q(su(2))$.

3. THE QUANTUM ALGEBRA $U_q(su(2))$

The classical Lie algebra $su(2)$ is generated by matrices H', E_+, E_-, which satisfy the commutation relations

$$ [H', E'_\pm] = \pm 2E'_\pm, \qquad [E'_+, E'_-] = H'. \tag{25} $$

We set $q = e^h$ and deform the relations (25) into

$$ \left. \begin{aligned} [H, E_+] = 2E_+, \quad [H, E_-] = -2E_-, \\ [E_+, E_-] = \frac{\sinh\frac{h}{2}H}{\sinh\frac{h}{2}} \equiv \frac{q^{H/2} - q^{-H/2}}{q^{1/2} - q^{-1/2}}, \end{aligned} \right\} \tag{26} $$

where $\sinh \frac{h}{2}H$ means the infinite formal series

$$ \sinh\frac{h}{2}H = \frac{h}{2}H + \frac{1}{3!}\left(\frac{h}{2}H\right)^3 + \frac{1}{5!}\left(\frac{h}{2}H\right)^5 + \dots . $$

The associative algebra, generated by the elements H, E_+, E_- which obey the relations (26), is called the quantum algebra $U_q(su(2))$. Its elements are finite, or infinite, series of H, E_+, E_-. In order to have a finite series one uses e^H instead of H. Somtimes $U_q(su(2))$ is also called the deformation of the universal enveloping algebra $U(su(2))$.

The element

$$ C = \left(\frac{q^{(H+1)/4} - q^{-(H+1)/4}}{q^{1/2} - q^{-1/2}}\right)^2 + E_-E_+ \in U_q(su(2)) \tag{27} $$

of $U_q(su(2))$ commutes with H, E_+, E_- and, consequently, with all elements of $U_q(su(2))$. It is called the Casimir element of $U_q(su(2))$.

The elements E_+^n, E_-^m, $m \geq n$, satisfy the relation

$$ [E_+^n, E_-^m] = E_-^{m-n} \frac{[m]!}{[m-n]!} \prod_{j=1}^{n} \frac{q^{\frac{m-j}{2}} q^{H/2} - q^{-\frac{m-j}{2}} q^{-H/2}}{q^{1/2} - q^{-1/2}} . \tag{28} $$

On $U_q(su(2))$ one introduces the structure of a Hopf algebra. We thus need to give the definition of a Hopf algebra. Let A be an associative algebra over C with unit

I. We introduce the operations $e : \mathbf{C} \to A$ and $m : A \otimes A \to A$ by means of the formulas

$$e(\alpha) = \alpha I, \qquad \alpha \in \mathbf{C}; \qquad m(a \otimes b) = ab, \qquad a, b \in A.$$

The mapping m is a homomorphism of the associative algebras. The algebra A with the operations e and m is called bialgebra if the homomorphisms $\epsilon : A \to \mathbf{C}$ and $\Delta : A \to A \otimes A$ are defined, such that the diagram

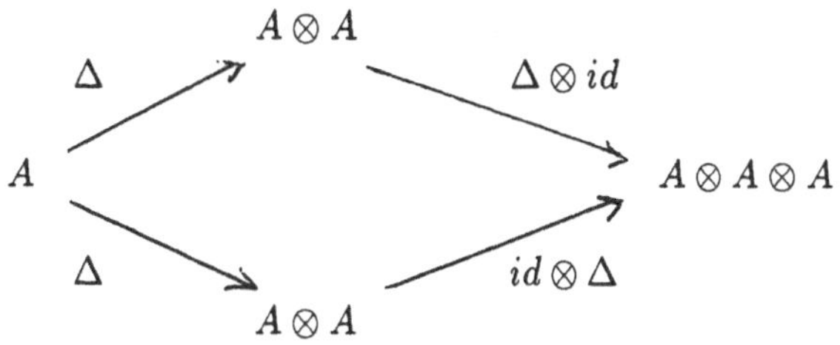

is commutative, where *id* is the identity mapping. That is

$$(\Delta \otimes id)\Delta = (id \otimes \Delta)\Delta.$$

A bialgebra A is a Hopf algebra if an antiauthomorphism S can be introduced into A such that the diagram

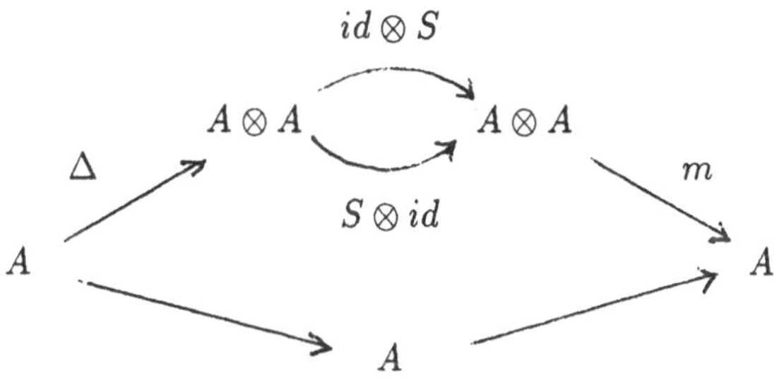

is commutative. The mapping S is called an antipode.

We introduce the operations Δ, ϵ and S by means of the formulas

$$\Delta(E_\pm) = E_\pm \otimes q^{H/4} + q^{-H/4} \otimes E_\pm, \tag{29}$$

$$\Delta(H) = H \otimes 1 + 1 \otimes H, \tag{29'}$$

$$S(E_\pm) = -q^{\pm 1/2} E_\pm, \qquad S(H) = -H, \tag{29''}$$

$$\epsilon(H) = \epsilon(E_\pm) = 0$$

and continue these relations onto the whole of the algebra $U_q(su(2))$ considering that Δ and ϵ are homomorphisms and S is an antiauthomorphism. These operations turn $U_q(su(2))$ into a Hopf algebra.

The mappings ϵ, $\Delta' = \sigma\Delta$, $S' = S^{-1}$, where the linear operation σ acts on $U_q(su(2))^2 \equiv U_q(su(2)) \otimes U_q(su(2))$ as a permutation: $\sigma(a \otimes b) = b \otimes a$, also defines a Hopf algebra on $U_q(su(2))$ which we denote by $U'_q(su(2))$. The formulas (29)–(29″) shows that

$$U'_q(su(2)) = U_{q^{-1}}(su(2)).$$

A simple (but cumbersome) verification shows that the operations Δ and Δ' are related by the formula

$$\Delta'(a) = R\Delta(a)R^{-1}, \qquad a \in U_q(su(2)),$$

where $R \in U_q(su(2))$ is defined as

$$R = \exp(\frac{h}{2}H \otimes H)\sum_{n=0}^{\infty}\frac{(1-q^{-1})^n}{[n]!} \times ((\exp\frac{h}{2}H)E_+)^n \otimes ((\exp(-\frac{h}{2}H)E_-)^n.$$

The elements R^{-1} and R are connected by the formula $(S \otimes id)R = R^{-1}$.

The element R is called the universal R-matrix. It satisfies the relations

$$(\Delta \otimes id)R = R_{13}R_{23}, \tag{30}$$

$$(id \otimes \Delta)R = R_{13}R_{12}, \tag{31}$$

where the indices i and j of R imply that R acts on i-th and j-th multipliers of the space

$$U_q(su(2))^3 \equiv U_q(su(2)) \otimes U_q(su(2)) \otimes U_q(su(2)).$$

It follows from (30) and (31) that

$$R_{12}R_{13}R_{23} = R_{23}R_{13}R_{12}.$$

This relation is called the Yang-Baxter equation. It is of great importance for the quantization theory.

4. FINITE DIMENSIONAL REPRESENTATIONS OF $U_q(su(2))$

We shall consider those finite dimensional representations T of $U_q(su(2))$ for which holds

$$T(E_\pm)^* = T(E_\mp), \qquad T(H)^* = T(H). \tag{32}$$

It has been proven that the finite dimensional representations of $U_q(su(2))$ are completely reducible, that is, that they decompose into a direct sum of irreducible representations.

In order to construct a finite dimensional representation of $U_q(su(2))$ it is sufficient to construct the operators $T(H)$, $T(E_\pm)$, which satisfy the relations (26). It follows from (32) that the operator $T(H)$ is Hermitian. Therefore we can take as basis of the carrier space $\mathcal{H}$ of the representation T the eigenvectors of $T(H)$. Using this fact we can solve the commutation relations (26) in the same way as in the classical case. As a consequence we obtain the following result. The finite dimensional irreducible representations of $U_q(su(2))$ are characterized (up to an isomorphism) by integral or half-integral numbers l. If an irreducible representation T_l corresponds to the number l, then there exists an orthogonal basis $\{e_m; m = -l, -l+1, \ldots, l\}$ of the carrier space $\mathcal{H}_l$, for which

$$T_l(E_+)e_m = \sqrt{[l-m][l+m+1]}e_{m+1},$$
$$T_l(E_-)e_m = \sqrt{[l+m][l-m+1]}e_{m-1}, \qquad (33)$$
$$T_l(H/2)e_m = me_m,$$

where $[m]$ is given by formula (2).

Thus, there is the one-to-one correspondence between the finite dimensional representations of $U_q(su(2))$ and those of the classical group $SU(2)$. Moreover, the corresponding representations act upon the same space.

The operator $T_l(C)$, where C is the Casimir operator, is a multiple of the identity operator,

$$T_l(C)e_m = \left(\frac{q^{(2l+1)/4} - q^{-(2l+1)/4}}{q^{1/2} - q^{-1/2}}\right)^2 e_m = [l + \frac{1}{2}]^2 e_m.$$

Using (33) we find that

$$T_l(E_\pm^n)e_m = \left(\frac{[l\mp m]![l\pm m+n]!}{[l\mp m-n]![l\pm n]!}\right)^{1/2} e_{m\pm n},$$

where $|m\pm n| \leq l$.

In the same manner as for the classical case, we find that

$$T_{l_1} \otimes T_{l_2} = \sum_l \oplus T_l, \qquad (34)$$

where the summation is over $l = |l_1 - l_2|, |l_1 - l_2| + 1, \ldots, l_1 + l_2$. However, the relation (29) means that

$$(T_{l_1} \otimes T_{l_2})(a) \neq (T_{l_2} \otimes T_{l_1})(a), \qquad a \in U_q(su(2)).$$

If t^l_{mn} denote the matrix elements of T_l in the basis $\{e_n\}$, then this inequality means that

$$t^l_{mn} t^{l'}_{m'n'} \neq t^{l'}_{m'n'} t^l_{mn}.$$

5. CLEBSCH-GORDAN COEFFICIENTS OF $U_q(su(2))$

Let $\{e_j\}$, $\{e'_k\}$, $\{e^l_m\}$ be the bases of the carrier spaces $\mathcal{H}_1$, $\mathcal{H}_2$, $\mathcal{H}_l$ for the representations of equation (34). Then the relation

$$e^l_m = \sum_{j,k} \begin{bmatrix} l_1 & l_2 & l \\ j & k & m \end{bmatrix}_q e_j \otimes e'_k$$

defines the Clebsch-Gordan coefficients (CGC's) for the tensor product $T_{l_1} \otimes T_{l_2}$. According to (3) a CGC is equal to zero if $m \neq j + k$. Therefore, below we assume that $j + k = m$. The CGC's of the algebra $U_q(su(2))$ satisfy orthogonality relations which have the same form as for the classical case.

The method of highest vectors [25] can be used for the evaluation of the CGC's. A detailed evaluation is given in [15]. We have

$$\begin{bmatrix} l_1 & l_2 & l \\ j & k & m \end{bmatrix}_q = (-1)^{l_1-j} q^B \frac{\Delta(l_1,l_2,l)[l+l_2-j]!}{[l_1-l_2+l]![l+l_2-l_1]![l_2-l+j]!}$$

$$\times \left(\frac{[l_1+j]![l_2-k]![l+m]![2l+1]}{[l_1-j]![l_2+k]![l-m]!}\right)^{1/2} {}_3\Phi_2\left(\begin{matrix} j-l_1,\ l_1+j+1,\ m-l \\ j-l+l_2+1,\ j-l-l_2 \end{matrix}\middle| q,q\right), \tag{35}$$

where

$$B = \frac{1}{4}\{l_2(l_2+1) - l_1(l_1+1) - l(l+1) + 2j(m+1)\},$$

$$\Delta(l_1,l_2,l) = \left(\frac{[l_1+l_2-l]![l_1-l_2+l]![l-l_1+l_2]!}{[l_1+l_2+l+1]!}\right)^{1/2}.$$

This is the q-analogue of Racah's formula for the CGC's of $SU(2)$. Applying the relation (23) to equation (35) we obtain the expressions for CGC's of $U_q(su(2))$. These are given in [15].

The CGC's of $U_q(su(2))$ satisfy the symmetry relations

$$\begin{bmatrix} l_1 & l_2 & l \\ j & k & m \end{bmatrix}_q = (-1)^{l_1+l_2-l} \begin{bmatrix} l_1 & l_2 & l \\ -j & -k & -m \end{bmatrix}_{q^{-1}},$$

$$\begin{bmatrix} l_1 & l_2 & l \\ j & k & m \end{bmatrix}_q = \begin{bmatrix} l_2 & l_1 & l \\ -k & -j & -m \end{bmatrix}_q,$$

$$\begin{bmatrix} l_1 & l_2 & l \\ j & k & m \end{bmatrix}_q = (-1)^{l-l_1-k} q^{-k/2} \left(\frac{[2l+1]}{[2l_1+1]}\right)^{1/2} \begin{bmatrix} l & l_2 & l_1 \\ m & -k & j \end{bmatrix}_q,$$

$$\begin{bmatrix} l_1 & l_2 & l \\ j & k & m \end{bmatrix}_q = \begin{bmatrix} \frac{1}{2}(l_1+l_2+m) & \frac{1}{2}(l_1+l_2-m) & l \\ \frac{1}{2}(l_1-l_2+j-k) & \frac{1}{2}(l_1-l_2-j+k) & l_1-l_2 \end{bmatrix}_q.$$

These relations generate the symmetry group for CGC's, analogous to the case of the classical CGC's [25].

Generating functions exist for the CGC's of $U_q(su(2))$ which are analogous to those for the classical CGC's. For example, we have

$$\left(\frac{[a-b+c]![a+\alpha]![2c+1]}{[b+c-a]![a+b-c]![a+b+c+1]![a-\alpha]!}\right)^{1/2} \frac{(1+x)^{(b+c-a)}}{[c-b+\alpha]!}$$

$$\times {}_2\Phi_1(c-b-a,\ -a+\alpha;\ c-b+\alpha+1;\ q,\ -q^{a-b-c}x)$$

$$= \sum_{\beta-\gamma=\alpha} \begin{bmatrix} a & b & c \\ \alpha & \beta & \gamma \end{bmatrix}_q \frac{q^A x^{b+\beta}}{([b+\beta]![b-\beta]![c+\gamma]![c-\gamma]!)^{1/2}}, \tag{36}$$

where $A = \frac{1}{4}\{b(2a-2\alpha-2c-3b-3)-(a-c)(a+c+1)\}$,

$$\frac{E_q(x_1t_2)E_q(x_2t_3)E_q(x_3t_1)}{E_q(x_1t_3)E_q(x_2t_1)E_q(x_3t_2q)} = \sum_{\substack{a,b,c\\ \alpha,\beta}} \begin{bmatrix} a & b & c \\ \alpha & \beta & \gamma \end{bmatrix}_q$$

$$\times \frac{x_1^{-a+b+c}x_2^{a-b+c}x_3^{a+b-c}t_1^{a+\alpha}t_2^{b+\beta}t_3^{c-\gamma}(1-q)^A q^B}{\Delta(a,b,c)I(a,b,c,\alpha,\beta)}, \tag{37}$$

where $\gamma = \alpha+\beta$, $A = a+b-2c+\beta-\alpha$,

$$B = -\frac{1}{2}\{a(3a-1)+b(3b-1)+c(3c+1)+\alpha(a+c-2b+1)$$
$$+\beta(b+c-2a+1)\} + ab+ac+bc,$$

$$I(a,b,c,\alpha,\beta) = ([a+\alpha]![a-\alpha]![b+\beta]![b-\beta]![c+\gamma]![c-\gamma]!)^{1/2}, \tag{38}$$

$$(t_1-qt_2)^{(a+b-c)}(t_2-t_3)^{(b+c-a)}(t_3-t_1)^{(a-b+c)} = (-1)^{a+b-2c}$$
$$\times \frac{\Delta(a,b,c)[a+b+c+1]!}{([2c+1])^{1/2}} \sum_{\alpha,\beta} \begin{bmatrix} a & b & c \\ \alpha & \beta & \gamma \end{bmatrix}_q q^D(-1)^{\beta-\alpha}\frac{t_1^{a+\alpha}t_2^{b+\beta}t_3^{c-\gamma}}{I(a,b,c,\alpha,\beta)}, \tag{39}$$

where $\gamma = \alpha+\beta$, $I(a,b,c,\alpha,\beta)$ is given by (38) and

$$D = \frac{1}{2}\{(c-a-\beta)(a-b-c-1)+(c-b-\alpha)(b-c-a-2)+\alpha b-\beta a\}-\frac{1}{4}(a+b-c)(a+b+c+1).$$

For CGC's of $U_q(su(2))$ we have the integral representation

$$E\begin{bmatrix} a & b & c \\ \alpha & \beta & \gamma \end{bmatrix}_q = \int_0^1 u^{b-\beta}(1-qu)^{(a-\alpha)}$$
$$\times\, {}_2\Phi_1(-c-\gamma, c-\gamma+1; b-a-\gamma+1; q, u)d_qu$$
$$= (1-q)\sum_{r=0}^{\infty} q^{r(b-\beta+1)}(1-q^{r+1})^{(a-\alpha)}$$
$$\times\, {}_2\Phi_1(-c-\gamma, c-\gamma+1; b-a-\gamma+1; q, q^r), \tag{40}$$

where

$$E = (-1)^{b+c+\beta+\gamma}q^{-A}\frac{[a-b+c]![a+b-\gamma+1]!I(a,b,c,\alpha,\beta)}{\Delta(a,b,c)[a+b+c+1]![c-\gamma]![2c+1]^{1/2}},$$

$$A = \frac{1}{4}\{a(a+1)-b(b+1)+c(c+1)+2\beta(\gamma+1)\}.$$

6. THE QUANTUM GROUP $SU_q(2)$

Since the quantum algebra $U_q(su(2))$ is not a Lie algebra in the usual sense we do not have an exponential map of $U_q(su(2))$ onto a Lie group or onto its analogue.

The Lie group $SU(2)$ can however *also* be *completely* characterized with the help of the *space of functions* on $SU(2)$, or the *space of functions* on its *universal enveloping algebra* $U(su(2))$. We thus understand the quantum *group* $SU_q(2)$ as the space of functions on $U_q(su(2))$ which satisfy additional conditions which turn it into a Hopf algebra. In order to introduce the structure of a Hopf algebra into the space of functions on $U_q(su(2))$ we make use of the matrix elements t^l_{mn} of the representations T_l of the quantum algebra $U_q(su(2))$. These matrix elements are functions on $U_q(su(2))$.

As in the classical case, the CGC's and the matrix elements t^l_{mn} are connected by the relations

$$\sum_l \begin{bmatrix} l_1 & l_2 & l \\ j_1 & k_1 & m_1 \end{bmatrix}_q \begin{bmatrix} l_1 & l_2 & l \\ j_2 & k_2 & m_2 \end{bmatrix}_q t^l_{m_1 m_2} = t^{l_1}_{j_1 j_2} t^{l_2}_{k_1 k_2}, \tag{41}$$

$$\sum_{j_1, j_2} \begin{bmatrix} l_1 & l_2 & l \\ j_1 & k_1 & m_1 \end{bmatrix}_q \begin{bmatrix} l_1 & l_2 & l \\ j_2 & k_2 & m_2 \end{bmatrix}_q t^{l_1}_{j_1 j_2} t^{l_2}_{k_1 k_2} = t^l_{m_1 m_2}, \tag{42}$$

(we consider the CGC's to be real).

Let t_{11}, t_{12}, t_{21}, t_{22} denote respectively the matrix elements $t^{1/2}_{1/2,1/2}$, $t^{1/2}_{1/2,-1/2}$, $t^{1/2}_{-1/2,1/2}$, $t^{1/2}_{-1/2,-1/2}$ of the simplest irreducible representation $T_{1/2}$ of $U_q(su(2))$ in the basis $\{e_{1/2}, e_{-1/2}\}$. The CGC's of the tensor product $T_{1/2} \otimes T_{-1/2}$ are given by the formulas

$$\begin{bmatrix} \frac{1}{2} & \frac{1}{2} & 1 \\ \frac{1}{2} & \frac{1}{2} & 1 \end{bmatrix}_q = \begin{bmatrix} \frac{1}{2} & \frac{1}{2} & 1 \\ -\frac{1}{2} & -\frac{1}{2} & -1 \end{bmatrix}_q = 1,$$

$$\begin{bmatrix} \frac{1}{2} & \frac{1}{2} & 1 \\ \frac{1}{2} & -\frac{1}{2} & 0 \end{bmatrix}_q = -\begin{bmatrix} \frac{1}{2} & \frac{1}{2} & 0 \\ \frac{1}{2} & -\frac{1}{2} & 0 \end{bmatrix}_q = \frac{q^{-1/4}}{[2]^{1/2}},$$

$$\begin{bmatrix} \frac{1}{2} & \frac{1}{2} & 1 \\ -\frac{1}{2} & \frac{1}{2} & 0 \end{bmatrix}_q = -\begin{bmatrix} \frac{1}{2} & \frac{1}{2} & 0 \\ \frac{1}{2} & -\frac{1}{2} & 0 \end{bmatrix}_q = \frac{q^{1/4}}{[2]^{1/2}}.$$

Substituting $l_1 = l_2 = \frac{1}{2}$ into (41) and using these CGC's we find that

$$t_{22}t_{11} = \frac{q^{1/2}}{[2]}(t^1_{00} + q^{-1}I), \qquad t_{11}t_{22} = \frac{q^{-1/2}}{[2]}(t^1_{00} + qI),$$

where $I = t^0_{00}$. Hence,

$$t_{22}t_{11} - qt_{11}t_{22} = (q-1)I. \tag{43a}$$

In the same manner we find that

$$\left.\begin{aligned} &t_{12}t_{21} = t_{21}t_{12}, \quad t_{12}t_{11} = \sqrt{q}t_{11}t_{12}, \quad t_{21}t_{11} = \sqrt{q}t_{11}t_{21}, \\ &\sqrt{q}t_{11}t_{22} - t_{21}t_{12} = \sqrt{q}I, \quad t_{22}t_{12} = \sqrt{q}t_{12}t_{22}, \\ &t_{22}t_{21} = \sqrt{q}t_{21}t_{22}, \quad t_{22}t_{11} - t_{11}t_{22} = (q^{1/2} - q^{-1/2})t_{21}t_{12}. \end{aligned}\right\} \tag{43b}$$

Using formula (34) it is easy to show that any finite dimensional irreducible representation of $U_q(su(2))$ can be obtained by successive tensor multiplication of representations $T_{1/2}$. If in the multiplication procedure we use CGC's, then we obtain the matrix elements for the representations T_l, $l = 1, 3/2, 2, \ldots$. Therefore, by taking products of the elements t_{11}, t_{12}, t_{21}, t_{22} and their linear combinations we obtain the algebra A which contains the matrix elements of the irreducible representations T_l of the quantum algebra $U_q(su(2))$. The algebra A is an associative algebra, generated by the elements $I \equiv t^0_{00}$, t_{11}, t_{12}, t_{21}, t_{22} and by the relations (43). The operation Δ is introduced in A by the formula

$$\Delta(t^l_{mn}) = \sum_r t^l_{mr} \otimes t^l_{rn},$$

while the other operations are natural. The algebra A as a Hopf algebra is called the quantum group $SU_q(2)$. The matrix elements t^l_{mn}, $l = 0, \frac{1}{2}, 1, \frac{3}{2}, \ldots; -l \le m, n \le l$, form a basis of $SU_q(2)$. The structure of a C^* algebra can be introduced into $SU_q(2)$ [10].

The explicit expressions for t^l_{mn} in terms of t_{11}, t_{12}, t_{21}, t_{22} are found with the help of formulas (41) and (42). We have

$$\begin{aligned} t^l_{mn} &= \left(\frac{[l-n]![l+m]!}{[l-m]![l+n]!} \right)^{1/2} \frac{t_{11}^{m+n} t_{12}^{m-n}}{[m-n]!} q^{(m+n)(m-n)/4} \\ &\quad \times {}_2\Phi_1(-l+m, l+m+1; m-n+1; -\sqrt{q} t_{12} t_{21}, \\ t^l_{nm} &= \left(\frac{[l-n]![l+m]!}{[l-m]![l+n]!} \right)^{1/2} \frac{t_{11}^{m+n} t_{21}^{m-n}}{[m-n]!} q^{(m+n)(m-n)/4} \\ &\quad \times {}_2\Phi_1(-l+m, l+m+1; m-n+1; -\sqrt{q} t_{12} t_{21}), \end{aligned}$$

if $m \ge n \ge -m$, and

$$\begin{aligned} t^l_{-m,n} &= \left(\frac{[l+n]![l+m]!}{[l-m]![l-n]!} \right)^{1/2} \frac{q^{-(m+n)(m-n)/4}}{[m+n]!} \\ &\quad \times {}_2\Phi_1(-l+m, l+m+1; m+n+1; -\sqrt{q} t_{12} t_{21}) t_{22}^{m-n} t_{21}^{m+n}, \\ t^l_{n,-m} &= \left(\frac{[l+n]![l+m]!}{[l-n]![l-m]!} \right)^{1/2} \frac{q^{-(m+n)(m-n)/4}}{[m+n]!} \\ &\quad \times {}_2\Phi_1(-l+m, l+m+1; m+n+1; -\sqrt{q} t_{12} t_{21}) t_{22}^{m-n} t_{12}^{m+n}, \end{aligned}$$

if $m \ge |n|$.

The structure of a $*$-Hopf algebra is introduced into the algebra A by the definition

$$\begin{pmatrix} t_{11} & t_{12} \\ t_{21} & t_{22} \end{pmatrix}^* = \begin{pmatrix} t^*_{11} & t^*_{21} \\ t^*_{12} & t^*_{22} \end{pmatrix} = \begin{pmatrix} t_{22} & -\sqrt{q} t_{12} \\ \frac{-1}{\sqrt{q}} t_{21} & t_{22} \end{pmatrix}$$

and is extended onto the entire algebra A by considering that $*$ is an antilinear antiautomorphism, that is

$$\begin{aligned} (\alpha a + \beta b)^* &= \overline{\alpha} a^* + \beta b^*, \quad \alpha, \beta \in C, \quad a, b \in A, \\ (ab)^* &= b^* a^*, \quad a, b \in A. \end{aligned}$$

There is an invariant linear functional (invariant integral) φ on A which is defined by the relation [8]

$$(id \otimes \varphi) \circ \Delta(a) = I \cdot \varphi(a),$$
$$(\varphi \otimes id) \circ \Delta(a) = \varphi(a) \cdot I, \quad a \in A.$$

The invariant functional is uniquely defined by the conditions

$$\varphi(t^0_{00}) \equiv \varphi(I) = 1, \qquad \varphi(t^l_{mn}) = 0, \qquad l > 0.$$

Using the functional φ, the scalar product

$$\langle a, b \rangle \equiv \varphi(ab^*)$$

is introduced into A. It has been demonstrated in [8] that

$$\left\langle t^l_{mn}, t^{l'}_{ks} \right\rangle = 0, \quad \text{if } (l, m, n) \neq (l', k, s), \tag{44a}$$

$$\left\langle t^l_{mn}, t^l_{mn} \right\rangle = \frac{1-q}{1-q^{2l-1}}\, q^{l-n} \equiv c_{ln}. \tag{44b}$$

7. THE WIGNER-ECKART THEOREM FOR $SU_q(2)$

Let $T = \sum_l \oplus T_l$ be a direct sum of irreducible representations of the quantum group $SU_q(2)$. For the sake of simplicity we consider that the multiplicities of representations T_l do not exceed 1. A set of operators R_n, $n = -l', -l'+1, \ldots, l'$, is called a tensor operator transforming according to the irreducible representation $T_{l'}$ of the quantum group $SU_q(2)$ if

$$TR_nT^* = \sum_{m=-l'}^{l'} t^{l'}_{mn} R_m, \tag{45}$$

where $(t^{l'}_{mn})$ is the matrix representation of $SU_q(2)$. The equality is understood as the equality of matrices, and $T^* = \sum_l \oplus T^*_l$ means the matrix which is the direct sum of matrices

$$T^*_l \equiv (t^l_{ij})^* = (t^{l\,*}_{ji}),$$

with the $*$-operation introduced above.

The equality $T^*_l T_l = T_l T^*_l = E$ holds, where E is the unit matrix (with the units I of the algebra A along its diagonal). Therefore, the formula (45) can be rewritten as

$$TR_n = \sum_{m=-l'}^{l'} t^{l'}_{mn} R_m T.$$

Writing down the basis elements e_m of the carrier space of the representation T_l in the form $|l,m\rangle$ we can represent this formula as

$$\sum_{j=-l}^{l} t^l_{ij} \langle l,j|\, R_n\, |l'',k\rangle = \sum_{m=-l'}^{l'} \sum_{s=-l''}^{l''} t^{l'}_{mn} \langle l,i|\, R_m\, |l'',s\rangle\, t^{l''}_{sk}. \tag{46}$$

This equality has to be understood as an equality in the algebra A.

Using the relations (44) we obtain from (46) that

$$c_{lj} \langle l,j|\, R_n\, |l'',k\rangle = \sum_{m=-l'}^{l'} \sum_{s=-l''}^{l''} \langle l,i|\, R_m\, |l'',s\rangle \left\langle t^{l'}_{mn} t^{l''}_{sk}, t^l_{ij} \right\rangle .$$

It follows from (41) that

$$\left\langle t^{l'}_{mn} t^{l''}_{sk}, t^l_{ij} \right\rangle = c_{lj} \begin{bmatrix} l' & l'' & l \\ m & s & i \end{bmatrix}_q \begin{bmatrix} l' & l'' & l \\ n & k & j \end{bmatrix}_q .$$

Therefore,

$$\langle l,j|\, R_n\, |l'',k\rangle = \sum_m \sum_s \langle l,i|\, R_m\, |l'',s\rangle \begin{bmatrix} l' & l'' & l \\ m & s & i \end{bmatrix}_q \begin{bmatrix} l' & l'' & l \\ n & k & j \end{bmatrix}_q .$$

As in the classical case, we obtain from here that

$$\langle l,j|\, R_n\, |l'',k\rangle = \langle l\|\, R^{l'}\, \|l''\rangle \begin{bmatrix} l' & l'' & l \\ n & k & j \end{bmatrix}_q \tag{47}$$

where the reduced matrix elements $\langle l\|\, R^{l'}\, \|l''\rangle$ do not depend upon the labels of the basis elements and

$$\langle l\|\, R^{l'}\, \|l''\rangle = \sum_{m,s} \langle l,i|\, R_m\, |l'',s\rangle \begin{bmatrix} l' & l'' & l \\ m & s & i \end{bmatrix}_q . \tag{48}$$

The formulas (47) and (48) are the Wigner-Eckart theorem for the quantum group $SU_q(2)$.

8. THE WIGNER-ECKART THEOREM FOR THE QUANTUM ALGEBRA $U_q(su(2))$

The quantum algebra $U_q(su(2))$ can be obtained from the algebra A. To demonstrate this we consider $q^{H/4}$, E_+, E_- as linear functionals on A which act on t_{11}, t_{12}, t_{21}, t_{22} according to the formulas

$$q^{H/4} \begin{pmatrix} t_{11} & t_{12} \\ t_{21} & t_{22} \end{pmatrix} = \begin{pmatrix} q & 0 \\ 0 & q^{-1} \end{pmatrix},$$

$$E_+ \begin{pmatrix} t_{11} & t_{12} \\ t_{21} & t_{22} \end{pmatrix} = \begin{pmatrix} 0 & 1 \\ 0 & 0 \end{pmatrix}, \qquad E_- \begin{pmatrix} t_{11} & t_{12} \\ t_{21} & t_{22} \end{pmatrix} = \begin{pmatrix} 0 & 0 \\ 1 & 0 \end{pmatrix}.$$

The action of $q^{H/4}$, E_+, E_- is extended upon other elements of A (that is, upon polynomials of t_{11}, t_{12}, t_{21}, t_{22}) with the help of the formulas

$$q^{H/4}(ab) = q^{H/4}(a)q^{H/4}(b),$$
$$E_\pm(ab) = E_\pm(a)q^{H/4}(b) + q^{-H/4}(a)E_\pm(b), \quad a,b \in A. \tag{49}$$

Products of elements $q^{H/4}$, E_+, E_- (considered as linear functionals on A) are defined in the same way as in the case of the product of usual functions f_1 and f_2: $(f_1f_2)(x) = f_1(x)f_2(x)$.

It has been shown in [8] that the elements H, E_+, E_-, as just defined above, satisfy the commutation relations (26). Using the results of Section 4.2 of ref. [8], one obtains the matrices for the representation T_l of the algebra $U_q(su(2))$ in the basis $|l,m\rangle$ in the following way. If $X \in U_q(su(2))$ then

$$T_l(x)_{mn} = X(t^l_{mn}), \qquad t^l_{mn} \in A.$$

In particular, it follows from (49) that

$$E_\pm(t^l_{mn}t^{l'}_{rs}) = T_l(E_\pm)_{mn}q^{T_{l_1}(H/4)_{rs}} + q^{-T_l(H/4)_{mn}}T_{l'}(E_\pm)_{rs}.$$

Applying this formula to both sides of relation (46) we have

$$\sum_{j=-l}^{l} (E_\pm)^l_{ij}\ \langle l,i|\ R_n\ |l'',k\rangle = \sum_{m=-l'}^{l'} (E_\pm)^{l'}_{mn}\ \langle l,i|\ R_m\ |l'',k\rangle\ q^{k/2} + \sum_{s=-l''}^{l''} q^{-n/2}\ \langle l,i|\ R_n\ |l'',s\rangle\ (E_\pm)^{l''}_{sk}$$

(here we have taken into account that $H\ |l,r\rangle\ = 2r\ |l,r\rangle$). This formula can be written as

$$\sum_j (E_\pm)^l_{ij}\ \langle l,j|\ R_n\ |l'',k\rangle\ q^{-k/2} - \sum_s q^{-(n+k)}\ \langle l,i|\ R_n\ |l'',s\rangle\ (E_\pm)^{l''}_{sk} = \sum_m (E_\pm)^{l'}_{mn}\ \langle l,i|\ R_m\ |l'',k\rangle\ ,$$

or in operator form as

$$E_\pm R_n q^{-H/4} - q^{\pm 1/2}q^{-H/4}R_nE_\pm = \sum_{m=-l'}^{l'} (E_\pm)^{l'}_{mn}R_m \tag{50}$$

(the upper and the lower signs have to be taken separately). Taking H instead of $E_\pm$, we obtain in the same manner that

$$HR_n - R_nH = \sum_{m=-l'}^{l'} (H)^l_{mn} R_m. \tag{50'}$$

(In fact, the sums in the formulas (50) and (50′) are reduced to one term.)

Thus, we can now define tensor operators for the quantum algebra $U_q(su(2))$. A set of operators R_n, $n = -l', -l'+1, \ldots, l'$, is called a tensor operator transforming according to the representation $T_{l'}$ of $U_q(su(2))$ if the relations (50) and (50′) are fulfilled. The formula (47) is valid for the matrix elements of these tensor operators, whereby the reduced matrix elements are given by the formula (48).

9. RACAH COEFFICIENTS FOR $U_q(su(2))$

Let T_{l_1}, T_{l_2}, T_{l_3} be three irreducible representations of the algebra $U_q(su(2))$ which act in the Hilbert spaces $\mathcal{H}_1$, $\mathcal{H}_2$, $\mathcal{H}_3$ respectively, and let $\{e_j\}$, $\{f_k\}$, $\{h_m\}$ denote the orthonormal bases of these spaces. The tensor product $\mathcal{H} \equiv \mathcal{H}_1 \otimes \mathcal{H}_2 \otimes \mathcal{H}_3$ can be considered as the product $(\mathcal{H}_1 \otimes \mathcal{H}_2) \otimes \mathcal{H}_3$ and as the product $\mathcal{H}_1 \otimes (\mathcal{H}_2 \otimes \mathcal{H}_3)$. As in the classical case [25], this leads to two orthonormal bases for the space $\mathcal{H}$, with basis vectors

$$e_p^{(l_1l_2)l_{12},l_3,l} = \sum_{i,j,k} \begin{bmatrix} l_1 & l_2 & l_{12} \\ i & j & m \end{bmatrix}_q \begin{bmatrix} l_{12} & l_3 & l \\ m & k & p \end{bmatrix}_q (e_i \otimes f_j) \otimes h_k, \tag{51}$$

and

$$e_p^{l_1(l_2l_3),l_{23},l} = \sum_{i,j,k} \begin{bmatrix} l_1 & l_2 & l_{12} \\ j & k & n \end{bmatrix}_q \begin{bmatrix} l_1 & l_{23} & l \\ i & n & p \end{bmatrix}_q e_i \otimes (f_j \otimes h_k), \tag{52}$$

where $i + j = m$, $m + k = p$, $j + k = n$. These basis vectors are related by the unitary matrix R

$$e_p^{(l_1l_2)l_{12},l_3,l} = \sum_{l_{23}} R(l_1l_2l_3, l_{12}l_{23}, l) e_p^{l_1(l_2l_3),l_{23},l}.$$

The elements $R(l_1l_2l_3, l_{12}l_{23}, l)$ of this matrix do not depend upon the indices i, j, m, k, n, p of (51) and (52) and are called Racah coefficients of $U_q(su(2))$. The unitarity of the matrix R gives the orthogonality relations for the Racah coefficients, which are of the same form as in the classical case. The formula

$$\begin{Bmatrix} l_1 & l_2 & l_{12} \\ l_3 & l & l_{23} \end{Bmatrix}_q = (-1)^{l_1+l_2+l_3+l}([2l_{12}+1][2l_{23}+1])^{-1/2} \times R(l_1l_2l_3, l_{12}l_{23}, l)$$

defines a q-analogue of Wigner $6j$ symbols.

The explicit expression for the Racah coefficients is derived in the same way as in the classical case [11]. Another method of their method derivation is given in [9]. We have

$$\begin{Bmatrix} l_1 & l_2 & l_{12} \\ l_2 & l & l_{23} \end{Bmatrix}_q = \frac{(-1)^{l_2+l_{12}+l+l_{23}}\Delta(l_1,l_2,l_{12})\Delta(l_{12},l_3,l)\Delta(l_1,l,l_{23})}{[l_2-l_1+l_{12}]![l_1+l_2-l_{12}]![l-l_3+l_{12}]![l_3-l+l_{12}]!}$$

$$\times \frac{\Delta(l_2,l_3,l_{23})[2l_2]![l_2+l_{12}-l+l_{23}]![l_2+l+l_{12}+l_{23}+1]!}{[l-l_1+l_{23}]![l_1-l+l_{23}]![l_2+l_3-l_{23}]![l_2-l_3+l_{23}]!}$$

$$\times {}_4\Phi_3\left(\begin{matrix} l_1-l_2-l_{12}, & l_3-l_2-l_{23}, & -l_1-l_2-l_{12}-1, & -l_2-l_3-l_{23}-1 \\ -2l_2, & -l_2-l_{12}+l-l_{23}, & -l_2-l_{12}-l-l_{23}-1 & \end{matrix}\middle| q,q\right).$$

Applying successively the transformation (24) we obtain different expressions for the Racah coefficients in terms of the function ${}_4\Phi_3$. In this manner we obtain the q-analogues for all known expressions of the classical Racah coefficients. These q-analogues differ from the classical expressions only by the replacement of the factorials $m!$ by the q-factorials $[m]!$ and the replacement of

$${}_4F_3(a,b,c,d;e,f,g;1) \quad \text{by} \quad {}_4\Phi_3(a,b,c,d;e,f,g;q,q).$$

Since $[n]! \equiv [n]_q! = [n]_{q^{-1}}!$, it follows

$$\begin{Bmatrix} a & b & c \\ d & e & f \end{Bmatrix}_q = \begin{Bmatrix} a & b & c \\ d & e & f \end{Bmatrix}_{q^{-1}}.$$

There are also q-analogues of the classical formulas for the asymptotic relations of the Racah coefficients [25]. The main relation is obtained as

$$\lim_{r\to\infty} R(a,b,d+r;c,f+r;e+r) = \begin{bmatrix} a & b & c \\ f-e & d-f & d-e \end{bmatrix}_q.$$

For $r, a, c \gg b$, $c-a=d=$ const we have

$$R(r-m,c-d,b;r-k,c;r) \approx (-1)^{b-d} q^{\{c(d+k)-b(c+m)\}/c}$$

$$\times \left([b-k]![b+k]![b-d]![b+d]!\right)^{1/2} \left(\frac{[c-m+1]}{[2c+1]}\right)^b \left(\frac{[c+m+1]}{[c-m+1]}\right)^{(d+k)/2}$$

$$\times \sum_s \frac{q^{cs}(-1)^s}{[s]![d+k+s]![b-k-s]![b-d-s]!}\left(\frac{[c+m+1]}{[c-m+1]}\right)^s.$$

For $a, b, c \gg m, n, f$ we obtain

$$\begin{Bmatrix} a & b & c \\ b+m & a+n & f \end{Bmatrix}_q \approx (-1)^{a+b+c-m-n}\left([f-m]![f+m]![f-n]![f+n]!\right)^{1/2}$$

$${}_2\Phi_1\left(-f+m,-f+n;m+n+1;q,\frac{[a+b-c][a+b+c+1]}{[a-b+c][b-a+c]}\right).$$

10. REPRESENTATIONS OF THE QUANTUM ALGEBRA $U_q(su(1,1))$

The universal enveloping algebras $U_q(su(2))$ and $U_q(su(1,1))$ of the classical Lie algebras $su(2)$ and $su(1,1)$ and the universal enveloping algebra $U(sl(2))$ of the complex Lie algebra $sl(2)$ are identical. The q-deformations $U_q(su(2))$, $U_q(su(1,1))$ and $U_q(sl(2))$ also define the same algebra. These quantum algebras have therefore the same representations. As for the classical case, the difference between these quantum algebras appears when we consider their "unitary" representations, since the "unitarity" conditions are different for them.

The quantum algebras $U_q(su(1,1))$ and $U_q(su(2))$ are generated by the elements E_+, E_-, H, which obey the commutation relations (26). The Casimir element is given by formula (27).

By a linear representation T of $U_q(su(1,1))$ we understand a homomorphism of $U_q(su(1,1))$ into the algebra of linear operators of a Hilbert space $\mathcal{H}$, which are defined on an everywhere dense subspace, such that $T(\exp(-2i\theta H))$, $0 \leq \theta < 2\pi$, gives a one-valued representation of the group $SO(2)$. The last condition means that the operator H is diagonal in an appropriate basis and has a discrete spectrum, which consists of integral and half-integral eigenvalues.

Let ϵ be equal to 0 or 1/2, and $\mathcal{H}_\epsilon$ be the complex Hilbert space with the orthonormal basis

$$\{e_m; m = n + \epsilon, n = 0, \pm 1, \pm 2, \dots\}.$$

Let us fix a complex number σ and denote by T_σ^ϵ the representation of the algebra $U_q(su(1,1))$ on $\mathcal{H}_\epsilon$, given by the formulas

$$\left.\begin{aligned} \frac{H}{2} e_m &= m e_m, \\ E_+ e_m &= [-\sigma + m] e_{m+1}, \\ e_- e_m &= [-\sigma - m] e_{m-1}, \end{aligned}\right\} \qquad (53)$$

where $[a]$ is defined by (2) (instead of $T_\sigma^\epsilon(H)$ and $T_\sigma^\epsilon(E_\pm)$ we write H, $E_\pm$ respectively). It is easily checked that the commutation relations (26) are satisfied.

First we consider the case $q = e^h$, $h \in R$. The function $f(z) = [z]$, $z \in \mathbf{C}$, has the period $4\pi i/h$. Therefore, the representations T_σ^ϵ and $T_{\sigma+4\pi i/h}^\epsilon$, $k \in Z$, coincide. For this reason we consider only those representations T_σ^ϵ for which $0 \leq \mathrm{Im}\ \sigma < 4\pi/h$.

For the eigenvalue of the Casimir operator C one obtains

$$C e_m = \left[\sigma + \frac{1}{2}\right]^2 e_m.$$

The irreducibility of the representations is studied in the same manner as for the classical case (see, for example, [26]). The following results are obtained:

<u>Theorem</u>. The representation T_σ^ϵ is irreducible if and only if 2σ is not an integer of the same evenness as 2ϵ.

If $\sigma \equiv l$ and $2l \equiv 2\epsilon \pmod 2$, then for $l \geq 0$ the invariant subspace

$$\mathcal{H}_\epsilon^f = \text{span}\{e_m;\ -l \leq m \leq l\}$$

exists in $\mathcal{H}_\epsilon$, which is a carrier space for the finite dimensional representation π_l of $U_q(su(1,1))$, equivalent to T_l. The quotient space $\mathcal{H}_\epsilon/\mathcal{H}_\epsilon^f$ decomposes into the orthogonal sum of two invariant irreducible subspaces

$$\mathcal{H}_\epsilon^{l+1} = \text{span}\{e_m;\ l < m\}, \qquad \mathcal{H}_\epsilon^{-l-1} = \text{span}\{e_m;\ m < -l\},$$

which carry irreducible representations of $U_q(su(1,1))$ (we denote these by T_{l+1}^+ and T_{-l-1}^- respectively). For $l = -1/2$ the space $\mathcal{H}_\epsilon$ decomposes into an orthogonal sum of two invariant subspaces

$$\mathcal{H}_\epsilon^{1/2} = \text{span}\{e_m;\ m \geq 1/2\}, \qquad \mathcal{H}_\epsilon^{1/2} = \text{span}\{e_m;\ m \leq -1/2\},$$

which form the carrier spaces for the irreducible representations $T_{1/2}^+, T_{1/2}^-$.

If $l < -1/2$, then two non-overlapping invariant subspaces

$$\widetilde{\mathcal{H}}_\epsilon^{-l} = \text{span}\{e_m;\ m \geq -l\}, \qquad \widetilde{\mathcal{H}}_\epsilon^{l} = \text{span}\{e_m;\ m \geq l\}$$

exist in $\mathcal{H}_\epsilon$, which carry irreducible representations of $U_q(su(1,1))$ (we denote these by $\widetilde{T}_{-l}^+$ and $\widetilde{T}_l^-$). The finite dimensional representation π_{-l-1} (which is equivalent to T_l) acts on the quotient space $\mathcal{H}_\epsilon/(\widetilde{\mathcal{H}}_\epsilon^{-l} + \widetilde{\mathcal{H}}_\epsilon^{l})$.

Equivalence relations exist between the irreducible representations constructed. The irreducible representations T_σ^ϵ and $T_{-\sigma-1}^\epsilon$ are equivalent. The operator which demonstrates the equivalence is evaluated in the same way as in the classical case (see, for example, [26]). The equivalence operator Q for T_σ^ϵ and $T_{-\sigma-1}^\epsilon$ is diagonal in the basis $\{e_m\}$ and its diagonal elements are

$$a_m = \frac{[\sigma+1][\sigma+2]\dots[\sigma+m]}{[-\sigma][-\sigma+1]\dots[-\sigma+m-1]}.$$

By evaluating the equivalence operators we can prove that the representations $T_{\pm l \pm 1}^\pm$ and $\widetilde{T}_{\mp l \pm 1}^\pm$ are equivalent. There are no other equivalence relations.

Thus the following classes of irreducible representations of $U_q(su(1,1))$ exist:

1) the representations T_σ^ϵ, $\epsilon \in \{0, 1/2\}$, $\text{Re}\,\sigma \geq -1/2$, $0 \leq \text{Im}\,\sigma < 2\pi/h$, $2\sigma \not\equiv 2\epsilon$ (mod 2),

2) the representations T_l^+, T_l^-, $l = 1/2, 1, 3/2, \dots$,

3) the finite dimensional representations T_l, $l = 0, 1/2, 1, 3/2, \dots$.

It has been proved in [20] that there are not other irreducible representations.

A representation of $U_q(su(1,1))$ is called "unitary" if, on the linear span of the basis vectors e_m, the relations

$$H^* = H, \qquad E_\pm^* = -E_\mp$$

are fulfilled. It is easy to prove that the representations T_σ^ϵ, $\sigma = i\rho - \frac{1}{2}$, $0 \le \rho < \frac{4\pi}{h}$, are "unitary" (the principal unitary series). Among the irreducible representations, constructed above, there are representations which are "unitary" in other topologies of the space $\mathcal{H}_\epsilon$, that is, which are equivalent to "unitary" representations. The following representations are equivalent to "unitary" representations:

1) the representations T_σ^0, $-\frac{1}{2} \le \sigma \le 0$ (the complimentary series),

2) all representations T_l^+, T_l^- (the discrete series),

3) the representations T_σ^ϵ, Im $\sigma = \frac{\pi}{h}$, Re $\sigma > -\frac{1}{2}$ (the strange series),

4) the identity representation T_0.

The strange series of unitary representations has no analogue in the classical case.

Now we consider the case $q = e^{ih}$, $h \in R$ and q not a root of unity.

Theorem. The representation T_σ^ϵ is irreducible if and only if $2\sigma \not\equiv 2\epsilon \pmod 2$. The irreducible representations T_σ^ϵ and $T_{\sigma+2k\pi/h}^\epsilon$, $k \in Z$, are identical. The irreducible representations T_σ^ϵ and $T_{-\sigma-1}^\epsilon$ (and only these) are equivalent.

If $2\sigma \equiv 2\epsilon \pmod 2$, then T_σ^ϵ is reducible. We have

$$T_l^\epsilon = \pi_l \Subset (T_{l+1}^+ \oplus T_{-l-1}^-), \quad l > 0,$$

where π_l is realized on the invariant subspace, and $T_{l+1}^+ \oplus T_{-l-1}^-$ on the quotient space. Moreover we have

$$T_{-1/2}^\epsilon = T_{1/2}^+ \oplus T_{-1/2}^-,$$

and

$$T_l^\epsilon = (\widetilde{T}_{-l}^+ \oplus \widetilde{T}_l^-) \Subset \pi_{-l-1}, \quad l < -1/2,$$

where $\widetilde{T}_{-l}^+ \oplus \widetilde{T}_l^-$ is realized on the invariant subspace. Studying equivalence relations for these representations we obtain the following classes of irreducible representations of $U_q(su(1,1))$:

1) the representations T_σ^ϵ, $\epsilon \in \{0, 1/2\}, 0 \le \text{Re } \sigma < 2\pi/h$, $2\sigma \not\equiv 2\epsilon \pmod 2$,

2) the representations T_l^+, T_l^-, $l = 1/2, 1, 3/2, 2, \ldots$,

3) the finite dimensional representations T_l, $l = 0, 1/2, 1, 3/2, \ldots$.

The following representations are "unitary" or equivalent to "unitary" ones:

1) the representations T_σ^ϵ, $\sigma = i\rho - \frac{1}{2}$, $\rho \in R$ (the principal unitary series),

2) the representations T_σ^ϵ, Im $\sigma = \pi/h$ (the strange series),

3) the identity representation.

For the case $q = \exp(h_1 + ih_2)$, $h_1, h_2 \in R\backslash\{0\}$, the representations T_σ^ϵ are studied in the same way. One obtains the following classes of irreducible representations:

1) the representations T_σ^ϵ, $\epsilon \in \{0, 1/2\}$, Re $\sigma \geq -\frac{1}{2}$, $2\sigma \not\equiv 2\epsilon \pmod 2$,

2) the representations T_l^+, T_l^-, $l = 1/2, 1, 3/2, 2, \dots$,

3) the finite dimensional representations T_l, $l = 0, 1/2, 1, 3/2, \dots$.

None of these representations (except for the identity representation) is equivalent to "unitary" representations.

11. MULTIPLICITY FREE REPRESENTATIONS OF THE QUANTUM ALGEBRA $U_q(su(1,1))$

If 2σ is an integer of the same evenness as 2ϵ, then the representation T_σ^ϵ of $U_q(su(1,1))$ of section 10 is indecomposable. The algebra $U_q(su(1,1))$ has however also other indecomposable representations. We describe in this section the multiplicity free indecomposable representations of $U_q(su(1,1))$.

As before, we consider the case that q is not a root of unity. For convenience we introduce in $U_q(su(1,1))$ the elements e, f, h (instead of E_+, E_-, H) which obey the commutative relations

$$[h, e] = e, \quad [h, f] = -f \quad [e, f] = \frac{q^h - q^{-h}}{q^{1/2} - q^{-1/2}} \equiv -[2h]. \tag{54}$$

Let $x_0, x_1, x_2 \dots$ denote a basis of an infinite dimensional vector space V and let Λ be a complex number [22]. Then the formulas

$$\left.\begin{aligned} &\pi_\Lambda(h)x_i = (\Lambda - i)x_i, && i = 0, 1, 2, \dots, \\ &\pi_\Lambda(f)x_i = x_{i+1}, && i = 0, 1, 2, \dots, \\ &\pi_\Lambda(e)x_0 = 0, && \\ &\pi_\Lambda(e)x_i = -[i][2\Lambda - (i-1)]x_{i-1} && \end{aligned}\right\} \tag{54'}$$

where $[a] = (q^{a/2} - q^{-a/2})/(q^{1/2} - q^{-1/2})$, define the representation π_Λ of $U_q(su(1,1))$ with highest weight Λ. (It is easy to verify that the commutation relations for the operators $\pi_\Lambda(h), \pi_\Lambda(f), \pi_\Lambda(e)$ are fulfilled.) In a similar way the representations with lowest weight can be introduced.

The representations π_Λ can be analysed in the same manner as for the case of representations T_σ^ϵ. We obtain the result that the representation π_Λ is reducible if and only if $\Lambda = \frac{1}{2}(k-1)$ for some integer $n, n \geq 0$. In this case π_Λ has an irreducible subrepresentation which is equivalent to π. The quotient representation π_Λ/π is irreducible and finite dimensional.

Now we introduce new multiplicity free indecomposable representations π of $U_q(su(1,1))$, for which the operator $\pi(h)$ is diagonal. Let $|\Lambda\rangle$ denote one of the eigenvectors of $\pi(h)$ such that $\pi(h)\,|\Lambda\rangle = \Lambda\,|\Lambda\rangle$. We create the vectors

$$\ldots, \pi(f^2)\,|\Lambda\rangle\,, \pi(f)\,|\Lambda\rangle\,,\,|\Lambda\rangle\,, \pi(e)\,|\Lambda\rangle\,, \pi(e^2)\,|\Lambda\rangle\,, \ldots \tag{55}$$

These vectors have the eigenvalues

$$\ldots, \Lambda-2, \Lambda-1, \Lambda, \Lambda+1, \Lambda+2, \ldots .$$

We denote these vectors by

$$\ldots, |\Lambda-2\rangle\,, |\Lambda-1\rangle\,, |\Lambda\rangle\,, |\Lambda+1\rangle\,, |\Lambda+2\rangle\,, \ldots \tag{56}$$

As in the classical case (see, [22]), we have the following possibilities.

Case 1a. None of the vectors $\pi(f^n)\,|\Lambda+m\rangle$, $n,m = 1,2,\ldots$, is equal to zero and no vector of the sequence $\pi(e)\,|\Lambda-n\rangle$, $n = 1,2,3,\ldots$, is equal to zero. Then the representation π is irreducible and equivalent to one of the representations T^ϵ_σ.

Case 1b. None of the vectors $\pi(f^n)\,|\Lambda+m\rangle$, $n,m = 1,2,\ldots$, is equal to zero and for some m, $m > 0$, it holds that $\pi(e)\,|\Lambda-m\rangle = 0$. In this case we relable the vectors by taking the vector $|\Lambda-m+1\rangle$ as the vector $|\Lambda\rangle$. Then we have $\pi(e)\,|\Lambda-1\rangle = 0$. Utilizing this relation it can be proved that the following equations hold

$$\pi(e)\,|\Lambda-n\rangle = [n-1][n-2\Lambda]\,|\Lambda-n+1\rangle\,, \qquad n>0, \tag{57}$$

$$\pi(f)\,|\Lambda+n'\rangle = [n'][2\Lambda+n'-1]\,|\Lambda+n'-1\rangle\,, \qquad n'>0, \tag{58}$$

where n and n' are integers and $[a]$ has the same meaning as before.

For $\Lambda = -(n'-1)/2$ the representation has two invariant subspaces with basis elements

$$|\Lambda-1\rangle\,, |\Lambda-2\rangle\,, |\Lambda-3\rangle\,, \ldots$$

and

$$|-\Lambda+1\rangle\,, |-\Lambda+2\rangle\,, |-\Lambda+3\rangle\,, \ldots .$$

Thus, this representation does not satisfy our conditions and $\Lambda \neq (n'-1)/2$, where $n' > 0$ and integral. If Λ is complex and $\Lambda \neq n/2$, where $n > 0$ and integral, then the formulas (55)–(58) define a representation σ_Λ which has only one invariant subspace with basis elements

$$|\Lambda-1\rangle\,, |\Lambda-2\rangle\,, |\Lambda-3\rangle\,, \ldots .$$

The corresponding subrepresentation of $U_q(su(1,1))$ is a representation with highest weight $\Lambda-1$. The quotient representation is a representation with lowest weight Λ. The representation σ_Λ can not be decomposed into a direct sum of these irreducible representations.

Now let $\Lambda = \frac{n}{2}$, where $n > 0$ and integer. Then the representation has two invariant subspaces. One of them is spanned by the basis vectors

$$|-\Lambda\rangle \ , \ |-\Lambda-1\rangle \ , \ |-\Lambda-2\rangle \ , \ldots$$

and the other one by the basis vectors

$$|\Lambda-1\rangle \ , \ |\Lambda-2\rangle \ , \ |\Lambda-3\rangle \ , \ldots .$$

The first subspace (we denote it by $V_{-\Lambda}$) is irreducible and is included into the second subspace (which is denoted by $V_{\Lambda-1}$. The representation on $V_{\Lambda-1}$ is indecomposable. The representation on $V_{\Lambda-1}/V_{-\Lambda}$ is finite dimensional. For this case, the representations σ_Λ cannot be represented as a direct sum of two or three representations. It is clear that the irreducible representations which are contained in σ_Λ are equivalent to the corresponding representations of section 10.

<u>Case 2</u>. No one of the vectors $\pi(e)\ |\Lambda - m\rangle$, $m = 1,2,3,\ldots$, is equal to zero, and for some m, where $m > 0$ and integral, it holds that $\pi(f)\ |\Lambda + m\rangle = 0$. Again we relabel the vectors by taking $|\Lambda + m - 1\rangle$ as the vector $|\Lambda\rangle$. In the new notation we have $\pi(f)\ |\Lambda + 1\rangle = 0$. Using this formula it is easy to show that

$$\pi(e)\ |\Lambda - n\rangle = [n][-2\Lambda + n - 1]\ |\Lambda - n + 1\rangle \ , \qquad n > 0, \tag{59}$$

$$\pi(f)\ |\Lambda + n'\rangle = [n' - 1][2\Lambda + n']\ |\Lambda + n' - 1\rangle \ , \qquad n' > 0, \tag{60}$$

where n and n' are integers.

If $\Lambda = \frac{n-1}{2}$, then the representation has two invariant subspaces. The first one has basis elements

$$|-\Lambda-1\rangle \ , \ |-\Lambda-2\rangle \ , \ |-\Lambda-3\rangle \ , \ldots$$

and the second one the basis elements

$$|\Lambda+1\rangle \ , \ |\Lambda+2\rangle \ , \ |\Lambda+3\rangle \ , \ldots .$$

Thus the representation does not satisfy the conditions imposed upon the case which we are considering.

Let $\Lambda \neq \frac{n-1}{2}$, where $n > 0$ and integer. If Λ is complex and $\Lambda \neq -n'/2$, where $n' > 0$ and integral, then the representation τ_Λ has only one invariant subspace and is indecomposable. The invariant subspace has the basis

$$|\Lambda+1\rangle \ , \ |\Lambda+2\rangle \ , \ |\Lambda+3\rangle \ , \ldots .$$

The corresponding subrepresentation is a representation with lowest weight $\Lambda+1$. The quotient representation is a representation with highest weight Λ.

Now let $\Lambda = -n'/2$, where $n' > 0$ and integer. Then we obtain a representation τ_Λ with two invariant subspaces. One of the invariant subspaces has the basis

$$|\Lambda+1\rangle \ , \ |\Lambda+2\rangle \ , \ |\Lambda+3\rangle \ , \ldots$$

and the other representation has the basis

$$|-\Lambda\rangle\ ,\ |-\Lambda+1\rangle\ ,\ |-\Lambda+2\rangle\ ,\dots .$$

The first subspace is irreducible and is contained in the second one. The second invariant subspace is indecomposable. All these representations are indecomposable.

The representation $\tau_\Lambda, \Lambda \neq \frac{n-1}{2}$, where $n > 0$ and integral, and the representation $\sigma_{\Lambda-1}$ yield the same set of irreducible representations of the quantum algebra $U_q(su(1,1))$.

12. REPRESENTATIONS OF $U_q(su(1,1))$ ON QUOTIENT SPACES OF $U_q(su(1,1))$

As in the classical case, (left) multiplication defines the representation of $U_q(su(1,1))$ on the space $U_q(su(1,1))$. And again as in the classical case [22], we want to consider quotient representations induced by this representation. Let I be the ideal of $U_q(su(1,1))$ generated by the element $h-\Lambda$, where Λ is a fixed complex number. We have $I = U_q(su(1,1))(h-\Lambda)$. Left multiplication generates a representation ρ_Λ of $U_q(su(1,1))$ on I. This representation can be described in the following manner. The elements

$$1, f^n e^m, m, n = 0,1,2,3,\dots , \qquad m \neq n = 0$$

can be taken as a basis of I. It is easy to evaluate that the operators $\rho_\Lambda(h)$, $\rho_\Lambda(f)$, $\rho_\Lambda(e)$ act on these basis elements as

$$\begin{aligned}
&\rho_\Lambda(h)1 = \Lambda 1, \quad \rho_\Lambda(e)1 = e, \quad \rho_\Lambda(f)1 = f,\\
&\rho_\Lambda(h)f^n e^m = (\Lambda - n + m)f^n e^m,\\
&\rho_\Lambda(f)f^n e^m = f^{n+1}e^m,\\
&\rho_\Lambda(e)f^n e^m = f^n e^{m+1} + [n][2\Lambda + 2m - n + 1]f^{n-1}e^m .
\end{aligned}$$

This action can be shown graphically in exactly the same manner as for the classical case (see Figure 8 in [22]).

For every fixed positive integer s the subspace I_s of I with basis elements

$$f^n e^m, \quad n = 0,1,2,\dots , \qquad m = s, s+1, s+2, \dots$$

is invariant under ρ_Λ. The corresponding subrepresentation is denoted by $\rho_{\Lambda,s}$. We use for the quotient representation $\rho_\Lambda/\rho_{\Lambda,s}$ the notation ρ_Λ^s. The representation ρ_Λ^1 is a representation d_Λ with highest weight Λ (as for the classical case we call it an elementary representation). On the quotient space I_s/I_{s+1} the elementary representation $d_{\Lambda+s}$ is realized. Thus, the representation ρ_Λ consists of the elementary representations

$$d_\Lambda, d_{\Lambda+1}, d_{\Lambda+2}, \dots . \tag{61}$$

Two possbilities exist. The first is for $\Lambda \neq -m/2$ where $m > 1$ and integer. For this case different representations of (61) have different eigenvalues for the Casimir operator. Therefore, the representation ρ_Λ^s can be decomposed into a direct sum of representations with highest weights. This decomposition is achieved by means of the following vectors of highest weight (that is, by means of vectors y for which $\rho_\Lambda(e)y = 0$):

$$x^0 = 1 + \sum_{i=1}^{s-1} \left\{ \prod_{j=1}^{i} [j][2\Lambda + j + 1] \right\}^{-1} f^i e^i, \tag{62}$$

$$x^m = e^m + \sum_{i=1}^{s-m-1} \left\{ \prod_{j=1}^{i} [j][2\Lambda + 2m + j + 1] \right\}^{-1} f^i e^{m+i}, \quad m = 1, 2, \dots s-1. \tag{63}$$

The s subspaces obtained from these s highest weight vectors, with the basis elements

$$x^0, fx^0, f^2x^0, \dots, \tag{64}$$

$$x^m, fx^m, f^2x^m, \dots, \quad m = 1, 2, 3, \dots, s-1, \tag{65}$$

are invariant with respect to ρ_Λ^s, as can be verified by direct evaluation. In fact, for each of the s distinct bases (64) and (65) the representation obtained is equivalent to the representation (54′).

Now let $\Lambda = -n/2$, where $n > 1$ and integer. For this case, pairs

$$(d_{-n/2}, d_{(n/2)-1}), (d_{(-n/2)+1}, d_{(n/2)-2}), \dots$$

of representations $d_{\Lambda+k}$ exist for which the Casimir operator has the same eigenvalue (the corresponding classical case is given in [22]). Thus, for this case the representation ρ_Λ^s can be decomposed into a direct sum of representations $d_\Lambda, d_{\Lambda+1}, d_{\Lambda+2}, \dots, d_{\Lambda+s-1}$ such that those representations which have the same eigenvalue for the Casimir operator occur as indecomposable pairs.

If the carrier space for ρ_Λ is extended to include (formal) infinite sums, then the representations ρ_Λ can be decomposed into a direct sum of representations with highest weights, and pairs of representations with highest weights which are coupled in an indecomposable manner. The action of ρ_Λ on these infinite sums is defined by linearity. The resultant representation is denoted by $\overline{\rho}_\Lambda$. If $\Lambda \neq -m/2$, $m > 0$ and integer, then $\overline{\rho}_\Lambda$ is decomposed into a direct sum of representations $d_\Lambda, d_{\Lambda+1}, d_{\Lambda+2}, \dots$. The vectors

$$x^0 = 1 + \sum_{i=1}^{\infty} \left\{ \prod_{j=1}^{i} [j][2\Lambda + j + 1] \right\}^{-1} f^i e^i, \tag{66}$$

$$x^m = e^m + \sum_{i=1}^{\infty} \left\{ \prod_{j=1}^{i} [j][2\Lambda + 2m + j + 1] \right\}^{-1} f^i e^{m+i}, \quad m = 1, 2, 3, \dots \tag{67}$$

are the highest weight vectors, belonging to the highest weights $\Lambda, \Lambda+1, \Lambda+2, \dots$ of these representations. Below we describe indecomposable pairs for representations with highest weights which appear in $\overline{\rho}_\Lambda$ when $\Lambda = -n/2$.

Fig. 1 shows an indecomposable pair $(d_{\Lambda'}, d_{\Lambda''})$ with highest weight vectors y and $x \equiv x^m$, $m \geq n-1$, which belong to the highest weights $\Lambda' = -n/2$, $\Lambda'' = n/2 - 1$, where n is a positive integer. The highest weight vector $x \equiv x^m$ is given by the formula (67). The vector y is given by the formula

$$y = 1 + \sum_{k=1}^{\infty} a_k f^k e^{k+m},$$

where

$$a_k = \prod_{j=1}^{k} \{[j][2\Lambda'' - j + 1]\}^{-1}, \qquad \text{if } k \leq 2\Lambda'',$$

$$a_{2\Lambda''+1} = 0,$$

$$a_k = g\{\prod_{j=1}^{r} [j][2\Lambda'' + j + 1]\}^{-1} \times \sum_{s=1}^{r} \frac{1}{[s][2\Lambda'' + s + 1]},$$

$$\text{if } k = 2\Lambda'' + r + 1, \quad r = 1, 2, 3, \ldots,$$

with

$$g = \frac{1}{[2\Lambda'']![2\Lambda'']!}.$$

Note that

$$\overline{\rho}_\Lambda(e) y = g f^{2\Lambda''} x.$$

REFERENCES

1. V.G. Drinfel'd, Proceedings of International Congress of Mathematics, Berkeley, Cal., 78–98 (1986).
2. P.P. Kulish, N.Yu. Reshetikhin, J. Sov. Math. 23:2435 (1983).
3. E.K. Skljanin, Funktcionalnij analis i ego pril., no. 3 40:34–48 (1983).
4. G. Moore, N. Seiberg, Commun. Math. Phys., 123:177–254 (1989).
5. L.L. Vaksman, Ja.S. Soibelmen, Funktcionalnij analis i ego pril., no 3 22:1–14 (1989).
6. T.H. Koornwinder, Nederl. Akad. Wetensch. Proc., Ser. A, 92:97–117 (1989).
7. T.H. Koornwinder, CWI Report AM-R8909, Amsterdam (1989).
8. T. Masuda, K. Mimachi, Y. Nakagami, M. Noumi, K. Ueno, C.R. Akad. Sci. Paris, Ser. Math., 307:559–664 (1988); J. Funkt. Anal., (to appear).
9. V.A. Groza, I.I. Kachurik, A.U. Klimyk, Preprint ITP-89-51E, Kiev (1989).
10. S.L. Woronowicz, Publ. RIMS, Kyoto Univ., 23:117–181 (1987).
11. A.N. Kirillov, N.Yu. Reshetikhin, LOMI preprint E-9-88, Leningrad (1988).
12. H.T. Koelink, T.H. Koornwinder, Nederl. Akad. Wetensch. Proc., Ser. A. (1990).
13. I.I. Kachurik, A.U. Klimyk, Preprint ITP-89-48E, Kiev (1989).
14. L.L. Vaksman, Dokl Akad. Nauk USRR, 306:269–271 (1989).
15. I.I. Kachurik, A.U. Klimyk, Preprint ITP-90-7E, Kiev (1990).

16. L.C. Biedenharn, J. Phys. A, 22:L873 (1989).
17. A.J. Macfarlane, Preprint DAMTP/88-29, Cambridge (1988).
18. L.C. Biedenharn, Preprint (1989).
19. L.L. Vaksman, Preprint (1989).
20. A.U. Klimyk, V.A. Groza, Preprint ITP-89-37P, Kiev (1989).
21. T. Masuda, K. Mimachi, Y. Nakagami, M. Noumi, Y. Saburi, K. Ueno, Unitary Representations of the Quantum Group $SU_q(1,1)$, parts I & II, Preprints (1989).
22. B. Gruber, A.U. Klimyk, J. Math. Phys., 19:2009–2017.
23. B. Gruber, A.U. Klimyk, J. Math. Phys., 25:755–764.
24. Ph. Feinsilver, Rocky Mountain J. Math., 25:755–764 (1982).
25. L.C. Biedenharn, J.D. Louck, Angular Momentum in Quantum Physics, Addison Wesley (1981).
26. S. Lang, $SL_2(R)$, Addison Wesley (1975).

FROM "QUANTUM GROUPS" TO "QUASI-QUANTUM GROUPS"

Y. Kosmann-Schwarzbach

UFR de Mathématiques, Université de Lille I
59655 Villeneuve d'Ascq, France

INTRODUCTION

The purpose of this lecture is not to review once more the theory of quantum groups, but to give a brief introduction to the recent work of Drinfeld[1], where he presented a new (quantum) object, the *quasi-Hopf algebras*, which leads to generalizations of quantum groups that, to my knowledge, had not yet been proposed, and, in an even shorter summary, the definition and properties of the *Lie quasi-bialgebras*[1] and *quasi-Poisson Lie groups*[2], which constitute the classical limit of Drinfeld's quantum objects.

The connections of this theory with physics are clear. Drinfeld developed the theory of quasi-Hopf algebras in 1989 in response to recent developements in *conformal field theory*, in particular the Wess-Zumino-Witten model, which required categories of modules more general than those of modules over a non-co-commutative Hopf algebra (Moore and Seiberg[3], and *see* Brustein, Ne'eman and Sternberg[4] for a review). The monodromy matrices of the Knizhnik-Zamolodchikov equations were an essential tool in the construction of these quasi-Hopf algebras. Takhtajan[5] claims that "relative versions of the quantum Clebsch-Gordan coefficients (3j-symbols) and quantum Wigner-Raccah coefficients (6j-symbols)" are obtained from the twisting and the nontrivial associativity constraint in quasi-quantum groups. The quasi-Hopf algebras could then serve to interpret recent results[6] on the quantization of the 6j-symbols.

How can one motivate the successive generalizations that we shall describe here, first from *Lie groups* – where the universal enveloping algebra of the associated Lie algebra is equipped with both a non-commutative multiplication (unless the group is Abelian) and a co-commutative co-multiplication (the 'diagonal'), *i. e.*, it is a co-commutative Hopf algebra –, to *quantum groups*, which are, by definition, a class of non-co-commutative Hopf algebras, then to the new *quasi-Hopf algebras*, which are no longer co-associative? It is natural to consider the category of modules (the finite-dimensional representations of the "group") over these algebras, and to require that these modules satisfy axioms which are familiar from the theory of categories, such as those of a *tensor category*, or those of a *braided monoïdal category*, or those of a *monoïdal category* in general. The *quasi-triangular* and the *triangular* cases, corresponding to various types of solutions of the quantum Yang-Baxter equation, are easy to define by this *categorical* approach, which is further justified by the various reconstruction theorems generalizing the Tannaka-Krein duality for compact groups.

Symmetries in Science V, Edited by B. Gruber *et al.*
Plenum Press, New York, 1991

When one relaxes the requirement of co-commutativity for the Hopf algebra co-multiplication, one obtains the Jimbo-Drinfeld approach to quantum groups[7,8]. It was Drinfeld's idea[1] to go further and to relax the Hopf algebra's requirement of co-associativity for the co-multiplication. See also Majid[9]. Drinfeld thus introduced objects that were still algebras but which were no longer Hopf algebras, and he called them *quasi-Hopf algebras*. Just as the term *quantum group* applies only to a special class of non-co-commutative Hopf algebras, the *quasi-quantum groups* are a special class of quasi-Hopf algebras, obtained by a deformation of a universal enveloping algebra, that we shall not fully describe here. We have chosen to put the words "quantum groups" and "quasi-quantum groups" in quotation marks in the title of this lecture, since we shall only outline some properties of the *categories of modules over Hopf and quasi-Hopf algebras*.

Table 1

	Algebra	*Category of modules*
LIE GROUPS	co-commutative Hopf algebra (universal enveloping algebra)	tensor category with trivial commutativity and associativity constraints
QUANTUM GROUPS	non-co-commutative Hopf algebra	monoïdal category with trivial associativity constraint
QUASI-TRIANGULAR QUANTUM GROUPS	non-co-commutative Hopf algebra with R-matrix	braided monoïdal category with trivial associativity constraint
TRIANGULAR QUANTUM GROUPS	unitary R-matrix	tensor category with trivial associativity constraint
QUASI-QUANTUM GROUPS	algebra with co-multiplication, non-co-associative with "Φ-matrix", non-co-commutative	monoïdal category
QUASI-TRIANGULAR QUASI-QUANTUM GROUPS	algebra with co-multiplication, non-co-associative with "Φ-matrix", non co-commutative with R-matrix	braided monoïdal category
TRIANGULAR QUASI-QUANTUM GROUPS	unitary R-matrix	tensor category

Table 1 above shows the properties of the algebras attached to the successive generalizations of the Lie groups, the *quantum groups* and the *quasi-quantum* groups, and, in the right-hand column, the various structures, which we are going to describe, of the *category of modules* over each of these algebras. We have listed the added features of the algebra and the category of modules in the *quasi-triangular* case, and in the even more particular *triangular* case, for both the quantum groups and the quasi-quantum groups.

Some remarks regarding terminology are in order. We have used the term *braided monoïdal categories* (to be defined below) in this table and in the rest of this lecture since they are monoïdal categories with an additional structure, which is why they give rise to representations of the *braid group* (see Yetter[10] and references therein). Braided monoïdal categories are also called *quasi-tensor categories*. *Tensor categories* are also called *symmetric monoïdal categories* because they give rise to representations of the symmetric group.

The prefix "quasi" is used in "quasi-triangular quantum group" because the associated category of modules is a "quasi-tensor category", the axioms of which are a weakening of those of a tensor category. When one considers the particular case of a triangular quantum group, the category becomes a tensor category. However, the use of the prefix "quasi" in "quasi-Hopf algebras" and "quasi-quantum groups" does not refer to quasi-tensor categories, but only indicates a weakening of the particular assumptions of "Hopf algebras" and "quantum groups".

There is a dual approach to the theory of quantum groups, which is due to Faddeev, Reshetikhin, Takhtajan (*see* ref. 11), Woronowicz[12], Manin[13] and others. Instead of the non co-commutative Hopf algebras arising as deformations of the co-commutative universal enveloping algebra of a Lie algebra, these authors consider the non commutative Hopf algebras arising as deformations of the commutative algebra of smooth functions over a Lie group. In that dual approach, one should consider the *category of co-modules* over each of these algebras. We shall say a few words about the idea behind the approach by deformation of the algebra of functions in sections 1.1-3 but shall only describe the generalizations of the universal enveloping algebras and their categories of *modules*, in the rest of the exposition.

The categorical approach to the theory of quantum groups is due to many authors, Lyubashenko[14], Drinfeld[7] (section 10), Manin[13], Reshetikhin and Turaev[15] and Majid, whose survey[16] we have used extensively. For the connections with the invariants of knots and links, see, *e. g.*, references 15 and 16. The extensions to the quasi-case are more recent. Besides Drinfeld's paper, one can read lectures by Cartier[17], Stasheff[18] and Majid[19]. Other references to the quasi-quantum groups include Takhtajan[5] and Reshetikhin[20].

1. QUANTUM GROUPS AND THEIR CATEGORIES OF MODULES

We show how a detour through the category of modules of a group leads to the generalization of groups to quantum groups.

1.1 *Commutative (resp., co-commutative) Hopf algebra associated with a group*

With a finite group, G, one classically associates an algebra of functions, $A = \mathcal{F}(G)$, with values in the field K of real or complex numbers. Similarly, with a compact matrix group, G , one associates an algebra, A , of representative functions. In both cases, the group structure on G induces a co-multiplication, Δ , in A , which is a linear map from A to $A \otimes A$, where $A \otimes A$ is identified with the algebra of functions over $G \times G$. Namely, to a function f on G, one associates the function, $\Delta(f)$, on $G \times G$ defined by

$$\Delta(f)(g,g') = f(gg') ,$$

or, in terms of matrix elements, t^i_j , for $(n \times n)$ matrices,

$$\Delta(t^i_j) = \sum_{k=1}^{n} t^i_k \otimes t^k_j \in A \otimes A \quad .$$

Thus A is equipped both with a commutative *multiplication*, $m: A \otimes A \longrightarrow A$, and a *co-multiplication*, $\Delta: A \longrightarrow A \otimes A$, which is non-commutative unless G is Abelian. In addition, the group structure on G induces both a *unit* on A , $f \in A \longrightarrow f(e) \in K$, where e is the unit element of G , and an *antipode*, $f \in A \longrightarrow f' \in A$, where $f'(g) = f(g^{-1})$. The structure of A is actually that of a *commutative Hopf algebra*, to be defined below.

With a finite group, G , one also associates a group algebra, $U = KG$, over K , or, when G is a Lie group, one associates with it the universal enveloping algebra, $U = U(\mathfrak{g})$, of its Lie algebra, $\mathfrak{g}$. In both cases, U is equipped with i) a *multiplication* with *unit*, $m : A \otimes A \longrightarrow A$, which is non-commutative unless G is Abelian, ii) a co-commutative *co-multiplication* with *co-unit*, $\Delta : U \longrightarrow U \otimes U$, also called the diagonal, and iii) an *antipode*, which together give U the structure of a *co-commutative Hopf algebra*. In the case where $U = KG$, Δ is the algebra morphism such that, for $a \in G$, $\Delta a = a \otimes a$. In the case where $U = U(\mathfrak{g})$, Δ is the algebra morphism such that, for $x \in \mathfrak{g}$, $\Delta(x) = x \otimes 1 + 1 \otimes x$.

When G is a finite group there is an obvious duality between the Hopf algebra of functions, A , and the group algebra, U . This is why the approaches to quantum groups that have been mentioned in the introduction are, roughly speaking, the *duals* of each other. For example, if G is a compact Lie group, identifying x in $\mathfrak{g}$ with the left-invariant vector field it defines, x can be viewed as a derivation of $A = \mathcal{F}(G)$, the smooth functions over G. Thus, for a, b in A,

$$x(a.b) = x(a).b + a.x(b) = (x \otimes 1 + 1 \otimes x)(a \otimes b),$$

i. e., we can identify $\mathfrak{g}$, and therefore $U(\mathfrak{g})$, with a subspace of $(\mathcal{F}(G))^*$, and denoting the natural duality by $\langle \ , \ \rangle$, the multiplication m on $\mathcal{F}(G)$ and the co-multiplication Δ on $U(\mathfrak{g})$ are dual to each other,

$$\langle \Delta x, a \otimes b \rangle = \langle x, m(a \otimes b) \rangle.$$

We shall now define *bialgebras* and *Hopf algebras* in general.

Definition. A *bialgebra* U over a field K is both an algebra, whose *multiplication* (or *product*) we denote by $m : U \otimes U \longrightarrow U$, with a unit that we denote by $e : K \longrightarrow U$, and a co-algebra, whose *co-multiplication* (or *co-product*) we denote by $\Delta : U \longrightarrow U \otimes U$, with a co-unit that we denote by $\eta : U \longrightarrow K$, such that Δ and η are morphisms of algebras (*compatibility of the algebra and co-algebra structures*).

By assumption the multiplication, m, is *associative*, *i. e.*, the following diagram commutes, where 1 denotes the identity mapping of U,

(Associativity)

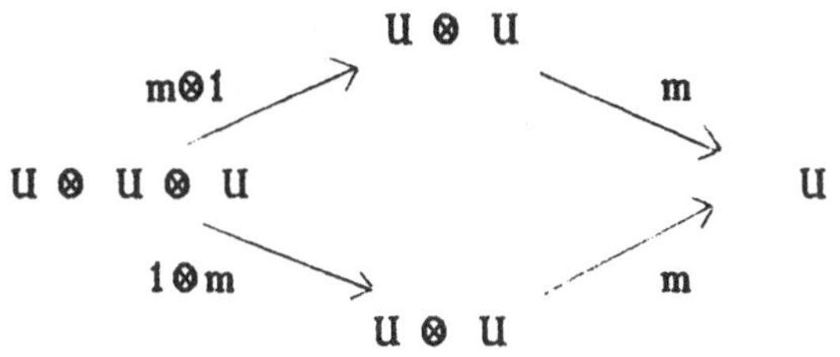

i. e., $$m \circ (1 \otimes m) = m \circ (m \otimes 1).$$

and the co-multiplication, Δ, is *co-associative*, *i. e.*, the following diagram also commutes,

(Co-associativity)

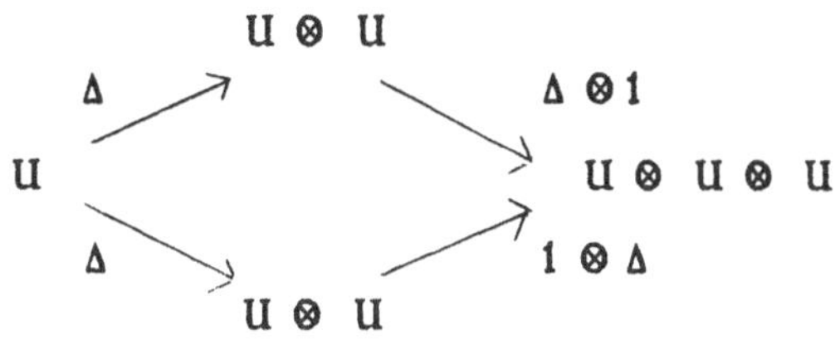

i. e., $$(1 \otimes \Delta) \circ \Delta = (\Delta \otimes 1) \circ \Delta.$$

The fact that e is a unit means

$$m \circ (1 \otimes e) = m \circ (e \otimes 1) = 1 .$$

The fact that η is a co-unit means

$$(1 \otimes \eta) \circ \Delta = (\eta \otimes 1) \circ \Delta = 1 .$$

Let $\sigma : U \otimes U \longrightarrow U \otimes U$ denote the usual *transposition* defined by

$$\sigma(x \otimes y) = y \otimes x .$$

A bialgebra is *commutative* if its multiplication is commutative, *i. e.*, if

$$m' = m , \quad \text{where} \quad m' = m \circ \sigma .$$

A bialgebra is *co-commutative* if its co-multiplication is co-commutative, *i. e*, if

$$\Delta' = \Delta , \quad \text{where} \quad \Delta' = \sigma \circ \Delta .$$

With the additional structure of an antipode, reflecting the existence of an inverse for each element in a group, a bialgebra becomes a Hopf algebra.

Definition. A *Hopf algebra* is a bialgebra, U , which is also equipped with a linear *antipode*, $S : U \longrightarrow U$, satisfying

$$m \circ (S \otimes 1) \circ \Delta = m \circ (1 \otimes S) \circ \Delta = \eta \circ e .$$

The antipode is an algebra anti-morphism, which is to say that $S(m(x \otimes y)) = m(S(y) \otimes S(x))$, and $S \circ e = e$. Some authors require that S be invertible, but S is not required to be involutive.

We shall sometimes use the standard notation for the image of an element $x \in U$ under the co-multiplication, Δ , $\Delta(x) = \sum (x_{(1)} \otimes x_{(2)})$, or simply $\Delta(x) = x_{(1)} \otimes x_{(2)}$. See Abe[21], Sweedler[22], Majid[16].

1.2 *Non-commutative and non-co-commutative Hopf algebras: "Quantum Groups"*

In the theory of quantum groups, one deals with Hopf algebras which are both non-commutative and non-co-commutative, and which are either deformations of commutative Hopf algebras (of functions) or deformations of co-commutative Hopf algebras (universal enveloping algebras).

In the approach by deformation of the algebra of functions, the R-*matrix* is the invertible element in $A \otimes A$ that measures the defect of commutativity of the multiplication in A. If $T = (t^i_j)$, and $T_1 = T \otimes 1$, $T_2 = 1 \otimes T$, then

$$R\, T_1 T_2 = T_2 T_1 R \,.$$

(If T is the $(n \times n)$ matrix of matrix elements for a quantum matrix group, then R is an invertible $(n^2 \times n^2)$ matrix.)

In the approach by deformation of the universal enveloping algebra, the R-matrix is the invertible element in $U \otimes U$ that measures the defect of co-commutativity of the co-multiplication in U . If x is in U ,

$$R\, \Delta(x) = \Delta'(x)\, R \,,$$

where $\Delta' = \sigma \circ \Delta$ is the opposite co-multiplication.

The important cases of quasitriangular Hopf algebras and triangular Hopf algebras will appear in a very natural fashion, once we consider the categories of modules of Hopf algebras. In fact, as we shall see *infra*, the requirement that the R-matrix be *quasi-triangular* corresponds to the case where the category of modules is a *braided monoïdal category*, which, as explained above is also called a *quasi-tensor category*, while the particular case where the R-matrix is *triangular* corresponds to the case of a *tensor category*. The relevant definitions are given in the next subsection and can be found in Mac Lane[23], Majid[16], Moore and Seiberg[3], Yetter[10].

1.3 *The category of modules of a co-commutative Hopf algebra*

Let us first say a few words about the classical case of a Lie group, where one can consider either the co-modules over its function algebra or the modules over the universal enveloping algebra of its Lie algebra.

When a Lie group, G , is modelled by the commutative Hopf algebra of smooth functions, $A = \mathcal{F}(G)$, one considers the category of A-*co-modules*. (See Abe[21], Sweedler[22] or Majid[16] for a definition.) In fact, if (V,ρ) is a finite-dimensional representation of a Lie group, G , then the linear space of polynomial functions over V is a co-module over the algebra, A , of smooth functions over G . For example, if $\mathbb{C}^n$ is the standard representation of $GL(n,\mathbb{C})$, then $\mathcal{F}(\mathbb{C}^n)$ is an $\mathcal{F}(GL(n,\mathbb{C}))$-co-module for the law

$$x^i \in \mathcal{F}(\mathbb{C}^n) \longrightarrow \sum a^i_k \otimes x^k \in \mathcal{F}(GL(n,\mathbb{C})) \otimes \mathcal{F}(\mathbb{C}^n) \,.$$

We remark that, when defining the tensor product of two A-co-modules, we use the multiplication in A, because the co-action of A on $V_1 \otimes V_2$ is defined by

$$v_1 \otimes v_2 \in V_1 \otimes V_2 \longrightarrow (a_1 \otimes v_1) \otimes (a_2 \otimes v_2) \in (A \otimes V_1) \otimes (A \otimes V_2)$$
$$\longrightarrow m(a_1 \otimes a_2) \otimes v_1 \otimes v_2 \in A \otimes V_1 \otimes V_2 \,.$$

When one relaxes the requirement that A be commutative, one deals with "non-commutative geometry". For example, the "quantum plane" is the standard co-module on the quantum group, $GL_q(2)$. See Manin[13] .

In the dual approach, if (V, ρ) is a finite-dimensional representation of a Lie algebra, $\mathfrak{g}$, then V is a U-module, where $U = U(\mathfrak{g})$. One can define direct sums, $(V_1 \oplus V_2, \rho_1 \oplus \rho_2)$, and tensor products, $(V_1 \otimes V_2, \rho_1 \otimes \rho_2)$, of U-modules, (V_1, ρ_1) and (V_2, ρ_2) . We remark that, when defining the tensor product of U-modules, we use the co-multiplication in U , because the action of U on $V_1 \otimes V_2$ is defined by

$$x \otimes v_1 \otimes v_2 \in U \otimes V_1 \otimes V_2 \longrightarrow x_{(1)} v_1 \otimes x_{(2)} v_2 \in V_1 \otimes V_2 ,$$

where $\Delta(x) = \sum (x_{(1)} \otimes x_{(2)})$.

The axioms satisfied by both the category of A-co-modules and that of U-modules are those of a *rigid Abelian tensor category*. In what follows, we shall only consider the universal enveloping algebra approach and its generalizations, following Drinfeld[7,1], and therefore all our discussions will concern the *categories of modules over Hopf algebras* or generalizations thereof.

In this subsection, we shall be concerned with a *co-commutative* Hopf algebra, U. Co-associativity of the co-multiplication is assumed, since it is one of the axioms of a bialgebra. We examine the properties of the category of U-modules which are, by definition, finite-dimensional modules over U , considered as an algebra. We shall list the properties of the category of U-modules, showing how the co-associativity of Δ implies the *associativity* of the tensor product of U-modules, while the co-commutativity of Δ implies the *commutativity* of the tensor product of U-modules. In the process, we shall use the vocabulary of the theory of categories that is relevant to the case at hand, in the order of decreasing generality, *monoïdal category, braided monoïdal category* or *quasi-tensor category, tensor category* or *symmetric monoïdal category, rigid Abelian tensor category.* The role of the *pentagon identity* and the two *hexagon identities* will be stressed. In the generalization to non-co-commutative Hopf algebras and, further, to quasi-Hopf algebras, some but not all of the properties of the category of modules will remain.

(i) *Consequences of co-associativity.* For each triple (V_1, V_2, V_3) of U-modules, consider the usual linear space isomorphism

$$\Phi_{V_1,V_2,V_3} : (V_1 \otimes V_2) \otimes V_3 \longrightarrow V_1 \otimes (V_2 \otimes V_3) ,$$

We first show that Φ_{V_1,V_2,V_3} is a U-*module-isomorphism, i. e.*, an invertible intertwining operator. In fact, the co-associativity of Δ means that

$$(x_{(1)(1)} \otimes x_{(1)(2)}) \otimes x_{(2)} = x_{(1)} \otimes (x_{(2)(1)} \otimes x_{(2)(2)}) ,$$

thus

$$\Phi_{V_1,V_2,V_3}(x \,.\, ((v_1 \otimes v_2) \otimes v_3)) = \Phi_{V_1,V_2,V_3}(x_{(1)} (v_1 \otimes v_2) \otimes x_{(2)} v_3)$$

$$= \Phi_{V_1,V_2,V_3}((x_{(1)(1)} v_1 \otimes x_{(1)(2)} v_2) \otimes x_{(2)} v_3)$$

$$= x_{(1)(1)} v_1 \otimes (x_{(1)(2)} v_2 \otimes x_{(2)} v_3) ,$$

while

$$x \,.\, \Phi_{V_1,V_2,V_3}((v_1 \otimes v_2) \otimes v_3) = x \,.\, (v_1 \otimes (v_2 \otimes v_3))$$

$$= x_{(1)} v_1 \otimes x_{(2)} (v_2 \otimes v_3) = x_{(1)} v_1 \otimes (x_{(2)(1)} v_2 \otimes x_{(2)(2)} v_3) .$$

By the co-associativity of Δ these two expressions coïncide.

Moreover, Φ is a *natural equivalence* from the functor $(\,.\, \otimes \,.\,) \otimes \,.\,$ to the functor $. \otimes (\,.\, \otimes \,.\,)$, which means that each Φ_{V_1, V_2, V_3} is a *functorial isomorphism*. This, in turn, means that the following diagram commutes, for any morphisms of U-modules $u_i : V_i \longrightarrow V_i'$, $i = 1, 2, 3$,

$$\begin{array}{ccc} (V_1 \otimes V_2) \otimes V_3 & \xrightarrow{\Phi_{V_1,V_2,V_3}} & V_1 \otimes (V_2 \otimes V_3) \\ {\scriptstyle (u_1 \otimes u_2) \otimes u_3}\downarrow & & \downarrow{\scriptstyle u_1 \otimes (u_2 \otimes u_3)} \\ (V'_1 \otimes V'_2) \otimes V'_3 & \xrightarrow{\Phi_{V'_1,V'_2,V'_3}} & V'_1 \otimes (V'_2 \otimes V'_3) \end{array}$$

Functoriality of Φ

It is clear that Φ satisfies the *pentagon identity, i. e.*, for any U-modules, V_1, V_2, V_3, V_4, the following *pentagonal diagram* commutes:

$$\begin{array}{ccccc} & & (V_1 \otimes V_2) \otimes (V_3 \otimes V_4) & & \\ & \overset{\Phi_{V_1 \otimes V_2, V_3, V_4}}{\nearrow} & & \overset{\Phi_{V_1, V_2, V_3 \otimes V_4}}{\searrow} & \\ ((V_1 \otimes V_2) \otimes V_3) \otimes V_4 & & & & V_1 \otimes (V_2 \otimes (V_3 \otimes V_4)) \\ {\scriptstyle \Phi_{V_1,V_2,V_3} \otimes 1_{V_4}}\downarrow & & & & \uparrow{\scriptstyle 1_{V_1} \otimes \Phi_{V_2,V_3,V_4}} \\ (V_1 \otimes (V_2 \otimes V_3)) \otimes V_4 & & \xrightarrow[\Phi_{V_1, V_2 \otimes V_3, V_4}]{} & & V_1 \otimes ((V_2 \otimes V_3) \otimes V_4) \end{array}$$

Pentagon

By definition, a natural equivalence Φ from the functor $(\,.\, \otimes \,.\,) \otimes \,.\,$ to the functor $. \otimes (\,.\, \otimes \,.\,)$ is an *associativity constraint* if it satisfies the pentagon identity. When *Mac Lane's coherence theorem*[24] is applied to Φ it shows that all of the various ways of decomposing the passage from an n-fold tensor product to any other n-fold tensor product lead to one and the same isomorphism. For instance, with $n = 5$, all the maps obtained from Φ and the identities from, for example, $\left(V_1 \otimes ((V_2 \otimes V_3) \otimes V_4) \otimes V_5\right)$ to, for example, $(V_1 \otimes V_2) \otimes ((V_3 \otimes V_4) \otimes V_5)$, are the same. In other words, the tensor product is associative up to well-defined isomorphisms.

There are also functorial isomorphisms, χ_V from $V \otimes K$ to V, and ξ_V from $K \otimes V$ to V (χ and ξ are natural transformations) such that the associativity constraint Φ is *compatible* with χ and ξ in the sense that all diagrams of the following form commute.

$$\begin{array}{ccc} (V_1 \otimes K) \otimes V_2 & \xrightarrow{\Phi_{V_1, K, V_2}} & V_1 \otimes (K \otimes V_2) \\ {\scriptstyle \chi_{V_1} \otimes 1_{V_2}} \searrow & & \nearrow {\scriptstyle 1_{V_1} \otimes \xi_{V_2}} \\ & V_1 \otimes V_2 & \end{array}$$

Compatibility of associativity and unit

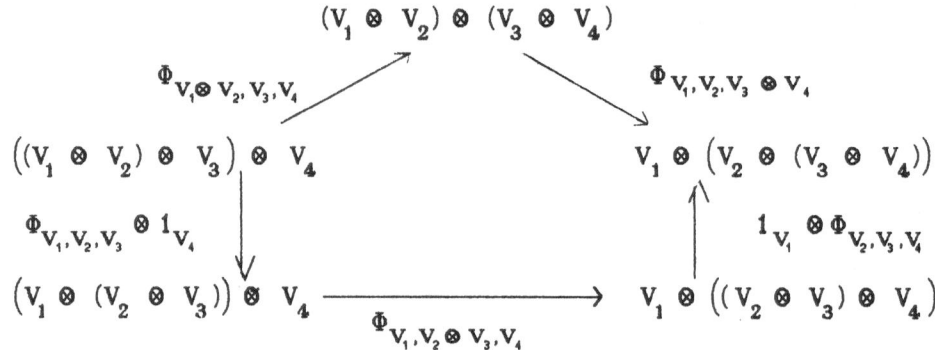

It then follows from another coherence theorem due to Kelly[25] that all diagrams involving both the two natural transformations, χ and ξ , and the associativity constraint, Φ , commute.

Therefore, when U is a Hopf algebra, the category of U-modules is a *monoïdal category* in the following sense:

Definition.- A *monoïdal category* is a category with tensor products, a unit object, an associativity constraint, and natural transformations, χ and ξ , compatible with the associativity constraint.

(ii) *Consequences of co-commutativity.* For each pair (V_1, V_2) of U-modules, consider the usual transposition,

$$\Psi_{V_1,V_2} : V_1 \otimes V_2 \longrightarrow V_2 \otimes V_1 .$$

We first show that Ψ_{V_1,V_2} is a U-*module isomorphism, i. e.*, an invertible intertwining map. In fact,

$$\Psi_{V_1,V_2}(x.(v_1 \otimes v_2)) = \Psi_{V_1,V_2}(x_{(1)}v_1 \otimes x_{(2)}v_2) = x_{(2)}v_2 \otimes x_{(1)}v_1$$

while

$$x.\Psi_{V_1,V_2}(v_1 \otimes v_2) = x.(v_2 \otimes v_1) = x_{(1)}v_2 \otimes x_{(2)}v_1 .$$

These two expressions are equal by the co-commutativity of the co-multiplication, Δ .

Moreover, Ψ is *functorial* in the sense that

$$\Psi_{V'_1,V'_2} \circ (u_1 \otimes u_2) = (u_2 \otimes u_1) \circ \Psi_{V_1,V_2} ,$$

where $u_i : V_i \longrightarrow V'_i$, $i = 1, 2$, are morphisms of U-modules.

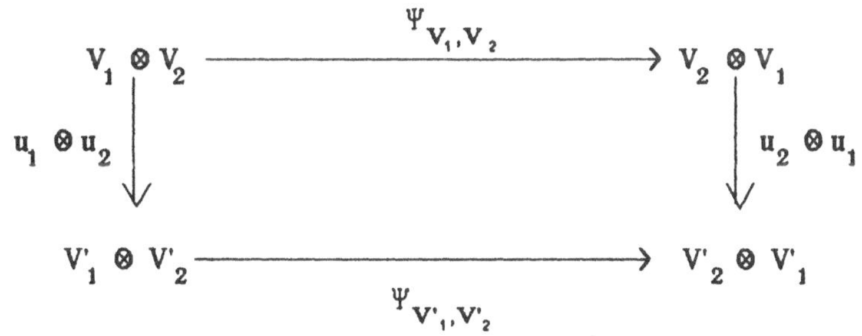

Functoriality of Ψ

The natural equivalence Ψ can be called the *commutativity constraint.* We remark that, in the older literature, this term is applied only to an *involutive* commutativity constraint. In the case where Ψ is the transposition, this additional condition is clearly satisfied, but it is not met by the natural equivalences obtained from, *e. g.*, quasi-triangular solutions of the quantum Yang-Baxter equation (see 1.4).

(iii) *Consequences of co-associativity and co-commutativity.* It is clear that the associativity constraint, Φ , and the commutativity constraint, Ψ , satisfy the two *hexagon identities, i. e.*, for any U-modules V_1, V_2, V_3 , the following *hexagonal diagrams* commute:

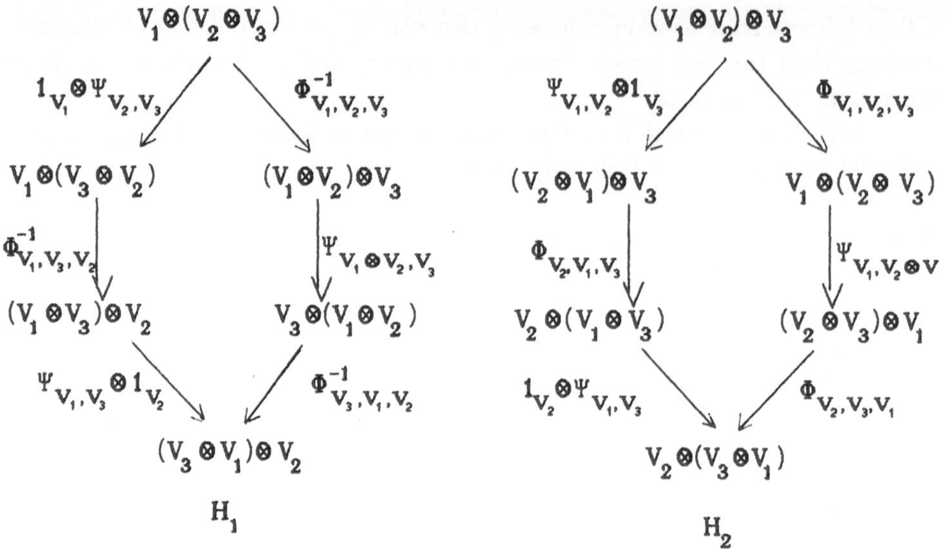

Hexagons

The fact that Ψ is functorial and that the two hexagon identities are satisfied is expressed by the statement, Ψ is a *commutativity constraint compatible with the associativity constraint*, Φ .

Therefore, when U is a co-commutative Hopf algebra, the category of U-modules is a *braided monoïdal category* in the following sense:

Definition: A *braided monoïdal category* is a monoïdal category with a commutativity constraint compatible with the associativity constraint.

There are additional *coherence theorems* (Joyal and Street[26], Freyd and Yetter[27]) which state that, in general, given an associativity constraint, Φ , and a commutativity constraint, Ψ, which is not necessarily involutive, the pentagon identity and the two hexagon identities imply that all diagrams constructed in the preceding way from Φ and Ψ commute.

From a braided monoïdal category, one can obtain representations of the *braid group* on n strings . See Majid[16], Cartier[17], Yetter[10].

Actually, in the case at hand, the category of modules over a co-commutative Hopf algebra, U , the commutativity isomorphism is involutive, and therefore either of the hexagon identities implies the other. In fact, the category of U-modules over a co-commutative Hopf algebra, U , satisfies stronger axioms, those of a *tensor category*, in the following sense:

Definition: A *tensor category* is a braided monoïdal category with an *involutive* commutativity constraint.

From a tensor category, one can obtain representations of the *symmetric group* on n elements.

The tensor category of U-modules over a co-commutative Hopf algebra is, in fact a *rigid Abelian tensor category*. We shall not give the technical definition of the rigid Abelian categories but we shall only mention that "rigidity" implies that each U-module has a dual – there is a contragredient representation for each representation – and that the biduality is the identity, while "Abelian" expresses the fact that direct sums are defined with the usual properties (there is a zero object and kernels and co-kernels exist), and, as we have seen, "tensor category" expresses the fact that tensor products are defined with the usual properties (associativity and commutativity).

In the next subsection, we shall examine the case of the quantum groups which correspond to Hopf algebras that are no longer co-commutative.

1.4 *The category of modules of a quantum group, the triangular and the quasi-triangular cases*

Since *quantum groups* are defined (see Drinfeld[7]) to be a certain class of Hopf algebras, and since the axioms of any Hopf algebra include that of *co-associativity* (see section 1.1), the category of U-modules, the representations of the quantum group, is *monoïdal* (see 1.3 (i)).

Let us now assume that there exists an R-matrix, *i. e.*, an element R in $U \otimes U$, that is invertible and satisfies

$$\Delta'(x) = R . \Delta(x) . R^{-1} ,$$

for x in U , where the dots indicate the multiplication in $U \otimes U$, and let us examine the consequences of the existence of the R-matrix for the category of U-modules. As we have already observed, R measures the extent to which the co-multiplication, Δ , in the Hopf algebra, U, is non-co-commutative. We can then define, for each pair (V_1, ρ_1), (V_2, ρ_2) of U-modules, an isomorphism of U-modules,

$$\hat{R}_{V_1,V_2} = \sigma \circ R_{V_1,V_2} : V_1 \otimes V_2 \longrightarrow V_2 \otimes V_1 ,$$

where σ is the usual transposition, and $R_{V_1,V_2} : V_1 \otimes V_2 \longrightarrow V_1 \otimes V_2$ is obtained from R by means of the U-module structure of $V_1 \otimes V_2$.

We shall set $\Psi_{V_1,V_2} = \hat{R}_{V_1,V_2}$. We can ask what are the additional conditions on R for Ψ to be a commutativity constraint compatible with the trivial associativity constraint, given by the natural isomorphisms of triple tensor products of U-modules. (In the next section, on quasi-Hopf algebras, we shall consider nontrivial associativity constraints.) In this way, we are led to introduce

Definition. A *quasi-triangular Hopf algebra* is a Hopf algebra, (U,m,Δ) , with an R-matrix, R , such that the two *hexagon identities* are satisfied when Ψ is the commutativity constraint defined by R , and Φ is the trivial associativity constraint.

Thus a quasi-triangular Hopf algebra is defined by an R-matrix, R , called *quasi-triangular*, such that

$$\begin{cases} \Delta_{12}(R) = R_{13} . R_{23} \\ \Delta_{23}(R) = R_{13} . R_{12} , \end{cases}$$

where we have introduced the notation, $R = \sum_i x_i \otimes y_i$, $R_{12} = \sum_i x_i \otimes y_i \otimes e$, $R_{13} = \sum_i x_i \otimes e \otimes y_i$, $R_{23} = \sum_i e \otimes x_i \otimes y_i$, where $\Delta_{12}(R) = (\Delta \otimes 1)(R)$ and $\Delta_{23}(R) = (1 \otimes \Delta)(R)$, and the dots denote the multiplication in $U \otimes U \otimes U$.

It follows from the definitions that *the category of modules over a quasi-triangular Hopf algebra* is a *braided monoïdal category*. We make a further

Definition. A triangular Hopf algebra is a *quasi-triangular Hopf algebra* such that R is involutive, *i. e.*,

$$R_{21} = (R_{12})^{-1} ,$$

where if $R = R_{12} = \sum_i x_i \otimes y_i \in U \otimes U$, then $R_{21} = \sum_i y_i \otimes x_i \in U \otimes U$.

With this definition, it is clear that *the category of modules over a triangular Hopf algebra* is a *tensor category*.

Any triangular or quasi-triangular R-matrix satisifies the *Yang-Baxter equation* or *quantum Yang-Baxter equation*,

(YBE) $$R_{12} \cdot R_{13} \cdot R_{23} = R_{23} \cdot R_{13} \cdot R_{12} .$$

This is proven using the identity $\Delta_{12}(R) = R_{12} \cdot R_{23}$ and the defining relation $\Delta'(x) . R = R . \Delta(x)$. An alternate proof is based on the following *dodecagon* diagram and it will be used in the more general situation of the quasi-triangular quasi-Hopf algebras in the next section, yielding the quasi-Yang-Baxter equation. In this diagram, we use an obvious shorthand notation, Φ_{123} for Φ_{V_1, V_2, V_3} etc..., and we have suppressed the tensor product signs.

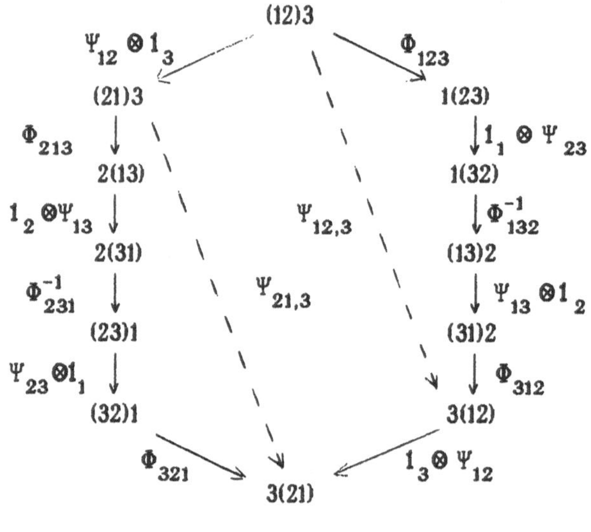

Dodecagon

When one inserts the dotted arrows, $\Psi_{12,3}$ and $\Psi_{21,3}$, the dodecagon is transformed into two hexagons, which commute by assumption, and a quadrilateral which also commutes because the commutativity constraint is functorial. The fact that the dodecagon, where Φ is the trivial associativity constraint, and Ψ is defined by a quasi-triangular R-matrix, commutes for all triples of U-modules expresses the fact that the Yang-Baxter equation is satisfied.

1.5 *Reconstruction theorems and Tannaka-Krein duality*

There are several *reconstruction theorems*, the statements of which I shall not reproduce in this lecture. The simplest of these theorems roughly states that for each rigid Abelian tensor category with trivial associativity and commutativity constraints that is equipped with what is called a *forgetful functor* that maps the objects and arrows of the category into the vector spaces and linear maps in a functorial way, there exists a co-commutative Hopf algebra, essentially corresponding to a *group*, whose category of modules is the given category. The name 'forgetful functor' comes from the fact that it 'forgets' the additional structure on the objects of the given category corresponding to their module structure over the Hopf algebra, *i. e.*, their structure as a representation space of the group. See Saavedra[28], Deligne and Milne[29].

The various generalizations of this theorem state that to more general categories, such as tensor (resp., braided monoïdal) categories with a trivial associativity constraint but a nontrivial (resp., and non-involutive) commutativity constraint, there corresponds a triangular (resp., quasi-triangular) Hopf algebra. See Lyubashenko[14], Majid[9], Deligne[30] and also Woronowicz[31] for precise statements and references, and also the surveys of Drinfeld[7](section 10), Manin[13] (section 12), Majid[16] (section 7), Moore and Seiberg[3] (appendix C), Cartier[17](section 3).

In the next section, we shall mention the extension of this approach (see Drinfeld[1], Majid[19]) to a tensor or quasi-tensor category whose commutativity and associativity constraints are both nontrivial. The corresponding algebra is no longer a Hopf algebra and this more general object is called a *quasi-Hopf algebra*. The motivation of the recent advances in this theory visibly comes from *conformal field theory*. The axioms obeyed by the representations of the chiral algebra are more general than those of the category of modules of a Hopf algebra. Thus reconstructing a CFT from the equations satisfied by its operators Ω and its fusion matrices (the 'bootstrap program') warrants the search for generalizations of Hopf algebras.

2. QUASI-QUANTUM GROUPS

2.1 *Quasi-Hopf algebras*

The quasi-Hopf algebras are a generalization of the Hopf algebras, in so far as the co-multiplication is no longer required to be co-associative. To ensure that their categories of modules are still monoïdal, one assumes that the defect in the co-associativity of the co-multiplication is measured by a "Φ-matrix". We first define the quasi-bialgebras which generalize the bialgebras.

Definition. A *quasi-bialgebra*, (U, m, Δ, Φ) , is an algebra with an associative multiplication, m, and a unit, e , equipped with a co-multiplication, Δ , and a co-unit η, such that the multiplication and the co-multiplication are *compatible*, *i. e.*, Δ and η are morphisms of algebras, and with an element Φ in $U \otimes U \otimes U$ such that, for each x in U ,

(A) $$(1 \otimes \Delta)(\Delta(x)) = \Phi.(\Delta \otimes 1)(\Delta(x)) . \Phi^{-1} ,$$

(P) $$\Delta_{34}(\Phi) . \Delta_{12}(\Phi) = \Phi_{234} . \Delta_{23}(\Phi) . \Phi_{123} ,$$

where $\Phi_{123} = \Phi \otimes e$, $\Phi_{234} = e \otimes \Phi$,

$\Delta_{12}(\Phi) = (\Delta \otimes 1 \otimes 1)(\Phi)$, $\Delta_{23}(\Phi) = (1 \otimes \Delta \otimes 1)(\Phi)$, $\Delta_{34}(\Phi) = (1 \otimes 1 \otimes \Delta)(\Phi)$,

and where the dots indicate the multiplication in $U \otimes U \otimes U$. Moreover, there is a co-unit, η,

$$(\eta \otimes 1) \circ \Delta = 1 = (1 \otimes \eta) \circ \Delta,$$

and Φ, η and the unit, e, satisfy

$$(1 \otimes \eta \otimes 1)(\Phi) = e \otimes e .$$

Thus, in a quasi-bialgebra, the commutative diagram expressing the co-associativity property that is valid in the case of a bialgebra, is replaced by the following commutative diagram,

$$\begin{array}{ccccc} & & U \otimes U & \xrightarrow{\Delta \otimes 1} & U \otimes U \otimes U \\ & \overset{\Delta}{\nearrow} & & & \downarrow \mathrm{Ad}(\Phi) \\ U & & & & \\ & \underset{\Delta}{\searrow} & & & \\ & & U \otimes U & \xrightarrow[1 \otimes \Delta]{} & U \otimes U \otimes U \end{array}$$

Diagram (A)

Obviously, the quasi-bialgebras generalize the bialgebras since, for $\Phi = 1$, (A) reduces to the co-associativity of Δ.

Let us show that (A) and (P) imply that the category of modules over a quasi-bialgebra is monoïdal. Let (U, m, Δ, Φ) be a quasi-bialgebra. Tensor products of U-modules are defined using the co-multiplication Δ, and we shall prove that, because of identities (A) and (P), Φ defines an *associativity constraint*.

For any triple of U-modules (V_1, V_2, V_3), the element Φ in $U \otimes U \otimes U$ defines a linear map, the composition of the multiplication by Φ with the usual associativity map,

$$\Phi_{V_1,V_2,V_3} : (V_1 \otimes V_2) \otimes V_3 \longrightarrow V_1 \otimes (V_2 \otimes V_3),$$

which is explicitly written, setting $\Phi = \sum (X_i \otimes Y_i \otimes Z_i)$ and omitting the summation signs,

$$\Phi_{V_1,V_2,V_3}((v_1 \otimes v_2) \otimes v_3) = X_i v_1 \otimes (Y_i v_2 \otimes Z_i v_3),$$

for $v_i \in V_i$, $i = 1, 2, 3$.

Identity (A) is equivalent to the fact that all maps Φ_{V_1,V_2,V_3} are U-module isomorphisms. (We remark that, unless we are dealing with a bi-algebra, where Φ is the trivial element $e \otimes e \otimes e$, the usual associativity maps are *not* U-module isomorphisms, they have to be composed with Φ to yield an associativity constraint.) In fact (A) means that, for all x in U,

$$x_{(1)} \otimes (x_{(2)(1)} \otimes x_{(2)(2)}) . \Phi = \Phi . ((x_{(1)(1)} \otimes x_{(1)(2)}) \otimes x_{(2)}),$$

which is

$$x_{(1)} X_i \otimes (x_{(2)(1)} Y_i \otimes x_{(2)(2)} Z_i) = (X_i x_{(1)(1)} \otimes Y_i x_{(1)(2)}) \otimes Z_i x_{(2)} .$$

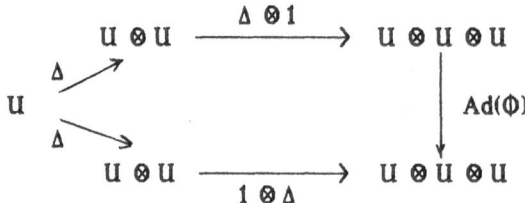

Now, for x in U,

$$\Phi_{V_1,V_2,V_3}\left(x \cdot ((v_1 \otimes v_2) \otimes v_3)\right) = \Phi_{V_1,V_2,V_3}(x_{(1)}\, (v_1 \otimes v_2\,) \otimes x_{(2)}\, v_3)$$

$$= \Phi_{V_1,V_2,V_3}\left((x_{(1)(1)}\, v_1 \otimes x_{(1)(2)}\, v_2) \otimes x_{(2)}\, v_3\right)$$

$$= X_i\, x_{(1)(1)}\, v_1 \otimes (Y_i\, x_{(1)(2)}\, v_2 \otimes Z_i\, x_{(2)}\, v_3)\,,$$

while

$$x \cdot \Phi_{V_1,V_2,V_3}((v_1 \otimes v_2) \otimes v_3) = x \cdot (X_i\, v_1 \otimes (Y_i v_2 \otimes Z_i v_3))$$

$$= x_{(1)}\, X_i\, v_1 \otimes x_{(2)}(Y_i v_2 \otimes Z_i v_3) = x_{(1)} X_i\, v_1 \otimes (x_{(2)(1)} Y_i v_2 \otimes x_{(2)(2)} Z_i v_3)\,.$$

Identity (A) implies that these two expressions coïncide, and, conversely, if these two expressions coïncide, (A) is satisfied. (Consider U as a U-module and take $v_i = e$, $i = 1, 2, 3$.) It is clear that Φ, thus defined, is functorial.

We shall see that identity (P) is the *pentagon identity*, *i. e.*, it is equivalent to the fact that the pentagonal diagram (see section 1.3) commutes for all triples of U-modules. In fact, identity (P) can be written

$$X_i X_{j(1)} \otimes \left(Y_i X_{j(2)} \otimes (Z_{i(1)} Y_j \otimes Z_{i(2)} Z_j)\right) = X_l X_h \otimes \left(X_k Y_{l(1)} Y_h \otimes (Y_k Y_{l(2)} Z_h \otimes Z_k Z_l)\right).$$

Now the pentagonal diagram commutes if, for $v_i \in V_i$, $i = 1, 2, 3, 4$,

$$X_i X_{j(1)} v_1 \otimes \left(Y_i X_{j(2)} v_2 \otimes (Z_{i(1)} Y_j v_3 \otimes Z_{i(2)} Z_j v_4)\right)$$

$$= X_l X_h v_1 \otimes \left(X_k Y_{l(1)} Y_h v_2 \otimes (Y_k Y_{l(2)} Z_h v_3 \otimes Z_k Z_l v_4)\right).$$

Therefore, identity (P) implies that all pentagonal diagrams commute, and conversely.

Thus, when (A) and (P) are satisfied, Φ is an *associativity constraint* in the technical sense that the *pentagonal diagram* commutes for every triple of U-modules. Moreover, the axiom involving Φ, the unit and the co-unit is written

$$X_i \otimes \eta(Y_i) \otimes Z_i = e \otimes e\,.$$

It expresses precisely the compatibility of the associativity constraint defined by Φ with the natural transformations χ and ξ. Thus the definition of the quasi-bialgebras implies that the category of modules over a quasi-bialgebra is a *monoïdal category*.

We remark that, contrary to the case of the bialgebras, the dual of a quasi-bialgebra is not a quasi-bialgebra. The situation is not self-dual.

We now define the quasi-Hopf algebras which generalize the Hopf algebras.

Definition: A *quasi-Hopf algebra* is a quasi-bialgebra, (U, m, Δ, Φ), with an antipode S, which is an algebra anti-morphism such that there exist elements a and b in U satisfying, for each x in U,

$$\sum S(x_{(1)}).\ a\ .\ x_{(2)} = \eta(x)\, a\,, \qquad \sum x_{(1)}.\ b.\ S(x_{(2)}) = \eta(x)\, b\,,$$

where $\Delta(x) = \sum x_{(1)} \otimes x_{(2)}$, and

$$\sum_i X_i\ .\ b\ .S(Y_i)\ .a\ .Z_i = e\,, \qquad \text{where} \quad \Phi = \sum_i X_i \otimes Y_i \otimes Z_i\,.$$

Following the general idea of the theory of quantum groups, it is natural to define the "quasi-quantum groups" as a class of quasi-Hopf algebras, depending on a continuous parameter. We shall consider this question in section 2.4. We first consider the case of the quasi-triangular quasi-Hopf algebras.

2.2 *Quasi-triangular and triangular quasi-Hopf algebras*

The quasi-triangular quasi–Hopf algebras generalize the quasi-triangular Hopf algebras, but their categories of modules remain braided monoïdal categories.

Definition. A *quasi-triangular quasi-Hopf algebra* is a quasi-Hopf algebra (U, m, Δ, Φ) with an R-matrix, *i. e.*, an element R in $U \otimes U$, that is invertible and satisfies

(C) $\Delta'(x) = R \,.\, \Delta(x) \,.\, R^{-1}$,

for x in U , such that the two *hexagon identities* are satisfied when Ψ is the commutativity constraint defined by R , and Φ is the associativity constraint defined by Φ .

A computation similar to the preceding ones involving identities (A) and (P) shows that a quasi-triangular quasi-Hopf algebra is defined by an R-matrix, R , such that

(H_1) $\Delta_{12}(R) = \Phi_{312} \cdot R_{13} \cdot (\Phi_{132})^{-1} \cdot R_{23} \cdot \Phi_{123}$,

(H_2) $\Delta_{23}(R) = (\Phi_{231})^{-1} \cdot R_{13} \cdot \Phi_{213} \cdot R_{12} \cdot (\Phi_{123})^{-1}$.

Identities (H_1) and (H_2) are equivalent to the fact that the two hexagonal diagrams commute for any triple of U-modules, *i. e.*, that the associativity constraint, Φ , defined by Φ , and the commutativity constraint, Ψ , defined by R , are *compatible* (see section 1.3), Therefore the category of modules of a quasi-triangular quasi-Hopf algebra is a *braided monoïdal category*.

Definition. A *triangular quasi-Hopf algebra* is a quasi-triangular quasi-Hopf algebra such that R is *involutive*, *i. e.*,

(U) $R_{21} = (R_{12})^{-1}$.

Condition (U) is called the *unitarity condition.*

In the case of a triangular Hopf algebra, the commutativity constraint is involutive and the category of modules of a triangular quasi-Hopf algebra is a *tensor category.*

2.3 *Quasi-Yang-Baxter equation*

The R-matrix of a quasi-triangular or triangular quasi-Hopf algebra satisifes a generalization of the Yang-Baxter equation, the *quasi-Yang-Baxter equation*,

(quasiYBE) $R_{12} \cdot \Phi_{312} \cdot R_{13} \cdot (\Phi_{132})^{-1} \cdot R_{23} \cdot \Phi_{123} = \Phi_{321} \cdot R_{23} \cdot (\Phi_{231})^{-1} \cdot R_{13} \cdot \Phi_{213} \cdot R_{12}$.

This identity is equivalent to the commutativity, for any triple of U-modules, of the *dodecagonal diagram* that we already introduced in section 1.3. As was shown in section 1.3, when one of the hexagonal diagrams commutes, the dodecagonal diagram commutes, so the quasi-Yang-Baxter equation is satisfied whenever (U, m, Δ, Φ, R) is a quasi-triangular quasi-Hopf algebra.

The defining axioms of the quasi-Hopf algebras and some of their consequences, the special case of the quasi-triangular quasi-Hopf algebras and the even more special case of the triangular quasi-Hopf algebras, are summarized in table 2, below, where, for simplicity, we have ommitted the properties that involve the unit or co-unit.

Table 2

QUASI-HOPF ALGEBRAS (U, m, Δ, Φ)

– *defect of co-associativity is measured by* $\Phi \in U \otimes U \otimes U$

(A) $\Delta_{12}(\Delta(x)) = \Phi . \Delta_{23}(\Delta(x)) . \Phi^{-1}$ (*generalizes co-associativity of* Δ, $\Delta_{12} \circ \Delta = \Delta_{12} \circ \Delta$)

– Φ *defines an associativity constraint*

(P) $\Delta_{34}(\Phi) . \Delta_{12}(\Phi) = \Phi_{234} . \Delta_{23}(\Phi) . \Phi_{123}$ (*the pentagon identity*)

Axioms (A) and (P) imply that the category of U-modules is *monoïdal.*

QUASI-TRIANGULAR QUASI-HOPF ALGEBRAS (U, m, Δ, Φ, R)

– (A) *and* (P)

– *defect of co-commutativity is measured by* $R \in U \otimes U$

(C) $\Delta'(x) = R . \Delta(x) . R^{-1}$ (*generalizes co-commutativity of* Δ, $\Delta' = \Delta$)

– R *defines a commutativity constraint compatible with* Φ

(H_1) $\Delta_{12}(R) = \Phi_{312} . R_{13} . (\Phi_{132})^{-1} . R_{23} . \Phi_{123}$

(H_2) $\Delta_{23}(R) = (\Phi_{231})^{-1} . R_{13} . \Phi_{213} . R_{12} . (\Phi_{123})^{-1}$

(*the hexagon identities ; they generalize the identities for a quasi-triangular Hopf algebra*, $\Delta_{12}(R) = R_{13} . R_{23}$, $\Delta_{23}(R) = R_{13} . R_{12}$)

Consequence (quasiYBE) (*generalizes* (YBE) $R_{12} R_{13} R_{23} = R_{23} R_{13} R_{12}$)

$$R_{12} \Phi_{312} R_{13} (\Phi_{132})^{-1} R_{23} \Phi_{123} = \Phi_{321} R_{23} (\Phi_{231})^{-1} R_{13} \Phi_{213} R_{12}$$

Axioms (A)(P)(C)(H_1) and (H_2) imply that the category of U-modules is a *braided monoïdal category.*

TRIANGULAR QUASI-HOPF ALGEBRAS (U, m, Δ, Φ, R)

– (A) (P) (C) (H_1) *and* (H_2)

– (U) $R_{21} = (R_{12})^{-1}$ (*unitarity*)

Axioms (A)(P)(C)(quasiYBE) and (U) imply that the category of U-modules is a *tensor category.*

2.4 *Properties of the quasi-quantum groups*

We briefly summarize the fundamental results concerning the quasi-Hopf algebras (Drinfeld[1], see also Cartier[17], Stasheff[18], Majid[19]) .

With additional conditions on the 'classical' limit of Φ as $\hbar$ tends to 0, quasi-Hopf algebras which are deformations of universal enveloping algebras become generalizations of the QUE-Hopf algebras introduced by Drinfeld[7], often called "quantum groups", so the more general objects, the QUE-quasi-Hopf algebras ought to be called "quasi-quantum groups".

The *quantum groups* (QUE-Hopf algebras, in the sense of Drinfeld[7]) are the Hopf algebras over the formal series in $\hbar$ which are deformations, in the parameter $\hbar$, of the universal enveloping algebra $U(g_0)$ of a complex Lie algebra, g_0 , and whose co-multiplication is, up to first-order terms in $\hbar$, the usual co-associative and co-commutative co-multiplication of $U(g_0)$, so that $\lim_{\hbar\to 0} 1/\hbar \left(\Delta'_\hbar(x) - \Delta_\hbar(x)\right)$ exists. Similarly, a *quasi-quantum group* (QUE-quasi-Hopf algebra, in the sense of Drinfeld[1]) is a deformation of a universal enveloping algebra such that, moreover, the element, Φ , that measures the defect of co-associativity is equal to the trivial element, $e \otimes e \otimes e$, up to terms of order $\hbar$, and the antisymmetrization of Φ vanishes up to terms of order $\hbar^2$.

The fundamental example of a quasi-triangular quasi-Hopf algebra, $U_{g,t}$, is constructed from a Lie algebra, g , over $\mathbb{C}[[\hbar]]$, the formal series in $\hbar$, such that, moreover, g is a *deformation* of $g_0 = g/(\hbar g)$, equipped with an *ad*-invariant symmetric tensor, t (scalar product), an element in the completed tensor product, $g \hat{\otimes} g$. Drinfeld proves the following

Theorem. Let $(U(g), m, \Delta)$ be the universal enveloping algebra of g, with the standard multiplication, m, and the standard co-multiplication, Δ. Let

$$R = e^{\hbar \frac{t}{2}} .$$

There exists $\Phi \in U \otimes U \otimes U$, such that $U_{g,t} = (U(g), m, \Delta, \Phi, R)$ is a *quasi-triangular quasi-Hopf algebra.*

The existence of Φ satisfying (A)(P)(H_1) and (H_2) is proved by considering the monodromy matrix of the *Knizhnik-Zamolodchikov equations*[32] for the 3-point functions in the Wess-Zumino-Witten case of conformal field theory.

In addition, an equivalence relation is defined on the quasi-Hopf algebras. If (U, m, Δ, Φ) is a quasi-Hopf algebra, then for any invertible $F \in U \otimes U$,

$$\begin{cases} \tilde{\Delta}(x) = F . \Delta(x) . F^{-1} \\ \tilde{\Phi} = F_{23} \cdot \Delta_{23}(F) . \Phi . \Delta_{12}(F^{-1}) . (F_{12})^{-1} , \end{cases}$$

defines a new quasi-Hopf algebra, $(U, m, \tilde{\Delta}, \tilde{\Phi})$, which is said to be obtained from (U, m, Δ, Φ) by *twisting* by F . If (U, m, Δ, Φ, R) is a *quasi-triangular quasi-Hopf algebra*, so is the twisted quasi-Hopf algebra, with

$$\tilde{R} = F_{21} . R . (F_{12})^{-1} .$$

This property, which is a novel feature that appears when the axioms of Hopf algebras are weakened into those of the quasi-Hopf algebras, has the following important consequence, discovered by Drinfeld[1].

Theorem. Let (U, m, Δ, Φ, R) be a quasi-triangular quasi-Hopf algebra over $\mathbb{C}[[\hbar]]$, which is a deformation of $U(g_0)$, where g_0 is a Lie algebra over $\mathbb{C}$, and whose co-multiplication is, up to first-order terms in $\hbar$, the usual co-multiplication of the universal envelopping algebra $U(g_0)$. Then (U, m, Δ, Φ, R) is obtained from $U_{g,t}$ by twisting, where g is a deformation of g_0 , and t is the *ad*-invariant symmetric tensor, t , in $g \hat{\otimes} g$,

$$t = \lim_{\hbar\to 0} \frac{1}{\hbar} (R_{21} R_{12} - e \otimes e).$$

The above theorem states that, up to twisting, all quasi-triangular quasi-quantum groups are of the form $U_{g,t}$. In particular, for a triangular quasi-quantum group, $t = 0$.

New results in the deformation theory of the quasi-Hopf algebras have been announced by Stasheff[18] and Shnider[33].

In the next section, we study the classical limit of the quasi-quantum groups (Lie quasi-bialgebras) and the corresponding Lie groups (quasi–Poisson Lie groups).

3. THE CLASSICAL LIMIT OF QUASI-QUANTUM GROUPS

3.1 *Lie quasi-bialgebras and quasi-Poisson Lie groups*

A *Lie quasi-bialgebra* is the classical analogue of a quasi-quantum group. It is a Lie algebra with an additional structure that generalizes that of a *Lie bialgebra.* As you may recall, a Lie bialgebra is the classical analogue of a QUE-Hopf algebra (Drinfeld[7]) and conversely, the quantum groups are deformations (*i. e.*, quantizations) of the enveloping algebra of a Lie bialgebra. A *Lie bialgebra*, $(\mathfrak{g}, \mu, \gamma)$, is a Lie algebra, $\mathfrak{g}$, whose Lie bracket operation we denote by $\mu : \wedge^2\mathfrak{g} \longrightarrow \mathfrak{g}$, with a co-bracket $\gamma : \mathfrak{g} \longrightarrow \wedge^2\mathfrak{g}$ (we have denoted the second exterior power of the vector space $\mathfrak{g}$ by $\wedge^2\mathfrak{g}$) such that

(1) γ and μ are *compatible*, which means that γ is a Lie-algebra 1-cocycle of $\mathfrak{g}$ with values in $\wedge^2\mathfrak{g}$ with respect to the adjoint action defined by μ, *i. e.*,

$$\delta_\mu \gamma = 0,$$

where δ_μ is the Lie algebra cohomology operator.

(2_0) γ is *Jacobian*, *i. e.*, it satisfies the identity,

$$[\gamma,\gamma] = 0 ,$$

where [,] is the Schouten bracket of linear multivector fields on $\mathfrak{g}$. (This means that, by transposition, γ defines a Lie algebra bracket, satisfying the Jacobi identity, on $\mathfrak{g}^*$. See, *e. g.*, Tulczyjew[34] for the Schouten bracket.)

In the generalization to Lie quasi-bialgebras, the bracket $[\gamma,\gamma]$, which is a linear map from $\mathfrak{g}$ to $\wedge^3\mathfrak{g}$, will no longer vanish, but we shall require that it be the coboundary of an element $\varphi \in \wedge^3\mathfrak{g}$ in the Lie-algebra cohomology defined by μ. Moreover, for reasons that will appear when we introduce the *double* of a Lie quasi-bialgebra, we shall assume that φ is a 3-cocycle on $\mathfrak{g}^*$ with values in the base field K with repect to the operator δ_γ which is defined formally in the same way as a Lie algebra cohomology, but is not of square 0 unless $[\gamma,\gamma] = 0$. Thus,

Definition. A *Lie quasi-bialgebra*, $(\mathfrak{g}, \mu, \gamma, \varphi)$, is a Lie algebra $(\mathfrak{g}, \mu)$ with a cobracket, $\gamma : \mathfrak{g} \longrightarrow \wedge^2\mathfrak{g}$, and an element $\varphi \in \wedge^3\mathfrak{g}$, such that

(1) $\delta_\mu\gamma = 0$,

(2) $[\gamma,\gamma] = \delta_\mu\varphi$,

(3) $\delta_\gamma\varphi = 0$.

This definition, in a slightly different language, is to be found in Drinfeld[1]. It was natural to inquire what are the *quasi-Poisson Lie groups*, the analogues for Lie quasi-bialgebras of the *Poisson Lie groups*, the global objects whose infinitesimal objects are the Lie bialgebras. In order to formulate and prove the fundamental result on the one-to-one correspondence between the quasi-Poisson Lie groups and the Lie quasi-bialgebras, we were led to reformulate Drinfeld's definition in the form given above, replace the study of the Lie quasi-bialgebras in the more general context of the *quasi-Lie quasi-bialgebras*, and give the following definition of quasi-Poisson Lie groups. (See Kosmann-Schwarzbach[2] for details.)

Recall that a *Poisson Lie group* (see *e. g.*, Verdier[35], Lu and Weinstein[36]) is a Lie group (G,π), where π denotes the multiplication in G, with a Poisson bracket, defined by a Poisson bivector field Λ, which is *multiplicative*, *i. e.*, for g, g' in G,

$$\Lambda(gg') = g.\Lambda(g') + \Lambda(g).g' .$$

Set $\rho(\Lambda)(g) = \Lambda(g).g^{-1} \in \bigwedge^2 \mathfrak{g}$. Obviously, Λ is multiplicative if and only if $\rho(\Lambda)(gg') = g.\rho(\Lambda)(g').g^{-1} + \rho(\Lambda)(g)$, *i. e.*, if $\rho(\Lambda)$ is a 1-cocycle of G with values in $\Lambda^2\mathfrak{g}$ with respect to the adjoint action of (G,π) . Thus (G,π,Λ) is a Poisson Lie group if and only if the bivector Λ satisfies

(I) $\rho(\Lambda)$ is a 1-cocycle of G with values in $\bigwedge^2\mathfrak{g}$ with respect to the adjoint action,

(II_0) Λ is *Poisson*, *i. e.*,

$$[\Lambda,\Lambda] = 0 ,$$

where $[\ ,\ ]$ denotes the Schouten bracket of multivector fields on G .

If (G, π, Λ) is a Poisson Lie group, then *the differential at the identity* of $\rho(\Lambda)$ is a Jacobian 1-cocycle, γ , of $(\mathfrak{g},\mu)$, the Lie algebra of (G,π) , so that $(\mathfrak{g}, \mu, \gamma)$ is a Lie bialgebra, and conversely, to each Lie bialgebra, there corresponds a unique connected and simply connected Poisson Lie group.

In the generalization to quasi-Poisson Lie groups, the bivector field is no longer required to be Poisson, but only to be such that $\rho([\Lambda,\Lambda])$, which is a mapping from G to $\bigwedge^3\mathfrak{g}$, be the coboundary of an element φ in $\bigwedge^3\mathfrak{g}$, with respect to the adjoint action of (G,π). Moreover, we shall require an additional condition corresponding to condition (3) in the definition of the Lie quasi-bialgebras.

Definition. A *quasi-Poisson Lie group* $(G, \pi, \Lambda, \varphi)$ is a Lie group (G,π) with a bivector field Λ and an element φ in $\bigwedge^3\mathfrak{g}$, such that

(I) Λ is multiplicative, *i. e.*, $\rho(\Lambda)$ is a 1-cocycle of (G,π),

(II) $[\Lambda,\Lambda] = \varphi^\lambda - \varphi^\rho$, *i. e.*, $\rho([\Lambda,\Lambda])$ is the coboundary of φ with respect to the adjoint action of (G,π) . (We have denoted φ^λ and φ^ρ the left- and right-invariant trivector fields defined by φ.)

(III) $[\Lambda,\varphi^\lambda] = 0$.

With these definitions, we obtain

Theorem. The Lie algebra of a quasi-Poisson Lie group $(G, \pi, \Lambda, \varphi)$ is the Lie quasi-bialgebra $(\mathfrak{g}, \mu, \gamma,- \varphi)$, where γ is the differential at the identity of $\rho(\Lambda)$. Conversely, to each Lie quasi-bialgebra, there corresponds a unique connected and simply connected quasi-Poisson Lie group.

The proof makes use of the relationships between Schouten brackets and Lie algebra cohomology. First, because Λ is multiplicative, one can show that

(a) $\Delta([\Lambda,\Lambda]) = -[\Delta\Lambda,\Delta\Lambda]$,

where $\Delta\Lambda$ (resp., $\Delta([\Lambda,\Lambda])$) is the differential at the identity of $\rho(\Lambda)$ (resp., $\rho([\Lambda,\Lambda])$). The bracket on the left-hand side is the Schouten bracket of bivector fields on G, while the bracket on the right-hand side is the Schouten bracket of linear fields of bivectors on $\mathfrak{g}$. (This formula is a particular case of a theorem that roughly states that the infinitesimal counterpart of the graded Lie algebra of 'multiplicative multivector fields' on G, equipped with the Schouten bracket and the derivation $d_\Lambda = [\Lambda, \ . \ ]$ that generalizes the Poisson cohomology operator, is the Lie algebra of linear multivector fields on $\mathfrak{g}$, equipped with the Schouten bracket and the derivation $d_\gamma = [\gamma, \ . \ ]$ that coïncides with δ_γ up to sign.) Formula (a) implies that, if G is connected, (2) is equivalent to (II) with φ changed into $-\varphi$. Then, the following relation holds,

(b) $[\Lambda, \varphi^\lambda](e) = -\delta_{\Delta\Lambda}\varphi$.

This is proved again using the fact that Λ is multiplicative. Since, for a multiplicative Λ , $[\Lambda,\varphi^\lambda]$ is left-invariant, we see that (3) is equivalent to (III) .

3.2 *Strictly cohomologous quasi-Poisson Lie groups and Lie quasi-bialgebras*

The important *twisting* operation on quasi-Hopf algebras that we have mentioned has the following classical analogue. We need to introduce a notation, $[r,r]^\mu$, for the *algebraic Schouten bracket* of an element r in $\bigwedge^2\mathfrak{g}$ with itself (Koszul[37]). It is the value at the identity of the Schouten bracket of the left-invariant bivector field defined by r with itself. This is the quantity, usually denoted $[r_{12}, r_{13}] + [r_{12}, r_{23}] + [r_{13}, r_{23}]$, that appears in the *classical Yang-Baxter equation*. (See, *e. g.*, Kosmann-Schwarzbach[38]).

Theorem. Let $(G, \pi, \Lambda, \varphi)$ be a quasi-Poisson Lie group. Let r be an element in $\bigwedge^2\mathfrak{g}$. Set

$$\text{(c)} \quad \begin{cases} \Lambda' = \Lambda + r^\lambda - r^\mu , \\ \varphi' = \varphi - 2\,\delta_{\Delta\Lambda} r + [r,r]^\mu . \end{cases}$$

Then $(G, \pi, \Lambda', \varphi')$ is a quasi-Poisson Lie group, and the most general quasi-Poisson Lie group of the form, $(G, \pi, \Lambda', \varphi'')$, is such that $\varphi'' = \varphi' + \varphi_1$, where φ_1 is Ad-invariant and satisfies $\delta_\gamma \varphi_1 + [r,\varphi_1]^\mu = 0$.

The proof of this theorem uses relation (b) above and the graded Jacobi identity for the Schouten bracket.

Two quasi-Poisson Lie groups, $(G, \pi, \Lambda, \varphi)$ and $(G, \pi, \Lambda', \varphi')$, are called *strictly cohomologous* if there exists an $r \in \bigwedge^2\mathfrak{g}$ such that (Λ, φ) and (Λ', φ') are related by relations (c).

If two strictly cohomologous quasi-Poisson Lie groups, $(G, \pi, \Lambda, \varphi)$ and $(G, \pi, \Lambda', \varphi')$, are related by relations (c), then their Lie quasi-bialgebras are $(\mathfrak{g}, \mu, \gamma, -\varphi)$ and $(\mathfrak{g}, \mu, \gamma', -\varphi')$ with

$$(\underline{c}) \quad \begin{cases} \gamma' = \gamma + \delta_\mu r \\ \varphi' = \varphi - 2\,\delta_{\Delta\Lambda} r + [r,r]^\mu \ . \end{cases}$$

In the same way, two Lie quasi bialgebras, $(\mathfrak{g}, \mu, \gamma, \varphi)$ and $(\mathfrak{g}, \mu, \gamma', \varphi')$, are said to be *strictly cohomologous* if there exists an $r \in \bigwedge^2\mathfrak{g}$ such that (γ,φ) and (γ',φ') are related by relations ($\underline{c}$).

3.3 *The double of a Lie quasi-bialgebra, marked Manin pairs and Manin pairs*

The construction of the double of a Lie bialgebra can be generalized to the case of a Lie quasi-bialgebra. More precisely,

Theorem. Each Lie quasi-bialgebra, $(\mathfrak{g}, \mu, \gamma, \varphi)$, defines a unique Lie-algebra structure, $[\ ,\ ]$, on $\mathfrak{g} \oplus \mathfrak{g}^*$, which prolongs the Lie algebra structure, μ, of $\mathfrak{g}$, and which leaves invariant the canonical scalar product. For $x \in \mathfrak{g}$, $\xi, \eta \in \mathfrak{g}^*$,

$$[x,\xi] = -\,\mathrm{ad}^{*\gamma}_{\xi}\, x + \mathrm{ad}^{*\mu}_{x}\, \xi ,$$

$$[\xi,\eta] = -\,\varphi(\xi,\eta) + \gamma(\xi,\eta) .$$

(Here, we have identified $\varphi \in \wedge^3\mathfrak{g}$ with a linear map from $\wedge^2\mathfrak{g}^*$ to $\mathfrak{g}$.) Conversely, a Lie algebra structure on $\mathfrak{g} \oplus \mathfrak{g}^*$ that leaves the canonical scalar product invariant and such that $(\mathfrak{g},\mu)$ is a Lie subalgebra, defines a Lie quasi-bialgebra structure on $(\mathfrak{g},\mu)$.

The vector space $\mathfrak{g} \oplus \mathfrak{g}^*$ equipped with this Lie algebra structure is called the *double* of the Lie quasi-bialgebra $\mathfrak{g}$.

The proof of this theorem can be worked out by calculations. It is immediate if one uses the *big bracket* of Kostant-Sternberg-Lecomte-Roger on $\wedge(\mathfrak{g}^* \oplus \mathfrak{g})$. (See Kostant and Sternberg[39], Lecomte et Roger[40] for a definition, and ref. 2 for a proof.)

In the theory of Lie bialgebras, it was shown by Drinfeld[7] that the Lie bialgebra structures on $(\mathfrak{g}, \mu)$ are in one-to-one correspondence with the *Manin triples*, $(\mathfrak{k}, \mathfrak{g}, \mathfrak{h})$, where $\mathfrak{k}$ is a Lie algebra with an invariant scalar product, and $\mathfrak{g}$ and $\mathfrak{h}$ are transversal isotropic Lie subalgebras of maximal dimension.

Following Drinfeld[1], we call $(\mathfrak{k}, \mathfrak{g})$ a *Manin pair* if $\mathfrak{k}$ is a Lie algebra with an invariant scalar product and $\mathfrak{g}$ is an isotropic Lie subalgebra of maximal dimension.

A *marked Manin pair* is a Manin pair $(\mathfrak{k}, \mathfrak{g})$ in which an isotropic complementary subspace, $\mathfrak{h}$, which is not necessarily a Lie subalgebra, of $\mathfrak{g}$ has been chosen. The choice of such a complementary subspace is always possible; the complements to $\mathfrak{g}$ are in one-to-one correspondence with the elements $r \in \wedge^2\mathfrak{g}$. It follows from the preceding theorem that the Lie quasi-bialgebra structures on $\mathfrak{g}$ are in one-to-one correspondence with the *marked Manin pairs*, $(\mathfrak{k}, \mathfrak{g}, \mathfrak{h})$.

Moreover, the equivalence classes of *strictly cohomologous* Lie quasi-bialgebras are in one-to-one correspondence with the *Manin pairs*.

3.4 *Outlook*

Underlying the results of section 3.3 are general properties that relate the various brackets (Schouten bracket, algebraic Schouten bracket on a Lie algebra, Nijenhuis-Richardson bracket of vector-valued forms) and the operators that generalize the Lie algebra and Poisson cohomologies. This is an unexpected by-product of the study of the Lie quasi-bialgebras. Moreover, the theory of quasi-Poisson Lie groups can be extended to include that of the *affine quasi-Poisson structures* which relies upon the properties of the affine multivector fields on a Lie group. These various questions are treated in ref. 2.

Though it is clear that the dual of a Lie quasi-bialgebra is a vector space equipped with both a Jacobian 1-cocycle cobracket and a bracket that satisfies the Jacobi identity only up to a coboundary — an object which we propose to call a *quasi-Lie bialgebra* —, it is not at all clear what the dual of a quasi-Poisson Lie group, the generalization of the dual of a Poisson-Lie group, is. This is a mathematical question that should lead to interesting comparisons with Batalin's article[41] on 'quasigroups' and with Karasev's preprint[42] on 'pseudogroups'.

The study of the 'quasi'-case, both quantum and classical, has just begun and I am confident that it will prove to be relevant to many developments in theoretical physics.

Acknowledgements. The author is very grateful to P. Cartier, whose lectures have been a great source of information and inspiration, to R. Gergondey who lectured on the category theory approach and Tannaka-Krein duality for quantum groups in Lille and at the Ecole Normale Supérieure in Paris in 1988, and several times since, to A. Weinstein for illuminating discussions on quasi-Poisson Lie groups in July 1990 and to J.-H. Lu for an important e.mail message concerning the Manin pairs.

References

[1] V. G. Drinfeld, Quasi-Hopf algebras, *Algebra i Analiz* 1(6), 114-148 (1989) and Quasi-Hopf algebras and Knizhnik-Zamolodchikov equations, Preprint ITP-89-43E, Acad. Sci. Ukr. SSR, Kiev (1989).

[2] Y. Kosmann-Schwarzbach, Grand crochet, crochets de Schouten et cohomologies d'algèbres de Lie, *Comptes rendus Acad. Sci. Paris*, to appear; Champs affines de multivecteurs sur les groupes de Lie, *ibid.*; Quasi-bigèbres de Lie et groupes de Lie quasi-Poisson, *ibid.*; Quasi-Poisson structures on Lie groups and Lie quasi-bialgebras, *in preparation.*

[3] G. Moore and and N. Seiberg, Classical and quantum conformal field theory, *Comm. Math. Phys.* 123, 177-254 (1989).

[4] R. Brustein, Yu. Neeman, S. Sternberg, Duality, crossing and Mac Lane's coherence, *Israel J. Math.* 71, - (1990).

[5] L. A. Takhtajan, Introduction to quantum groups, *in* "Proceedings of the Summer Workshop on Quantum Groups, Clausthal, 1989", Springer-Verlag, to appear.

[6] L. C. Biedenharn, A q-boson realization of the quantum group $SU_q(2)$ and the theory of q-tensor operators, *in* "Proceedings of the Summer Workshop on Quantum Groups, Clausthal, 1989", Springer-Verlag, to appear, and L. C. Biedenharn and M. Tarlani, On q-tensor operators for quantum groups, *Lett. Math. Phys.* 20, 271-278 (1990).

[7] V. G. Drinfeld, Quantum Groups, *in* "Proceedings of the International Congress of Mathematicians, Berkeley, California, 1986", Amer. Math. Soc. (1987).

[8] M. Jimbo, A q-difference analogue of U(g) and the Yang-Baxter equation, *Lett. Math. Phys.* 10, 63-69 (1985).

[9] S. Majid, Reconstruction theorems and rational conformal field theories, *Preprint*, Cambridge (1990).

[10] D. N. Yetter, Quantum groups and representations of monoidal categories, *Math. Proc. Cambridge Phil. Soc.* 108, 261-290 (1990).

[11]L. D. Faddeev, N. Yu. Reshetikhin and L. A. Takhtajan, Quantization of Lie groups and Lie algebras, *in* "Algebraic Analysis", M. Kashiwara and T. Kawai eds., Academic Press, 129-139 (1989).

[12]S. L. Woronowicz, Compact matrix pseudo-groups, *Comm. Math. Phys.* 111, 613-665 (1987).

[13]Yu. I. Manin, "Quantum Groups and Noncommutative Geometry", Publ. Centre Rech. Math., Univ. Montréal (1988).

[14]V. V. Lyubashenko, Hopf algebras and vector symmetries, *Russian Math. Surveys* 41(5), 153-154 (1986); Tensor categories and RCFT. I. Hopf algebras in rigid categories, Preprint, ITP-90-30E, Acad. Sci. Ukr. SSR, Kiev (1990).

[15]N. Yu. Reshetikhin and V. G. Turaev, Ribbon graphs and their invariants derived from quantum groups, *Comm. Math. Physics* 127, 1-26 (1990).

[16]S. Majid, Quasitriangular Hopf algebras, *Int. J. Modern Physics* A5(1), 1-91 (1990).

[17]P . Cartier, Développements récents sur les groupes de tresses. Applications à la topologie et à l'algèbre, exposé n° 716 *in* Séminaire Bourbaki (Novembre 1989).

[18]J. Stasheff, Drinfeld's quasi-Hopf algebras and beyond, *in* "Proceedings of the Conference on Deformation Theory with Applications to Physics, Amherst, MA, June 1990", M. Gerstenhaber and J. Stasheff eds., *Contemporary Math.*, Amer. Math. Soc., to appear.

[19]S. Majid, Tannaka-Krein theorem for quasiHopf algebras and other results, *in* "Proceedings of the Conference on Deformation Theory with Applications to Physics, Amherst, MA, June 1990", M. Gerstenhaber and J. Stasheff eds., *Contemporary Math.*, Amer. Math. Soc., to appear, and Quasi-quantum groups as internal symmetries of topological quantum field theories, *Preprint*, Cambridge (1990).

[20]N. Reshetikhin, Multiparameter quantum groups and twisted quasitriangular Hopf algebras, *Lett. Math. Phys.* 20, 331-335 (1990).

[21]E. Abe, "Hopf Algebras", Cambridge Univ. Press (1969).

[22]M. E. Sweedler, "Hopf Algebras", Benjamin (1969).

[23]S. Mac Lane, "Categories for the Working Mathematician", Springer-Verlag (1971).

[24]S. Mac Lane, Natural associativity and commutativity, *Rice University Studies* 49, 28-46 (1963).

[25]G. M. Kelly, On MacLane's conditions for coherence of natural associativities, *J. of Algebra* 1, 397-402 (1964).

[26]A. Joyal and R. Street, Braided tensor categories, *Preprint* (1986).

[27]P. J. Freyd and D. N. Yetter, Braided compact closed categories with applications to low-dimensional topology, *Adv. in Math.* 77, 156-182 (1989).

[28]N. Saavedra Rivano, "Catégories tannakiennes", *Lect. Notes in Math.* 265, Springer-Verlag (1972).

[29]P. Deligne and J. Milne, Tannakian Categories, *in* "Hodge cycles, motives and Shimura varieties", *Lect. Notes in Math.* 900, Springer-Verlag (1982).

[30]P. Deligne, Catégories tannakiennes, *in* "Festschrift in honor of Grothendieck", P. Cartier *et al.* eds., *Progress in Math.*, Birkhäuser, to appear.

[31]S. L. Woronowicz, Tannaka-Krein duality for compact matrix pseudo-groups. Twisted SU(n), *Invent. Math.* 93, 35-76 (1988).

[32]V. G. Knizhnik and A. B. Zamolodchikov, Current algebras and Wess-Zumino models in two dimensions, *Nuclear Phys.* B 247, 83-103 (1984).

[33] S. Shnider, Deformations of quasi-Hopf algebras, in preparation.

[34] W. M. Tulczyjew, The graded Lie algebra of multivector fields and the generalized Lie derivative of forms, *Bull. Acad. Pol. Sci.*, Sci. math., 22, 937-942 (1974).

[35] J.-L. Verdier, Groupes quantiques, *in* "Séminaire Bourbaki 1986-87", Société mathématique de France, *Astérisque* 152-153, 305-319 (1987).

[36] J.-H. Lu and A. Weinstein, Poisson Lie groups, dressing transformations and Bruhat decompositions, *Journal Diff. Geom.* 31, 501-526 (1990).

[37] J.-L. Koszul, Crochet de Schouten-Nijenhuis et cohomologie, *in* "Elie Cartan et les mathématiques d'aujourd'hui", Société mathématique de france, *Astérisque*, hors série, 257-271 (1985)

[38] Y. Kosmann-Schwarzbach, Quantum and classical Yang-Baxter equations, *Modern Phys. Lett.* A5(13), 981-990 (1990).

[39] B. Kostant and S. Sternberg, Symplectic reduction, BRS cohomology, and infinite- dimensional Clifford algebras, *Ann. Phys.* 176, 49-113 (1987)

[40] P. Lecomte et C. Roger, Modules et cohomologie des bigèbres de Lie, *Comptes rendus Acad. Sci. Paris* 310 série I, 405-410 (1990).

[41] I. A. Batalin, Quasigroup construction and first class constraints, *J. Math. Phys.* 22, 1837-1850 (1981).

[42] M. V. Karasev, Poisson manifolds with connections, their pseudogroups, pseudobrackets and Dirac type brackets, *Preprint* (1989).

Symmetries of Icosahedral Quasicrystals

P. Kramer, Z. Papadopolos, D. Zeidler
Institut für Theoretische Physik der Universität
D-7400 Tübingen, FRG

Introduction

The icosahedral group $A(5)$ as the point symmetry group of quasicrystals is by now well established for a variety of materials. Quasicrystals have been grown up to macroscopic scale and display polyhedral shapes with this symmetry, compare the review by Guyot, Kramer and de Boissieu 1990 [1]. The non-crystallographic icosahedral point group, considered in crystallography for local sites only, is defined more precisely through its three-dimensional (3D) faithful representation, the symmetry group of the regular icosahedron and dodecahedron. This representation requires, for the embedding into a periodic lattice, at least the dimension 6D. The embedding into a lattice and the study of 3D sections through the embedding periodic structure is one of the main theoretical tools for the analysis of ideal icosahedral quasicrystals. In this survey we shall develop some aspects of this embedding for centered hypercubic 6D lattices. The primitive hypercubic lattice has been studied before, but will be considered for comparison. A detailed analysis is given for the root lattice D_6. We hope to show that the 3D sections of this lattice display a rich geometric structure which we expect to encounter in the geometry and physics of the corresponding quasicrystals.

In section 1 we describe the lattices, point out the particular role played by the root lattice D_6, and introduce inflation symmetries. The geometry of Voronoi domains and their duals is developed in section 2. Tiles and tilings arising from the 3D projection are described in section 3. Vertex configurations for a particular tiling are given in section 4. In section 5 it is shown that this tiling is dissectable, with planar sections arranged in a quasiperiodic triangular pattern. Two different tilings are transformed into one another in section 6. In section 7 we compare two tilings which arise from different hypercubic lattices. Most of the results given here are new, part of them have been communicated before [2]. For a recent analysis of icosahedral tilings along different lines compare Danzer 1990 [3].

1 Lattices in 6D with icosahedral point symmetry

The icosahedral group, denoted in what follows by $A(5)$, may be generated from its typical elements g_2, g_3 of order 2 and 3 respectively with the relations

$$(g_2)^2 = (g_3)^3 = (g_2 g_3)^5 = e. \tag{1}$$

Symmetries in Science V, Edited by B. Gruber *et al.*
Plenum Press, New York, 1991

Table 1. Characters for irreps of $A(5)$

		Irrep	[5]	[41]	[32]	$[31^2_+]$	$[31^2_-]$
Class	$\|Class\|$						
$C(e)$	1		1	4	5	3	3
$C(g_5)$	12		1	-1	0	τ	$1-\tau$
$C(g_5^2)$	12		1	-1	0	$1-\tau$	τ
$C(g_3)$	20		1	1	-1	0	0
$C(g_2)$	15		1	0	1	-1	-1

Its class structure and orthogonal irreps are given in Table 1. There are two irreps, denoted in a partition notation by $[31^2_+], [31^2_-]$ of dimension 3. The first one of these two irreps is identical with the symmetry group of the icosahedron and dodecahedron and is interpreted in 3D crystallography as a non-crystallographic point group, compare Hahn and Klapper 1983 [4].

The description of 3D quasicrystals in terms of quasiperiodic functions, based on the work of H. Bohr 1925 [5], 1926 [6], employs 3D sections through nD lattices where the appropriate section is often determined by the requirement of point symmetry in the 3D section.

For quasicrystals with icosahedral non-crystallographic point symmetry, one is led to nD lattices whose point group admits A(5) as a subgroup and contains the irrep $[31^2_+]$. We shall focus on the family of hypercubic lattices in 6D which admit this embedding. The simplest way of getting this embedding is by inducing from the non-trivial one-dimensional irrep of the dihedral subgroup D_5. Clearly this yields the dimension $|A(5)| : |D_5| = 6$ and, by reciprocity, compare Haase et al. 1987 [7], the two 3D irreps of $A(5)$. Induction from a real one-dimensional representation provides a representation in terms of signed permutation matrices, hence an embedding into the defining representation of the hyperoctahedral group $\Omega(n), n = 6$. The group $\Omega(n)$ is the semidirect product

$$\Omega(n) = (Z/2Z)^n \times_s S(n). \tag{2}$$

Upon subducing the defining representation D of $\Omega(6)$ to $A(5)$, the reduction to irreducible form is obtained as

$$MD(g)M^{-1} = D^{[31^2_+]}(g) \oplus D^{[31^2_-]}(g). \tag{3}$$

An explicit choice of the embedding $A(5) < \Omega(6)$, compare Kramer and Zeidler 1989 [8], leads to the reducing matrix

$$M = \sqrt{\frac{1}{2}} \begin{bmatrix} 0 & s & \bar{s} & \bar{c} & 0 & c \\ s & c & c & 0 & \bar{s} & 0 \\ c & 0 & 0 & s & c & s \\ 0 & c & \bar{c} & s & 0 & \bar{s} \\ c & \bar{s} & \bar{s} & 0 & \bar{c} & 0 \\ \bar{s} & 0 & 0 & c & \bar{s} & c \end{bmatrix} \tag{4}$$

where $s = \sin\beta, c = \cos\beta$ and $\tan\beta = \tau^{-1}$. We shall denote the two orthogonal 3D spaces which carry the irreps $[31^2_+], [31^2_-]$ by $E_\parallel, E_\perp$ respectively. The space $E_\parallel$ will carry the geometry of the quasicrystal while the space $E_\perp$ will be needed for the constructions. The position of the 2-, 3-, and 5-fold axes in the two irrep

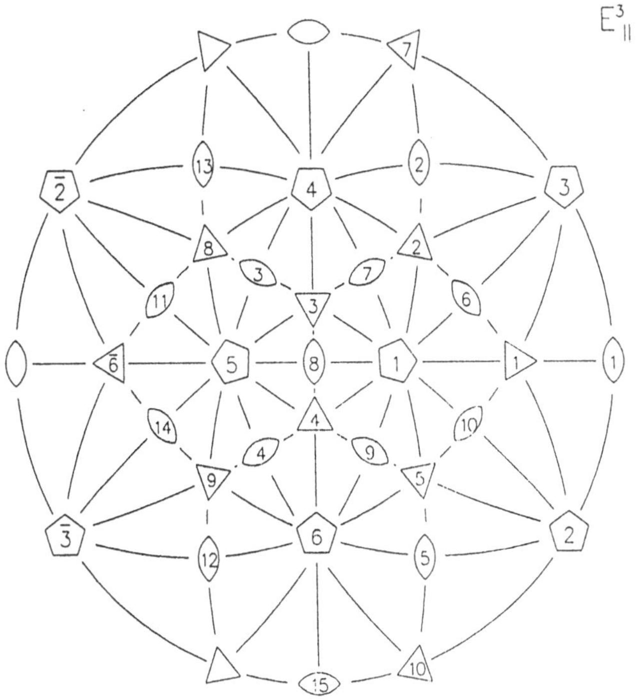

Figure 1. Stereographic projection of icosahedral symmetry operations in $E_{\|}$.

spaces $E_{\|}, E_{\perp}$ in a stereographic notation are given in Figs. 1, 2. The axes are denoted by numbers. Any element of the icosahedral group can be denoted by a signed permutation of the six 5-fold axes. This notation fixes the embedding of $A(5)$ into the hyperoctahedral group. The 6D inversion element from $\Omega(6)$, i, together with $A(5)$, generates the point group $A(5) \times \{e, i\}$.

The group $\Omega(n)$ is the holohedry of the hypercubic lattice $Z^n, n = 6$, and so one has the required lattice embedding. A basis of this lattice is spanned by the orthogonal unit vectors $e_1, \ldots, e_6$. For general dimension n, the hypercubic lattice admits two centerings denoted, in analogy with the 3D case, by the letters F and I with the typical positions $F : \frac{1}{2}(110000)$, $I : \frac{1}{2}(111111)$ respectively. The lattice $2F$, scaled by a factor 2, is known as the checkerboard lattice, i.e. the sublattice of $P : (100000)$ with $n_1 + \ldots + n_6 = even$. The reciprocals of the lattices obey $P^R = P, (2F)^R = I, I^R = 2F$. The two centerings may be described in terms of centering matrices $Z^{(2F)}, Z^{(2I)}$ with integer coefficients and rational inverses as

$$Z^{(2F)} = \begin{bmatrix} 1 & 0 & 0 & 0 & 0 & 0 \\ 0 & 1 & 0 & 0 & 0 & 0 \\ 0 & 0 & 1 & 0 & 0 & 0 \\ 0 & 0 & 0 & 1 & 0 & 0 \\ 0 & 0 & 0 & 0 & 1 & 0 \\ \bar{1} & \bar{1} & \bar{1} & \bar{1} & \bar{1} & 2 \end{bmatrix}, Z^{(2I)} = \begin{bmatrix} 2 & 0 & 0 & 0 & 0 & 1 \\ 0 & 2 & 0 & 0 & 0 & 1 \\ 0 & 0 & 2 & 0 & 0 & 1 \\ 0 & 0 & 0 & 2 & 0 & 1 \\ 0 & 0 & 0 & 0 & 2 & 1 \\ 0 & 0 & 0 & 0 & 0 & 1 \end{bmatrix} \quad (5)$$

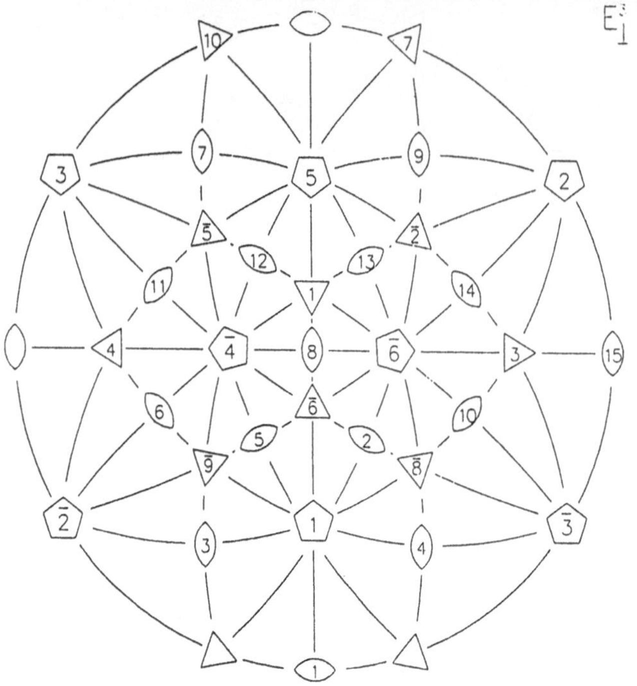

Figure 2. Stereographic projection of icosahedral symmetry operations in $E_{\perp}$.

The centered lattices preserve the hyperoctahedral and hence the icosahedral point group, expressed now in the form of Q-equivalent representations

$$(Z^{(2F)})^{-1}D(g)(Z^{(2F)}),\ (Z^{(2I)})^{-1}D(g)(Z^{(2I)}). \tag{6}$$

The lattice $2F$ is identical with the root lattice D_6, compare Conway and Sloane 1988 [9]. The role of root lattices for quasicrystals is pointed out by Baake et

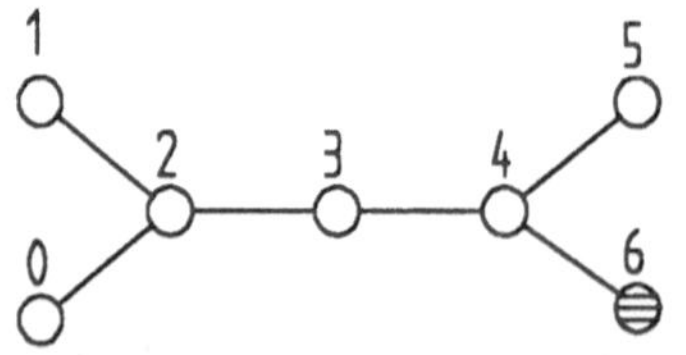

Figure 3. The extended Dynkin diagram for the lattice D_6. The nodes $0 \ldots 5$ mark the simple roots, 6 the additional (negative maximal) root vector. Two root vectors marked by nodes and connected directly with no or with a single branch are at angle $\frac{\pi}{2}$ or $\frac{2\pi}{3}$ respectively.

al. 1990 [10]. A root lattice is based on a root system which in turn is a vector star with specific symmetries, length and angles. The root systems appear both in the classification of finite-dimensional semisimple Lie algebras, Humphreys 1972 [11], and in the classification of crystallographic reflection groups, Coxeter 1973 [12]. Root lattices are generated by integral linear combinations of the root vectors. For them one can derive, Conway and Sloane 1988 [9], the list $A_n(n \geq 1), Z^n(n \geq 2), D_n(n \geq 4), E_6, E_7, E_8$. The geometry of root systems and root lattices is condensed in the corresponding Dynkin and extended Dynkin diagram. The nodes of the Dynkin diagram mark the simple root vectors, the lines the corresponding angles. The extended Dynkin diagram for the lattice D_6 is given in Fig. 3. The Weyl reflection group of the root system is the semidirect product $(Z/2Z)^5 \times_s S(6)$ (restriction to even reflections), and the full automorphism group is $\Omega(6)$. With the choice of the point group $A(5)$, we shall restrict the 6D space group from the holohedry $Z^6 \times_s \Omega(6)$ to $Z^6 \times_s A(5)$.

We turn now to the *inflation symmetry* of the hypercubic lattices. An inflation symmetry g is a real transformation with the properties:

(i) it is a general lattice symmetry contained in $Gl(n, Z)$,

(ii) it commutes with the chosen point group of the lattice.

One does not require orthogonality or finite order of g. Since the 6D representation of $\Omega(6)$ is irreducible, the only inflation symmetry with respect to this point group is, by Schur's Lemma, the trivial scaling by ± 1. So inflation symmetry requires the choice of a point group with a reducible representation in the lattice space. Consider now the hypercubic lattice with the icosahedral point group. Let $P_\parallel$, $P_\perp$ denote the projectors for the two irreps of $A(5)$ with respect to the orthogonal basis. These projectors are given by the scalar products of the projections of the 6D unit vectors to the two 3D subspaces $E_\parallel$, $E_\perp$ respectively,

$$P_{\parallel ij} = e_{i\parallel} \cdot e_{j\parallel}, \; P_{\perp ij} = e_{i\perp} \cdot e_{j\perp}. \tag{7}$$

By application of Schur's Lemma, any inflation symmetry from (ii) must be of the form $D(g) = \alpha P_\parallel + \beta P_\perp$. The condition (i) must be checked for each one of the three lattices and reads

$$D(g) = Z^{-1}(\alpha P_\parallel + \beta P_\perp) Z \in Gl(6, Z), \; Z = I, \; Z^{(2F)}, \; Z^{(2I)}. \tag{8}$$

The solutions of these equations are generated by the coefficients

$$\begin{array}{cccc} & P & 2F & 2I \\ \alpha & \tau^3 & \tau & \tau \\ \beta & -\tau^{-3} & -\tau^{-1} & -\tau^{-1} \end{array} \tag{9}$$

Clearly the matrices are not orthogonal in the basis where the irreps of $A(5)$ are explicitly reduced. In this basis, they represent scalings of the two irrep spaces. In all three cases, the inflation symmetries form infinite discrete abelian groups generated from a single element.

2 Voronoi domains and their duals

In crystallography, one associates to a lattice its unit cell which plays, with respect to the action of the translation group, the role of a *fundamental domain*. Instead of the usual primitive unit cell we shall use the *Voronoi domain* V. This domain is the set of points with a distance closer to a fixed lattice point, say $x = 0$, than to any other lattice point. The domain $V(0)$ is invariant under the holohedry of the lattice, and it is bounded by hyperplanes of the form $x \cdot b_i = \frac{1}{2} b_i \cdot b_i$, where the set $(b_i,\ i = 1, \ldots, \nu)$ is the set of *Voronoi vectors*. Naturally, a Voronoi domain and the same set of Voronoi vectors are associated to any lattice point. The full set of these Voronoi domains and boundaries, translated to all lattice points, will be called the *Voronoi complex* $\mathcal{V}$ of the lattice. The boundaries are convex and in face-to-face position, and any intersection of two boundaries is a boundary, thus the name cell complex is justified, compare Munkres 1984 [13].

Given a fixed boundary P of a Voronoi domain, there is a finite set of lattice points q whose Voronoi domain have P as a proper boundary. We define the *dual* P^* of P as the convex hull of this set. It can be shown, compare Kramer and Schlottmann 1989 [14] and Schlottmann 1990 [15], that the polytopes P and P^* have complementary dimension and are orthogonal to one another. By translation of the duals to all lattice points one obtains the *dual complex* $\mathcal{V}^*$. The boundaries of the two complexes respect the bounding relations,

$$P_1 \subseteq P_2 \leftrightarrow P_1^* \supseteq P_2^*. \tag{10}$$

For root lattices, the Voronoi domain has a particularly simple form, compare Conway and Sloane 1988 [9]: it is generated from a fundamental simplex S by application of the automorphism group of the root system. The fundamental simplex S is coded by the Dynkin diagram and by the corresponding simple root vectors, compare Fig. 4. The simple root vectors of the lattice D_6 are given in Table 2, along with the additional root vector which is coded in the extended Dynkin diagram. The automorphism group is $\Omega(6)$. The Voronoi vectors are all

Table 2. Simple roots and vertices of the fundamental simplex S for D_6

m	simple root	vertex v_m
0	$-e_1 - e_2$	$\frac{1}{2}(111111)$
1	$e_1 - e_2$	$\frac{1}{2}(\bar{1}11111)$
2	$e_2 - e_3$	$\frac{1}{2}(001111)$
3	$e_3 - e_4$	$\frac{1}{2}(000111)$
4	$e_4 - e_5$	$\frac{1}{2}(000011)$
5	$e_5 - e_6$	(000001)
6	$(e_5 + e_6)$	(000000)

root vectors, they are of the form $\pm e_i \pm e_j, i \neq j$. All boundaries $P(n)$ of $V(0)$ can be generated by application of coset representatives with respect to a subgroup $H < \Omega(6)$ to subsimplices S. The corresponding polytopes are (hyper-) cubes and single or double pyramids with (hyper-) cubic basis. For cubes we introduce the notation

$$P(0 \ldots 0\epsilon_{j+1} \ldots \epsilon_6) = \frac{1}{2}\left(\sum_{i=1}^{j} \lambda_i e_i + \sum_{i=j+1}^{6} \epsilon_i e_i\right),\ -1 \leq \lambda_i \leq 1,\ \epsilon_i = \pm 1. \tag{11}$$

A join of a polytope P to an extra point e is denoted by

$$P \circ e :\ x = \mu P + (1 - \mu) e,\ 0 \leq \mu \leq 1. \tag{12}$$

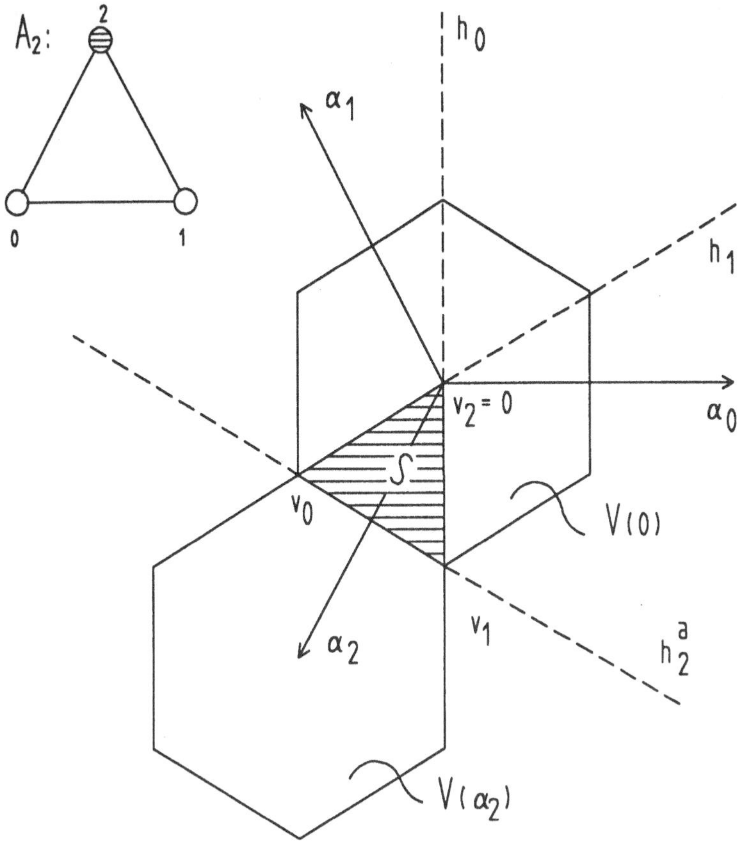

Figure 4. Simple example: the root lattice A_2. α_0, α_1 are simple roots, α_2 is the additional root, $\alpha_i \cdot \alpha_i = 2$. Reflections in hyperplanes $h_i : x \cdot \alpha_i = 0, i = 0, 1$ generate $V(0)$ from the fundmental simplex S, the reflection in $h_2^a : x \cdot \alpha_2 = 1$ generates the lattice A_2.

The dual boundaries are denoted by the corresponding sets of lattice points. In Table 4 one finds in subsequent columns the dimension n of the boundary, the boundary P, its dual P^*, the orbit length, the subsimplex S, and the stability group H.

3 Tiles and tilings associated with the icosahedral group

In this and in the following sections, the point symmetry group of the lattice Z^6 will be restricted to $A(5)$ and the space group to $Z^6 \times_s A(5)$. The inclusion of the inversion i in 6D generates $A(5) \times \{e, i\}$ and, from this point group, mirror planes which will be mentioned in some places. The tiles for quasiperiodic tilings associated with the icosahedral group will be constructed from projections of boundaries $P(3)$ or $P^*(3)$ to the subspace $E_\parallel$. Since a translation does not change the shape and orientation of a projection, we may restrict the analysis to boundaries obtained from the first Voronoi domain $V(0)$. To classify the shapes and orientations of the (projected) boundaries, it suffices to find representatives for the orbits in 6D under the icosahedral group. All boundaries on the same orbit must yield projections

Table 3. Space group representatives of boundaries under $D_6 \times_s A(5)$

n	shape	P(n)	P*(6 − n)
5		$P(100010) \circ \begin{cases} e_1 \\ e_5 \end{cases}$	$0, e_5 + e_1$
		$P(1000\bar{1}0) \circ \begin{cases} e_1 \\ -e_5 \end{cases}$	$0, -e_5 + e_1$
4		$P(110010) \circ e_1$	$0, e_5 + e_1, e_2 + e_1$
		$P(1100\bar{1}0) \circ e_1$	$0, -e_5 + e_1, e_2 + e_1$
		$P(100110) \circ e_1$	$0, e_5 + e_1, e_4 + e_1$
		$P(100\bar{1}\bar{1}0) \circ e_1$	$0, -e_5 + e_1, -e_4 + e_1$
3	G	$P(0001\bar{1}1)$	$0, e_6 - e_5, e_6 + e_4, -e_5 + e_4$
	F	$P(111000)$	$0, e_3 + e_2, e_3 + e_1, e_2 + e_1$
	A	$P(111010) \circ e_1$	$0, e_5 + e_1, e_3 + e_1, e_2 + e_1$
	B	$P(1\bar{1}\bar{1}010) \circ e_1$	$0, e_5 + e_1, -e_3 + e_1, -e_2 + e_1$
	C	$P(1\bar{1}1010) \circ e_1$	$0, e_5 + e_1, e_3 + e_1, -e_2 + e_1$
	D	$P(1110\bar{1}0) \circ e_1$	$0, -e_5 + e_1, e_3 + e_1, e_2 + e_1$
	E	$P(1\bar{1}10\bar{1}0) \circ e_1$	$0, -e_5 + e_1, e_3 + e_1, -e_2 + e_1$ †

† degenerate in $E_{\parallel}$ and $E_{\perp}$

of the same shape. The orbit analysis is identical for dual pairs of boundaries of $V_{\perp}(0)$.

We shall not go into the details of this orbit analysis. It turns out that there are seven orbits and seven shapes for $P(3), P^*(3)$ whose representatives are given in Table 3. One of these polytopes has a degenerate projection, the others are denoted by the letters A, B, C, D, F, G and their duals. The projections of A, B, C, D to the subspaces $E_{\parallel}, E_{\perp}$ are pyramids on a rhombus base. The edges point along 3- and 5-fold directions with respect to the icosahedral group. Their length scales in powers of τ. Upon inclusion of the inversion, A and B have two, C and D one mirror plane. The projections of G, F are the two rhombohedra known from the first icosahedral tiling, compare Kramer and Neri 1984 [16]. G and F have 3-fold symmetry axes and three mirror planes. The projections of $A^*, B^*, C^*, D^*, F^*, G^*$ are tetrahedra. They have the same local point symmetries as their duals. All edges of these tetrahedra point along 2-fold axes of $A(5)$ with a length corresponding to the short or long diagonal of the rhombus. This result follows from the fact that the boundaries $P^*(1)$, which form the boundaries of the tetrahedra, are lines along Voronoi vectors of the type $\pm e_i \pm e_j$, $i \neq j$, all of which point along 2-fold axes in the projection to $E_{\parallel}, E_{\perp}$. The faces of all these polytopes can be seen by unfolding the polytopes into a plane. The different axes are coded by different types of lines, compare Figs. 5, 6.

For the construction of the tiling we require the projection of the first Voronoi domain and its boundaries to $E_{\perp}$. This projection $V_{\perp}(0)$ is the well-known triacontahedron which is also the projection of the 6D hypercube. So the projections of the Voronoi domains for the lattices P and $2F$ coincide under icosahedral projection. The polytope was given by Kepler 1611 [17], compare also Kowalewski 1938 [18] and Kramer 1986 [19]. All other projections of boundaries are polytopes

Table 4. Representative boundaries P and duals P^* of the Voronoi domain under $\Omega(6)$, orbit length, subsimplex S, and stability group H of the boundary.

n	P(n)	P*(6 – n)	orbit length	subsimplex S	H
5	$P(000011) \circ \left\{ \begin{matrix} e_6 \\ e_5 \end{matrix} \right.$	$0, e_5 + e_6$	60	$v_0v_1v_2v_3v_4v_5$	$\Omega(4) \times S(2)$
4	$P(000111) \circ e_6$	$0, e_i + e_6, 4 \le i \le 5$	480	$v_0v_1v_2v_3v_5$	$\Omega(3) \times S(2) \times S(1)$
3	$P(000111)$	$0, e_i + e_j, 4 \le i < j \le 6$	160	$v_0v_1v_2v_3$	$\Omega(3) \times S(3)$
	$P(001111) \circ e_6$	$0, e_i + e_6, 3 \le i \le 5$	960	$v_0v_1v_2v_5$	$\Omega(2) \times S(3) \times S(1)$
2	$P(001111)$	$0, e_i + e_j, 3 \le i < j \le 6$	240	$v_0v_1v_2$	$\Omega(2) \times S(4)$
	$P(011111) \circ e_6$	$0, e_i + e_6, 2 \le i \le 5$	960	$v_0v_1v_5$	$\Omega(1) \times S(4) \times S(1)$
1	$P(011111)$	$0, e_i + e_j, 2 \le i < j \le 6$ $e_2 + e_3 + e_4 + e_5 + e_6 - e_i,$ $2 \le i \le 5$	192	v_0v_1	$\Omega(1) \times S(5)$
	$P(111111) \circ e_6$	$0, e_i + e_6, 1 \le i \le 5$	384	v_0v_5	$S(5) \times S(1)$
C	$P(111111)$	$0, e_1 + e_2 + e_3 + e_4 + e_5 + e_6,$ $e_i + e_j, e_1 + e_2 + e_3$ $+e_4 + e_5 + e_6 - e_i - e_j,$ $1 \le i < j \le 6$	64	v_0	$S(6)$
	e_6	$\pm e_i + e_6, 1 \le i \le 6$	12	v_5	$\Omega(5) \times S(1)$

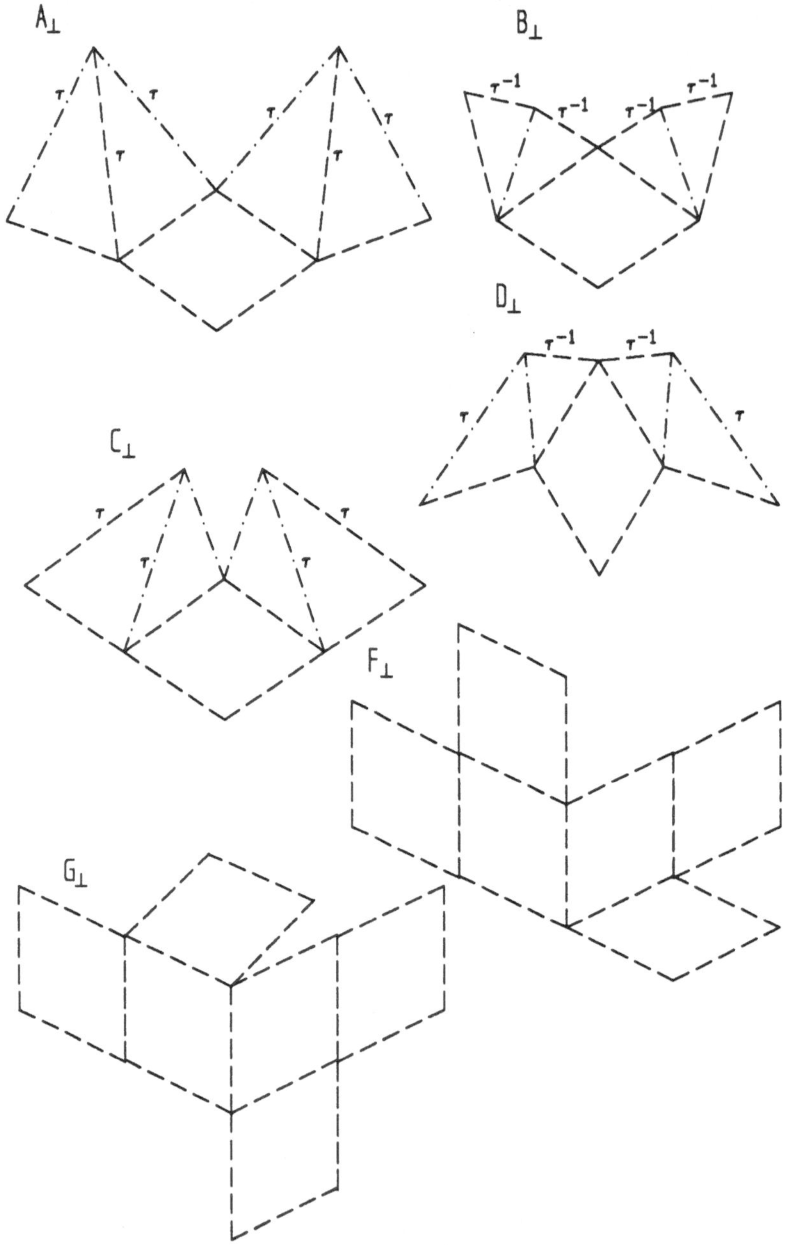

Figure 5. The projections of A, B, C, D, F and G to $E_{\perp}$. Edges run along 5-fold (− − −) or 3-fold (− · −) axes. Scalings by powers of τ w.r.t. a standard lenght are marked for each type of axes.

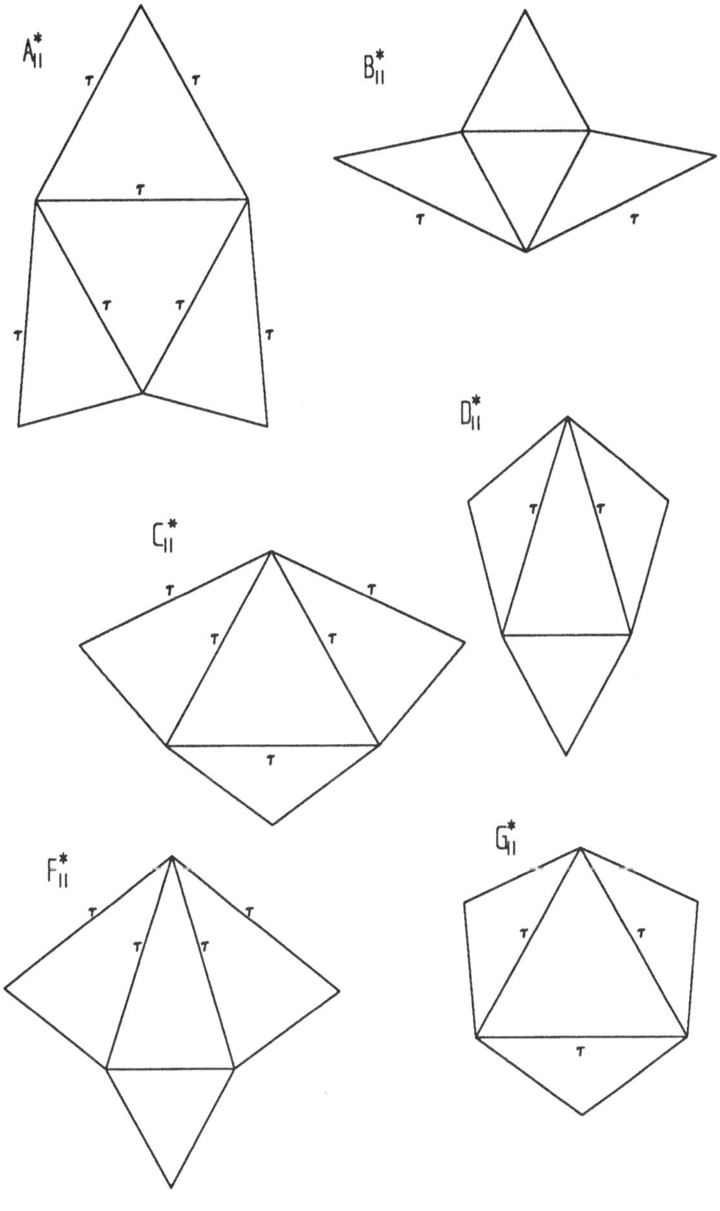

Figure 6. The projections of A^*, B^*, C^*, D^*, F^* and G^* to $E_{\|}$. The edges represented by full lines are along 2-fold axes.

inside of this triacontahedron. The boundaries $P(5)$ are double pyramids on a 4D cubic base. In the projection to $E_\perp$, they fall into two distinct orbits under $A(5)$, Fig. 7. On the first orbit, the two extra points appear outside, on the second orbit inside of the projection image of the 4D cube. The total of 60 projections share in pairs one of the outer face, and a 2-fold axis perpendicular to this face, with the triacontahedron. The dual boundaries $P_\perp^*(1)$ in $E_\perp$ are the projections of the two Voronoi vectors pointing outward along this 2-fold axis.

The projections $P_\perp(3)$ overlap inside of $V_\perp(0)$ in various ways. A boundary $P_\perp(3)$ of fixed shape and orientation may occur in several positions which cannot be related by icosahedral rotations and hence belong to different orbits under the point group. It turns out that all these positions can be shifted into one another by projections of Voronoi vectors and hence, in the 6D analysis, by translations. So when we refer to the representatives of seven orbits for the boundaries $P(3), P^*(3)$, we choose the representatives from the intersection of the orbit of $P(3), P^*(3)$ under the space group with $V(0)$. This orbit analysis is seen in the projections to $E_\perp$.

Consider pairs of dual boundaries $P(3), P^*(3)$ and their projections to $E_\parallel$ and $E_\perp$ respectively. From each such pair one may form a new 6D klotz polytope

$$K = P_\perp(3) + P_\parallel^*(3). \tag{13}$$

Clearly these polytopes have all their boundaries perpendicular or parallel to the subspaces $E_\perp, E_\parallel$, and their projections to these spaces coincide with the projections of the boundaries $P(3), P^*(3)$. The set of all non-degenerate klotz polytopes, called the *klotz construction* $\mathcal{K}$, can be shown to yield a periodic space-filling of 6D, compare Kramer and Schlottmann 1989 [14] and Schlottmann 1990 [15]. The intersection of the 3D plane with points $x = c_\perp + E_\parallel$ with the klotz construction clearly yields a *tiling* $\mathcal{T}^*$ whose tiles are the projections $P_\parallel^*(3)$. For the lattice $2F$ we denote this tiling by $\mathcal{T}^{*(2F)}$, its tiles are tetrahedra of six shapes. The interchange of the complexes $\mathcal{V}$ and $\mathcal{V}^*$ in the definition of the klotz construction yields a second, dual construction whose tiling we denote by $\mathcal{T}^{(2F)}$, with tiles given by the four pyramids and the two rhombohedra.

There is a constructive algorithm which effectively produces the tiling $\mathcal{T}^*$, compare Kramer 1986 [19], Baake, Kramer, Schlottmann and Zeidler 1990 [20]. We describe it here for general tilings $\mathcal{T}^*$ from a lattice in 6D:

1) Choose in $E_\perp$ an initial point $c_\perp^\alpha \in V_\perp(0)$,

2) draw in $E_\parallel$ an initial vertex v^α.

3) determine in $E_\perp$ all boundaries $P_\perp(3) \subset V_\perp(0)$ which have $c_\perp^\alpha$ as an inner point,

4) adjoin in $E_\parallel$ all the projections of duals $P_\parallel^*(3)$ to these boundaries to the vertex v^α to obtain a vertex configuration,

5) choose in $E_\parallel$ a new vertex $v^\beta = v^\alpha + q_\parallel$ from the vertex configuration, where q is a Voronoi vector,

6) construct in $E_\perp$ the point $c_\perp^\beta = c_\perp^\alpha - q_\perp \in V_\perp(0)$,

7) repeat step (1) with the new point $c_\perp^\beta$. By this algorithm, one proceeds in the tiling along a sequence of edges, and inside of $V_\perp(0)$ along a graph formed by projections $-q_\perp$ of Voronoi vectors.

Notice that the step along the projection $-q_\perp$ is allowed if and only if the point $c_\perp^\alpha$ is an interior point of the projection of the boundary $P(5)$ dual to q. The tiling is produced tile by tile up to an arbitrary size. Cases where a point $c_\perp^\omega$ is on the boundary of a projection require an extra analysis.

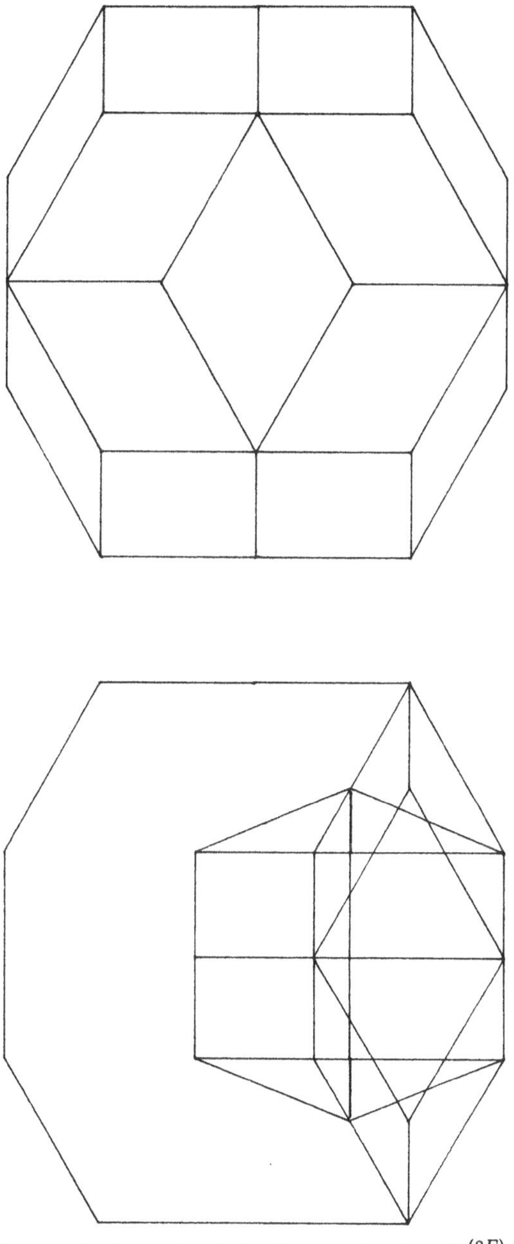

Figure 7. The triacontahedron (top) is the projection $V_{\perp}^{(2F)}(0)$ of the central Voronoi domain for the lattice $2F$. The projected boundaries $P_{\perp}(5)$ are two types of double pyramids which share in pairs the outer rhombus faces (bottom) of the triacontahedron.

One purpose of the graph algorithm is to generate the tiling. More important is its role as a fundamental tool for exploring structure properties of the tiling. In this sense it will be used in the following sections.

4 Vertex configurations of the tiling $\mathcal{T}^{*(2F)}$

A *vertex configuration* in a tiling is a set of tiles which share a vertex and fill the solid angle around it. By the graph algorithm described in the last section, a vertex configuration is determined by a set of tiles in $E_{\parallel}$ whose duals in $E_{\perp}$ share an interior point of $V_{\perp}(0)$. Hence the study of vertex configurations reduces to the study of intersections of boundaries, which may be written as

$$p_{\perp\alpha} = \cap_i P_{\perp\alpha_i}(3). \tag{14}$$

This study may be implemented by first finding the intersections of the boundaries of type $F_{\perp}, G_{\perp}$. The projections $V_{\perp}(0)$ of the lattices P and $2F$ coincide. Moreover, the positions of these rhombohedra agree with the ones occuring in the algorithm for the lattice P, which was studied by Baake, Ben-Abraham et al. 1990 [22]. The algorithm then implies that any vertex configuration of $\mathcal{T}^{*(2F)}$ branches off from one of the 24 vertex configurations of $\mathcal{T}^{*(P)}$. Due to the presence of the other tiles of pyramide shape, several vertex configurations of $\mathcal{T}^{*(2F)}$ correspond to a single configuration of $\mathcal{T}^{*(P)}$.

For the infinite tiling, the graph underlying the tiling construction covers $V_{\perp}$ with a uniform density. It follows from this observation that the relative frequency of the vertex configuration α is given by

$$f_{\alpha} = vol(p_{\perp\alpha})/vol(V_{\perp}(0)). \tag{15}$$

In Table 5, we list the 36 vertex configurations of the tiling $\mathcal{T}^{*(2F)}$. Given are the shapes in terms of the letters A, B, C, D, F, G, and the edge vectors of the dual boundaries. For the unit vectors we use the short-hand notation $e_j \rightarrow j, -e_j \rightarrow \bar{j}$.

5 The tiling $\mathcal{T}^{*(2F)}$ is dissectable

In a periodic lattice, there is a well-known countable infinite set of net planes occupied by lattice points. Parallel sets of net planes are perpendicular to a fixed vector from the reciprocal lattice. Thus, periodic lattices are dissectable along these net planes. This appearance of net planes with a simple geometric description in general is absent in quasicrystals. The quasilattice obtained from the tiling $\mathcal{T}^{*(2F)}$ has the remarkable property that quasiperiodic sets of vertex points occur in parallel planes perpendicular to any one of the fivefold axes.

The construction of the tiling $\mathcal{T}^{*(2F)}$ follows from the graph algorithm given in section 4. If a point inside of $V_{\perp}(0)$ is an interior point of a boundary $P_{\perp}(5)$, then the corresponding vertex of the tiling belongs to an edge which is the projection to $E_{\parallel}$ of the Voronoi vector dual to $P(5)$. The graph must then also have a continuation along an edge given by the projection to $E_{\perp}$ of the inverse Voronoi vector. The position of the boundaries $P_{\perp}(5)$ was described in section 4. For a point of the graph inside a given $P_{\perp}(5)$, the graph must always continue along the twofold direction associated with $P_{\perp}(5)$.

Consider now an equatorial ring of ten rhombus faces of the triacontahedron with their 2-fold normals perpendicular to a fixed 5-fold axis. An equatorial 2D decagonal cut through this configuration is shown in Fig. 8. The 2D cuts of the 10 pairs of boundaries have the shape of two types of hexagons. This decagon and these hexagons are in one-to-one correspondence to the 2D projection of the 4D Voronoi domain for the root lattice A_4, which was analyzed in great detail in Baake, Kramer, Schlottmann and Zeidler 1990 [20], [21]. Consider now an interior point on the equatorial cut of $V_\perp(0)$. If it is interior to a fixed hexagon, the graph must continue along a 2-fold direction, which by definition is in the equatorial plane. Since the equatorial cut is completely covered by cuts through the 20 selected boundaries $P_\perp(5)$, any initial point on the cut produces an infinite continuation of the 2D part of the 3D graph. It turns out that the length of the edges of this graph matches precisely the corresponding length in the 2D construction of the triangle pattern, Fig. 9, obtained from the root lattice A_4. This implies that any initial point of the graph on the equatorial cut produces in the 3D tiling $\mathcal{T}^{*(2F)}$ an infinite planar set of vertices which form a triangle pattern. Inspection of a parallel shift of the cutting plane through the triacontahedron, Fig. 8 , shows that there is a full equatorial zone which produces exactly the same decagon and hexagons from the boundaries $P_\perp(5)$.Any point of the graph which is in this zone will produce a triangular pattern.

When a path on the 3D graph, along edges which are not in the plane, leaves and returns to this equatorial zone, it will generate in the tiling parallel planes with triangular vertex patterns.

A similar analysis applies to all equatorial zones perpendicular to the six five-fold directions. We conclude that the infinite 3D tiling $\mathcal{T}^{*(2F)}$ is *dissectable* along parallel sets of planes perpendicular to the six 5-fold directions, and that its vertex points on these dissections are connected in a triangle pattern.

6 Local and global transformation of the tiling $\mathcal{T}^{*(2F)}$

The six tetrahedral tiles of the tiling $\mathcal{T}^{*(2F)}$, whose edges are formed by the projections of Voronoi vectors, have been considered before from an entirely different point of view by Mosseri and Sadoc 1982 [23]. These authors have shown, moreover, that these tiles arise from the tetrahedral decomposition of another set of four tiles H, S, Z, A. The set H, S, Z, A in turn was constructed by Mosseri and Sadoc from an earlier set of seven tiles proposed by Kramer 1982 [24]. We shall use lower case letters $x_\parallel^* = h_\parallel^*, s_\parallel^*, z_\parallel^*, a_\parallel^*$ for these tiles, compare Fig. 10 , in order to distinguish them from the ones considered in section 3. A remarkable property of the four tiles is that locally they can be packed into four new tiles $h_\parallel^*(\tau), s_\parallel^*(\tau), z_\parallel^*(\tau), a_\parallel^*(\tau)$, which are copies of the original set scaled by a linear factor τ. This inflation is represented by the scheme

$$\begin{array}{ccccc} & h_\parallel^* & s_\parallel^* & z_\parallel^* & a_\parallel^* \\ h_\parallel^*(\tau) & 1 & 2 & 2 & 2 \\ s_\parallel^*(\tau) & 1 & 1 & 1 & 2 \\ z_\parallel^*(\tau) & 1 & 1 & 1 & 1 \\ a_\parallel^*(\tau) & 0 & 1 & 0 & 2 \end{array} \tag{16}$$

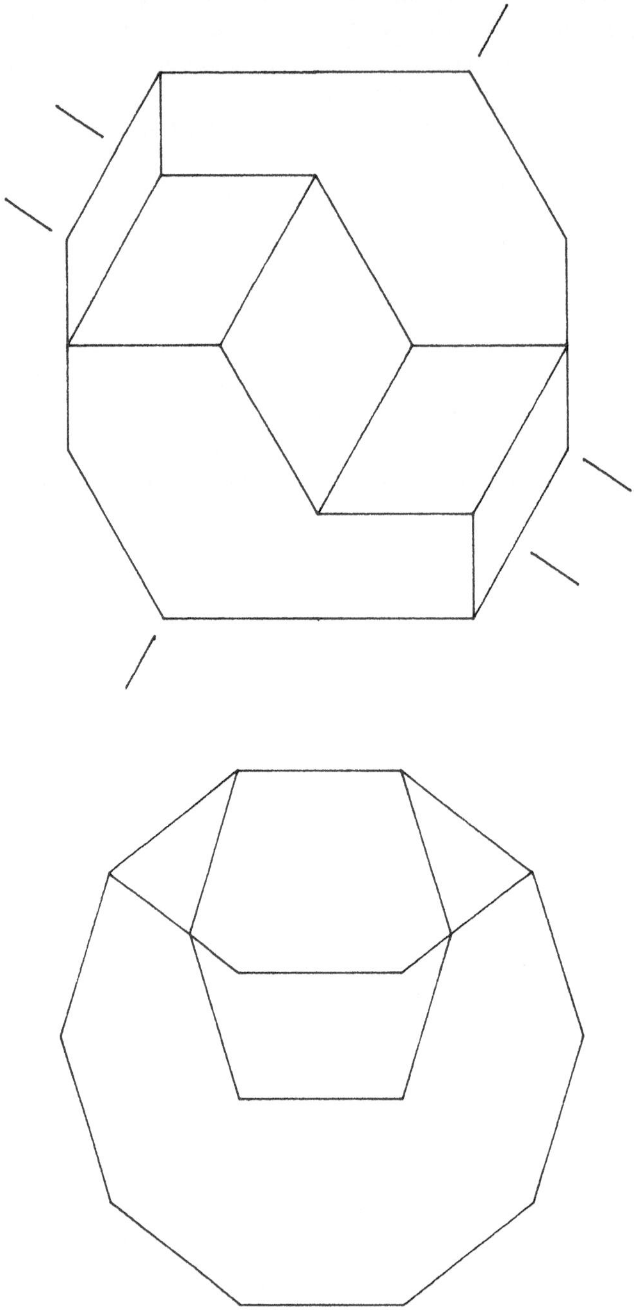

Figure 8. 10 rhombus faces with their 2-fold normals perpendicular to a 5-fold axis form an equatorial ring of the triacontahedron (top). A planar cut through an equatorial zone of this ring (bottom) yields, for each pair of boundaries $P_{\perp}(5)$ from one of the rhombus faces, a pair of hexagons sharing a decagon edge.

Figure 9. The triangle pattern is obtained from the 4D root lattice A_4 by projection into a 2D plane of 5-fold symmetry. Its tiles are two types of triangles.

By use of this scaling, one can generate a 3D tilings from a series of local inflations of an initial tile. This local linear scaling is in line with the inflation symmetry of the lattice $2F$ given in section 2. In contrast, inspection of the tetrahedral tiles shows that one cannot pack them into the same tetrahedral tiles scaled by a factor τ. We shall now explore the relation between the two tilings locally and globally by use of the construction based on the lattice $2F$.

Consider the local packings of tetrahedra which may be represented by the scheme

$$\begin{array}{ccccccc} & A^*_{\|} & B^*_{\|} & C^*_{\|} & D^*_{\|} & F^*_{\|} & G^*_{\|} \\ h^*_{\|} & 1 & 1 & 0 & 0 & 2 & 2 \\ s^*_{\|} & 1 & 0 & 2 & 0 & 0 & 0 \\ z^*_{\|} & 1 & 0 & 1 & 0 & 0 & 1 \\ a^*_{\|} & 0 & 0 & 0 & 1 & 1 & 0 \end{array} \qquad (17)$$

First of all it can be shown that these local packings of the tetrahedra $A^*_{\|}, \ldots G^*_{\|}$ of the shapes $h^*_{\|}, \ldots a^*_{\|}$ appear in the tiling $\mathcal{T}^{*(2F)}$. This can be seen by first choosing a vertex of the new tiles, shared by a maximum number of tetrahedra. A vertex

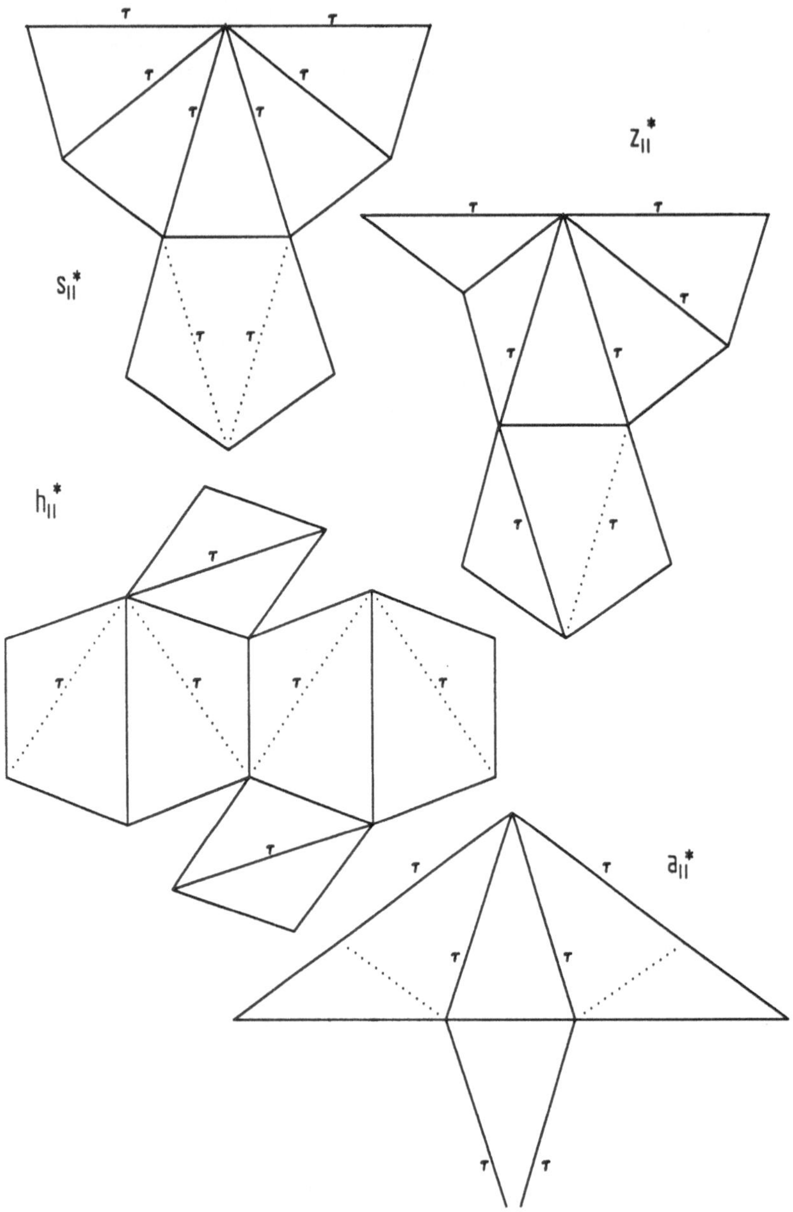

Figure 10. The tiles $h_{\|}^{*}$, $s_{\|}^{*}$, $z_{\|}^{*}$, and $a_{\|}^{*}$. All edges are along 2-fold axes. No folding along dotted axes.

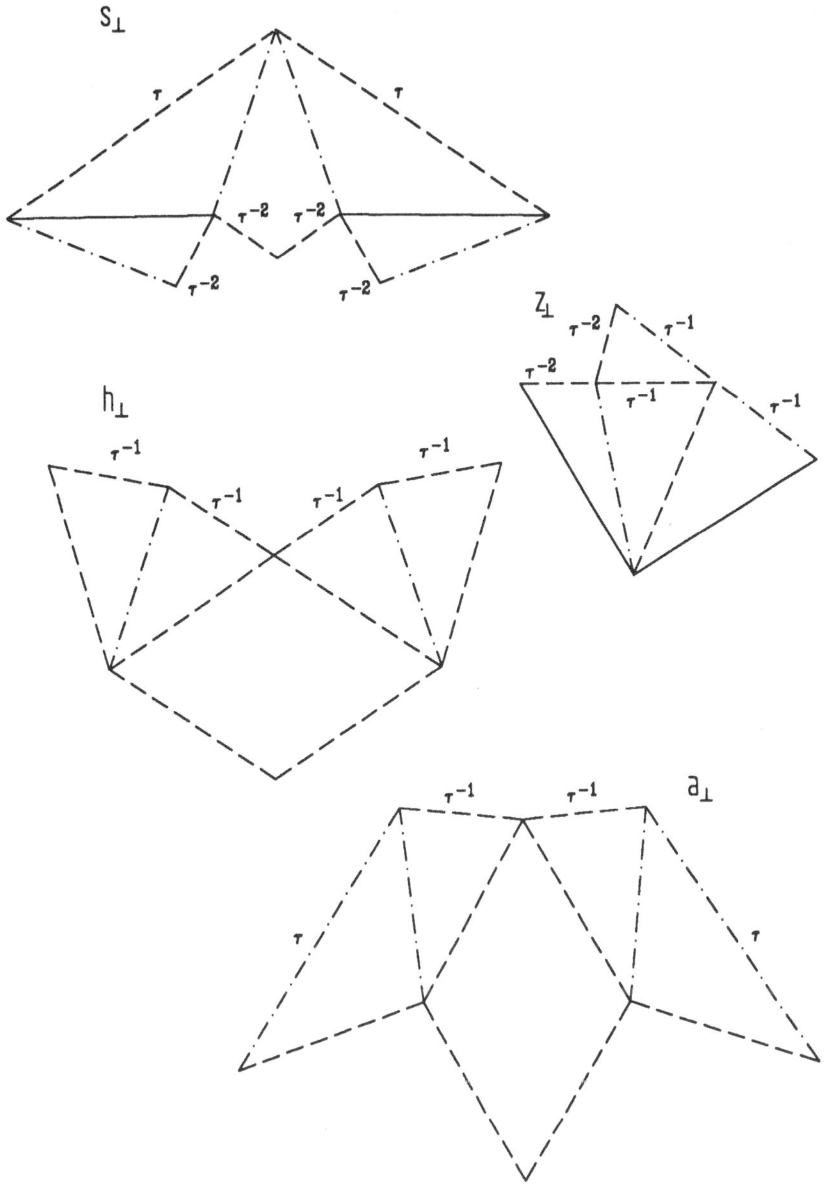

Figure 11. The polytopes $h_\perp, s_\perp, z_\perp, a_\perp$. The edges are denoted as in Figs. 5, 6 but all objects are enlarged by τ w.r.t. the objects in Figs. 5, 6 and 10.

shared by all the tetrahedra exists for $s_{\|}^*$, $z_{\|}^*$, $a_{\|}^*$, for $h_{\|}^*$ there is a vertex shared by 4 of the 6 tetrahedra. If the duals of these sets of tetrahedra are constructed in $V_\perp(0)$, one finds that the dual boundaries intersect in nondegenerate polytopes which will be denoted by $x_\perp = h_\perp$, $s_\perp$, $z_\perp$, $a_\perp$. The shapes of these polytopes are shown in Fig. 11 by unfolding them into the plane. Moreover in the case of $h_\perp$, the additional tetrahedra are enforced by the appearance of the first 4 tetrahedra and by the graph algorithm. We stress that the tiles $x_{\|}^*$ must be considered as objects with additional fixed edges on some of their faces, indicated by dotted lines in Fig. 10, and, in case of $a_{\parallel}^*$, along with a fixed vertex on the longest edge. These additional edges and vertices reflect the packing from tetrahedra and lower the local symmetry of the outer shape. With this restriction and upon inclusion of the inversion symmetry, the tile $h_{\parallel}^*$ has two mirror planes, $s_{\parallel}^*, a_{\parallel}^*$ have one mirror plane, and the tile $z_{\parallel}^*$ appears in two mirror copies. All these packings appear as parts of vertex configurations in the tiling $\mathcal{T}^{*(2F)}$. Clearly the construction in $E_{\perp}$ admits the point symmetry group $A(5)$, and so the packings in $E_{\parallel}$ appear in all icosahedral orientations.

Each packing of tetrahedra into a new tile $x_{\parallel}^*$ is now coded in $V_{\perp}(0)$ by the set of interior points of $x_{\perp}$. This coding includes the additional edges or vertices. Now one can move along edge vectors of $x_{\parallel}^*$ to any other vertex and consider the packing of tetrahedra, seen from this new vertex. In $V_{\perp}(0)$, this implies moving the coding polytope $x_{\perp}$ along a path built from the inverse Voronoi vectors projected to $E_{\perp}$. So each coding polytope appears on the intersection of an orbit under translations with the Voronoi domain. This part of the orbit is determined by the vertex set of the tile $x_{\parallel}^*$, and by the Voronoi vectors whose projections connect these vertices.

Consider now the polytopes $A_{\perp}, \ldots, G_{\perp}$ and $h_{\perp}, \ldots, a_{\perp}$ in $V_{\perp}(0)$, along with their orbits under the icosahedral point group and under the translation group, but restricted to $V_{\perp}(0)$. By inspection of these orbits it can be shown that, for any $P_{\perp}(3)$, there exists a unique set of polytopes $x_{\perp}$ which intersect at most in 2D faces and whose union is $P_{\perp}(3)$. Moreover, translated copies of $P_{\perp}(0)$ have identical decompositions. These unions follow the scheme

	$h_{\perp}$	$s_{\perp}$	$z_{\perp}$	$a_{\perp}$
$A_{\perp}$	1	2	4	0
$B_{\perp}$	1	0	0	0
$C_{\perp}$	0	2	2	0
$D_{\perp}$	0	0	0	1
$F_{\perp}$	3	0	0	3
$G_{\perp}$	3	0	6	0

(18)

This result is easily interpreted in terms of the graph algorithm: any interior point of a fixed $P_{\perp}(3)$ on the graph generates the tile $P_{\parallel}^*(3)$ in the tiling. This point will in general be an interior point of one and only one polytope $x_{\perp} \subset P_{\perp}(3)$. Then the graph generates the tile $P_{\parallel}^*(3)$ as part of the new tile $x_{\parallel}^*$, and only of this new tile.

The implications for the tiling $\mathcal{T}^{*(2F)}$ are as follows: The tetrahedral tiling can be locally and globally transformed into a tiling whose tiles are $h_{\parallel}^*, \ldots a_{\parallel}^*$. The converse is also true since any one of the new tiles, provided that all additional edges and vertices are given, has a unique tetrahedral decomposition. Given the transformation of the two tilings, we have paved the ground for an analysis of the local and global inflation symmetry of the tiling $\mathcal{T}^{*(2F)}$. We expect this inflation symmetry to follow the general inflation scheme for the new tiles, but additional restrictions will arise from the projection scheme associated with the lattice D_6.

7 Relations between the tilings $\mathcal{T}^{*(P)}$ and $\mathcal{T}^{*(2F)}$

The primitive hypercubic lattice $P \sim Z^6$, with the standard embedding of $A(5)$

into its hypercubic holohedry $\Omega(6)$, has been extensively studied since 1984 for use with icosahedral quasicrystals. The cell complex $\mathcal{V}^{(P)}$ for this lattice simply consists of a hierarchy of (hyper-)cubes, translated copies of the unit hypercube $V(0)^{(P)}$ by all linear combinations of the unit vectors $\pm e_i, i = 1 \dots 6$, which are the Voronoi vectors of this lattice. The dual complex $\mathcal{V}^{*(P)}$ is a second such hierarchy, but now with a vertex at $x = 0$, compare Kramer 1986 [19].

In contrast to the usual hexagrid method used with hypercubic and related lattices, Kramer and Neri 1984 [16], the algorithm given in section 3 works for general lattices and can be used to compare different tilings.

It was pointed out in section 2 that the projected Voronoi domains $V_\perp^{(2F)}(0)$ and $V_\perp^{(P)}(0)$ coincide, so that the algorithm for the lattices $P, 2F$ runs within the same polytope. One can choose the same initial point $c_\perp^a$ for the tilings. The Voronoi vectors of the lattice $2F$ are of the form $\pm e_i \pm e_j, i \neq j$. The boundaries $P^{(P)}(4)$ are precisely the 4D hypercubes which occur in the boundaries $P^{(2F)}(5)$. In the projection to $E_\perp$ this implies that any interior point of $P_\perp^{(P)}(4)$ is also an interior point of the corresponding boundary $P^{(2F)}(5)$. Interior points of a fixed boundary $P_\perp^{(2F)}(5)$, not contained in the corresponding $P_\perp^{(P)}(4)$, occur only in the first type of orbit described in section 3. The difference polytope consists of two disconnected polytopes $B_\perp$. Any point interior to this difference polytope produces in the tiling $\mathcal{T}^{*(2F)}$ the only long edge line of the tile $B_\parallel^*$.

The interpretation of these geometric relations in terms of the tilings $\mathcal{T}^{*(2F)}$ and $\mathcal{T}^{*(P)}$ yields for vertices, edges and faces the following results:

(i) Any edge line of $\mathcal{T}^{*(2F)}$, except for the long edge of all tiles $B_\parallel^*$ described above, can be converted into a rhombus face of a tiling $\mathcal{T}^{*(P)}$. Short and long edges along the same 2-fold direction become the diagonals of rhombus faces perpendicular to one another. The long edge of the tile $B_\parallel^*$ must be removed from the tiling.

(ii) The vertices of the tiling $\mathcal{T}^{*(2F)}$ form the even vertices of the tiling $\mathcal{T}^{*(P)}$.

The projections $F_\perp, G_\perp$ occur in $V_\perp^{(2F)}(0)$ with the same shape and positions as the two types of boundaries $P_\perp^{(P)}(3)$ in $V_\perp^{(P)}(0)$. The duals to these objects are two tetrahedra, in $\mathcal{T}^{*(2F)}$ and two rhombohedra in $\mathcal{T}^{*(P)}$ respectively. Upon replacing the edges of the tetrahedra by rhombus faces according to (i), the two tetrahedra are transformed into the two rhombohedra. Since the algorithm for $\mathcal{T}^{(P)}$ yields already a non-periodic space-filling, the partial information on $F_\perp, G_\perp$ and their duals $F_\parallel^*, G_\parallel^*$ in the tiling $\mathcal{T}^{*(2F)}$ suffices to determine by local transformations a tiling $\mathcal{T}^{*(P)}$.

(iii) The tiling $\mathcal{T}^{*(2F)}$ admits a unique and local derivation of a tiling $\mathcal{T}^{*(P)}$ through the transformation of the tetrahedra $F_\parallel^*, G_\parallel^*$ into the corresponding rhombohedra.

The converse is not true: Given the tiling $\mathcal{T}^{*(P)}$, the even vertices still yield the vertices of a tiling $\mathcal{T}^{*(2F)}$. The tetrahedra $F_\parallel^*, G_\parallel^*$ could also be locally obtained from the corresponding rhombohedra. This information does not suffice for fixing the remaining tetrahedra $A_\parallel^*, \dots D_\parallel^*$ which would be needed to derive a tiling $\mathcal{T}^{*(2F)}$.

We cannot expect an equivalence between the two tilings for another reason: The inflation symmetry of the two lattices considered in section 1 leads to a different scaling. The scaling by τ^3 for the tiling $\mathcal{T}^{(P)}$ is discussed by Lück 1990 [25], and a scaling by τ is expected for the tiling $\mathcal{T}^{*(2F)}$.

Conclusion

In the study of non-periodic quasicrystals, many results stem from the replacement of 3D by nD crystallography. The geometry of quasicrystals arises from sections through the nD periodic lattices. This geometry is richer than 3D crystallography. Sections through a given 6D icosahedral lattice like the root lattice D_6 generate several distinct tilings which can serve as cell models for the atomic structure. In contrast to 3D crystals, the local atomic environments vary within a class of different vertex configurations, with a broad range of relative frequencies. The root lattices D_6 and A_4 can generate dissectable tilings. New symmetries, not considered for 3D crystals, appear in the form of inflation/deflation transformations. Eventually they give rise to local and global inflation/deflation symmetries of the tiles and cells.

All these symmetries are geometric in nature. We expect part of these geometric symmetries to be represented on the level of observables and quantum processes in quasicrystals. The new inflation symmetries appear in the kinematical diffraction, compare Baake et al. 1990 [2]. From 1D studies it is known that inflation symmetry governs the states and the spectrum of electrons propagating in quasicrystals. A systematic analysis of nD lattice embeddings should set the geometric ground for the study of the physics of quasicrystals.

Acknowledgments

The present work is part of a study, done at Tübingen University together with M. Baake, R.W. Haase, D. Joseph and M. Schlottmann, on the structure and physics of quasicrystals. Many general and specific discussions with these colleagues are gratefully acknowledged. This work is supported by the Deutsche Forschungsgemeinschaft.

References

[1] P. Guyot, P. Kramer and M. de Boissieu, *Quasicrystals*, Reports on Progress in Physics (1990)

[2] M. Baake, P. Kramer, Z. Papadopolos and D.Zeidler, *Icosahedral Dissectable Tilings from the Root Lattice D_6* to appear in Proc XVIII Int. Coll. on Group Theoretical Methods in Physics, Moscow 1990, Lecture Notes in Physics, ed. by V .V. Dodonov et al., Springer

[3] L. Danzer, *Quasiperiodicity: Local and Global Aspects* to appear in Proc XVIII Int. Coll. on Group Theoretical Methods in Physics, Moscow 1990, Lecture Notes in Physics, ed. by V .V. Dodonov et al., Springer

[4] Th. Hahn and H. Klapper, *Point groups and Crystal Classes*, in: Int. Tables for Crystallography, Vol. A, ed. Th. Hahn, 745-786 (1983)

[5] H. Bohr Acta Math. **45** 29-127 (1925) , Acta Math. **46** 101-214

[6] H. Bohr Acta Math. **47** 237-281 (1926)

[7] R. W. Haase, L. Kramer, P. Kramer and H. Lalvani, *Polyhedra of Three Quasilattices Associated with the Icosahedral Group* Acta Cryst. **A43**, 574-587 (1987)

[8] P. Kramer and D. Zeidler, *Structure Factors for Icosahedral Quasicrystals* Acta Cryst. **A45**, 524-533 (1989)

[9] J. H. Conway and N. J. A. Sloane, *Sphere Packings, Lattices and Groups*, Springer, New York 1988

[10] M. Baake, D. Joseph, P. Kramer and M. Schlottmann, *Root Lattices and Quasicrystals* J. Phys. A in press

[11] J. E. Humphreys, *Introduction to Lie Algebras and Representation Theory*, Springer, New York 1972

[12] H.S.M Coxeter, *Regular Polytopes*, 3rd ed., Dover, New York 1973

[13] J. R. Munkres, *Elements of Algebraic Topology*, Menlo Park: Addison-Wesley 1984

[14] P. Kramer and M. Schlottmann, *Dualization of Voronoi Domains and Klotz Construction: A General Method for the Generation of proper Space Fillings* J. Phys. A 22 L1097-102 (1989)

[15] M. Schlottmann, *Quasiperiodic Tilings and Periodic Structures*, preprint TPT-QC-90-04-2

[16] P. Kramer and R. Neri *On Periodic and Non-Periodic Space Fillings of E^m Obtained by Projection* Acta Cryst. **A40** 580-587 (1984)

[17] J. Kepler 1611, *Strena seu de Nive Sexangula*, in: Ges. Werke vol. 4, ed. M. Caspar and F. Hammer, Munich 1941

[18] G. Kowalewski *Der Keplersche Körper und andere Bauspiele*, Köhlers Antiquarium, Leipzig 1938

[19] P. Kramer, *On the Theory of a Non-periodic Quasilattice Associated with the Icosahedral Group*, Z. Naturf. **41a** 897-911 (1986)

[20] M. Baake, P. Kramer, M. Schlottmann and D. Zeidler *Planar Patterns with Fivefold Symmetry as Sections of Periodic Structures in 4-Space* Int. J. Mod. Phys. B (1990) to appear

[21] M. Baake, P. Kramer, M. Schlottmann and D. Zeidler, *The Triangle Pattern - a New Quasiperiodic Tiling with Fivefold Symmetry*, Mod. Phys. Lett. **B4** 249-258 (1990)

[22] M. Baake, S. I. Ben-Abraham, P. Kramer and M. Schlottmann, *The Vertices of the Ideal 3-D Icosahedral Quasicrystal*, in: *Quasicrystals and Incommensurate Structures in Condensed Matter*, eds. M J. Yacaman et al., 85-95, World Scientific, Singapore 1990

[23] R. Mosseri and J. F. Sadoc 1982, *Two and Three Dimensional Non-periodic Networks Obtained from Self-similar Tiling*, in: The Physics of Quasicrystals, eds. P.J. Steinhardt et al., 720-34 , World Scientific, Singapore 1987

[24] P. Kramer, *Non-periodic Central Space Fillings with Icosahedral Symmetry using Copies of Seven Elementary Cells*, Acta Cryst. **A38** 712-9 (1982)

[25] R. Lück, *Matching Rules for the Icosahedral Penrose Tiling Derived by Inflation-Deflation Procedures*, J. Non-Cryst. Solids **117/118** 820 (1990)

Table 5. Representative Vertex Configurations

Common Tetrahedra of the Vertex Types 1.1 1.2			
$(\bar{2}+3,\bar{2}+4,3+4)_G$	$(\bar{1}+\bar{4},\bar{1}+6,\bar{4}+6)_G$	$(\bar{1}+\bar{2},\bar{1}+5,\bar{2}+5)_G$	$(\bar{1}+3,\bar{1}+\bar{5},3+\bar{5})_G$
$(2+3,2+6,3+6)_G$	$(3+4,3+5,4+5)_G$	$(\bar{2}+4,\bar{2}+5,4+5)_F$	$(\bar{1}+\bar{4},\bar{1}+\bar{5},\bar{4}+\bar{5})_F$
$(\bar{1}+\bar{2},\bar{1}+3,\bar{2}+3)_F$	$(\bar{1}+5,\bar{1}+6,5+6)_F$	$(2+\bar{4},2+\bar{5},\bar{4}+\bar{5})_F$	$(2+\bar{4},2+6,\bar{4}+6)_F$
$(2+3,2+\bar{5},3+\bar{5})_F$	$(3+5,3+6,5+6)_F$		
$(\bar{2}+\bar{1},\bar{4}+\bar{1},\bar{5}+\bar{1})_A$	$(\bar{1}+\bar{5},3+\bar{5},\bar{4}+\bar{5})_A$	$(\bar{1}+\bar{2},4+\bar{2},5+\bar{2})_A$	$(\bar{1}+\bar{4},2+\bar{4},6+\bar{4})_A$
$(\bar{2}+5,4+5,6+5)_A$	$(2+6,\bar{4}+6,5+6)_A$	$(3+2,\bar{4}+2,6+2)_A$	$(2+3,4+3,\bar{5}+3)_A$
$(\bar{2}+4,3+4,5+4)_A$	$(\bar{2}+\bar{1},3+\bar{1},\bar{5}+\bar{1})_C$	$(3+5,4+5,6+5)_C$	$(\bar{1}+6,\bar{4}+6,5+6)_C$
$(\bar{2}+\bar{1},\bar{4}+\bar{1},6+\bar{1})_C$	$(\bar{1}+\bar{2},3+\bar{2},4+\bar{2})_C$	$(\bar{1}+5,\bar{2}+5,6+5)_C$	$(2+6,3+6,5+6)_C$
$(2+\bar{5},3+\bar{5},\bar{4}+\bar{5})_C$	$(\bar{1}+\bar{4},2+\bar{4},\bar{5}+\bar{4})_C$	$(3+2,\bar{4}+2,\bar{5}+2)_C$	$(\bar{2}+\bar{1},5+\bar{1},6+\bar{1})_D$
Vertex type 1.1	orientations: 120	frequency: $1/\tau^{12}$	
$(1+3,5+3,6+3)_B$	$(\bar{1}+3,\bar{2}+3,\bar{6}+3)_B$	$(1+3,2+3,4+3)_C$	$(4+3,\bar{5}+3,\bar{6}+3)_C$
$(\bar{1}+3,\bar{5}+3,\bar{6}+3)_D$	$(1+3,2+3,6+3)_D$	$(1+3,4+3,5+3)_D$	$(\bar{2}+3,4+3,\bar{6}+3)_D$
Vertex type 1.2	orientations: 120	frequency: $1/\tau^{9}$	
$(2+3,4+3,6+3)_C$	$(\bar{1}+3,4+3,\bar{5}+3)_C$	$(4+3,5+3,6+3)_D$	$(\bar{1}+3,\bar{2}+3,4+3)_D$
Common Tetrahedra of the Vertex Types 2.1 2.2			
$(\bar{2}+3,\bar{2}+4,3+4)_G$	$(\bar{1}+\bar{4},\bar{1}+6,\bar{4}+6)_G$	$(\bar{1}+\bar{2},\bar{1}+5,\bar{2}+5)_G$	$(\bar{1}+3,\bar{1}+\bar{5},3+\bar{5})_G$
$(2+3,2+\bar{4},3+\bar{4})_G$	$(2+3,2+6,3+6)_G$	$(3+\bar{4},3+\bar{5},\bar{4}+\bar{5})_G$	$(3+4,3+5,4+5)_G$
$(\bar{2}+4,\bar{2}+5,4+5)_F$	$(\bar{1}+\bar{4},\bar{1}+\bar{5},\bar{4}+\bar{5})_F$	$(\bar{1}+\bar{2},\bar{1}+3,\bar{2}+3)_F$	$(\bar{1}+5,\bar{1}+6,5+6)_F$
$(2+\bar{4},2+6,\bar{4}+6)_F$	$(3+5,3+6,5+6)_F$		
$(\bar{2}+\bar{1},\bar{4}+\bar{1},\bar{5}+\bar{1})_A$	$(\bar{1}+\bar{5},3+\bar{5},\bar{4}+\bar{5})_A$	$(\bar{1}+\bar{2},4+\bar{2},5+\bar{2})_A$	$(\bar{1}+\bar{4},2+\bar{4},6+\bar{4})_A$
$(\bar{2}+5,4+5,6+5)_A$	$(2+6,\bar{4}+6,5+6)_A$	$(3+2,\bar{4}+2,6+2)_A$	$(2+3,4+3,\bar{5}+3)_A$
$(\bar{2}+4,3+4,5+4)_A$	$(\bar{2}+\bar{1},3+\bar{1},\bar{5}+\bar{1})_C$	$(3+5,4+5,6+5)_C$	$(\bar{1}+6,\bar{4}+6,5+6)_C$
$(\bar{2}+\bar{1},\bar{4}+\bar{1},6+\bar{1})_C$	$(\bar{1}+\bar{2},3+\bar{2},4+\bar{2})_C$	$(\bar{1}+5,\bar{2}+5,6+5)_C$	$(2+6,3+6,5+6)_C$
$(\bar{1}+\bar{4},2+\bar{4},\bar{5}+\bar{4})_C$	$(2+\bar{4},3+\bar{4},\bar{5}+\bar{4})_D$	$(2+3,\bar{4}+3,\bar{5}+3)_D$	$(\bar{2}+\bar{1},5+\bar{1},6+\bar{1})_D$
Vertex type 2.1	orientations: 120	frequency: $1/\tau^{9}$	
$(1+3,5+3,6+3)_B$	$(\bar{1}+3,\bar{2}+3,\bar{6}+3)_B$	$(1+3,2+3,4+3)_C$	$(4+3,\bar{5}+3,\bar{6}+3)_C$
$(\bar{1}+3,\bar{5}+3,\bar{6}+3)_D$	$(1+3,2+3,6+3)_D$	$(1+3,4+3,5+3)_D$	$(\bar{2}+3,4+3,\bar{6}+3)_D$
Vertex type 2.2	orientations: 120	frequency: $1/\tau^{9}$	
$(2+3,4+3,6+3)_C$	$(\bar{1}+3,4+3,\bar{5}+3)_C$	$(4+3,5+3,6+3)_D$	$(\bar{1}+3,\bar{2}+3,4+3)_D$
Vertex type 3.1	orientations: 120	frequency: $2/\tau^{13}$	
$(\bar{2}+3,\bar{2}+\bar{6},3+\bar{6})_G$	$(\bar{2}+3,\bar{2}+4,3+4)_G$	$(\bar{1}+\bar{4},\bar{1}+6,\bar{4}+6)_G$	$(\bar{1}+\bar{2},\bar{1}+5,\bar{2}+5)_G$
$(\bar{1}+3,\bar{1}+\bar{6},3+\bar{6})_G$	$(\bar{1}+3,\bar{1}+\bar{5},3+\bar{5})_G$	$(2+3,2+6,3+6)_G$	$(3+4,3+5,4+5)_G$
$(\bar{2}+4,\bar{2}+5,4+5)_F$	$(\bar{1}+\bar{4},\bar{1}+\bar{5},\bar{4}+\bar{5})_F$	$(\bar{1}+\bar{2},\bar{1}+\bar{6},\bar{2}+\bar{6})_F$	$(\bar{1}+5,\bar{1}+6,5+6)_F$
$(2+\bar{4},2+\bar{5},\bar{4}+\bar{5})_F$	$(2+\bar{4},2+6,\bar{4}+6)_F$	$(2+3,2+\bar{5},3+\bar{5})_F$	$(3+5,3+6,5+6)_F$
$(\bar{2}+\bar{1},\bar{4}+\bar{1},\bar{5}+\bar{1})_A$	$(\bar{1}+\bar{5},3+\bar{5},\bar{4}+\bar{5})_A$	$(\bar{1}+\bar{6},\bar{2}+\bar{6},3+\bar{6})_A$	$(\bar{1}+\bar{2},4+\bar{2},5+\bar{2})_A$
$(\bar{1}+\bar{4},2+\bar{4},6+\bar{4})_A$	$(\bar{2}+5,4+5,6+5)_A$	$(2+6,\bar{4}+6,5+6)_A$	$(3+2,\bar{4}+2,6+2)_A$
$(2+3,4+3,\bar{5}+3)_A$	$(\bar{2}+4,3+4,5+4)_A$	$(\bar{3}+\bar{1},5+\bar{1},6+\bar{1})_B$	$(1+3,5+3,6+3)_B$
$(3+5,4+5,6+5)_C$	$(\bar{1}+6,\bar{4}+6,5+6)_C$	$(\bar{1}+5,\bar{2}+5,6+5)_C$	$(2+6,3+6,5+6)_C$
$(\bar{2}+\bar{1},\bar{5}+\bar{1},\bar{6}+\bar{1})_C$	$(\bar{2}+\bar{1},\bar{3}+\bar{1},\bar{4}+\bar{1})_C$	$(2+\bar{5},3+\bar{5},\bar{4}+\bar{5})_C$	$(\bar{1}+\bar{2},4+\bar{2},\bar{6}+\bar{2})_C$

Table 5. Cont.

$(\bar1+\bar4,2+\bar4,\bar5+\bar4)_C$	$(3+2,\bar4+2,\bar5+2)_C$	$(1+3,2+3,4+3)_C$	$(4+3,\bar5+3,\bar6+3)_C$
$(\bar2+\bar1,\bar3+\bar1,5+\bar1)_D$	$(\bar3+\bar1,\bar4+\bar1,6+\bar1)_D$	$(3+\bar1,\bar5+\bar1,\bar6+\bar1)_D$	$(3+\bar2,4+\bar2,\bar6+\bar2)_D$
$(\bar1+3,\bar5+3,\bar6+3)_D$	$(1+3,2+3,6+3)_D$	$(1+3,4+3,5+3)_D$	$(\bar2+3,4+3,\bar6+3)_D$
Vertex type 4.1	orientations: 120	frequency: $1/\tau^{12}$	
$(\bar1+\bar4,\bar1+6,\bar4+6)_G$	$(\bar1+\bar2,\bar1+5,\bar2+5)_G$	$(\bar1+3,\bar1+\bar6,3+\bar6)_G$	$(\bar1+3,\bar1+\bar5,3+\bar5)_G$
$(2+3,2+6,3+6)_G$	$(3+4,3+5,4+5)_G$	$(\bar2+4,\bar2+\bar6,4+\bar6)_F$	$(\bar2+4,\bar2+5,4+5)_F$
$(\bar1+\bar4,\bar1+\bar5,\bar4+\bar5)_F$	$(\bar1+\bar2,\bar1+\bar6,\bar2+\bar6)_F$	$(\bar1+5,\bar1+6,5+6)_F$	$(2+\bar4,2+\bar5,\bar4+\bar5)_F$
$(2+\bar4,2+6,\bar4+6)_F$	$(2+3,2+\bar5,3+\bar5)_F$	$(3+4,3+\bar6,4+\bar6)_F$	$(3+5,3+6,5+6)_F$
$(\bar2+\bar1,\bar4+\bar1,\bar5+\bar1)_A$	$(\bar1+\bar5,2+\bar5,3+\bar5)_A$	$(\bar1+\bar6,3+\bar6,4+\bar6)_A$	$(\bar1+\bar2,4+\bar2,5+\bar2)_A$
$(\bar1+\bar4,2+\bar4,6+\bar4)_A$	$(\bar2+5,4+5,6+5)_A$	$(2+6,\bar4+6,5+6)_A$	$(3+2,\bar4+2,6+2)_A$
$(2+3,4+3,\bar5+3)_A$	$(\bar2+4,3+4,5+4)_A$	$(\bar3+\bar1,5+\bar1,6+\bar1)_B$	$(1+3,5+3,6+3)_B$
$(3+5,4+5,6+5)_C$	$(\bar1+6,\bar4+6,5+6)_C$	$(\bar1+5,\bar2+5,6+5)_C$	$(2+6,3+6,5+6)_C$
$(\bar2+\bar1,\bar5+\bar1,\bar6+\bar1)_C$	$(\bar2+\bar1,\bar3+\bar1,\bar4+\bar1)_C$	$(\bar1+\bar5,2+\bar5,\bar4+\bar5)_C$	$(\bar1+\bar6,\bar2+\bar6,4+\bar6)_C$
$(\bar1+\bar2,4+\bar2,\bar6+\bar2)_C$	$(\bar1+\bar4,2+\bar4,\bar5+\bar4)_C$	$(3+2,\bar4+2,\bar5+2)_C$	$(1+3,2+3,4+3)_C$
$(4+3,\bar5+3,\bar6+3)_C$	$(\bar2+4,3+4,\bar6+4)_C$	$(\bar2+\bar1,\bar3+\bar1,5+\bar1)_D$	$(\bar3+\bar1,\bar4+\bar1,6+\bar1)_D$
$(3+\bar1,\bar5+\bar1,\bar6+\bar1)_D$	$(\bar1+3,\bar5+3,\bar6+3)_D$	$(1+3,2+3,6+3)_D$	$(1+3,4+3,5+3)_D$
Vertex type 5.1	orientations: 120	frequency: $2/\tau^{12}$	
$(\bar2+3,\bar2+\bar6,3+\bar6)_G$	$(\bar2+3,\bar2+4,3+4)_G$	$(\bar1+\bar4,\bar1+6,\bar4+6)_G$	$(\bar1+\bar2,\bar1+5,\bar2+5)_G$
$(\bar1+3,\bar1+\bar6,3+\bar6)_G$	$(\bar1+3,\bar1+\bar5,3+\bar5)_G$	$(2+3,2+\bar4,3+\bar4)_G$	$(2+3,2+6,3+6)_G$
$(3+\bar4,3+\bar5,\bar4+\bar5)_G$	$(3+4,3+5,4+5)_G$	$(\bar2+4,\bar2+5,4+5)_F$	$(\bar1+\bar4,\bar1+\bar5,\bar4+\bar5)_F$
$(\bar1+\bar2,\bar1+\bar6,\bar2+\bar6)_F$	$(\bar1+5,\bar1+6,5+6)_F$	$(2+\bar4,2+6,\bar4+6)_F$	$(3+5,3+6,5+6)_F$
$(\bar2+\bar1,\bar4+\bar1,\bar5+\bar1)_A$	$(\bar1+\bar5,3+\bar5,\bar4+\bar5)_A$	$(\bar1+\bar6,\bar2+\bar6,3+\bar6)_A$	$(\bar1+\bar2,4+\bar2,5+\bar2)_A$
$(\bar1+\bar4,2+\bar4,6+\bar4)_A$	$(\bar2+5,4+5,6+5)_A$	$(2+6,\bar4+6,5+6)_A$	$(3+2,\bar4+2,6+2)_A$
$(2+3,4+3,\bar5+3)_A$	$(\bar2+4,3+4,5+4)_A$	$(\bar3+\bar1,5+\bar1,6+\bar1)_B$	$(1+3,5+3,6+3)_B$
$(3+5,4+5,6+5)_C$	$(\bar1+6,\bar4+6,5+6)_C$	$(\bar1+5,\bar2+5,6+5)_C$	$(2+6,3+6,5+6)_C$
$(\bar2+\bar1,\bar5+\bar1,\bar6+\bar1)_C$	$(\bar2+\bar1,\bar3+\bar1,\bar4+\bar1)_C$	$(\bar1+\bar2,4+\bar2,\bar6+\bar2)_C$	$(\bar1+\bar4,2+\bar4,\bar5+\bar4)_C$
$(1+3,2+3,4+3)_C$	$(4+3,\bar5+3,\bar6+3)_C$	$(\bar2+\bar1,\bar3+\bar1,5+\bar1)_D$	$(\bar3+\bar1,\bar4+\bar1,6+\bar1)_D$
$(3+\bar1,\bar5+\bar1,\bar6+\bar1)_D$	$(3+\bar2,4+\bar2,\bar6+\bar2)_D$	$(2+\bar4,3+\bar4,\bar5+\bar4)_D$	$(\bar1+3,\bar5+3,\bar6+3)_D$
$(2+3,\bar4+3,\bar5+3)_D$	$(1+3,2+3,6+3)_D$	$(1+3,4+3,5+3)_D$	$(\bar2+3,4+3,\bar6+3)_D$
Vertex type 6.1	orientations: 30	frequency: $1/\tau^{3}$	
$(\bar1+3,\bar1+5,3+5)_G$	$(\bar1+3,\bar1+6,3+6)_G$	$(\bar1+5,\bar1+6,5+6)_F$	$(3+5,3+6,5+6)_F$
$(\bar1+5,3+5,6+5)_C$	$(\bar1+6,3+6,5+6)_C$	$(3+\bar1,5+\bar1,6+\bar1)_D$	$(\bar1+3,5+3,6+3)_D$
Common Tetrahedra of the Vertex Types 7.1 7.2			
$(\bar1+\bar2,\bar1+5,\bar2+5)_G$	$(\bar1+3,\bar1+6,3+6)_G$	$(\bar2+3,\bar2+5,3+5)_F$	$(\bar1+\bar2,\bar1+3,\bar2+3)_F$
$(\bar1+5,\bar1+6,5+6)_F$	$(3+5,3+6,5+6)_F$		
$(\bar1+\bar2,3+\bar2,5+\bar2)_A$	$(\bar2+5,3+5,6+5)_A$	$(\bar1+5,\bar2+5,6+5)_C$	$(\bar2+\bar1,3+\bar1,6+\bar1)_C$
$(\bar1+6,3+6,5+6)_C$	$(\bar2+\bar1,5+\bar1,6+\bar1)_D$		
Vertex type 7.1	orientations: 120	frequency: $1/\tau^{6}$	
$(\bar2+3,4+3,5+3)_B$	$(\bar1+3,4+3,6+3)_C$	$(4+3,5+3,6+3)_D$	$(\bar1+3,\bar2+3,4+3)_D$
Vertex type 7.2	orientations: 120	frequency: $2/\tau^{5}$	

Table 5. Cont.

$(\bar{1}+3,\bar{2}+3,5+3)_B$	$(\bar{1}+3,5+3,6+3)_D$		
Vertex type 8.1	orientations: 120	frequency: $1/\tau^6$	
$(\bar{2}+3,\bar{2}+4,3+4)_G$	$(\bar{1}+\bar{2},\bar{1}+5,\bar{2}+5)_G$	$(\bar{1}+3,\bar{1}+6,3+6)_G$	$(3+4,3+5,4+5)_G$
$(\bar{2}+4,\bar{2}+5,4+5)_F$	$(\bar{1}+\bar{2},\bar{1}+3,\bar{2}+3)_F$	$(\bar{1}+5,\bar{1}+6,5+6)_F$	$(3+5,3+6,5+6)_F$
$(\bar{1}+\bar{2},4+\bar{2},5+\bar{2})_A$	$(\bar{2}+5,4+5,6+5)_A$	$(\bar{2}+4,3+4,5+4)_A$	$(3+5,4+5,6+5)_C$
$(\bar{1}+\bar{2},3+\bar{2},4+\bar{2})_C$	$(\bar{1}+5,\bar{2}+5,6+5)_C$	$(\bar{2}+\bar{1},3+\bar{1},6+\bar{1})_C$	$(\bar{1}+6,3+6,5+6)_C$
$(\bar{1}+3,4+3,6+3)_C$	$(4+3,5+3,6+3)_D$	$(\bar{2}+\bar{1},5+\bar{1},6+\bar{1})_D$	$(\bar{1}+3,\bar{2}+3,4+3)_D$
Common Tetrahedra of the Vertex Types 9.1 9.2 9.3			
$(\bar{2}+3,\bar{2}+4,3+4)_G$	$(\bar{1}+\bar{4},\bar{1}+6,\bar{4}+6)_G$	$(\bar{1}+\bar{2},\bar{1}+5,\bar{2}+5)_G$	$(2+3,2+\bar{4},3+\bar{4})_G$
$(2+3,2+6,3+6)_G$	$(3+4,3+5,4+5)_G$	$(\bar{2}+4,\bar{2}+5,4+5)_F$	$(\bar{1}+\bar{2},\bar{1}+3,\bar{2}+3)_F$
$(\bar{1}+3,\bar{1}+\bar{4},3+\bar{4})_F$	$(\bar{1}+5,\bar{1}+6,5+6)_F$	$(2+\bar{4},2+6,\bar{4}+6)_F$	$(3+5,3+6,5+6)_F$
$(\bar{1}+\bar{2},4+\bar{2},5+\bar{2})_A$	$(\bar{1}+\bar{4},2+\bar{4},6+\bar{4})_A$	$(\bar{2}+5,4+5,6+5)_A$	$(2+6,\bar{4}+6,5+6)_A$
$(3+2,\bar{4}+2,6+2)_A$	$(\bar{2}+4,3+4,5+4)_A$	$(\bar{2}+\bar{1},3+\bar{1},\bar{4}+\bar{1})_A$	$(\bar{1}+\bar{4},2+\bar{4},3+\bar{4})_C$
$(3+5,4+5,6+5)_C$	$(\bar{1}+6,\bar{4}+6,5+6)_C$	$(\bar{1}+\bar{2},3+\bar{2},4+\bar{2})_C$	$(\bar{1}+5,\bar{2}+5,6+5)_C$
$(2+6,3+6,5+6)_C$			
Vertex type 9.1	orientations: 120	frequency: $2/\tau^{10}$	
$(2+3,4+3,\bar{6}+3)_A$	$(\bar{1}+3,\bar{2}+3,\bar{6}+3)_B$	$(\bar{2}+\bar{1},\bar{4}+\bar{1},5+\bar{1})_C$	$(\bar{1}+3,2+3,\bar{6}+3)_C$
$(2+3,4+3,5+3)_C$	$(\bar{2}+3,4+3,\bar{6}+3)_D$	$(\bar{4}+\bar{1},5+\bar{1},6+\bar{1})_D$	$(\bar{1}+3,2+3,\bar{4}+3)_D$
$(2+3,5+3,6+3)_D$			
Vertex type 9.2	orientations: 120	frequency: $2/\tau^{11}$	
$(2+3,4+3,\bar{5}+3)_A$	$(1+3,5+3,6+3)_B$	$(\bar{1}+3,\bar{2}+3,\bar{6}+3)_B$	$(\bar{1}+3,\bar{4}+3,\bar{5}+3)_B$
$(\bar{2}+\bar{1},\bar{4}+\bar{1},6+\bar{1})_C$	$(1+3,2+3,4+3)_C$	$(4+3,\bar{5}+3,\bar{6}+3)_C$	$(\bar{1}+3,\bar{5}+3,\bar{6}+3)_D$
$(2+3,\bar{4}+3,\bar{5}+3)_D$	$(1+3,2+3,6+3)_D$	$(1+3,4+3,5+3)_D$	$(\bar{2}+3,4+3,\bar{6}+3)_D$
$(\bar{2}+\bar{1},5+\bar{1},6+\bar{1})_D$			
Vertex type 9.3	orientations: 120	frequency: $2/\tau^8$	
$(\bar{1}+3,2+3,4+3)_A$	$(\bar{2}+\bar{1},\bar{4}+\bar{1},5+\bar{1})_C$	$(2+3,4+3,5+3)_C$	$(\bar{4}+\bar{1},5+\bar{1},6+\bar{1})_D$
$(\bar{1}+3,2+3,\bar{4}+3)_D$	$(\bar{1}+3,\bar{2}+3,4+3)_D$	$(2+3,5+3,6+3)_D$	
Common Tetrahedra of the Vertex Types 10.1 10.2 10.3 10.4			
$(\bar{2}+3,\bar{2}+4,3+4)_G$	$(\bar{1}+\bar{4},\bar{1}+6,\bar{4}+6)_G$	$(\bar{1}+\bar{2},\bar{1}+5,\bar{2}+5)_G$	$(3+4,3+5,4+5)_G$
$(\bar{2}+4,\bar{2}+5,4+5)_F$	$(\bar{1}+\bar{2},\bar{1}+3,\bar{2}+3)_F$	$(\bar{1}+3,\bar{1}+\bar{4},3+\bar{4})_F$	$(\bar{1}+5,\bar{1}+6,5+6)_F$
$(3+\bar{4},3+6,\bar{4}+6)_F$	$(3+5,3+6,5+6)_F$		
$(\bar{1}+\bar{2},4+\bar{2},5+\bar{2})_A$	$(\bar{2}+5,4+5,6+5)_A$	$(\bar{2}+4,3+4,5+4)_A$	$(\bar{2}+\bar{1},3+\bar{1},\bar{4}+\bar{1})_A$
$(\bar{1}+\bar{4},3+\bar{4},6+\bar{4})_A$	$(3+6,\bar{4}+6,5+6)_A$	$(3+5,4+5,6+5)_C$	$(\bar{1}+6,\bar{4}+6,5+6)_C$
$(\bar{2}+\bar{1},\bar{4}+\bar{1},6+\bar{1})_C$	$(\bar{1}+\bar{2},3+\bar{2},4+\bar{2})_C$	$(\bar{1}+5,\bar{2}+5,6+5)_C$	$(\bar{2}+\bar{1},5+\bar{1},6+\bar{1})_D$
Vertex type 10.1	orientations: 120	frequency: $2/\tau^9$	
$(\bar{1}+3,2+3,4+3)_A$	$(2+3,\bar{4}+3,6+3)_B$	$(2+3,4+3,6+3)_C$	$(\bar{1}+3,2+3,\bar{4}+3)_D$
$(4+3,5+3,6+3)_D$	$(\bar{1}+3,\bar{2}+3,4+3)_D$		
Vertex type 10.2	orientations: 120	frequency: $1/\tau^{12}$	
$(2+3,4+3,\bar{5}+3)_A$	$(1+3,5+3,6+3)_B$	$(\bar{1}+3,\bar{2}+3,\bar{6}+3)_B$	$(\bar{1}+3,\bar{4}+3,\bar{5}+3)_B$
$(2+3,\bar{4}+3,6+3)_B$	$(1+3,2+3,4+3)_C$	$(4+3,\bar{5}+3,\bar{6}+3)_C$	$(\bar{1}+3,\bar{5}+3,\bar{6}+3)_D$

Table 5. Cont.

$(2+3,\bar{4}+3,\bar{5}+3)_D$	$(1+3,2+3,6+3)_D$	$(1+3,4+3,5+3)_D$	$(\bar{2}+3,4+3,\bar{6}+3)_D$
Vertex type 10.3	orientations: 120	frequency: $1/\tau^9$	
$(\bar{1}+3,\bar{4}+3,6+3)_B$	$(\bar{1}+3,4+3,6+3)_C$	$(4+3,5+3,6+3)_D$	$(\bar{1}+3,\bar{2}+3,4+3)_D$
Vertex type 10.4	orientations: 120	frequency: $1/\tau^9$	
$(2+3,4+3,\bar{5}+3)_A$	$(\bar{1}+3,\bar{4}+3,\bar{5}+3)_B$	$(2+3,\bar{4}+3,6+3)_B$	$(2+3,4+3,6+3)_C$
$(\bar{1}+3,4+3,\bar{5}+3)_C$	$(2+3,\bar{4}+3,\bar{5}+3)_D$	$(4+3,5+3,6+3)_D$	$(\bar{1}+3,\bar{2}+3,4+3)_D$
Common Tetrahedra of the Vertex Types 11.1 11.2 11.3 11.4 11.5			
$(\bar{1}+\bar{4},\bar{1}+6,\bar{4}+6)_G$	$(\bar{1}+\bar{2},\bar{1}+5,\bar{2}+5)_G$	$(\bar{2}+3,\bar{2}+5,3+5)_F$	$(\bar{1}+\bar{2},\bar{1}+3,\bar{2}+3)_F$
$(\bar{1}+3,\bar{1}+\bar{4},3+\bar{4})_F$	$(\bar{1}+5,\bar{1}+6,5+6)_F$	$(3+\bar{4},3+6,\bar{4}+6)_F$	$(3+5,3+6,5+6)_F$
$(\bar{2}+\bar{1},3+\bar{1},\bar{4}+\bar{1})_A$	$(\bar{1}+\bar{2},3+\bar{2},5+\bar{2})_A$	$(\bar{1}+\bar{4},3+\bar{4},6+\bar{4})_A$	$(\bar{2}+5,3+5,6+5)_A$
$(3+6,\bar{4}+6,5+6)_A$	$(\bar{1}+6,\bar{4}+6,5+6)_C$	$(\bar{1}+5,\bar{2}+5,6+5)_C$	
Vertex type 11.1	orientations: 120	frequency: $1/\tau^9$	
$(2+3,4+3,\bar{6}+3)_A$	$(\bar{2}+3,4+3,5+3)_B$	$(\bar{1}+3,\bar{2}+3,\bar{6}+3)_B$	$(2+3,\bar{4}+3,6+3)_B$
$(\bar{2}+\bar{1},\bar{4}+\bar{1},5+\bar{1})_C$	$(\bar{1}+3,2+3,\bar{6}+3)_C$	$(2+3,4+3,5+3)_C$	$(\bar{2}+3,4+3,\bar{6}+3)_D$
$(\bar{4}+\bar{1},5+\bar{1},6+\bar{1})_D$	$(\bar{1}+3,2+3,\bar{4}+3)_D$	$(2+3,5+3,6+3)_D$	
Vertex type 11.2	orientations: 120	frequency: $1/\tau^9$	
$(2+3,4+3,\bar{5}+3)_A$	$(1+3,5+3,6+3)_B$	$(\bar{2}+3,4+3,5+3)_B$	$(\bar{1}+3,\bar{2}+3,\bar{6}+3)_B$
$(\bar{1}+3,\bar{4}+3,\bar{5}+3)_B$	$(2+3,\bar{4}+3,6+3)_B$	$(\bar{2}+\bar{1},\bar{4}+\bar{1},6+\bar{1})_C$	$(1+3,2+3,4+3)_C$
$(4+3,\bar{5}+3,\bar{6}+3)_C$	$(\bar{1}+3,\bar{5}+3,\bar{6}+3)_D$	$(2+3,\bar{4}+3,\bar{5}+3)_D$	$(1+3,2+3,6+3)_D$
$(1+3,4+3,5+3)_D$	$(\bar{2}+3,4+3,\bar{6}+3)_D$	$(\bar{2}+\bar{1},5+\bar{1},6+\bar{1})_D$	
Vertex type 11.3	orientations: 120	frequency: $1/\tau^6$	
$(\bar{2}+3,4+3,5+3)_B$	$(\bar{1}+3,\bar{4}+3,6+3)_B$	$(\bar{2}+\bar{1},\bar{4}+\bar{1},6+\bar{1})_C$	$(\bar{1}+3,4+3,6+3)_C$
$(4+3,5+3,6+3)_D$	$(\bar{2}+\bar{1},5+\bar{1},6+\bar{1})_D$	$(\bar{1}+3,\bar{2}+3,4+3)_D$	
Vertex type 11.4	orientations: 120	frequency: $2/\tau^8$	
$(\bar{1}+3,2+3,4+3)_A$	$(\bar{2}+3,4+3,5+3)_B$	$(2+3,\bar{4}+3,6+3)_B$	$(\bar{2}+\bar{1},\bar{4}+\bar{1},5+\bar{1})_C$
$(2+3,4+3,5+3)_C$	$(\bar{4}+\bar{1},5+\bar{1},6+\bar{1})_D$	$(\bar{1}+3,2+3,\bar{4}+3)_D$	$(\bar{1}+3,\bar{2}+3,4+3)_D$
$(2+3,5+3,6+3)_D$			
Vertex type 11.5	orientations: 120	frequency: $2/\tau^6$	
$(\bar{1}+3,\bar{4}+3,6+3)_B$	$(\bar{1}+3,\bar{2}+3,5+3)_B$	$(\bar{2}+\bar{1},\bar{4}+\bar{1},5+\bar{1})_C$	$(\bar{1}+3,5+3,6+3)_D$
$(\bar{4}+\bar{1},5+\bar{1},6+\bar{1})_D$			
Vertex type 12.1	orientations: 120	frequency: $2/\tau^{13}$	
$(\bar{1}+\bar{4},\bar{1}+6,\bar{4}+6)_G$	$(\bar{1}+\bar{3},\bar{1}+5,\bar{3}+5)_G$	$(\bar{1}+\bar{3},\bar{1}+6,\bar{3}+6)_G$	$(\bar{1}+\bar{2},\bar{1}+4,\bar{2}+4)_G$
$(\bar{1}+\bar{2},\bar{1}+5,\bar{2}+5)_G$	$(\bar{1}+2,\bar{1}+\bar{5},2+\bar{5})_G$	$(\bar{1}+2,\bar{1}+\bar{4},2+\bar{4})_G$	$(\bar{1}+3,\bar{1}+\bar{6},3+\bar{6})_G$
$(\bar{1}+3,\bar{1}+\bar{5},3+\bar{5})_G$	$(\bar{1}+4,\bar{1}+\bar{6},4+\bar{6})_G$	$(1+3,1+5,3+5)_G$	$(1+3,1+6,3+6)_G$
$(2+3,2+6,3+6)_G$	$(3+4,3+5,4+5)_G$	$(\bar{3}+5,\bar{3}+6,5+6)_F$	$(\bar{2}+4,\bar{2}+5,4+5)_F$
$(1+5,1+6,5+6)_F$	$(2+\bar{4},2+6,\bar{4}+6)_F$	$(2+3,2+\bar{5},3+\bar{5})_F$	$(3+4,3+\bar{6},4+\bar{6})_F$
$(3+1,5+1,6+1)_A$	$(\bar{2}+\bar{1},\bar{4}+\bar{1},\bar{5}+\bar{1})_A$	$(\bar{1}+\bar{5},2+\bar{5},3+\bar{5})_A$	$(\bar{1}+\bar{6},3+\bar{6},4+\bar{6})_A$
$(\bar{1}+\bar{2},4+\bar{2},5+\bar{2})_A$	$(\bar{1}+\bar{3},5+\bar{3},6+\bar{3})_A$	$(\bar{1}+\bar{4},2+\bar{4},6+\bar{4})_A$	$(\bar{2}+5,4+5,6+5)_A$
$(2+6,\bar{4}+6,5+6)_A$	$(3+2,\bar{4}+2,6+2)_A$	$(2+3,4+3,\bar{5}+3)_A$	$(\bar{2}+4,3+4,5+4)_A$
$(\bar{2}+\bar{1},\bar{5}+\bar{1},\bar{6}+\bar{1})_C$	$(\bar{2}+\bar{1},\bar{3}+\bar{1},\bar{4}+\bar{1})_C$	$(1+5,4+5,6+5)_C$	$(\bar{2}+5,\bar{3}+5,6+5)_C$

Table 5. Cont.

$(1+6,2+6,5+6)_C$	$(\bar{3}+6,\bar{4}+6,5+6)_C$	$(3+2,\bar{4}+2,\bar{5}+2)_C$	$(1+3,2+3,4+3)_C$
$(4+3,\bar{5}+3,\bar{6}+3)_C$	$(\bar{2}+4,3+4,\bar{6}+4)_C$	$(\bar{2}+\bar{1},\bar{3}+\bar{1},5+\bar{1})_D$	$(\bar{3}+\bar{1},\bar{4}+\bar{1},6+\bar{1})_D$
$(2+\bar{1},\bar{4}+\bar{1},\bar{5}+\bar{1})_D$	$(3+\bar{1},\bar{5}+\bar{1},\bar{6}+\bar{1})_D$	$(\bar{2}+\bar{1},4+\bar{1},\bar{6}+\bar{1})_D$	$(\bar{1}+5,\bar{2}+5,\bar{3}+5)_D$
$(1+5,3+5,4+5)_D$	$(\bar{1}+6,\bar{3}+6,\bar{4}+6)_D$	$(1+6,2+6,3+6)_D$	$(\bar{1}+2,\bar{4}+2,\bar{5}+2)_D$
$(\bar{1}+3,\bar{5}+3,\bar{6}+3)_D$	$(1+3,2+3,6+3)_D$	$(1+3,4+3,5+3)_D$	$(\bar{1}+4,\bar{2}+4,\bar{6}+4)_D$
Vertex type 13.1	orientations: 120	frequency: $2/\tau^{14}$	
$(\bar{1}+\bar{4},\bar{1}+6,\bar{4}+6)_G$	$(\bar{1}+\bar{3},\bar{1}+5,\bar{3}+5)_G$	$(\bar{1}+\bar{3},\bar{1}+6,\bar{3}+6)_G$	$(\bar{1}+\bar{2},\bar{1}+4,\bar{2}+4)_G$
$(\bar{1}+\bar{2},\bar{1}+5,\bar{2}+5)_G$	$(\bar{1}+3,\bar{1}+\bar{6},3+\bar{6})_G$	$(\bar{1}+3,\bar{1}+\bar{5},3+\bar{5})_G$	$(\bar{1}+4,\bar{1}+\bar{6},4+\bar{6})_G$
$(1+3,1+5,3+5)_G$	$(1+3,1+6,3+6)_G$	$(2+3,2+6,3+6)_G$	$(3+4,3+5,4+5)_G$
$(\bar{3}+5,\bar{3}+6,5+6)_F$	$(\bar{2}+4,\bar{2}+5,4+5)_F$	$(\bar{1}+\bar{4},\bar{1}+\bar{5},\bar{4}+\bar{5})_F$	$(1+5,1+6,5+6)_F$
$(2+\bar{4},2+\bar{5},\bar{4}+\bar{5})_F$	$(2+\bar{4},2+6,\bar{4}+6)_F$	$(2+3,2+\bar{5},3+\bar{5})_F$	$(3+4,3+\bar{6},4+\bar{6})_F$
$(3+1,5+1,6+1)_A$	$(\bar{2}+\bar{1},\bar{4}+\bar{1},\bar{5}+\bar{1})_A$	$(\bar{1}+\bar{5},2+\bar{5},3+\bar{5})_A$	$(\bar{1}+\bar{6},3+\bar{6},4+\bar{6})_A$
$(\bar{1}+\bar{2},4+\bar{2},5+\bar{2})_A$	$(\bar{1}+\bar{3},5+\bar{3},6+\bar{3})_A$	$(\bar{1}+\bar{4},2+\bar{4},6+\bar{4})_A$	$(\bar{2}+5,4+5,6+5)_A$
$(2+6,\bar{4}+6,5+6)_A$	$(3+2,\bar{4}+2,6+2)_A$	$(2+3,4+3,\bar{5}+3)_A$	$(\bar{2}+4,3+4,5+4)_A$
$(\bar{2}+\bar{1},\bar{5}+\bar{1},\bar{6}+\bar{1})_C$	$(\bar{2}+\bar{1},\bar{3}+\bar{1},\bar{4}+\bar{1})_C$	$(\bar{1}+\bar{5},2+\bar{5},\bar{4}+\bar{5})_C$	$(\bar{1}+\bar{4},2+\bar{4},\bar{5}+\bar{4})_C$
$(1+5,4+5,6+5)_C$	$(\bar{2}+5,\bar{3}+5,6+5)_C$	$(1+6,2+6,5+6)_C$	$(\bar{3}+6,\bar{4}+6,5+6)_C$
$(3+2,\bar{4}+2,\bar{5}+2)_C$	$(1+3,2+3,4+3)_C$	$(4+3,\bar{5}+3,\bar{6}+3)_C$	$(\bar{2}+4,3+4,\bar{6}+4)_C$
$(\bar{2}+\bar{1},\bar{3}+\bar{1},5+\bar{1})_D$	$(\bar{3}+\bar{1},\bar{4}+\bar{1},6+\bar{1})_D$	$(3+\bar{1},\bar{5}+\bar{1},\bar{6}+\bar{1})_D$	$(\bar{2}+\bar{1},4+\bar{1},\bar{6}+\bar{1})_D$
$(\bar{1}+5,\bar{2}+5,\bar{3}+5)_D$	$(1+5,3+5,4+5)_D$	$(\bar{1}+6,\bar{3}+6,\bar{4}+6)_D$	$(1+6,2+6,3+6)_D$
$(\bar{1}+3,\bar{5}+3,\bar{6}+3)_D$	$(1+3,2+3,6+3)_D$	$(1+3,4+3,5+3)_D$	$(\bar{1}+4,\bar{2}+4,\bar{6}+4)_D$
Vertex type 14.1	orientations: 120	frequency: $(\tau^2+1)/\tau^{13}$	
$(\bar{1}+\bar{4},\bar{1}+6,\bar{4}+6)_G$	$(\bar{1}+\bar{3},\bar{1}+5,\bar{3}+5)_G$	$(\bar{1}+\bar{3},\bar{1}+6,\bar{3}+6)_G$	$(\bar{1}+\bar{2},\bar{1}+5,\bar{2}+5)_G$
$(\bar{1}+3,\bar{1}+\bar{6},3+\bar{6})_G$	$(\bar{1}+3,\bar{1}+\bar{5},3+\bar{5})_G$	$(1+3,1+5,3+5)_G$	$(1+3,1+6,3+6)_G$
$(2+3,2+6,3+6)_G$	$(3+4,3+5,4+5)_G$	$(\bar{3}+5,\bar{3}+6,5+6)_F$	$(\bar{2}+4,\bar{2}+\bar{6},4+\bar{6})_F$
$(\bar{2}+4,\bar{2}+5,4+5)_F$	$(\bar{1}+\bar{4},\bar{1}+\bar{5},\bar{4}+\bar{5})_F$	$(\bar{1}+\bar{2},\bar{1}+\bar{6},\bar{2}+\bar{6})_F$	$(1+5,1+6,5+6)_F$
$(2+\bar{4},2+\bar{5},\bar{4}+\bar{5})_F$	$(2+\bar{4},2+6,\bar{4}+6)_F$	$(2+3,2+\bar{5},3+\bar{5})_F$	$(3+4,3+\bar{6},4+\bar{6})_F$
$(3+1,5+1,6+1)_A$	$(\bar{2}+\bar{1},\bar{4}+\bar{1},\bar{5}+\bar{1})_A$	$(\bar{1}+\bar{5},2+\bar{5},3+\bar{5})_A$	$(\bar{1}+\bar{6},3+\bar{6},4+\bar{6})_A$
$(\bar{1}+\bar{2},4+\bar{2},5+\bar{2})_A$	$(\bar{1}+\bar{3},5+\bar{3},6+\bar{3})_A$	$(\bar{1}+\bar{4},2+\bar{4},6+\bar{4})_A$	$(\bar{2}+5,4+5,6+5)_A$
$(2+6,\bar{4}+6,5+6)_A$	$(3+2,\bar{4}+2,6+2)_A$	$(2+3,4+3,\bar{5}+3)_A$	$(\bar{2}+4,3+4,5+4)_A$
$(\bar{2}+\bar{1},\bar{5}+\bar{1},\bar{6}+\bar{1})_C$	$(\bar{2}+\bar{1},\bar{3}+\bar{1},\bar{4}+\bar{1})_C$	$(\bar{1}+\bar{5},2+\bar{5},\bar{4}+\bar{5})_C$	$(\bar{1}+\bar{6},\bar{2}+\bar{6},4+\bar{6})_C$
$(\bar{1}+\bar{2},4+\bar{2},\bar{6}+\bar{2})_C$	$(\bar{1}+\bar{4},2+\bar{4},\bar{5}+\bar{4})_C$	$(1+5,4+5,6+5)_C$	$(\bar{2}+5,\bar{3}+5,6+5)_C$
$(1+6,2+6,5+6)_C$	$(\bar{3}+6,\bar{4}+6,5+6)_C$	$(3+2,\bar{4}+2,\bar{5}+2)_C$	$(1+3,2+3,4+3)_C$
$(4+3,\bar{5}+3,\bar{6}+3)_C$	$(\bar{2}+4,3+4,\bar{6}+4)_C$	$(\bar{2}+\bar{1},\bar{3}+\bar{1},5+\bar{1})_D$	$(\bar{3}+\bar{1},\bar{4}+\bar{1},6+\bar{1})_D$
$(3+\bar{1},\bar{5}+\bar{1},\bar{6}+\bar{1})_D$	$(\bar{1}+5,\bar{2}+5,\bar{3}+5)_D$	$(1+5,3+5,4+5)_D$	$(\bar{1}+6,\bar{3}+6,\bar{4}+6)_D$
$(1+6,2+6,3+6)_D$	$(\bar{1}+3,\bar{5}+3,\bar{6}+3)_D$	$(1+3,2+3,6+3)_D$	$(1+3,4+3,5+3)_D$
Vertex type 15.1	orientations: 120	frequency: $1/\tau^{12}$	
$(\bar{1}+\bar{4},\bar{1}+6,\bar{4}+6)_G$	$(\bar{1}+\bar{3},\bar{1}+5,\bar{3}+5)_G$	$(\bar{1}+\bar{3},\bar{1}+6,\bar{3}+6)_G$	$(\bar{1}+\bar{2},\bar{1}+4,\bar{2}+4)_G$
$(\bar{1}+\bar{2},\bar{1}+5,\bar{2}+5)_G$	$(\bar{1}+2,\bar{1}+\bar{5},2+\bar{5})_G$	$(\bar{1}+2,\bar{1}+\bar{4},2+\bar{4})_G$	$(\bar{1}+3,\bar{1}+\bar{6},3+\bar{6})_G$
$(\bar{1}+3,\bar{1}+\bar{5},3+\bar{5})_G$	$(\bar{1}+4,\bar{1}+\bar{6},4+\bar{6})_G$	$(1+3,1+6,3+6)_G$	$(2+3,2+6,3+6)_G$
$(\bar{3}+5,\bar{3}+6,5+6)_F$	$(\bar{2}+4,\bar{2}+5,4+5)_F$	$(1+3,1+4,3+4)_F$	$(1+4,1+5,4+5)_F$

Table 5. Cont.

$(1+5,1+6,5+6)_F$	$(2+\bar{4},2+6,\bar{4}+6)_F$	$(2+3,2+\bar{5},3+\bar{5})_F$	$(3+4,3+\bar{6},4+\bar{6})_F$
$(3+1,4+1,6+1)_A$	$(\bar{2}+\bar{1},\bar{4}+\bar{1},\bar{5}+\bar{1})_A$	$(\bar{1}+\bar{5},2+\bar{5},3+\bar{5})_A$	$(\bar{1}+\bar{6},3+\bar{6},4+\bar{6})_A$
$(\bar{1}+\bar{2},4+\bar{2},5+\bar{2})_A$	$(\bar{1}+\bar{3},5+\bar{3},6+\bar{3})_A$	$(\bar{1}+\bar{4},2+\bar{4},6+\bar{4})_A$	$(\bar{2}+5,4+5,6+5)_A$
$(2+6,\bar{4}+6,5+6)_A$	$(3+2,\bar{5}+2,6+2)_A$	$(2+3,4+3,\bar{5}+3)_A$	$(\bar{2}+4,3+4,5+4)_A$
$(4+1,5+1,6+1)_C$	$(\bar{2}+\bar{1},\bar{5}+\bar{1},\bar{6}+\bar{1})_C$	$(\bar{2}+\bar{1},\bar{3}+\bar{1},\bar{4}+\bar{1})_C$	$(1+5,4+5,6+5)_C$
$(\bar{2}+5,\bar{3}+5,6+5)_C$	$(1+6,2+6,5+6)_C$	$(\bar{3}+6,\bar{4}+6,5+6)_C$	$(\bar{4}+2,\bar{5}+2,6+2)_C$
$(1+3,2+3,4+3)_C$	$(4+3,\bar{5}+3,\bar{6}+3)_C$	$(1+4,3+4,5+4)_C$	$(\bar{2}+4,3+4,\bar{6}+4)_C$
$(\bar{2}+\bar{1},\bar{3}+\bar{1},5+\bar{1})_D$	$(\bar{3}+\bar{1},\bar{4}+\bar{1},6+\bar{1})_D$	$(2+\bar{1},\bar{4}+\bar{1},5+\bar{1})_D$	$(3+\bar{1},\bar{5}+\bar{1},\bar{6}+\bar{1})_D$
$(\bar{2}+\bar{1},4+\bar{1},\bar{6}+\bar{1})_D$	$(\bar{1}+5,\bar{2}+5,\bar{3}+5)_D$	$(\bar{1}+6,\bar{3}+6,\bar{4}+6)_D$	$(1+6,2+6,3+6)_D$
$(\bar{1}+2,\bar{4}+2,\bar{5}+2)_D$	$(\bar{1}+3,\bar{5}+3,\bar{6}+3)_D$	$(1+3,2+3,6+3)_D$	$(\bar{1}+4,\bar{2}+4,\bar{6}+4)_D$
Vertex type 16.1	orientations: 120	frequency: $1/\tau^{12}$	
$(\bar{1}+\bar{4},\bar{1}+6,\bar{4}+6)_G$	$(\bar{1}+\bar{3},\bar{1}+5,\bar{3}+5)_G$	$(\bar{1}+\bar{3},\bar{1}+6,\bar{3}+6)_G$	$(\bar{1}+\bar{2},\bar{1}+4,\bar{2}+4)_G$
$(\bar{1}+\bar{2},\bar{1}+5,\bar{2}+5)_G$	$(\bar{1}+3,\bar{1}+\bar{6},3+\bar{6})_G$	$(\bar{1}+3,\bar{1}+\bar{5},3+\bar{5})_G$	$(\bar{1}+4,\bar{1}+\bar{6},4+\bar{6})_G$
$(1+3,1+6,3+6)_G$	$(2+3,2+6,3+6)_G$	$(\bar{3}+5,\bar{3}+6,5+6)_F$	$(\bar{2}+4,\bar{2}+5,4+5)_F$
$(\bar{1}+\bar{4},\bar{1}+\bar{5},\bar{4}+\bar{5})_F$	$(1+3,1+4,3+4)_F$	$(1+4,1+5,4+5)_F$	$(1+5,1+6,5+6)_F$
$(2+\bar{4},2+\bar{5},\bar{4}+\bar{5})_F$	$(2+\bar{4},2+6,\bar{4}+6)_F$	$(2+3,2+\bar{5},3+\bar{5})_F$	$(3+4,3+\bar{6},4+\bar{6})_F$
$(3+1,4+1,6+1)_A$	$(\bar{2}+\bar{1},\bar{4}+\bar{1},\bar{5}+\bar{1})_A$	$(\bar{1}+\bar{5},2+\bar{5},3+\bar{5})_A$	$(\bar{1}+\bar{6},3+\bar{6},4+\bar{6})_A$
$(\bar{1}+\bar{2},4+\bar{2},5+\bar{2})_A$	$(\bar{1}+\bar{3},5+\bar{3},6+\bar{3})_A$	$(\bar{1}+\bar{4},2+\bar{4},6+\bar{4})_A$	$(\bar{2}+5,4+5,6+5)_A$
$(2+6,\bar{4}+6,5+6)_A$	$(3+2,\bar{5}+2,6+2)_A$	$(2+3,4+3,\bar{5}+3)_A$	$(\bar{2}+4,3+4,5+4)_A$
$(4+1,5+1,6+1)_C$	$(\bar{2}+\bar{1},\bar{5}+\bar{1},\bar{6}+\bar{1})_C$	$(\bar{2}+\bar{1},\bar{3}+\bar{1},\bar{4}+\bar{1})_C$	$(\bar{1}+\bar{5},2+\bar{5},\bar{4}+\bar{5})_C$
$(\bar{1}+\bar{4},2+\bar{4},\bar{5}+\bar{4})_C$	$(1+5,4+5,6+5)_C$	$(\bar{2}+5,\bar{3}+5,6+5)_C$	$(1+6,2+6,5+6)_C$
$(\bar{3}+6,\bar{4}+6,5+6)_C$	$(\bar{4}+2,\bar{5}+2,6+2)_C$	$(1+3,2+3,4+3)_C$	$(4+3,\bar{5}+3,\bar{6}+3)_C$
$(1+4,3+4,5+4)_C$	$(\bar{2}+4,3+4,\bar{6}+4)_C$	$(\bar{2}+\bar{1},\bar{3}+\bar{1},5+\bar{1})_D$	$(\bar{3}+\bar{1},\bar{4}+\bar{1},6+\bar{1})_D$
$(3+\bar{1},\bar{5}+\bar{1},\bar{6}+\bar{1})_D$	$(\bar{2}+\bar{1},4+\bar{1},\bar{6}+\bar{1})_D$	$(\bar{1}+5,\bar{2}+5,\bar{3}+5)_D$	$(\bar{1}+6,\bar{3}+6,\bar{4}+6)_D$
$(1+6,2+6,3+6)_D$	$(\bar{1}+3,\bar{5}+3,\bar{6}+3)_D$	$(1+3,2+3,6+3)_D$	$(\bar{1}+4,\bar{2}+4,\bar{6}+4)_D$
Vertex type 17.1	orientations: 120	frequency: $2/\tau^{13}$	
$(\bar{1}+\bar{4},\bar{1}+6,\bar{4}+6)_G$	$(\bar{1}+\bar{3},\bar{1}+5,\bar{3}+5)_G$	$(\bar{1}+\bar{3},\bar{1}+6,\bar{3}+6)_G$	$(\bar{1}+\bar{2},\bar{1}+5,\bar{2}+5)_G$
$(\bar{1}+3,\bar{1}+\bar{6},3+\bar{6})_G$	$(\bar{1}+3,\bar{1}+\bar{5},3+\bar{5})_G$	$(1+3,1+6,3+6)_G$	$(2+3,2+6,3+6)_G$
$(\bar{3}+5,\bar{3}+6,5+6)_F$	$(\bar{2}+4,\bar{2}+\bar{6},4+\bar{6})_F$	$(\bar{2}+4,\bar{2}+5,4+5)_F$	$(\bar{1}+\bar{4},\bar{1}+\bar{5},\bar{4}+\bar{5})_F$
$(\bar{1}+\bar{2},\bar{1}+\bar{6},\bar{2}+\bar{6})_F$	$(1+3,1+4,3+4)_F$	$(1+4,1+5,4+5)_F$	$(1+5,1+6,5+6)_F$
$(2+\bar{4},2+\bar{5},\bar{4}+\bar{5})_F$	$(2+\bar{4},2+6,\bar{4}+6)_F$	$(2+3,2+\bar{5},3+\bar{5})_F$	$(3+4,3+\bar{6},4+\bar{6})_F$
$(3+1,5+1,6+1)_A$	$(\bar{2}+\bar{1},\bar{4}+\bar{1},\bar{5}+\bar{1})_A$	$(\bar{1}+\bar{5},2+\bar{5},3+\bar{5})_A$	$(\bar{1}+\bar{6},3+\bar{6},4+\bar{6})_A$
$(\bar{1}+\bar{2},4+\bar{2},5+\bar{2})_A$	$(\bar{1}+\bar{3},5+\bar{3},6+\bar{3})_A$	$(\bar{1}+\bar{4},2+\bar{4},6+\bar{4})_A$	$(\bar{2}+5,4+5,6+5)_A$
$(2+6,\bar{4}+6,5+6)_A$	$(3+2,\bar{4}+2,6+2)_A$	$(2+3,4+3,\bar{5}+3)_A$	$(\bar{2}+4,3+4,5+4)_A$
$(3+1,4+1,5+1)_C$	$(\bar{2}+\bar{1},\bar{5}+\bar{1},\bar{6}+\bar{1})_C$	$(\bar{2}+\bar{1},\bar{3}+\bar{1},\bar{4}+\bar{1})_C$	$(\bar{1}+\bar{5},2+\bar{5},\bar{4}+\bar{5})_C$
$(\bar{1}+\bar{6},\bar{2}+\bar{6},4+\bar{6})_C$	$(\bar{1}+\bar{2},4+\bar{2},\bar{6}+\bar{2})_C$	$(\bar{1}+\bar{4},2+\bar{4},\bar{5}+\bar{4})_C$	$(1+5,4+5,6+5)_C$
$(\bar{2}+5,\bar{3}+5,6+5)_C$	$(1+6,2+6,5+6)_C$	$(\bar{3}+6,\bar{4}+6,5+6)_C$	$(3+2,\bar{4}+2,\bar{5}+2)_C$
$(1+3,2+3,4+3)_C$	$(4+3,\bar{5}+3,\bar{6}+3)_C$	$(1+4,3+4,5+4)_C$	$(\bar{2}+4,3+4,\bar{6}+4)_C$
$(\bar{2}+\bar{1},\bar{3}+\bar{1},5+\bar{1})_D$	$(\bar{3}+\bar{1},\bar{4}+\bar{1},6+\bar{1})_D$	$(3+\bar{1},\bar{5}+\bar{1},\bar{6}+\bar{1})_D$	$(\bar{1}+5,\bar{2}+5,\bar{3}+5)_D$
$(\bar{1}+6,\bar{3}+6,\bar{4}+6)_D$	$(1+6,2+6,3+6)_D$	$(\bar{1}+3,\bar{5}+3,\bar{6}+3)_D$	$(1+3,2+3,6+3)_D$

Table 5. Cont.

Vertex type 18.1	orientations: 120	frequency: $2/\tau^{12}$	
$(\bar{1}+\bar{4},\bar{1}+6,\bar{4}+6)_G$	$(\bar{1}+\bar{3},\bar{1}+6,\bar{3}+6)_G$	$(\bar{1}+3,\bar{1}+\bar{6},3+\bar{6})_G$	$(\bar{1}+3,\bar{1}+\bar{5},3+\bar{5})_G$
$(1+3,1+6,3+6)_G$	$(2+3,2+6,3+6)_G$	$(\bar{3}+5,\bar{3}+6,5+6)_F$	$(\bar{2}+\bar{3},\bar{2}+5,\bar{3}+5)_F$
$(\bar{2}+4,\bar{2}+\bar{6},4+\bar{6})_F$	$(\bar{2}+4,\bar{2}+5,4+5)_F$	$(\bar{1}+\bar{4},\bar{1}+\bar{5},\bar{4}+\bar{5})_F$	$(\bar{1}+\bar{2},\bar{1}+\bar{6},\bar{2}+\bar{6})_F$
$(\bar{1}+\bar{2},\bar{1}+\bar{3},\bar{2}+\bar{3})_F$	$(1+3,1+4,3+4)_F$	$(1+4,1+5,4+5)_F$	$(1+5,1+6,5+6)_F$
$(2+\bar{4},2+\bar{5},\bar{4}+\bar{5})_F$	$(2+\bar{4},2+6,\bar{4}+6)_F$	$(2+3,2+\bar{5},3+\bar{5})_F$	$(3+4,3+\bar{6},4+\bar{6})_F$
$(3+1,4+1,6+1)_A$	$(\bar{2}+\bar{1},\bar{4}+\bar{1},\bar{5}+\bar{1})_A$	$(\bar{1}+\bar{5},3+\bar{5},\bar{4}+\bar{5})_A$	$(\bar{1}+\bar{6},\bar{2}+\bar{6},3+\bar{6})_A$
$(\bar{1}+\bar{2},4+\bar{2},5+\bar{2})_A$	$(\bar{1}+\bar{3},\bar{2}+\bar{3},6+\bar{3})_A$	$(\bar{1}+\bar{4},\bar{5}+\bar{4},6+\bar{4})_A$	$(\bar{2}+5,4+5,6+5)_A$
$(2+6,\bar{4}+6,5+6)_A$	$(3+2,\bar{5}+2,6+2)_A$	$(2+3,4+3,\bar{5}+3)_A$	$(\bar{2}+4,3+4,5+4)_A$
$(4+1,5+1,6+1)_C$	$(\bar{2}+\bar{1},\bar{5}+\bar{1},\bar{6}+\bar{1})_C$	$(\bar{2}+\bar{1},\bar{3}+\bar{1},\bar{4}+\bar{1})_C$	$(2+\bar{5},3+\bar{5},\bar{4}+\bar{5})_C$
$(\bar{2}+\bar{6},3+\bar{6},4+\bar{6})_C$	$(\bar{1}+\bar{2},\bar{3}+\bar{2},5+\bar{2})_C$	$(\bar{1}+\bar{2},4+\bar{2},\bar{6}+\bar{2})_C$	$(\bar{2}+\bar{3},5+\bar{3},6+\bar{3})_C$
$(2+\bar{4},\bar{5}+\bar{4},6+\bar{4})_C$	$(1+5,4+5,6+5)_C$	$(\bar{2}+5,\bar{3}+5,6+5)_C$	$(1+6,2+6,5+6)_C$
$(\bar{3}+6,\bar{4}+6,5+6)_C$	$(\bar{4}+2,\bar{5}+2,6+2)_C$	$(1+3,2+3,4+3)_C$	$(4+3,\bar{5}+3,\bar{6}+3)_C$
$(1+4,3+4,5+4)_C$	$(\bar{2}+4,3+4,\bar{6}+4)_C$	$(\bar{3}+\bar{1},\bar{4}+\bar{1},6+\bar{1})_D$	$(3+\bar{1},\bar{5}+\bar{1},\bar{6}+\bar{1})_D$
$(\bar{1}+6,\bar{3}+6,\bar{4}+6)_D$	$(1+6,2+6,3+6)_D$	$(\bar{1}+3,\bar{5}+3,\bar{6}+3)_D$	$(1+3,2+3,6+3)_D$
Vertex type 19.1	**orientations: 120**	**frequency: $2/\tau^{14}$**	
$(\bar{1}+\bar{4},\bar{1}+6,\bar{4}+6)_G$	$(\bar{1}+\bar{3},\bar{1}+5,\bar{3}+5)_G$	$(\bar{1}+\bar{3},\bar{1}+6,\bar{3}+6)_G$	$(\bar{1}+\bar{2},\bar{1}+5,\bar{2}+5)_G$
$(\bar{1}+3,\bar{1}+\bar{6},3+\bar{6})_G$	$(\bar{1}+3,\bar{1}+\bar{5},3+\bar{5})_G$	$(\bar{3}+5,\bar{3}+6,5+6)_F$	$(\bar{2}+4,\bar{2}+\bar{6},4+\bar{6})_F$
$(\bar{2}+4,\bar{2}+5,4+5)_F$	$(\bar{1}+\bar{4},\bar{1}+\bar{5},\bar{4}+\bar{5})_F$	$(\bar{1}+\bar{2},\bar{1}+\bar{6},\bar{2}+\bar{6})_F$	$(1+2,1+3,2+3)_F$
$(1+2,1+6,2+6)_F$	$(1+3,1+4,3+4)_F$	$(1+4,1+5,4+5)_F$	$(1+5,1+6,5+6)_F$
$(2+\bar{4},2+\bar{5},\bar{4}+\bar{5})_F$	$(2+\bar{4},2+6,\bar{4}+6)_F$	$(2+3,2+\bar{5},3+\bar{5})_F$	$(3+4,3+\bar{6},4+\bar{6})_F$
$(3+1,5+1,6+1)_A$	$(\bar{2}+\bar{1},\bar{4}+\bar{1},\bar{5}+\bar{1})_A$	$(\bar{1}+\bar{5},2+\bar{5},3+\bar{5})_A$	$(\bar{1}+\bar{6},3+\bar{6},4+\bar{6})_A$
$(\bar{1}+\bar{2},4+\bar{2},5+\bar{2})_A$	$(\bar{1}+\bar{3},5+\bar{3},6+\bar{3})_A$	$(\bar{1}+\bar{4},2+\bar{4},6+\bar{4})_A$	$(\bar{2}+5,4+5,6+5)_A$
$(2+6,\bar{4}+6,5+6)_A$	$(3+2,\bar{4}+2,6+2)_A$	$(2+3,4+3,\bar{5}+3)_A$	$(\bar{2}+4,3+4,5+4)_A$
$(2+1,3+1,6+1)_C$	$(3+1,4+1,5+1)_C$	$(\bar{2}+\bar{1},\bar{5}+\bar{1},\bar{6}+\bar{1})_C$	$(\bar{2}+\bar{1},\bar{3}+\bar{1},\bar{4}+\bar{1})_C$
$(\bar{1}+\bar{5},2+\bar{5},\bar{4}+\bar{5})_C$	$(\bar{1}+\bar{6},\bar{2}+\bar{6},4+\bar{6})_C$	$(\bar{1}+\bar{2},4+\bar{2},\bar{6}+\bar{2})_C$	$(\bar{1}+\bar{4},2+\bar{4},\bar{5}+\bar{4})_C$
$(1+5,4+5,6+5)_C$	$(\bar{2}+5,\bar{3}+5,6+5)_C$	$(1+6,2+6,5+6)_C$	$(\bar{3}+6,\bar{4}+6,5+6)_C$
$(1+2,3+2,6+2)_C$	$(3+2,\bar{4}+2,\bar{5}+2)_C$	$(1+3,2+3,4+3)_C$	$(4+3,\bar{5}+3,\bar{6}+3)_C$
$(1+4,3+4,5+4)_C$	$(\bar{2}+4,3+4,\bar{6}+4)_C$	$(\bar{2}+\bar{1},\bar{3}+\bar{1},5+\bar{1})_D$	$(\bar{3}+\bar{1},\bar{4}+\bar{1},6+\bar{1})_D$
$(3+\bar{1},\bar{5}+\bar{1},\bar{6}+\bar{1})_D$	$(\bar{1}+5,\bar{2}+5,\bar{3}+5)_D$	$(\bar{1}+6,\bar{3}+6,\bar{4}+6)_D$	$(\bar{1}+3,\bar{5}+3,\bar{6}+3)_D$
Vertex type 20.1	**orientations: 120**	**frequency: $1/\tau^{12}$**	
$(\bar{1}+3,\bar{1}+\bar{6},3+\bar{6})_G$	$(\bar{1}+3,\bar{1}+\bar{5},3+\bar{5})_G$	$(\bar{3}+\bar{4},\bar{3}+6,\bar{4}+6)_F$	$(\bar{3}+5,\bar{3}+6,5+6)_F$
$(\bar{2}+\bar{3},\bar{2}+5,\bar{3}+5)_F$	$(\bar{2}+4,\bar{2}+\bar{6},4+\bar{6})_F$	$(\bar{2}+4,\bar{2}+5,4+5)_F$	$(\bar{1}+\bar{4},\bar{1}+\bar{5},\bar{4}+\bar{5})_F$
$(\bar{1}+\bar{3},\bar{1}+\bar{4},\bar{3}+\bar{4})_F$	$(\bar{1}+\bar{2},\bar{1}+\bar{6},\bar{2}+\bar{6})_F$	$(\bar{1}+\bar{2},\bar{1}+\bar{3},\bar{2}+\bar{3})_F$	$(1+2,1+3,2+3)_F$
$(1+2,1+6,2+6)_F$	$(1+3,1+4,3+4)_F$	$(1+4,1+5,4+5)_F$	$(1+5,1+6,5+6)_F$
$(2+\bar{4},2+\bar{5},\bar{4}+\bar{5})_F$	$(2+\bar{4},2+6,\bar{4}+6)_F$	$(2+3,2+\bar{5},3+\bar{5})_F$	$(3+4,3+\bar{6},4+\bar{6})_F$
$(3+1,5+1,6+1)_A$	$(\bar{2}+\bar{1},\bar{4}+\bar{1},\bar{5}+\bar{1})_A$	$(\bar{1}+\bar{5},2+\bar{5},3+\bar{5})_A$	$(\bar{1}+\bar{6},3+\bar{6},4+\bar{6})_A$
$(\bar{1}+\bar{2},4+\bar{2},5+\bar{2})_A$	$(\bar{1}+\bar{3},5+\bar{3},6+\bar{3})_A$	$(\bar{1}+\bar{4},2+\bar{4},6+\bar{4})_A$	$(\bar{2}+5,4+5,6+5)_A$
$(2+6,\bar{4}+6,5+6)_A$	$(3+2,\bar{4}+2,6+2)_A$	$(2+3,4+3,\bar{5}+3)_A$	$(\bar{2}+4,3+4,5+4)_A$
$(2+1,3+1,6+1)_C$	$(3+1,4+1,5+1)_C$	$(\bar{2}+\bar{1},\bar{5}+\bar{1},\bar{6}+\bar{1})_C$	$(\bar{2}+\bar{1},\bar{3}+\bar{1},\bar{4}+\bar{1})_C$

Table 5. Cont.

$(\bar{1}+\bar{5},2+\bar{5},\bar{4}+\bar{5})_C$	$(\bar{1}+\bar{6},\bar{2}+\bar{6},4+\bar{6})_C$	$(\bar{1}+\bar{2},\bar{3}+\bar{2},5+\bar{2})_C$	$(\bar{1}+\bar{2},4+\bar{2},\bar{6}+\bar{2})_C$
$(\bar{1}+\bar{3},\bar{4}+\bar{3},6+\bar{3})_C$	$(\bar{1}+\bar{3},\bar{2}+\bar{3},5+\bar{3})_C$	$(\bar{1}+\bar{4},2+\bar{4},\bar{5}+\bar{4})_C$	$(\bar{1}+\bar{4},\bar{3}+\bar{4},6+\bar{4})_C$
$(1+5,4+5,6+5)_C$	$(\bar{2}+5,\bar{3}+5,6+5)_C$	$(1+6,2+6,5+6)_C$	$(\bar{3}+6,\bar{4}+6,5+6)_C$
$(1+2,3+2,6+2)_C$	$(3+2,\bar{4}+2,\bar{5}+2)_C$	$(1+3,2+3,4+3)_C$	$(4+3,\bar{5}+3,\bar{6}+3)_C$
$(1+4,3+4,5+4)_C$	$(\bar{2}+4,3+4,\bar{6}+4)_C$	$(3+\bar{1},\bar{5}+\bar{1},\bar{6}+\bar{1})_D$	$(\bar{1}+3,\bar{5}+3,\bar{6}+3)_D$
Vertex type 21.1	orientations: 120	frequency: $2/\tau^{13}$	
$(\bar{1}+\bar{4},\bar{1}+6,\bar{4}+6)_G$	$(\bar{1}+\bar{3},\bar{1}+6,\bar{3}+6)_G$	$(\bar{1}+3,\bar{1}+\bar{6},3+\bar{6})_G$	$(\bar{1}+3,\bar{1}+\bar{5},3+\bar{5})_G$
$(\bar{3}+5,\bar{3}+6,5+6)_F$	$(\bar{2}+\bar{3},\bar{2}+5,\bar{3}+5)_F$	$(\bar{2}+4,\bar{2}+\bar{6},4+\bar{6})_F$	$(\bar{2}+4,\bar{2}+5,4+5)_F$
$(\bar{1}+\bar{4},\bar{1}+\bar{5},\bar{4}+\bar{5})_F$	$(\bar{1}+\bar{2},\bar{1}+\bar{6},\bar{2}+\bar{6})_F$	$(\bar{1}+\bar{2},\bar{1}+\bar{3},\bar{2}+\bar{3})_F$	$(1+2,1+3,2+3)_F$
$(1+2,1+6,2+6)_F$	$(1+3,1+4,3+4)_F$	$(1+4,1+5,4+5)_F$	$(1+5,1+6,5+6)_F$
$(2+\bar{4},2+\bar{5},\bar{4}+\bar{5})_F$	$(2+\bar{4},2+6,\bar{4}+6)_F$	$(2+3,2+\bar{5},3+\bar{5})_F$	$(3+4,3+\bar{6},4+\bar{6})_F$
$(3+1,4+1,6+1)_A$	$(\bar{2}+\bar{1},\bar{4}+\bar{1},\bar{5}+\bar{1})_A$	$(\bar{1}+\bar{5},2+\bar{5},3+\bar{5})_A$	$(\bar{1}+\bar{6},3+\bar{6},4+\bar{6})_A$
$(\bar{1}+\bar{2},4+\bar{2},5+\bar{2})_A$	$(\bar{1}+\bar{3},5+\bar{3},6+\bar{3})_A$	$(\bar{1}+\bar{4},2+\bar{4},6+\bar{4})_A$	$(\bar{2}+5,4+5,6+5)_A$
$(2+6,\bar{4}+6,5+6)_A$	$(3+2,\bar{5}+2,6+2)_A$	$(2+3,4+3,\bar{5}+3)_A$	$(\bar{2}+4,3+4,5+4)_A$
$(4+1,5+1,6+1)_C$	$(2+1,3+1,6+1)_C$	$(\bar{2}+\bar{1},\bar{5}+\bar{1},\bar{6}+\bar{1})_C$	$(\bar{2}+\bar{1},\bar{3}+\bar{1},\bar{4}+\bar{1})_C$
$(\bar{1}+\bar{5},2+\bar{5},\bar{4}+\bar{5})_C$	$(\bar{1}+\bar{6},\bar{2}+\bar{6},4+\bar{6})_C$	$(\bar{1}+\bar{2},\bar{3}+\bar{2},5+\bar{2})_C$	$(\bar{1}+\bar{2},4+\bar{2},\bar{6}+\bar{2})_C$
$(\bar{1}+\bar{3},\bar{2}+\bar{3},5+\bar{3})_C$	$(\bar{1}+\bar{4},2+\bar{4},\bar{5}+\bar{4})_C$	$(1+5,4+5,6+5)_C$	$(\bar{2}+5,\bar{3}+5,6+5)_C$
$(1+6,2+6,5+6)_C$	$(\bar{3}+6,\bar{4}+6,5+6)_C$	$(1+2,3+2,6+2)_C$	$(\bar{4}+2,\bar{5}+2,6+2)_C$
$(1+3,2+3,4+3)_C$	$(4+3,\bar{5}+3,\bar{6}+3)_C$	$(1+4,3+4,5+4)_C$	$(\bar{2}+4,3+4,\bar{6}+4)_C$
$(\bar{3}+\bar{1},\bar{4}+\bar{1},6+\bar{1})_D$	$(3+\bar{1},\bar{5}+\bar{1},\bar{6}+\bar{1})_D$	$(\bar{1}+6,\bar{3}+6,\bar{4}+6)_D$	$(\bar{1}+3,\bar{5}+3,\bar{6}+3)_D$
Vertex type 22.1	orientations: 120	frequency: $1/\tau^{15}$	
$(\bar{1}+\bar{4},\bar{1}+6,\bar{4}+6)_G$	$(\bar{1}+\bar{3},\bar{1}+5,\bar{3}+5)_G$	$(\bar{1}+\bar{3},\bar{1}+6,\bar{3}+6)_G$	$(\bar{1}+\bar{2},\bar{1}+4,\bar{2}+4)_G$
$(\bar{1}+\bar{2},\bar{1}+5,\bar{2}+5)_G$	$(\bar{1}+3,\bar{1}+\bar{6},3+\bar{6})_G$	$(\bar{1}+3,\bar{1}+\bar{5},3+\bar{5})_G$	$(\bar{1}+4,\bar{1}+\bar{6},4+\bar{6})_G$
$(\bar{3}+5,\bar{3}+6,5+6)_F$	$(\bar{2}+4,\bar{2}+5,4+5)_F$	$(\bar{1}+\bar{4},\bar{1}+\bar{5},\bar{4}+\bar{5})_F$	$(1+2,1+3,2+3)_F$
$(1+2,1+6,2+6)_F$	$(1+3,1+4,3+4)_F$	$(1+4,1+5,4+5)_F$	$(1+5,1+6,5+6)_F$
$(2+\bar{4},2+\bar{5},\bar{4}+\bar{5})_F$	$(2+\bar{4},2+6,\bar{4}+6)_F$	$(2+3,2+\bar{5},3+\bar{5})_F$	$(3+4,3+\bar{6},4+\bar{6})_F$
$(3+1,4+1,6+1)_A$	$(\bar{2}+\bar{1},\bar{4}+\bar{1},\bar{5}+\bar{1})_A$	$(\bar{1}+\bar{5},2+\bar{5},3+\bar{5})_A$	$(\bar{1}+\bar{6},3+\bar{6},4+\bar{6})_A$
$(\bar{1}+\bar{2},4+\bar{2},5+\bar{2})_A$	$(\bar{1}+\bar{3},5+\bar{3},6+\bar{3})_A$	$(\bar{1}+\bar{4},2+\bar{4},6+\bar{4})_A$	$(\bar{2}+5,4+5,6+5)_A$
$(2+6,\bar{4}+6,5+6)_A$	$(3+2,\bar{5}+2,6+2)_A$	$(2+3,4+3,\bar{5}+3)_A$	$(\bar{2}+4,3+4,5+4)_A$
$(4+1,5+1,6+1)_C$	$(2+1,3+1,6+1)_C$	$(\bar{2}+\bar{1},\bar{5}+\bar{1},\bar{6}+\bar{1})_C$	$(\bar{2}+\bar{1},\bar{3}+\bar{1},\bar{4}+\bar{1})_C$
$(\bar{1}+\bar{5},2+\bar{5},\bar{4}+\bar{5})_C$	$(\bar{1}+\bar{4},2+\bar{4},\bar{5}+\bar{4})_C$	$(1+5,4+5,6+5)_C$	$(\bar{2}+5,\bar{3}+5,6+5)_C$
$(1+6,2+6,5+6)_C$	$(\bar{3}+6,\bar{4}+6,5+6)_C$	$(1+2,3+2,6+2)_C$	$(\bar{4}+2,\bar{5}+2,6+2)_C$
$(1+3,2+3,4+3)_C$	$(4+3,\bar{5}+3,\bar{6}+3)_C$	$(1+4,3+4,5+4)_C$	$(\bar{2}+4,3+4,\bar{6}+4)_C$
$(\bar{2}+\bar{1},\bar{3}+\bar{1},5+\bar{1})_D$	$(\bar{3}+\bar{1},\bar{4}+\bar{1},6+\bar{1})_D$	$(3+\bar{1},\bar{5}+\bar{1},\bar{6}+\bar{1})_D$	$(\bar{2}+\bar{1},4+\bar{1},\bar{6}+\bar{1})_D$
$(\bar{1}+5,\bar{2}+5,\bar{3}+5)_D$	$(\bar{1}+6,\bar{3}+6,\bar{4}+6)_D$	$(\bar{1}+3,\bar{5}+3,\bar{6}+3)_D$	$(\bar{1}+4,\bar{2}+4,\bar{6}+4)_D$
Vertex type 23.1	orientations: 120	frequency: $1/\tau^{12}$	
$(\bar{1}+\bar{4},\bar{1}+6,\bar{4}+6)_G$	$(\bar{1}+\bar{3},\bar{1}+5,\bar{3}+5)_G$	$(\bar{1}+\bar{3},\bar{1}+6,\bar{3}+6)_G$	$(\bar{1}+\bar{2},\bar{1}+4,\bar{2}+4)_G$
$(\bar{1}+\bar{2},\bar{1}+5,\bar{2}+5)_G$	$(\bar{1}+2,\bar{1}+\bar{5},2+\bar{5})_G$	$(\bar{1}+2,\bar{1}+\bar{4},2+\bar{4})_G$	$(\bar{1}+3,\bar{1}+\bar{6},3+\bar{6})_G$
$(\bar{1}+3,\bar{1}+\bar{5},3+\bar{5})_G$	$(\bar{1}+4,\bar{1}+\bar{6},4+\bar{6})_G$	$(\bar{3}+5,\bar{3}+6,5+6)_F$	$(\bar{2}+4,\bar{2}+5,4+5)_F$
$(1+2,1+3,2+3)_F$	$(1+2,1+6,2+6)_F$	$(1+3,1+4,3+4)_F$	$(1+4,1+5,4+5)_F$

Table 5. Cont.

$(1+5,1+6,5+6)_F$	$(2+\bar{4},2+6,\bar{4}+6)_F$	$(2+3,2+\bar{5},3+\bar{5})_F$	$(3+4,3+\bar{6},4+\bar{6})_F$
$(2+1,3+1,5+1)_A$	$(\bar{3}+\bar{1},\bar{4}+\bar{1},\bar{6}+\bar{1})_A$	$(\bar{1}+\bar{5},2+\bar{5},3+\bar{5})_A$	$(\bar{1}+\bar{6},3+\bar{6},4+\bar{6})_A$
$(\bar{1}+\bar{2},4+\bar{2},5+\bar{2})_A$	$(\bar{1}+\bar{3},5+\bar{3},6+\bar{3})_A$	$(\bar{1}+\bar{4},2+\bar{4},6+\bar{4})_A$	$(\bar{3}+5,4+5,6+5)_A$
$(2+6,\bar{4}+6,5+6)_A$	$(3+2,\bar{4}+2,6+2)_A$	$(2+3,4+3,\bar{6}+3)_A$	$(3+4,5+4,\bar{6}+4)_A$
$(2+1,5+1,6+1)_C$	$(3+1,4+1,5+1)_C$	$(\bar{4}+\bar{1},\bar{5}+\bar{1},\bar{6}+\bar{1})_C$	$(\bar{2}+\bar{1},\bar{3}+\bar{1},\bar{6}+\bar{1})_C$
$(1+5,4+5,6+5)_C$	$(\bar{2}+5,\bar{3}+5,4+5)_C$	$(1+6,2+6,5+6)_C$	$(\bar{3}+6,\bar{4}+6,5+6)_C$
$(1+2,3+2,6+2)_C$	$(3+2,\bar{4}+2,\bar{5}+2)_C$	$(1+3,2+3,4+3)_C$	$(2+3,\bar{5}+3,\bar{6}+3)_C$
$(1+4,3+4,5+4)_C$	$(\bar{2}+4,5+4,\bar{6}+4)_C$	$(\bar{2}+\bar{1},\bar{3}+\bar{1},5+\bar{1})_D$	$(\bar{3}+\bar{1},\bar{4}+\bar{1},6+\bar{1})_D$
$(2+\bar{1},\bar{4}+\bar{1},\bar{5}+\bar{1})_D$	$(3+\bar{1},\bar{5}+\bar{1},\bar{6}+\bar{1})_D$	$(\bar{2}+\bar{1},4+\bar{1},\bar{6}+\bar{1})_D$	$(\bar{1}+5,\bar{2}+5,\bar{3}+5)_D$
$(\bar{1}+6,\bar{3}+6,\bar{4}+6)_D$	$(\bar{1}+2,\bar{4}+2,\bar{5}+2)_D$	$(\bar{1}+3,\bar{5}+3,\bar{6}+3)_D$	$(\bar{1}+4,\bar{2}+4,\bar{6}+4)_D$
Vertex type 24.1	orientations: 120	frequency: $1/\tau^9$	
$(\bar{3}+\bar{4},\bar{3}+6,\bar{4}+6)_F$	$(\bar{3}+5,\bar{3}+6,5+6)_F$	$(\bar{2}+\bar{3},\bar{2}+5,\bar{3}+5)_F$	$(\bar{2}+4,\bar{2}+\bar{6},4+\bar{6})_F$
$(\bar{2}+4,\bar{2}+5,4+5)_F$	$(\bar{1}+\bar{5},\bar{1}+\bar{6},\bar{5}+\bar{6})_F$	$(\bar{1}+\bar{4},\bar{1}+\bar{5},\bar{4}+\bar{5})_F$	$(\bar{1}+\bar{3},\bar{1}+\bar{4},\bar{3}+\bar{4})_F$
$(\bar{1}+\bar{2},\bar{1}+\bar{6},\bar{2}+\bar{6})_F$	$(\bar{1}+\bar{2},\bar{1}+\bar{3},\bar{2}+\bar{3})_F$	$(1+2,1+3,2+3)_F$	$(1+2,1+6,2+6)_F$
$(1+3,1+4,3+4)_F$	$(1+4,1+5,4+5)_F$	$(1+5,1+6,5+6)_F$	$(2+\bar{4},2+\bar{5},\bar{4}+\bar{5})_F$
$(2+\bar{4},2+6,\bar{4}+6)_F$	$(2+3,2+\bar{5},3+\bar{5})_F$	$(3+\bar{5},3+\bar{6},\bar{5}+\bar{6})_F$	$(3+4,3+\bar{6},4+\bar{6})_F$
$(2+1,4+1,6+1)_A$	$(\bar{3}+\bar{1},\bar{5}+\bar{1},\bar{6}+\bar{1})_A$	$(\bar{1}+\bar{5},2+\bar{5},\bar{6}+\bar{5})_A$	$(\bar{1}+\bar{6},4+\bar{6},\bar{5}+\bar{6})_A$
$(\bar{3}+\bar{2},4+\bar{2},\bar{6}+\bar{2})_A$	$(\bar{1}+\bar{3},5+\bar{3},6+\bar{3})_A$	$(2+\bar{4},\bar{3}+\bar{4},\bar{5}+\bar{4})_A$	$(1+5,\bar{3}+5,4+5)_A$
$(1+6,2+6,\bar{3}+6)_A$	$(1+2,\bar{5}+2,6+2)_A$	$(2+3,4+3,\bar{5}+3)_A$	$(1+4,5+4,\bar{6}+4)_A$
$(4+1,5+1,6+1)_C$	$(2+1,3+1,4+1)_C$	$(\bar{2}+\bar{1},\bar{3}+\bar{1},\bar{6}+\bar{1})_C$	$(\bar{3}+\bar{1},\bar{4}+\bar{1},\bar{5}+\bar{1})_C$
$(2+\bar{5},3+\bar{5},\bar{6}+\bar{5})_C$	$(\bar{1}+\bar{5},2+\bar{5},\bar{4}+\bar{5})_C$	$(\bar{1}+\bar{6},\bar{2}+\bar{6},4+\bar{6})_C$	$(3+\bar{6},4+\bar{6},\bar{5}+\bar{6})_C$
$(\bar{1}+\bar{2},\bar{3}+\bar{2},\bar{6}+\bar{2})_C$	$(\bar{3}+\bar{2},4+\bar{2},5+\bar{2})_C$	$(\bar{1}+\bar{3},\bar{4}+\bar{3},6+\bar{3})_C$	$(\bar{1}+\bar{3},\bar{2}+\bar{3},5+\bar{3})_C$
$(\bar{1}+\bar{4},\bar{3}+\bar{4},\bar{5}+\bar{4})_C$	$(2+\bar{4},\bar{3}+\bar{4},6+\bar{4})_C$	$(1+5,\bar{3}+5,6+5)_C$	$(\bar{2}+5,\bar{3}+5,4+5)_C$
$(2+6,\bar{3}+6,\bar{4}+6)_C$	$(1+6,\bar{3}+6,5+6)_C$	$(1+2,3+2,\bar{5}+2)_C$	$(\bar{4}+2,\bar{5}+2,6+2)_C$
$(1+3,2+3,4+3)_C$	$(4+3,\bar{5}+3,\bar{6}+3)_C$	$(\bar{2}+4,5+4,\bar{6}+4)_C$	$(1+4,3+4,\bar{6}+4)_C$

MOLECULAR SYMMETRY ADAPTED BASES IN THE BORN-OPPENHEIMER APPROXIMATION +

R. Lemus and A. Frank

Instituto de Ciencias Nucleares, UNAM
Apartado Postal 70-543
México, D.F., 04510 México

We present a procedure to obtain molecular symmetry adapted bases in the Born-Oppenheimer approximation within the framework of the second quantization picture, by constructing a suitable group chain which either carries the irreducible representations of the point symmetry group of the molecule or those of a subgroup. In the latter case an induction procedure is required to establish the basis.

1. INTRODUCTION

Adaptation of basis to the point symmetry group of a molecule has proved to be very useful. Since its first applications,[1,2] symmetry adaptation has been increasingly employed and at the present time it is a very well known tool in Quantum Chemistry. Most of the applications, however, have been restricted to configuration space.

Later on, the group theoretical point of view was combined with the second quantization formalism,[3,4] to produce an approach which allows to express the Hamiltonian and other operators in terms of the generators of a chain of groups and which provides an efficient procedure for the evaluation of matrix elements. Special care should be taken in the case of molecules, however, since in this case, because of the many center nature of the system, the set of single particle states (atomic orbitals) is not orthogonal. When dealing with nonorthogonal bases, the creation and annihilation operators no longer satisfy the usual anticommutation relations[5] and it is not possible to apply the usual group theoretical procedures.[1-4] There are two possible ways to surmount this problem. One possibility is to orthonormalize the states from the beginning with the help of Schmidt's or Löwdins'[6] procedures and then use the techniques developed for unitary groups.[7,8] The second alternative, suggested by Moshinsky,[5] deals with sets of nonorthogonal states by means of the introduction of a dual basis. In the latter case the general linear groups must be employed.[5]

Since for molecular systems (in the Born-Oppenheimer approximation) the angular momentum is not conserved, the three-dimensional orthogonal group does not appear as a subgroup. Thus, the canonical chain is suitable for the classification procedure of the basis states, whatever the method involved. The symmetry adapted basis must be obtained by projection techniques, since the canonical chain does not contain any information on the point symmetry of the molecule.

In this paper we present an approach based on the construction of a suitable chain of groups, which already contains information on the discrete symmetry of the

+This work was supported in part by DGAPA-UNAM under project IN-101889

molelcule. We start our discussion by introducing the set of orthogonal single particle states in configuration space.

2. ORTHOGONAL SINGLE PARTICLE STATES

In the separated atoms limit the basis which spans the molecular states corresponds to the atomic orbitals. We may characterize these atomic spin-orbitals by

$$\Phi^{\rho,i}_{nlm\sigma}(\mathbf{r},s)=\phi^{\rho,i}_{nlm}(\mathbf{r})\delta_{\sigma s}\ , \tag{2.1}$$

where $\mathbf{r}$ is the position vector and s is the spin variable of the particle. The indices $\{nlm\}$ and σ characterize the orbital and spin part of the single particle states, respectively, while the superindex ρ characterizes the set $\{r_1, r_2, \ldots, r_i, \ldots, r_t\}$ of equivalent atoms.

The set of atomic orbitals (2.1) constitutes a non-orthogonal basis. In order to obtain an orthogonal set we follow a procedure consisting of two steps. We first project over the atomic orbitals to obtain functions which carry irreducible representations of the point group $\mathcal{G}$. We then orthogonalize the functions belonging to the same irreducible representation, when necessary.

The action of an operator $\hat{O}_R$, associated to an element $R \in \mathcal{G}$, has the following effect over the set (2.1):

$$\hat{O}_R\Phi^{\rho,i}_{nlm\sigma}(\mathbf{r},s)=\sum_{m',j}\Phi^{\rho,j}_{nlm'\sigma}(\mathbf{r},s)\Delta^{(l)}(R)^{ji}_{m'm}\ , \tag{2.2}$$

where $\Delta^{(l)}(R)$ is a reducible representation of $\mathcal{G}$ given by the direct product

$$\Delta^{(l)}(R)=\delta(R)\otimes\mathbf{D}^{(l)}(R)\ ,$$

where

$$\delta(R)=\begin{cases}1 & \text{if}\quad \hat{O}_R\mathbf{R}_i=\mathbf{R}_j\ ,\\ 0 & \text{if}\quad \hat{O}_R\mathbf{R}_i\neq\mathbf{R}_j\ .\end{cases}$$

The parameters $\mathbf{R}_r$ are the position vectors of the nuclei, while the matrix $\mathbf{D}^l(R)$ corresponds to an irreducible representation of the three-dimensional orthogonal group.

The representation $\Delta(R)$ can be reduced by a similarity transformation which brings the matrices $\Delta(R)$ into the direct sum

$$\mathbf{S}\Delta^{(l)}(R)\mathbf{S}^{-1}=\sum_{\Gamma}a_\Gamma D^\Gamma(R);\quad a_\Gamma\quad \text{integers}\ , \tag{2.3}$$

where the $\mathbf{D}^\Gamma(R)$ are unitary irreducible representations. One can then construct, by means of $\mathbf{S}$, linear combinations of the atomic orbitals (2.1) which transform under $\hat{O}_R$ according to (2.3) and which belong to the various rows of the irreducible representations $\mathbf{D}^{\Gamma_1}(R)$, $\mathbf{D}^{\Gamma_2}(R)$,.... Explicitly, these linear combinations are given by

$${}_{a_\Gamma}F^{\Gamma\gamma}_{\rho nl\sigma}(\mathbf{r},s)=\sum_{im}\mathbf{S}^{a_\Gamma\Gamma\gamma}_{\rho rlnm}\Phi^{\rho,i}_{nlm\sigma}(\mathbf{r},s)\ , \tag{2.4}$$

where the left subindex a_Γ takes into account the possible repetition of irreducible representations in the reduction (2.3). The functions ${}_{a_\Gamma}F^{\Gamma\gamma}_{\rho nl\sigma}(\mathbf{r},s)$ form a basis for the Γ-th irreducible representation of the point group $\mathcal{G}$. Consequently

$$\hat{O}_R\ \ {}_{a_\Gamma}F^{\Gamma\gamma}_{\rho nl\sigma}(\mathbf{r},s)=\sum_{\gamma'}\mathbf{D}^\Gamma_{\gamma'\gamma}(R)\,{}_{a_\Gamma}F^{\Gamma\gamma'}_{\rho nl\sigma}(\mathbf{r},s)\ . \tag{2.5}$$

Basis functions belonging to different irreducible representations are orthogonal to each other, so we only need to orthogonalize the $a_\Gamma = 1, 2, \ldots$ functions belonging to the same irreducible representations Γ and γ. The orthonormalization can be achieved by means of Löwdins' procedure.[6] Its application provides us with an orthonormalized set of single particle states, which carry irreducible representations of the point group. In terms of the basis functions (2.4), these single particle states are given by

$$_\alpha\Psi^\Gamma_{\gamma\sigma}(\mathbf{r}, s) = \sum_{a_\Gamma, \rho nl} C^{a_\Gamma \Gamma\gamma}_{\rho nl} \; {}_{a_\Gamma}F^{\Gamma\gamma}_{\rho nl\sigma}(\mathbf{r}, s) = {}_\alpha \Psi^\Gamma_\gamma(\mathbf{r})\delta_{s\sigma} \quad , \tag{2.6}$$

where the C are orthonormalization coefficients.

The form of the basis functions (2.4) depends on the representation employed to carry out the projection. In order to fix this representation, we assume (with the exception of linear molecules, where $\mathcal{G} = 0(2)$ or $0(2) \otimes C_i$) that the quantum numbers Γ and γ are associated with the group chain

$$\begin{array}{ccc} \mathcal{G} & \supset & C_n \\ \downarrow & & \downarrow \\ |\Gamma\;, & & \gamma>\;, \end{array} \tag{2.7}$$

where C_n is an axial subgroup of $\mathcal{G}$, the generator of which corresponds to a pure rotation C_n of maximum order.

In the next section we transform the basis (2.6) to the second quantization scheme and establish the Hamiltonian.

3. SECOND QUANTIZATION FORMALISM

Once the set of single particle states has been determined we introduce anticommuting creation operators $a^\dagger$[3]

$$_\alpha\Psi^\Gamma_{\gamma\sigma}(\mathbf{r}, s) \rightarrow a^\dagger_{\alpha\Gamma\gamma\sigma}|0> \;, \tag{3.1}$$

where $|0>$ is the vacumm state. The Slater determinant n-particle state is then given by

$$|\alpha_1\Gamma_1\gamma_1\sigma_1, \ldots, \alpha_n\Gamma_n\gamma_n\sigma_n >= a^\dagger_{\alpha_1\Gamma_1\gamma_1\sigma_1} \cdots a^\dagger_{\alpha_n\Gamma_n\gamma_n\sigma_n}|0> \;. \tag{3.2}$$

In addition to the usual annihilation operators $a_{\alpha\Gamma\gamma\sigma}$, we define the covariant operators

$$\tilde{a}_{\alpha\Gamma\gamma\sigma} = (-)^{\frac{1}{2}-\sigma} a_{\alpha,\Gamma,-\gamma,-\sigma} \;, \tag{3.3}$$

which transform under $SU(2)$ and $\mathcal{G}$ in the same way as the $a^\dagger_{\alpha\Gamma\gamma\sigma}$. The minus sign in γ denotes the conjugate representation of C_n.

We next turn our attention to the Hamiltonian. The nonrelativistic electronic Hamiltonian for a molecule with N nuclei and n electrons in the Born-Oppenheimer approximation (in atomic units) is given by

$$\hat{H} = \sum_{i=1}^{n} W(i) + \sum_{i>j}^{n} V(i, j) \quad , \tag{3.4}$$

where the one and two body operators are, respectively,

$$W(i) = -\frac{1}{2}\nabla^2_i - \sum_{\alpha=1}^{N} \frac{Z_\alpha}{|\mathbf{r}_i - \mathbf{R}_\alpha|} \;,$$

and

$$V(i,j) = \frac{1}{|\mathbf{r}_i - \mathbf{r}_j|} ,$$

where Z_α and $\mathbf{R}_\alpha$ are the charges and vector positions of the nuclei. The $\mathbf{r}_i$ are the vector positions of the electrons refered to the nuclei' center of mass.

In the second quantization scheme the Hamiltonian (3.4), refered to the basis (3.1), takes the form

$$\hat{\mathcal{H}} = \hat{\mathcal{W}} + \hat{\mathcal{V}} , \tag{3.5}$$

where

$$\hat{\mathcal{W}} = \sum_{\alpha\Gamma\gamma}\sum_{\alpha'\Gamma'\gamma'} < {}_\alpha\Psi_\gamma^\Gamma(1)|W(1)|{}_{\alpha'}\Psi_{\gamma'}^{\Gamma'}(1) > \sum_\sigma a^\dagger_{\alpha\Gamma\gamma\sigma} a_{\alpha'\Gamma'\gamma'\sigma} , \tag{3.6}$$

and

$$\hat{\mathcal{V}} = \sum_{\alpha_1\Gamma_1\gamma_1}\sum_{\alpha_2\Gamma_2\gamma_2}\sum_{\alpha_1'\Gamma_1'\gamma_1'}\sum_{\alpha_2'\Gamma_2'\gamma_2'} < {}_{\alpha_1}\Psi_{\gamma_1}^{\Gamma_1}(1){}_{\alpha_2}\Psi_{\gamma_2}^{\Gamma_2}(2)|V(1,2)|{}_{\alpha_1'}\Psi_{\gamma_1'}^{\Gamma_1'}(1){}_{\alpha_2'}\Psi_{\gamma_2'}^{\Gamma_2'}(2) >$$
$$\times \sum_{\sigma_1}\sum_{\sigma_2} a^\dagger_{\alpha_1\Gamma_1\gamma_1\sigma_1} a^\dagger_{\alpha_2\Gamma_2\gamma_2\sigma_2} a_{\alpha_1'\Gamma_1'\gamma_1'\sigma_1} a_{\alpha_2'\Gamma_2'\gamma_2'\sigma_2} . \tag{3.7}$$

In (3.6) and (3.7) we have taken into account that the Hamiltonian (3.4) is spin-independent.

The matrix elements of $W(1)$ and $V(1,2)$ are different from zero only if the integrands belong to the totally symmetric irreducible representation A_1 of $\mathcal{G}$. Since $\mathcal{H}$ is, by definition, invariant under $\mathcal{G}$, the product of wave functions has to be transformed to an A_1 coupling. We thus rewrite $\hat{\mathcal{H}}$ in terms of the coupled basis Ψ_l^Γ whose relation with the ${}_\alpha\Psi_\gamma^\Gamma$ is given by

$${}_{\alpha_1}\Psi_{\gamma_1}^{\Gamma_1}\ {}_{\alpha_2}\Psi_{\gamma_2}^{\Gamma_2} = \sum_{\Gamma\gamma} C(\Gamma_1\Gamma_2\Gamma;\gamma_1\gamma_2 l)({}_{\alpha_1}\Psi^{\Gamma_1} \times {}_{\alpha_2}\Psi^{\Gamma_2})_l^\Gamma , \tag{3.8}$$

where the $C(;)$ are Clebsch-Gordan coefficients associated to the group chain (2.7).

Introduction of the coupled basis (3.8) in the one and two body operators (3.6) and (3.7), leads to the expressions

$$\hat{\mathcal{W}} = \sum_\alpha\sum_\Gamma \int [{}_\alpha\Psi^{\Gamma^*}(1) \times W(1){}_\alpha\Psi^\Gamma(1)]^{A_1} d\mathbf{r}_1 (a^\dagger_{\alpha\Gamma} \times \tilde{a}_{\alpha\Gamma})^{(A_1,0)} , \tag{3.9}$$

$$\hat{\mathcal{V}} = \frac{1}{2}\sum_{\alpha_1\Gamma_1}\sum_{\alpha_2\Gamma_2}\sum_{\alpha_1'\Gamma_1'}\sum_{\alpha_2'\Gamma_2'}\sum_{\Gamma'\Gamma''}\int\int [({}_{\alpha_1}\Psi^{\Gamma_1^*}(1) \times {}_{\alpha_2}\Psi^{\Gamma_2^*}(2))^{\Gamma'}$$
$$\times V(1,2)({}_{\alpha_1'}\Psi^{\Gamma_1'}(1) \times {}_{\alpha_2'}\Psi^{\Gamma_2'}(2))^{\Gamma''}]^{A_1} d\mathbf{r}_1 d\mathbf{r}_2$$
$$\{[(a^\dagger_{\alpha_1\Gamma_1} \times a^\dagger_{\alpha_2\Gamma_2})^{(\Gamma',0)} \times (\tilde{a}_{\alpha_1'\Gamma_1'} \times \tilde{a}_{\alpha_1'\Gamma_1'})^{(\Gamma'',0)}]^{(A_1,0)}$$
$$+ \sqrt{3}[(a^\dagger_{\alpha_1\Gamma_1} \times a^\dagger_{\alpha_2\Gamma_2})^{(\Gamma',1)} \times (\tilde{a}_{\alpha_1'\Gamma_1'} \times \tilde{a}_{\alpha_2'\Gamma_2'})^{(\Gamma'',1)}]^{(A_1,0)}\} , \tag{3.10}$$

where the notation $(a^\dagger_{\alpha_1\Gamma_1} \times a^\dagger_{\alpha_2\Gamma_2})^{(\Gamma,S)}$ indicates $\mathcal{G}$-coupling to Γ and spin angular momentum coupling to s.

Having established the general form of the Hamiltonian, we need to construct an appropriate basis to carry out its diagonalization.

4. SYMMETRY ADAPTED BASIS STATES

The set of creation $a^{\dagger}_{\alpha\Gamma\gamma\sigma}$ and annihilation $a_{\alpha\Gamma\gamma\sigma}$ operators satisfy the familiar anticommutation relations for fermions

$$\{a^{\dagger}_{\alpha\Gamma\gamma\sigma}, a^{\dagger}_{\alpha'\Gamma'\gamma'\sigma'}\} = \{a_{\alpha\Gamma\gamma\sigma}, a_{\alpha'\Gamma'\gamma'\sigma'}\} = 0,$$
$$\{a^{\dagger}_{\alpha\Gamma\gamma\sigma}, a_{\alpha'\Gamma'\gamma'\sigma'}\} = \delta_{\alpha\alpha'}\delta_{\Gamma\Gamma'}\delta_{\gamma\gamma'}\delta_{\sigma\sigma'} \ , \tag{4.1}$$

while the operators

$$C^{\alpha'\Gamma'\gamma'\sigma'}_{\alpha\Gamma\gamma\sigma} \equiv a^{\dagger}_{\alpha\Gamma\gamma\sigma} a_{\alpha'\Gamma'\gamma'\sigma'} \ , \tag{4.2}$$

have de commutation relation[3]

$$[C^{\alpha_1'\Gamma_1'\gamma_1'\sigma_1'}_{\alpha_1\Gamma_1\gamma_1\sigma_1}, \ C^{\alpha_2'\Gamma_2'\gamma_2'\sigma_2'}_{\alpha_2\Gamma_2\gamma_2\sigma_2}] = C^{\alpha_2'\Gamma_2'\gamma_2'\sigma_2'}_{\alpha_1\Gamma_1\gamma_1\sigma_1}\delta_{\alpha_2\alpha_1'}\delta_{\Gamma_2\Gamma_1'}\delta_{\gamma_2\gamma_1'}\delta_{\sigma_2\sigma_1'}$$
$$- C^{\alpha_1'\Gamma_1'\gamma_1'\sigma_1'}_{\alpha_2\Gamma_2\gamma_2\sigma_2}\delta_{\alpha_1\alpha_2'}\delta_{\Gamma_1\Gamma_2'}\delta_{\gamma_1\gamma_2'}\delta_{\sigma_1\sigma_2'} \ , \tag{4.3}$$

and therefore are generators of a unitary group $U(2N)$ where N corresponds to the number of wave functions ${}_{\alpha}\Psi^{\Gamma}_{\gamma}$ and the factor 2 accounts for the spin of the electron.

Since the Hamiltonian (3.5) and any other physical operator can be written in terms of these generators, the unitary group $U(2N)$ is the dynamical group for the system.

The usual procedure to construct a basis involves decomposing the dynamical algebra of the system into its subalgebras. Associated to each possible decomposition there is a corresponding basis. Since we are interested in characterizing the states with irreducible representations of $\mathcal{G}$, the decomposition must include $\mathcal{G}$ as a subgroup.

We may construct generators of subgroups of $U(2N)$ if we carry out contractions either with respect to σ or $\{\alpha\Gamma\gamma\}$. In the first case we obtain the N^2 operators

$$C^{\alpha'\Gamma'\gamma'}_{\alpha\Gamma\gamma} \equiv \sum_{\sigma} C^{\alpha'\Gamma'\gamma'\sigma}_{\alpha\Gamma\gamma\sigma} \ , \tag{4.4}$$

which are generators of a $U(N)$ group. In the second case we find the four operators

$$\mathrm{C}^{\sigma'}_{\sigma} = \sum_{\alpha\Gamma\gamma} C^{\alpha\Gamma\gamma\sigma'}_{\alpha\Gamma\gamma\sigma} \ , \tag{4.5}$$

which are generators of a $U(2)$ group. Since

$$[C^{\alpha'\Gamma'\gamma'}_{\alpha\Gamma\gamma} \ , \ \mathrm{C}^{\sigma'}_{\sigma}] = 0 \ , \tag{4.6}$$

we have obtained the generators of the chain of groups[3]

$$U(2N) \supset U(N) \times U(2) \ . \tag{4.7}$$

The spin operator $\hat{S}$ is given by

$$\hat{S}_{\Sigma} = \sum_{\sigma\sigma'} < \frac{1}{2}\sigma|S_{\Sigma}|\frac{1}{2}\sigma' > \sum_{\alpha\Gamma\gamma} a^{\dagger}_{\alpha\Gamma\gamma\sigma} a_{\alpha\Gamma\gamma\sigma'} \ , \tag{4.8}$$

and from the explicit matrix elements we get

$$\hat{S}_{+1} = \mathrm{C}^{+\frac{1}{2}}_{-\frac{1}{2}} \, , \, \hat{S}_0 = \frac{1}{2}(\mathrm{C}^{\frac{1}{2}}_{\frac{1}{2}} - \mathrm{C}^{-\frac{1}{2}}_{-\frac{1}{2}}) \, , \, \hat{S}_{-1} = \mathrm{C}^{\frac{1}{2}}_{-\frac{1}{2}} \ ,$$

which are the generators of the unimodular subgroup $SU(2) \subset U(2)$. Instead of (4.7) we may use the chain

$$U(2N) \supset U(N) \times SU(2) \ . \tag{4.9}$$

The further reduction of the $U(N)$ group is not unique, i.e., it depends on the molecule we are interested in as well as the single particle basis included. There are four possibilities, which are:

I.-) The molecular point group is C_1.
In this case the basis functions (2.4) belong to the totally symmetric representation and therefore the canonical chain

$$U(N) \supset U(N-1) \supset \ldots \supset U(1) \ , \tag{4.10}$$

provides an appropriate basis (Gelfands' states) to diagonalize the Hamiltonian (3.5).

II.-) The number of basis functions spanning a given irreducible representation Γ (with maximum value of $\alpha = a$) is the same for all representations Γ of $\mathcal{G}$ to which the basis functions (2.4) belong and the chain contains a symmetric group S_n which is isomorphic to $\mathcal{G}$.
In this case the generators of the group $U(N)$ can be contracted either with respect to $\{\Gamma\gamma\}$ or α. In the former case we obtain the a^2 operators

$$C_{\alpha}^{\alpha'} = \sum_{\Gamma\gamma} C_{\alpha\Gamma\gamma}^{\alpha'\Gamma\gamma} \ , \ \alpha, \alpha' = 1, 2, \ldots, a \quad , \tag{4.11}$$

which generate the unitary group $U^{\alpha}(a)$. In the second, we obtain the b^2 operators

$$C_{\Gamma\gamma}^{\Gamma'\gamma'} = \sum_{\alpha} C_{\alpha\Gamma\gamma}^{\alpha\Gamma'\gamma'} \ , \ \Gamma\gamma = (\Gamma\gamma)_1, (\Gamma\gamma)_2, \ldots, (\Gamma\gamma)_b \quad , \tag{4.12}$$

which constitute the genertors of a $U^{\Gamma}(b)$ group. Since the generators (4.11) and (4.12) commute, we obtain the chain

$$U(N) \supset U^{\alpha}(a) \times U^{\Gamma}(b) \, , \ ab = N \ . \tag{4.13}$$

Since α is a multiplicity index, we can define for the group $U^{\alpha}(a)$ the canonical chain

$$U^{\alpha}(a) \supset U^{\alpha}(a-1) \supset \ldots \supset U^{\alpha}(1) \ . \tag{4.14}$$

On the other hand, the group $U^{\Gamma}(b)$ contains the information on the ireducible representations of $\mathcal{G}$. We expect to obtain from this group an orthogonal subgroup $0(n-1)$, which contains the symmetric group S_n, isomorphic to $\mathcal{G}$. Before doing this we note that single particle states belonging to A_1 of $\mathcal{G}$ do not affect the many body state, since

$$A_1 \otimes \Gamma = \Gamma \quad \text{for all } \Gamma \ .$$

This fact suggest that we can exclude, if necessary, the A_1 irreducible representation from the set (4.12), to obtain the generators of $U^{\Gamma}(b-1)$

$$C_{\Gamma\gamma}^{\Gamma'\gamma'} = \sum_{\alpha}{}' C_{\alpha\Gamma\gamma}^{\alpha\Gamma'\gamma'} \ , \quad \text{except } A_1 \ . \tag{4.15}$$

The inclusion of the subgroup $U^{\Gamma}(b-1)$ depends on the dimension $n-1$ of the orthogonal group contained in the chain. The generators of the orthogonal group are obtained by taking the antisymmetric part of (4.12) or (4.15):

$$\Lambda_{\Gamma\gamma}^{\Gamma'\gamma'} = \frac{1}{2}(C_{\Gamma\gamma}^{\Gamma'-\gamma'} - C_{\Gamma'\gamma'}^{\Gamma-\gamma}) \ . \tag{4.16}$$

We thus obtain the full chain

$$
\begin{array}{ccccc}
U(N) \supset & U^{\alpha}(a) & \times & U^{\Gamma}(b) & \\
& \cup & & \cup & \\
& U^{\alpha}(a-1) & & U^{\Gamma}(b-1)\} & \text{if necessary} \\
& \cup & & \cup & \\
& \vdots & & O(n-1) & \\
& \cup & & \cup & \\
& U^{\alpha}(1) & & S_n \approx \mathcal{G} \supset C_n \, . &
\end{array}
\tag{4.17}
$$

III.-) The same conditions as in II, but now the chain contains the symmetric group S_n which is isomorphic to a subgroup of $\mathcal{G}$.

Sometimes the symmetry conditions in II over the multiplicity index α are not fulfiled, unless we restrict ourselves to a subgroup H of $\mathcal{G}$. In this case the associated chain is similar to (14.17), but with

$$S_n \approx H \ , \ H \subset \mathcal{G} \, . \tag{4.18}$$

This chian labels the states according to the irreducible representations of H (instead of $\mathcal{G}$). It is possible, however, to recover the labeling of $\mathcal{G}$ by inducting from H to $\mathcal{G}$.[9] We briefly explain the procedure below.

The group $\mathcal{G}$ can be written in terms of H by means of a left coset expansion

$$\mathcal{G} = \sum_{\sigma}^{|\sigma|} S_\sigma H \ , \ S_\sigma \in \mathcal{G} \ , \ S_\sigma \notin H \, .$$

The set $\{S_\sigma H\}$ of left cosets of H in $\mathcal{G}$ is a basis for a representation of $\mathcal{G}$ called left ground representation of $\mathcal{G}$ by H, ${}^{\gamma}\mathcal{G}_H(g), g \in \mathcal{G}$. Therefore

$$g \ S_\lambda H = \sum_{\sigma}^{|\sigma|} S_\sigma H \ {}^{\gamma}\mathcal{G}(g)_{\sigma\lambda} \, .$$

Given the set of functions $|\alpha_r>, r = 1, 2, \ldots, \mu$, carrying the μ-th irreducible representation $\mathbf{D}^{\mu}(H)$ of H, the space

$$S_\lambda|\alpha_r> \equiv |\alpha_{\lambda r}> \ , \ \lambda = 1, 2, \ldots, \frac{|\mathcal{G}|}{|H|} \ ,$$

is an invariant space of $\mathcal{G}$ and is therefore a basis for a representation of $\mathcal{G}$, given by

$$g|\alpha_{\lambda r}> = \sum_{\sigma t} |\alpha_{\sigma t}> \mathbf{D}^{\mu}(h_\lambda(g))_{tr} \ {}^{\gamma}\mathcal{G}(g)_{\sigma\lambda} \ , \tag{4.19}$$

where $h_\lambda(g)$ is called the subelement of g in H and is defined by

$$gS_\lambda = S_\tau h_\lambda(g) \, .$$

The chain (4.18) provides sets of $|\alpha_r>$ basis functions. Since we know the action of S_τ on the states (3.1), we can construct the basis $|\alpha_{\lambda r}>$. On the other hand, the operators S_τ do not change the multiplicity index nor the label Γ. Thus the set $|\alpha_{\lambda r}>$ corresponds to linear combinations of states associated to the same number of electrons in the unitary groups and to the irreducible representations of H which are consistent with the Frobenius reciprocity theorem.[9] The states $|\alpha_{\lambda r}>$ provide the representations (4.19), which in case of being reducible, can be taken to block diagonal form to obtain the states which span irreducible representations of $\mathcal{G}$. It is worth mentioning that usually the dimension of the representation (4.19) is two.

IV.-) The group $\mathcal{G}$ can be written in terms of a direct product

$$\begin{array}{ccccc} \mathcal{G} & = & G_1 & \otimes & G_2 \\ \downarrow & & \downarrow & & \downarrow \\ |\Gamma\gamma> & = & |\Gamma & \otimes & \gamma>, \end{array} \tag{4.20}$$

where γ not necessarily labels an uniaxial group C_n, and the multiplicity index α has the same values for all γ.

As for case III, the group $U(N)$ can be contracted with respect to γ

$$C_{\alpha\Gamma}^{\alpha'\Gamma'} = \sum_{\gamma} C_{\alpha\Gamma\gamma}^{\alpha'\Gamma'\gamma} \quad \alpha\Gamma = (\alpha\Gamma)_1, \dots, (\alpha\Gamma)_a \tag{4.21}$$

or $\alpha\Gamma$

$$C_{\gamma}^{\gamma'} = \sum_{\alpha\Gamma} C_{\alpha\Gamma\gamma}^{\alpha\Gamma\gamma'}, \quad \gamma, \gamma' = 1, 2, \dots, g_2 \quad , \tag{4.22}$$

where the number of functions belonging to the γ representation should be the same for all γ. In the former case we obtain a^2 operators generating the group $U^{\alpha\Gamma}(a)$. In the latter, we obtain g_2^2 operators which generate the group $U^{\gamma}(g_2)$. Since these generators commute,

$$U(N) \supset U^{\Gamma}(a) \times U^{\gamma}(g_2) \; . \tag{4.23}$$

In analogy to the reduction of $U^{\Gamma}(b)$ in (4.17), both $U^{\Gamma}(a)$ and $U^{\gamma}(g_2)$ should contain orthogonal groups which in turn contain symmetric groups $S_{n_1}^{\Gamma}$ and $S_{n_2}^{\gamma}$, isomorphic to G_1 and G_2, respectively. We then couple $S_{n_1}^{\Gamma}$ and $S_{n_2}^{\gamma}$ to the symmetric group S_n, isomorphic to $\mathcal{G}$. Just as in the case II and III, the totally symmetric representation A_1 may be substracted from $U(c)$, $c = a, g_2$, to obtain a group $U(c-1)$, so the A_1 representations may be excluded from $U^{\Gamma}(a)$ or $U^{\gamma}(g_2)$ to obtain the groups $U^{\Gamma}(a-1)$, $U^{\Gamma}(a-2), \dots$ and $U^{\gamma}(g_2-1)$, $U^{\gamma}(g_2-2), \dots$. The corresponding group chain is thus given by

$$\begin{array}{ccccc} U(N) & \supset & U^{\Gamma}(a) & \times & U^{\gamma}(g_2) \\ & & \cup & & \cup \\ & & U^{\Gamma}(a-1) & & U^{\gamma}(g_2-1) \\ & \text{excluding} & \cup & & \cup \\ & \text{all } A_1 \text{ irred. rep.} & \vdots & & \vdots \\ & & \cup & & \cup \\ & & U^{\gamma}(n_1-1) & & U^{\gamma}(n_2-1) \\ & & \cup & & \cup \\ & & O(n_1-1) & & O(n_2-1) \\ & & \cup & & \cup \\ & & (S_{n_1}^{\Gamma} \approx G_1) & \times & (S_{n_2}^{\gamma} \approx G_2) \\ & & & \| & \\ & & & (S_n \approx \mathcal{G}) \; . & \end{array} \tag{4.24}$$

We have found that there is an alternative chain which is associated to orthogonal groups. Instead of excluding all A_1 irreducible representations we may exclude only one in $U^{\Gamma}(a)$ (and in $U^{\gamma}(g_2)$ if it is possible) and then continue the decomposition with orthogonal groups. We thus have the chain

$$\begin{array}{ccccc} U(N) & \supset & U^{\Gamma}(a) & \times & U^{\gamma}(g_2) \\ & & \cup & & \cup \\ & & U^{\Gamma}(a-1) & & U^{\gamma}(g_2-1)\} \text{ if possible} \\ & & \cup & & \cup \\ & & O(a-1) & & O(g_2-1) \\ & & \cup & & \cup \\ & & \vdots & & \vdots \\ & & \cup & & \cup \\ & & O(n_1-1) & & O(n_2-1) \\ & & \cup & & \cup \\ & & (S^{\Gamma}_{n_1} \approx G_1) & \times & (S^{\gamma}_{n_2} \approx G_2) \\ & & & \| & \\ & & & (S_n \approx \mathcal{G}) \ . & \end{array} \tag{4.25}$$

The possibility of the decomposition $U^{\gamma}(g_2) \supset U^{\gamma}(g_2-1)$ depends on the dimension of the group G_2. The generators of the orthogonal groups are obtained as in (4.16).

In the next section we consider some examples of the four cases analyzed in this section.

5. EXAMPLES

We now present examples of molecules whose valence electrons occupy either the $1s$ and/or the $2sp$ shell. The next table indicates the case which applies for these molecules.

Molecule(s)	Case *I*	Case *II*	Case *III*	Case *IV*	Point Group
$H-A$				×	$C_{\infty v} \approx O(2)$
$A-A$				×	$D_{\infty h} = C_{\infty v} \otimes C_i$
H_2O			×		C_{2v}
$H_3(1s)$		×			D_3
$H_3(1s+2sp)$			×		D_{3h}
H_2O_2		×			C_2
CH_4		×			T_d
$CH\ F\ Cl\ Br$	×				C_1

$$A = O, N, C, B, Be, Li \ .$$

We present details for the H_3 molecule, including its Hamiltonian and the explicit expressions for the wave functions. For the other molecules we shall only indicate the appropriate group chain.

<u>H_3 (1s shell)</u>

The H_3 molecule has an equilibrium framework of three hydrogen atoms at the corners of an equilateral triangle. The first basis we consider consists of an atomic orbital with angular momentum $l = 0$ ($1s$ shell), located at each of the hydrogen atoms. The atomic spin-orbitals (2.1) are then

$$\Phi^{r_1}_{100\sigma}(\mathbf{r}, s) \equiv \Phi_1(\mathbf{r})\delta_{\sigma s} \ , \quad \Phi^{r_2}_{100\sigma}(\mathbf{r}, s) \equiv \Phi_2(\mathbf{r})\delta_{\sigma s} \ ,$$

$$\Phi^{r_3}_{100\sigma}(\mathbf{r},s) \equiv \Phi_3(\mathbf{r})\delta_{\sigma s} \,. \tag{5.1}$$

The point group of othe system is formally D_{3h}, but since we are dealing with the 1s shell the point group D_3 is enough to describe the system.

The set of atomic orbitals carries a reducible representation $\Delta(R)$ of D_3 whose reduction leads to

$$\Delta(R) = \mathbf{D}^{A_1}(R) \oplus \mathbf{D}^{E}(R) .$$

By applying the corresponding projection operators to (5.1) we obtain the basis functions (2.5), which in this case coincide with the orthonormalized functions (2.6). The single particle states are thus given by

$$\begin{aligned}
\Psi_{A_1\sigma}(\mathbf{r},s) &= N_{A_1}[\Phi_1(\mathbf{r}) + \Phi_2(\mathbf{r}) + \Phi_3(\mathbf{r})]\delta_{\sigma s} \,, \\
\Psi_{E_1\sigma}(\mathbf{r},s) &= N_{E_1}[\Phi_1(\mathbf{r}) + \epsilon^*\Phi_2(\mathbf{r}) + \epsilon\Phi_3(\mathbf{r})]\delta_{\sigma s} \,, \\
\Psi_{E_2\sigma}(\mathbf{r},s) &= N_{E_2}[\Phi_1(\mathbf{r}) + \epsilon\Phi_2(\mathbf{r}) + \epsilon^*\Phi_3(\mathbf{r})]\delta_{\sigma s} \,,
\end{aligned} \tag{5.2}$$

where the N_{Γ_γ} are normalization constants

$$N_{A_1} = \frac{1}{\sqrt{3}}\frac{1}{\sqrt{1+2s_{ij}}} \quad , \quad N_{E_1} = N_{E_2} = \frac{1}{\sqrt{3}}\frac{1}{\sqrt{1-s_{ij}}}$$

and

$$s_{ij} = \int \Phi_i(\mathbf{r})\Phi_j(\mathbf{r})d\mathbf{r} \,,$$

$$\epsilon = \exp\left(\frac{i2\pi}{3}\right) ,$$

the latter phase arising from the irreducible representations of C_3 in the group chain

$$\begin{array}{ccc}
D_3 & \supset & C_3 \\
\downarrow & & \downarrow \\
|\,\Gamma, & & \gamma > \,.
\end{array}$$

According to the basis (5.2), the creation (annihilation) operators are

$$a^\dagger_{A_1\sigma}(a_{A_1\sigma}) \quad , \quad a^\dagger_{E_1\sigma}(a_{E_1\sigma}) \quad , \quad a^\dagger_{E_2\sigma}(a_{E_2\sigma}) \quad ,$$

and the second quantized Hamiltonian in the coupled basis is given by

$$\hat{\mathcal{H}} = \hat{\mathcal{W}} + \hat{\mathcal{V}} \quad , \tag{5.3}$$

where

$$\begin{aligned}
\hat{\mathcal{W}} &= \omega_1(a^\dagger_{A_1} \times \tilde{a}_{A_1})^{(A_1,0)} + \omega_2(a^\dagger_E \times \tilde{a}_E)^{(A_1,0)} \quad , \\
2\hat{\mathcal{V}} &= v_1(a^\dagger_{A_1} \times a^\dagger_{A_1})^{(A_1,0)}(\tilde{a}_{A_1} \times \tilde{a}_{A_1})^{(A_1,0)} \\
&\quad + v_2[(a^\dagger_{A_1} \times a^\dagger_{A_1})^{(A_1,0)}(\tilde{a}_E \times \tilde{a}_E)^{(A_1,0)} + (a^\dagger_E \times a^\dagger_E)^{(A_1,0)}(\tilde{a}_{A_1} \times \tilde{a}_{A_1})^{(A_1,0)}] \\
&\quad + v_3(a^\dagger_E \times a^\dagger_E)^{(A_1,0)}(\tilde{a}_E \times \tilde{a}_E)^{(A_1,0)} \\
&\quad + v_4[(a^\dagger_E \times a^\dagger_E)^{(A_2,1)} \times (\tilde{a}_E \times \tilde{a}_E)^{(A_2,1)}]^{(A_1,0)} \\
&\quad + v_5[(a^\dagger_E \times a^\dagger_E)^{(E,0)} \times (\tilde{a}_E \times \tilde{a}_E)^{(E,0)} + \sqrt{3}(a^\dagger_E \times a^\dagger_E)^{(E,1)} \times (\tilde{a}_E \times \tilde{a}_E)^{(E,1)}]^{(A_1,0)} \\
&\quad + v_6[(a^\dagger_{A_1} \times a^\dagger_E)^{(E,0)} \times (\tilde{a}_{A_1} \times \tilde{a}_E)^{(E,0)} \\
&\qquad + \sqrt{3}(a^\dagger_{A_1} \times a^\dagger_E)^{(E,1)} \times (\tilde{a}_{A_1} \times \tilde{a}_E)^{(E,1)}]^{(A_1,0)} \\
&\quad + v_7[(a^\dagger_{A_1} \times a^\dagger_E)^{(E,0)} \times (\tilde{a}_E \times \tilde{a}_E)^{(E,0)} + (a^\dagger_E \times a^\dagger_E)^{(E,0)} \times (\tilde{a}_{A_1} \times \tilde{a}_E)^{(E,0)} \\
&\qquad + \sqrt{3}\{(a^\dagger_{A_1} \times a^\dagger_E)^{(E,1)} \times (\tilde{a}_E \times \tilde{a}_E)^{(E,1)} \\
&\qquad + (a^\dagger_E \times a^\dagger_E)^{(E,1)} \times (\tilde{a}_{A_1} \times \tilde{a}_E)^{(E,1)}\}]^{(A_1,0)}
\end{aligned}$$

and $\{\omega_i\}$ and $\{v_i\}$ are the one and two body interaction matrix elements. Explicitly,

$$\omega_1 = \int \Psi^*_{A_1}(1)W(1)\Psi_{A_1}(1)d\mathbf{r}_1 \ ,$$

$$\omega_2 = \int \{\Psi^*_{E_1}(1)W(1)\Psi_{E_1}(1) + \Psi^*_{E_2}(1)W(1)\Psi_{E_2}(1)\}d\mathbf{r}_1 \ ,$$

$$\begin{aligned}
v_1 &= \int\int \Psi^*_{A_1}(1)\Psi^*_{A_1}(2)V(1,2)\Psi_{A_1}(1)\Psi_{A_1}(2)d\mathbf{r}_1 d\mathbf{r}_2 \ , \\
v_2 &= \int\int \Psi^*_{A_1}(1)\Psi^*_{A_1}(2)V(1,2)[\Psi_{E_1}(1)\Psi_{E_2}(2) + \Psi_{E_2}(1)\Psi_{E_1}(2)]d\mathbf{r}_1 d\mathbf{r}_2 \ , \\
v_3 &= \int\int [\Psi^*_{E_1}(1)\Psi^*_{E_2}(2) + \Psi^*_{E_2}(1)\Psi^*_{E_1}(2)]V(1,2)[\Psi_{E_1}(1)\Psi_{E_2}(2) \\
&\quad + \Psi_{E_2}(1)\Psi_{E_1}(2)]d\mathbf{r}_1 d\mathbf{r}_2 \ , \\
v_4 &= \sqrt{3}\int\int [\Psi^*_{E_1}(1)\Psi^*_{E_2}(2) - \Psi^*_{E_2}(1)\Psi^*_{E_1}(2)]V(1,2)[\Psi_{E_1}(1)\Psi_{E_2}(2) \\
&\quad - \Psi_{E_2}(1)\Psi_{E_1}(2)]d\mathbf{r}_1 d\mathbf{r}_2 \ , \\
v_5 &= \int\int [\Psi^*_{E_1}(1)\Psi^*_{E_1}(2)V(1,2)\Psi_{E_2}(1)\Psi_{E_2}(2) \\
&\quad + \Psi^*_{E_2}(1)\Psi^*_{E_2}(2)V(1,2)\Psi_{E_1}(1)\Psi_{E_1}(2)]d\mathbf{r}_1 d\mathbf{r}_2 \ , \\
v_6 &= \int\int [\Psi^*_{A_1}(1)\Psi^*_{E_1}(2)V(1,2)\Psi_{A_1}(1)\Psi_{E_2}(2) \\
&\quad + \Psi^*_{A_1}(1)\Psi^*_{E_2}(2)V(1,2)\Psi_{A_1}(1)\Psi_{E_1}(2)]d\mathbf{r}_1 d\mathbf{r}_2 \ , \\
v_7 &= \int\int [\Psi^*_{A_1}(1)\Psi^*_{E_1}(2)V(1,2)\Psi_{E_2}(1)\Psi_{E_2}(2) \\
&\quad + \Psi^*_{A_1}(1)\Psi^*_{E_2}(2)V(1,2)\Psi_{E_1}(1)\Psi_{E_1}(2)]d\mathbf{r}_1 d\mathbf{r}_2 \ .
\end{aligned}$$

In this case the bilinear products (4.2) generate the dynamical group $U(6)$ and the appropriate chain of subgroups corresponds to that of case II. We then have that the basis is labeled by the group chain

$$\begin{array}{ccccccccc}
U(6) \supset U^{\Gamma}(3) & \times & SU(2) \supset & U^{E}(2) & \times & SU(1) \supset & O(2) & \times & SU(1) \supset \\
\downarrow & & \downarrow & \downarrow & & \downarrow & \downarrow & & \\
\|[1^m] \ , & & S \ , & [h_1 h_2] \ , & & \Sigma \ , & \tau \ , & &
\end{array}$$

$$\begin{array}{ccccc}
S_3 & \times & SU(1) \supset C_3 & \times & SU(1) \\
\downarrow & & \downarrow & & \\
\Gamma \ , & & \gamma & & > \ ,
\end{array} \tag{5.4}$$

where we indicate below each group the quantum numbers that label their representations. The index m can take values of 1,2 or 3, corresponding to the molecules H_3^{++},

H_3^+ and H_3, respectively. The Casimir operators associated to these groups are

Group	Casimir Operator	Eigenvalue
$U(6)$	$\hat{m} = \sum_{\Gamma\gamma\sigma} C_{\Gamma\gamma\sigma}^{\Gamma\gamma\sigma}$	m
	$\hat{C}_{2u(6)} = \hat{m}(7-\hat{m})$	$m(7-m)$
$U(3)$	$\hat{m} = \sum_{\Gamma\gamma} C_{\Gamma\gamma}^{\Gamma\gamma}$	m
	$\hat{C}_{2u(3)} = \sum_{\Gamma\gamma}\sum_{\Gamma'\gamma'} C_{\Gamma\gamma}^{\Gamma'\gamma'} C_{\Gamma'\gamma'}^{\Gamma\gamma}$	$5m - \frac{1}{2}m^2 - 2S(S+1)$
$U(2)$	$m_E = C_{E_1}^{E_1} + C_{E_2}^{E_2}$	m_E
	$\hat{C}_{2u(2)} = \sum_{\gamma\gamma'} C_{E\gamma}^{E\gamma'} C_{E\gamma'}^{E\gamma}$	$h_1^2 + h_2^2 + h_1 - h_2$
$O(2)$	$J^2 = (C_{E_1}^{E_1} - C_{E_2}^{E_2})^2$	τ^2

where we note that the group $U^E(2)$ has been obtained by deleting the A_1 representation from $U^\Gamma(3)$.

We write below explicit expressions for the wave functions with maximum weight in the spin quantum number.

$U(6)$	$U(3)$	$U(2)$	$O(2)$	S_3	C_3	$SU(2)$	$SU(1)$	
H_3^{++}								
$\|[1]$	$[1]$	$[0]$	0	A_1		$;\ \frac{1}{2}$	$\frac{1}{2}>$	$= a^\dagger_{A_1+}\|0>$
$\|[1]$	$[1]$	$[1]$	1	E	γ	$;\ \frac{1}{2}$	$\frac{1}{2}>$	$= a^\dagger_{E_\gamma+}\|0>$

$U(6)$	$U(3)$	$U(2)$	$O(2)$	S_3	C_3	$SU(2)$	$SU(1)$	
H_3^{+}								
$\|[1^2]$	$[2]$	$[0]$	0	A_1		$;\ 0$	$0>$	$= a^\dagger_{A_1+}a^\dagger_{A_1-}\|0>$
$\|[1^2]$	$[2]$	$[1]$	1	E	γ	$;\ 0$	$0>$	$= \frac{1}{\sqrt{2}}(a^\dagger_{A_1+}a^\dagger_{E_\gamma-} - a^\dagger_{A_1-}a^\dagger_{E_\gamma+})\|0>$
$\|[1^2]$	$[2]$	$[2]$	0	A_1		$;\ 0$	$0>$	$= \frac{1}{\sqrt{2}}(a^\dagger_{E_1+}a^\dagger_{E_2-} - a^\dagger_{E_1-}a^\dagger_{E_2+})\|0>$
$\|[1^2]$	$[2]$	$[2]$	2	E	γ	$;\ 0$	$0>$	$= a^\dagger_{E_\gamma+}a^\dagger_{E_\gamma-}\|0>$
$\|[1^2]$	$[11]$	$[1]$	1	E	γ	$;\ 1$	$1>$	$= a^\dagger_{A_1+}a^\dagger_{E_\gamma+}\|0>$
$\|[1^2]$	$[11]$	$[11]$	0^-	A_2		$;\ 1$	$1>$	$= a^\dagger_{E_1+}a^\dagger_{E_2+}\|0>$

$U(6)$ $U(3)$ $U(2)$ $O(2)$ S_3 C_3 $SU(2)$ $SU(1)$

H_3

$|[1^3]$ $[21]$ $[1]$ 1 E γ $;\frac{1}{2}$ $\frac{1}{2}>$ $=$
$a^\dagger_{A_1+}a^\dagger_{A_1-}a^\dagger_{E_\gamma+}|0>$

$|[1^3]$ $[21]$ $[11]$ 0^- A_2 $;\frac{1}{2}$ $\frac{1}{2}>$ $=$
$\frac{1}{\sqrt{6}}\{a^\dagger_{A_1+}[a^\dagger_{E_1+}a^\dagger_{E_2-} + a^\dagger_{E_1-}a^\dagger_{E_2+}] - 2a^\dagger_{A_1-}a^\dagger_{E_1+}a^\dagger_{E_2+}\}|0>$

$|[1^3]$ $[21]$ $[2]$ 0 A_1 $;\frac{1}{2}$ $\frac{1}{2}>$ $=$
$\frac{1}{\sqrt{2}}(a^\dagger_{A_1+}[a^\dagger_{E_1+}a^\dagger_{E_2-} - a^\dagger_{E_1-}a^\dagger_{E_2+}]|0>$

$|[1^3]$ $[21]$ $[2]$ 2 E γ $;\frac{1}{2}$ $\frac{1}{2}>$ $=$
$a^\dagger_{A_1+}a^\dagger_{E_\gamma+}a^\dagger_{E_\gamma-}|0>$

$|[1^3]$ $[21]$ $[21]$ 1 E 1 $;\frac{1}{2}$ $\frac{1}{2}>$ $=$
$a^\dagger_{E_1+}a^\dagger_{E_1-}a^\dagger_{E_2+}|0>$

$|[1^3]$ $[21]$ $[21]$ 1 E 2 $;\frac{1}{2}$ $\frac{1}{2}>$ $=$
$a^\dagger_{E_2+}a^\dagger_{E_2-}a^\dagger_{E_1+}|0>$

$|[1^3]$ $[1^3]$ $[11]$ 0^- A_2 $;\frac{3}{2}$ $\frac{3}{2}>$ $=$
$a^\dagger_{A_1+}a^\dagger_{E_1+}a^\dagger_{E_2+}|0>$

These wave functions constitute an appropriate set of basis states to carry out the diagonalization of the Hamiltonian (5.3).

Finally, we point out that there is a dynamical symmetry associated to chain (5.4), which is given by

$$\hat{H} = E_0 + f_1\hat{m} + f_2\hat{m}_E + f_3\hat{m}^2 + f_4\hat{m}_E^2 + f_5\hat{C}_{2u(3)} + f_6\hat{C}_{2u(2)} + f_7J^2 ,$$

where the parameters $\{f_i\}$ are linearly related with the parameters of the Hamiltonian (5.3).

We next discuss the appropriate chains of subgroups for the other molecules in the table.

$\underline{H-A}$

This kind of structure corresponds to heteronuclear diatomic molecules. We take as a basis the $1s$ shell for H and the $2sp$ shell for A, corresponding to the first series of diatomic hydride molecules. This atomic basis leads to three Σ^+ basis functions and one Π basis function, which carry irreducible representations of the point group

$C_{\infty v}$. Orthonormalizing the Σ^+ functions we obtain the set of single particle states in the second quantized scheme

$$a^\dagger_{1\Sigma^+\sigma}(a_{1\Sigma^+\sigma}) \ , \ a^\dagger_{2\Sigma^+\sigma}(a_{2\Sigma^+\sigma}) \ , \ a^\dagger_{3\Sigma^+\sigma}(a_{3\Sigma^+\sigma}) \ , \ a^\dagger_{\Pi^\pm\sigma}(a_{\Pi^\pm\sigma}) \ ,$$

whose bilinear products $a^\dagger_{\alpha\Gamma\sigma}a_{\alpha'\Gamma'\sigma'}$ generate a $U(10)$ group. For these molecules the group chain corresponds to case IV, in the particular reduction where there is no γ label, that is, the chain results from the reduction of the $U^\Gamma(a)$ group. In this case there are two possible chains corresponding to (4.25)

$$U(10) \supset U^\Gamma(5) \times SU(2) \supset U^\Gamma(4) \times SU(2) \supset O(4) \times SU(2) \supset \\ O(3) \times SU(2) \supset O(2) \times SU(1) \ ,$$

and to (4.24)

$$U(10) \supset U^\Gamma(5) \times SU(2) \supset U^\Gamma(4) \times SU(2) \supset U^\Gamma(3) \times SU(2) \supset \\ U^\Gamma(2) \times SU(2) \supset O(2) \times SU(1) \ ,$$

where the $O(2)$ group is isomorphic to $C_{\infty v}$. In both chains the reduction of the unitary groups is achieved by successive deletions of Σ^+ states. The $O(2)$ group is labeled by the angular momentum projection M along the axis of the molecule; the Σ^+ states correspond to $M = 0$ and the $\Pi^\pm$ states to $M = 1$. Since the $\Pi^\pm$ states effectively carry the total projection M, it is possible to delete the Σ^+ states in these reductions.

$\underline{A - A}$

This structure corresponds to homonuclear diatomic molecules. Considering again the $2sp$ shell, we obtain the following molecular functions: two Σ_g^+, two Σ_u^+, one Π_g and one Π_u. These functions span irreducible representations of the point group $D_{\infty v} \otimes C_i$. The corresponding creation (annihilation) operators are

$$a^\dagger_{1\Sigma_g^+}(a_{1\Sigma_g^+}) \ , \ a^\dagger_{2\Sigma_g^+}(a_{2\Sigma_g^+}) \ , \ a^\dagger_{1\Sigma_u^+}(a_{1\Sigma_u^+}) \ , \ a^\dagger_{2\Sigma_u^+}(a_{2\Sigma_u^+}) \ ,$$

$$a^\dagger_{\Pi_g}(a_{\Pi_g}) \ , \ a^\dagger_{\Pi_u}(a_{\Pi_u}) \ ,$$

whose bilinear products generate a $U(16)$ group. The appropriate group chain corresponds to case IV, where chain (4.25) is given by

$$U(16) \supset U(8) \times SU(2) \supset U^\Gamma(4) \times U^\gamma(2) \times SU(2) \supset O^\Gamma(4) \times O^\gamma(2) \times SU(2) \supset \\ O^\Gamma(3) \times SO^\gamma(2) \times SU(2) \supset O^\Gamma(2) \times S_2 \times SU(2) \approx D_{\infty h} \times SU(2) \ ,$$

and chain (4.24) is

$$U(16) \supset U(8) \times SU(2) \supset U^\Gamma(4) \times U^\gamma(2) \times SU(2) \supset U^\Gamma(3) \times O^\gamma(2) \times SU(2) \supset \\ U^\Gamma(2) \times SO^\gamma(2) \times SU(2) \supset O(2) \times S_2 \times SU(2) \approx D_{\infty h} \times SU(2) \ .$$

We note that these reductions are possible because the number of gerade and ungerade functions in the single particle case is the same.

$\underline{H_2O_2}$

This molecule has the following geometric structure[10]

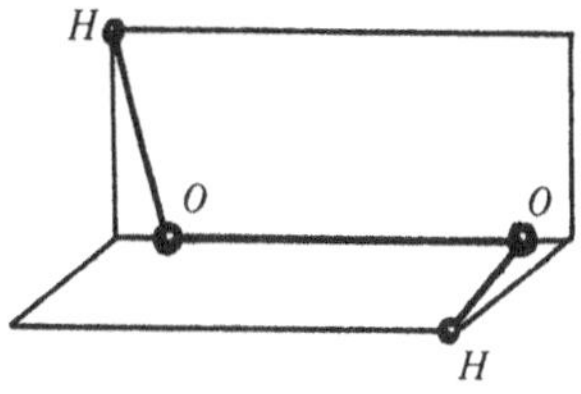

In this case we take as atomic basis for the hydrogen atoms the 1s shell and for the oxygen atoms the 2sp shell. We obtain four A functions and four B functions which constitute a basis for the C_2 point group. The creation (annihilation) operators are then given by

$$a^{\dagger}_{\alpha A\sigma}(a_{\alpha A\sigma}) \quad , a^{\dagger}_{\alpha B\sigma}(a_{\alpha B\sigma}) \quad , \quad \alpha = 1,2,3,4.$$

and the dynamical group is $U(20)$. The associated chain belongs to case II and is given by (4.17)

$$\begin{array}{ccccccccc} U(20) \supset U(10) & \times & SU(2) \supset U^{\alpha}(5) & \times & U(2) & \times & SU(2) \\ & & \cup & & \cup & & \\ & & U^{\alpha}(4) & & O(2) & & \\ & & \cup & & \cup & & \\ & & U^{\alpha}(3) & & SO(2) & & \\ & & \cup & & \cup & & \\ & & U^{\alpha}(2) & & S_2 \approx C_2 & & \\ & & \cup & & & & \\ & & U^{\alpha}(1) & & & & \end{array}$$

$\underline{CH_4}$

Methane has a central carbon and four hydrogen atoms located at the corners of a tetrahedron. Again, the basis for the hydrogen atoms is the 1s shell and for the carbon atom the 2sp shell. This atomic basis leads to two A_1 functions and two F_2 functions, which carry irreducible representations of the T_d point group. The creation (annihilation) operators are

$$a^{\dagger}_{\alpha A_1\sigma}(a_{\alpha A_1\sigma}) \quad , \quad a^{\dagger}_{\alpha F_2\gamma\sigma}(a_{\alpha F_2\gamma\sigma}) \quad , \quad \alpha = 1,2.,$$

and the dynamical group is $U(16)$. The corresponding chain belongs to case II and is given by

$$\begin{array}{ccccccc} U(16) \supset U(8) & \times & SU(2) \supset U^{\alpha}(2) & \times & U^{\Gamma}(4) & \times & SU(2) \\ & & \cup & & \cup & & \\ & & U^{\alpha}(1) & & U^{\Gamma}(3) & & \\ & & & & \cup & & \\ & & & & O(3) & & \\ & & & & \cup & & \\ & & & & S_4 \approx T_d \supset C_3 \ . & & \end{array}$$

H_2O

The water molecule belongs to the C_{2v} point group. The minimal basis needed to describe H_2O corresponds to the $1s$ shell for the hydrogen atoms and the $2sp$ shell for oxygen. The projection of these atomic orbitals leads to the following distribution of basis functions

C_{2v}	α		
A_1	×	×	×
A_2			
B_1	×	×	
B_2	×		

This unsymmetrical distribution of basis functions does not permit us to construct the group chain by the methods of cases II or IV. However, if we consider the distribution of the basis functions with respect to the sybgroup C_2

C_2	α		
A	×	×	×
B	×	×	×

we obtain the required symmetry in α and can apply case III. Accordingly, the appropriate chain is

$$\begin{array}{ccccccc} U(12) \supset U(6) & \times & SU(2) \supset U^{\alpha}(3) & \times & U^{\Gamma}(2) & \times & SU(2) \\ & & \cup & & \cup & & \\ & & U^{\alpha}(2) & & O(2) & & \\ & & \cup & & \cup & & \\ & & U^{\alpha}(1) & & S_2 \approx C_2 \ . & & \end{array} \tag{5.5}$$

This group chain labels the states with respect to the subgroup C_2 of the point group C_{2v}. In order to obtain the states carrying the irreducible representations of C_{2v} we need to induce from C_2,

$$C_{2v} = C_2 + \sigma_v C_2 \ ,$$

so we apply the operator $\hat{\sigma}_v$ to the wave functions associated to chain (5.5) to obtain the space

$$\{|\Gamma> \ , \ \hat{\sigma}_v|\Gamma>\} \ ,$$

which is invariant under C_{2v} and has to be reduced to obtain the irreducible representations of the point group. Note that the new linear combinations only mix the $O(2)$ label in (5.5).

H_3 (1s and 2sp shell)

The treatment of this molecule is similar to that of water. Taking the $1s$ and $2sp$ shell as atomic basis, we obtain the following distribution of molecular single particle states according to the D_{3h} point group and C_3 subgroup:

D_{3h}	α			
A_1	×	×	×	
A_2	×			
A_1'				
A_2'	×			
E	×	×	×	×
E'	×			

C_3	α				
A_1	×	×	×	×	×
E_1	×	×	×	×	×
E_2	×	×	×	×	×

We note again the symmetrical distribution for the multiplicity index α in the subgrup C_3. In this case the dynamical group is $U(30)$ and the corresponding chain is

$$\begin{array}{ccccccc} U(30) \supset U(15) & \times & SU(2) \supset U^{\alpha}(5) & \times & U^{\Gamma}(3) & \times & SU(2) \\ & & \cup & & \cup & & \\ & & U^{\alpha}(4) & & U^{\Gamma}(2) & & \\ & & \cup & & \cup & & \\ & & U^{\alpha}(3) & & O(2) & & , \\ & & \cup & & \cup & & \\ & & U^{\alpha}(2) & & SO(2) & & \\ & & \cup & & \cup & & \\ & & U^{\alpha}(1) & & \mathcal{A}_3 \approx C_3 & & \end{array} \tag{5.6}$$

where $\mathcal{A}_3$ is the alternating group. This chain carries the irreducible representations of the C_3 subgroup. The appropriate basis is again obtained by induction to the D_{3h} point group. We may carry out the induction either in one step

$$C_3 \uparrow D_{3h} \ ,$$

or in two steps

$$C_3 \uparrow D_3 \uparrow D_{3h} \ .$$

In the former case, we have the left coset expansion of D_{3h}

$$D_{3h} = C_3 + C_2 C_3 + \sigma_n C_3 + \sigma_v C_3 \ ,$$

so we need to apply the operators $\hat{C}_2$, $\hat{\sigma}_n$ and $\hat{\sigma}_v$ to the functions (5.6) to arrive at the D_{3h} invariant space

$$\{|\Gamma> \ , \ \hat{C}_2|\Gamma> \ , \ \hat{\sigma}_n|\Gamma> \ , \ \hat{\sigma}_v|\Gamma>\} \ ,$$

which must be reduced to obtain the D_{3h} basis states.

The second possibility is to first induce to D_3

$$D_3 = C_3 + C_2 C_3$$

and then from D_3 to D_{3h}

$$D_{3h} = D_3 + \sigma_n D_3 \ ,$$

which simplifies the procedure, since we then diagonalize two bidimensional spaces instead of a four-dimensional one.

<u>*CH F Cl Br*</u>

This molecule has a central carbon atom and the Bromine, Hydrogen, Clorine and Fluorine located at the corners of a tetrahedron. This structure belongs to the C_1 point group. The basis functions span the totally symmetric representation and the canonical chain of Gelfands' states constitutes a suitable basis.

6. CONCLUSIONS

We have presented a systematic procedure to construct symmetry adapted bases for molecular systems in the Born-Oppenheimer approximation.

From a non-orthogonal set of atomic orbitals a set of orthonormal single particle states is constructed, which carry irreducible representations of the molecular point group. Because of the orthonormality of the set, the dynamical group of the system turns out to be a unitary group $U(2N)$ which is then separated into a spatial part $U(N)$ and a spin part $SU(2)$. The new aspect in our analysis concerns the reduction of the $U(N)$ group. Instead of using the canonical chain as a basis, we analyze the symmetries contained in the distribution of the basis functions and obtain alternative group chains for $U(N)$ containing a symmetric group S_n, which is isomorphic to the point group or to a subgroup of it.

We have found that the procedure to follow in the reduction of $U(N)$ is not unique. It depends on the molecular system as well as on the single particle basis employed. We have, however, classified in four cases our method of establishing the suitable group chains. Although we have only given examples of molecules in which 1*s* and/or 2*sp* shell basis functions were considered, we expect that other cases (e.g. systems where the 3*spd* shell is taken into account) will be essentially included either in one of the four cases presented or in a combination of them.

A symmetry adapted basis permits not only to simplify the diagonalization of the Hamiltonian but also provides analytic solutions obtained from the dynamical symmetries associated to the group chains. We believe that it is possible to take advantage of this fact in the study of electronic excited states.

REFERENCES

1. Racah G., Phys. Rev. 51 (1942) 186; 62 (1942) 438; 63 (1943) 367; 76 (1949) 1352.
2. Racah G., Group Theory and Spectroscopy in Springer Tracts in Mod. Phys. 37, Springer, Heidelberg, 1965.
3. Moshinsky M., "Group Theory and the Many Body Problem", Gordon & Breach, New York, 1967.
4. Judd B., "Second Quantization and Atomic Spectroscopy", John Hopkins Univ. Press, Baltimore, 1967.
5. Moshinsky M. and Seligman T.H., Annals of Physics 66 (1971) 311.
6. Lõwdin P.O., J. Chem. Phys. 18 (1950) 365.
7. Matsen F.A. and Pauncz R., "The Unitary Group in Quantum Chemistry", Studies in Physical and Theoretical Chemistry 44. Elsevier Science Publishing Company, Inc., 1986.
8. Lectures Notes in Chemistry. The Unitary Group. Spring-Verlag. Edited by Jürgen Hinze. 1981.
9. Altmann L.S., "Induced Representations in Crystals and Molecules. Point, space and non-rigid molecule group". Academic Press. 1977.
10. Herzberg G., "Molecular Spectra and Molecular Structure. II.- Infrared and Raman Spectra of Polyatomic Molecules". Van Nostrand. 1945.

ON CERTAIN SUBMODULES OF THE ENVELOPING FIELDS

Romuald Lenczewski

Institute of Mathematics
Technical University of Wroclaw
50-370 Wroclaw, Poland

1. Introduction

In [1]-[3] we studied indecomposable modules of certain Lie algebras. They were obtained as certain submodules, quotient modules or subquotient modules of the enveloping algebra of a Lie algebra g, $U(g)$. In particular, Verma modules of highest (lowest) weight Λ were obtained. It turned out that the formulas obtained in [1]-[3] for the left g action on the Poincare-Birkchoff-Witt basis could be extended to negative exponents in the standard monomials. Moreover, what is noteworthy, that is how the infinite dimensional indecomposable modules of the Lorentz algebra so(3,1) were recovered in [1]. In this paper an algebraic construction of modules of that type via the enveloping field of g, $K(g)$ is given. In order to define the enveloping fields one has to start with some ring theory and define a ring of fractions in an appropriate way (see, for example, [4],[10]). When considering the enveloping algebras modulo certain two-sided ideals one can obtain rings of fractions for the corresponding quotients (in the case of primitive ideals see, for example [7] and [8]). However, if one wants to give an algebraic construction for the extensions of Verma modules comprising monomials with the inverses of the usual generators another approach seems to be helpful.

The submodules of $K(g)$, denoted by $\tilde{U}(g)$, spanned by monomials with exponents extended to all integers are considered in this paper. They are extensions of the enveloping algebras $U(g)$. On their quotient modules (viewed as left g modules) the modules considered in [1]-[3] are obtained.

The concept is illustrated on the example of complex su(2). However, a similar construction is possible for some other Lie algebras. In particular, the Heisenberg-Weyl algebra is listed as a second example, although a much more complete treatment and a realization for its modules of the type considered here is given in [6]. Also, the Lorentz algebra can be treated along the same lines.

Basic definitions and results on the enveloping fields are taken from [4]. The structure of the enveloping fields was first studied in [9]. Also, the results from [5] are used.

2. Enveloping fields $K(g)$

Let A be a ring, S a subset of A. S is said to allow of an arithmetic of fractions if the following conditions are satisfied:

Symmetries in Science V, Edited by B. Gruber *et al.*
Plenum Press, New York, 1991

(i) $1 \in S$

(ii) $sr \in S \quad if \quad s, r \in S$

(iii) the elements of S are not divisors of zero in A

(iv) for $s \in S$ and $a \in A$ there exists $t \in S$, $\quad b \in A$ such that $at = sb$

(v) for $s \in S$ and $a \in A$ there exists $r \in S$, $\quad c \in A$ such that $ra = cs$

Let now $(a, s), (b, t) \in A \times S$. We write $(a, s) \sim (b, t)$ if there exists $c, d \in A$ such that $ac = bd$, $sc = td \in S$. It is an equivalence relation (see [4], p.117). The corresponding equivalence class will be denoted by a/s. On the quotient set of $A \times S$ under $\sim$, denoted by B, we define addition and multiplication as follows:

$$(A) \qquad (a/s) + (b/t) = (ac + bd)/e$$

where $sc = td = e \in S$, $c, d \in A$,

$$(M) \qquad (a/s)(b/t) = (ac)/(tu)$$

where $bu = sc$, $c \in A$,$u \in S$. For details see [4].

One can verify that (A) and (M) are well defined operations. Moreover, the addition is associative and commutative, $0/1$ is the zero element and the additive inverse of a/s is $(-a)/s$.

Likewise, the multiplication is associative , $1/1$ is the unity and

$$a/s = (a/1)(1/s) = (a/1)(s/1)^{-1}$$

and the distributive law holds. Thus, B is a ring. The mapping $a \mapsto (a/1)$ of A into B is an injective ring homomorphism under which we can identify A with its image in B. Then every element of S is invertible and $a/s = as^{-1}$.

The ring B is called the ring of fractions of A defined by S. It can be shown that if A is integral and $S = A - \{0\}$ allows of an arithmetic of fractions, then the ring of fractions becomes a skew field called the field of fractions of A (see [4], p 121).

Finally, since the enveloping algebra $U(g)$ of a Lie algebra g is Noetherian and integral, it possesses a field of fractions called the enveloping field of g. We denote it by $K(g)$.

$K(g)$ becomes a left g - module under the left action of g. Namely, we define

$$x(a/s) = (xa)/s$$

where $x \in g$. The module $\tilde{U}(g)$ spanned by extended monomials $x_1^{n_1}...x_k^{n_k}$ where $n_1, ..., n_k$ integers, $x_1, ..., x_k \in g$ and are ordered as in the Poincare-Birkhoff-Witt basis forms a submodule of $K(g)$. The demonstration of that fact amounts to showing that the brackets $[g, U(g)]$ can be extended to $[g, \tilde{U}(g)]$ by allowing of negative exponents. It will be demonstrated on the example of complex su(2).

3. Complex rotation algebra

Let $g = \text{su}(2)$. A basis of su(2) can be chosen as h_+, h_-, h_3 satisfying the following

$$[h_+, h_-] = 2h_3 \quad [h_3, h_+] = h_+ \quad [h_3, h_-] = -h_-$$

One can easily show that in $U(g)$ the following relations hold:

$$a) \qquad h_3 h_+^n = h_+^n h_3 + n h_+^n$$

$$b) \qquad h_3 h_-^m = h_-^m h_3 - m h_-^m$$

$$c) \qquad h_- h_+^n = h_+^n h_- \quad 2n h_+^{n-1} h_3 \quad n(n \quad 1) h_+^{n-1}$$

where n, m are natural numbers. We will show that they remain valid if n, m are integers and the elements with negative exponents are understood to come from the enveloping field $K(g)$. We thus have to show that

$$\begin{aligned} a^*) \quad & (h_3/1)(1/h_+^n) = (1/h_+^n)(h_3/1) - (n/h_+^n) \\ b^*) \quad & (h_3/1)(1/h_-^m) = (1/h_-^m)(h_3/1) + (m/h_-^m) \\ c^*) \quad & (h_-/1)(1/h_+^n) = (1/h_+^n)(h_-/1) + (2n/h_+^{n+1})(h_3/1) - (n(n+1))/h_+^{n+1} \end{aligned}$$

To prove that we are going to use the definitions of (A) and (M). Namely, let $a = 1$, $b = h_3$, $s = h_+^n$, $t = 1$. Then $h_3 u = h_+^n c$ holds if $c = h_3 + n$ by a). Therefore,

$$(1/h_+^n)(h_3/1) = (h_3 + n)/h_+^n$$

which proves a^*). Similarly, let now $a = 1$, $b = h_3$, $s = h_-^m$, $t = 1$. Then $h_3 u = h_-^m c$ holds for $u = h_-^m$ and $c = h_3 - m$ by b). Hence,

$$(1/h_-^m)(h_3/1) = (h_3 - m)/h_-^m$$

which proves b^*). Finally, to prove c^*) let $a = 1$, $b = h_-$, $s = h_+^n$, $t = 1$. Then $h_- u = h_+^n c$ holds for $u = h_+^{n+1}$ and $c = h_+ h_- - 2(n+1)h_3 - n(n+1)$ by c) and hence

$$(1/h_+^n)(h_-/1) = (h_+ h_- - 2(n+1)h_3 - n(n+1))/h_+^{n+1}$$

Also, let $a = 2n$, $b = h_3$, $s = h_+^{n+1}$, $t = 1$ to get $h_3 u = h_+^{n+1} c$ with $u = h_+^{n+1}$ and $c = h_3 + n + 1$ by a) and hence

$$(2n/h_+^{n+1})(h_3/1) = (2nh_3 + 2n(n+1))/h_+^{n+1}$$

which ends the proof.

Thus, the extension of the exponents to all integers gives a submodule of the enveloping field of su(2). Hence, the following relations hold:

$$\begin{aligned} & h_3 X(n,m,r) = X(n,m,r+1) + (n-m)X(n,m,r) \\ & h_+ X(n,m,r) = X(n+1,m,r) \\ & h_- X(n,m,r) = X(n,m+1,r) - 2nX(n-1,m,r+1) + n(2m-n+1)X(n-1,m,r) \end{aligned}$$

where the set of $X(n,m,r) = h_+^n h_-^m h_3^r$ for m, n, r integers can be shown to be linearly independent. Clearly, the enveloping algebra $U(g)$ is isomorphic to the submodule spanned by $X(n,m,r)$ where n, m, r are natural numbers.

Let us consider now the following vector space

$$\begin{aligned} \tilde{V}_{\Lambda\mu} = {} & \tilde{U}(h_+)\tilde{U}(h_-)U(h_3)(h_3 - \Lambda) + \tilde{U}(h_+)\tilde{U}(h_-)U(h_3^{-1})(h_3^{-1} - \Lambda^{-1}) + \\ & + \tilde{U}(h_+)U(h_-)\tilde{U}(h_3)(h_- - \mu) + \tilde{U}(h_+)U(h_-^{-1})\tilde{U}(h_3)(h_-^{-1} - \mu^{-1}) \end{aligned}$$

where $\Lambda \neq 0, \mu \neq 0$ and Λ, μ are complex numbers. It is easy to show that $\tilde{V}_{\Lambda\mu}$ is left-g-invariant, thus a left-g-submodule, hence we can consider the quotient module $\tilde{d}_{\Lambda,\mu} = \tilde{U}(g)/\tilde{V}_{\Lambda,\mu}$ with the basis consisiting of extended h_+- monomials, i.e., $X(n) = h_+^n$, n an integer. The left g-action on this module is given by:

$$\begin{aligned} & h_3 X(n) = (\Lambda + n)X(n) \\ & h_+ X(n) = X(n+1) \\ & h_- X(n) = \mu X(n) \quad n(2\Lambda + n + 1)X(n \quad 1) \end{aligned}$$

In the limit $\mu \to 0$ in the matrix elements we obtain an extension $\tilde{d}_\Lambda$ of the Verma module d_Λ of lowest weight Λ (In a similar manner we can obtain an extension of the Verma module of highest weight Λ). Thus,

$$\begin{aligned} h_3X(n) &= (\Lambda + n)X(n) \\ h_+X(n) &= X(n+1) \\ h_-X(n) &= -n(2\Lambda + n - 1)X(n-1) \end{aligned}$$

Clearly, $\tilde{d}_\Lambda$ does not have a lowest or highest weight. It is always indecomposable since d_Λ is its proper submodule. Besides, it may have another proper submodule depending on the value of Λ. Namely, if 2Λ is an integer, then the vector space

$$f_\Lambda = span\{X(n), \quad n \geq -2\Lambda + 1\}$$

is an infinite dimensional submodule. Three cases may be distinguished:

a) If $\Lambda = 1/2$, then it coincides with d_Λ ,

b) If $\Lambda < 1/2$, then we obtain $f_\Lambda \subset d_\Lambda \subset \tilde{d}_\Lambda$,

c) If $\Lambda > 1/2$ then we obtain $d_\Lambda \subset f_\Lambda \subset \tilde{d}_\Lambda$.

In Case a) both $f_\Lambda = d_\Lambda$ as well as the quotient $\tilde{d}_\Lambda/d_\Lambda$ are infinite dimensional and irreducible modules. In Case b) $\tilde{d}_\Lambda/d_\Lambda$ is an infinite dimensional irreducible module and d_Λ/f_Λ is a finite dimensional irreducible module. In Case c) $\tilde{d}_\Lambda/f_\Lambda$ is an infinite dimensional irreducible module and f_Λ/d_Λ is a finite dimensional irreducible module.

4. Heisenberg-Weyl algebra

The Heisenberg-Weyl algebra is defined by the following Lie brackets:

$$[a^+, a] = -E \qquad [a^+, E] = [a, E] = 0$$

The following relations can be derived on its enveloping algebra (see [5]):

$$\begin{aligned} a^+a^m &= a^na^+ - na^{n-1}E \\ a(a^+)^m &= (a^+)^m a + m(a^+)^{m-1}E \end{aligned}$$

where n, m are natural numbers. However, one can show as was demonstrated in Section 3 for su(2), that such formulas can be derived also on $\tilde{U}(g)$, where g is the Heisenberg-Weyl algebra and $\tilde{U}(g)$ is defined also in an analogous way, i.e. as a linear span of extended standard monomials (with integral exponents). Thus, the following relations are obtained for $\tilde{U}(g)$ viewed as a left g-module:

$$\begin{aligned} a^+X(n,m,r) &= X(m+1,n,r) \\ aX(m,n,r) &= X(m,n+1,r) + mX(m-1,n,r+1) \\ EX(m,n,r) &= X(m,n,r+1) \end{aligned}$$

where $X(m,n,r) = (a^+)^m a^n E^r$, $X(0,0,0) = 1$ and m, n, r are integers. $\tilde{U}(g)$ is indecomposable and $U(g)$ is its proper submodule. Now, let us introduce the following vector space:

$$\begin{aligned} V_{\sigma\mu} = \tilde{U}(a^+)\tilde{U}(a)U(E)(E - \mu) + \tilde{U}(a^+)\tilde{U}(a)U(E^{-1})(E^{-1} - \mu^{-1}) + \\ + \tilde{U}(a^+)U(a)\tilde{U}(E)(a \quad \sigma) + \tilde{U}(a^+)U(a^{-1})\tilde{U}(E)(a^{-1} \quad \sigma^{-1}) \end{aligned}$$

where σ and μ are complex numbers. It is a submodule of $\tilde{U}(g)$ and the corresponding quotient module has a basis in the form of $X(m) = (a^+)^m$ where m is an integer. The left g-action is given by:

$$\begin{aligned} a^+X(m) &= X(m+1) \\ aX(m) &= \sigma X(m) + m\mu X(m-1) \\ EX(m) &= \mu X(m) \end{aligned}$$

This is an infinite dimensional indecomposable module. Its submodule spanned by $X(m)$, where m is a natural number, gives the extended harmonic oscillator Fock space.

Acknowledgements

I would like to thank Professor Philip Feinsilver for helpful remarks and references.

References

[1] Gruber B.,Lenczewski R., " *Indecomposable Representations of the Lorentz Algebra in an Angular Momentum Basis*", J.Phys.A **16** (1983),3703-22.

[2] Lenczewski R.,Gruber B., "*Indecomposable Representations of the Poincare Algebra*", J.Phys.A **19** (1986),1-20.

[3] Lenczewski R., "*Indecomposable Representations of the Poincare Algebra in an Energy-cyclic Angular Momentum Basis*" , J.Math.Phys. **28** (1987), 1237-42.

[4] Dixmier J., "*Enveloping Algebras*" , North-Holland, Amsterdam, 1978.

[5] Gruber B.,Doebner H.D.,Feinsilver Ph., "*Representations of the Heisenberg Weyl Algebra and Group*", Kinam Vol.4 (1982), 241-278.

[6] Loeb D.E., Rota G.C., "*Formal Power Series of Logarithmic Type*" , Adv.Math. **75** (1989), 1-118.

[7] Jantzen J.C., "*Primitive Ideals in the Enveloping Algebra of a Semisimple Lie Algebra*", Mathematical Surveys and Monographs No. **24** (1987), AMS, Providence.

[8] Rentschler R., "*Primitive Ideals in Enveloping Algebras (General Case)*" , Mathematical Surveys and Monographs, No. **24** (1987), AMS, Providence.

[9] Gelfand I.M., Kirillov A.A., "*Sur les corps lies aux algebres enveloppantes des algebres de Lie*", Publ. Inst. Hautes Etudes Sci., **31** (1966), 5-19.

[10] Goldie A.W., "*The Structure of Noetherian Rings*", Lect. Notes in Math. **246**, Springer, Berlin, 213-321.

INVARIANTS AND STATES GENERATING SYMMETRY

OF NONSTATIONARY SYSTEMS

V. Man'ko

Lebedev Physics Institute, 117924 Moscow
Leninsky prospect 53

1. INTRODUCTION

The aim of this work is to consider, following Ref. [1–3], the relation of spectrum generating algebras [4] and dynamical groups [5] (or noninvariance groups [6]) with integrals of motion and symmetries of the Schrödinger equation. We consider these concepts for examples of stationary and nonstationary quantum systems. The nonstationary systems have no energy levels. For this reason, there are no spectrum generating algebras for these systems. On the other hand, using the integrals of motion, it is possible to construct a generalization of the spectrum generating algebra for nonstationary systems as well. This construction can be called state generating symmetry of nonstationary quantum system. This construction was used in [2], [7] for quantum parametric oscillators. For stationary quantum systems nonstationary integrals of motion as elements of a spectrum generating algebra have been introduced by Dothan [8]. The spectrum generating algebra $O(4,2)$ for the hydrogen atom has been studied in Ref. [1], [9]–[12]. Symmetry of potential in quantum mechanics is considered to be responsible for level degeneration. So, the degeneration of hydrogen atom discrete levels was explained by Fock [13] as a consequence of $O(4)$ hidden symmetry.

In this article we will consider the example of a conventional stationary oscillator on the basis of time-dependent invariants. We will also discuss the notion of a symmetry of a mathematical equation that gives a possibility to apply this notion to the Schrödinger equation for quantum mechanics.

SYMMETRY OF EQUATIONS AND SOLUTIONS

We will analyze the concept of the symmetry of mathematical equations, the notion of symmetry of a physical system and the connection of these symmetries with integrals of motion. This analysis is essential for nonstationary quantum systems.

The following expressions are employed in the physics literature: symmetry, a group of symmetries, dynamic symmetry, hidden symmetry, broken symmetry, intrinsic symmetry, internal symmetry algebra, spontaneous symmetry braking, geometric symmetry, Lie symmetry, non-Lie symmetry, an invariance group, a spectrum-generating algebra, a nonvariance group, and symmetry algebra. The abundance of terms reflects, on the one hand, the various aspects of symmetry that are important in different applications, and, on the other hand, the increasing importance of exact definitions (in the mathematical sense) for reducing the possibility of different interpretations arising in approaches using the concept of symmetry.

We will follow the approach of studies [1–3] in discussing our concept of the words "symmetry of equations". Let there be a function $\phi(x)$. This function may be a finite- or infinite-component column, or a matrix with elements depending on the variables $x_1, \dots, x_N$. We will consider the relation

$$\hat{A}\phi = 0 \tag{1}$$

where the operator $\hat{A}$ may be linear, nonlinear, differential, integrodifferential, i.e., in any form. Expression (1) shows the symbolic mathematical relation corresponding to the equation under discussion. This may be a system of nonlinear equations from a Liouville system, a Korteweg-de Vries equation, a linear Schrödinger or Dirac equation, or Maxwell equations. Normally the symmetry of equation (1) is understood to mean the set of transformations of the coordinates $x_1, \dots, x_N$ and the functions ϕ (substitution of variables) that form a group and do not change the form of the equation. We will understand the symmetry of an equation in a somewhat broader sense of this word based on the following concept. In all the cases when we speak of the symmetry of equation (1) we are implying that if a certain solution of this equation is found, symmetry transformations may be used to obtain a different solution of this same equation. Hence we will provide a definition of the symmetry of equation (1).

Definition of the symmetry of an equation. We will call the symmetry of equation (1) the set of operators $\hat{B}_\alpha$ (α is a set of indices labeling the operators) such that they convert solutions α of equation (1) into solutions of the same equation, i.e. the following equality is satisfied

$$\hat{A}(\hat{B}_\alpha\phi) = 0. \tag{2}$$

The operators may be linear, nonlinear, differential, integrodifferential, or may simply indicate a substitution of variables. Normally the symmetry of equation (1) is understood to mean the set of operators that do not change the form of the equation, while we will call the substitution of dependent and independent variables a geometric symmetry. This may be understood as follows. If we map the numbers $\phi, x_1, x_2, \dots, x_N$, corresponding to the solutions of (1) by points in space, the operations that are geometrically descriptive are the substitution of variables that in this case convert certain points (solutions) into other points (solutions).

We should point out at the outset that it also makes sense to consider the symmetry of sets of equations (systems of equations) rather than a single equation (equation system (1))

$$\hat{A}_\mu\phi_\mu = 0, \quad \mu = 1, 2, \dots . \tag{3}$$

Here the symmetry of a set of these equations may be a set of operations $\hat{B}_{\alpha}^{k}$ that convert solutions of a single μ-th equation from the entire set into a solution of another different μ'-th equation from the entire set of equations, i.e.,

$$\hat{A}_{\mu'}(\hat{B}_{\alpha}^{k}\phi_{\mu}) = 0. \tag{4}$$

This definition essentially refers to what today is called a Backlund transformation.

We will also give a definition of the symmetry of a physical system. Since there are no other exact methods of describing physical systems other than in the language of mathematical equations, the symmetry of a physical system will be understood as the symmetry of the equation describing this physical system.

We discussed the concept of the symmetry of equations and the symmetry of physical systems. Now we will discuss one additional important concept: the "symmetry of solutions". Thus, by definition, this symmetry operation of an equation converts one solution to another solution of the same equation. However, for a certain set of symmetry operations of an equation, solutions exist which, under the action of these symmetry operations, do not go over to other solutions, but rather stay the same. Naturally different sets of unchanging solutions correspond to different subsets of such equation-symmetry operations.

We will call the "symmetry of a solution" the subset of equation operation symmetries that leave the solution the same. The symmetry of a solution does not coincide with the symmetry of an equation; the former contains significantly fewer operations than the symmetry of an equation. Different solutions have different symmetries. Thus, for the case of a hydrogen atom whose Schrödinger equation has spherical symmetry, only solutions with a zero orbital angular momentum will have this symmetry. In a particular physical system, namely, a harmonic three-dimensional oscillator, only the ground state will have unitary symmetry; the remaining states will go over to other states with the same energy under the action of unitary rotations.

One additional important point is that for many years the concept of symmetry has been identified with a symmetry group, in the belief that the words "the symmetry of a physical system" and "the symmetry group of the physical system" denote the same idea. Recently there has been a trend to differentiate these concepts and to identify a "symmetry group of a physical system" as a broader concept than the concept of the "symmetry of a physical system". From a mathematical point of view this means that in symmetry approaches in physics it is necessary to use both group representation theory and to use other mathematical schemes as well (algebras, superalgebras, supergroups, etc.).

The symmetry of the Hamiltonian (defined below) of a physical system is closely related to the integrals of motion of the physical system. In recent years there has been interest in the study (particularly within the scope of quantum mechanics) of quantum integrals of motion of nonstationary systems for which energy is not conserved. It is believed that the broader the symmetry of a physical system the larger the number of integrals of motion of the system (conservation of all three components of the angular momentum for a spherically-symmetric Hamiltonian in the classical quantum cases, and conservation of a single component of the angular momentum in axially-symmetric potentials). Accounting for the possible explicit time-dependent of the physical quantities which are conserved along the trajectories realized in classical and quantum mechanics has made it possible to analyze the relationship between the symmetry of a Hamiltonian and the number of integrals of motion, which has turned out to be different — specifically: the number of invariants is identical for any symmetry of the physical system. In what follows we will limit our examination to quantum mechanics within the scope of linear equations.

2. QUANTUM INTEGRALS OF MOTION

In this section we will discuss the concepts of the "symmetry of a physical system" and the "integrals of motion of quantum systems" and the relation between these concepts in greater detail. We will incorporate this examination into quantum systems described by a Schrödinger equation, although the discussion below is also valid for classical systems and relativistic systems described by, for example, a Dirac equation. We will examine a finite-dimensional quantum system described by the Schrödinger equation $(i\hbar\partial/\partial t - \hat{H})\psi = 0$. In (1) the operator $\hat{A} = i\hbar\partial/\partial t - \hat{H}$, and the function ϕ is the wave function ψ.

We note that the existence of a Hamiltonian implies the existence of an evolution operator $\hat{U}(t_2, t_1)$, $|\psi(t_2)\rangle = \hat{U}(t_2, t_1)\,|\psi(t_1)\rangle$, and vice versa, the concept of a wave function $\psi(t)$ itself implies the existence of an evolution operator $\hat{U}(t_2, t_1)$ given for all times $-\infty < t_1 \leq t_2 < \infty$. Thus, a quantum system is understood to be a physical system with a unitary (this is not mandatory, and is used for simplicity) evolution operator $\hat{U}(t_2, t_1)$. Other systems are never considered within the framework of quantum mechanics.

One question may be asked: how many independent integrals of motion does a quantum system have, and does the number of such integrals depend on the form of the potential and the symmetry of this Hamiltonian? At first glance, as we noted above, the answer is positive. Thus, if we use a harmonic three-dimensional oscillator with identical frequencies along all axes, all three angular momentum components will be conserved due to the spherical symmetry. If the frequency along one of the axes is changed, only axial symmetry remains and only the component of the angular momentum along the axis is conserved. It seems that the number of integrals of motion has been reduced. Now if all three frequencies are made equal, there is no conservation of the angular momentum. It may be demonstrated that the number of integrals of motion has been reduced to one: the conserved quantity is the energy of the anisotropic harmonic oscillator. Now if the various oscillator frequencies are made to be time-dependent (the energy is no longer conserved) it would seem that no integral of motion (ignoring the parity operators) remains. However, this is not the case: the number of independent integrals of motion is the same for all cases discussed here.

One additional important aspect requires discussion. We have used the word "symmetry of the Hamiltonian" and for definiteness we are referring to quantum mechanics here. The concept seems obvious as though no additional explanation is necessary and the symmetry of the Hamiltonian is understood to be the symmetry of the physical system. However, there are many interpretations in precisely this area: different authors impart different meanings to the words "symmetry of a physical system", "symmetry of a Hamiltonian", and "symmetry of equations", describing a physical system, which was discussed above. When we speak of the spherical symmetry of a three-dimensional isotropic harmonic oscillator, what we have in mind is that the obvious geometric property of the potential does not change with any rotation of the measuring system. However, the concept of symmetry may be far more comprehensive than rotations in coordinate space even in this particular case.

We will call the symmetry of a quantum system the set of operators that convert a state (a solution of the Schrödinger equation) into other states (solutions).

As noted above, the concept of the symmetry of a physical system is often identified with a symmetry group. Thus, a set of symmetry operators is often considered a group. This is obligatory, however. From the viewpoint of the definition above, the set of symmetry operators may be another mathematical object rather than a group (it is not obligatory to require the existence of inverse operators; it is possible to use a Lie algebra, a superalgebra, etc.).

Symmetry is a rather broad and all-encompassing concept. Hence, we will differentiate between the concepts of a symmetry group and the symmetry of a physical system. A symmetry group is only a portion of the symmetry operators of a physical system.

One additional area requiring explanation is the form of the symmetry operators. Up to this point we have considered the symmetry of equations describing a physical system limiting the discussion to first-order derivative operators (symmetry in the Lie sense). We have already discussed the terms "Lie symmetry" and "non-Lie symmetry". The "Lie symmetry" of an equation (in this case the Schrödinger equation) contains, for the case of operators forming a Lie algebra, operators that are in linear derivative form. The words "non-Lie symmetry" refer to sets of symmetry operators containing more than linear derivative expressions. The set of first order derivative symmetry operators comprises only a small portion of the symmetry operators which may also be second and higher order derivatives as well as integral and integrodifferential operators. We only require that the symmetry operators convert any solution of an equation into a solution of the same equation.

We will discuss in some detail the properties of the integrals of motion of a dynamical system. We will consider dynamical systems as all physical and mathematical tasks that finally reduce to a solution of the equation

$$\partial\psi/\partial\tau = \hat{A}\psi, \tag{5}$$

the function ψ may be a vector from a finite-dimensional space (for example, a spin problem), a scalar wave function, or a finite- and infinite-component wave function. The parameter τ may have a variable physical meaning, as may the operator $\hat{A}$. If $\tau = -it/\hbar$ and $\hat{A} = \hat{H}$, where t is time, $\hbar$ is Planck's constant and $\hat{H}$ is the Hamiltonian, we have the Schrödinger equation. However, many other problems of physics and mathematics are described by this relation. The discussion below holds for all such dynamical systems. Specifically, any Lie group may be considered as dynamical system.

A quantum integral of motion is an operator acting in the state space of the physical system whose average value does not change during the evolution of the system. We will carry out our examination in Schrödinger representation and make some statements.

As invariant $\hat{I}$ for a system possessing a Hamiltonian satisfies the condition

$$[i\hbar\partial/\partial t - \hat{H}, \hat{I}]\,|\psi\rangle = 0, \tag{6}$$

where the vector $|\psi\rangle$ satisfies the Schrödinger equation, if $\hat{H}$ is a Hermitian operator.

The average value of the $\hat{I}$ invariant is time-independent (by definition)

$$\frac{d}{dt}\langle\psi(t)|\,\hat{I}(t)\,|\psi(t)\rangle = 0 \tag{7}$$

due to equation (6). Condition (7) is normally used as the definition of the operator integral of motion. If a system with a Hamiltonian has a set of invariants, arbitrary functions of these, specifically their commutators or anticommutators, are also invariant.

The evolution operator $\hat{U}(t_2, t_1)$ of a dynamical system by definition changes the $|\psi(t_1)\rangle$ state into the $|\psi(t_2)\rangle$ state:

$$|\psi(t_2)\rangle = \hat{U}(t_2, t_1)\,|\psi(t_1)\rangle\,. \tag{8}$$

Its matrix element in the x-representation $\langle x|\,\hat{U}(t_2,t_1)\,|x'\rangle$ is the Feynman amplitude of $G(x, x', t_2, t_1)$ (the Green's function). Often it is convenient to set $t_1 = 0$ and $t_2 = t$, $\langle x|\,\hat{U}(t)\,|x'\rangle = G(x, x', t)$. It is possible to examine the Green's function in any other representation, which corresponds to an examination of the matrix elements of the $\hat{U}(t)$ operator in all possible bases, including an "oblique" basis: $\psi(x,t) = \langle x|\,\hat{U}(t)\,|\psi\rangle$ is the wave function in the coordinate representation.

The integrals of motion and the Green's function are related by a number of relations. The Green's function in the coordinate representation is a solution of an eigenvalue system of equations.

We will prove the statements above.

Any operator $\hat{I}(t) = \hat{U}(t)\hat{I}(0)\hat{U}^{-1}(t)$ is an operator-integral of motion.

The proof of the theorem is trivial and is reduced to testing for the time-dependence of the mean value of the operator $\hat{I}(t)$. Indeed, the mean value of the operator $\hat{I}(t)$ takes the form

$$\langle\hat{I}(t)\rangle = \langle\psi(0)|\,\hat{U}^{+}(t)\hat{U}(t)\hat{I}(0)\hat{U}^{-1}(t)\hat{U}(t)\,|\psi(0)\rangle\,. \tag{9}$$

We will use the properties $\langle\psi(t)| = \langle\psi(0)|\hat{U}^{+}(t)$, $\hat{U}^{-1}\hat{U} = \hat{E}$. Since the operator $\hat{U}(t)$ is unitary, we have the equality

$$\langle\hat{I}(t)\rangle = \langle\psi(0)|\,\hat{I}(0)\,|\psi(0)\rangle\,, \tag{10}$$

i.e. the mean value of the $\hat{I}(t)$ operator is time-independent. According to the definition of the operator-integral of motion, the $\hat{I}(t)$ operator is an integral of motion. Let the $\hat{I}(0)$ operators be selected in the following form. We will use n operators $\hat{x}_i$ as the $\hat{I}(0)$ operator for a system with n degrees of freedom. Then consistent with the proven statement, the operators we have designated as $\hat{x}_{0i} = \hat{U}(t)\hat{x}_i\hat{U}^{-1}(t)$, $i = 1, 2, \dots, n$, are integrals of motion. These operators commute, since $[\hat{x}_{0i}, \hat{x}_{0j}] = [\hat{U}\hat{x}_i\hat{U}^{-1}, \hat{U}\hat{x}_j\hat{U}^{-1} = 0$. What is the physical meaning of the operators $\hat{x}_{0i}$? If we measure the mean value of the coordinate $\hat{x}(0) = \hat{x}$ in the state $|\psi(0)\rangle$ at the time $t = 0$, we obtain the number $\langle\psi(0)|\,\hat{x}\,|\psi(0)\rangle$. If at time t we measure the mean of the quantity described by the Hermitian operator $\hat{x}_0(t)$ (the Hermiticity derives from the Hermiticity of the $\hat{x}$ operator and the unitary of the $\hat{U}(t)$ operator), we obtain the same number. Thus the integral of motion $\hat{x}_0(t)$ describes the initial value of the system coordinate. Analogously we may formulate n integrals of motion selecting the momentum operators $\hat{p}$ ($\hat{p} = -i\hbar\partial/\partial x$ in the coordinate representation) as the $\hat{I}(0)$ operators. Then the n operators $\hat{p}_{0i} = i\hbar\hat{U}(t)\partial/\partial x_i\hat{U}^{-1}(t)$ are integrals of motion, where $\langle\hat{p}_{0i}\rangle = \langle\psi(t)|\,\hat{p}_{0i}(t)\,|\psi(t)\rangle$ and $\langle\hat{p}_{0i}\rangle = \langle\psi(0)|\,-i\hbar\partial/\partial x_i\,|\psi(0)\rangle$. The physical meaning of the integrals of motion $\hat{p}_{0i}(t)$ is that they describe the initial mean values of the i-th momentum component. Thus any system with n degrees of freedom always has $2n$ integrals of motion $\hat{x}_{0i}(t)$, $p_{0i}(t)$, $\hat{x}_0 = \hat{U}(t)\hat{x}\hat{U}^{-1}(t)$, $\hat{p}_0 = \hat{U}(t)\hat{p}\hat{U}^{-1}(t)$ describing the initial mean coordinates of the system in its phase space. We note that all operators are examined in the Schrödinger representation and the operators $\hat{x}_0(t)$ and $\hat{p}_0(t)$ are not the same as the coordinate and momentum operators in the Heisenberg representation.

Any operator-integral of motion $\hat{I}(t)$ always takes the form $\hat{I}(t) = \hat{U}(t)\hat{I}(0)\hat{U}^{-1}(t)$.

We will carry out the proof in the following manner. The fact that $\hat{I}(t)$ is an integral of motion indicates, by definition of the integral of motion, the existence of the equality

$$\langle\psi(t)|\,\hat{I}(t)\,|\psi(t)\rangle = \langle\psi(0)|\,\hat{I}(0)\,|\psi(0)\rangle\,. \tag{11}$$

Replacing the vector $|\psi(t)\rangle$ by the vector $|\psi(t)\rangle = \hat{U}(t)\,|\psi(0)\rangle$, we obtain from (11) the equality

$$\langle\psi(0)|\,\hat{U}^{+}(t)\hat{I}(t)\hat{U}(t)\,|\psi(0)\rangle = \langle\psi(0)|\,\hat{I}(0)\,|\psi(0)\rangle\,. \tag{12}$$

Then, since (12) is valid for any vector $|\psi(0)\rangle$ from the state space of the system, the operator equality below derives from this equality for the matrix elements

$$\hat{U}^{+}(t)\hat{I}(t)\hat{U}(t) = \hat{I}(0), \tag{13}$$

or the necessary equality

$$\hat{I}(t) = \hat{U}(t)\hat{I}(0)\hat{U}^{-1}(t). \tag{14}$$

Here we use the unitary of the evolution operator $\hat{U}^{+}(t) = \hat{U}^{-1}(t)$ and the well-known property of the complex state space where the bilinear form $\langle\phi|\,\hat{I}\,|\psi\rangle$ is determined by its quadratic form.

The eigenvalues of the operators-integral of motion are time-independent.

From the viewpoint of physics, this statement is obvious. However, if we do not incorporate physical concepts, this statement must be proven, which we shall do using the preceding theorems.

We will consider the operator $\hat{I}(0)$, the integral of motion $\hat{I}(t)$ at the time $t = 0$, and will find its eigenfunctions $|\phi_\lambda(0)\rangle$ and the eigenvalues λ which, naturally, are time-independent

$$\hat{I}(0)\,|\phi_\lambda(0)\rangle = \lambda\,|\phi_\lambda(0)\rangle\,. \tag{15}$$

We act on both sides of equality (15) by the evolution operator $\hat{U}(t)$ from the left and represent the vector $\hat{U}(t)\,|\phi_\lambda(0)\rangle$ in the form $|\phi_\lambda(t)\rangle$. Then we have the equality

$$\hat{U}(t)\hat{I}(0)\hat{E}\,|\phi_\lambda(0)\rangle = \lambda\,|\phi_\lambda(t)\rangle\,. \tag{16}$$

Since the operator $\hat{E} = \hat{U}^{-1}(t)\hat{U}(t)$ and $\hat{U}(t)\hat{I}(0)\hat{U}^{-1}(t) = \hat{I}(t)$, the equality (16) can be rewritten in the form

$$\hat{I}(t)\,|\phi_\lambda(t)\rangle = \lambda\,|\phi_\lambda(t)\rangle \tag{17}$$

where the eigenvalues λ of the $\hat{I}(t)$ operator are time-independent by formulation. The operators $\hat{I}(t)$ are related through a unitary evolution operator; hence they have an identical spectrum. Thus, the $\hat{I}(t)$ operator has no other eigenvalues aside from the numbers λ, and all these terms, as we have shown, are time-independent.

The square of the operator-integral of motion is an integral of motion.

Indeed, an evolution operator exists in a system with a Hamiltonian; we will examine the operator $\hat{J}(t) = \hat{I}(t)\hat{I}(t) = \hat{I}^2(t)$, where $\hat{I}(t)$ is an integral of motion. This means that, consistent with the preceding theorems, the operator $\hat{I}(t) = \hat{U}(t)\hat{I}(0)\hat{U}^{-1}(t)$. However, then $\hat{J}(t) = \hat{U}(t)\hat{I}(0)\hat{U}^{-1}(t)\hat{U}(t)\hat{I}(0)\hat{U}^{-1}(t) = \hat{U}(t)\hat{J}(0)\hat{U}^{-1}(t)$. This relation demonstrates that the operator $\hat{J}(t) = \hat{U}(t) \times \hat{J}(0)\hat{U}^{-1}(t)$, where the operator $\hat{J}(t) = \hat{I}^2(t)$ is an integral of motion. It is easy to prove by induction that any power of the operator-integral of motion is an integral of motion. Hence, any function of the operator-integral of motion for systems with a Hamiltonian (evolution operator) is an integral of motion. We will represent this function in a power series expansion of the operator; each term of the series is an integral of motion. If a system with a Hamiltonian has several different integrals of motion, their products are also integrals of motion. This is proven by the same method we used above to prove that the square of an integral of motion is an integral of motion.

One important thing we utilized in the proof of the fact that the square of an integral of motion is an integral of motion is the presence of a Hamiltonian in the system (although physically this statement seems obvious). Nonetheless, if a quantum system is not described by a Hamiltonian (open systems) but rather by a density matrix subject to a general kinetic equation, then generally speaking, the square of the integral is not an integral of motion for such systems. The operator-integrals of motion $\hat{x}_{0i}(t)$, $\hat{p}_{0i}(t)$ and $\hat{E}$ form the basis of the Lie algebra of a Heisenberg group. The enveloping algebra of this Lie group contains the entire infinite set of integrals of motion of any Hamiltonian quantum system. The integrals of motion $\hat{x}_{0i}(t)$ and $\hat{p}_{0i}(t)$ are independent integrals of motion.

The integral of motion changes a solution of the Schrödinger equation into a solution of the same equation.

The proof of this important statement is reduced to its test. Let $\hat{I}(t)$ be an integral of motion. We will then represent it in the form $\hat{I}(t) = \hat{U}(t)\hat{I}(0)\hat{U}^{-1}(t)$. The solution of the Schrödinger equation $|\psi(t)\rangle$ can be written in the form $|\psi(t)\rangle = \hat{U}(t)\,|\psi(0)\rangle$. We will also use the operator $\hat{I}(t)$ to act on the solution $|\psi(t)\rangle$. We will obtain a new function $|\phi(t)\rangle$ with the form

$$|\phi(t)\rangle = \hat{I}(t)\,|\psi(t)\rangle = \hat{U}(t)\hat{I}(0)\hat{U}^{-1}(t)\hat{U}(t)\,|\psi(t)\rangle .$$

However, since the function $|\phi(t)\rangle$ takes the form $|\phi(t)\rangle = \hat{U}(t)\,|\phi(0)\rangle$ where $|\phi(0)\rangle = \hat{I}(0)\,|\psi(0)\rangle$, it satisfies the Schrödinger equation due to the fact that the evolution operator satisfies the equation $i\hbar\partial\hat{U}/\partial t = \hat{H}\hat{U}$.

An important element is the fact that by using the operator-integrals of motion we may obtain the solutions of the Schrödinger equation that are not only found in the unity-normalized or δ-function normalized class of solutions describing the state space of the quantum system.

Green's function is an eigenfunction of an integral of motion.

Let us explain one additional aspect of integrals of motion. It turns out that Green's function is a solution of a new equation defined by the invariants $\hat{x}_{0i}(t)$, $\hat{p}_{0i}(t)$. Thus, it follows from the physical meaning of the Green's function as the amplitude of the transition from an initial point to a final point and the physical meaning of the integral of motion $\hat{x}_{0i}(t)$, as the initial coordinate operator that

$$\hat{x}_{0i}(t)G(x_i, x_t', t) = x_i' G(x_i, x_t', t). \tag{18}$$

It is easy to prove the validity of one additional relation, specifically,

$$\hat{p}_{0i}(t)G(x_i, x_t', t) = i\hbar\frac{\partial}{\partial x_i'}G(x_i, x_i', t). \tag{19}$$

Evidently, these relations are valid for arbitrary dynamical systems, i.e. systems described by equations of the Schrödinger type with evolution operators.

Relations (18), (19) determine the Green's function accurate to within a phase multiplier which is time-dependent and determined from the Schrödinger equation. We often see that it is much easier to solve the set of equations (18), (19) than a normal equation for the Green's function if, of course, we independently know beforehand the explicit form of all integrals of motion. This approach is analogous to that in classical mechanics where knowledge of all invariants is equivalent to finding a solution to the equations of motion or the action as a generating function of a canonical transformation representing the transition from initial points x_{0i}, p_{0i} to the current points $x_i(t)$, $p_i(t)$. It is the same in quantum mechanics: knowledge of all integrals of motion makes it possible to determine the evolution of the system, i.e., find the Green's function. According to a discussed general definition of the symmetry operators (2) of the equation (1) applied to the Schrödinger equation the integrals of motion of a quantum system form a set of symmetry operators of this system and this is an important link between the integrals of motion of the system and its symmetry.

The integrals of motion $\hat{x}_{0i}(t)$ and $\hat{p}_{0i}(t)$ form a Lie algebra of a Heisenberg group. These may be used to construct the generators of a symplectic group (a set of quadratic forms for the operators $\hat{x}_{0i}(t)$, $\hat{p}_{0i}(t)$). Hence it may be stated that for any physical system there exists a symmetry described by a symplectic group whose generators relate all solutions of Schrödinger's equation. It can also be said that the symmetry groups of quantum and classical systems are identical since there is a unique correlation between the classical and quantum integrals of motion: the initial points of the system in its phase space. We may also state that there exists in any physical system a preassigned symmetry in the sense noted above, since the integrals of motion $\hat{x}_{0i}(t)$ and $\hat{p}_{0i}(t)$ may be used to construct generators (representations) of any groups; we have the right to limit our discussion not only to the generators but to any functions of the generators, expanding to an enveloping algebra. Thus it may be said that a hydrogen atom has a symmetry group such as $SU(10)$. Technically the construction of the action of the generators of this group involves mapping the space of bound states of the hydrogen atom (having a countable base) onto the state space of a ten-dimensional oscillator and the action of the generators is then represented by the normal formulae. Although it may be demonstrated that such a statement of the existence of an arbitrary symmetry in a given physical system is not significant, nonetheless if it is possible to explicitly determine the symmetry operators which on one occasion form a single Lie algebra and on another occasion form another Lie algebra for the same system, this may be a useful tool since it will make it possible to use available mathematical developments to obtain new physical results. In the familiar example of the motion of a charge in a magnetic field, we may state that the symmetry of this task is described by both groups $U(2,1)$ and $ISP(r,R)$. In this problem we can find explicitly, the generators — the integrals of motion, constructed from the invariants $\hat{x}_{0i}(t)$ and $\hat{p}_{0i}(t)$ and forming Lie algebras of these two different groups. The matrix elements of the evolution operator can be related to the matrix elements of the representations of any of these groups. Again we emphasize that the problem of finding the evolution operator $\hat{U}(t)$ satisfying the Schrödinger equation

$$i\hbar\partial\hat{U}/\partial t = \hat{H}\hat{U} \tag{20}$$

is equivalent to finding all independent integrals of motion $\hat{I}(t)$ satisfying the equation

$$\frac{\partial\hat{I}}{\partial t} + \frac{i}{\hbar}[\hat{H}\cdot\hat{I}] = 0. \tag{21}$$

If the operator $\hat{U}(t)$ is found, we construct all integrals of motion by the formula (14) selecting any independent operators as $\hat{I}(0)$. And vice versa, if all independent integrals of motion are found, we can find the matrix element of the evolution operator.

We note that we can assume the integrals of motion $\hat{I}(t)$ are Heisenberg operators. However, the relation of Heisenberg operators $I_H(t)$ to the integrals of motion $\hat{I}(t)$ is as follows

$$I_H(t) = \hat{U}^{-2}(t)\hat{I}(t)\hat{U}^{2}(t). \tag{22}$$

When the Hamiltonian of the system H is time-independent, the value of the Heisenberg operator makes it possible to immediately find the integral of motion consistent with the formula

$$\hat{I}(t) = \hat{I}_H(-t). \tag{24}$$

In this case if you know the solutions of the Heisenberg equations, the integrals of motion are also known.

3. INTEGRALS OF MOTION FOR THE 'OSCILLATOR

In this section we will demonstrate time-dependent integrals of motion for a basic quantum system — the 3-dimensional oscillator. Simultaneously, we will discuss the connection of the symmetry of a potential with the quantity and the form of time-dependent integrals of the motion. In fact, the concept introduced for the symmetry of equations demands a new consideration of Noether's theorem.

We will now examine how the familiar integrals of motion — namely the angular momentum projections — change for a three-dimensional harmonic oscillator when the symmetry is gradually broken. First we will select an isotropic oscillator (spherical symmetry of the potential), then we will select one frequency different from the other two frequencies (axial symmetry), and finally we will select an oscillator with three different frequencies. Thus, we may first write the Hamiltonian of the isotropic oscillator in the form

$$\hat{H}_s = \frac{\hat{p}_x^2 + \hat{p}_y^2 + \hat{p}_z^2}{2m} + \frac{m\omega^2}{2}(\hat{x}^2 + \hat{y}^2 + \hat{z}^2). \tag{25}$$

Consistent with Noether's theorem, when there is symmetry of the Hamiltonian with respect to the rotation group (in both the classical and quantum cases) three quantities are preserved $\hat{L}_x$, $\hat{L}_y$, $\hat{L}_z$: projections of the angular momentum $L = \hat{r} \times \hat{p}$

$$\hat{L}_x = \hat{y}\hat{p}_z - \hat{z}\hat{p}_y, \quad \hat{L}_y = \hat{z}\hat{p}_x - \hat{x}\hat{p}_z, \quad \hat{L}_z = \hat{x}\hat{p}_y - \hat{y}\hat{p}_x. \tag{26}$$

The commutation relations take on the form

$$[\hat{L}_x, \hat{L}_y] = i\hbar\hat{L}_z, \quad [\hat{L}_y, \hat{L}_z] = i\hbar\hat{L}_x, \quad [\hat{L}_z, \hat{L}_x] = i\hbar\hat{L}_y. \tag{27}$$

All three projections of the momentum $\hat{L}_x$, $\hat{L}_y$, $\hat{L}_z$ commute with the Hamiltonian $\hat{H}_s$

$$[\hat{L}_x, \hat{H}_s] = 0, \quad [\hat{L}_y, \hat{H}_s] = 0, \quad [\hat{L}_z, \hat{H}_s] = 0, \tag{28}$$

resulting in the preservation of the mean values of $\langle\hat{L}_x\rangle$, $\langle\hat{L}_y\rangle$, and $\langle\hat{L}_z\rangle$ over time. We will now examine the Hamiltonian of a less symmetry system, specifically

$$\hat{H}_\alpha = \frac{\hat{p}_x^2 + \hat{p}_y^2 + \hat{p}_z^2}{2m} + \frac{m\omega^2}{2}(\hat{x}^2 + \hat{y}^2)\frac{m\Omega^2}{2}\hat{z}^2. \tag{29}$$

The potential for this oscillator has axial symmetry, and according to Noether's theorem, the $\hat{L}_z$-component of the angular momentum is preserved, since

$$[\hat{L}_z, \hat{H}_\alpha] = 0. \tag{30}$$

However, now $[\hat{L}_x, \hat{H}_\alpha] \neq 0$, $[\hat{L}_y, \hat{H}_\alpha] \neq 0$. It is obvious that these projections of the angular momentum are no longer integrals of motion. We remember, by referring back to the preceding examination of n-dimensional quantum system, that, since the Hamiltonian $\hat{H}_\alpha$ is the sum of single-dimensional Hamiltonians, we may formulate six integrals of motion:

$$\begin{aligned} \hat{x}_0(t) &= \hat{x}\cos\omega t - \hat{p}_x \sin\omega t/m\omega, \quad \hat{p}_{x0}(t) = m\omega\hat{x}\sin\omega t + \hat{p}_x\cos\omega t, \\ \hat{y}_0(t) &= \hat{y}\cos\omega t - \hat{p}_y \sin\omega t/m\omega, \quad \hat{p}_{y0}(t) = m\omega\hat{y}\sin\omega t + \hat{p}_y\cos\omega t, \\ \hat{z}_0(t) &= \hat{z}\cos\Omega t - \hat{p}_z \sin\Omega t/m\Omega, \quad \hat{p}_{z0}(t) = m\Omega\hat{z}\sin\Omega t + \hat{p}_z\cos\Omega t. \end{aligned} \tag{31}$$

Here $\hat{p}_x = -i\hbar\partial/\partial x$, $\hat{p}_y = -i\hbar\partial/\partial y$, $\hat{p}_z = -i\hbar\partial/\partial z$. These six quantities satisfy the preservation condition of the mean quantities over time. With $t = 0$ we have $\hat{y}_0(0) = \hat{y}$, $\hat{p}_{y0}(0) = \hat{p}_y$, $\hat{z}_0(0) = \hat{z}$, $\hat{p}_{z0}(0) = \hat{p}_z$. Since according to the theorem formulated above (see subsection (2)) any function of the integrals of motion (for systems with a Hamiltonian) is an integral of motion, we will draft two preserved quantities $\hat{L}_x(t)$, $\hat{L}_y(t)$ from the six invariants in the following manner:

$$\hat{\hat{L}}_x(t) = \hat{y}_0(t)\hat{p}_{z0}(t) - \hat{z}_0(t)\hat{p}_{y0}(t), \quad \hat{\hat{L}}_y(t) = \hat{z}_0(t)\hat{p}_{x0}(t) - \hat{x}_0(t)\hat{p}_{z0}(t). \tag{32}$$

In explicit form these preserved quantities are written through the coordinate and momenta operators

$$\begin{aligned} \hat{\hat{L}}_x(t) &= m\hat{y}\hat{z}(\Omega\sin\Omega t\cdot\cos\omega t - \omega\sin\omega t\cdot\cos\Omega t) \\ &+ \frac{\hat{p}_y\hat{p}_z}{m}\left(\frac{\sin\omega t}{\omega}\cos\Omega t - \frac{\sin\Omega t}{\Omega}\cos\omega t\right) \\ &+ \hat{y}\hat{p}_z\left(\cos\omega t\cdot\cos\Omega t + \frac{\omega}{\Omega}\sin\omega t\cdot\sin\Omega t\right) \\ &- \hat{z}\hat{p}_y\left(\frac{\Omega}{\omega}\sin\omega t\cdot\sin\Omega t + \cos\omega t\cdot\cos\Omega t\right), \\ \hat{\hat{L}}_y(t) &= m\hat{z}\hat{x}(\omega\sin\omega t\cdot\cos\Omega t - \Omega\sin\Omega t\cdot\cos\omega t) \\ &+ \frac{\hat{p}_z\hat{p}_x}{m}\left(\frac{\sin\omega t}{\omega}\cos\Omega t - \frac{\sin\Omega t}{\Omega}\cos\omega t\right) \\ &+ \hat{z}\hat{p}_x(\cos\omega t\cdot\cos\Omega t + \frac{\Omega}{\omega}\sin\Omega t\cdot\sin\omega t) \\ &- \hat{x}\hat{p}_z(\cos\omega t\cdot\cos\Omega t + \frac{\omega}{\Omega}\sin\Omega t\cdot\sin\omega t). \end{aligned} \tag{33}$$

The commutators, as functions of the integrals of motion, are integrals of motion and take the form

$$[\hat{\hat{L}}_x, \hat{\hat{L}}_y] = i\hbar\hat{L}_z, \quad [\hat{\hat{L}}_y, \hat{L}_z] = i\hbar\hat{\hat{L}}_x, \quad [\hat{L}_z, \hat{\hat{L}}_x] = i\hbar\hat{\hat{L}}_y. \tag{34}$$

The integrals of motion are also "Hamiltonians"

$$\hat{\hat{H}} = \frac{p_0^2}{2m} + m\omega^2\frac{\hat{r}_0^2}{2}, \quad \hat{\hat{H}} = \frac{p_0^2}{2m} + m\Omega^2\frac{\hat{r}_0^2}{2}. \tag{35}$$

The integral of motion is also the "square of the momentum"

$$\hat{\hat{L}}^2 = \hat{\hat{L}}_x^2 + \hat{\hat{L}}_y^2 + \hat{L}_z^2. \tag{36}$$

This operator commutes with the other three "momentum projections" $\hat{\hat{L}}_x$, $\hat{\hat{L}}_y$, and $\hat{L}_z$. In reconstructing the symmetry of the potential, i.e., with $\omega = \Omega$, the $\hat{\hat{L}}_x$ and $\hat{\hat{L}}_y$ operators go over to projections of the angular momentum $\hat{L}_x$ and $\hat{L}_y$ while the operator $\hat{\hat{L}}^2$ goes over to the operator of the square of the momentum. Since the integrals of motion $\hat{\hat{H}}$ (35), $\hat{L}_z$ and $\hat{\hat{L}}^2$ (36) commute, it is possible to formulate solutions Schrödinger equation of an axially-symmetric oscillator with a Hamiltonian $\hat{\hat{H}}$, diagonalizing these three invariants:

$$\hat{\hat{H}}(t)\,|nlm\rangle = \hbar\omega(n+3/2)\,|nlm\rangle\,, \quad \hat{\hat{L}}^2\,|nlm\rangle = \hbar^2 l(l+1)\,|nlm\rangle\,, \qquad \hat{L}_z\,|nml\rangle = m\,|nlm\rangle\,. \tag{37}$$

In accordance with the theorem of the eigenvalues of the integrals of motion, the numbers, n, l, and m are time-independent, n is a nonnegative integer: the "principle quantum number", l is a nonnegative integer, the "orbital momentum", m is the magnetic quantum number (already without quotation marks). The states of $|nlm\rangle$ are not stationary (energy is not represented). It is obvious that even though there is no spherical symmetry in our problem, we have found the solutions of $|nlm\rangle$ (they may be written in the coordinate representation explicitly, i.e. it is possible to find the functions $(x, y, z\,|nlm\rangle$), given by the same quantum numbers as the stationary states of a spherically-symmetric isotropic harmonic oscillator.

We will now consider an oscillator whose potential has no axial symmetry; specifically, an oscillator described by the Hamiltonian

$$\hat{H}_n = \frac{\hat{p}^2}{2m} + \frac{m\omega_x^2\hat{x}^2}{2} + \frac{m\omega_y^2\hat{y}^2}{2} + \frac{m\Omega^2\hat{z}^2}{2}\,. \tag{38}$$

The projection of the angular momentum onto the z axis will produce a condition where operator $\hat{L}_z$ will not commute with this Hamiltonian and will no longer be an integral of motion. However, for an asymmetric oscillator it is also possible to formulate six quantities — invariants $\hat{r}_0(t)$ and $\hat{p}_0(t)$:

$$\begin{aligned}
\hat{x}_0(t) &= \hat{x}\cos\omega_x t - \hat{p}_x\sin\omega_x t/m\omega_x, & \hat{p}_{x0}(t) &= \hat{x}m\omega_x\sin\omega_x t + \hat{p}_x\cos\omega_x t,\\
\hat{y}_0(t) &= \hat{y}\cos\omega_y t - \hat{p}_y\sin\omega_y t/m\omega_y, & \hat{p}_{y0}(t) &= \hat{y}m\omega_y\sin\omega_y t + \hat{p}_y\cos\omega_y t,\\
\hat{z}_0(t) &= \hat{z}\cos\Omega t - \hat{p}_z\sin\Omega t/m\Omega, & \hat{p}_{z0}(t) &= \hat{z}m\Omega\sin\Omega t + \hat{p}_z\cos\Omega t.
\end{aligned} \tag{39}$$

From these invariants we may formulate a vector integral of motion $\hat{\hat{L}} = [\hat{r}(t) \times \hat{p}_0(t)]$ with the components

$$\begin{aligned}
\hat{\hat{L}}_x(t) &= m\hat{y}\hat{z}(\Omega\sin\Omega t\cos\omega_y t - \omega_y\sin\omega_y t\cdot\cos\Omega t)\\
&+ \frac{\hat{p}_y\hat{p}_z}{m}\Big(\frac{1}{\omega_y}\sin\omega_y t\cdot\cos\Omega t - \frac{1}{\Omega}\cos\omega_y t\cdot\sin\Omega t\Big)\\
&+ \hat{y}\hat{p}_z\Big(\cos\omega_y t\cdot\cos\Omega t + \frac{\omega_y}{\Omega}\sin\omega_y t\cdot\sin\Omega t\Big)\\
&- \hat{z}\hat{p}_y\Big(\cos\omega_y t\cdot\cos\Omega t + \frac{\Omega}{\omega_y}\sin\omega_y t\cdot\sin\Omega t\Big),
\end{aligned}$$

$$
\begin{aligned}
\hat{\hat{L}}_y(t) &= m\hat{z}\hat{x}(\omega_x \cos\Omega t \cdot \sin\omega_x t - \Omega \sin\Omega t \cdot \cos\omega_x t) \\
&+ \frac{\hat{p}_z\hat{p}_x}{m}\left(\frac{1}{\omega_x}\sin\omega_x t \cdot \cos\Omega t - \frac{1}{\Omega}\sin\Omega t \cdot \cos\omega_x t\right) \\
&+ \hat{z}\hat{p}_x(\cos\Omega t \cdot \cos\omega_x t + \frac{\Omega}{\omega_x}\sin\Omega t \cdot \sin\omega_x t) \\
&- \hat{x}\hat{p}_z(\cos\omega_x t \cdot \cos\Omega t + \frac{\omega_x}{\Omega}\sin\Omega t \cdot \sin\omega_x t); \\
\hat{\hat{L}}_z(t) &= \hat{m}\hat{y}\hat{x}(\omega_y \sin\omega_y t \cdot \cos\omega_x t - \omega_x \sin\omega_x t \cdot \cos\omega_y t) \\
&+ \frac{\hat{p}_x\hat{p}_y}{m}\left(\frac{1}{\omega_y}\sin\omega_y t \cdot \cos\omega_x t - \frac{1}{\omega_x}\sin\omega_x t \cdot \cos\omega_y t\right) \\
&+ \hat{x}\hat{p}_y\big(\cos\omega_x t \cdot \cos\omega_y t + \frac{\omega_x}{\omega_y}\sin\omega_x t \cdot \sin\omega_y t\big) \\
&- \hat{y}\hat{p}_x\big(\cos\omega_y t \cdot \cos\omega_x t + \frac{\omega_y}{\omega_x}\sin\omega_x t \cdot \sin\omega_y t\big).
\end{aligned}
\tag{40}
$$

These three integrals of motion (explicitly dependent on time) commute in the following manner:

$$
|\hat{\hat{L}}_x, \hat{\hat{L}}_y| = i\hbar\hat{\hat{L}}_z, \quad |\hat{\hat{L}}_y, \hat{\hat{L}}_z| = i\hbar\hat{\hat{L}}_x, \quad |\hat{\hat{L}}_z, \hat{\hat{L}}_x| = i\hbar\hat{\hat{L}}_y. \tag{41}
$$

The square of the "momentum" $\hat{\hat{L}}^2 = \hat{\hat{L}}_x^2 + \hat{\hat{L}}_y^2 + \hat{\hat{L}}_z^2$ is also an integral of motion and commutes with all projections $\tilde{L}_x, \tilde{L}_y, \tilde{L}_z$ and with the Hamiltonian

$$
\hat{H} = \frac{\hat{p}_0^2}{2m} + \frac{m\Omega^2}{2}(\hat{x}_0^2(t) + \hat{y}_0^2(t) + \hat{z}_0^2(t)). \tag{42}
$$

For an "asymmetric" oscillator (asymmetric in the sense of an absence of geometric symmetry of its potential with respect to the rotations of the axes of the x, y and z coordinates in space), the "spherical" symmetry with respect to the "rotations represented by the infinitesimal transformations of $\hat{\hat{L}}_x$, $\hat{\hat{L}}_y$, $\hat{\hat{L}}_z$ is nonetheless preserved. Evidently the Schrödinger equation $H\psi = E\psi$ does not possess symmetry, although the temporal Schrödinger equation $i\hbar\dot{\Psi} = H\psi$ has symmetry with respect to these rotations. We may classify all states of an asymmetric oscillator based on the values of the "principle quantum number" n ($n = 0, 1, 2, \dots$) — the preserved eigenvalue of the "Hamiltonian" of the auxiliary oscillator (42), the eigenvalue $l\,(l+1)$ ($l = 0, 1, 2, \dots$) of the "square of the angular momentum" $\tilde{L}^2$ and the "projection" of the angular momentum l_z ($l_z = -l, -l+1, \dots, l-1, l$) — the eigenvalues of the integral of motion $\hat{\hat{L}}_z(t)$ (40). Thus although the geometric spherical symmetry of the potential of an initially isotropic harmonic oscillator (both quantum and classical) was not broken, and the number of integrals of motion was not reduced, their forms nonetheless changed, obtaining a time dependence. At the same time, the symmetry of the anisotropic oscillator also coincides with the symmetry of an isotropic oscillator; as before, it remains "spherical", although this must be understood in accordance with the definition of the symmetry of a physical system as the symmetry of the equation describing this system. This means that the solution of the temporal Schrödinger equation is converted by the symmetry operators (integrals of motion) $\hat{\hat{L}}_x(t)$, $\hat{\hat{L}}_y(t)$,

$\hat{\hat{L}}_z(t)$ into solutions of the same equation. This means that the state of an asymmetric oscillator is converted by these operators into other states of the same oscillator.

At a first glance it may seem that there is an inconsistency with the common understanding of Noether's theorem. Indeed, the spherical symmetry of the potential vanished in the transition from an isotropic oscillator to an anisotropic oscillator, while the number of integrals of motion was not reduced; they only changed their explicit forms. However, again, if we understand the symmetry of the anisotropic oscillator in a more general sense (not as the geometric symmetry of the potential with respect to the rotations in real space, but rather as symmetry in the sense of a set of rotation operators given by the integrals of motion $\hat{\hat{L}}_x$, $\hat{\hat{L}}_y$, $\hat{\hat{L}}_z$ and converting solutions of Schrödinger's equation into other solutions of the same equation), it is possible to treat Noether's theorem in a more general sense as well. The integrals of motion (time-dependent) understood in a more general sense correspond to precisely this symmetry of a system understood in a broader sense; these integrals of motion are nothing other than transformations of a set of symmetry operations. This statement is in some sense a tautology, since, as we have demonstrated, for a quantum system, the set of symmetry operators and the set of integrals of motion coincide. Hence, if a system has symmetry this indicates the existence of integrals of motion, and if there are integrals of motion, this automatically means that the system also has symmetry. What can be said for a classical system regarding the behaviour of an action when using symmetry operators for symmetry understood in the sense described above? These operators convert classical trajectories — solutions of the same equations of motion – into other trajectories – solutions of equations of motion. However, these transformations do not leave a functional: the invariant action. They may convert it to another functional whose extremals coincide with the extremals of the initial action. Thus, if possible to expand the understanding of Noether's theorem to encompass the more broadly understood symmetry of a physical system and the existence of integrals of motion that are time-dependent.

We will again return to a discussion of an anisotropic oscillator with different oscillation frequencies along three coordinate axes. As we demonstrated, three "angular momentum" operators exist, $\ddot{L}_x(t)$, $\ddot{L}_y(t)$, and $\ddot{L}_z(t)$, which are symmetry operators of the temporal Schrödinger equation and at the same time are integrals of motion. However, the symmetry of the system given by the rotation operators-exponents of the infinitesimal operators $\hat{\hat{L}}_x$, $\hat{\hat{L}}_y$ and $\hat{\hat{L}}_z$ is a generalization of only that portion of the symmetry of the isotropic oscillator that describes the degeneration of the isotropic oscillator in terms of the magnetic quantum number (for the case of an anisotropic oscillator, the states with the "principle quantum number" — the eigenvalue of "Hamiltonian" (42) — are also degenerated in terms of "magnetic quantum number" — the eigenvalue of the integral of motion $\hat{\hat{L}}_z(t)$ (40). However, dynamic symmetry operators exist for the isotropic oscillator that, when used, not only change states with a fixed angular momentum, but generally speaking, all states of the oscillator are bound in a single multiplet. The dynamical group of an isotropic oscillator is also completely converted to an anisotropic oscillator. For this it is necessary to utilize integrals of motion $\hat{r}_0(t)$ and $\hat{p}_0(t)$ (39) in formulating the dynamical symmetry generators of the anisotropic oscillator. Dynamical symmetry may be described, for example, as an inhomogeneous symplectic group $ISP(6, R)$ whose generators (integrals of motion) are all linear and quadratic expressions formulated from the integrals of motion $\hat{r}(t)$ and $\hat{p}(t)$. It is clear that in the transition to the isotropic case (i.e. with all oscillation frequencies equal along the coordinate axes) in this uniform manner, the dynamical symmetry generators of the anisotropic oscillator convert to dynamical symmetry generators of an isotropic oscillator.

4. STATES GENERATING SYMMETRY

In this section we want to discuss the relation of the concept of a dynamical symmetry of a stationary quantum systems with respect to the introduced notion of a symmetry of the Schrödinger equation, even with a nonstationary Hamiltonian. The notion of symmetry explains the levels degeneracy, for example, of the systems with rotationally symmetry potentials with respect to magnetic quantum numbers. The hydrogen atom has, in addition, an extra degeneracy.

It was shown by Fock that an "accidental" degeneracy of levels of the discrete spectrum of the hydrogen atom with respect to the orbital angular momentum is described by a "hidden" symmetry with respect to the group of four-dimensional rotations. The "hidden" symmetry of the hydrogen atom becomes explicit by going over from the Schrödinger equation in the momentum representation to a stereographic projection. Then the functions corresponding to the bound states of the hydrogen atom satisfy, in the new variables, a four-dimensional Laplace equation whose symmetry with respect to rotations is obvious. In 1935 Fock also showed that the infinite degeneracy of the states of the continuous spectrum of the hydrogen atom is described by a "hidden" symmetry of the Schrödinger equation with respect to $O(3, 1)$.

In 1965 the next important step generalizing Fock's approach was taken; the essence of this step follows. Before 1965 symmetry operators were assumed to transform states corresponding to some energy level of system to other states with the same energy. Then the set (family or multiplet) of states of an energy level constitute the basis for a representation of the symmetry group of the problem. The idea was to unify in one family (multiplet) states belonging to different energy levels. Idealizing, one irreducible representation of the dynamic group should embrace all the stationary states of a physical system. Thus applying dynamic symmetry operators to the unique stationary states would generate the other stationary states of a quantum system. Thus, adding to Fock's generators of $O(4)$ the group generators complementing it to $O(4, 2)$, we could generate all the excited states of the hydrogen atom's discrete spectrum from the ground state of the hydrogen atom. In this sense $O(4, 2)$, considered as the symmetry group of the hydrogen atom, generalizes the Fock symmetry $O(4)$ description as an "accidental" degeneracy of its energy levels and extends its symmetry so that one irreducible degenerate representation of this (non-compact) group contains all the states of the discrete spectrum of the hydrogen atom. From what follows, it is clear that the dynamic symmetry algebra of a stationary quantum system contains generators which can either commute, or not commute, with the Hamiltonian, i.e., corresponding to interactions causing transitions between states of different energy levels. Note that for the one-dimensional stationary quantum oscillator the symmetry group $O(2, 1)$ which contains generators which do not commute with the Hamiltonian was considered in study [4] in 1959, but it did not begin a new trend.

Thus, let us summarize our understanding of the dynamic symmetry of a stationary quantum system with Hamiltonian $\hat{H}$ in the Schrödinger representation, whose states (wave functions $\psi_{n,i}$) satisfy the stationary Schrödinger equation

$$\hat{H}\psi_{n,i} = E_n\psi_{n,i}. \tag{43}$$

In formula (43) the principal quantum number n labels the energy levels, index i denotes all other quantum numbers that distinguish different states with the same energy. Operators $\hat{S}_\alpha$, such that

$$[\hat{S}_\alpha, \hat{H}] = 0 \tag{44}$$

(they can form a Lie algebra) transform stationary states with the same n into each other:

$$\hat{S}_\alpha\psi_{n,i} = \sum_{i'} S_\alpha^{ii'}\psi_{n,i'} \tag{45}$$

thus they are operators of the symmetry describing the degeneracy of the energy levels of the system. Suppose that $\hat{S}_\alpha$ are complemented by operators $\hat{C}_l$ not commuting with $\hat{H}$ such that

$$\hat{C}_l\psi_{n,i} = \sum_{n'}\sum_{i'} C_l^{ii',nn'}\psi_{n',i'}. \tag{46}$$

If the set of operators $\hat{C}_l$ and $\hat{S}_\alpha$ form a finite-dimensional Lie algebra such that its complete system of wave functions $\psi_{n,i}$ is the basis of one irreducible representation of this Lie algebra, this algebra is called the spectrum generating algebra [4] or noninvariance algebra [6], or dynamic symmetry (algebra) of the system with Hamiltonian $\hat{H}$ [1,5]. To discuss the dynamic symmetry group we clearly need the existence of exponents of $\hat{S}_\alpha$ and $\hat{C}_l$, but frequently in the literature no distinction is made between noninvariance groups and noninvariance algebras.

Usually a dynamic symmetry is connected with the existence of a finite-dimensional algebra, but it is clear from the above that all $\hat{S}_\alpha$ and $\hat{C}_l$ must do is to generate all the states of the system from one state. They can be generators of an infinite dimensional Lie algebra or superalgebra, etc. The tendency is to classify all such cases as "dynamic symmetry". The analysis of this concept shows that if the generators of a hidden symmetry $\hat{S}_\alpha$ commuting with the Hamiltonian (the generators themselves do not depend on time) are integrals of motion, then the additional generators $\hat{C}_l$, not commuting with $\hat{H}$ are not integrals of motion. Thus the dynamic symmetry of a quantum system determined above is constructed by generators which are not integrals of motion. Is it possible to modify the definition of dynamic symmetry to recover the relation of the symmetry with the integrals of motion, i.e., to make the symmetry operators integrals of motion? This question is most significant when we try to generalize the idea of dynamic symmetry or a spectrum generating algebra to the case of systems with time dependent Hamiltonians $\hat{H}(t)$. Such Hamiltonians describe for example, parametric oscillators, charge motion in a variable electromagnetic field, and in general, any system if we take into account that our universe is not stationary.

For nonstationary quantum systems the concept of a dynamic symmetry group connected with the stationary Schrödinger equation and the presence of energy levels can not be generalized directly since this system itself has no energy levels. The problem of whether or not it is possible to construct the dynamic group of a stationary system from generators — quantum integrals — was solved in studies [7,8]. The generalization of spectrum generating algebra to the nonstationary Hamiltonian case, as well as with generators — integrals of motion — was proposed in studies [7,12]. This generalization has been called the state generating algebra of the quantum system [3].

To change the idea of the state generating algebra of a quantum system it is necessary to change from a stationary Schrödinger equation to a Schrödinger equation with time (we use dimensional variables)

$$[i\hbar\partial/\partial t - \hat{H}(t)]\Psi_{n,i}(t) = 0. \tag{47}$$

Here $\psi_{n,i}(t)$ is a time-dependent wave function of the system in Schrödinger representation. It is important that the quantum numbers of a nonstationary system are the same as for a stationary system but their physical meaning is somewhat different. Thus, n does not label the energy levels (they do not currently exist) or the time-dependent eigenvalue $E_n(t)$ of the Hamiltonian, but rather it is an eigenvalue of an operator — integral of motion. Therefore we will discuss the value of equation (47) rather than equation (43). Now if we have operators $\hat{S}_\alpha(t)$ and $\hat{C}_l(t)$ satisfying

$$\begin{aligned}(i\hbar\frac{\partial}{\partial t} - \hat{H}(t))\hat{S}_\alpha(t)\psi_{n,i}(t) = 0,\\ (i\hbar\frac{\partial}{\partial t} - \hat{H}(t))\hat{C}_l(t)\psi_{n,i}(t) = 0,\end{aligned} \tag{48}$$

i.e., transforming the solutions of (47) into solutions we call elements of the state generating algebra or operators of the dynamic symmetry of the nonstationary (or stationary) quantum system. The operators $\hat{S}_\alpha(t)$, $\hat{C}_l(t)$ can form a (finite or infinite-dimensional) Lie algebra, superalgebra, etc. It is important that these operators are integrals of motion of a nonstationary quantum system. Thus we have

$$\hat{S}_\alpha(t)\psi_{n,i}(t) = \sum_{i'} S_\alpha^{ii'}\psi_{n,i'}(t) \tag{49}$$

$$\hat{C}_l(t)\psi_{n,i}(t) = \sum_{n'}\sum_{i'} C_l^{ii',nn'}\psi_{n',i'}(t). \tag{50}$$

We say that we have found the dynamic symmetry of the system, if the set of operators — integral of motion — enables us to recover all the states of the (nonstationary) system from a unique (nonstationary) state. If $\hat{S}_\alpha(t)$ and $\hat{C}_l(t)$ generate a Lie algebra, then the state space of the quantum system is the space of an irreducible representation of this Lie algebra.

Therefore the route from Fock's work to the present is, it could be said, almost trivial (though it required half a century). First operators, called symmetry operators, transformed wave functions of stationary states of the same energy level into each other. Thirty years after the concept of symmetry had been generalized and operators transforming wave functions of stationary states of any energy levels into each other, wave functions of stationary states of any energy levels were attributed to symmetry operators.

Five years after that, the scheme was generalized once more to include nonstationary systems. Symmetry (dynamic symmetry) operators was the name given to operators transforming wave functions which describe all the nonstationary states of the same quantum system. The dynamic symmetry operators depend on Schrödinger's representation on the time and are quantum integrals of motion.

A possible method for further generalization is to study the totality of several systems with different Hamiltonians and construct operators transforming the wave functions of the states of all these systems into each other. This is akin to considering matrix Hamiltonians in supersymmetric quantum mechanics. Knowing such operators and one state of the quantum system we should be able to generate all the states of all other systems. For instance, applying the generators of such a large dynamic symmetry to the ground state of the hydrogen atom we could construct all the states and, for example, the states of a quantum oscillator. This example can be investigated in detail with the use of the known relation between the hydrogen atom problem and that of the four-dimensional harmonic oscillator.

Let us give examples of Hamiltonians describing systems for which a more detailed investigation of symmetry (as in any of the above definitions) and the explicit search of integrals of motion is of interest. For instance, let

$$\hat{H}(t) = \sum_i c_i(t)\hat{L}_i \tag{51}$$

be a well-known Hamiltonian in Schrödinger representation expressed in terms of Lie algebra generators $\hat{L}_i$.

Another example is an operator with the form

$$\hat{H}(t) = \sum_n A_n P^n(\hat{L}_i, t) \tag{52}$$

where P^n is an n-th degree polynomial in $\hat{L}_i$ with scalar coefficients depending on time. In particular, the stationary Hamiltonian can be a Casimir operator.

We have a more complicated example when we have a set of (finite or infinite-dimensional) Lie algebras G_α ($\alpha = 1, \dots, N$) with generators $\hat{L}_i^\alpha$ of some of their representations. We construct a Hamiltonian in the form

$$\hat{H}(t) = \sum_{m,n,\alpha,\beta} P_\alpha^n(\hat{L}_i^\alpha, t) P_\beta^m(\hat{L}_i^\beta, t). \tag{53}$$

The cases (50–52) can be generalized considering superalgebras and replacing polynomials by other functions (exponents, etc.).

Another type of Hamiltonians are those where the energy of interaction is described by

$$\hat{H}(t) = \sum_{m,n,\alpha,\beta} P_\alpha^n(\hat{L}_i^\alpha, t) P_\beta^m(\hat{L}_i^\beta, t) \tag{54}$$

where $\hat{L}_i^\beta$ are superalgebra generators.

Note that almost all the problems of quantum mechanics are described by Hamiltonians of the above type. Thus Hamiltonians of nonstationary quadratic systems [12] belong to type (53) with degrees of polynomials $n + m$ no greater than two for Heisenberg-Weyl algebras (quadratic expressions in Bose creation-annihilation operators). These Hamiltonians can be considered as operators of the form (51) with $\hat{L}_i$ generating a representation of $ISP(2N, R)$ (with $N = \infty$ for field theory). The problems of quantum field theory and solid state physics are also described by Hamiltonians (53), (54) and the nontrivial results, not obtained with the help of perturbation theory, are obtained considering 4th degree polynomials in creation-annihilation operators.

5. CONCLUDING REMARKS

As we have discussed in previous sections the notion of the symmetry of quantum systems based on the definition introduced for the symmetry of the Schrödinger equation gives a possibility to identify this symmetry with the set of quantum time-dependent integrals of motion of the same system. On the other hand, this set may be considered as a state generating symmetry of this system. Time-dependent invariants, which do not coincide with the Heisenberg operators, may be used, for example, to measure the influence of an external force on the system, because the distribution function for these integrals of motion is invariant too. The discussed linear time-dependent invariants for the oscillator may be measured experimentally with a gravitational wave detector in order to find the influence of gravitational perturbation without quantum limitation of the detector sensitivity [3]. The approach to the symmetry and invariants of quantum systems gives a possibility not to distinguish the stationary and nonstationary quantum systems. Thus, the concept of state generating symmetry generalizes the concept of spectrum generating algebra to the cases of nonstationary quantum systems (and it may be applied, of course, to stationary systems too). The states generating symmetry is used to combine into one representation of the symmetry algebra all the states of the quantum system independent of whether this system is stationary or nonstationary. The concrete nonstationary quantum systems as charge in time-dependent electromagnetic field or polydimensional oscillator are considered from the discussed point of view in Ref. [1-3, 7, 15, 16].

REFERENCES

1. I.A. Malkin and V.I. Man'ko, Sov. Phys. JETP Lett. 2:230 (1956), (in Russian).
2. I.A. Malkin, V.I. Man'ko and D.A. Trifonov. Phys. Lett. 30A:414 (1969).
3. V.V. Dodonov and V.I. Man'ko In: *Invariants and the evolution of nonstationary quantum systems*, Ed. by M.A. Markov, Proceedings of Lebedev Physics Institute, 183:103–263, Nova Science Publishers, Commack, 1987.

4. Y. Dothan, M. Gell'mann and Y. Neeman. Phys. Lett. 15:148 (1965).
5. A.O. Barut and A. Bohm. Phys. Rev. B139:1107 (1965).
6. N.O. Mukunda, L. O'Raifeartaigh and E. Sudarshan, Phys. Rev. Lett. 15:1041 (1965).
7. I.A. Malkin, V.I. Man'ko and D.A. Trifonov, Nuovo Cim. 4A:773 (1971).
8. Y. Dothan, Phys. Rev. 2d:2944 (1970).
9. A.O. Barut and H. Kleinert, Phys. Rev. 156:1541 (1967).
10. Y. Namba, Phys. Rev. 160:1171 (1967).
11. Y. Neeman, *Algebraic theory of elementary particles*, Plenum Press, 1967.
12. I.A. Malkin and V.I. Man'ko, *Dynamical symmetries and coherent states of quantum systems*, Moscow, Nauka Publishers, 1979 (in Russian).
13. V.A. Fock, Z. Phys. 98:145 (1935).
14. S. Gohen and H.J. Lipkin, Ann. Phys. 6:301 (1959).
15. I.A. Malkin, V.I. Man'ko and D.A. Trifonov, Phys. Rev. 2D:1371 (1970).
16. I.A. Malkin, V.I. Man'ko and D.A. Trifonov, J. Math. Phys. 14:576 (1973).

ORIGINS OF NUCLEAR AND HADRON SYMMETRIES

Y. Ne'eman

Sackler Faculty of Exact Sciences, Tel-Aviv Univ.*¶
Tel-Aviv, 69978, Israel; and
Center for Particle Theory #
University of Texas
Austin, Texas 78712, USA

Dj. Sijacki

Institute of Physics
P.O.B. 57,+ Belgrade
11001, Yugoslavia; and
Tel-Aviv University*
Tel-Aviv, 69978, Israel

One component of the strongly-coupled IR region in QCD consists of exchange of a phenomenological field $G\mu\nu(x)$ formed by a color-neutral pair of gluons. The $G_{\mu\nu}$ acts formally as a Riemannian metric with $J^P = 0^+, 2^+$ quanta coupled symmetrically to nuclear matter. This is the origin of the IBM paradigm and several features of hadrons.

I. U(6) DYNAMICAL SYMMETRY IN NUCLEI (IBM)

The Interacting Boson Model[1] has been very successful as a dynamicalsymmetry of nuclei. The model's point of departure is the observation that in most even-even nuclei, the lowest excitations are 0^+ and 2^+, with relatively close energies, realized by proton or neutron pairs. The model postula-tes a phenomenological U(6) symmetry between the six states in $(0^+, 2^+)$. The $(0^+, 2^+)$ excitations already appear in the presence of a single nucleon pair (above a closed shell), i.e. their origin may be due to the fundamental forces involved, rather than to collective motion: it seems plausible to look for an explanation in QCD itself.. It is easy to verify that the conventional view of QCD, as involving the exchange of mesons (quark-anti-quark pairs with masses, in the direct channel, between .14 to 1.3 Gev) as its main effective contribution to the force between zero-color (hadron) states, does not fit with an excitation spectrum that skips the 1^- dipoles altogether. Rather, the study of potentials appears to indicate a gravity-like effect[2] - a somewhat surprising feature.

Another system of symmetries characterizing nuclear states has been shown to involve quadrupolar excitations[3]. The groups invoked are either a compact[3] SU(3) or non-compact[4,5] SL(3,R) - or the inhomogeneous[6] T(5)(σ)SO(3) - depending upon the commutation relations between the quadrupole moments. The entire system is contained in[7] Sp(3,R). The role

* Supported in part by the USA-Israel BNSF, Contract 87-00009/1 and by the FDR-Israel GIF, Contract I-52.212.7/87.
¶ Wolfson Chair Extraordinary in Theoretical Physics.
+ Supported in part by Science Foundation (Belgrade).
# Supported in part by the USA DOE Grant DE-FG05-85 ER 40200.

played by a basically gravitational feature in a strong interaction represents yet another indication that these interactions involve a gravity-like component intrinsic to strong interactions.

II. GRAVITY-LIKE FEATURES IN HADRON STRONG INTERACTIONS

The non-perturbative features of QCD have made it difficult, in almost any situation, to apply the theory exactly - except for some vacuum configurations studied especially to prove confinement[8] (yet without completely conclusive results to date) and for the (perturbative) asymptotic domain where QCD was originally identified. Various approximation schemes have been tried, ranging from the non-relativistic quark model (NRQM) to the Bag (BM) and to the Skyrme-Witten models (SWM). Their validity is generally restricted to one or two individual features of the theory, missing most others by a wide range[9].

Before the rise of QCD and throughout the earlier stages in the evolution of the theory, an ad hoc "strong gravity" hypothesis[10] was tried, in which the f^0 meson (with $J=2^+$ and a mass of 1270 MeV) was given a central role as the "strong graviton". It was assumed that at the fundamental level there are two massless 2^+ gauge graviton fields ("f/g gravity") and that their mixing results in one massless (Einstein's) and one massive (f^0) state. However, in view of the f^0's presently acknowledged quark-antiquark structure, its postulated gauge-field nature can now at most be regarded as "phenomenological"[11]. Moreover, the results of f/g gravity were inconclusive and the model was abandoned with the establishment of QCD.

Other features of hadrons and strong interactions bear a certain resemblance to gravity. In particular, it was this resemblance which made it possible for Yoneya and for Scherk and Schwarz[12] to reinterpret the String: turning Dual Models, a highly suggestive description of the hadrons' strong interactions, into a theory of gravity.

Another seemingly gravity-induced feature is the system of Regge trajectories. These exhibit $\Delta j = 2$ intervals in angular momentum, a fact which led Dothan et al.[4] to identify the trajectories with unitary irreducible representations of SL(3,R) and of its double-covering group $\widetilde{SL}(3,R)$[13]. Physically, sl(3,R) is generated by the time-derivatives of gravitational quadrupoles[4] (i.e. pulsation rates), thus involving moments of inertia. The emergence of the inertial features of gravity (related to the gravitational interaction through the Principle of Equivalence) in an essentially strong "nuclear" interaction is yet another piece of this puzzle.

III. QCD EXPLAINS HADRON FLAVOR SYMMETRIES

Iachello has often compared the IBM to the 1961 postulation of SU(3) in particle physics (a "flavor" symmetry in the present dynamical picture) from purely phenomenological considerations and from the identification of the observed patterns. In that example, the true "strong" interaction is now known to be generated by Quantum Chromodynamics (QCD), a force induced by "color" SU(3). It is flavor-invariant, and thus explains the approximate flavor-SU(3) invariance at the historical departure point, while the flavor degrees of freedom themselves stem from other origins.

As a matter of fact, the postulation of SU(3) as an approximate symmetry of the strong interactions met with great resistance originally[14], since it seemed paradoxical that an essentially non-perturbative interaction should lend itself smoothly to a perturbative treatment of the symmetry-breaking, as demonstrated by the success of the Gell-Mann/Okubo, or of the Coleman/Glashow mass formulae etc. The dynamical answer[15] was provided in terms of a separation of the phenomenological strong interactions into an SU(3) invariant part (namely QCD in the present view) carrying the non-perturbative features, and a "Fifth Interaction" (the name was later plagiarized and used to describe the supposed departure from the Principle of Equivalence in the Eötvös experiment). The "Fifth" is a perturbative feature linked with the existence of generations (distinguishing the s quark from the u,d set and the μ and its neutrino from the electronic set). It is presently represented by the ad hoc postulation of heavy quark masses, numerous Yukawa Higgs couplings etc. An understanding of its "fundamental" content will probably have to wait for the results of experiments in the TeV region[16].

In particle physics, the emergence of the flavor SU(3) symmetry is thus understood presently as resulting from the fact that color and flavor SU(3) commute. With QCD induced by color-SU(3), this component of the strong interactions has to be flavor-SU(3) invariant. In various studies[17], we have similarly tried to present dynamical models which could explain the other hadronic algebraic feature[4], namely the Regge sequences.

In what follows, we present a QCD-generated derivation of all the above gravity-like algebraic features, including both the Arima-Iachello "IBM" and the Sp(3,R) quadrupole-related symmetries in nuclei - together with the string-like dynamics, confinement, scaling and Regge systematics in hadrons [18]. At the same time, all results of asymptotic freedom (the UV end) remain unhindered, i.e. the fit to the NRQM stands.

IV. QCD-INDUCED "EFFECTIVE" STRONG GRAVITY

Our basic ansatz is weaker than the full QCD "dogma" that hadron observable states are color-SU(3) singlets. We only assume the proven saturation properties, i.e. that the color-singlet configurations are the lowest lying ones.

Since the hadron lowest ground states are colorless (our assumption) and in the approximation of an external QCD potential (in analogy to the treatment of the hydrogen atom in the Schroedinger equation), the hadron spectrum above these levels will be generated by color-singlet quanta, whether made of dressed 2-gluon configurations, 3-gluons,.. Every possible configuration will appear. No matter what the mechanism responsible for a given flavor state, the next vibrational, rotational or pulsed excitation will correspond to the "addition" of one such collective color-singlet multigluon quantum superposition.

In the fully relativistic QCD theory, these contributions have to come from summations of appropriate Feynman diagrams, in which dressed n-gluon configurations are exchanged. We rearrange the sum by lumping together contributions from n-gluon irreducible parts, $n = 2, 3,..\infty$ and with the same Lorentz quantum numbers. The simplest such system will have the quantum numbers of gluonium, i.e. $n = 2$. The color singlet external field can thus be constructed from the QCD gluon field as a sum (η_{ab} is the SU(3) Killing-Cartan metric, d_{abc} is the totally symmetric $8 \times 8 \times 8 \to 1$ coefficient)

$$B^a{}_\mu B^b{}_\nu \eta_{ab} + B^a{}_\mu B^b{}_\nu B^c{}_\sigma d_{abc} + \ldots \qquad (1)$$

In the above, $B^a{}_\mu$ is the dressed gluon field. It will be useful for the applications to separate the "flat connection" $N^a{}_\mu$, i.e. the zero-mode of the field. Writing for the curvature or field strength) $F^a{}_{\mu\nu} = \partial_\mu B^a{}_\nu - \partial_\nu B^a{}_\mu - i f^a{}_{bc} B^b{}_\mu B^c{}_\nu$, we define

$$B^a{}_\mu = N^a{}_\mu + A^a{}_\mu \ , \text{ so that } F(N) = 0 \tag{2}$$

i.e. $\partial_\mu N^a{}_\nu - \partial_\nu N^a{}_\mu = i f^a{}_{bc} N^b{}_\mu N^c{}_\nu$, or, in form language,

$$dN = N \wedge N \tag{3}$$

Note that (3) implies that $N^a{}_\mu$ is the Cartan left-invariant form of SU(3) in a Soft-Group-Manifold[19] version of the 8+4 dimensional $\{SU(3) \otimes M\}$, M - Minkowski space, after spontaneous fibration[20] and contraction of the holonomic SU(3) indices. Eq. (3) is therefore also the BRS equation for the ghosts C, when we replace d → s and N → C.

We can now rewrite the 2-gluon configuration as

$$G_{\mu\nu}(x) = B^a{}_\mu B^b{}_\nu \eta_{ab} \tag{4}$$

which looks very much like a spacetime metric. It is an <u>effective</u> spacetime metric representing some of the geometric features induced in the spacetime base-manifold of the color-SU(3) Principal Bundle. We assume that (4) is the dominating configuration in the excitation systematics and note that Lorentz invariance forces further assumption that this metric to obey a Riemannian constraint

$$D_\sigma G_{\mu\nu} = 0, \tag{5}$$

where D_σ is the covariant derivative of the "effective gravity", (the connection will be given by a Christoffel symbol constructed with the metric (4)). The separation of the flat part of $B^a{}_\mu$ in Eq. (2) reproduces the separation of a tetrad $e^a{}_\mu(x) = \delta^a{}_\mu + f^a{}_\mu(x)$ into the flat background piece and the quantum gravitational contribution. As a result, $G_{\mu\nu}(x)$ itself can be separated similarly. We note two points:

(1) Out of the 10 components of $G_{\mu\nu}$ in (4), the 6 that survive after the 4 Riemannian constraints (5) have spin/parity assignments $J^P = 0^+, 2^+$. This suggests a relationship with the IBM model systematics[1], in which the fundamental excitation was selected with these quantum numbers, to fit the phenomenology. We noted in §I that the absence of dipolar excitations is in itself an indication that a gravity-like force is involved[2].

(2) An effective Riemannian metric induces Einsteinian dynamics. However, <u>our correspondence is between low-energy (IR) QCD, with its strong coupling, and the high-energy (UV) strong coupling region of our effective gravity</u> (and not with the weak coupling Newtonian limit). <u>Gravity becomes a strong force in the quantum regime</u> (two Planck mass particles attract each other gravitationally $10^{19} \times 10^{19} \sim 10^{38}$ times stronger than two nucleons). The quantum gravity lagrangian includes curvature-quadratic counterterms generated by the renormalization procedure and corresponds to the (effective) invariant action,

$$I_{inv} = - \int d^4x \sqrt{-G}\, (\alpha\, R_{\mu\nu} R^{\mu\nu} - \beta R^2 + \gamma \kappa^{-2} R) \tag{6}$$

This is the Stelle[22] action, when used for true gravity. It was shown by Stelle to be renormalizable, a feature befitting our present application, since QCD is renormalizable and any piece of it should preserve the finiteness feature; but (6) is not unitary, which makes it unutilizable for true gravity - but which befits the present application: a "piece" of QCD should not be unitary, QCD being an irreducible theory. Stelle's main result, however, was to show that renormalizability is caused by $1/p^4$ propagators. But $1/p^4$ propagators are dynamically eqivalent to confinement![23]. Such propagators were generally introduced in QCD ad hoc. Here, they stem from our basic premise. Note that although we only assumed that the lowest states are color-singlets, the $1/p^4$ propagators will cause any colored state to be bound and confined; adding a quark or gluon to a color-singlet hadron will polarize the vacuum, creating pairs until the configuration becomes color-neutral.

In recent years, quadratic lagrangians like (6), with, in addition, torsion-squared terms, have been investigated classically in the context of the Poincaré Gauge Theory of gravity[24]. The exact solutions display, aside from the Newtonian potential M/r, a component behaving $\sim r^2$, dominating the strong-field limit and originating in the curvature-squared terms as in (6).

There is one more feature that relates to Eq. (6). It has been shown[25] that the String, as a gravitational theory, is equivalent to an action such as (6), i.e. with quadratic counterterms. From a different viewpoint, the embedding of the String in curved "target" spacetimes has been interpreted[26] as a series of constraints on the manifold's geometry due to the necessity of preserving the cancellation of the conformal anomaly. Such constraints are regarded in String Theory as replacing Einstein's equation in fixing the geometry of the target space - and their lowest terms are also those of (6). As a result, our ansatz explains the good fit of String Theory in its original hadron version, in reproducing the IR region features: color confinement, string flux-tubes etc.

One more comment relates to "f/g gravity"[10] and the f^0. In that hypothesis, it was necessary to assign the mass of the f^0 to some unknown Higgs-type effect, in order to forego writing a geometric Einsteinian equation for a massive graviton, a doubtful procedure. In the present "effective" picture, the $G_{\mu\nu}$ "effective" QCD-induced metric field is massless in the ansatz' approximation, because of the gauge invariance generated by the Covariance Group (the diffeomorphisms) of the "effective" Einsteinian theory.

At this point, it is important to note that the exchange of two gluons according to Eq. (1) does not correspond to the Yukawa-type exchange of "gluonium", a possible bound state of two gluons whose mass is estimated (from lattice calculations) to be of the order of 1.5 GeV. The correspondence between a pole in the s-channel and an exchange in the t-channel is based on the Klein-Gordon nature of the usual $1/p^2$ propagator. But this does not hold for a $1/p^4$ propagator! At most, such a propagator would describe the difference between two such exchanges!

V. CLASSICAL AND QUANTUM ALGEBRAS FOR HADRONS AND NUCLEI

The external field $G_{\mu\nu}(x)$ transforms under Lorentz transformations as a (reducible) second rank symmetric tensor field, with Abelian components: $[G_{\mu\nu},G_{\rho\sigma}] = 0$. Algebraically, $G_{\mu\nu}$ and the Lorentz generators form the algebra of $T_{10}(\sigma)SO(1,3)$, an inhomogeneous Lorentz group with tensor "translations". This is a classical relativistic algebra. For the quantum case, when the gluon field is expanded in creation and annihilation operators, we can write, $G_{\mu\nu} = T_{\mu\nu} + U_{\mu\nu}$, where the quadrupolar excitation-rate is given by

$$T_{\mu\nu} = \eta_{ab} \int d\tilde{k} \, [\alpha^a_\mu{}^+(k) \, \alpha^b_\nu{}^+(k) \, e^{2ikx} + \alpha^a_\mu(k) \, \alpha^b_\nu(k) \, e^{-2ikx}] \tag{7}$$

for (infinite) gl(3,R) non-compact excitation bands[*], whereas

$$U_{\mu\nu} = \eta_{ab} \int d\tilde{k} \, [\alpha^a_\mu{}^+(k) \, \alpha^b_\nu(k) + \alpha^a_\mu(k) \, \alpha^b_\nu{}^+(k)]. \tag{8}$$

generates finite u(3) spectral multiplets. We have made use of the canonical transformation:

$$[\alpha^a_\mu(k) \;+ 1/2 \, N^a_\mu \, e^{ikx} \;] \rightarrow \alpha^a_\mu(k) \;; \tag{9a}$$

$$[\alpha^a_\mu{}^+(k) + 1/2 \, N^a_\mu \, e^{-ikx}] \rightarrow \alpha^a_\mu{}^+(k). \tag{9b}$$

Using $[\alpha^a_\mu(k),\alpha^b_\nu{}^+(k')] = \delta^{ab} \, \delta_{\mu\nu} \, \delta(k-k')$, one verifies that the operators $T_{\mu\nu}$ and $U_{\mu\nu}$, together with the operators

$$S_{\mu\nu} = \eta_{ab} \int d\tilde{k} \, [\alpha^a_\mu{}^+(k) \, \alpha^b_\nu(k) - \alpha^a_\mu(k) \, \alpha^b_\nu{}^+(k)] \tag{10}$$

close respectively on the gl(4;R) and u(1,3) algebras. Note that the largest (linearly realized) algebra with generators quadratic in the $\alpha_\mu{}^+$, α_μ operators is sp(1,3;R), where the notation "1,3" implies a definition over Minkowski space.

$$sp(1,3;R) \rightarrow \left| \begin{array}{l} \rightarrow u(1,3) \rightarrow\rightarrow\rightarrow\rightarrow su(1,3) \rightarrow\rightarrow\rightarrow \\ \rightarrow gl(4;R) \rightarrow\rightarrow\rightarrow\rightarrow sl(4;R) \rightarrow\rightarrow\rightarrow \\ \rightarrow t_{10}(\sigma)so(1,3) \rightarrow t_9(\sigma)so(1,3) \rightarrow \end{array} \right| \rightarrow so(1,3) \tag{11}$$

The gl(4;R) algebra represents a Spectrum-Generating Algebra for the set of states of a given flavor[17].

We now return to the expansion in Eq. (1). The sl(4;R) is generated by the 2-gluon configuration. What about 3-gluon and n-nucleon exchanges? The corresponding algebras do not close and generate the full Ogievetsky algebra[27] of the Diffeomorphisms in Minkowski space, the 4-dimensional analog of the Virasoro algebra (the algebra of Diffeomorphisms on the circle).

Had we considered the entire (infinite) sequence when writing Eq. (1), we would have generated this diff(4;R). The maximal linear subalgebra of diff(4;R) is gl(4;R). The remaining generators (i.e. the quotient diff(4;R)/gl(4,R)) can be explicitly realized in terms of the gl(4,R) generators[28], both for tensors and for spinors. In our case, this would involve functions as matrix elements of the representation of our generators $T_{\mu\nu}$ and $S_{\mu\nu}$ in Eq. (7) and (9). As in General Relativity, the entire "$G_{\mu\nu}$-covariance" can be realized in terms of the invariant action given in Eq. (6).

But diff(3,1;R) can also be represented linearly. It will then involve infinite, ever more massive, repetitions of the representations of sl(4;R). In either way, we find that using sl(4;R) takes care of the entire sequence in Eq. (1).

The inhomogeneous versions of the algebras in Eq. (11), i.e. their semi-direct product with the translations t_4, are relevant to the Hilbert space spectrum of states. In the case of u(1,3) in Eq. (11), when selecting a time-like vector (for massive states), the stability subgroup is the compact u(3) with finite representations - as against the non-compact gl(3;R) for sl(4;R). This fits with the situation in nuclei, as we shall see in the coming sections.

VI. HADRON DYNAMICAL SYMMETRIES

Dynamically, we have discussed the role of 2-gluon excitations in generating the transition to a hadronic excited level, from any given hadron state. To this we can now add **scaling symmetry**: the general linear group GL(4,R) decomposes into its unimodular SL(4,R) subgroup, the R^+ of scale transformations and Parity Π: GL(4,R) = [Π(σ)SL(4,R)](x)R^+. This scaling symmetry corresponds to the observations in deep-inelastic photon-nucleon scattering experiments. **Color confinement** too manifests itself algebraically at several levels. The sl(4;R) subalgebra **preserves the 4-dimensional measure**, a geometric realization of confinement as a dynamical **4-volume-preserving rotation-deformation-vibration pulsation mechanism.**

When dealing with the hadron Hilbert space states, momenta come in and the translations thus have to be algebraically adjoined. Thus sa(4;R) = t_4(σ)sl(4;R). **The massive states of the hadron spectrum are then classified according to the stability subalgebra, here** sa´(3;R) = t_3´(σ)sl(3;R). The t_3´ quantum numbers are trivialized, as is done with the formal translations t_2´of the Euclidean 2-dimensional stability subalgebra for the massless states in the Poincaré group representations. Hadron states are then characterized by the sl(3;R) subalgebra, whose infinite representations correspond to Regge trajectories[4]. These preserve the measure in 3-space: a Regge trajectory described by such a representation corresponds to a given "bag". Spinors span the infinite representations of the double-covering groups $\tilde{SA}(4;R) = T_4(\sigma)\tilde{SL}(4;R)$.

We collect all hadronic field configurations obtained by successive application of the quantum 2-gluon field's sl(4;R) dynamical algebra, into an infinite-component field (manifield). These manifields are subject to the following constraints:

(i) Owing to Eq. (5), the wave-equations have to be Lorentz-covariant;
(ii) The lowest manifield components should fit the basic quark system field configuration in its Lorentz content;
(iii) They should transform according to non-unitary representations of the $\tilde{SO}(1,3) \subset \tilde{SL}(4;R)$ in order to meet the experimental fact that a boosted particle keeps its spin quantum number. These natural requirements of our 2-gluon picture determine uniquely the selection of manifields and their equations[17], while the "non-unitarity" condition on $\tilde{SO}(1,3)$ representations is achieved by making use of "Å-deunitarized" $\tilde{SL}(4;R)$ unitary irreducible representations[28].

For mesons, take a manifield Φ, $(\Box + M^2)\Phi = 0$, transforming according to the ladder representation $D(1/2,1/2)^{Å}$. For the 3-quark octet configuration we use the manifield Ψ, $(i\chi^\mu\partial_\mu - M)\Psi = 0$ transforming according to the spinorial multiplicity-free SL(4;R) representation $[D(1/2,0)(+)D(0,1/2)]^{Å}$. For the decuplet 3-quark configuration we take a manifield Ψ_ρ fulfilling $(i\chi^\mu\partial_\mu - M)\Psi_\rho = 0$. It transforms according to the spinorial multiplicity-free representation $[D(1/2,0)(+)D(0,1/2)]^{Å}(x)(1/2,1/2)$.

We find a good fit with the experimental data. Note that this classification was originally suggested and realized phenomenologically[17] and formulated using field theory and QCD as an ad hoc ansatz. The present work purports to supply solid dynamical foundations deriving from what appears to be an extremely versatile approximation for QCD in the IR region.

VII. THE ARIMA-IACHELLO U(6) IN NUCLEI

Returning to nuclei, we look again at the ansatz defining the IBM model[1,29] U(6) symmetry described in section II: a fundamental 0^+ and 2^+ quasi-degenerate set of excitations. Such a set would never arise from the exchange of the lighter quark-antiquark pairs (i.e. mesons with spins 0 and 1). This would not generate quadrupole excitations. Skipping the 1^- dipole is generally intimately connected with the (tensor) gravitational potential[2]. The 2^+ mesons such as the $f^0(1270)$ are represented by tensor fields and could thus do it[10], but their range is too short. Our QCD-induced effective gravity as described by Eq. (1-6) appears to supply the correct answer, as we shall now see.

Out of the 10 components of $G_{\mu\nu}$ the 6 that survive the 4 Riemannian constraints (5) have spin/parity assignments $J^P = 0^+, 2^+$. It is mainly this feature, suggesting a relationship with the IBM systematics, that we apply here.

Algebraically, $G_{\mu\nu}(x)$ carries the 10-dimensional (non-unitary) irreducible representation of GL(4,R). In true gravity, this is a geometric group, the linear subgroup of the Covariance Group Diff(4,R). Here it is a dynamical construct, except for the geometric Lorentz subgroup. The non-relativistic subgroup of SL(4,R) (the traceless piece, including shears, aside from Lorentz transformations) is SL(3,R). Under this group, the 0^+ and 2^+ states span together one irreducible 6-dimensional representation. The couplings to the "effective" gravity are given by the SL(4,R) group; they will thus be SL(3,R) invariant.

There is thus full justification, in this picture, for the IBM postulate of a U(6) symmetry between the defining states! Note also that this SL(3,R), obtained from the basic QCD fields, takes on a geometric interpretation, once we use $G_{\mu\nu}$ as a formal metric field. In that picture, SL(3,R) predicts the conservation of 3-volume, i.e. incompressibility, for nuclei and hadrons (where it justifies the "bag" model as an approximation of hadron dynamics).

When applying "effective gravity" to nuclei, it is natural to assume that closed shells assume the role of "vacua", as rigid structures. "Graviton" excitations should then be searched for in the valence nucleon systematics. In this sector of even-even nuclei the $G_{\mu\nu}$ quanta can indeed excite nucleon pairs; the overwhelming preponderance of proton-proton and neutron-neutron over proton-neutron pairs can be fully explained in terms of the Clebsch-Gordan coefficients in the direct channel. Dynamically for one pair, we assume that the pairing force itself is due to the exchange of a "strong graviton" between the two nucleons. The paired system then displays further excited states with the absorption of additional such quanta. The picture now is of an external field supplying these quanta, perhaps like the role of the electromagnetic field in the hydrogen atom in the Schroedinger equation treatment. It is thus natural that proton pairs and neutron pairs should have the same energy difference between 0^+ and 2^+, since these are due to the same flavor-independent component of QCD - precisely for the same reason that the Eightfold Way (flavor) SU(3) invariance is due to the flavor-independence of QCD.

In estimating the amplitude for such pair excitations, we note that the 2-gluon exchange here is dominated by a pole in the direct channel at $k^2 = 4\ m^2$, m the nucleon mass. Remembering that "effective strong graviton" excitations are also seen in hadrons, where they cause the $\Delta J = 2$ sequences of resonances along Regge trajectories, we get a value for the effective $1/\kappa^2$ in Eq. (6). It is given by the slope of the Regge trajectories, roughly $1/\kappa^2 \sim 1\ (\mathrm{GeV})^2$.

We now return to the Hamiltonian corresponding to Eq. (6), as translated into our effective dynamics. The curvature R corresponds in true gravity to terms $\partial\Gamma + \Gamma.\Gamma$; the Christoffel formula gives $\Gamma \sim G^{-1}\partial G$ and $R \sim \partial(G^{-1}\partial G) + (G^{-1}\partial G)(G^{-1}\partial G)$. Here G stands for $G_{\mu\nu}$, $\underline{b}$ and $\underline{b}^+$ represent the destruction and creation of a 6-dimensional "strong graviton" quantum,

$$H = (1/M^3)\int dk\ \{C_1\ (k^2/\kappa^2)\ (\underline{b}^+.\underline{b}) + C_2\ (k^2/\kappa^2)\ (\underline{b}+.\underline{b})(\underline{b}+.\underline{b}) +$$

$$+ A_1\ k^4\ (\underline{b}^+.\underline{b})(\underline{b}^+.\underline{b}) + A_2\ k^4\ (\underline{b}^+.\underline{b})(\underline{b}^+.\underline{b})(\underline{b}^+.\underline{b}) +$$

$$+ A_3\ k^4\ (\underline{b}^+.\underline{b})(\underline{b}^+.\underline{b})(\underline{b}^+.\underline{b})(\underline{b}^+.\underline{b})\}. \quad (12)$$

The coefficients C_i and A_i respectively contain dynamical information relating to the γ and α,β terms in (6), an approximation for the non-linear effect of the $\sqrt{-G}$, the reduced matrix element for the coupling to the nucleon pair, etc. M is a mass parameter that takes care of the dimensionality.

We select M to be of the order of the impacted system, i.e. the valence nucleons, $M \sim 20$ GeV. For k^2 we use the dispersion relation result mentioned above, i.e. $\langle k^2\rangle \sim 4\ (\mathrm{GeV})^2$. Using our string-Regge result $(1/\kappa^2) \sim 1\ (\mathrm{GeV})^2$, and assuming the C coefficients to be of the order of unity, we get for the first terms $\langle\Delta E\rangle \sim .5$ MeV . This is of course by far no more than an order of magnitude check, but it seems the values are roughly in the right ballpark. The values will decrease for larger M. Our (12) is of course equivalent to the IBM Hamiltonian with higher order terms.

VIII. SYMMETRIES OF DEFORMED NUCLEI

In nuclei the relevant symmetries in Eq. (11) correspond to the non-relativistic subgroups, i.e. to the Sp(3,R) and the related $T_5(\sigma)SO(3)$, SL(3,R) and SU(3). Moreover, averaging over the color SU(3) degrees of freedom, which are summed over anyhow, in Eq. (7-10) and integrating over the $d\hat{k}$, we get the Sp(3,R) generators of Rowe[7], as expressed in terms of shell-model harmonic oscillators.

It is interesting that this picture in which the system of creation and annihilation operators of the Arima-Iachello U(6) and those of the Sp(3,R) of deformed nuclei are constructed as, respectively, quadratic and linear expressions involving the same dynamical elements validates a "guess" expressed by Lipkin some time ago[30].

REFERENCES

1. A. Arima and F. Iachello, Phys. Rev. Lett. 35, 1069 (1975); same authors, Ann. Phys. 99, 253 (1976); 111, 201 (1978); 123, 468 (1979).

2. see for example C.W. Misner, K.S. Thorne and J.A. Wheeler,

Gravitation, W.H. Freeman and Co Pub., San Francisco (1973), §36.1.

3. J.P. Elliott, Proc. Roy. Soc. A245 (1958) 128, 562; J.P. Elliott and M. Harvey, Proc. Roy. Soc. A272 (1963) 557.

4. Y. Dothan, M. Gell-Mann and Y. Ne'eman, Phys. Lett. 17 (1965) 148.

5. L. Weaver and L.C. Biedenharn, Phys. Lett, 32B (1970) 326; Nucl. Phys. A185 (1972) 1.

6. H. Ui, Prog. Theor. Phys. 44 (1970) 153; L. Weaver, L.C. Biedenharn and R.Y. Cusson, Ann. Phys. 77 (1973) 250.

7. S. Goshen and H.J. Lipkin, Ann. Phys. 6 (1959) 301; D.J. Rowe, in Dynamical Groups and Spectrum Generating Algebras, A. Bohm, Y. Ne'eman and A.O. Barut editors, World Scientific, Singapore (1989), p. 287; G. Rosensteel and D.J. Rowe, Phys. Rev. Lett. 47, 223 (1981); J.P. Draayer and K.J. Weeks, Phys. Rev. Lett. 51, 1422 (1983).

8. C.G. Callan, R.E. Dashen and D. Gross, Phys. Rev. D17 (1978) 2717; ibid. D19 (1978) 1826.

9. D.B. Kaplan, Phys. Lett. 235B (1990) 163.

10. C.J. Isham, A. Salam and J. Strathdee, Phys. Rev. D8 (1973) 2600; ibid. D9 (1974) 1702; ibid. Lett. Nuo. Cim. 5 (1972) 969.

11. M. Gell-Mann, Phys. Rev. 125 (1962) 1067, footn. 38; P.G.O. Freund, Phys. Lett. 2 (1962) 136.

12. T. Yoneya, Prog. Theoret. Phys. 51 (1974) 1907; J. Scherk and J.H. Schwarz, Nucl. Phys. B81 (1974) 118.

13. D.W. Joseph, Un. of Nebraska prep., unpub. (1969); Y. Ne'eman, Ann. Inst. H. Poincaré 28 (1978) 639; Y. Ne'eman and Dj. Šijački, Int. J. Mod. Phys. A2 (1987) 1655.

14. C.N. Yang and R.J. Oakes, Phys. Rev. Lett. 125 (1963) 1067.

15. Y. Ne'eman, Phys. Rev. 134 (1965) B1355.

16. Y. Ne'eman, Nucl. Phys. (Proc. Supp. Sec.) B13 (1990) 582.

17. Y. Ne'eman and Dj. Sijacki, Phys. Lett. 157B, 267 (1985) and Phys. Rev. D37, 3267 (1988).

18. Dj. Šijački and Y. Ne'eman, Physics Lett. B, to be published.

19. Y. Ne'eman and T. Regge, Riv. Nuo. Cim. Ser. 3, 1 (1978) #5.

20. J. Thierry-Mieg and Y. Ne'eman, Ann. Phys.(NY) 123 (1979) 247.

21. see for example S.L. Adler, Rev. Mod. Phys. 54 (1982) 729.

22. K.S. Stelle, Phys. Rev. D16 (1977) 953.

23. see for example J. Kiskis, Phys. Rev. D11 (1975) 2178; G. B. West, Phys. Lett. 115B (1982) 468.

24. F.W. Hehl, Y. Ne'eman, J. Nisch and P. v.d. Heyde, P. Baeckler and F.W. Hehl, in From SU(3) to Gravity, E. Gotsman and G. Tauber editors, Cambridge University Press, Cambridge (1985), p. 341.

25. S. Deser and A.N. Redlich, Phys. Lett. 176B (1986) 350.

26. E.S. Fradkin and A.A. Tseytlin, Phys. Lett. B158 (1985) 316.

27. V.I. Ogievetsky, Lett. Nuo. Cim. 8 (1973) 988.

28. Dj. Ŝijaĉki and Y. Ne'eman, J. Math. Phys. 16 (1985) 2457.

29. F. Iachello and I. Talmi, Rev. Mod. Phys. 54 (1987) 339.

30. H.J. Lipkin, Nucl. Phys. A350 (1980) 16.

PROJECTION OPERATOR METHOD AND Q-ANALOG OF ANGULAR MOMENTUM THEORY

Yu.F. Smirnov, V.N. Tolstoy and Yu.I. Kharitonov

Nuclear Physics Institute, Moscow State University

Moscow 119899, USSR

1. INTRODUCTION

Quantum algebras were introduced at first in Refs.[1,2]. Then this concept was developed in details in Refs.[3,4] and in the papers of other authors (see for example [5-11] and the papers cited there). Because of deep analogy consisting between quantum and usual Lie algebras which is reflected in the fact that the quantum algebra $A_q(l,r)$ of order l and rank r transforms into usual Lie algebra $A(l,r)$ in the limit $q\to 1$ a number of notations and theorems of the theory of Lie algebra representations can be transferred onto quantum algebras. In particular as it was shown in Refs [5-17] the q-analogs of well known quantities of Wigner-Racah algebra (WRA) (3j, 6j, 9j-symbols etc.) can be introduced. The detail investigation of the representation of quantum algebras was begun with the simplest quantum algebra $SU_q(2)$ that is a q-analog of the angular momentum theory (AMT) [18-21].

In this paper we apply to this problem an original approach namely the projection operator method that was developed by us for usual Lie algebras in Refs [22,23] and appears rather effective as well in AMT as for higher Lie algebras. The important advantage of this method is the fact that for the calculation of quantities of WRA any explicit realization of algebra generators is unnecessary. Only commutation rules for generators, their Hermitian properties and the existence of the highest vector are enough for the development of q-algebra unitary representation theory. Below it will be shown that most part of AMT formulae will conserve their shape in the case of $SU_q(2)$ algebra except for exchange of usual numbers (x) by so called q-numbers $[x]$:

$$[x] \equiv \frac{q^{x}-q^{-x}}{q-q^{-1}} \qquad (1.1)$$

where $q=e^{\hbar}$. Often the exponent $\hbar$ is referred as "Planck's

Symmetries in Science V, Edited by B. Gruber *et al.*
Plenum Press, New York, 1991

constant" since the classical Lie algebra is the $\hbar=0$ limit of corresponding quantum algebra similarly to transition from quantum to classical mechanics at vanishing Planck's constant. In the limit $q=1$ (or $\hbar=0$) q-numbers coincide with usual numbers

$$\lim_{q\to 1}[x] = (x). \tag{1.2}$$

Obviously $[0]=0$, $[1]=1$, $[2]=q+q^{-1}$, $[3]=q^2+1+q^{-2}$ etc. In general case of nonnegative integer n we have

$$[n] = q^{n-1}+ q^{n-3}+\ldots+ q^{-n+3}+ q^{-n+1}. \tag{1.3}$$

Also we obtain

$$[-x] = -[x]. \tag{1.4}$$

Generally speaking the notation q-number must include the value of q-parameter as subscript $[x] = [x]_q$. However we shall omit it because of invariance of $[x]$ with respect to substitution $q \to \bar{q} \equiv q^{-1}$

$$[x] = [x]_q = [x]_{\bar{q}} . \tag{1.5}$$

Below we shall use the so called q-factorial

$$[n]! \equiv [n][n-1]\ldots[2][1] \tag{1.6}$$

for nonnegative integer n. As usually we assume $[0]! = 1$ and $[-n]! = \infty$ at $n = 1,2,\ldots$. The operator expression

$$[A] \equiv \frac{q^A- q^{-A}}{q - q^{-1}} = \frac{\mathrm{sh}\ \hbar A}{\mathrm{sh}\ \hbar} \tag{1.7}$$

will be used for Cartan generator A. The Planck's constant $\hbar$ will be assumed real in our calculation. The special case when q is equal to the root of unit was considered in Refs.[12,13].

2. $SU_q(2)$ ALGEBRA AND ITS IRREDUCIBLE REPRESENTATIONS

The q-analog of $SU(2)$ algebra is defined by three generators J_0, J_+, J_- with following properties

$$[J_0,J_\pm] = \pm J_\pm , \tag{2.1}$$

$$[J_+,J_-] = [2J_0] \equiv \frac{\mathrm{sh}\ 2\hbar J_0}{\mathrm{sh}\ \hbar} = \frac{q^{2J_0}- q^{-2J_0}}{q - q^{-1}} , \tag{2.2}$$

$$J_0^+ = J_0 , \tag{2.3}$$

$$J_\pm^+ = J_\pm . \tag{2.4}$$

The irreducible (IR) D^J of highest weight J contains the highest vector $|JJ\rangle$ satisfying the equations

$$J_0|JJ\rangle = J|JJ\rangle \ , \tag{2.5}$$

$$J_{\pm}|JJ\rangle = 0 \ , \tag{2.6}$$

$$\langle JJ|JJ\rangle = 1 \ . \tag{2.7}$$

The general basis vector of this IR having the weight m can be constructed using the lowering generator J_-

$$|Jm) = J_-^{\,n}|JJ\rangle \ , \quad m = J-n \ . \tag{2.8}$$

The squared norm of this vector

$$(Jm|Jm) = N^2(n) = \langle JJ|J_+^{\,n}J_-^{\,n}|JJ\rangle \tag{2.9}$$

can be calculated by using the relation

$$J_+J_-^{\,n} = J_-^{\,n}J_+ + [n]J_-^{\,n-1}[2J_0-n+1] \tag{2.10}$$

that can be proven by induction on n. Similarly more general permutation expression can be obtained

$$J_+^{\,a}J_-^{\,b} = \sum_{c=0}^{min(a,b)} \frac{[a]![b]!}{[c]![a-c]![b-c]!} J_-^{\,b-c}J_+^{\,a-c} \frac{[2J_0-b+a]!}{[2J_0-b+a-c]!} \ . \tag{2.11}$$

Using (2.10) and (2.5) we find

$$N^2(n) = [n][2J-n+1] \, N^2(n-1), \quad n = 1,2,\dots \ . \tag{2.12}$$

For the finite dimensional IR D^J this norm must vanish at some integer $n=n_0$. Such vanishing of the norm is possible only at integer $n_0=2J+1$ i.e. at integer or half integer values of the "angular momentum" J:

$$J = 0,1/2,1,\dots \ . \tag{2.13}$$

For its projection m the following values are admissible

$$m = J,J-1,\dots,-J \ . \tag{2.14}$$

Thus the weight structure of $SU_q(2)$ IR's is similar to the IR's of usual $SU(2)$ algebra and the dimension of IR's are the same in both of cases

$$dim \ D^J = 2J+1 \ . \tag{2.15}$$

This analogy between IR's of these two algebras is rather deep.

Iterating the Eq. (2.12) we obtain

$$N^2(n) = \frac{[n]![2J]!}{[2J-n]!} \ . \tag{2.16}$$

Therefore the basis vectors of IR D^J are of the form

$$|jm\rangle = \sqrt{\frac{[j+m]!}{[2j]![j-m]!}}\, J_-^{j-m}|jj\rangle = \qquad (2.17a)$$

$$= \sqrt{\frac{[j-m]!}{[2j]![j+m]!}}\, J_+^{j+m}|j,-j\rangle. \qquad (2.17b)$$

The last part gives the expression of basis vector in term of the lowest weight vector $|j,-j\rangle$ that will be used in Section 6. The basis (2.17) is ortonormalized

$$\langle jm'|jm\rangle = \delta_{m,m'}\ . \qquad (2.18)$$

Acting by generator J_- on vector (2.17a) we obtain

$$J_-|jm\rangle = \sqrt{[j+m][j-m+1]}\ |j,m-1\rangle\ . \qquad (2.19)$$

The application of J_+ to (2.17b) gives the result

$$J_+|jm\rangle = \sqrt{[j-m][j+m+1]}\ |j,m+1\rangle\ . \qquad (2.20)$$

Thus the explicit form of D^j IR for $SU_q(2)$

$$\langle jm'|J_0|jm\rangle = m\ \delta_{m,m'}, \qquad (2.21a)$$

$$\langle jm'|J_-|jm\rangle = \delta_{m',m-1}\sqrt{[j+m][j-m+1]}\ , \qquad (2.21a)$$

$$\langle jm'|J_+|jm\rangle = \delta_{m',m+1}\sqrt{[j-m][j+m+1]} \qquad (2.21a)$$

coincides with corresponding formulae for $SU(2)$ except for the substitution of usual number $(j\pm m)$ and $(j\mp m+1)$ by q-numbers in two last rows. As for the powers of generators we have

$$\langle jm'|J_+^{\,r}|jm\rangle = \delta_{m',m+r}\sqrt{\frac{[j-m]![j+m+r]!}{[j+m]![j-m-r]!}}\ , \qquad (2.22a)$$

$$\langle jm'|J_-^{\,r}|jm\rangle = \delta_{m',m-r}\sqrt{\frac{[j+m]![j-m+r]!}{[j-m]![j+m-r]!}}\ . \qquad (2.22b)$$

In the theory of $SU_q(2)$ IR the important role plays the Casimir operator

$$\mathbf{C}_2 = J_-J_+ + [J_0+1/2]^2 = J_+J_- + [J_0-1/2]^2, \qquad (2.23a)$$

which is Hermitian one

$$\mathbf{C}_2^+ = \mathbf{C}_2\ . \qquad (2.23b)$$

As usually it commutes with each $SU_q(2)$ generator

$$[\mathbf{C}_2, J_0] = 0 \ , \tag{2.24a}$$

$$[\mathbf{C}_2, J_\pm] = 0 \ . \tag{2.24b}$$

The vectors (2.17) are its eigenvectors

$$\mathbf{C}_2 |jm\rangle = \Lambda |jm\rangle \tag{2.25}$$

corresponding to eigenvalue

$$\Lambda = [j+1/2]^2 \ . \tag{2.26}$$

In the limit $q \to 1$ the Casimir operator transforms into the squared angular momentum operator

$$\mathbf{C}_2 = \mathbf{J}^2 + 1/4 \ , \tag{2.27}$$

$$\Lambda = j(j+1) + 1/4 \ . \tag{2.28}$$

3. PROJECTION OPERATORS FOR $SU_q(2)$ ALGEBRA

At first we are interesting in the projection operator (PO) $P^j_{j,j}=P^j$ having the property

$$P^j |j' m=j\rangle = \delta_{j,j'} |jj\rangle \ , \tag{3.1}$$

i.e. acting on an arbitrary vector $|j)$ of weight m

$$|m=j) = \sum_{j' \geqslant j} B_{j'} |j' j\rangle \tag{3.2}$$

the operator P^j projects the component $|jj\rangle$ being the highest weight vector of IR D^j

$$P^j |j) = B_j |jj\rangle \ . \tag{3.3}$$

Similarly to Refs.[22-25] we seek this PO as a power series of generators J_+ and J_-

$$P^j = \sum_{r=0}^{\infty} C_r \ J_-^r J_+^r \ . \tag{3.4}$$

The exponents of these generators are the same due to condition

$$[P^j, J_0] = 0 \ . \tag{3.5}$$

Since

$$P^j |jj\rangle = |jj\rangle \ , \tag{3.6}$$

we obtain because of (2.6) that

$$C_0 = 1 \tag{3.7}$$

and

$$J_+ P^j |j) = J_+ \sum_{r=0}^{\infty} C_r \, J_-^r J_+^r |j) = 0 \,. \tag{3.8}$$

By using of Eq. (2.10) the following recurrent relation for C_r coefficients can be found

$$C_{r-1} + [r][2j+r+1]C_r = 0 \,. \tag{3.9}$$

Solving it we have

$$C_r = (-1)^r \frac{[2j+1]!}{[r]![2j+r+1]!} \,. \tag{3.10}$$

Obviously the PO is Hermitian one

$$(P^j)^+ = P^j \,. \tag{3.11}$$

The problem of convergency of the formal series (3.4) is not essential because we shall apply it to vectors $|j)$ containing in its expansion (3.2) only components $|j'j\rangle$ with $j' \leqslant j_{max}$. It means that only finite numbers of terms of (3.4) series will give nonvanishing contribution into the expression $P^j|j)$. Namely the terms of (3.4) with $r > j_{max} - j$ are vanishing and can be omitted. Such situation takes place in Clebsch-Gordan coefficients (CGC), Racah coefficients (RC) calculations etc. By Hermitian conjugation of Eq.(3.8) we obtain an important property of PO

$$P \, J_- = 0 \,. \tag{3.12}$$

In similar manner other extremal projector

$$P^j_{-j,-j} = \sum_{r=0}^{\infty} C_r \, J_+^r J_-^r \tag{3.13}$$

projecting the vector $|-j)$ on the lowest weight of IR D^j can be constructed. It satisfies the equations

$$J_- \, P^j_{-j,-j} = P^j_{-j,-j} \, J_+ = 0 \tag{3.14}$$

with the same coefficients C_r as in Eq.(3.10).

In conclusion of this section it should be noted that more general PO of the form

$$P^j_{m,m'} = \sqrt{\frac{[j+m]!}{[2j]![j-m]!}} \, J_-^{\,j-m} \, P^j \, J_+^{\,j-m'} \sqrt{\frac{[j+m']!}{[2j]![j-m']!}} \tag{3.15}$$

will need in further calculations. These POs have the properties

$$(P^j_{m,m'})^+ = P^j_{m',m} \,, \tag{3.16}$$

$$P^{j}_{m,m'}\,|j'm'\rangle = \delta_{j,j'}\,|jm\rangle\,, \tag{3.17}$$

i.e.acting on the general vector of weight m'

$$|m')=\sum_{j'} B_{j'm'}\,|j'm'\rangle\,, \tag{3.18}$$

such PO projects the component $|jm'\rangle$ and transfers it onto $|jm\rangle$ component:

$$P^{j}_{m,m'}\,|m')=B_{j,m'}\,|jm\rangle\,. \tag{3.19}$$

Because of this projecting property

$$P^{j}_{m,m'}P^{j}_{m',m} = \delta_{j,j'}P^{j}_{m,m'} \tag{3.20}$$

such transfer operators will be referred below as projection operators too, although the exact term PO can be used strictly speaking only for idempotents $P^{j}_{m,m}$. The structure of PO depends of the value of q-parameter. When it will be necessary (for example in sections 5 and 6) q will be indicated as upper index of PO.

4. "VECTOR COUPLING" OF Q-ANGULAR MOMENTA

It is known that the expansion of the tensor product of $SU(2)$ IRs $D^{j_1}\otimes D^{j_2}$ into irreducible components D^{j}

$$D^{j_1}\otimes D^{j_2} = \sum_{j=|j_1-j_2|}^{j_1+j_2} D^{j} \tag{4.1}$$

is interpreted in the AMT as a rule of vector coupling of angular momentum. Namely a summary momentum

$$\mathbf{j} = \mathbf{j}_1 + \mathbf{j}_2 \tag{4.2}$$

for the quantum system consisting from two subsystems 1 and 2 with partial angular momenta j_1 and j_2 respectively can take the following values

$$j = |j_1-j_2|,|j_1-j_2|+1,|j_1-j_2|+2,\ldots,j_1+j_2\,. \tag{4.3}$$

In this case vector equation (4.2) means

$$J_0(1,2) = J_0(1) + J_0(2)\,, \tag{4.4a}$$

$$J_\pm(1,2) = J_\pm(1) + J_\pm(2) \tag{4.4b}$$

where generators $J_{0,\pm}(i)$ act on the variables of i-th subsystem.

Now we turn to the of "vector coupling" of angular momenta in the case of $SU_q(2)$ algebra .It should be noted that equation (4.4) must be modified in the following manner

$$J_0^q(1,2) = J_0(1) + J_0(2) \,, \tag{4.5a}$$

$$J_\pm^q(1,2) = J_\pm(1)\, q^{J_0(2)} + q^{-J_0(1)} J_\pm(2) \,. \tag{4.5b}$$

This modification is the consequence of the fact that in the theory of quantum algebras the so called coproduct of IRs is introduced. In standard notation for Hopf algebras the relations (4.5) must be written in the following form

$$J_0^q = J_0 \otimes I + I \otimes J_0 \,, \tag{4.6a}$$

$$J_\pm^q = J_\pm \otimes q^{J_0} + q^{-J_0} \otimes J_\pm \,. \tag{4.6b}$$

However we shall use below notation (4.5) in order to conserve the maximal possible similarity with usual AMT.

It is easy to prove that the operators (4.5) are satisfying to commutation relations (2.1) and (2.2) while usual operators (4.4) are violating these expressions.

The action of generators (4.5) on basis vectors $|J_1m_1\rangle \times |J_2m_2\rangle$ of the tensor product of IRs is given by formulae

$$J_0^q(1,2)|J_1m_1\rangle|J_2m_2\rangle = (m_1+m_2)|J_1m_1\rangle|J_2m_2\rangle \,, \tag{4.7a}$$

$$J_\pm^q(1,2)|J_1m_1\rangle|J_2m_2\rangle = q^{m_2}\langle J_1,m_1\pm 1|J_\pm|J_1m_1\rangle|J_1m_1\pm 1\rangle|J_2m_2\rangle +$$
$$+ q^{-m_1}\langle J_2,m_2\pm 1|J_\pm|J_2m_2\rangle|J_1m_1\rangle|J_2m_2\pm 1\rangle \,. \tag{4.7b}$$

It should be remarked that the q-analog of the binomial expansion formula is valid

$$\left[J_\pm^q(1,2)\right]^r = \left[\, J_\pm(1)q^{J_0(2)} + q^{-J_0(1)} J_\pm(2)\right]^r =$$
$$= \sum_s \frac{[r]!}{[s]![r-s]!} J_\pm^s(1)\, J_\pm^{r-s}(2)\, q^{sJ_0(2)-(r-s)J_0(1)} \,. \tag{4.8}$$

The summary Casimir operator is of the form

$$\mathbf{C}_2^q(1,2) = J_-^q(1,2)J_+^q(1,2) + [J_0^q(1,2) + 1/2]^2 =$$
$$= J_-(1)J_+(2)q^{-J_0(1)+J_0(2)+1} + J_+(1)J_-(2)q^{-J_0(1)+J_0(2)-1} +$$
$$+(\mathbf{C}_2(1)-[J_0(1)+1/2]^2)q^{2J_0(2)} + (\mathbf{C}_2(2)-[J_0(2)+1/2]^2)q^{-2J_0(1)} +$$
$$+ [J_0(1)+J_0(2)+1/2]^2 \,. \tag{4.9}$$

The generalization of the vector coupling procedure on the case of $SU_q(2)$ consists in the following. It is necessary to construct from the tensor product basis vectors $|j_1m_1\rangle|j_2m_2\rangle$ such linear combinations

$$|j_1j_2:jm\rangle_q = \sum_{m_1,m_2} \langle j_1m_1;j_2m_2|jm\rangle_q \; |j_1m_1\rangle|j_2m_2\rangle \tag{4.10}$$

which belong to the IR D^j of $SU_q(2)$, i.e. they are eigenvectors of the Casimir operator (4.9) with eigenvalues $\Lambda=[j+\frac{1}{2}]^2$:

$$C_2^q(1,2)|j_1j_2;jm\rangle_q = [j+1/2]^2|j_1j_2;jm\rangle_q \; . \tag{4.11}$$

It is clear that

$$J_0^q(1,2)|j_1j_2;jm\rangle_q = m|j_1j_2;jm\rangle_q \; . \tag{4.12}$$

The coefficient $\langle j_1m_1;j_2m_2|jm\rangle_q$ in linear combinations (4.10) are called as Clebsch-Gordan coefficients (q-CGC) for $SU_q(2)$ quantum algebra. Multiplying both sides of Eq. (4.10) by vector $\langle j_1m_1|\langle j_2m_2|$ and using (4.9), (2.19) and (2.20) we obtain the recurrent relation for q-CGC

$$q^{-m_1+m_2-1}\sqrt{[j_1-m_1][j_1+m_1+1][j_2+m_2][j_2-m_2+1]}\langle j_1m_1+1;j_2m_2-1|jm\rangle_q +$$
$$+q^{-m_1+m_2+1}\sqrt{[j_1+m_1][j_1-m_1+1][j_2-m_2][j_2+m_2+1]}\langle j_1m_1-1;j_2m_2+1|jm\rangle_q +$$
$$+\Big(q^{2m_2}[j_1-m_1][j_1+m_1+1] + q^{-2m_1}[j_2-m_2][j_2+m_2+1] +$$
$$+ [m+1/2]^2 - \Lambda\Big)\langle j_1m_1;j_2m_2|jm\rangle_q = 0 \; . \tag{4.13}$$

It was used in Ref.[11] for the derivation of the general analytical formula for q-CGCs in manner of the Racah paper [26].However we use the PO approach for this aim. Simultaneously the structure of Clebsch-Gordan series for $SU_q(2)$ will be found or more correctly it will be confirmed that the Clebsch-Gordan series for $SU_q(2)$ coincides with the Eq.(4.1). However before to turn to this point it is pertinent to list the orthormality relations for the q-CGCs

$$\sum_{m_1,m_2} \langle j_1m_1;j_2m_2|jm\rangle_q \; \langle j_1m_1;j_2m_2|j'm'\rangle_q = \delta_{j,j'}\delta_{m,m'} , \tag{4.14}$$

$$\sum_{j,m} \langle j_1m_1;j_2m_2|jm\rangle_q \; \langle j_1m_1';j_2m_2'|jm\rangle_q = \delta_{m_1,m_1'}\delta_{m_2,m_2'} . \tag{4.15}$$

The q-CGCs form an orthogonal matrix and the following equation which is inverse with respect to transformation (4.10) is valid

$$|j_1m_1\rangle|j_2m_2\rangle = \sum_{j,m} \langle j_1m_1;j_2m_2|jm\rangle_q \; |j_1j_2;jm\rangle_q \; . \tag{4.16}$$

5. Q-ANALOGS OF CLEBSCH-GORDAN COEFFICIENTS

Using PO we can write the vector (4.10) characterizing by the summary angular momentum J in a form

$$|J_1J_2;Jm\rangle_q = (Q'_q)^{-1}P^{J,q}_{m,m'}(1,2)|J_1m'_1(1)\rangle|J_2m'_2(2)\rangle, \quad (5.1)$$

where $m'=m'_1+m'_2$. Thus the q-CGC can be reduced to the matrix element of PO

$$\langle J_1m_1,J_2m_2|Jm\rangle_q = \quad (5.2)$$

$$= (Q'_q)^{-1}\langle J_1m_1(1)|\langle J_2m_2(2)|P^{J,q}_{m,m'}(1,2)|J_1m'_1(1)\rangle|J_2m'_2(2)\rangle$$

where Q'_q is a normalizing factor. As for values of m'_1 and m'_2 in Eqs. (5.1) and (5.2) they can be chosen in arbitrary manner but for simplification of calculations it is convenient to take $m'_1=J_1$ and $m'_2=J-J_1$. Then the Eq.(5.2) can be rewritten in the form

$$\langle J_1m_1;J_2m_2|Jm\rangle_q = \quad (5.3)$$

$$= (Q_q)^{-1}\langle J_1m_1(1)|\langle J_2m_2(2)|P^{J,q}_{m,J}(1,2)|J_1J_1(1)\rangle|J_2,J-J_1(2)\rangle,$$

where

$$Q^2_q = \langle J_1J_1(1)|\langle J_2,J-J_1(2)|P^{J,q}_{J,J}(1,2)|J_1J_1(1)\rangle|J_2,J-J_1(2)\rangle. \quad (5.4)$$

Since $|J_1J_1(1)\rangle$ is a highest weight vector the generators $J_+(1)$ in PO $P^{J,q}_{m,J}(1,2)$ can be omitted and for the normalizing factor Q_q we obtain

$$Q_q^{\,2} = \langle J_1J_1;J_2,J-J_1|JJ\rangle^2 =$$

$$= \langle J_2,J-J_1(2)|P^{J,q}_{(J_1)}(2)|J_2,J-J_1(2)\rangle, \quad (5.5)$$

where

$$P^{J,q}_{(J_1)}(2) = \sum_{r\geqslant 0} \frac{(-1)^r[2J+1]!}{[r]![2J+r+1]!}\, q^{-2rJ_1} J_-^{\,r}(2)J_+^{\,r}(2). \quad (5.6)$$

Let's adopt an usual phase convention for Q_q being positive (arithmetic) square root of $Q_q^{\,2}$. As a results all q-CGCs will be real.

It is clear from Eq.(5.3) that only values of summary angular momentum J satisfying the conditions $-J_2\leqslant J-J_1\leqslant J_2$ are possible.It means the following restriction

$$J_1-J_2 \leqslant J \leqslant J_1+J_2 \,. \quad (5.7)$$

Since angular momenta j_1 and j_2 are "equal in rights" the restriction

$$j_2-j_1 \leqslant j \leqslant j_1+j_2 \tag{5.8}$$

is valid too. Combining two these conditions we obtain for $SU_q(2)$ the same rule of vector coupling of angular momenta (4.3) as for usual $SU(2)$ algebra. Thus the Clebsch-Gordan series for $SU_q(2)$ coincides with (4.1).

Finally the general expression for q-CGCs can be written in form

$$\langle j_1m_1;j_2m_2|jm\rangle_q = \tag{5.9}$$

$$= \frac{\langle j_1m_1(1)|\langle j_2m_2(2)|P^{j,q}_{m,j}(1,2)|j_1j_1(1)\rangle|j_2,j-j_1(2)\rangle}{\langle j_1j_1(1)|\langle j_2,j-j_1(2)|P^{j,q}(1,2)|j_1j_1(1)\rangle|j_2,j-j_1(2)\rangle^{1/2}}.$$

To calculate the numerator of this expression it is necessary to express PO $P^{j,q}_{m,j}(1,2)$ in terms of generators $J_\pm(1,2)$, then to do the binomial expansion of their powers in terms of $J_\pm(1)$ and $J_\pm(2)$ using the Eq.(4.8). The last step is a calculation of matrix elements of $J_\pm{}^s(i)$ using formulas (2.22a,b). The resulting expression takes a form

$$\langle j_1m_1(1)|\langle j_2m_2(2)|P^{j,q}_{m,j}(1,2)|j_1j_1(1)\rangle|j_2,j-j_1(2)\rangle =$$

$$= [2j+1]\sqrt{\frac{[2j]![j+m]![2j_1]![j_2-m_2]![j_1+j_2-j]!}{[j-m]![j_1-m_1]![j_1+m_1]![j_2+m_2]![j-j_1+j_2]!}}\times$$

$$\times\sum_r \frac{(-1)^r[j-j_1+j_2+r]![j-m+r]!\;q^{-r(j_1+m_1)-m_1j+mj_1}}{[r]![2j+r+1]![j_1+j_2-j-r]![j-j_1-m_2+r]!}. \tag{5.10}$$

The denominator of the Eq.(5.9) can be found using the Eq.(5.5)

$$\langle j_1j_1(1)|\langle j_2,j-j_1(2)|P^{j,q}_{m,j}(1,2)|j_1j_1(1)\rangle|j_2,j-j_1(2)\rangle =$$

$$= \frac{[2j+1]![j_1+j_2-j]!}{[j-j_1+j_2]!}\sum_r \frac{(-1)^r[j-j_1+j_2+r]!\;q^{-2rj_1}}{[r]![2j+r+1]![j_1+j_2-j-r]!} =$$

$$= \frac{[2j+1]![j_1+j_2-j]!}{[j-j_1-j_2]!}\sum_s \frac{(-1)^{j_1+j_2-j-s}[2j_2-s]!\,q^{2(j_1+j_2-j-s)j_1}}{[s]![j_1+j_2+j+1-s]![j_1+j_2-j-s]!}. \tag{5.11}$$

This sum may be calculated using one of Vandermonde formulae [27]. As a result we find

$$Q_q^2 = q^{(j_1+j_2-j)(j-j_1+j_2+1)} \frac{[2j_1]![2j+1]!}{[j+j_1-j_2]![j_1+j_2+j+1]!} . \quad (5.12)$$

It should be noted that the matrix element (5.11) can be calculated also in recurrent manner [28]. Thus we obtain the following explicit analytical formula for q-CGCs

$$\langle j_1m_1;j_2m_2|jm\rangle_q =$$

$$= \delta_{m,m_1+m_2} q^{-\frac{1}{2}(j_1+j_2-j)(j+j_1+j_2+1)+j_1m_2-j_2m_1} \times$$

$$\times \sqrt{\frac{[2j+1][j+m]![j_2-m_2]![j_1+j_2+j+1]![j_1+j_2-j]![j_1-j_2+j]!}{[j-m]![j_1-m_1]![j_1+m_1]![j_2+m_2]![j-j_1+j_2]!}}$$

$$\times \sum_z \frac{(-1)^{j_1+j_2-j-z}[2j_2-z]![j_1+j_2-m-z]!q^{z(j_1+m_1)}}{[z]![j_1+j_2-j-z]![j_2-m_2-z]![j_1+j_2+j+1-z]!} . \quad (5.13)$$

In a "classical" limit $q=1$ it coincides with the general formula for CGCs obtained in Ref.[18]. It is once more version of q-CGCs formulae alternative to ones derived in Refs.[9-11,14-17].

Simple analytical formulae can be found for important particular cases [28]

$$\langle 00;jm|j'm'\rangle_q = \langle jm;00|j'm'\rangle_q = \delta_{j,j'}\,\delta_{m,m'}, \quad (5.14)$$

$$\langle jm,j'm'|00\rangle_q = \delta_{j',j}\,\delta_{m,-m'} \frac{(-1)^{j-m}}{\sqrt{[2j+1]}} q^m, \quad (5.15)$$

$$\langle j_1m_1;j_2m_2|jj\rangle_q = \quad (5.16)$$

$$= \delta_{j,m_1+m_2}(-1)^{j_1-m_1} q^{\frac{1}{2}(j_1+j_2-j)(j-j_1+j_2+1)-(j+1)(j_1-m_1)} \times$$

$$\times \sqrt{\frac{[2j+1]![j_1+m_1]![j_2+m_2]![j_1+j_2-j]!}{[j_1-m_1]![j_2-m_2]![j_1-j_2+j]![j-j_1+j_2]![j_1+j_2+j+1]!}} ,$$

$$\langle j_1j_1;j_2m_2|jm\rangle_q = \quad (5.17)$$

$$= \delta_{m,j_1-m_2}\, q^{\frac{1}{2}(j_1+j_2-j)(j-j_1+j_2+1)-j_1(j-m)} \times$$

$$\times \sqrt{\frac{[2j+1][j+m]![2j_1]![j_2-m_2]![j-j_1+j_2]!}{[j-m]![j_2+m_2]![j_1-j_2+j]![j_1+j_2-j]![j_1+j_2+j+1]!}} .$$

6. SYMMETRY PROPERTIES OF CLEBSCH-GORDAN COEFFICIENTS

First of all we are interesting in a symmetry property of q-CGC with respect to permutation of angular momenta j_1 and j_2. It should be noted that the operators (4.5b) of summary angular momentum are noninvariant with respect to permutation p_{12} of $\mathbf{j}_1$ and $\mathbf{j}_2$. In fact

$$J^q_\pm(1,2) \to J^q_\pm(2,1) = J^{\bar{q}}_\pm(1,2) \qquad (6.1)$$

at substitution $J_\pm(1) \leftrightarrow J_\pm(2)$. Also the Casimir operator is not conserving at permutation p_{12}

$$\mathbf{C}^q_2(1,2) = \mathbf{C}^{\bar{q}}_2(2,1) \neq \mathbf{C}^q_2(2,1) \ . \qquad (6.2)$$

Therefore the vectors $|j_1 j_2 : jm\rangle_q$ and $|j_2 j_1 : jm\rangle_q$ are eigenvectors of <u>different</u> Casimir operators $\mathbf{C}^q_2(1,2)$ and $\mathbf{C}^q_2(2,1)$ respectively. Thus the inner product of these vectors is not diagonal on quantum number j although it is diagonal on m

$$_q\langle j_2 j_1 : j'm | j_1 j_2 : jm\rangle_q \neq 0 \ \text{ at } j' \neq j \ . \qquad (6.3)$$

Because of such nonconservation of summary angular momenta j at the permutation of momenta j_1 and j_2 it is reasonable to compare the vectors $|j_1 j_2 : jm\rangle_q$ and $|j_2 j_1 : jm\rangle_{\bar{q}}$ because they are eigenvectors of the same Casimir operator $\mathbf{C}^q_2(1,2)$. Thus to find the connection between CGCs $(j_1 m_1 ; j_2 m_2 | jm)_q$ and $(j_2 m_2 ; j_1 m_1 | jm)_{\bar{q}}$ it should to compare vectors

$$|j_1 j_2 ; jm\rangle_q = (Q_q)^{-1} P^{j,q}_{m,j}(1,2) | j_1 j_1(1)\rangle | j_2, j - j_1(2)\rangle, \qquad (6.4)$$

and

$$|j_2 j_1 ; jm\rangle_{\bar{q}} = (Q_{\bar{q}})^{-1} P^{j,\bar{q}}_{m,j}(2,1) | j_2 j_2(2)\rangle | j_1, j - j_2(1)\rangle. \qquad (6.5)$$

The POs $P^j_{m,j}$ in both of this expression are identical. As for positive normalizing factors they are described by expressions

$$Q_q = q^{\frac{1}{2}(j_1+j_2-j)(j-j_1+j_2+1)} \sqrt{\frac{[2j+1]!\,[2j_1]!}{[j+j_1-j_2]!\,[j_1+j_2+j+1]!}} \ , \qquad (6.6)$$

$$Q_{\bar{q}} = q^{-\frac{1}{2}(j_1+j_2-j)(j+j_1-j_2+1)} \sqrt{\frac{[2j+1]!\,[2j_2]!}{[j-j_1+j_2]!\,[j_1+j_2+j+1]!}} \ . \qquad (6.7)$$

In order to reduce the vector (6.5) to the form (6.4) it is necessary to rewrite the vector $|j_1\; j-j_2\rangle$ as following

$$|j_1, j-j_2(1)\rangle \sim \left[J_-(1)\right]^{j_1+j_2-j} |j_1 j_1(1)\rangle \sim \tag{6.8}$$

$$\sim q^{-(j_1+j_2-j)J_0(2)} \left[J_-(1,2) - q^{-J_0(1)} J_-(2)\right]^{j_1+j_2-j} |j_1 j_1(1)\rangle.$$

Using the binomial expansion

$$\left[J^q_\pm(1,2) - q^{-J_0(1)} J_\pm(2)\right]^s = \tag{6.9}$$

$$= \sum_{s=0}^{k} \frac{(-1)[k]!}{[s]![k-s]!} q^{\mp s(k-1)} \left[J^q_\pm(1,2)\right]^{k-s} \left[J^q_\pm(2)\right]^s q^{-sJ_0(1)}$$

and taking account the property (3.8) we see that nonvanishing contribution can give only first term of this expression (not containing $J^q_\pm(1,2)$ generator) with phase factor $(-1)^{j_1+j_2-j}$. Rest factors appearing in this expression are canceling the difference of normalizing factors Q_q and $Q_{\bar{q}}$. As a result we have

$$|j_1 j_2; jm\rangle_q = (-1)^{j_1+j_2-j} |j_2 j_1; jm\rangle_{\bar{q}} \;. \tag{6.10}$$

From this equation the symmetry property of q-CGCs follows

$$\langle j_1 m_1; j_2 m_2 | jm\rangle_q = (-1)^{j_1+j_2-j} \langle j_2 m_2; j_1 m_1 | jm\rangle_{\bar{q}} \tag{6.11}$$

i.e. the same phase factor as in AMT appears at simultaneous substitution q by $\bar{q}\equiv q^{-1}$. In similar way another symmetry properties of q-CGCs can be proven

$$\langle j_1 m_1; j_2 m_2 | jm\rangle_q =$$

$$= (-1)^{j_2+m_2} q^{-m_2} \sqrt{\frac{[2j+1]}{[2j_1+1]}} \langle j_2 -m_2; jm | j_1 m_1\rangle_{\bar{q}} \tag{6.12}$$

$$= (-1)^{j_1+m_1} q^{m_1} \sqrt{\frac{[2j+1]}{[2j_2+1]}} \langle jm, j_1 -m_1 | j_2 m_2\rangle_{\bar{q}} \tag{6.13}$$

$$\langle j_1 m_1, j_2 m_2 | jm\rangle_q = (-1)^{j_1+j_2-j} \langle j_2 -m_2, j_1 -m_1 | j -m\rangle_{\bar{q}} \;. \tag{6.14}$$

It is important to underline the Regge symmetry holds for q-CGCs. In fact if one expresses the q-CGC

$\langle j'_1m'_1;j'_2m'_2|j'm'\rangle_q$ with

$$j'_1 = \tfrac{1}{2}(j_1+j_2+m_1+m_2), \qquad m'_1 = \tfrac{1}{2}(j_1-j_2+m_1-m_2),$$

$$j'_2 = \tfrac{1}{2}(j_1+j_2-m_1-m_2), \qquad m'_2 = \tfrac{1}{2}(j_1-j_2-m_1+m_2), \qquad (6.15)$$

$$j' = j\,, \qquad m' = j_1-j_2$$

in explicit form using the formula (5.13) then the resulting expression will coincide with Eq.(5.13) again because most part of q-factorials in (5.13) is invariant with respect the substitution (6.15) and in four pairs of factorials there is an exchange of terms of each pair, for example $j_1-j_2+j \leftrightarrow j+m$, $j_1-m_1 \leftrightarrow j_2+m_2$ etc. It means that

$$\langle j_1m_1;j_2m_2|jm\rangle_q = \langle j'_1m'_1;j'_2m'_2|j'm'\rangle_q\,. \qquad (6.16)$$

This Regge symmetry property together with (6.11)-(6.14) generate total set of 144 symmetry properties of q-CGCs.

In Ref.[11] the q-analog of 3j-symbols was defined as a follows

$$\begin{pmatrix} j_1 & j_2 & j_3 \\ m_1 & m_2 & m_3 \end{pmatrix}_q = \frac{(-1)^{j_1-j_2-m_3}}{\sqrt{[2j_3+1]}}\, q^{-\frac{1}{3}(m_1-m_2)} \langle j_1m_1;j_2m_2|jm\rangle_q\,. \qquad (6.17)$$

3j-Symbols are characterized by developed symmetry properties:

(i) they are invariant with respect to even permutations of columns;

(ii) at odd permutations of columns the phase factor $(-1)^{j_1+j_2+j_3}$ appears and the substitution of q by $\bar{q}\equiv q^{-1}$ takes place;

(iii) the reflection of all projections $m_i \to -m_i$ gives

$$\begin{pmatrix} j_1 & j_2 & j_3 \\ m_1 & m_2 & m_3 \end{pmatrix}_q = (-1)^{j_1+j_2+j_3} \begin{pmatrix} j_1 & j_2 & j_3 \\ -m_1 & -m_2 & -m_3 \end{pmatrix}_{\bar{q}}\,. \qquad (6.18)$$

In order to include the Regge symmetry properties of 3j-symbols it is convenient to introduce the Regge symbol

$$\begin{pmatrix} j_1 & j_2 & j_3 \\ m_1 & m_2 & m_3 \end{pmatrix}_q \equiv \begin{bmatrix} j_1+m_1 & j_2+m_2 & j_3+m_3 \\ j_1-m_1 & j_2-m_2 & j_3-m_3 \\ j_3-j_1+j_2 & j_3+j_1-j_2 & -j_3+j_1+j_2 \end{bmatrix}_q \qquad (6.19)$$

that is invariant with respect transposition and even permutations of rows and columns. At odd permutations of rows and columns the phase factor $(-1)^{j_1+j_2+j_3}$ appears and the substitution $q \to \bar{q}$ takes place.

7. TENSOR OPERATORS, WIGNER-ECKART THEOREM

As a q-analog of rank k tensor operator we shall consider a set of $2k+1$ operators $T^k_{æ}(q)$ ($æ=k,k-1,\dots,-k+1,-k$) satisfying the following commutation relations with generators of $SU_q(2)$ algebra

$$[J_0, T^k_{æ}(q)] = æ\, T^k_{æ}(q) , \tag{7.1}$$

$$J_\pm T^k_{æ}(q) = q^{æ} T^k_{æ}(q) J_\pm + \sqrt{[k\mp æ][k\pm æ+1]}\; T^k_{æ\pm 1}(q) q^{-J_0}. \tag{7.2}$$

Acting by generators $J_{0,\pm}$ on vectors $\Phi^{j,k}_{m,æ}(q) \equiv T^k_{æ}(q)|jm\rangle$ and taking account (7.1) and (7.2) we obtain

$$J_0\left[T^k_{æ}(q)|jm\rangle\right] = (m+æ) T^k_{æ}(q)|jm\rangle , \tag{7.3}$$

$$J_\pm\left[T^k_{æ}(q)|jm\rangle\right] = q^{æ}\sqrt{[j\mp m][j\pm m+1]}\; T^k_{æ}(q)|j,m\pm 1\rangle +$$

$$+ q^{-m}\sqrt{[k\mp æ][k\pm æ+1]}\; T^k_{æ\pm 1}(q)|jm\rangle . \tag{7.4}$$

From the comparison of these expressions with (4.7) it is clear that vectors $\Phi^{j,k}_{m,æ}(q)$ are transforming as basis vectors of the tensor product $D^j \otimes D^k$ of IRs of $SU_q(2)$ algebra. Therefore it is possible to expand these vectors on components $\Phi^{j,k:j''}_{m''}(q)$ belong to the IR $D^{j''}$ of $SU_q(2)$

$$\Phi^{j,k}_{m,æ}(q) = \sum_{j''m''} \langle jm;kæ|j''m''\rangle_q\, \Phi^{j,k:j''}_{m''}(q) . \tag{7.5}$$

Multiplying both sides of this equation by vector $\langle j'm'|$ and taking account the orthogonality property

$$\left\langle j'm' \middle| \Phi^{j,k:j''}_{m''}(q)\right\rangle = \delta_{j'j''}\, \delta_{m'm''}\, f^{j,k}_{j'}(q) \tag{7.6}$$

where $f^{j,k}_{j'}(q)$ is independent on m', m, $æ$ we obtain the q-analog of well known Wigner-Eckart theorem:

$$\langle j'm'|T^k_{æ}(q)|jm\rangle = \langle jm;kæ|j'm'\rangle_q\, (j'\|T^k(q)\|j) \tag{7.7}$$

or in more standard form

$$\langle j'm'|T^k_{æ}(q)|jm\rangle = \frac{\langle jm;kæ|j'm'\rangle_q}{\sqrt{[2j+1]}}\, (-1)^{2k}\langle j'\|T^k(q)\|j\rangle. \tag{7.8}$$

As an example the tensor operator the first rank $J^1_{æ}(q)$

($\varkappa=0,\pm 1$) is constructed by us from generators $J_{0,\pm}$ in explicit form:

$$J^1_{\pm 1}(q) = \frac{\mp 1}{\sqrt{[2]}} q^{-J_0} J_\pm , \qquad (7.9a)$$

$$J^1_0(q) = \frac{1}{\sqrt{[2]}} \left[q^{-1}[2J_0] + (q-q^{-1})J_+J_-\right] =$$

$$= \frac{1}{\sqrt{[2]}} \left[q^{-1}[2J_0] + (q-q^{-1})(\mathbf{C}_2-[J_0-1/2]^2)\right] . \qquad (7.9b)$$

It is clear that these expressions are more complicate then in $SU(2)$ case but in the limit $q=1$ they coincide with standard cyclic components of angular momentum. Calculating necessary CGCs we find the following expression for the reduced matrix elements of the tensor (7.9)

$$\langle J' \| J^1(q) \| J\rangle = \delta_{JJ'} \frac{1}{[2]} \sqrt{[2J][2J+1][2J+2]} . \qquad (7.10)$$

For the unit operator we have

$$(J' \| I \| J) = \delta_{JJ'} \sqrt{[2J+1]} . \qquad (7.11)$$

On the base of the definition (7.1) and Wigner-Eckart theorem the total q-analog of the tensor operator algebra can be formulated.

In conclusion of this section it should be noted that the derivation of Wigner-Eckart theorem in terms of quantum group $SU_q(2)$ was given by A.Klimyk.The definition of tensor operators close to our one was given and actively exploited in the frame of q-boson calculus by L.Biedenharn [6].

8. RECOUPLING OF ANGULAR MOMENTA, 6J-SYMBOL

The vector coupling of three angular momenta can be realized in two ways: $(\mathbf{j}_1+\mathbf{j}_2)+\mathbf{j}_3$ and $\mathbf{j}_1+(\mathbf{j}_2+\mathbf{j}_3)$. The vectors corresponding to the fist coupling scheme are of the form:

$$|J_1J_2(J_{12}),J_3:Jm\rangle_q = \qquad (8.1)$$

$$= \sum_{m_i m_{12}} \langle J_1m_1;J_2m_2|J_{12}m_{12}\rangle_q \langle J_{12}m_{12};J_3m_3|Jm\rangle_q |J_1m_1\rangle|J_2m_2\rangle|J_3m_3\rangle .$$

For the second coupling scheme we have

$$|J_1,J_2J_3(J_{23}):Jm\rangle_q = \qquad (8.2)$$

$$= \sum_{m_i m_{23}} \langle J_2m_2;J_3m_3|J_{23}m_{23}\rangle_q \langle J_1m_1;J_{23}m_{23}|Jm\rangle_q |J_1m_1\rangle|J_2m_2\rangle|J_3m_3\rangle .$$

The transition from the set of vectors (8.1) to vectors (8.2) can be done using Racah coefficients

$$|j_1 j_2(j_{12}).j_3 : jm\rangle_q =$$

$$= \sum_{j_{23}} U(j_1 j_2 j j_3 ; j_{12} j_{23})_q |j_1, j_2 j_3(j_{23}) : jm\rangle_q . \qquad (8.3)$$

They form the orthogonal matrix

$$\sum_{j_{23}} U(j_1 j_2 j j_3 ; j_{12} j_{23})_q \, U(j_1 j_2 j j_3 ; j'_{12} j_{23})_q = \delta_{j_{12} j'_{12}} \qquad (8.4)$$

and can be expressed in terms of q-CGCs

$$U(j_1 j_2 j j_3 ; j_{12} j_{23})_q = \sum_{m_i m_{lk}} \langle j_1 m_1 ; j_2 m_2 | j_{12} m_{12}\rangle_q \times \qquad (8.5)$$

$$\times \langle j_{12} m_{12} ; j_3 m_3 | jm\rangle_q \langle j_2 m_2 ; j_3 m_3 | j_{23} m_{23}\rangle_q \langle j_1 m_1 ; j_{23} m_{23} | jm\rangle_q .$$

These recoupling coefficients do not depend on the projection m of summary angular momentum j. Therefore it is possible to multiply both sides of (8.5) by q^m and to do the summation on m

$$\sum_{m=-j}^{j} q^m = [2j+1] .$$

As a result a new expression for the Racah coefficients $U(\ldots)$ will be obtained that coincides with right side of (8.5) except for the additional sum on index m and the factor $[2j+1]^{-1}$. It is useful to introduce a q-analog of 6j-symbol instead of Racah coefficients

$$U(abcd;ef)_q = \sqrt{[2e+1][2f+1]} \, (-1)^{a+b+c+d} \begin{Bmatrix} a & b & e \\ d & c & f \end{Bmatrix}_q . \qquad (8.6)$$

If to express it in terms of 3j-symbols

$$\begin{Bmatrix} j_1 & j_2 & j_3 \\ l_1 & l_2 & l_3 \end{Bmatrix}_q = \qquad (8.7)$$

$$= \sum_{m_i n_i} (-q)^{-m_1-m_2-m_3-n_1-n_2-n_3} (-1)^{j_1+j_2+j_3+l_1+l_2+l_3}$$

$$\times \begin{pmatrix} j_1 & j_2 & j_3 \\ m_1 & m_2 & m_3 \end{pmatrix}_q \begin{pmatrix} j_1 & l_2 & l_3 \\ -m_1 & n_2 & -n_3 \end{pmatrix}_q \begin{pmatrix} l_1 & j_2 & l_3 \\ -n_1 & -m_2 & n_3 \end{pmatrix}_q \begin{pmatrix} l_1 & l_2 & j_3 \\ n_1 & -n_2 & -m_3 \end{pmatrix}_q$$

then the symmetry properties of 6j-symbols can be easy found. Namely the 6j-symbols are invariant with respect to permutations of columns

$$\begin{Bmatrix} j_1 & j_2 & j_3 \\ l_1 & l_2 & l_3 \end{Bmatrix}_q = \begin{Bmatrix} j_2 & j_1 & j_3 \\ l_2 & l_1 & l_3 \end{Bmatrix}_q = \dots \quad . \tag{8.8}$$

Also they are invariant with respect to substitution two arbitrary momenta in the first row by corresponding momenta from the second row

$$\begin{Bmatrix} j_1 & j_2 & j_3 \\ l_1 & l_2 & l_3 \end{Bmatrix}_q = \begin{Bmatrix} l_1 & l_2 & j_3 \\ j_1 & j_2 & l_3 \end{Bmatrix}_q = \dots \quad . \tag{8.9}$$

Finally Racah coefficients $U(\dots)$ and 6j-symbols are invariant with respect to substitution $q\to\bar{q}$ as it can be seen from the general analytical formula (8.14) for Racah coefficients. The last one may be derived using the projection operators in the following manner.

Let us consider the matrix element

$$X=\langle j_1m_1|\langle j_2m_2|\langle j_3m_3|P^{j_{23}}P^{j}P^{j_{12}}|j_1m_1'\rangle|j_2m_2'\rangle|j_3m_3'\rangle \tag{8.10}$$

with $m_1=j-j_{23}$, $m_2=m_2'=j_2$, $m_3=j_{23}-j_2$, $m_1'=j_{12}-j_2$, $m_3'=j-j_{12}$. Here the projection operators $P^{j_{12}}$, $P^{j_{23}}$ and P^{j} constructed from generators

$$J_\pm^q(1,2) = J_\pm(1)\, q^{J_0(2)} + q^{-J_0(1)} J_\pm(2) \ ,$$

$$J_\pm^q(2,3) = J_\pm(2)\, q^{J_0(3)} + q^{-J_0(2)} J_\pm(3) \ ,$$

and

$$J_\pm^q(1,2,3) = J_\pm^q(1,2)\, q^{J_0(3)} + q^{-J_0(1,2)} J_\pm(3) =$$

$$= J_\pm(1)\, q^{J_0(2,3)} + q^{-J_0(1)} J_\pm(2,3) = \tag{8.11}$$

$$= J_\pm(1) q^{J_0(2)+J_0(3)} + q^{-J_0(1)} J_\pm(2) q^{J_0(3)} + q^{-J_0(1)-J_0(2)} J_\pm(3)$$

respectively. Acting by $P^{j_{12}}$ on the right side of the matrix element (8.10) and by $P^{j_{23}}$ on its left side we obtain

$$X = \langle j_1m_1, j_2m_2|j_{12}m_{12}\rangle_q \langle j_{12}m_{12}, j_3m_3|jm\rangle_q U(j_1j_2jj_3;j_{12}j_{23})_q$$

$$\times \langle j_2m_2', j_3m_3'|j_{23}m_{23}'\rangle_q \langle j_1m_1', j_{23}m_{23}'|jm\rangle_q \ . \tag{8.12}$$

At particular values of projections m_i, m_i' taken in

(8.10) all q-CGCs do not contain any internal sums and are simple root squares from the combinations of q-factorials. This matrix element can be calculated also in direct manner

$$X = \sum_{n_i n'_i} \langle J_2 m_2 | \langle J_3 m_3 | P^{J_{23}} | J_2 n_2 \rangle | J_3 n_3 \rangle \times$$

$$\times \langle J_1 n_1 | \langle J_2 n_2 | \langle J_3 n_3 | P^J | J_1 n'_1 \rangle | J_2 n'_2 \rangle | J_3 n'_3 \rangle \times$$

$$\times \langle J_1 n'_1 | \langle J_2 n'_2 | P^{J_{12}} | J_1 m'_1 \rangle | J_2 m'_2 \rangle . \qquad (8.13)$$

Taking account the properties of the vector $|J_2 m_2\rangle = |J_2 m'_2\rangle = |J_2 J_2\rangle$ as a highest weight vector it is possible to omit generator $J_+(2)$ in the third factor of r.h.s. (8.13), and $J_-(2)$ generator in the first factor. Because of corresponding properties of projectors $P^{J_{12}}$ and $P^{J_{23}}$ the operators $J^q_+(1,2)$ and $J^q_-(2,3)$ in the summary momentum operators (8.11) may be omitted in the projector P^J. After such simplifications it becomes clear that the summation on n'_i and n''_i in (8.13) is absent in fact since $n_1 = J - J_{23}$, $n_2 = n'_2 = J_{12} + J_{23} - J - r$, $n_3 = J - J_{12} + r$, $n'_1 = J - J_{23}$, $n'_3 = J - J_{12}$. Here r is a summation index in the explicit expression of the projector P^J. Calculating its matrix element in (8.13) in a direct manner and taking account that the matrix elements of $P^{J_{12}}$ and $P^{J_{23}}$ can be expressed in terms of simple q-CGCs

$$\langle J_2 J_2, J_3 J_{23} - J_2 | J_{23} J_{23} \rangle_q \, \langle J_2 J_{12} + J_{23} - J - r, J_3 J - J_{12} + r | J_{23} J_{23} \rangle_q \times$$

$$\times \langle J_1 J - J_{23} + r, J_2 J_{12} + J_{23} - J - r | J_{12} J_{12} \rangle_q \, \langle J_1 J_{12} - J_2 J_2 J_2 | J_{12} J_{12} \rangle_q$$

we find from Eqs. (8.12) and (8.13) the following explicit expression for the Racah coefficients

$$U(J_1 J_2 J J_3; J_{12} J_{23})_q = \sqrt{[2J_{12}+1][2J_{23}+1]}\, (-1)^{J - J_{23} - J_{12} + J_2}$$

$$\times \frac{\Delta(J_1 J_2 J_{12})\, \Delta(J_2 J_3 J_{23})\, \Delta(J_{12} J_3 J)\, \Delta(J_1 J_{23} J)}{[J_1 - J_2 + J_{12}]!\,[-J_1 + J_2 + J_{12}]!\,[J_2 - J_3 + J_{23}]!\,[-J_2 + J_3 + J_{23}]!} \times$$

$$\times \frac{[J_{12} + J_3 + J + 1]!\,[J_1 + J_{23} + J + 1]!}{[J_1 - J_{23} + J]!\,[-J_{12} + J_3 + J]!} \times \qquad (8.14a)$$

$$\times \sum_r \frac{(-1)^r [J_1 + J - J_{23} + r]!\,[J_3 + J - J_{12} + r]!\,[J_2 - J + J_{12} + J_{23} - r]!}{[r]!\,[2J + r + 1]!\,[J_1 - J + J_{23} - r]!\,[J_3 + J_{12} - J - r]!\,[J_2 + J - J_{12} - J_{23} + r]!} .$$

Here

$$\Delta(abc) = \sqrt{\frac{[a+b-c]![a-b+c]![-a+b+c]!}{[a+b+c+1]!}} . \qquad (8.14b)$$

In the particular case of one vanishing momentum in the 6j-symbol we have

$$\begin{Bmatrix} J_1 & J_2 & J_3 \\ 0 & J_3 & J_2 \end{Bmatrix}_q = \frac{(-1)^{J_1+J_2+J_3}}{\sqrt{[2J_2+1][2J_3+1]}} . \qquad (8.15)$$

As an example of recoupling of four angular momenta $\mathbf{l}_1+\mathbf{l}_2+\mathbf{l}'_2+\mathbf{l}'_3=\mathbf{J}_2$ the derivation of the Elliott-Biedenharn identity can be considered. If the transition between two coupling schemes

$$(((\mathbf{l}_1+\mathbf{l}_2)_{J_1}+\mathbf{l}'_2)_{l'_1}+\mathbf{l}'_3)_{J_2} \text{ and } (\mathbf{l}_1+(\mathbf{l}_2+(\mathbf{l}'_2+\mathbf{l}'_3)_{J_1})_{l_3})_{J_2} \qquad (8.16)$$

is under consideration it can be done in two ways. The first one is realized through intermediate coupling scheme

$$(\mathbf{l}_1+\mathbf{l}_2)_{J_3}+(\mathbf{l}'_2+\mathbf{l}'_3)_{J_1} \qquad (8.17)$$

with repeated using of Racah coefficients

$$Y = U(J_3l'_2J_2l'_3;l'_1J_1)_q U(l_1l_2J_2J_1\ J_3l_3)_q . \qquad (8.18)$$

The second way consists in using of two intermediate coupling schemes

$$((\mathbf{l}_1+(\mathbf{l}_2+\mathbf{l}'_2)_k)_{l'_1}+\mathbf{l}'_3)_{J_2} \text{ and } (\mathbf{l}_1+((\mathbf{l}_2+\mathbf{l}'_2)_k+\mathbf{l}'_3)_{l_3})_{J_2}. \qquad (8.19)$$

In this case the total recoupling coefficient Y can be expressed in terms of product of three Racah coefficients with summation on one intermediate momentum k

$$Y = \sum_k U(l_1l_2l'_1l'_2;J_3k)_q \times$$

$$\times U(l_1kJ_2l'_3;l'_1l_3)_q\, U(l_2l'_2l_3l'_3;kJ_1)_q . \qquad (8.20)$$

Substituting Racah coefficients in Eq.(8.18) and (8.20) by 6j-symbols we obtain the q-analog of Elliott-Biedenharn identity

$$\begin{Bmatrix} J_1 & J_2 & J_3 \\ l_1 & l_2 & l_3 \end{Bmatrix}_q \begin{Bmatrix} J_1 & J_2 & J_3 \\ l'_1 & l'_2 & l'_3 \end{Bmatrix}_q =$$

$$= \sum_k (-1)^{k+J_1+J_2+J_3+l_1+l_2+l_3+l'_1+l'_2+l'_3} \times$$

$$\times [2k+1] \begin{Bmatrix} l_1 & k & l'_1 \\ l'_3 & J_2 & l_3 \end{Bmatrix}_q \begin{Bmatrix} l_2 & k & l'_2 \\ l'_1 & J_3 & l_1 \end{Bmatrix}_q \begin{Bmatrix} l_3 & k & l'_3 \\ l'_2 & J_1 & l_2 \end{Bmatrix}_q . \qquad (8.21)$$

The important feature of the transition (8.16) was that in all coupling schemes (8.16),(8.17) and (8.19) the order of recoupled angular momenta was the same one. All these coupling schemes are different each other only by position of brackets symbolizing the vector coupling of a pair angular momenta. Namely due to this fact each step of this transition (8.16) could be reduced to standard Racah coefficient. If the recoupling of angular momenta $J_1, J_2, \dots, J_n$ is connected with a change of order of these momenta the so called R-matrix [10,11] should be used. In particular the recoupling of angular momenta

$$((\mathbf{j}_1+\mathbf{j}_2)_{j_{12}}+(\mathbf{j}_3+\mathbf{j}_4)_{j_{34}})_j \rightarrow ((\mathbf{j}_1+\mathbf{j}_3)_{j_{13}}+(\mathbf{j}_2+\mathbf{j}_4)_{j_{24}})_j \quad (8.22)$$

connected with 9j-symbols in the usual AMT requires the using of R-matrix in the case of $SU_q(2)$.

9. UNIVERSAL R-MATRIX

It is clear from the Eq.(6.10) that transposition of two coupled angular momenta transforms the vector $|J_1J_2:Jm\rangle_q$ into similar vector $|J_2J_1:Jm\rangle_{\bar{q}}$ with parameter $\bar{q}\equiv q^{-1}$. If we want to connect vectors $|J_1J_2:Jm\rangle_q$ and $|J_2J_1:Jm\rangle_q$ with the same parameter q it is necessary to use the R-matrix determined by the equation

$$R_q^{12}|J_1J_2:Jm\rangle_q \equiv R_q^{j_1j_2}|J_1J_2:Jm\rangle_q = \lambda_j^{j_1j_2}(q)|J_2J_1:Jm\rangle_q \quad (9.1)$$

where the factor $\lambda_j^{j_1j_2}(q)$ will be given below. Since vectors in the l.h.s. and r.h.s. of (9.1) are belonging to the eigenvalue $[J+1/2]^2$ of the <u>different</u> Casimir operators $\mathbf{C}_2^q(1,2)$ and $\mathbf{C}_2^{\bar{q}}(1,2)$ respectively the R-matrix should to transform also the generators $J_\pm^q(1,2)$ into $J_\pm^{\bar{q}}(1,2)$

$$R_q^{12}\, J_\pm^q(1,2) = J_\pm^{\bar{q}}(1,2)\, R_q^{12} \,. \quad (9.2)$$

According to Refs [5,10] it is reasonable to seek an R-matrix in form of power series

$$R_q^{12} = \sum_{n=0}^{\infty} B_n\, J_+^{\,n}(1)\, J_-^{\,n}(2) \times$$

$$\times\, q^{2J_0(1)J_0(2)-nJ_0(1)+nJ_0(2)-\frac{n(n+1)}{2}} \,. \quad (9.3)$$

From the equations (9.3) the recurrent relation for coefficients B_n follows

$$(q-q^{-1})B_{n-1} - [n]B_n = 0 \,. \quad (9.4)$$

Assuming $B_0 = 1$ we obtain the result

$$B_n = \frac{(q-q^{-1})^n}{[n]!} \tag{9.5}$$

that means $R_q^{12}=1$ at $q=1$.

The alternative form of R-matrix exists

$$\tilde{R}_q^{12} = \sum_{n=0}^{\infty} (-1)^n B_n J_+^n(1) J_-^n(2) \times$$

$$\times q^{-2J_0(1)J_0(2)-nJ_0(1)+nJ_0(2)+\frac{n(n+1)}{2}} . \tag{9.6}$$

However we shall use below only the form of R-matrix given by (9.3) and (9.5).

If we apply the R-matrix to the vector $|j_1 j_2 : jm\rangle_q$ then taking account the relation

$$R_q^{12} P^{j,q}(1,2) = P^{j,\bar{q}}(1,2) R_q^{12} \tag{9.7}$$

which is a consequence of (9.2) we obtain

$$R_q^{12}|j_1 j_2 : jm\rangle_q = Q_q^{-1} P^{j,\bar{q}}(1,2) R_q^{12}|j_1 j_1\rangle|j_2 j-j_1\rangle =$$

$$= (Q_q)^{-1} Q_{\bar{q}}\, q^{2j_1(j-j_1)} |j_1 j_2 : jm\rangle_{\bar{q}} . \tag{9.8}$$

Substituting here the explicit expression (6.6) for the normalizing factor Q_q and comparing the result with the definition (9.2) we find

$$\lambda_j^{j_1 j_2}(q) = q^{c(j)-c(j_1)-c(j_2)} (-1)^{j-j_1-j_2}, \tag{9.9}$$

where $c(j) = j(j+1)$ is the eigenvalue of the "classical" Casimir operator $\mathbf{J}^2$. Obviously,

$$(R_q^{12})^{-1} = R_{\bar{q}}^{12} , \tag{9.10}$$

$$\tilde{R}_q^{12} = R_{\bar{q}}^{21} . \tag{9.11}$$

The last relation means that for the operator (9.6) satisfying the equation (9.2) the relation

$$\tilde{R}_q^{12}|j_1 j_2 : jm\rangle_q = \lambda_j^{j_1 j_2}(\bar{q})|j_2 j_1 : jm\rangle_q \tag{9.12}$$

is valid.

The important property of R-matrices

$$R_q^{13}\, R_q^{23} = R_q^{12,3} \qquad (9.13)$$

can be proven by direct substitution of expression (9.3) to the l.h.s. of this equation and using the binomial formula (4.8) for the generators $(J_\pm^q(1,2))^n$ occurring into r.h.s. of (9.13). Similarly we have

$$R_q^{13}\, R_q^{12} = R_q^{1,23}\ . \qquad (9.14)$$

Combining (9.13) and (9.14) it is possible show that R-matrices (9.3) are satisfying the so called Yang-Baxter equation

$$R_q^{12}\, R_q^{13}\, R_q^{23} = R_q^{23}\, R_q^{13}\, R_q^{12}\ . \qquad (9.15)$$

Since the solution of Yang-Baxter equation is referred as R-matrix the operator determined by Eqs.(9.3) and (9.4) is called the universal R-matrix.

Concluding the section we calculate the matrix elements of the R-operator in uncoupled basis $|J_1 m_1\rangle |J_2 m_2\rangle$. It is evident [10] that

$$\langle J_1 m_1+n, J_2 m_2-n | R_q^{12} | J_1 m_1 J_2 m_2\rangle = q^{2m_1 m_2 - n m_1 + n m_2 - \frac{n(n+1)}{2}} \times$$

$$\times \frac{(q-q^{-1})^n}{[n]!} \left\{\frac{[J_1+m_1+n]!\,[J_1-m_1]!\,[J_2+m_2]!\,[J_2-m_2+n]!}{[J_1+m_1]!\,[J_1-m_1-n]!\,[J_2+m_2-n]!\,[J_2-m_2]!}\right\}^{\frac{1}{2}} . \qquad (9.16)$$

10. Q-ANALOG OF 9J-SYMBOL

We shall connect a q-analog of 9j-symbol with the transformation bracket between two sets of vectors:

$$R_q^{J_3 J_2} | J_1 J_3 (J_{13}), J_2 J_4 (J_{24}) : Jm\rangle_q \qquad (10.1)$$

and

$$| J_1 J_2 (J_{12}), J_3 J_4 (J_{34}) : Jm\rangle_q \qquad (10.2)$$

that are eigenvectors corresponding to eigenvalue $[J+\frac{1}{2}]^2$ of the same Casimir operator $C_2^q(1,2,3,4)$ constructed from the summary angular momentum operators

$$J_\pm^q(1,2,3,4) =$$

$$= J_\pm^q(1,2)\, q^{J_0(3)+J_0(4)} + q^{-J_0(1)-J_0(2)} J_\pm^q(3,4). \qquad (10.3)$$

Therefore the matrix of the transformation

$$R_q^{j_3 j_2} |j_1 j_3(j_{13}), j_2 j_4(j_{24}) : jm\rangle_q =$$

$$= \sum_{j_{12} j_{34}} \begin{bmatrix} j_1 & j_2 & j_{12} \\ j_3 & j_4 & j_{34} \\ j_{13} & j_{24} & j \end{bmatrix}_q |j_1 j_2(j_{12}), j_3 j_4(j_{34}) : jm\rangle_q \qquad (10.4)$$

is diagonal with respect quantum numbers j and m. Its value does not depend on the projection m. This transformation bracket can be expressed in terms of q-CGCs

$$\begin{bmatrix} j_1 & j_2 & j_{12} \\ j_3 & j_4 & j_{34} \\ j_{13} & j_{24} & j \end{bmatrix}_q =$$

$$= \langle j_1 j_2(j_{12}), j_3 j_4(j_{34}) : jm | R_q^{j_3 j_2} | j_1 j_3(j_{13}), j_2 j_4(j_{24}) : jm\rangle_q =$$

$$= \sum_{m_i m'_k m_{rs}} \langle j_1 m_1, j_2 m_2 | j_{12} m_{12}\rangle_q \langle j_3 m_3, j_4 m_4 | j_{34} m_{34}\rangle_q \times$$

$$\times \langle j_1 m_1, j_3 m'_3 | j_{13} m_{13}\rangle_q \langle j_2 m'_2, j_4 m_4 | j_{24} j_{24}\rangle_q \times$$

$$\times \langle j_2 m_2 | \langle j_3 m_3 | R_q^{j_3 j_2} | j_3 m_3{}'\rangle | j_2 m'_2\rangle. \qquad (10.5)$$

For the calculation of this bracket its expression in terms of Racah coefficients is more convenient

$$\begin{bmatrix} j_1 & j_2 & j_{12} \\ j_3 & j_4 & j_{34} \\ j_{13} & j_{24} & j \end{bmatrix}_q =$$

$$= \sum_{j_{123}} U(j_{13} j_2 j j_4 ; j_{123} j_{24})_q \, U(j_2 j_1 j_{123} j_3 ; j_{12} j_{13})_q \times$$

$$\times U(j_{12} j_3 j j_4 ; j_{123} j_{34})_q \, \lambda_{j_{123}}^{j_{13} j_2}(q) \, \lambda_{j_{12}}^{j_1 j_2}(\bar{q}). \qquad (10.6)$$

This formula can be obtained if the transition (10.4) is fulfilled in three steps

$$(((\mathbf{j}_1 + \mathbf{j}_3)_{j_{13}} + (\mathbf{j}_2 + \mathbf{j}_4)_{j_{24}})_j \rightarrow (((\mathbf{j}_1 + \mathbf{j}_3)_{j_{13}} + \mathbf{j}_2)_{j_{123}} + \mathbf{j}_4)_j \rightarrow$$

$$\rightarrow \left[R_q^{32} = R_{\bar q}^{12}\, R_q^{13,2}\right] \rightarrow \qquad (10.7)$$

$$(((\mathbf{j}_2+\mathbf{j}_1)_{j_{12}}+\mathbf{j}_2)_{j_{123}}+\mathbf{j}_4)_j \rightarrow ((\mathbf{j}_1+\mathbf{j}_2)_{j_{13}}+(\mathbf{j}_3+\mathbf{j}_4)_{j_{34}})_j \,.$$

Each this step generates one Racah coefficient in (10.6). The application of R-matrices $R_q^{13,2}$ and $R_{\bar q}^{12}$ to corresponding intermediate vectors gives the factors $\lambda_{j_{123}}^{j_{13}j_2}(q)$ and $\lambda_{j_{12}}^{j_1 j_2}(\bar q)$ in r.h.s. of (10.6) respectively. The transition inverse to (10.4) is of the form

$$|j_1 j_2(j_{12}), j_3 j_4(j_{34}): jm\rangle_q =$$

$$= \sum_{j_{13}\, j_{24}} \begin{bmatrix} j_1 & j_2 & j_{12} \\ j_3 & j_4 & j_{34} \\ j_{13} & j_{24} & j \end{bmatrix}_{\bar q} R_q^{23}\, |j_1 j_2(j_{12}), j_3 j_4(j_{34}): jm\rangle_q \,. \qquad (10.8)$$

Therefore the orthogonality relation for transformation brackets (10.6) can be written as follows

$$\sum_{j_{13}\, j_{24}} \begin{bmatrix} j_1 & j_2 & j_{12} \\ j_3 & j_4 & j_{34} \\ j_{13} & j_{24} & j \end{bmatrix}_{\bar q} \begin{bmatrix} j_1 & j_2 & j_{12} \\ j_3 & j_4 & j_{34} \\ j'_{13} & j'_{24} & j \end{bmatrix}_q = \delta_{j_{13} j'_{13}}\, \delta_{j_{24} j'_{24}} \,. \qquad (10.9)$$

The symmetry properties of the transformation bracket (10.6) can be found using symmetry properties of Racah coefficient. However the so called 9j-symbol [17] is more symmetrical. It is defined by the equation

$$\begin{bmatrix} j_1 & j_2 & j_{12} \\ j_3 & j_4 & j_{34} \\ j_{13} & j_{24} & j \end{bmatrix}_q = \sqrt{[2j_{12}+1][2j_{34}+1][2j_{13}+1][2j_{24}+1]}\,\times$$

$$\times\, q^{-c(j_{12})-c(j_{13})-c(j_{24})-c(j_{34})} \begin{Bmatrix} j_1 & j_2 & j_{12} \\ j_3 & j_4 & j_{34} \\ j_{13} & j_{24} & j \end{Bmatrix}_q \,. \qquad (10.10)$$

Using the r.h.s. of (10.6) we can express the q-analog of 9j-symbol also in terms of 6j-symbols

$$\begin{Bmatrix} j_1 & j_2 & j_{12} \\ j_3 & j_4 & j_{34} \\ j_{13} & j_{24} & j \end{Bmatrix}_q = \sum_z (-1)^{2z}\, q^{c(z)+c(j_1)+c(j_{24})+c(j_{34})} \times$$

$$\times\, [2z+1] \begin{Bmatrix} j_1 & j_2 & j_{12} \\ z & j_3 & j_{13} \end{Bmatrix}_q \begin{Bmatrix} j_3 & j_4 & j_{34} \\ j & j_{12} & z \end{Bmatrix}_q \begin{Bmatrix} j_{13} & j_{24} & j \\ j_4 & z & j_2 \end{Bmatrix}_q \,. \qquad (10.11)$$

On the base of 6j-symbol symmetry properties it is possible to prove [17] that

(i) the 9j-symbols are invariant with respect to tranposition;

(ii) they are invariant with respect to even permutations of columns or rows;

(iii) at odd permutations of columns or rows the 9j-symbol must be multiplied by the factor $(-1)^X q^Y$, where $X = \sum j_i$ and $Y = \sum c(j_i)$.

In limit $q=1$ these symmetry properties coincide with well known relations for standard 9j-symbols of AMT.

The symmetry properties of 9j-symbols are connected with the Yang-Baxter equation. If we act by both sides of Eq.(9.15) on the vector $|j_4 j_3(j_{34}), j_2(j_{234}), j_1 : jm\rangle_q$ and expand the resulting vectors on a set of orthonormalized vectors $|j_4 j_1(j_{14}), j_2(j_{124}), j_3 : jm\rangle_q$ the following relation will be obtained [10,11,17]

$$\sum_z (-1)^{2z} [2z+1]\, q^{c(z)+c(j)+c(j_{14})+c(j_{34})} \times$$

$$\times \begin{Bmatrix} j_4 & j_1 & j_{14} \\ j_{124} & j_2 & z \end{Bmatrix}_q \begin{Bmatrix} j_3 & j & j_{124} \\ j_1 & z & j_{234} \end{Bmatrix}_q \begin{Bmatrix} j_{34} & j_{234} & j_2 \\ z & j_4 & j_3 \end{Bmatrix}_q =$$

$$= \sum_z (-1)^{2z} [2z+1]\, q^{c(z)+c(j_4)+c(j_{124})+c(j_{234})} \times$$

$$\times \begin{Bmatrix} j_3 & j & j_{124} \\ j_2 & j_{14} & z \end{Bmatrix}_q \begin{Bmatrix} j_{34} & j_{234} & j_2 \\ j & z & j_1 \end{Bmatrix}_q \begin{Bmatrix} j_4 & j_1 & j_{14} \\ z & j_3 & j_{34} \end{Bmatrix}_q \qquad (10.12)$$

that corresponds to the symmetry of 9j-symbol

$$\begin{Bmatrix} j_4 & j_1 & j_{14} \\ j_3 & j & j_{124} \\ j_{34} & j_{234} & j_2 \end{Bmatrix}_q = \begin{Bmatrix} j_3 & j & j_{124} \\ j_{34} & j_{234} & j_2 \\ j_4 & j_1 & j_{14} \end{Bmatrix}_q \qquad (10.13)$$

with respect to even permutation of upper row into low position

As for the 9j-symbol with one vanishing momentum the following result is valid due to (8.15)

$$\begin{Bmatrix} f & f & 0 \\ d & g & e \\ b & a & e \end{Bmatrix}_q = \frac{(-1)^{b+g+e+f}\, q^{c(d)+c(e)+c(a)+c(f)}}{\sqrt{[2e+1][2f+1]}} \begin{Bmatrix} a & b & e \\ d & g & f \end{Bmatrix}_q . \qquad (10.14)$$

In conclusion of this section the useful equation will be given

$$R_q^{j_{13},j_2} |j_1 j_3(j_{13}), j_2 j_4(j_{24}) : jm\rangle_q =$$

$$= \sum_{j_{12} j_{34}} \begin{bmatrix} j_1 & j_2 & j_{12} \\ j_3 & j_4 & j_{34} \\ j_{13} & j_{24} & j \end{bmatrix}_q \lambda_{j_{12}}^{j_1 j_2}(q) |j_1 j_2(j_{12}), j_3 j_4(j_{34}) : jm\rangle_q \tag{10.15}$$

that is a consequence of (10.4) and (9.14).

11. ALGEBRA OF TENSOR OPERATORS

In analogy with usual algebra of tensor operators in AMT the tensor product of two tensor operators can be determined

$$\left[U^{k_1}(q_1) \otimes V^{k_2}(q_2)\right]_{\varkappa}^{k}(q) =$$

$$= \sum_{\varkappa_1 \varkappa_2} \langle k_1 \varkappa_1 ; k_2 \varkappa_2 | k \varkappa\rangle_q \, U_{\varkappa_1}^{k_1}(q_1) \, V_{\varkappa_2}^{k_2}(q_2) \tag{11.1}$$

where values of parameters q_1, q_2 that coincide with q or q^{-1} will be fixed below. The scalar product of tensor operators is defined as follows

$$\left(U^k(q_1) \cdot V^k(q_2)\right)(q) = (-1)^{-k} \sqrt{[2k+1]} \left[U^k(q_1) \otimes V^k(q_2)\right]_0^0(q) =$$

$$= \sum_{\varkappa} (-1)^{\varkappa} q^{\varkappa} \, U_{\varkappa}^k(q_1) \, V_{-\varkappa}^k(q_2) \quad . \tag{11.2}$$

At first we consider the case when both of operators U and V act on the same variables. Using the formulae of the section 7 and just the same reasons as in usual AMT [18-21] we obtain the following result

$$\langle \alpha' j' \| \left[U^{k_1}(q) \otimes V^{k_2}(q)\right]^{k}(\bar{q}) \| \alpha j\rangle = \sqrt{[2k+1]} \times \tag{11.3}$$

$$\times \sum_{\alpha'' j''} (-1)^{j+k+j'} \begin{Bmatrix} j & k & j' \\ k_1 & j_2 & k_2 \end{Bmatrix}_q \langle \alpha' j' \| U^{k_1}(q) \| \alpha'' j''\rangle \langle \alpha'' j'' \| V^{k_2}(q) \| \alpha j\rangle$$

that coincides with corresponding formula of usual AMT except for substitution of usual dimension $(2k+1)$ of D^k IR by "quantized" dimension $[2k+1]$ of this IR and using of q-analog instead of usual 6j-symbol. For the scalar product we have

$$\langle\alpha' J' \| \left[U^k(q)\cdot V^k(q)\right](\bar{q}) \| \alpha J\rangle = \delta_{JJ'}\delta_{mm'} \frac{1}{[2J+1]} \times$$

$$\times \sum_{\alpha'' J''} (-1)^{J''-J} \langle\alpha' J' \| U^k(q) \| \alpha'' J''\rangle \langle\alpha'' J'' \| V^k(q) \| \alpha J\rangle \qquad (11.4)$$

If operators $U^{k_1}(q_1;1)$ and $V^{k_2}(q_2;2)$ are acting on two different variables 1 and 2 then the matrix element of their tensor product will be expressed in terms of 9j-symbols as it is well known from usual AMT. In order to obtain the q-analog of a formula it is necessary to use R-matrix as it is clear from the preceding section. In this connection we shall define the adjoint action of R-matrix on the combination of tensor operator and the vector to which it is applied. Namely similarly to (9.16) we shall write

$$R_q^{j,k} \circ \left\{T^k_{\varkappa}(q_1)\,|Jm\rangle\right\} = \sum_n \frac{(q-q^{-1})^n}{[n]!} q^{2\varkappa m - nm + n\varkappa - \frac{n(n+1)}{2}} \times$$

$$\times \langle J\; m+n | J_+^{\;n} | Jm\rangle \langle k\; \varkappa-n | J_-^{\;n} | k\varkappa\rangle\; T^k_{\varkappa-n}(q_1)\,| J\; m+n\rangle\;. \qquad (11.5)$$

For the vector coupled combination

$$\left\{T^k(q_1)\,|J\rangle\right\}^{J'}_{m'}(q) = \sum_{m\varkappa} \langle Jm;k\varkappa | J'm'\rangle_q\; T^k(q_1)\,|Jm\rangle \qquad (11.6)$$

it means

$$R_q^{j,k} \circ \left\{T^k(q_1)\,|J\rangle\right\}^{J'}_{m'}(q) = q^{c(J')-c(J)-c(k)} \left\{T^k(q_1)\,|J\rangle\right\}^{J'}_{m'}(\bar{q})$$

(11.7)

in correspondence with (9.1). If $q_1 = q^{-1}$ the r.h.s. of this equation can be expressed in terms of reduced matrix element of T-operator using the Wigner-Eckart. From (11.5) and (11.6) is clear that the following generalization of the Wigner-Eckart theorem can be formulated

$$\langle J'm' | R_q^{j,k} \circ T^k_{\varkappa}(q_1)\,| Jm\rangle =$$

$$= (-1)^{2k} \frac{\langle Jm;k\varkappa | J'm'\rangle_q}{\sqrt{[2J'+1]}} \langle J' \| R_q^{j,k} \circ T^k(\bar{q}) \,| J\rangle\;, \qquad (11.8)$$

where the generalized reduced matrix element can be expressed in terms of usual reduced matrix element as follows

$$\langle j' \| R_q^{j,k} \circ T^k(\bar{q}) \| j\rangle = q^{c(j')-c(j)-c(k)} \langle j' \| T^k(\bar{q}) \| j\rangle . \quad (11.9)$$

In this connection it is reasonable to consider the following matrix elements of tensor product of two operators acting on different variables

$$\langle j_1' j_2' : j' m' | R_q^{j,k_1} \circ \left[U^{k_1}(\bar{q};1) \otimes V^{k_2}(q;2) \right]_{\varkappa}^{k}(q) | j_1 j_2 : jm\rangle =$$

$$= (-1)^{2k} \frac{\langle jm; k\varkappa | j'm'\rangle_q}{\sqrt{[2j'+1]}} \times \quad (11.10)$$

$$\times \langle j_1' j_2' : j' \| R_q^{j,k_1} \circ \left[U^{k_1}(\bar{q};1) \otimes V^{k_2}(q;2) \right]^{k}(q) \| j_1 j_2 : j\rangle .$$

The calculation of this matrix element in terms of reduced matrix elements of individual operators U and V can be done using the formula (10.15). As a result the following expression will be obtained

$$\langle j_1' j_2' : j' \| R_q^{j,k_1} \circ \left[U^{k_1}(\bar{q};1) \otimes V^{k_2}(q;2) \right]^{k}(q) \| j_1 j_2 : j\rangle =$$

$$= \sqrt{[2j+1][2j'+1][2k+1]}\; q^{-c(j)-c(k)-c(j_2')-c(j_1)-c(k_1)} \times$$

$$\times \begin{Bmatrix} j_1 & j_2 & j \\ k_1 & k_2 & k \\ j_1' & j_2' & j' \end{Bmatrix}_q \langle j' \| U^{k_1}(\bar{q}) \| j_1\rangle \langle j_2' \| V^{k_2}(q) \| j_2\rangle . \quad (11.11)$$

For the scalar product it gives

$$_q\langle j_1' j_2' : j' m' | R_q^{j,k} \circ \left(U^k(\bar{q};1) \cdot V^k(q;2) \right)(q) | j_1 j_2 : jm\rangle_q =$$

$$= \delta_{jj'} \delta_{mm'} (-1)^{j_1+j_2'+j} q^{c(j_1')-c(j_1)-c(j_2')+c(j_2)} \times$$

$$\times \begin{Bmatrix} j & j_2' & j_1' \\ k & j_1 & j_2 \end{Bmatrix}_q \langle j_1' \| U^k(\bar{q}) \| j_1\rangle \langle j_2' \| V^k(q) \| j_2\rangle . \quad (11.12)$$

In particular case $U=I$ (i.e. $k=0$ and $R_q^{j,k}=I$) the following relation can be obtained from (11.11)

$$ {}_q\langle j_1' j_2' : j' \| V^k(q;2) \| j_1 j_2 : j \rangle_q = \delta_{j_1 j_1'} (-1)^{j_1+j_2+k+j'} \times $$

$$ \times \sqrt{[2j+1][2j'+1]} \begin{Bmatrix} j' & k & j \\ j_2 & j_1 & j_2' \end{Bmatrix} \langle j_2' \| V^k(q) \| j_2 \rangle . \qquad (11.13) $$

Assuming $V=I$ in (11.11) we can go to the relation

$$ {}_q\langle j_1' j_2' : j' \| U^k(\bar{q};1) \| j_1 j_2 : j \rangle_q = \delta_{j_2 j_2'} (-1)^{j_1'+j_2+k+j} \times $$

$$ \times \sqrt{[2j+1][2j'+1]} \begin{Bmatrix} j' & k & j \\ j_1 & j_2' & j_1' \end{Bmatrix}_q \langle j_1' \| U^k(\bar{q}) \| j_1 \rangle . \qquad (11.14) $$

It is evident that in the limit $q=1$ all formulae of this section coincide with corresponding relations of standard algebra of tensor operators in AMT. As an example of application of these formulae let us consider the scalar operator

$$ (J^1(q) \circ J^1(q))(\bar{q}) = \sum_{\varkappa} (-1)^{\varkappa} q^{\varkappa} J^1_{\varkappa}(q) J^1_{-\varkappa}(q) = $$

$$ = \frac{1}{[2][2]} \left\{ (q-q^{-1})^2 (C_2^q)^2 + 4C_2^q - 1 \right\} . \qquad (11.15) $$

Using the relations (11.4) and (7.10) we find

$$ \langle jm | (J^1(q) \circ J^1(q))(\bar{q}) | jm \rangle = \frac{[2j][2j+2]}{[2][2]} . \qquad (11.16) $$

Obviously the same result will be obtained if the eigenvalue $[j+1/2]^2$ of Casimir operator $\mathbf{C}_2$ is substituted in the r.h.s. of (11.15). In the limit $q=1$ the Eq. (11.16) gives the correct eigenvalue $j(j+1)$ of the square of angular momentum operator $\mathbf{J}^2$. The above developed q-analog of tensor operator algebra will be useful for the derivation of various relations between elements of the Wigner-Racah algebra for $SU_q(2)$ (for example the recurrent relations for 3j,6j-symbols etc). Such applications and connection of q-analogs of 3j and 6j-symbols with orthogonal polynomials of discrete variable on nonuniform grids [29] will be described in future publications.

In conclusion it should be noted that results of this paper can be easily extended on noncompact $SU_q(1,1)$ algebra at least on positive discrete series tensor products.

ACKNOWLEDEMENTS

Authors are thankful to A.N. Kirillov, A.U. Klimyk and Ya. Soibelman for illuminating discussions.

REFERENCES

1. E.K. Sklyanin, Funct.Anal.Appl., **16**,263 (1982).
2. P.P. Kulish, N.Yu. Reshetikhin, Zapiski Nauch. Semin. LOMI, **101** 112 (1980).
3. V.G. Drinfeld, Doklady AN SSSR, **32**,254 (1985).
4. M. Jimbo, Lett.Math.Phys.,11,247 (1986).
5. N.Yu. Reshetikhin, LOMI preprints E-4-87, E-17-87 (1988).
6. L.C. Biedenharn, J.Phys.A: Math.Gen., **22**,L873 (1989).
7. A.J. Macfarlane, J.Phys.A: Math.Gen.,**22**,4581 (1989).
8. M. Rosso, Comm.Math.Phys., **117**,581 (1988).
9. L.L. Vaksman, Doklady AN SSSR, **306**,269 (1989).
10. A.N. Kirillov, N.Yu. Reshetikhin, LOMI preprint E-9-88 (1988).
11. Zhong-Qi-Ma, Preprint ICTP, Trieste, IC/89/162 (1989).
12. A.Ch. Ganchev, V.B. Petkova, Preprint ICTP, Trieste, IC/89/158 (1989).
13. R. Roche, D. Arnaudon, Lett.Math.Phys.,**17**,295 (1989).
14. I.I. Kachurik, A.U. Klimyk, Preprint ITP-89-48E (1989).
15. H. Ruegg, Preprint, University de Geneva, UGVA-DPT, 1989/08-625 (1989).
16. H.T. Koelink, T.H. Kornwinder, Preprint of Mathematical Institute, University of Leiden, W-88-12 (1988).
17. M. Nomura, J.Math.Phys., **30**,2397 (1989).
18. D.T. Sviridov, Yu.F. Smirnov, Theory of optical spectra of transition metal ions, Moscow, Nauka, 1977(in Russian).
19. D.A. Varshalovich, A.N. Moskalev, V.K. Khersonsky, Quantum Theory of Angular Momentum, Leningrad, Nauka, 1975 (in Russian).
20. A.P. Yutsis, A.A.Bandzaitis, Theory of angular momentum in quantum mechanics, Vilnjus, Mintis, 1965 (in Russian).
21. L.C. Biedenharn, J.D.Louck, Angular Momentum in Quantum Physics Addison-Wesley, 1981.
22. R.M. Asherova, Yu.F. Smirnov, V.N. Tolstoy, Theor. Math. Fiz., **8**,255 (1971).
23. R.M. Asherova, Yu.F. Smirnov, V.N. Tolstoy, Matem. Zametki, **36**,15 (1979).
24. J. Shapiro, J.Math.Phys., **6**,1680 (1965).
25. D.P. Zelobenko, Doklady AN SSSR, **4**,317 (1984).
26. G. Racah, Phys.Rev., **62**,438 (1942).
27. G. Gasper and M.Rahman,Basic Hypergeometric series, Cambridge University Press,1989,287p..
28. Yu.F. Smirnov, V.N. Tolstoy, Yu.I. Kharitonov, Leningrad Institute of Nuclear Physics, Preprint №1607, 1990.
29. A.F. Nikiforov, S.K. Suslov, V.B. Uvarov, Classic orthonal polynomials of discrete variable, Moscow, Nauka, 1985 (in Russian).

SYMMETRY BREAKING AND FRACTIONAL QUANTIZATION OF QUANTUM SYSTEMS

R. Tao

Department of Physics
Southern Illinois University
Carbondale, IL 62901, USA

I. INTRODUCTION

It is well established now that integral quantization of a class of quantum systems, such as the integral quantum Hall effect and the quantization of charge transport, is a topological invariant related to the first Chern class.[1] However, the relationship between fractional quantization of quantum systems and the first Chern class is still an open question.[2]

From the experiments of quantum Hall effect[3,4], we have learned that fractional quantization bears many similarities to integral one. For example, both of them have a very high precision of the quantized value. Even the patterns in development of longitudinal resistivity and Hall conductance in both cases are similar. It is, therefore, naturally expected that there must also be a deep relationship between fractional quantization and the topological invariant of the first Chern class.

In this paper, we will first derive a general formula for response coefficients and demonstrate that all quantities appeared in recent discussion of integral and fractional quantization can be expressed as response coefficients. In Section IV, we discuss the quantization of response coefficients. Especially, we will establish that fractional quantization of a quantum system is related to the degeneracy of ground state which comes from some symmetry breaking. Fractional quantization is then shown to be also a topological invariant of the first Chern Class. Section V is devoted to the application of our theory to the quantum Hall effect.

II. RESPONSE COEFFICIENTS

Let us consider a quantum system whose Hamiltonian contains a set of variables $\vec{\alpha} = (\alpha_1, \cdots, \alpha_n)$ $(n \geq 2)$. The force acting on the system by a change of variable α_i is given by $P_i = \partial H/\partial\alpha_i$. The flow is defined as the time rate of change of these variables $\dot{\alpha}_j$ $(j = 1, 2, \ldots n)$. When $\dot{\alpha}_i$ varies by $\delta\dot{\alpha}_i$, the system responds with forces $\delta\vec{P}$. The response coefficient η_{ji} is defined by[5]

$$\delta P_j = \sum_i \eta_{ji}\delta\dot{\alpha}_i. \tag{1}$$

The kinetic coefficient L_{ji} is the inverse of η_{ji} and satisfies

$$\delta\dot{\alpha}_j = \sum_i L_{ji}\delta P_i. \tag{2}$$

Symmetries in Science V, Edited by B. Gruber *et al.*
Plenum Press, New York, 1991

For every point in the $\vec{\alpha}$-space, there is a Hamiltonian $H(\vec{\alpha})$ which characterizes the quantum system. Generally, at two different points in the $\vec{\alpha}$-space, the system is in different physical states, but we are interested in a periodic case in which α_i has period T_i $(i = 1, \cdots, n)$. This implies that any such two points $(\alpha_1, \cdots, \alpha_i, \cdots)$ and $(\alpha_1, \cdots, \alpha_i + T_i, \cdots)$ are equivalent. Then the whole $\vec{\alpha}$-space can be represented by a single unit cell $0 \leq \alpha_i \leq T_i$ $(i = 1, \cdots, n)$.

To see the importance of response coefficients, we want to show that most quantities, found to have quantization in experiments or theories recently, can be expressed as response coefficients. For example, in the quantum Hall effect (see Fig.1) α_1 and α_2 are two solenoid fluxes. Both α_1 and α_2 are periodic with a period of h/e due to gauge invariance. We have $\dot{\alpha}_1 = -V$, the voltage applied in the x-direction, and $\partial H/\partial \alpha_2 = -I$, the Hall current produced in the y-direction. The investigated Hall conductance is just the response coefficient η_{21} now.[6–8] In a two-dimensional lattice, two lattice momenta k_1 and k_2 are periodic variables with periodicity equal to the reciprocal lattice vectors. Then $\dot{k}_1 = -eE_x/\hbar$, related to the applied electric field in the x-direction, while $\partial H/\partial k_2$ is the group velocity of electrons in the y-direction, associated to the current density j_y. The response coefficient η_{21} is the well-known Hall conductance of electrons in a periodic potential.[9] When a periodic force $f(t)$ of period T acts on electrons in a one-dimensional crystal, two periodic variables are t and the reciprocal vector k. Now $\partial H/\partial k = -\hbar j/(en_e)$ where n_e is the electron density. The flow $\dot{t} = 1$ is a constant. The total charge[10] transported during one period T is connected to the response coefficient through the equation $Q = jT = Ten_e\eta_{21}/\hbar$.

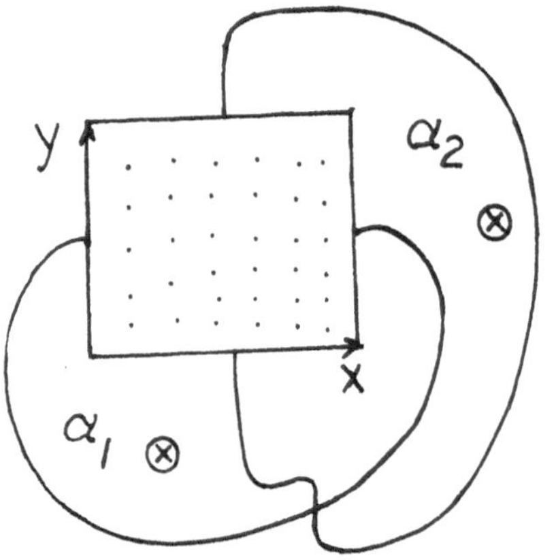

Fig.1. The quantum Hall system. The square is a two-dimensional electron system under a magnetic field. The solenoid α_1 produces the applied voltage in the x-direction $V = -\dot{\alpha}_1$. The Hall current is related to $\partial H/\partial \alpha_2$. When the periodic boundary condition in both directions is used, the system becomes a torus. The two solenoids are along the two toroidal axes.

III. GENERAL EXPRESSION FOR RESPONSE COEFFICIENTS

We derive an expression for η_{ji} now. As α_i varies by $\delta\alpha_i$. The Hamiltonian varies from H to $H + P_i\delta\alpha_i$. We consider $H_t = H + \delta H$ with

$$\delta H = -\delta\dot{\alpha}_i \Pi, \tag{3}$$

and $\Pi = \int^t P_i dt$. Since H_t differs from $H + P_i\delta\alpha_i$ only by a full time derivative

$d(\Pi_i\delta\alpha_i)/dt$, they are equivalent. Let ρ be the density operator, satisfying the equation

$$i\hbar\partial\rho/\partial t = [H, \rho]. \tag{4}$$

When H has a deviation δH, ρ responds with a deviation $\delta\rho$. After keeping the linear term, we obtain

$$i\hbar\partial\delta\rho/\partial t = [H, \delta\rho] + [\delta H, \rho]. \tag{5}$$

The solution is given by

$$\delta\rho = \frac{1}{i\hbar}\int_{-\infty}^{t} \exp[iH(\tau - t)/\hbar][\delta H, \rho]\exp[-iH(\tau - t)/\hbar]d\tau \tag{6}$$

The response force $\delta P_j = \mathrm{Tr}(P_i\delta\rho)$ reads

$$\delta P_j(t) = \frac{1}{i\hbar}\int_{-\infty}^{t} \mathrm{Tr}[\rho, \Pi_i]P_j(t-\tau)\delta\alpha_i(\tau)d\tau, \tag{7}$$

where $P_j(t) = e^{iHt/\hbar}P_je^{-iHt/\hbar}$. If we take the original density operator as canonical one, $\rho = e^{-\beta H}/Z$ where $\beta = 1/k_BT$ and $Z = \mathrm{Tr}(e^{-\beta H})$, the canonical partition function, then

$$[\rho, \Pi_i] = \int_0^\beta \rho e^{\lambda H}[\Pi_i, H]e^{-\lambda H}d\lambda = i\hbar\int_0^\beta \rho P_i(-i\hbar\lambda)d\lambda. \tag{8}$$

Equation (7) now can be written as

$$\delta P_j(t) = \int_{-\infty}^{t} \phi_{ji}(t-\tau)\delta\dot{\alpha}_i(\tau)d\tau, \tag{9}$$

where the response function is given by

$$\phi_{ji}(t) = \int_0^\beta d\lambda \mathrm{Tr}[\rho P_i(-i\hbar\lambda)P_j(t)]. \tag{10}$$

In the static case, $\delta\dot{\alpha}$ is independent of time, the response coefficient is given by

$$\eta_{ji} = \int_0^\infty dt\int_0^\beta \mathrm{Tr}[\rho P_i(-i\hbar\lambda)P_j(t)]. \tag{11}$$

Using the set of eigenfunction of $H(\vec{\alpha})$, $\{\Psi_s\}$, we expand Eq.(11),

$$\eta_{ji} = \frac{1}{Z}\sum_{s\neq l}\int_0^\infty\int_0^\beta d\lambda e^{-\beta E_s + (\lambda - it/\hbar)(E_s - E_l)}\left\langle\Psi_s|\frac{\partial H}{\partial\alpha_i}|\Psi_l\right\rangle\left\langle\Psi_l|\frac{\partial H}{\partial\alpha_j}|\Psi_s\right\rangle. \tag{12}$$

The formula $\int_0^\infty dte^{ikt} = i/k + \pi\delta(k)$ enables us to simplify Eq.(12) further. We divide η_{ji} into two parts: an anti-symmetric part $\eta^a_{ji} = -\eta^a_{ij}$ and a symmetric part $\eta^s_{ji} = \eta^s_{ij}$, which are given by

$$\eta_{ji}^{a} = \frac{i\hbar}{Z}\sum_{s} e^{-\beta E_s}\left[\left\langle\frac{\partial\Psi_s}{\partial\alpha_i}|\frac{\partial\Psi_s}{\partial\alpha_j}\right\rangle - \left\langle\frac{\partial\Psi_s}{\partial\alpha_j}|\frac{\partial\Psi_s}{\partial\alpha_i}\right\rangle\right], \tag{13}$$

$$\eta_{ji}^{s} = \frac{\hbar\pi\beta}{Z}\sum_{s\neq l} e^{-\beta E_s}\delta(E_s - E_l)\left\langle\Psi_s|\frac{\partial H}{\partial\alpha_i}|\Psi_l\right\rangle\left\langle\Psi_l|\frac{\partial H}{\partial\alpha_j}|\Psi_s\right\rangle. \tag{14}$$

Now consider the low temperature case and assume that there is a gap between the ground state and the excited states. Then from Eq.(14) η_{ji}^{s} vanishes and $\eta_{ji} = \eta_{ji}^{a}$. When the $\vec{\alpha}$-space is periodic and represented by a unit cell, $0 \leq \alpha_i \leq T_i$ $(i = 1, \cdots, n)$, the meaningful η_{ji} should be the average value on this unit cell. If the ground state is non-degenerate, we write

$$\eta_{ji} = 2\pi\hbar(\gamma_{ji})/T_i T_j \tag{15}$$

where γ_{ji} is given by

$$\gamma_{ji} = \frac{i}{2\pi}\int_0^{T_i}\int_0^{T_j}\left[\left\langle\frac{\partial\Psi_0}{\partial\alpha_j}|\frac{\partial\Psi_0}{\partial\alpha_i}\right\rangle - \left\langle\frac{\partial\Psi_0}{\partial\alpha_i}|\frac{\partial\Psi_0}{\partial\alpha_j}\right\rangle\right] d\alpha_i d\alpha_j \tag{16}$$

Ψ_0 is the normalized ground state. Using Stoke's theorem, we can rewrite Eq.(16) into the form

$$\gamma_{ji} = \frac{i}{2\pi}\oint\left\langle\Psi_0|\,\frac{\partial\Psi_0}{\partial\vec{\alpha}}\right\rangle d\vec{\alpha}. \tag{17}$$

The integration in Eq.(17) is along the contour of the unit cell $0 \leq \alpha_i \leq T_i$ and $0 \leq \alpha_j \leq T_j$. As shown by Berry[11], after one adiabatic move along a closed loop, the wave function Ψ_0 gains a phase factor $\theta = \oint\left\langle\Psi_0|\,\frac{\partial\Psi_0}{\partial\vec{\alpha}}\right\rangle d\vec{\alpha}$. Equation (17) directly relates γ_{ji} to the Berry phase,

$$\gamma_{ji} = \frac{\theta}{2\pi}. \tag{18}$$

IV. FRACTIONAL QUANTIZATION

Assume that the quantum system is initially in the ground state Ψ_0. In an adiabatic move, the quantum system will remain in the ground state. Since the $\vec{\alpha}$–space is periodic, after one circular move along the contour of the unit cell, the new arriving ground state must be the same as Ψ_0 if the ground state is non-degenerate. This implies that the Berry phase is 2π times an integer. Then we have the well-known result of integral quantization: in the case of non-degenerate ground state, γ_{ji} is an integer. This integer is a holonomy, related to the first Chern class.[1]

The above discussion clarifies that a non-degenerate ground state cannot produce fractional quantization of γ_{ji}. If the ground state is always q–fold degenerate throughout the whole $\vec{\alpha}$–space, and there is an energy gap separating these ground states from the excited states, at low temperature we can still express η_{ji} by Eq.(15), but

$$\gamma_{ji} = \frac{i}{2q\pi}\sum_{k=1}^{q}\oint\left\langle\Psi_0^{(k)}|\,\frac{\partial\Psi_0^{(k)}}{\partial\vec{\alpha}}\right\rangle d\vec{\alpha}, \tag{19}$$

where q in the denominator is from the partition function Z and $\Psi_0^{(k)}$ $(k = 1, 2, \cdots, q)$ are orthonormal ground states. Now it is clear that γ_{ji} is fractionally quantized with a denominator q.

We must distinguish degeneracy produced by some symmetry breaking from accidental degeneracy. An accidental degeneracy can be easily removed by a small change of some parameter. Therefore, the accidental degeneracy is not relevant to fractional quantization of γ_{ji}. Equation (19) requires that the ground state is always $q-$fold degenerate throughout the whole $\vec{\alpha}-$space. Therefore, this degeneracy must come from some symmetry.

For example, we can assume that there are two operators $\hat{O}_1$ and $\hat{O}_2$. Both of them commute with H, but do not commute each other. We first diagonalize H and $\hat{O}_2$ simultaneously, say, to obtain a ground state Ψ_0. Then we use $\hat{O}_1$ to act on Ψ_0. $\hat{O}_1\psi_0$ is also a ground state. Since $[\hat{O}_1, \hat{O}_2] \neq 0$, $\hat{O}_2\Psi_0$ and $\hat{O}_2(\hat{O}_1\Psi_0)$ are different, $\hat{O}_1\Psi_0$ is a different ground state. Now consider $\hat{O}_1^2\Psi_0$. If this new ground state is still different from Ψ_0, we will continue to use $\hat{O}_1$ to act on it. In this way, we produce a sequence of ground states, $\Psi_0, \hat{O}_1\Psi_0, \hat{O}_1^2\Psi_0, \cdots$ The assumption of a gap separating the ground state from the excited states assures that the degeneracy of ground state cannot be infinite. Then the above sequence must be periodic with a finite periodicity, say, q, $\hat{O}_1^q\Psi_0 = \Psi_0$. When the ground state is degenerate, if we start from one ground state Ψ_0 to do an adiabatic move, the system will shift from Ψ_0 to other ground states. Therefore, after one closed adiabatic move along the contour of the unit cell, the system usually does not return to the initial starting state. But since the degeneracy is q-fold, q circles along the contour of the unit cell will bring the system back to its initial state. Therefore, we can rewrite equation (19) into the form

$$\gamma_{ji} = \frac{i}{2q\pi} \oint_q \left\langle \Psi_0 \right| \left. \frac{\partial \Psi_0}{\partial \vec{\alpha}} \right\rangle d\vec{\alpha}. \tag{20}$$

The notation $\oint_q$ indicates an integration along a loop which circles the contour of the unit cell $0 \leq \alpha_i \leq T_i$ and $0 \leq \alpha_j \leq T_j$ q times. Equation (20) clearly indicates the relationship between the fractionally quantized γ_{ji} and the first Chern class.

V. APPLICATIONS

We consider a two-dimensional electron gas in a magnetic field which is perpendicular to the plane. In Landau gauge, the Hamiltonian of the system is given by

$$H = \frac{1}{2m}\sum_{j=1}^{N}\left[(-i\hbar\frac{\partial}{\partial x_j} + e\alpha_1/L_1)^2 + (-i\hbar\frac{\partial}{\partial y_j} + eBx_j + e\alpha_2/L_2)^2\right] + \sum_{j<l}\frac{e^2}{\epsilon|\vec{r}_j - \vec{r}_l|}, \tag{21}$$

We use electron charge $-e$. B is the applied magnetic field, ϵ is a dielectric constant, and L_1 and L_2 are the lengths of the system in the x- and y-direction respectively. The system has area L_1L_2. The total orbits in one Landau level is L_1L_2Be/h. The system is assumed to have a fractional filling $\nu = N/(L_1L_2Be/h = p/q$ where p and q are two mutual primes. Two parameters α_j ($j = 1, 2$) can be considered as two solenoid fluxes (see Fig.1). If we use the periodic boundary condition in both x- and y-directions, the two-dimensional surface becomes a toroidal surface. Then α_1 and α_2 are two magnetic solenoids along the axes of the torus. Because of gauge invariant, both α_1 and α_2 are periodic parameters with periodicity h/e. We introduce two operators

$$\pi_{jx} = -i\hbar\frac{\partial}{\partial x_j} + eBy_j, \quad \pi_{jy} = -i\hbar\frac{\partial}{\partial y_j} \tag{22}$$

The commutator between π_{jx} and π_{jy} is given

$$[\pi_{jx}, \pi_{ly}] = i\hbar eB\delta_{jl}. \tag{23}$$

We further introduce two operators

$$\hat{T}_1 = \exp\left[i\sum_j^N \frac{2\pi}{L_2 eB}\pi_{jx}\right] \quad \hat{T}_2 = \exp\left[i\sum_j^N \frac{2\pi}{L_1 eB}\pi_{jy}\right]. \tag{24}$$

When $\hat{T}_1$ and $\hat{T}_2$ act on any Landau orbit, they produce a new Landau orbit within the same Landau level. It is easy to check that

$$[\hat{T}_1, H] = [\hat{T}_2, H] = 0, \quad \text{but} \quad [\hat{T}_1, \hat{T}_2] \neq 0, \hat{T}_1\hat{T}_2 = \hat{T}_2\hat{T}_1 e^{-i2\pi p/q}. \tag{25}$$

Since p and q are mutual primes, among the sequence of $\hat{T}_1, \hat{T}_1^2, \hat{T}_1^3, \cdots$, the first one commuting with $\hat{T}_2$ is $\hat{T}_1^q$. The ground state is q−fold degenerate. If we diagonalize H and $\hat{T}_2$ simultaneously and assume one ground state is Ψ_0. The other degenerate ground states are $\hat{T}_1^j \Psi_0$ ($j = 1, 2, \cdots q-1$).

Now let us calculate the Hall conductance. From the Hamiltonian in Eq.(21), if we adiabatically increase the solenoid α_1 by one unit h/e, the system moves from the initial ground state Ψ_0 to a new ground state $\hat{T}_1\Psi_0$. If we increase α_1 by $2h/e$, we move to $\hat{T}_1^2\Psi_0, \cdots$. As α_1 increases by qh/e, the system moves back into its initial state. On the other hand, adiabatic increase of α_2 by one unit h/e is equivalent to use $\hat{T}_2$ to act on the wave function. One move along the contour of $0 \leq \alpha_j \leq h/e$ ($j = 1, 2$) is the operator

$$\hat{T}_2^{-1}\hat{T}_1^{-1}\hat{T}_2\hat{T}_1 = \exp(i2\pi p/q) \tag{26}$$

where equation (25) has been used. Then either from Eq.(19) or Eq.(20), we have fractional quantization of $\gamma_{21} = -p/q$. The Hall conductance is given by

$$\sigma_H = \frac{e^2}{h} p/q. \tag{27}$$

ACKNOWLEDGEMENTS

I wish to thank Prof. Gruber for his invitation and hospitality. This research is supported in part by the Office of Naval Research grant N00014-90-J-4041, Illinois ENR grant SWSC-14, and grants from MTC and ORDA of SIUC.

REFERENCES

1. E. Avron, R. Seiler, and B. Simon, Phys.Rev. Lett. **51**,51 (1983); B. Simon, *ibid*,**51**, 2167 (1983); E. Avron and R. Seiler, *ibid*, 54, 259 (1985); E. Avron, A. Raveh, and B. Zur, Rev. Mod. Phys.**60**, 873 (1988).
2. R. Tao, Phys. Rev. B **35**, 9853 (1987).
3. K. von Klitzing, G. Dorda, and M. Pepper, Phys. Rev. Lett.**45**, 494 (1980); B. I. Halperin, Phys. Rev. B **25**, 2185 (1982).
4. D.C. Tsui, H. L. Sörmer, and A.C.Gossard, Phys. Rev. Lett.**48**, 1559 (1982); R. B. Laughlin, *ibid*.**50**, 1395 (1982); F. D. M. Haldane, *ibid*.**51**,605 (1983).
5. R. Kubo, J. Phys. Soc. Jpn. **12**, 570 (1957).

6. R. Tao and Y.S. Wu, Phys. Rev. B**30**, 1097 (1984).
7. R. Tao, J. Phys. C **19**, 175 (1986).
8. R. Tao and F. M. D. Haldane, Phys. Rev. B**33**, 3844 (1986).
9. D. J. Thouless, M. Kohmoto, M. P. Nightingale, M. den Nijs, Phy. Rev. Lett. **49**, 405 (1982).
10. D. J. Thouless, Phys. Rev. B**27**, 6083 (1983); Q. Niu and D. J. Thouless, J.Phys. A **17**, 2453 (1984).
11. M. V. Berry, Proc. R. Soc. London A**392**, 45 (1984).

NEW PHASES OF D ≥ 2 CURRENT AND DIFFEOMORPHISM ALGEBRAS IN PARTICLE PHYSICS

Chia-Hsiung Tze

Institute of High Energy Physics and Physics Department,
Virginia Polytechnic Institute and State University
Blacksburg, VA 24061, U.S.A.

INTRODUCTION

As one who thought deeply about all aspects of symmetries Hermann Weyl [1] had traced their origin in nature to the mathematical character of physical laws. In the last thirty years, the developments in particle physics have been dominated by one single theme, the exploitation of symmetries. They can be either exact or approximate, ultimately fundamental or effective[2]. With its unqualified successes the use of symmetries has become synonymous with that of Lie algebras and groups. In the early seventies the mathematician, Jean Dieudonné [3] wrote : " Les groupes de Lie sont devenus le centre des mathematiques; on ne peut rien faire de sérieux sans eux" . In this era of gauge and string theories, we may, without much exaggeration, assert the preeminent role at the frontiers of physics of infinite dimensional Lie group theory by replacing the words "des mathematiques" above by " de la physique théorique" .

In a broader perspective, with the coming of age of gauge theories, string theories and D=2 conformal field theories, the range of applicable mathematics seems limited only by one's ingenuity and imagination . It spans the gamut of all major branches of 19th century and modern mathematics, from Riemann surfaces to hyperkähler manifolds, from infinite Lie groups to non-commutative geometry, from knots and links to p-adic numbers and analysis. As will be illustrated below, all of these apparently disparate structures are often brought together through the intermediary of one set of physical phenomena. This linkage reflects both the unity of mathematics as well as its unreasonable effectiveness in accounting for the physical world.

We have certainly gone a long way from the Young tableaux and Clebch-Gordan series in finite parameter Lie algebras applied to global (flavor) then local (gauge) symmetries of point particles to the full use of the representation theory of infinite parameter (super-)Virasoro-Kac-Moody algebras in string, 2-dimensional conformal and integrable field theories. Indeed solving for two dimensional quantum field theories is almost equivalent to solving for the representation theories of the loop and/or Virasoro groups. Though the task is rather difficult we could dream of a parallel outcome in four dimensions. So while we started out by often invoking symmetries as substitutes for dynamics we have ended up fulfilling the old Einsteinian dictum " symmetry dictates dynamics".

In his instructions to the speakers at this Symposium, Professor Gruber commissioned comprehensive reviews aimed not just at physicists using symmetries in their research but also at experts in other areas of sciences. This criterion has partly guided my choice of topics. My special interests are in algebraic and topological structures in particle physics. Accordingly, I shall take as my main and unifying theme, a few global aspects of Kac-Moody-Virasoro typed algebras, the representation theory of their current and diffeomorphism groups, seen in the context of a few concrete semi-topological field theories with solitons in D≥2 spacetime dimensions.

In these alotted pages, a truly comprehensive review is admittedly out of the question. I shall therefore not dwell on the numerous well established and extensively reviewed results of two dimensional. Rather using the latter as standards, I will focus on a few basic developments in D≥ 2 dimensions and discuss their open problems. My threefold emphasis will be on a) the question of fermi-boson equivalence or D≥3 bosonization, b) the possibility of anyonic transmutation and c) the role of complex and hypercomplex analyticity. These topics best illustrate some natural directions toward a nonperturbative, algebraic understanding of D≥2 dimensional quantum field theories. Four related topics are singled out for discussion:

1) To introduce the basic concepts and notations, a brief review of 2-cocycles as central extensions of D=2 current algebras, its equivalent fermionic and bosonic representations, via the Wess-Zumino- Novikov-Witten (WZNW) model. The complex analytic structure of the Kac-Moody-Virasoro algebras.

2) Going behond affine Lie algebras, D=4 current algebra with it q-number, non central, Abelian extension, its canonical realization in a Skyrme model with a Wess-Zumino term. Attempts at constructing vertex operators and a representation theory. Generalized fermi-bose correspondence and comments on hypercomplex analyticity of generalized Kac-Moody-Virasoro algebras.

3) An realization of D=3 current and diffeomorphism algebras in the CP_1 σ-model with a Chern-Simons-Hopf term. An anyonic vertex operator construction. A generalized spin and statistics connection by way of the Gauss-Bonnet theorem . Its relation to self-linking, twisting,writhing numbers of Feynman paths.

4) Going behond D=3 anyons, exceptional D= 7, 15 anyonic Hopf 2- and 3-membranes and their connection to division algebras via Adams' theorem. Comments on their current, diffeomorphism algebras.

These topics will be covered respectively in sections 1 to 4 , section 5 encloses some parting remarks. Our treatment of established results will be brief and primarily conceptual . For proofs and greater details we refer the interested reader to our long, though incomplete list of references.

1. D=2 KAC-MOODY GROUPS , FERMIONIZATION AND COMPLEX ANALYTICITY

1.1 CURRENT ALGEBRAS AND COCYCLES

For those unfamilar with current algebras,a few brief historical remarks may be in order. Comprehensive accounts of current algebras are to be found in the classic books by Adler and Dashen[4] and Ne'eman [5] and in a modern update by Treiman, Jackiw, Zumino and Witten [6]. Before the advent of gauge theories, amids the profusion of hadronic states the introduction of current algebras was motivated by the unifying idea that the basic objects for strong interaction physics should be the observable currents rather than the then still elusive fundamental fields. Thus, while the electromagnetic interactions among all charged particles are governed by the interaction Hamiltonian

$$H_e = e \int d^3x\, j_\mu(x)\, A^\mu(x) \tag{1.1}$$

e being the electric charge , j_μ the electric current and A_μ the electromagnetic potential, the leptonic and non-leptonic weak decays of hadrons are effectively accounted for by the interaction Hamiltonians

$$H_l = G \int d^3x\, j^h_\mu(x)\, j^l_\mu(x)\,, \quad H_{nl} = G \int d^3x\, j^h_\mu(x)\, j^h_\mu(x) \quad , \tag{1.2}$$

G is the Fermi-coupling constant, j^h_μ (x) and $j^l_\mu(x)$ denote the weak currents of the hadrons and the leptons respectively. At a fixed time t , the currents $j_\mu(x)$ are mappings from physical space into some internal space of a symmetry group G .

From these $j_\mu^\alpha(x)$'s , one compute the corresponding charge operators

$$Q^\alpha = \int d^3x \, j_0^\alpha(x) \,. \tag{1.3}$$

Essentially the fundamental hypothesis of current algebra was that , irrespective of the details (or even of the existence) of an underlying quantum field theory, the charges and current close under an algebra of equal- time canonical commutation relations. In order of their reliability, the postulated relations are of the generic forms of

a) a charge-charge algebra

$$\left[Q^\alpha(t) \,,\; Q^\beta(t)\right] = if^{\alpha\beta\gamma}Q^\gamma(t) \,, \tag{1.4}$$

b) a mixed charge -current algebra

$$\left[Q^\alpha(t) \,,\; j_\mu^\beta(\mathbf{x}, t)\right] = if^{\alpha\beta\gamma}j_\mu^\gamma(\mathbf{x},t) \,, \tag{1.5}$$

c) a current-current algebra

$$\left[j_0^\alpha(\mathbf{x}) \,,\; j_0^\beta(\mathbf{y})\right] = if^{\alpha\beta\gamma}j_0^\gamma(\mathbf{x}) \, \delta^3(\,\mathbf{x} - \mathbf{y}\,), \tag{1.6}$$

$$\left[j_0^\alpha(\mathbf{x}) \,,\; j_i^\beta(\mathbf{y})\right] = if^{\alpha\beta\gamma}j_i^\gamma(\mathbf{x}) \, \delta^3(\,\mathbf{x} - \mathbf{y}\,) + S_{0i}^{\alpha\beta}(\,\mathbf{x},\mathbf{y}) \,, \tag{1.7}$$

where $f^{\alpha\beta\gamma}$ are the structure constants of the algebra of the symmetry group G , e.g. G ≈ SU(N), SU(N)xSU(N) , N= 2 , 3. The additional matrix valued term $S(\mathbf{x}, \mathbf{x}')$ in (1.7) is the celebrated singular Schwinger term.

One particular feature must be noted. In the old current algebras, the Schwinger terms are highly model dependent and occur only in the space-time current commutator (1.7) while their modern cousins, being of topological origin, have more restricted forms and appear in the time-time current commutator or *local charge algebra* (1.6) . As such these topological extension terms, being the repositories of "good" or "bad" anomalies, have important implications on new physical effects or on the over all quantum consistency of the associated (gauge) field theory.

Clearly (a) and (b) are but special integrated form of (c) . We recall that (1.4) and (1.5) were widely and successfully used. Similarly several sum rules were derived from (1.6) and agreed reasonably well with experiments. Today the above algebras are seen to arise from the underlying dynamics of quantum chromodynamics and the standard model of electroweak interactions.

Looking back ,what physicists missed during the 60's was the possible topological significance of the Schwinger term(s). At one time these terms were even banned by decree from an universal current algebra. Yet they are necessarily present for consistency with Lorentz invariance and energy positivity. Its form is constrained by the associativity of the algebra, i.e. the Jacobi identity . It turns out that these singular terms are the residual local signatures of nontrivial 2-cocycles or projective representations of quantum systems with an infinite numbers of degrees of freedom and with topologically nontrivial configuration spaces. This fact was realized early on by I.M. Gelfand and his followers [7]who pioneered the representation theory of current groups in arbitrary dimensions. Unfortunately the relevance of their works musts await the coming of age of affine and loop algebras, ushered in by the advent of superstring and conformal field theories.

In the title of this review, by "new phases" we mean the nontrivial phases of the projective representations of infinite dimensional algebras, the 2-cocycles or Berry's phases[8]. We next recall their mathematical and physical meanings.

It is commonly said that the phrases of the complex Schrödinger wave function Ψ of a quantum system do not matter since physically observable effects are determined by the real norm $|\Psi|^2$. This statement is true provided Ψ is the wave function for the whole system, which is seldom the case in practice or even in principle[9]. As the Aharonov-Bohm effect [10]and the rich theoretical analyses and experimental confirmations of the Berry phase recently show, the relative phases of wave functions describing a part of the entire system are most relevant and physically detectable. Moreover they often have drastic and stunning effects on the properties of the subsystem(s) in question. These *anomaly phenomena* to be illustrated here by the fermi-bose equivalence in D=2 quantum field theory, the emergence of the D=4 baryonic topological soliton, the Skyrmion, from QCD, by the anyonic membrane excitations in odd dimensional , semi-topological field theories etc...

What are cocycles ?[9]Consider a quantum system Σ with a symmetry group G of transformations T(g) . For each fixed g , T(g) can be represented up to a phase factor $e^{i\omega_1(q,g)}$ by an (anti-) unitary operator U(g) in a Hilbert space *H*. Let q be the dynamical variable(s) on which g acts : $q \rightarrow q^g$, A wave function $\Psi(q)$ transforms as

$$U(g)\,\Psi(q) = e^{i\omega_1(q,g)}\,\Psi(q^g) \tag{1.8}$$

Consistency with the group composition law

$$U(g_1)\,U(g_2) = U(g_1 g_2) = U(g_{12}) \tag{1.9}$$

implies

$$\omega_1(q^{g_1}; g_2) - \omega_1(q; g_{12}) + \omega_1(q; g_1) = 0 \quad (\bmod Z)\,. \tag{1.10}$$

The real phase $\omega_1(q;g)$, a 1-cocycle, depends generally on both g and q . Specifically if Σ is a non-Abelian gauge (chiral field) theory , then in the Hamiltonian formalism, the g's, q's and $\Psi(q)$ correspond respectively to local gauge (global chiral) transformations, the spatial components of the gauge potential **A** (chiral current J) and the Schrödinger wavefunctional Ψ(**A**) .

Similarly, to the group relation (1.9) corresponds the composition law

$$U(g_1)\,U(g_2) = e^{-2\pi i\,\omega_2(q;g_1,g_2)}U(g_{12}) \tag{1.11}$$

Associativity of (1.11) leads to the consistency condition

$$\omega_2(q^{g_1}; g_2, g_1) - \omega_2(q; g_{12}, g_3) + \omega_2(q; g_1, g_2) - \omega_2(q; g_1, g_2) = 0 \quad (\bmod Z)\,. \tag{1.12}$$

Such a phase is a 2-cocycle and the unitary representations bearing it are the *ray or projective* representations of frequent occurence in quantum theory.

One could continue this process and abstractly define higher cocycles. Thus the 3-cocycle is given through

$$\big(U(g_1)\,U(g_2)\big)\,U(g_3) = e^{-2\pi i\,\omega_3(q;g_1,g_2,g_3)}\,U(g_1)\,\big(U(g_2)U(g_3)\big)\,. \tag{1.13}$$

However it violates associativity; moreover nonassociative entities cannot be represented by linear operators in a Hilbert space. As of now no physical effect is attributable to 3 - or higher cocycles. So we shall limit ourselves here to 2-cocycles as we consider next the global aspects of the affine Kac-Moody algebras in 2-spacetime dimensions.

1.2 D= 2 KAC-MOODY GROUPS

There exist several approaches to construct D=2 Kac-Moody groups. We adopt the simple and instructive construction of Mickelsson [11] as it readily generalizes to higher dimensions; specifically for the case of D=4. The latter's papers should be consulted for greater details. Consider (suitable smooth) mappings where the target space is a finite dimensional Lie group G and the base space, the unit circle $S^1 \approx \partial D = \{ z \in \mathbf{C} \mid |z| = 1 \}$. This S^1 is seen as a boundary of an unit disc D = { z ε C | |z| ≤ 1)} . Let LG be the space of loops $f : S^1 \to G$ and $\Omega G \approx \{ f \in LG \mid f(1) = 1 \}$ the space of based loops. While LG and ΩG both has a natural group structure under point-wise multiplication ,namely given two maps γ_1 and $\gamma_2 : S^1 \to G$, their product composition is $\gamma_1 \cdot \gamma_2 : S^1 \to G$ such that $\gamma_1 \cdot \gamma_2(z) = \gamma_1(z) \cdot \gamma_2(z)$, only ΩG is a C^∞- manifold . Now let $DG = \{ f : D \to G \mid f(1)=1 \}$, the space of based smooth maps from D into G and let $\pi : DG \to \Omega G$ be the natural projection, $\pi(f) \equiv f|_{S^1}$. Then the triple (DG , π , ΩG) is a principal fibre bundle with as its structure group $\mathcal{G} = \{ f : D \to G \mid f(1)=1 , x \in S^1 \}$ acting on DG from the right..

By contracting $S^1 \approx \partial D$ to a single point , the North pole of S^2, $\mathcal{G}$ can alternatively be the space { f : D → G | f(North pole of S^2) =1 , x ε S^1 } . Then for f ε DG , g ε $\mathcal{G}$ define the 1-cocycle ω_1

$$\omega_1(f,g) = \frac{\psi^2}{16\,\pi^2}\int_D \langle f^{-1}\partial_\alpha f \,,\, \partial g_\beta\, g^{-1}\rangle\, dx_\alpha \wedge dx_\beta + C(g) \tag{1.14}$$

with

$$C(g) = \frac{-\psi^2}{48\,\pi^2}\int_{D_3} \left\langle \tilde{g}^{-1}\partial_\alpha \tilde{g} \,,\, \frac{1}{2}\left[\tilde{g}^{-1}\partial \tilde{g}_\beta ,\, \tilde{g}^{-1}\partial_\gamma \tilde{g}\right]\right\rangle dx_\alpha \wedge dx_\beta \wedge dx_\gamma \tag{1.15}$$

where < ... > denotes the Kiling form on **g** , the Lie algebra of G . In (1.15) the map $\tilde{g} : D_3 \to G$, D_3 being a 3-dimensional unit ball , is now an arbitary extension of $g : S^2 \to G$. ψ^2 is the length squared of the longest root of **g** . If $\tilde{g}_1$ and $\tilde{g}_2$ are two extensions of the same g , then $C(\tilde{g}_1) - C(\tilde{g}_2) \in Z$, so that the phase exp($2\pi i\, \omega_1$) is well defined. The 1-cocycle ω_1 allows one to define for g ε $\mathcal{G}$ in the foregoing extension DG x U(1) the following equivalence relation "~ "

$$(f , \lambda) \sim (f g , \lambda \exp\{ 2\pi i\, \omega_1(f, g)) \tag{1.16}$$

whose transitivity property is but the 1-cocycle consistency condition satisfied by ω_1 (1.10) . The Kac-Moody group $\mathcal{G}$ is then a principal U(1) bundle P over the loop apace ΩL ≈ Map(S^1 , G) , P ≈ {DG x U(1) / ~} . The right action of U(1) in DG x U(1) commutes with the g-action; U(1) acts on P . So the KM group $\mathcal{G}$ can be defined by the pairs (f, λ) , f ε Map(S^1 , G) , with a multiplication law

$$(f, \lambda) (f',\lambda') = (f f' , \lambda \lambda' \exp\{ 2\pi i\, \omega_2(f , f') \} , \tag{1.17}$$

where

$$\omega_2(f, f') = \frac{\psi^2}{16\pi^2}\int_D \langle f^{-1}\partial_\alpha f \,,\, \partial_\beta f'\, f'^{-1}\rangle\, dx_\alpha \wedge dx_\beta \tag{1.18}$$

ω_2 (f, f') satisfies (1.12) so exp ($2\pi i\, \omega_2$) is then a U(1) valued 2-cocycle in DG . The bundle P is a group; a central extension of ΩG by U(1) .

So the Kac-Moody algebra **g** of $\mathcal{G}$ is a 1-dimensional central extension of the loop algebra Map (S^1 , **g**) . Given f_1 and f_2 ε Map (S^1 ,G) the Lie algebra cocycle corresponding to (1.18) is simply given by Map (S^1 , **g**) with the commutator is defined point-wise as

$$[f_1 , f_2] (\theta) \equiv [f_1(\theta) , f_2(\theta)] . \tag{1.19}$$

The central extension is given by the Lie algebra 2-cocycle $c(f_1, f_2)$ corresponding to the group cocycle ω_2

$$c(f_1, f_2) = 4\pi \frac{d^2}{d\sigma\, d\tau} \omega_2(e^{\sigma f_1}, e^{\tau f_2}) |_{\sigma=\tau=0}$$

$$= \frac{\psi^2}{4\pi} \int_D \langle \partial_\alpha \tilde{f}, \partial_\beta \tilde{f} \rangle \, d x_\alpha \wedge dx_\beta = = \frac{\psi^2}{4\pi} \int_0^{2\pi} \left\langle f_1(\theta) \frac{d}{d\theta} f_2(\theta) \right\rangle d\theta \, . \tag{1.20}$$

So if $G = SU(N)$, $< X, Y > = Tr (XY)$ and $\psi^2 = 2$. Then (1.20) , which defines a symplectic, nondegenerate and closed Kirillov 2-form on ΩG leads to a modified commutator

$$[f_1(\theta) , f_2(\theta)] + i\, x\, c(f_1, f_2) \, . \tag{1.21}$$

Alternatively it takes the more familiar form of

$$[T^a_n , T^b_m] = f^{abc} T^c_{n+m} + \frac{i x \psi^2}{4\pi} m\, \delta^{ab} \delta_{m,-n} \tag{1.22}$$

Here $T^a_n = T^a e^{in\theta}$ are the Fourier components of f near the identity map in an orthonormal basis $\{ T^a \}$ (a = 1, ...dim **g**) of **g** with structure constants f_{abc} and where $x\, \psi^2 \in Z$. (1.22) shows the 1-dimensionality of the central extension. It is called a level k =1 Kac-Moody algebra (KMA) . A level k KMA is simply gotten by multiplying ω_2 in (1.18) by $k \in Z$.

The pervasive phenomenon of fermi-bose equivalence in 2-dimensions is best illustrated by two equivalent representations , one fermionic, the other bosonic of the same untwisted affine KMA (1.21) . Witten[12] considered a conformal invariant system of N free Majorana fermions with a non-Abelian chiral symmetry $G \approx O(N)$. In light cone or conformal coordianates $z \equiv z_+ = x + i\, y$ and $\bar{z} \equiv z_- = x - i\, y$, the Euclidean action reads

$$S(\psi , \bar{\psi}) = \frac{1}{2} \int d^2x \sum_{i=1}^{N} \left[\psi^i \partial_{\bar{z}} \psi^i + \bar{\psi}^i \partial_z \bar{\psi}^i \right] . \tag{1.23}$$

From second quantization the anticommutation relations read $\{ \psi^i , \psi^j \} = \hbar\, \delta(x-y)\, \delta^{ij}$ .Then equivalent to the Dirac equations for $\psi_i = \psi^i = \begin{pmatrix} \psi^i_+ \\ \psi^i_- \end{pmatrix}$, are the conservation laws

$$\partial_{\bar{z}} J^a_+ = \partial_z J^a_- = 0 \tag{1.24}$$

for the chiral currents

$$J^a_\pm = \frac{1}{2} :\psi^T_\pm M^a \psi_\pm: \quad , \tag{1.25}$$

M^a (a= 1,2,...N) are real skew symmetric NxN O(N) representation matrices. Consequently J^a_+ is only a function of z and J^a_- a function of $\bar{z}$, they also mutually commute , so they can be taken as independent . This shows the theory to be invariant under a much larger infinite invariance group $G(z) \times G(\bar{z})$ whose generators are J^a_+ and J^a_- . The resulting two commuting ∞-dimensional Kac-Moody algebras are

$$\left[J^a_\pm (z_\pm) \ , J^b_\pm(w_\pm) \right] = i\hbar\, f^{ab}{}_c\, J^c_\pm(w_\pm)\, \delta (z_\pm - w_\pm) + \frac{i\, \kappa_\lambda}{4\pi} \hbar^2 \delta^{ab}\, \delta'(z_\pm - w_\pm) \tag{1.26}$$

where κ_λ is , up to a representation free normalization, the Dynkin index of the representation:

$-\kappa_\lambda\,\delta_{ab} = tr(M^a M^b)$. In fact $c_\lambda d_\lambda = \kappa_\lambda \dim G$; $-c_\lambda = (M^a)^2$ is the value of the quadratic Casimir in the representation λ and $d_\lambda = N$ is the dimension of λ .

The existence and physical origin of the Schwinger term were in fact known to P. Jordan[13, 14] long before the works of Goto, Imamura [15] and Schwinger [16]. Indeed the validity of (1.26) presupposes a Dirac vacuum (i.e, 2nd quantization) ; specifically the condition for the global existence of such a fermion ground state is encoded in a local " deformation" of the algebra of currents by the addition of a Schwinger term. Furthermore, Jordan et al pointed out that the current commutator derives from their D=2 quantum massless spinor field

$$[J_1\,,J_0] = \frac{i}{\pi}\,\partial_1\delta(x-y) \tag{1.27}$$

is reproduced exactly by the commutator $[\partial_1\phi\,,\partial_0\phi] = i\,\hbar\,\partial_1\delta(x-y)$ resulting from the Heisenberg relation $[\phi\,,\partial_0\phi] = i\,\hbar\,\delta(x-y)$ for a Bose field ϕ , provided one sets

$$J_\mu = :\bar{\psi}\,\gamma_5\gamma_\mu\psi: = \left(\frac{\hbar}{\pi}\right)^{1/2}\partial_\mu\phi\,. \tag{1.28}$$

This mapping is the first example of fermi-bose equivalence or abelian*bosonization* . The $\hbar$ dependent factor in (1.28) testifies to its purely quantum character . Later on, another canonical example was established by Coleman[17]: the equivalence between the fermion of the massive Thirring model and the quantum soliton of Sine-Gordon model . The corresponding lagrangian densities are

$$L_{MT} = \frac{i}{2}\,\bar{\psi}\partial\!\!\!/\psi - m\,\bar{\psi}\,\psi - \frac{1}{2}g\,(\bar{\psi}\gamma_\mu\psi)^2 \tag{1.29}$$

$$L_{SG} = :\frac{1}{2}\partial_\mu\phi\partial^\mu\phi - \alpha\beta^{-1}(\,1-\cos(\beta\phi):\,. \tag{1.30}$$

For subsequent comparison we only write down the "vertex operator" $\psi = \begin{pmatrix}\psi_1\\ \psi_2\end{pmatrix}$ for creating a point-like fermion as a topologically nontrivial bose field coherent state excitation. Pioneered by Skyrme ,its construction was completed by Mandelstam[18]

$$\psi_1 = N:\exp\left\{-2i\,\beta^{-1}\int_{-\infty}^{x} d\xi\,\dot{\phi} - \frac{1}{2}\,i\beta\phi\right\}: \tag{1.31a}$$

$$\psi_2 = -iN:\exp\left\{-2i\,\beta^{-1}\int_{-\infty}^{x} d\xi\,\dot{\phi} + \frac{1}{2}\,i\beta\phi\right\}: \tag{1.31b}$$

Characteristic features can be inferred from these explicit expressions. Here we observe that while the fermionic currents are local, the fermion fields themselves are nonlocal in terms of the field ϕ with nontrivial topology. Actually this short (local) to long (global) distance connection reflects a quite general a trademark of anomalies or quantum symmetry breaking. For an elaboration of this intriguing phenomenon, we recommend the excellent reviews of Morozov [19] and Shifman [20].

Subsequently Witten[12, 21] put forth a non-abelian extension of the above fermi-bose equivalence . His model is governed by the following semi-topological action [22]

$$S_{\lambda,k}(g) = \frac{1}{4\lambda^2}\int d^2\xi(\partial_\mu g^{-1}\partial_\mu g) + k\,\Gamma(g)\,. \tag{1.32}$$

$$\Gamma(g) = \frac{1}{24\pi^2}\int_D d^3Y\,\varepsilon^{\alpha\beta\gamma}\,\mathrm{Tr}\left(\tilde{g}^{-1}\partial_\alpha\tilde{g}\tilde{g}^{-1}\partial_\beta\tilde{g}\tilde{g}^{-1}\partial_\gamma\tilde{g}\right), \tag{1.33}$$

with $g \in G$, namely a $G_L \times G_R$ (say $G \approx O(N)$) D=2 invariant chiral model made up of the sum of the standard geometrical nonlinear σ-model and a topological action $\Gamma(g)$. This added Wess-Zumino term is defined over a 3-dimensional ball D (with coordinates Y^α) whose boundary is 2-spacetime. The boundary values of $g(\xi)$ determine (1.23) modulo 2π. Γ is an example of a multiplevalued action; the singlevaluedness of the Feynman action $\exp(iS_{\lambda, k})$ implies that the quantization of $k = n\hbar$.

A renormalization group analysis shows that the Wess-Zumino-Novikov-Witten (WZNW) model (1.32) has an infrared fixed point when $\lambda = \frac{4\pi}{k}$, it then reads

$$S_k(g) = \frac{k}{16\pi}\left\{\int d^2\xi(\partial_\mu g^{-1}\partial_\mu g) + \Gamma(g)\right\} \tag{1.34}$$

which is now invariant, exactly like the system (1.23), under the infinite dimensional Kac-Moody group $G(z)_L \times G(\bar{z})_R$, namely under the transformation $g(\xi) \to \Omega(z)\, g(\xi)\bar{\Omega}^{-1}(\bar{z})$

Indeed the equations of motions for (1.34) are the same as (1.24) if the $J^a_\pm$ are defined as

$$T^a J^a_+ = -i\sqrt{2}\frac{n\hbar}{4\pi} g^{-1}(\partial_+ g) \tag{1.35}$$

$$T^a J^a_- = -i\sqrt{2}\frac{n\hbar}{4\pi}(\partial_- g)g^{-1} \tag{1.36}$$

The T^a are generators of G . Then the obtained canonical Poisson brackets promoted to Dirac brackets yield in the case of n =1 the same KMA (1.17) of the massless O(N) fermion theory. The nonabelian bosonization rules are given by equating the $T^a J^a_\pm$ from (1.35) with (1.25) . It can on fact be proved that the two theories (1.23) and (1.34) are dynamical identical.

Yet the translation dictionary for this generalized fermi-bose equivalence is still to our knowledge incomplete. Despite attempts, we still do not have the non-abelian counterparts of the vertex operators (1.31) giving the fermionic field in term of exponential of the non-abelian currents. What we have is the powerful Frenkel-Kac construction to be recalled subsequently.

1.3 VIRASORO -KAC-MOODY ALGEBRA : REPRESENTATIONS

Examples of conformally invariant field theories(CFT) are statistical mechanical systems at their critical points[23, 24]. The representation theory of the conformal group places contraints on the critical exponents and on correlation functions[25]. Since the two dimensional conformal group , Vir , the Virasoro-Botts group of diffeomorphisms [26, 27] of the circle is infinite dimensional, it has a very rich and powerful in structure [24, 28]. One could actually realized[25] the conformal bootstrap program of Polyakov[29] in two dimensionns. It amounts to solving for the representation theory of the Dirac-Schwinger algebra of the energy momentum tensor components, the generators of Vir . This will give a complete classification of all possible D=2 conformal field theories.

The basic objects of a CFT are the primary fields $\phi(z, \bar{z})$. They transform as tensors

$$\phi(z, \bar{z}) \to \phi(z, \bar{z})' = (\partial_z z')^h (\partial_{\bar{z}} \bar{z}')^{\bar{h}}\, \phi(z', \bar{z}') \tag{1.37}$$

under conformal transformations $z \to z' = f(z)$, $\bar{z} \to \bar{z}' = \bar{f}(\bar{z})$. h and $\bar{h}$ are the conformal weights. Since under rescaling $z \to \lambda z$, λ real , and under a rotation $z \to \exp\{-i\theta\}z$, $\phi \to \lambda^{h+\bar{h}}\phi$ and $\phi \to \exp\{-i(h-\bar{h})\theta\}\,\phi$, $d = h + \bar{h}$ and $s = h - \bar{h}$ are called the scaling dimension and the conformal spin of ϕ respectively.

The tracelessness and conservation of the energy momentum tensor $T_{\mu\nu}$ of a CFT imply that

$$\partial_{\bar z} T = 0 \quad , \quad \partial_z \bar T = 0 \tag{1.38}$$

namely the two nonzero components of $T_{\mu\nu}$, $T(z) \equiv T_{zz}(z)$ and $\bar T(\bar z) \equiv T_{\bar z\bar z}(\bar z)$ are holomorphic and anti-holomorphic functions respectively.

Now any primary field ϕ has the following operator product expansion (OPE) with $T(z)$

$$T(z)\,\phi(\xi) = \frac{h\,\phi(\xi)}{(z-\xi)^2} + \frac{\partial_\xi \phi(\xi)}{(z-\xi)} + \text{finite terms} \tag{1.39}$$

as for the OPE of T with itself

$$T(z)\,T(\xi) = \frac{\frac{c}{2}}{(z-\xi)^4} + \frac{2\,T(\xi)}{(z-\xi)^2} + \frac{\partial_\xi T(\xi)}{(z-\xi)} + \text{finite terms} \tag{1.40}$$

The anomalous first term is due to the famous nonvanishing D=2 trace anomaly . For example c = n , if the field theory is a free massless theory of n scalar field. Indeed (1.40) and its barred counterpart are together another expression of the Virasoro algebra of Vir realized quantum mechanically by the central extension of the algebra of the diffeomorphisms of the circle S^1. It is given by the product of two commuting Virasoro algebras $Vir_L \times Vir_R$, the first of which is

$$[L_n\,,\,L_m] = (n-m)\,L_{n+m} + \frac{c}{12}\,n(n^2-1)\,\delta_{m,-n}\,; \tag{1.41}$$

the second obtains by the mere replacement $L_n \to \bar L_n$ in (1.41) with the same c since $T + \bar T$ is real. The L_n and $\bar L_n$, $n \in Z$, are respectively the hermitian $(L_n^\dagger = L_{-n})$ moments of $T(z)$ and $\bar T(\bar z)$: $T(z) = \sum_n z^{-n-2} L_n$.

While (1.41) describes the infinitesimal transformation $\delta z = z^{n+1}$, $T(z)$ obeys the following composition law for finite transformations $z \to z' = f(z)$:

$$T(z)\,dz^2 = T(f)\,df^2 + \frac{c}{12}\{f, z\}\,dz^2 \tag{1.42}$$

where

$$\{f, z\}\,dz^2 \equiv d^3f\,df^{-1} - \frac{3}{2}(d^2f\,df^{-1})^2 \tag{1.43}$$

is the Schwarzian quadratic differential . Its properties, unique for a weight 2 conformal object, are

$$\{f, z\} = 0 \quad \text{if} \quad f = \frac{\alpha z + \beta}{\gamma z + \delta} \in SL(2, \mathbf{R}) \,, \tag{1.44}$$

$$\{f, z\}\,(dz)^2 = \{f, \xi\}(d\xi)^2 + \{\xi, z\}\,(dz)^2 \,, \tag{1.45}$$

$SL(2, \mathbf{R})$ is the maximal subalgebra of Vir and generated by L_0 and $L_{\pm 1}$.

There are numerous reviews of the representation theory of the Virasoro algebra are [24, 30, 31]. For later reference, we only mention the following facts. In a conformal field theory such as (1.34), the Hilbert space must be partioned into irreducible representations of the Virasoro algebras. The dictate of physics, i.e. energy positivity, requires the representations to be of highest weight i.e. such that

$$L_0 |h\rangle = h |h\rangle \quad , L_n|h\rangle = 0 \quad , n > 0 . \tag{1.46}$$

Such a Verma module V(c,h) is spanned by the linear independent vectors

$$L_{-1}^{n_1} L_{-2}^{n_2} \ldots L_{-nr}^{n_r} |h\rangle \tag{1.47}$$

and is graded by the level $\sum_j j\, n_j$. For unitary representations it is necessary that either

$$c \geq 1 \text{ and } h \geq 0 \tag{1.48}$$

or

$$c = 1 - \frac{6}{(m+2)\,(m+3)} \quad \text{and} \quad h = \frac{[(m+3)p - (m+2)q]^2 - 1}{4(m+2)\,(m+1)} \tag{1.49}$$

where m = 0,2,2,...; p = 1,2,..., m+1 ; q =1,2,...p.

A conformal field theory is thus characterized by the value of its central charge and the set of highest weights { h , $\bar{h}$ } of its irreducible representations. In addition the Wilson operator product algebra for these fields should also be specified. Having in mind a WZNW theory at its critical point , the possible values of h and $\bar{h}$ can be determined and formulae for the characters in a Kac Moody highest weight representation have been computed by Kac and Petersen[32].

To every Kac-Moody algebra is associated a Virasoro algebra as a derivation algebra. Thus since $\lambda = \frac{4\pi}{k}$ corresponds to a conformal invariant fixed point , the Wess-Zumino-Novikov-Witten model is also invariant under the Virasoro-Bott conformal group. The L_n 's are given through the generic Sugawara-Sommerfield form[33, 34] of the system energy momentum tensor

$$T(z) = \frac{1}{2\,K + C_2} \sum_m :J^a\, J^a: \tag{1.50}$$

whence

$$L_n = \frac{1}{2\,K + C_2} \sum_m :J^a_{n-m}\, J^a_m: \tag{1.51}$$

J^a_m are the moments of the current $J^a_+ = \sum_m J^a_m z^{-(m+1)}$ and the KMA (1.21) i.e. (1.26) reads

$$\left[J^a_n , J^b_m\right] = f^{ab}{}_c\, J^c_{n+m} + \frac{K}{2}\, n\, \delta_{m,-n} . \tag{1.52}$$

K , a real constant in each representation in general, is called the *level* of the KMA . The central charge c of (1.41) is given by

$$c = \frac{2K \dim G}{2K + C_2} \tag{1.53}$$

where C_2 is the quadratic Casimir operator for the adjoint representation of G. So when (1.52) and (1.41) are combined together with $\left[L_n , J^a_n\right] = -\, m\, J^a_{n+m}$, and similar relations for the barred counterparts, the full invariance algebra of the WZNW model is the semi-direct products (Vir_L x KMA_L) x (Vir_R x KMA_R) .

If G is simply -laced and of rank n, then Witten's result (K =1) (1.34) implies that for level K=1 , the corresponding c = n , an integer , and hence that the level 1 KMA currents of G should be reproducable from n free bosonic fields. This bose field realization is the Frenkel-Kac construction. Its results are as follows:

In a Cartan-Weyl basis and self-explained standard notations, a KM algebra (1.52) of a simply-laced G with rank n reads

$$\left[H^i_n, H^j_m\right] = m\, \delta^{ij}\delta_{m+n} \qquad , \left[H^i_m, E^\alpha_n\right] = \alpha^i E^\alpha_{m+n}$$

$$\begin{aligned}\left[E^\alpha_m, E^\beta_n\right] &= \varepsilon(\alpha,\beta)E^{\alpha+\beta}_{m+n} \qquad \alpha\cdot\beta = -1\\ &= \alpha\cdot H_{m+n} + K\, m\, \delta_{m+n} \quad \alpha\cdot\beta = -2\\ &= 0 \qquad \alpha\cdot\beta \geq 0\end{aligned} \tag{1.54}$$

i, j = 1,2,... rank G , by hermiticity $H^{i\dagger}_n = H^i_{-n}$, $E^{\alpha\dagger}_n = E^{-\alpha}_{-n}$. it admits an explicit realization from n free bosonic fields

$$X^i(z) \equiv q^i - ip^i \ln z + i \sum_{n\neq 0} \frac{1}{n}\alpha^i_n z^{-n}, \tag{1.55}$$

namely

$$H^i_n = \oint_C \frac{dz}{2\pi i} z^n H^i(z) \quad \text{where} \quad H^i(z) = i\, \partial_z X^i(z), \tag{1.56}$$

$$E^\alpha_n = \oint_C \frac{dz}{2\pi i} z^n E^\alpha(z) \quad \text{where} \quad E^\alpha(z) = c_\alpha\!: \exp\{i\, \alpha\cdot\mathbf{X}(z)\}, \tag{1.57}$$

c_α are Klein factors or cocycles obeying $c_\alpha\cdot c_\beta = (-i)^{\alpha\cdot\beta}\, c_\beta\cdot c_\alpha$ and $c_\alpha\cdot c_\beta = \varepsilon(\alpha,\beta)c_{\alpha+\beta}$.

Since a D=2 field theory of a free Majorana-Weyl (i.e real chiral) fermion corresponds to a CFT with c= 1/2 , we expect to be able to realize a K =1 KMA for 2n real fermions in a vector representation say of S0(2n) by a CFT of n scalar fields with momentum vector being the vectors of SO(2n). Thus using the complex basis $\Psi^{\pm a} \equiv \frac{1}{\sqrt{2}}\left(\psi^{2a-1} \pm i\, \psi^{2a}\right)$

one has for the Cartan subalgebra S0(2n) currents

$$J^{a,-a}(z) = :\Psi^a\, \Psi^{-a}: = i\, \partial_z X^a(z) \qquad (a < b) \tag{1.58}$$

while the other non-commuting SO(2n) currents read

$$J^{\pm a,\pm b}(z) = c_{\pm a,\pm b}\!: \exp\{\, i\, (\pm X^a \pm X^b)\!: \qquad (a < b) \tag{1.59}$$

In fact one has a generalized fermi-bose equivalence , a generalized Mandelstam -Halpern, vertex construction in

$$\Psi^{\pm a}(z) = c_{\pm a} : \exp\{\, \pm i\, X^a\} \tag{1.60}$$

This Frenkel-Kac bosonization[35] is key to the incorporation of Yang-Mills symmetries in the heterotic string and allows for enormous simplifications in handling vertex operators of fermionic CFT's. The key question is how and how much can the above sampling of the rich representation and analyticity structures be generalized to four and higher dimensions. We survey the various excursions toward higher dimensional worlds next.

2. BEYOND THE AFFINE AND DIFFEOMORPHISM ALGEBRAS OF THE CIRCLE

2.1 D=4 GAUGE AND CURRENT GROUPS

We have recalled the tremendous successes of affine Lie algebras realized as loop algebras in D=2 quantum field theories. A natural next question concerns how much of these structures carries over to a four dimensional setting by replacing the circle $S^1$by a higher dimensional arbitrary Riemannian manifold M[36]. Indeed the group Map(M;G) of smooth maps $M \rightarrow G$ is an infinite dimensional Lie group and appears almost as simple as the loop group Map(S^1, G). There had been results on the representation theory of these algebras and groups; they were reviewed in 1983 by R.S. Ismagilov[37]. However it is a remarkable fact of loop algebras that all positive energy irreducible representations are both unitary and necessarily projective. It would therefore be most interesting to seek their higher dimensional analogs in D≥2 counterpart of affine Lie algebras, namely algebras with nontrivial extensions. From the standpoint of physics where $M \approx S^3$, the compactified physical space, such groups are of primary importance in quantum field theory as the "gauge groups" and their special cases, the"current groups' [6]. They are the algebraic structures underlying current gauge theories and effective chiral theories of strong interactions at the Gev[38]as well as the Tev energy scales[39]. Unfortunately not much is known after several ongoing efforts. Here we assess the results and mention the novel directions some have been undertaking to make further progress.

As shown by the works of Bars[40] and of Bruce and Bose[41] it is an easy matter as far as obtaining the algebras with extensions, say of the sphere group Map (S^d, G) . As illustrations we summarize the results for the simplest case of d = 2 and 3 respectively. Generically the current algebra reads:

$$\left[J^a(x)\ ,\ J^b(x')\right] = f^{abc}J^c(x)\ \delta(x - x') + S^{ab} \tag{2.1}$$

It is a very straightforward matter to find the most general Schwinger terms S^{ab}consistent with the Jacobi identity . They are

For $M = S^2$ parametrized by the Euler angle θ and φ , z = cos θ

$$S^{ab} = \delta^{ab}\,[\, f_1(\theta,\phi)\ \delta(z - z')\ \delta'(\phi - \phi') + f_2(\theta,\phi)\ \delta'(z - z')\delta(\phi - \phi')\,] \tag{2.2}$$

where

$$f_1 = \frac{\partial h(\theta,\phi)}{\partial z}\ ,\ f_2 = -\frac{\partial h(\theta,\phi)}{\partial \phi}\ , \tag{2.3}$$

h(θ, φ) is an arbitary function on S^2.

For $M = S^3$

$$S^{ab} = \delta^{ab}\,[\, f_1\delta(z - z')\ \delta(\gamma-\gamma')\delta'(\alpha-\alpha') + f_2\delta'(z - z')\delta(\gamma-\gamma')\delta(\alpha-\alpha')$$

$$+ f_3\delta(z - z')\ \delta'(\gamma-\gamma')\ \delta(\alpha-\alpha')] \tag{2.4}$$

with

$$f_1 = \frac{\partial h_3}{\partial z} - \frac{\partial h_2}{\partial \gamma}\ ,\quad f_2 = \frac{\partial h_1}{\partial \gamma} - \frac{\partial h_3}{\partial \alpha}\ ,\quad f_3 = \frac{\partial h_2}{\partial \alpha} - \frac{\partial h_1}{\partial z}\ . \tag{2.5}$$

f_i or h_i (i = 1,2 3) are three arbitrary functions of z = cosβ and the Euler angles α,β, γ . These arbitrary functions h and h_i are in fact identifiable with components of closed 1-forms on S^2 and S^3 respectively .

Paralleling the algebra of the Fourier moments $J^a{}_m$ of Map (S^1 , **g**) , one expands the G-algebra valued currents J^a (θ, φ) on S^2 and J^a(α,β, γ) on S^3 in spherical harmonics $Y_{l,m}$ and Wigner's $D_{l,m,m'}$ functions

$$J^a(\theta,\phi) = \sum_{l,m} J^a_{l,m} Y_{l,m}(\theta,\phi) \tag{2.6}$$

$$J^a(\alpha,\beta,\gamma) = \sum_{l,m,m'} J^a_{l,m,m'} D_{l,m,m'}(\alpha,\beta,\gamma) \tag{2.7}$$

The notable features distinguishing the above algebras from the affine Lie algebras are the following :

a) the resulting algebras of moments will clearly have an infinite number of central elements corresponding to the number of components of the function h for S^2, and h_i (i= 1,2, 3) . So for d > 1 , the central extension is no longer one dimensional ; there are an infinite number of central extensions. This new phenomenon for Dim M > 1 agrees with a general cohomological theorem of Feigin[42]. The latter states that if g^M is the Lie algebra of Map(M;G) , then the second cohomology group $H^2(g^M)$ is infinite dimensional for M with Dim M > 1 . One can interpret the space $H^2(g^M)$ as an infinite set of classes of independent 1-dimensional central extensions.

b) Focusing of the moments $J^a_{l,m}$ and $J^a_{l,m,m'}$, it was shown that while grading operators can be constructed by the indices m and m' , none exists for the index l . This feature implies that in contrast to the D=2 affine case it is not possible to associate with these sphere algebras a root vector system in a finite dimensional root vector space .

To construct the corresponding KM groups, their representation theory and make contact with physics ,we return to Mickelsson's bundle formulation . To be specific we restrict to D=3+1 dimensions ; extension to higher dimensions being straightforward[43]. Let us consider the case of the "gauge group" Map($S^3 \to G$) , specifically with $G \approx SU(N)$, $N \geq 3$. Let $\mathcal{A}$ be the space of gauge connections on S^4 and $\mathcal{G}$ be the gauge group of point based maps $f : S^4 \to G$ with f(p) =1 for some fixed p εS^4 . Let $D = \{ x \,\varepsilon\, R^4 \mid |x| \leq 1 \}$be the unit disk so that $S^3 \approx \partial D$ .and let $DG = \{ f : D \to G \mid f(p) = 1 \}$ for some fixed p ε ∂D. Now the space $\Omega_3 G = \{ f : S^3 \to G \mid f(p) = 1 \}$ is infinitely connected since $\pi_3(\Omega_3 G) \approx \pi_3(G) \approx Z$, its connected components $\Omega_3^n G$ are labelled by the instanton number n . As in the two dimensional case, each connected component of the bundle $\mathcal{A} \to \mathcal{A}/\mathcal{G}$ on S^4 is homotopically equivalent to $DG \to \Omega_3^0 G$ where the restriction to the zero instanton sector is of no consequence as all sectors are homeomorphic.

Now there is a D=4 analog of the principle bundle P discussed earlier. This bundle P_3 on $\Omega_3 G$ consists of equivalence classes (f,λ) in DGx U(1) w.r.t.

$$(f, \lambda) \sim (f g, \lambda \exp\{ 2\pi i\, \omega_1(f,g) \}) \tag{2.8}$$

$$\omega_1(f,g) = \overline{\omega_1}(f^{-1}df, g) \tag{2.9}$$

for g ε $\mathcal{G}$,

$$\overline{\omega_1}(A, g) = \frac{i}{24\pi^3} \int \mathrm{Tr} \left\{ -dg\, g^{-1} \left(\frac{1}{2} \left(\frac{1}{2} AdA + dAA + \frac{1}{2} A^3 \right) \right. \right.$$

$$\left. + \frac{1}{4} (dg\, g^{-1} A)^2 + \frac{1}{2} (dg\, g^{-1})^3 A \right\} + C_5(g) \tag{2.10}$$

and

$$C_5(g) \equiv \frac{i}{240\, \pi^3} \int_{D_5} \mathrm{Tr}(dg\, g^{-1})^5 \tag{2.11}$$

Here $A = f^{-1}df$. $\overline{\omega_1}$ is a 1-cocycle of the group DG.

Now unlike the bundle P in the 2-dimensional case, P_3 has no natural group structure. To see this, define a 2-cocycle ω_2 for the group DG as

$$\omega_2(A\,;\,g_1,g_2) \equiv -\frac{i}{48\pi^3}\int_{S^3} \mathrm{Tr}\Big\{[(dg_2\, g_2^{-1})\,(g_1^{-1} A\, g_1)\,(g_1^{-1} dg_1)$$

$$- (dg_2\, g_2^{-1})\,(g_1^{-1} dg_1)\,(g_1^{-1} A\, g_1)]\Big\} + R_3(g_1, g_2) \qquad (2.12)$$

where the A independent term R_3 is of no importance to our subsequent discussion. We defer the details of the cohomological derivation of (2.10) to a large body of literature. We note that ω_2 differs from its two dimensional counterpart (1.17) by being a function of the gauge potential A . Consequently a proper extension of DG is not simply a U(1) but rather the infinite Abelian group (by point wise multiplication) Map ($\mathcal{A}_3$, U(1)) where $\mathcal{A}_3$ is the space of g-valued vector potential in S^3 . We have then a non central , operator valued extension by an abelian ideal, Map ($\mathcal{A}_3$, U(1). This is in accord with the cited Feigin's theorem .The group composition rule reads

$$(f, \lambda)\cdot(f', \lambda') = (f f', \lambda\, \lambda'_f \exp\{2\pi i\, \omega_2(A\,;\,f\,,f'\,\}\,) \qquad (2.13)$$

where $\lambda'_f(A) \equiv \lambda\,(f^{-1}A\,f + f^{-1}d\,f)$. The associativity of this product is guaranteed by the 2-cocycle nature of ω_2 (A,; g_1, g_2) . The gauge transformation for A ε $\mathcal{A}_3$ is defined by the restriction of f to $S^3 \approx \partial D$, the boundary of the unit 4-disk. As in the D=2 case, one can now define a group Q_3 by way of the abelian extension mod out the equivalence relation " ~ " (2.13) . Q_3 = (DG x Map($\mathcal{A}_3$, U(1)) / ~) , the obtained set of equivalence classes , is thus the principal bundle on $\Omega_3 G$ with as structure group Map($\mathcal{A}_3$, U(1)) . Q_3 is seen as an associated bundle to P_3 through the natural action of U(1) in the space Map($\mathcal{A}_3$, U(1)), its group structure being inherited from that of DG x Map($\mathcal{A}_3$, U(1)).

The Lie algebra 2-cocycle c_3 in Map(D, g) corresponding to ω_2 (2.12) can be computed to be

$$c_3(A\,;\,f_1, f_2) = \frac{1}{12\,\pi^2}\int \mathrm{Tr}\,\{A\,(\,df_1 df_2 - df_2 df_1)\} \qquad (2.14)$$

where $A = A_i^a T^a dx^i$, df_1 and df_2 are three matrix-valued 1-forms. As we are solely interested here in current groups arising from chiral σ-models, we see that (1.14) defines the extension of the current algebra Map(S^3 , G) by the ableian ideal Map($\mathcal{A}_3$, U(1)) . If we define our smeared current J (f), in an evident generic notation

$$J(f) \equiv \int dx\, f^a(x)\, J^a(x) \quad , \quad f = f^a T^a , \qquad (2.15)$$

then the D=4 Kac-Moody current algebra reads

$$[J(f_1)\,,J(f_2)\,] = i\,J\big([f_1, f_2]\big) + c_3(\,A;\, f_1, f_3) \qquad (2.16)$$

where the integration in (2.15) is over S^3 and A reduces to the flat connection $\omega = U^{-1}dU$, U ε G as we will illustrate next.

2.2. CANONICAL REALIZATION , SOLITON OPERATOR AND REPRESENTATION THEORY

We saw that a salient and powerful feature of affine Lie algebras is the existence of equivalent fermionic and bosonic representations. The existence of the Skyrmion testifies to the existence of a similar phenomenon in four dimensions. A question of great theoretical and phenomenological interest is the full extent and exact mathematical nature of this analogy , namely its proper place in the representation theory of D=4 Kac-Moody algebras.

Any physically motivated current algebra A has too many representations only a subset of which is of physical relevance. So to better single out these physical representations, which must be adopted to the dynamics at hand, one is led in practice to assume some concrete dynamics underlying A such as an effective field theory. After all the vertex operator representation of the Virasoro algebra was first discovered by physicists in the dual resonance model. A concrete point starting for looking at the representation problem is the D=4 analog of the D=2 WZNW model , which may emerge from a large N , low energy limit of QCD. Its manifestly SU(N)xSU(N) chiral invariant action reads

$$S = S_{\sigma m} + S_{WZ} \tag{2.17}$$

where

$$S_{\sigma m} = \frac{-1}{16 f_\pi^2} \int_M d^4x \mathrm{Tr} \left(\partial_\mu U^{-1} \partial_\mu U\right) \tag{2.18}$$

$$S_{WZ} = \frac{-i N_c}{240 \pi^2} \int_\Omega \mathrm{Tr} \, (U^{-1} dU)^5 . \tag{2.19}$$

The SU(N) matrix field U parametrizes the field space $G \approx \frac{SU(N)x\, SU(N)}{SU(N)} \approx SU(N)$. With strong interactions in mind, N is the number of flavor. $U^{-1}dU \equiv \omega = \omega^a T^a$ is the Maurer-Cartan, left-invariant current 1-form. The { T^a} denotes an anti-Hermitian basis of SU(N). $S_{\sigma m}$ is the standard nonlinear σ-model action . S_{WZ} is the D=4 Wess-Zumino term where for the same reason as its D=2 counterpart , the integration is over Ω , a 5-dimensional disk with as its boundary the spacetime M., N_c is similarly quantized .

With the boundary condition that $U(x) \rightarrow I$ at spatial infinity, 3-space is effectively compactified onto a 3-sphere S^3. The configuration space of our model is then the infinite Lie group Map (S^3 ; G≈SU(N)) . From here we can obtain the corresponding current algebra of (2.17) either by the cohomological or the canonical field theory method . By way of the field equations of (2.17) expressed as a current conservation law

$$\partial_\mu J^a_\mu = 0 \tag{2.20}$$

for the current

$$J^a_\mu = \frac{1}{8 f_\pi^2} \mathrm{Tr} \, (T^a \omega_\mu) + 5 \, \lambda \, \varepsilon_{\mu\nu\rho\sigma} \mathrm{Tr}\{ \, T^a \omega_\nu \omega_\rho \omega_\sigma \}. \tag{2.21}$$

Its first term is of the usual Sugawara-Sommerfield form, its second term derives from the Wess-Zumino anomaly, $\lambda \equiv \frac{- i N_c}{240 \pi^2}$. In particular the resulting *local charge density algebra* reads

$$[J^a_o(\mathbf{x}) , J^b_o(\mathbf{x})] = i f^{abc} J^c_o(\mathbf{x}) \, \delta^3(\mathbf{x} - \mathbf{y}) + S^{ab}(A \, ; \mathbf{x}, \mathbf{y}) \tag{2.22}$$

where

$$S^{ab} = 10 \, i \, \lambda \, \varepsilon_{ijk} \mathrm{Tr} \left(\left\{T^a, T^b\right\} \omega_i \omega_j\right) \partial_k \, \delta^3(\mathbf{x} - \mathbf{y}) \tag{2.23}$$

(2.22) is an exact realization of the commutator (2.16) given previously. The operator-valued Schwinger term or Abelian extension S^{ab} , originates from the Wess-Zumino action. It is the flat connection ($\omega = U^{-1}dU$) limit of the anomalous gauge generator algebra of Faddeev and

Shatashvili [44] for a quantum theory of left-handed fermions coupled to an external gauge field In the case of global chiral symmetry, the presence of S^{ab} signals the possibility of projective representations, new sectors in the model's Hilbert of physical states. It is thus called a "good" anomaly. In the gauge theory case on the other hand, we have an inconsistent quantum gauge theory since a nonvanishing S^{ab} for $N \geq 3$ is a topological obstruction to the implementation of Gauss's law or local gauge invariance. We have here a "bad" anomaly.

What is most remarkable about the algebra (2.22) and its canonical bosonic realization (2.17) is that the later admit a fermionic soliton. It is well known that a D=4 dimensional σ-model i.e. the action $S_{\sigma m}$ augmented by suitable stabilizing higher field derivative terms admits topological $S^3 \rightarrow G$ solitons, the Skyrmions[45]. Since the works of Balachandran et al [46, 47, 48], there has been an explosion of phenomenological applications to hadronic physics[38, 49]. But of interest to us is Witten's proof that the added topological Wess-Zumino action induces the realization of a projective fermionic representation of the current algebra. It confirmed Skyrme's conjecture existence of a D=4 bosonization .

We recall in a nutshell Witten's semi-classical argument . Take the static classical 1-Skyrmion map $U_S(\mathbf{x}) : S^3 \rightarrow SU(2)$, seen as a suitable SU(2) embedding in $G \approx SU(3)$ with topological charge B

$$B = \frac{1}{24\pi^2} \int_{S^3} d^3x \, \mathrm{Tr}\,(U^{-1}dU)^3 = 1 \tag{2.24}$$

the generator of $\pi_3 (G) \approx Z$. Using the time dependent ansatz $U(\mathbf{x},t) = g(t)\, U_S\, g(t)^{-1}$, $g(t) \varepsilon SU(3)$ being the collecfive coordinate matrix, we now adiabatically rotate the skyrmion by an angle of 2π around some axis . The resulting contribution coming solely from the Wess-Zumino term is $(i \pi N_C)$, giving a geometrical phase factor of $\exp (i \pi N_C) = (- 1)^{N_C}$, the spin phase , to the quantum mechanical wavefunctional . So the soliton is a fermion for N_C = odd integer. N_C is identified as the number of colors by matching the flavor anomalies of the effective chiral model (2.17) with those of its underlying gauge theory, quantum chromodynamics.

To go beyond the semi-classical description of the Skyrmion, one would like to obtain the D-=4 counterpart of the local Skyrme-Mandelstam[18, 50, 51] fermionic operator (1.31) for creating a point like soliton out of the vacuum. The first effort was due to Skyrme himself[52]. It is at least of conceptual interest to sketch the essential elements of his construction . For the Sine-Gordon model with field denoted by α , he showed that the operator

$$S = FK = \exp(\pm \frac{i}{2}\alpha(x_0)) \exp\left[\frac{i}{2}\int_{x_0}^{\infty} \frac{\partial \alpha}{\partial t}\, dx\right] \tag{2.25}$$

obeys a massless Dirac equation. After normal ordering or renormalization , (2.25) was to become the Mandelstam operator . It later generalized to the vertex operator representation of affine Lie algebras. In analogy to (2.25) , Skyrme argued that the corresponding Weyl-liked operator S should also be made of two factors

$$S = S_2 S_1 = \exp\left\{i\int I_\alpha \frac{e_{\alpha\, i}(x-x_0)_i}{r}\, \omega(r)\, d^3x\right\} \exp\{i t_\alpha \theta_\alpha\} \tag{2.26}$$

with $r = |\vec{x} - \vec{x}_0|$, $I_\alpha(x)$ is the time -component of the isospin current , $e_{\alpha\, i}$ is a proper orthogonal matrix interlocking spatial and isospin directions and ω a suitable angular function of r . S_2 identifies the auxiliary momentum p_0 with a suitable field expression and $e_{\alpha\, i}$ with a matrix characterizing the field orientation. In S_1 , t_α is a rotation operator conjugate to $e_{\alpha\, i}$ relating to the internal symmetry index α, the θ_α(α =1,2,3) are functions of the soliton map . By applying the collective coordinate method to a static point like soliton and naively manipulating with S seen as canonical transformation, Skyrme partially diagonalized the nonlinear field system turning it into an effective, rotator Hamiltonian $H_{eff} = At_{\alpha}t_{\alpha} + B +$ interactions, with its isobaric spectrum so typical of old strong coupling theories. Then by a rather unconvincing argument he projected out the spin 1/2 state with H_{eff} leading to a Dirac hamiltonian of a free point particle plus interactions.

Much later Rajeev[53] took as operator which would create a soliton state from the vacuum the unitary operator

$$U(g_1) = \exp\{i \int d^3x I_0^a \theta^a(x)\} \tag{2.27}$$

where $g(x) = e^{i\lambda^a\theta^a(X)}$. U is to implement a projective or 2-cocycle representation of the 3-sphere current group $\Gamma \approx$ Map(S^3, G) in that $\widehat{U}(g_1)\,\widehat{U}(g_2) = e^{-2\pi i\,\omega_2(q;g_1,g_2)}\widehat{U}(g_{12})$. As is well known [54], there exist nontrivial projective representations of Γ provided its 2nd cohomology $H^2(\Gamma)$ is nontrivial. This is the case for $G \approx SU(N)$, $N \geq 3$ as π^5 (SU(N)) =Z for N≥3 due to the isomorphism $H^2(\Gamma) \approx H^5(G) \approx \pi^5(G)$. Nontrivial $H^5(G)$ is [55]precisely the condition for the existence of the Wess-Zumino anomaly. In a rather sketchy analysis, it was argued that for largely separated Skyrmions, two U (2.27)at different spatial points anticommute. In any case it is clear that to obtain a true local soliton operator obeying some spinor wave equation etc...greater kinematical and dynamical inputs through a canonical quantization of a definite model need to be brought to bear on Rajeev 's program. In our opinion, it may be more fruitful to try extracting such a soliton operator from a Skyrmion wavefunctional seen as a section of a Dirac determinant bundle.

The attempts by Skyrme and Rajeev while embodying the necessary central ideas are at best heuristic and incomplete. A technically rigorous construction has yet to be performed .One must face such ignored yet crucial and difficult issues such as regularizations, the meaning of exponentials of D>2 non-abelian field operators and other new topological ingredients. Surely one could profit from the recent experience (see Sect.C.1) in establishing certain exact operatorial boson fermi correspondence in three spacetime dimensions.

From an algebraic viewpoint, the existence of a kind of D=4 quantum fermi-bose correspondence has provided a strong inducement to attack a larger problem, that of the representation theory of the D=4 current algebra (2.16). Indeed there is also an pressing phenomenological need to do so. With the profusion of Gev hadrons, the possibility of a strongly coupled Higgs sector at SSC energies and the still intractable infrared structure of QCD, we may revive and seek to further advance the old program of current algebra , this time with an added topological twist, the Wess-Zumino chiral anomaly. The hope would be that , knowing the physical representations of such extended current algebra based on QCD, such an approach would provide a systematic nonperturbative (albeit effective) handle to portray strong interactions.

Compared to the rich developments of the representation theory of affine algebras what , if anything, is known about the representations of Map(M; G) ? The answer is " rather surprisingly little" . Till recently, there was only one irreducible representation due Gelfand , Graev and Vershik [7],but it has no apparent physical relevance. Another physically well founded attempt is by Mickelson and Rajeev[56]. The goal is to construct a suitable (3+1) dimensional generalization of the fermionic Fock space of D=2 current algebra with a Schwinger term (say (2.22).). To do so they generalized to the case of a larger linear group modelled by rank p ($0 < p < \infty$) Schatten classes the methods of Pressley and Segal [57]for constructing cocycle representations of the infinite dimensional restricted general linear groups. This is no place for involved technical details, we can only sum up their results.

Working in a D =odd spatial manifold M such as S^d, they consider the system of a Dirac spinor coupled to an external Yang-Mills field with gauge group G and its corresponding algebra **g**. Its current algebra is just Γ = Map (S^d ; **g**) is just (2.26) without the extension term C_3. Let f

: $S^d \rightarrow \mathbf{g}$ be Lie algebra valued functions , λ^i the representation matrices of $\mathbf{g}$. Since the Dirac Hamiltonian is unbounded below, the "1st quantized" unitary representation of Γ given by

$$[J(f) \, , \psi_\alpha(x)\,] = \lambda_\alpha^{i\beta} f^i(x)\, \psi_\beta(x) \tag{2.28}$$

is physically unsuitable as it has no vacuum state or highest weight vector . Second quantization cures this instability i.e. by constructing a Dirac vacuum as the highest weight state. The operator product $J^i(x) = \psi^\dagger\lambda^i\psi$ needs a short distance say a point splitting regularization, which involves substracting the VEV of $J\{f\}$. As a result, one no longer has a representation of Γ but a central extension of it .

In contrast to the situation in one spatial dimension where normal ordering is enough to a well-defined quantum theory , for d >1 further renormalizations are required. Thus for d >1 $J\{f\}^2$ are still not well defined after substraction . For d=3 , $J\{f\}$ requires an additional multiplicative renormalizaton, implying that such an operator is meaningless within the purely fermionic Fock space as it creates out states of infinite norm of the vacuum . A larger Hilbert space is then introduced . It include the fermionic states which no longer form a complete set and new bosonic states created from the vacuum by $J\{f\}$ and having the quantum numbers of a two fermion states. In this manner Mickelsson and Rajeev [56] found a nonunitary representation of an Abelian extension of Map(S^d ; $\mathbf{g}$) i.e. (2.12) . Their procedure illustrates the flip "local " side of the anomaly or of the ray representations of the KM group Map (S^d ; $\mathbf{g}$) . Specifically they found a linear representation with highest weight vector, essentially including these bosonic states . Very recently [56] they did manage to construct unitary representations in certain special cases of a 3-parameter family of deformations of the abelain extension $\widehat{\mathbf{gl}_2}$ of the general linear algebra $\mathbf{gl}_2$. It would be of great interest to see physical applications of such results.

We note that the necessity to include bosonic states along with fermionic ones, say to implement unitarity, seems consistent with the more recent conclusion on D=3 bosonization . In fact while D=3, 4 purely bosonic field theories do admit fermionic solitons, only in D=2 are such theories exactly equivalent to a local fermion model. Luscher [58] has shown that there exists an analogous exact quantum correspondence between certain D=3 interacting field theories, but this equivalence is between a purely bosonic model and one involving not only a basic local fermion field but also other bosonic fields. Similarly an illustration of exact of D=4 fermi-bose corespondence was put forth by Mickelsson[59]. He showed that on a spatial manifold with topology $S^2 \times S^1$ a D=4 Yang-Mills system coupled to a U(1) monopole and the mixed field system of a 4-component fermion coupled to a U(1) monopole have identical KMA. These results should motivated further work on physical representations of D>2 current algebras.

2.3 D > 2 DIFFEOMORPHISMS

As noted before, solving for quantum field theory is often equivalent to knowing all the unitary representations of its invariance groups. In two dimensions CFT testifies to the truth of the above assertion. In higher dimensions the place of the Virasoro-Bott conformal group is taken by the group of diffeomorphisms of a given manifold. In particular the diffeomorphism invariant topological quantum field theories should naturally take the place of CFT 's.

In the 60's diffeomorphism groups were considered in the motion of incompressible fluids by V. Arnold[60]. Subsequently applications of representations of the group SDiff(R^n) of volume preserving diffeomorphisms of R^n (n=2, 3) have been made in classical and quantum fluids, specifically to vortex filaments and other topological defects [61, 62]. In particle physics, recent attempts to quantize relativistic closed p-(super)branes[63], which generalize (super)strings, have led to the analysis of the algebras of SDiff(S^2) , of SDiff(M_g), M_g being a Riemann surface of genus g , of SDiff(S^3) etc.... and their possible central extensions. We briefly survey the status of these algebras as the D>2 analogs of the Virasoro algebra [64].

A p-brane M_p is a bosonic extended object of p-spatial dimensions propagating in d-spacetime dimensions according to the Polyakov action

$$S_p = \int d^{p+1}\sigma\sqrt{g}\left[g^{ij}\partial_i X\cdot\partial_j X - (p-1)\right] \tag{2.29}$$

$\sigma^i = (\sigma^a , \tau)$, σ^a (a= 1,...,p) are the coordinates on the p-brane, τ parametrizes the latter's time evolution. Working in the light cone gauge, well tested in string theory, means imposing first the following condition on the p-brane metric g_{ij}

$$g_{0a}= 0 , \qquad g_{00} = - \det h_{ab} \equiv h \tag{2.30}$$

where h_{ab} is the spatial metric on the p-brane. One can then choose the light cone gauge

$$X^+ = p^+\tau \ , \tag{2.31}$$

$X^\pm \equiv \frac{1}{\sqrt{2}}(X^0 \pm X^{d-1})$. What remains of the (p+1)-dimensional general coordinate invariance group is its subgroup which preserves (2.30) and (2.32) and consists of reparametrizations of the spatial variables $\sigma^a \rightarrow \sigma^{a'} (\sigma^b)$ with the restriction that they preserve the volume

$$\det\left(\frac{\partial\sigma^{a'}}{\partial\sigma^b}\right) = 1 \ . \tag{2.32}$$

The existence of this reparametrization invariance relates to that of the constraint

$$\partial_{[a}X^m\partial_{b]}P^m \equiv \phi_{ab}\approx 0 , \tag{2.33}$$

$$m = 1,2,...d-2 , \ P^m(\sigma) = \frac{\partial X^m}{\partial\tau} \ .$$

In mathematics the above subgroup is called the group of volume-preserving diffeomorphisms which we denote by ($VPDiffM_p$). Its general classical algebra has been computed, valid for any topologies and geometries. Infinitesimally (2.33) is equivalent to the variation $\delta\sigma^a = \eta^a(\sigma^b)$ where $\nabla_a(\alpha^{-1}\eta^a) =0$. The latter is solved by

$$\eta^a = \xi^a + \sum_{r=1}^{b_1} c_{(r)}\omega^a_{(r)} \tag{2.34}$$

in terms of coexact ($\xi^a d\sigma^a$) and harmonic ($\omega^a_{(r)}d\sigma^a$) 1-forms on M_p . b_1 denotes the 1st Betti number of M_p , $c_{(r)}$ are constant coefficients,

$$\xi^a \equiv \frac{1}{(p-2)!}\varepsilon^{an_2n_3...n_p}\partial_{n_2}\Lambda_{n_3...n_p} \ . \tag{2.35}$$

The classical algebra reads

$$\left[L_{\Lambda_1} , L_{\Lambda_2}\right] = L_{\Lambda_{12}} , \tag{2.36a}$$

$$\left[P_{(r)} , L_{\Lambda_2}\right] = L_{\Lambda_{(\phi)}}, \tag{2.36b}$$

$$\left[P_{(r)} , P_{(s)}\right] = L_{\Lambda_{(\phi\phi)}} \tag{2.36c}$$

where the generators are $L_\Lambda = \xi^a\partial_a$, $P_{(r)}= \omega^a_{(r)}\partial_a$ and the parameters on the RHS of (2.36 a-c) are given by

$$(\Lambda_{12})_{a_3...a_p} = - \varepsilon_{aba_3...a_p} \xi_1^a \xi_2^b \tag{2.37}$$

$$(\Lambda_{(r)})_{a_3...a_p} = - \varepsilon_{aba_3...a_p} \omega_{(r)}^a \xi^b \tag{2.38}$$

$$(\Lambda_{(r)(s)})_{a_3...a_p} = - \varepsilon_{aba_3...a_p} \omega_{(r)}^a \omega_{(s)}^b \tag{2.39}$$

In the quantum theory of membranes (p=2) , the constraints $\phi^{ab} \approx 0$ become $\hat{L}_\Lambda$|phys> = $\hat{P}_{(r)}$|phys> 0 on physical states through the operators

$$\hat{L}_\Lambda = -i \int d^2\sigma \xi^a \partial_a X^m P^m \qquad , \hat{P}_{(r)} = -i \int d^2\sigma\, \omega_{(r)}^a\, \partial_a X^m P^m \tag{2.40}$$

They satisfy at the level Poisson brackets the classical algebra (2.36) . For an arbitrary closed bosonic membrane $\Sigma = M_{2,\,g}$, namely a 2-sphere with g handles , the area-preserving diffeomorphism algebra was found with the most general central extension consistent with the Jacobi identy. It reads

$$\left[\hat{L}_{\Lambda_1}, \hat{L}_{\Lambda_2}\right] = \hat{L}_{\Lambda_{12}} + \int_\Sigma d^2\sigma V^a \Lambda_1 \overleftrightarrow{\partial}_a \Lambda_2 , \tag{2.41a}$$

$$\left[\hat{P}_{(r)}, \hat{L}_{\Lambda}\right] = \hat{L}_{\Lambda_{(r)}} - 2\int_\Sigma d^2\sigma\ V^a \varepsilon_{ab} \omega_{(r)}^b \Lambda , \tag{2.41b}$$

$$\left[\hat{P}_{(r)}, \hat{P}_{(s)}\right] = \hat{L}_{\Lambda_{(rs)}} + \int_\Sigma d^2\sigma\ \varepsilon_{ab} \omega_{(r)}^a \omega_{(s)}^b W . \tag{2.41c}$$

where W is any scalar and $\nabla_a(\alpha^{-1}V^a) = 0$, namely $\alpha^{-1}V^a$ is any divergenceless vector field on Σ . Consequently , from (2.42a-c) we see that the most general central extension is specified by one arbitrary scalar function and by 2g arbitarry constant coefficients of the harmonic forms on Σ [65] . This is so since the dimension of the space of harmonic 1-forms on Σ , dim $H^1(M_{2,g}, R)$ = dim R^{2g} = 2g [66].

From the above analysis it follows that , for the 2-torus (indeed the n-torus) and the 2-sphere, there are no nontrivial central extensions . The algebra for the torus was one found long ago by Arnold in hydrodynamics[67]. It is also remarkable that the Lie algebra of SU(∞) as well as the large N limit W_∞ of the operator algebra W_N generated by primary fields with integer spin 1,2,...N are isomorphic to the L- subalgebra (2.41a) of the area-preserving diffeomorphisms algebra of of the 2-plane ,whether or not it is compactified i.e. to S^2 [68].

As far as explicit representations, Figueirido and Ramos[69] have constructed Fock space representations of the algebra of vector fields of the n-torus, i.e of Diff(T^n). They used a generalization of the infinite wedge representation of Kac, Petersen, of Feigin and Fuchs for Diff (S^1)[32, 70]. Further renormalizations beside normal ordering were necessary to make everything well defined. Their representations are generated not by linear operators but bilinear forms so they do not arise from a Verma module. The unitarity question remains to be answered before physical applications. Of course, if one could manage to quantize membranes, this would amount to finding interesting unitary representations of their corresponding diffeomorphism groups with or without central extensions.

In two dimensions one can form the semi-direct sum of the Kac-Moody and Virasoro algebras where the first appears as an ideal. The join structure underlies conformal field theories with internal symmetries such as the WZNW model. A natural question is whether Map(S^n, g) and Diff(S^n) can be so combined into a larger algebra. This question was investigated for n=2,3 . It was shown [41] on the basis of consistency with the Jacobi identity that a) there exist no such a larger algebra containing Diff(S^n) and the centrally extended S^n- algebra e.g. (2.2) for n=2 and (2.4) for n=3 , b) however such a structure does exist if the S^n- algebra is not centrally extended . An example of (b) could be the current algebra Map(S^3; su (N)) of D=4 WZNW (2.17) non centrally extended by an abelian ideal.

From the standpoints both of physics and mathematics, the representation theory of higher dimensional analogs of Kac-Moody-Virasoro algebras is an object of great mathematical fascination and much potential physical importance. Since the D> 2 conformal groups are finite dimensional, to analyze D> 2 system at their critical points one should target infinite dimensional subgroups of the general D-dimensional diffeomorphism groups, ones which have the conformal group as a subgroup . In four dimensions several attempts to define and study four dimensional structures endowed with the richness of D=2 CFT. One group [71, 72] studies representations of infinite self-dual and anti-self dual subalgebras of the Diff(R^4) making use of quaternionic Schwartzian derivative and Fueter quaternionic analyticity. Inspired by the connection between D=3 Chern-Simons theory and D=2 CFT ,another group [73] seeks by descending from a D=5 Chern-Simons theory to find D=4 analogs of 2d CFT . Another group[74, 75] has undertaken a more radical approach à la Penrose. They propose replacing the Riemann surfaces of D=2 CTF by twistor spaces and complex analyticity by holomorphic sheaf cohomology. In this manner a classical infinite algebra has been found, one which has as its subalgebras both the D=4 SU(2,2) conformal and D=2 conformal algebras. The hope is to apply the corresponding quantum algebra with extension to classify fields and field theories reformulated in terms of twistors. If any of these programs succeeds, very enticing mathematical vistas will surely lie ahead.

3 . ODD PHENOMENA IN ODD DIMENSIONS

3.1 ANYONS REVISITED : CURRENT ALGEBRA AND VERTEX OPERATOR

Two overlapping topics in field theory have attracted a great deal of attention. They are the topological quantum field theories (TQFT) in D ≥ 3 spacetime dimensions[76], ushered in by the mathematical works of Donaldson and Jones and field theories[77] with excitations bearing*any* spin and statistics, *anyons*, which may well account for the fractional quantum Hall effect and high temperature superconductivity.

In his trail brazing analysis of D=3 Chern-Simons theories, Witten [78] showed the beautiful correspondence between the expectation values of the Wilson loops traced by 'colored' *point* sources in spacetime and Jone's polynomials for knots. To obtain the fundamental Skein relation of knot theory , he had to regularize or *frame* the Wilson loops inaddition to performing the standard regularization. A different form of such a regularization had been discovered by Polyakov[79, 80] in his proof of the fermi-bose transmutation of D=3 "baby Skyrmions" in their geometric *point-like limit.* It is at that juncture that an interesting overlap with the theory of anyonsoccurs . Our work[81, 82] on which this section is based takes off at this intersection between physics, the biology of DNA molecules, differential geometry, topology, representation theory of current algebras and division algebras.

To be specific we shall choose without loss of generality the model par excellence for anyons ,one governed by the action [77]

$$A = \int d^3x \left\{ | D_\mu Z |^2 + \frac{\theta}{8\pi^2} \varepsilon^{\mu\nu\lambda} A_\mu F_{\nu\lambda} + A_\mu J^\mu \right\} . \tag{3.1}$$

the CP_1 σ- model with a Chern-Simons term. Its basic field is a two component complex spinor $Z^T = (Z_1, Z_2)$ with $|Z|^2 = 1$, consequently it lives on S^3. The more familiar unit normed Wegner vector field **n** is given by the complex Hopf projection map taking $Z \in S^3$ to $\mathbf{n} = Z^+\sigma Z \in S^2 \approx CP_1$. D_μ is the covariant derivative w.r.t the composite U(1) gauge field $A_\mu = iZ^\dagger\partial_\mu Z$, $F_{\mu\nu}$ being the associated field strength. θ is a free parameter $(0 \le \theta \le \pi)$. This model should best be viewed as the low energy limit ($e^2 \to \infty$) of a system with a Maxwellian kinetic term $\frac{1}{2e^2}F_{\mu\nu}F_{\mu\nu}$ added on to (3.1).

It is well known that (3.1) (with θ =0) to admit exact $S^2 \to S^2$ general solitons. While the standard third term in (3.1) is the Aharonov-Bohm term, the second or Chern-Simons (C-S) term also reads for θ= π as $S_H = \frac{-1}{2}\int d^3x\, A_\mu J^\mu$, namely as an interaction between the field A_μ and the conserved topological current

$$J_\mu = \frac{1}{8\pi}\varepsilon_{\mu\nu\lambda}\varepsilon_{abc}n^a\partial^\nu n^b\partial^\lambda n^c \cdot \tag{3.2}$$

The soliton (electric) charge $Q = \int d^2x\, J_0$ is an integer labelling the elements of $\pi_2(S^2) = Z$. The field boundary condition is such that spacetime is $R^3 \cup (\infty) \approx S^3$; the C-S action is in fact the Whitehead form of the Hopf invariant for the maps $\mathbf{n} : S^3 \to CP_1 \approx S^2$, classified by the generators of $\pi_3(S^2) \approx Z$. The configuration space of fields is the infinte Lie group of 2-sphere base preserving smooth mappings $\Gamma \approx \{\mathbf{n}: S^2 \to S^2\}$. The homotopic relations $\pi_0(\Gamma) \approx \pi_2(S^2) \approx Z$ and $\pi_1(\Gamma) \approx \pi_3(S^2) \approx Z$ allow the topological possibilities of solitons and exotic spin -statistics connection respectively[84]. The latter option is implemented dynamically by a topological Chern-Simons action.

What effect does the Chern-Simons term have on the chiral soliton ? Wilczek and Zee[83] showed that either the interchange of two Q=1-solitons or a 2π rotation of one of them around the other gives a statistical (alias spin) phase factor $e^{i\theta} = (-1)^{2s}$ to the wave function. Hence the soliton has spin $s = \frac{\theta}{2\pi}$ and intermediate statistics, it is an *anyon*. A key ingredient for their proof is that the soliton map giving raise to this phase be of Hopf invariant 1. It will be a guiding criterion in our subsequent $D \ge 3$ generalization of the θ-spin and statistics connection.

Being a σ-model, (3.1) with θ = 0 has a canonical realization [84]of the following pseudo-Sugawara-Sommerfield equal time algebra of currents

$$\left[I_0^a(x), I_0^b(y)\right] = -i\,\varepsilon^{abc}I_0^c(x)\delta^2(\vec{x}-\vec{y}), \tag{3.3a}$$

$$\left[I_0^a(x), I_i^b(y)\right] = -i\,\varepsilon^{abc}I_i^c(x)\delta^2(\vec{x}-\vec{y}) + if^{-1}(\delta^{ab} - n^a(y)n^b(y))\,\partial_{x_i}\delta^2(\vec{x}-\vec{y}), \tag{3.3b}$$

$$\left[I_0^a(x), I_0^b(y)\right] = 0. \tag{3.3c}$$

where $I_\mu \equiv f^{-1}\varepsilon^{abc}\partial_\mu n^b n^c$. It is a pseudo-current algebra in it does not close since the operator-valued Schwinger term in (3.3b) is explicitly a function of the field **n**. Actually a larger close algebra can be obtained from (3.3a)-(3.3c) by introducing a new rank two symmetric tensor operator $S^{ab}(y)$ which is realized as $S^{ab} =(\delta^{ab} - n^a(y)n^b(y))$ by the model (3.1) . However one feature which distinguishes even and odd -dimensional current algebras seem to be the following . In contrast to the 4-dimensional σ-model with Wess-Zumino term leading to a noncentral extension of its algebra of currents (vis 2.22) by an abelian ideal, the model (3.1) with its Chern-Simons terms in fact leads to exactly the same algebra (3.3) where, say in the local charge density algebra (3.3a) we replace I_0^a by the new canonical momentum

$$L^a = I_0^a + \frac{\theta}{2\pi}\varepsilon_{ij}A_i(x)\partial_i n^a(x). \tag{3.4}$$

In other words the current algebra is without extension and is independent of the Chern-Simons term; it is θ-independent. While in odd dimension locally there is no signature of the Chern-Simon anomaly, globally at the level of representations, Semenoff ,Sodano [85]and Karabali [86]have shown that at the level of the wavefunctional ,when one exponentiates the algebra and boundary conditions then enter, it does make a big difference in whether one uses à la Skyrme as static vertex soliton creation operator

$$U_I(\vec{x}) = \exp\left\{i\int d^2y\, I_0^a(y)\,\omega^a(\vec{x}-\vec{y})\right\} \tag{3.5}$$

or

$$\widehat{U}_L(\vec{x}) = \exp\left\{i\int d^2y\, L^a(y)\,\omega^a(\vec{x}-\vec{y})\right\} \tag{3.6}$$

where the soliton profile map

$$\omega(\vec{x}) = g(r)\begin{pmatrix}\sin\phi\\ -\cos\phi\\ 0\end{pmatrix} \tag{3.7}$$

twists the vacuum configuration $n^a = (0, 0,1)$ into a soliton with charge $Q \neq 0$. Indeed while (3.6) does not transform like a scalar under rotation (3.7) provide a proper representations of the rotation group SO(2). In fact it can be shown that one has the graded commutation relations

$$\widehat{U}(\vec{x})\,\widehat{U}(\vec{y}) = e^{i\frac{\theta}{\pi}\Delta}\,\widehat{U}(\vec{y})\widehat{U}(\vec{x}) \tag{3.8}$$

where the multi-valued phase $\Delta(x, y) \equiv \Theta(x,y) - \Theta(y,x) = \pi \bmod 2\pi n$, $\Theta(\vec{x},\vec{y}) = \tan^{-1}\frac{(x_2-y_2)}{(x_1-y_1)}$.

(3.8) is the signature of exotic statistics. As with all such very ill-defined vertex operators, (3.6) is a topologically nontrivial coherent state operator with its classical soliton profile, it does not create a state of definite spin. As Skyrme was already keenly aware, it is in general not an easy task to use collective coordinates and proper regularization (!) procedures to project out from (3.6) operators with definite spin. All this and more remain to be done if vertex operators are to be useful entities in formulating effective Hamiltonian theories where anyonic excitations are the basic quasi-particles.

3.2 GEOMETRY OF A PHASE: LINKING THE SOLITON'S TWIST AND WRITHE TO EXOTIC SPIN AND STATISTICS

In the last section, we discussed the problem of zero in on some workable, important nonperturbative states or degrees of freedom of a highly nonlinear theory e.g. projecting out anyons of definite spin. It is clear that by going to the pointlike limit one can get to the lowest energy and spin states of a extended object. It is very much in this spirit that Polyakov[79] pioneered a tractable Wilson loop approach to the low energy behavior of soliton Green functions of model (3.1). To study the effects induced by the long range Chern-Simons interactions, he approximated the partition function **Z** by

$$Z = \sum_{(P)}^{\text{all closed paths}} e^{-mL(P)} \left\langle \exp\left(i \int_P dx^\mu A_\mu \right) \right\rangle \tag{3.9}$$

P is a Feynman path of a pointlike soliton , hence a curve, in spacetime R^3, L(P) is the total path length .

The first exponential in (3.9) is just the action of the path P of a free relativistic point soliton of mass m. Let

$$\Phi(P) = \left\langle \exp(i \oint_P A^\mu \, dx_\mu) \right\rangle , \tag{3.10}$$

the bracket < ...> denotes functional averaging w.r.t. the Hopf action. It embodies the Aharonov-Bohm effect , characteristic of topologically massive gauge theories: namely the Chern-Simons-Hopf action induces magnetic flux on electric charges and vice versa, thus producing dyonic objects. Being Gaussian, this phase Φ(P) is exactly calculable, thereby the analytic appeal of the point soliton approximation. By direct integration of the equation of motion , Φ(P) is given by

exponentiating the effective action :

$$\Phi(P) = \frac{1}{N} \exp\left\{ iS_0 + i \int d^3x \left(\frac{\theta}{4\pi^2} \varepsilon^{\mu\nu\rho} A_\mu \partial_\nu A_\rho + A_\mu J^\mu \right) \right\} . \tag{3.11}$$

S_0 is the free point particle action , N a suitable normalization. The conserved current of a Q=1 point source $J_\mu(x) = \int d\tau \, \delta^3(x-y(\tau)) \frac{dy_\mu(\tau)}{d\tau}$ is given geometrically. From (3.1) the equation for J_μ reads

$$J_\mu(x) = - \frac{\theta}{2\pi^2} \varepsilon_{\mu\nu\rho} \partial^\nu A^\rho \tag{3.12}$$

which on substitution in (3.11) with θ = π gives $\Phi(P) = \exp\left(\frac{i}{2} \int d^3x \, A_\mu J^\mu \right)$. Then for the above *given*point current and in the gauge $\partial^\alpha A_\alpha = 0$, A_μ can be solved to give :

$$\Phi(P) = \exp\{ i\pi \, I_G(P) \} , \tag{3.13}$$

where

$$I_G(C_\alpha \rightarrow C_\beta) = \frac{1}{4\pi} \oint_{C_\alpha} dx^\mu \oint_{C_\beta} dy^\nu \frac{\varepsilon_{\mu\nu\lambda}(x^\lambda - y^\lambda)}{|x-y|^3} \tag{3.14}$$

in the limit where the two smooth closed 3-space curves C_α and C_β <u>coincide</u>, namely $C_\alpha = C_\beta = P$, the soliton worldline.

Were C_1 and C_2 in R^3 (or S^3) *disjoint* curves , (3.14) would be their Gauss linking coefficient. If we denote by Ω (M_2) the solid angle subtended by C_1 at the point M_2 of C_2, Stoke's theorem gives $I_G = \frac{1}{4} \int_{C_2} d\Omega(M_2)$, which measures the variation of this solid angle divided 4π as M_2 runs along C_2 ; it is an integer, the algebraic number of loops of one curve around the other.

However though the integrand in (3.14) is that of Gauss' invariant , the integration is over <u>one</u> and the same curve. $I_G(P)$ is therefore *undetermined.* This artifact of the point-limit approximation must be amended by a proper definition or regularization of $I_G(P)$.

Polyakov's regularization consists in trading the δ-function in $\int_P dx^\mu \iint d^2y_\mu \delta(x-y)$, an equivalent expression for $4\pi I_G$, for the Gaussian $\delta_\varepsilon(x-y)=(2\pi\varepsilon)^{\frac{-3}{2}}\exp\left(\frac{-|x-y|^2}{\varepsilon}\right)$. He found that $I_G(P)_{Reg} = - T(P)$, the total torsion or twist of the curve P in spacetime with $T(P) = \frac{1}{2\pi}\oint_P dx\cdot(\mathbf{n} \times \frac{\partial \mathbf{n}}{\partial s}) \equiv \frac{1}{2\pi}\int_P \tau(s)\, ds$. s and **n** denote the arc length and the principal normal vector to P at the point **x**(s).

What is the meaning of this regularization ? By substituting the Gaussian, the dominant contribution to the surface integral comes from an infinitesimal strip Σ_P ; so this procedure effectively turns a spacetime curve into a *ribbon* . Precisely in 1961 such a process was used by Calugareanu [87]in his search for new invariants of the knot. The entity $I_G(C_\alpha \rightarrow C_\beta)$ turns out to be perfectly well defined and gives a new topological invariant SL , the self–linking number for a simple closed *ribbon.* SL is in fact the linking number of C_β with a twin curve C_α moved an infinitesimally small distance ε along the principle normal vector field to C_β. As disjoint curves

they can be linked and unlinked exactly the strands of a circular supercoiled DNA molecule[88]. In modern knot theory this construction is termed the *framing* of a curve C_β . Most notable is the existence of the "conservation law":

$$SL = T + W \tag{3.15}$$

explicitly

$$SL(P) = \frac{1}{4\pi}\int_{P\times P} d\Omega_2 + \frac{1}{2\pi}\int_P \tau\, ds \tag{3.16}$$

whereby SL, is the algebraic sum of two *differential geometric* characteristics of a closed ribbon, its total torsion or twisting number T and its *writhing number* or *writhe* W.. W is also the Gauss integral for the map $\phi : S^1\times S^1 \rightarrow S^2$, is the element solid angle, the pullback volume 2–form $d\Omega_2$ of S^2 under φ. While their sum SL is a topological invariant so must be an integer, T and W are metrical properties of the ribbon and its "axis" respectively, they can take a continuum of values. A coiled phone cord best illustrate the relation W + T = SL for a ribbon . When unstressed with its axis curling like a helix in space, its writhe is large while its twist is small. When stretched with its axis almost straight, its twist is large while its writhe is small.

By way of the dilatation invariance of W and the map **e**(s,u) ($\mathbf{e}^2 = 1$), a local Frenet-Serret frame vector attached to the curve, we can write $W = \frac{1}{4\pi}\int_0^L ds \int_0^1 du\, \varepsilon_{abc}\, e^a \partial_s e^b \partial_u e^c$, a,b,c = (1,2,3) and $\partial_s = \partial/\partial s$, $\partial_u = \partial/\partial u$. A conformally invariant action for the frame field **e**, W is manifestly a WZNW term as well as a Berry phase upon exponentiation[8]. Since W = – T (mod Z) , (3.15) explains Polyakov's double integral representation (modulo an integer) for the torsion T(P) .

By way of (3.15) the alternate form $\Phi(P) = \exp(- i\pi\, T(P)) \exp(+i\pi\, n)$ is the "spin " phase factor, essential to Polyakov 's proof that the 1-solitons of model (3.1) with $\theta=\pi$ are fermions by obeying a Dirac equation in their point like limit [80]. For arbitrary θ, we go over to the more general case of pointlike anyons . So we see the relation W = -T + SL as the very mathematical expression of the connection between statistics and spin in the geometric point soliton limit.

We now recall that in the geometry of 2-surfaces, a form of the Gauss-Bonnet theorem says $K = 2\pi\chi$. Like (3.15) , it relates a topological entity such as the Euler characteristics χ of a closed surface M to a metrical entity such as the total Gaussian curvature K for M. In applying (3.15) to supercoiled DNA, Fuller[89] in fact showed (3.15) to follow from the Gauss-Bonnet formula, one of the simplest examples of an index theorem. Thus it is pleasing to see how a fundamental physics principle, the relation between spin and statistics is mirrored by such a fundamental theorem of geometry, indeed the simplest of index theorem.

4. ANYONIC MEMBRANES

4.1 HOPF'S ESSENTIAL FIBRATIONS AND DIVISION ALGEBRAS

By 1935, Hopf [90, 91]discovered an unique link between topology and the four division algebras **K**= **R**, **C**, **H** and **Ω** , namely the real, complex numbers, the quaternions and octonions by connecting the latter and the fibrations of S^{2n-1} by a great S^{n-1}-sphere, n=1, 2, 4 and 8 respectively.

Hopf's construction [91],[90]of his maps is highly instructive. It can be directly inferred from Hurwitz's theorem which states: the only dimensions n of R^n with a multiplication $R^n \times R^n \rightarrow R^n$, denoted by $F(X,Y) = X\overline{Y}$ with $X\overline{Y}=0 \leftrightarrow X=0$ or $Y=0$ are n=1, 2, 4, 8. Namely these multiplications can be realized by the four division algebras over the reals **R**, **K** ≈ **R** , **C** ,**H** and **Ω** ,the real, complex numbers, quaternions and octonions respectively, X , Y ε **K** i.e. $R^n \approx$ **K**.

Next by a linear identification of the product space **K**x **K** with R^{2n}, the product F(X,Y) , X, Y ε **K**, defines a bilinear map, the <u>Hopf map</u>

$$H: R^{2n} \rightarrow S^{n+1} \tag{4.1}$$

with

$$H(X,Y) = (|X|^2 - |Y|^2, 2\,F(X,Y)) = (|X|^2 - |Y|^2, 2\,X\overline{Y}). \tag{4.2}$$

It follows that for $|X|^2 + |Y|^2 = 1$, $|H(X,Y)|^2 = (|X|^2 - |Y|^2)^2 + 4\,|XY|^2 = 1$. Considers two spheres, S^{2n-1} as the space of pairs (X, Y) of **K** with $|X|^2 + |Y|^2 = 1$ and S^n as the space of all pairs (s, **k**) of a real number $s = |X|^2 - |Y|^2$ and $\mathbf{k} = 2\,X\overline{Y}$ ε **K**. Thus H restricts to the map H: $S^{2n-1} \rightarrow S^n$ with S^{2n} and S^{n-1} as base space and fiber respectvely and S^{2n-1} as the fibre space, .

We parametrize S^n by a unit (n+1)-vector parametrizing $\vec{N}$, $\vec{N}^2=1$. Let $K^T = (K_1, K_2)$, K_1, K_2 ε **K**, $K^\tau K=1$, be a unit normed **K**-valued 2-spinor parametrizing S^{2n-1}. The Hopf map (4.1) then reads

$$\vec{N} = Sc\left(K^\dagger \vec{\gamma}\, K\right) \tag{4.3}$$

with $K^\dagger = \left(\overline{K}_1, \overline{K}_2\right)$ and $\gamma_\mu = \begin{pmatrix} 0 & e_\mu \\ \overline{e_\mu} & 0 \end{pmatrix}$, μ= 0, 1,...m-1 and $\gamma_m = \begin{pmatrix} 1 & 0 \\ 0 & -1 \end{pmatrix}$, m =1, 2, 4 and 8.

Alternatively, with S^{2n-1} in **K**x**K** and S^n = **K** ∪ {∞}, the Hopf projection map, $\pi : S^{2n-1} \rightarrow S^n$ also reads

$$\pi(X,Y) = \begin{cases} X/Y & \text{provided } Y \neq 0 \\ \infty & \text{if } Y=0 \end{cases} \tag{4.4}$$

where $|X|^2 + |Y|^2 = 1$, X, Y ε **K**. $\pi^{-1}(X,Y)$, the pre-image (or inverse) of this Hopf map, is the geometric intersection of S^{2n-1} with an n-subspace of **K**x **K**, namely a great (n-1) sphere S^{n-1} or a (n-1) *cycle.* So the image of any point on S^n is a S^{n-1}-sphere on S^{2n-1}. This is apparent since $\vec{N}$ or X/Y is invariant under the phase transformation $K \to KU$ ($X \to XU$, $Y \to YU$). $U = \bar{U}$, $|U|^2 = 1$, is a unit normed, pure imaginary **K**-number, i.e. $U \in S^0 \approx Z_2$, $S^1 \doteq U(1)$, $S^3 \doteq SU(2)$ and S^7, an (n-1)-cycle for n = 1, 2, 4 and 8 respectively.

(4.2) THE HOPF INVARIANTS AND ITS DISGUISES

The Hopf invariant $\gamma(\Phi)$ classifies the maps $\Phi : S^{2n-1} \to S^n$. As an added topological action in the model (3.1) (n=2) it is essential to a dynamical realization of exotic spin and statistics. Our work[81] is in essence about the many faces of $\gamma(\Phi)$, its mathematically different and physically telling expressions[92]. First there is the connection to the abelian Chern-Simons invariant.

Let $V^{(p)}(M)$ be the space of p-forms on a manifold M, $p \leq \dim M$. On S^n, we select a normalized n-form area element ω_n, $\int_{S^n} \omega_n = 1$. On S^{2n-1}, by pulling back the Hopf map F, we define a second induced n-form $\tilde{F}_n = \Phi^*\omega_n \in V^{(n)}(S^{2n-1})$ which is closed ($d\tilde{F}_n = 0$) since $d(F^*\omega_n) = F^*(d\omega_n) = 0$ and $d\omega_n = 0$. By de Rham's 2nd theorem $H^n(S^{2n-1}) \approx 0$, all closed n-forms on S^{2n-1} are exact, there is a non-unique (n-1)-form $\tilde{A}_{n-1} \in V^{(n-1)}(S^{2n-1})$ such that $d\tilde{A}_{n-1} = \tilde{F}_n$. So the integral

$$\gamma(\Phi) = \oint_{S^{2n-1}} \tilde{A}_{n-1} \wedge \tilde{F}_n \tag{4.5}$$

is defined. The following features hold:

a) $\gamma(\Phi)$ is independent of the choice of either $\tilde{A}_{n-1}$ ($d\tilde{A}_{n-1} = \tilde{F}_n$) or of ω_n, b) $\gamma(\Phi) = 0$ for all maps $\Phi : S^{2n-1} \to S^n$ with n odd, c) $\gamma(\Phi)$ is invariant for any two smooth and homotopic maps $S^{2n-1} \to S^n$.

(4.5) is the Whitehead form of the Hopf invariant $\gamma(\Phi)$. In physics this form is the Chern-Simons term for the Kalb-Ramond field $A_{n-1} \equiv 2\pi \tilde{A}_{n-1}$ and property (a) translates into the gauge invariance of this antisymmetric Abelian gauge field F.

There are variants of the Hopf invariant. Let us first parametrize the map $F: S^{2n-1} \to S^n$ by a (n+1) component unit vector $\vec{N} \in S^n$, ($\vec{N}^2 = 1$). If $\vec{N}_0$ is an arbitrary fixed point on S^n, then as in the case of the complex Hopf fibration , $\vec{N}(x) = \vec{N}_0$ is thus the equation of a closed hypercurve $C_0 \approx S^{n-1}$on S^{2n-1}. Equivalently, the preimage of $C_0 = F^{-1}(\vec{N}_0)$ of $\vec{N}_0$ is an (n-1)-cycle in S^{2n-1}. If S_0 is some n-dimensional closed connected submanifold on S^{2n-1} with, as its boundary δS_0, C_0, then $\vec{N}(x)$ maps S_0, a Seifert surface, onto the whole n-sphere. The Hopf invariant $\gamma(\vec{N})$ can be defined as the number of times $\vec{N}$ maps S_0 onto S^n. It is the mapping degree of $\vec{N}(x)$ restricted to S_0, from S_0 to S^n, $\vec{N}(x) : \Sigma_0 \to S^n$ and is independent of the point $\vec{N}_0$ of S^n. With $\pi_n(S^n) \approx Z$, the Hopf invariant is then an integer . By a theorem of Eilenberg and Niven, representative maps $S^n \to S^n$ for n=2,4 and 8 with winding number m are given simply by X^m with $X \in \underline{C}$, H and $\underline{\Omega}$ respectively. So we also have a generalized flux and loop integral representation of $\gamma(\vec{N})$

$$\gamma(\vec{N}) = \oint_{\Sigma^0 \sim S^n} \tilde{F}_n = \oint_{C_0 \sim \partial\Sigma^0} \tilde{A}_{n-1} \tag{4.6}$$

where $\tilde{F}_n = d\tilde{A}_{n-1}$ is the area element n-form of S^n mapped by $\vec{N}$ into S^{2n-1}. As it should be, these $\tilde{F}_n$ and $\tilde{A}_{n-1}$ are the same ones occuring in the Whitehead form of $\gamma(\vec{N})$. The Hopf invariant gives, upon exponentiation, a generalized Aharonov-Bohm-Berry phase factor associated with its antisymmetric U(1) gauge field.

The connection, due to Hopf himself, between his invariant and Gauss' linking number cannot be simpler : $\gamma(\Phi)$ was originally defined as a linking number ! The map **N** represents an element in $\pi_{2n-1}(S^n)$. Pick two distinct points N_1 and N_2 on S^n, then their pre-images $F(N_a) = C_a$ (a=1,2) are (n-1)-manifolds in S^{2n-1}. After assigning a natural orientation to these hypercurves

we get two (n-1)-spheres in S^{2n-1} or (n-1)-cycles C_1 and C_2. They can be linked or unlinked ; $\gamma(\alpha)$ is just the linking numbers $Lk(\alpha_1,\alpha_2)$ of C_1 and C_2 and depends only on α. $\gamma(\alpha)$ is thus a homomorphism :

$$H : \pi_{2n-1}(S^n) \to Z \tag{4.7}$$

with the generalized Gauss linking coefficient to be given.

We finally list some useful properties of the Hopf invariant :

a) For n odd, H is zero in consequence of the anticommutativity of linking numbers

b) For n even, Hopf proved that maps of an *even* H always exist.

c) If the map $\Gamma : S^{2-1} \to S^{2n-1}$ has degree p the $\gamma(\Phi \circ \Gamma) = p\,\gamma(\Phi)$.

d) If the map $\Psi : S^n \to S^n$ has degree q then $\gamma(\Psi \circ \Phi) = q^2 \gamma(\Phi)$,

where the degree of the map $S^n \to S^n$ is an element of $\pi_n(S^n)$.

4.3 COMBINING WHITE'S AND ADAMS' THEOREMS

In his 1969 thesis the mathematician White [93] derived the D>3 version of Calugareanu's formula as a byproduct of a reformulation of the Gauss-Bonnet-Chern theorem for Riemannian manifolds. In view of the established connection in section 3.2 it was natural to for us to extend Polyakov's approach to the D>3 counterparts of Wilcek-Zee σ-model (3.1). First we need to generalize Gauss linking number to higher dimensional manifolds.

Extending to D>3 manifolds the procedure for linking 3-space curves, we consider two continuous maps f(M) and g(N) from two smooth, oriented, non intersecting manifolds M and N, dim(M) = m and dim(N) = n, into R^{m+n+1}. Let S^{m+n} be a unit (m+n)–sphere centered at the origin of R^{m+n+1}. Let $d\Omega_m$ be the pull–back S^{m+n} volume form under the map $e : M \times N \to S^{n+m}$ where to each pair of points $(m,n) \in M \times N$ is associated the unit vector e in R^{m+n+1} : $e(m,n) = \frac{g(n)-f(m)}{|g(n)-f(m)|}$. The degree of this map

$$L(f(M),g(N)) \equiv L(M,N) = \frac{1}{\Omega_{n+m}} \int_{M \times N} d\Omega_{n+m}. \tag{4.8}$$

is the Gauss linking number of M and N. Ω_n $(=2\pi^{(n+1)/2}/\Gamma((n+1)/2))$ is the volume of S^n. Due to

The four rows reflect the one to one correspondence between the four division algebras over **R** and the real (**R**), complex (**C**), quaternionic (**H**) and octonionic (**W**) Hopf bundles (displayed in bold letters). The first three principal bundles are actually the simplest members of the three infinite sequences of the **K**= **R**, **C**, **H** universal Stiefel bundles over Grassmannian manifolds . The fourth bundle stands alone, a fact connected to the non-associativity of octonions.

The spheres S^p, p = 0, 1, 3 and 7 are the fibers,the first three are Lie groups while S^7 is a very special coset space, the space of the unit octonions. The latter has been an exotic object of fascination and discoveries in mathematics and in the Kaluza-Klein compactification of D=11 supergravity and supermembrane theories . An n-sphere S^n is parallelizable if there is a continous family of n orthonormal vectors at each its points. The fact that S^1, S^3 and S^7 are the only parallelizable spheres is yet another corollary to Adams' theorem. S^r, r = 1, 3, 7, 15 constitute. the corresponding fibre spaces. Finally the sequence of base spaces S^n, n = 1, 2, 4, 8 are equally interesting as **K**-projective lines, as is clear from their coset forms. With their holonomy groups Z_2, SO(2), SO(4) and SO(8) being the norm groups of **R**, **C**, **H** and **Ω** they can be said to have a real, complex, quaternionic and octonionic Kähler structures .

The Hopf maps $f : S^{2n-1} \to S^n$, n=1, 2, 4, 8 with *Hopf invariant one* have found important physical applications in condensed matter physics and in quantum field theory. Even the connection between Hopf maps and nonstandard spin and statistics had been lurking in the background. Thus, in the D=2 ϕ^4 field theory , it was shown that the n=1 real Hopf map realizes the 1-kink soliton , which carries intermediate spin and admits exotic statistics. Besides being the Dirac 1-monopole bundle, the n=2 complex Hopf map underlies the θ spin and statistics of D=3 CP(1) model. The n=4 quaternionic Hopf map is the embedding map for the SO(4) invariant, D=4 SU(2) BPST 1-instanton or the SO(5) invariant, D=5 SU(2) Yang monopole with eg=1/2. The n=8 octonionic Hopf map appears as a SO(9) invariant, D=8 SO(8) 1-instanton. The latter two maps admit further realizations in terms of U(1) tensor gauge fields associated with extended Dirac monopoles with eg = 1/2 in p-form Maxwellian electrodynamics.

We have noted that the field theory realizations of the **R**- and **C**- Hopf fiberings both admit exotic spin and statistics. It is then only natural to ask whether this pattern persist in suitable theories built on the two remaining Hopf fibrations, $S^{2n-1} \to S^n$ for n= 4, 8. Clearly the answer should be sought within the quaternionic D=7 HP(1) ($\approx S^4$) and the octonionic D=15 ΩP(1)≈ (S^8) σ-models augmented with their respective Hopf invariant term. We consider them next.

4.4 DIVISION ALGEBRAS AND SIGMA-MODELS WITH A HOPF TERM

In mathematics, the standard nonlinear σ-models are well known as *harmonic maps.* One associates with the map $\Psi : M \to N$ between two Riemannian manifolds M and N an action $S = \frac{1}{2}\int_M |d\Psi(x)|^2 \, d^m x$. $d\Psi(x)$ is the differential of Ψ at the point x ⊖ M and $d^m x$, the volume element of M. In a coordinate patch, $|d\Psi|^2 = g^{ij}\frac{\partial\Psi^\alpha}{\partial x^i}\frac{\partial\Psi^\beta}{\partial x^j}h_{\alpha\beta}$ is the pullback on M of the metric $ds^2 = h_{\alpha\beta}d\Psi^\alpha d\Psi^\beta$ on N. Ψ is called harmonic if it leads to a vanishing Euler-Lagrange operator (or tension field) $\mathrm{div}(d\Psi) \equiv 0$. The quadratic Hopf map $\Psi(X,Y) : S^{2n-1} \to S^n$ (n= 2, 4, 8) is in fact a harmonic polynomial map, with constant Lagrangian density $|d\Psi(x)|^2 = 2n$. As such it is the simplest harmonic representative of maps with Hopf invariant 1 .

While the D=3 CP(1) σ-model [83]) admits exact finite energy static solitons, the corresponding D=7 HP(1)($\approx S^4$) and the D=15 ΩP(1)($\approx S^8$) σ-models do not . This is clear from the Hobart-Derrick scaling argument . In practice, as in the Skyrme model , dynamical stability can be insured either by coupling the model to a gauge field or by adding to the standard KP(1) σ-model action with suitable chiral invariant terms of higher order in the field derivatives.

the non-commutativity property $L(M, N) = (-1)^{(m-1)(n-1)} L(N, M)$, $L(M,N)$ vanishes for even dimensional submanifolds M and N .

White's main theorem states: Let $f: M^n \rightarrow R^{D=2n+1}$ be an smooth embedding of a closed oriented differentable manifold into Euclidean (2n+1) space. Let v be an unit vector along the mean curvature vector of M^n . If n is odd (i.e. D=3, 7, 11,15 etc...) then

$$SL(f,f_\varepsilon) = \frac{1}{\Omega_{2n}} \int_{M \times M} d\Omega_{2n} + \frac{1}{\Omega_n} \int_M \tau^* \, dV \tag{4.9}$$

is the self-linking number of a hyper-ribbon. The latter consists of M^n and the same manifold deformed a small distance ε along v. The two terms on the RHS of (4.9) are respectively the generalized writhing and twisting numbers, W and T , of the hyper-ribbon. The cases of even n (D= 1,5,9,...) are of no interest to us since both W and T are zero and hence also SL =0.

The universality of the formula SL = W + T (3.15) mirrors that of Gauss-Bonnet-Chern theorem. As a possible physical application, we expect that for solitons in suitable D> 3 models White's general formula T = -W (mod Z) similarly links their spin and statistical phases . It would define and relate the twisting and writhing of odd dimensional closed S^3-, S^5-, S^7-... hyper-ribbons, the world volumes of topological S^2 -, S^4-, S^6- membranes solitons in D=7, 11, 15... dimensional spacetime respectively. The first problem is to cut down this infinity of choices ? What are the natural D >3 σ-model counterparts of (3.1) which may admit solitons with exotic spin and statistics ?

In seeking for exact analogs of θ-spin and statistics among D>3 extended objects, at least three key features of the CP(1) model (3.1) should be maintained: 1) the existence of topological solitons, 2) the presence in the action of an Abelian Chern-Simons-Hopf invariant, 3) the associated Hopf mappings $S^{2n-1} \rightarrow S^n$ include ones with Hopf invariant 1. The first two requirements are embodied in the time component of the key equation (3.12) . Upon integration over all of space of both members of this equation, one obtains the topological charge-magnetic flux coupling which is at the very basis of the fractional statistics phenomenon in (2+1) dimensions. As to the third requirement, essential to the proof of the fractional spin and statistics for one soliton, the following striking feature holds true for these Hopf mappings. While for any n even there always exists a map $f: S^{2n-1} \rightarrow S^n$ with only even integer Hopf invariant $\gamma(f)$, the existence question of Hopf maps of invariant 1 received the final answer in the celebrated theorem of Adams[94]:

If there exists a Hopf map Φ: $S^D \rightarrow S^{(D+1)/2}$ of invariant $\gamma(\Phi) = 1$, in fact of $\gamma(\Phi)$ = any integer, then D must equal 1, 3, 7 and 15 ($m = (D+1)/2 = 1, 2, 4$ and 8)

So there can only be four and only four classes of Hopf maps with $\gamma(\Phi) = 1$.
They along with their associated hidden (or holonomic) gauge field structures are best displayed through the following diagram of spheres bundles over spheres:

$$U(1) \approx SO(2)$$
$$\|$$
$$Z_2 = O(1) \approx S^0 \rightarrow S^1 \rightarrow S^1/Z_2 \approx RP(1) \approx SO(2)/Z_2$$
$$\|$$
$$SO(2) \approx U(1) \approx S^1 \rightarrow S^3 \rightarrow S^2 == CP(1) \approx SU(2)/U(1)$$
$$\|$$
$$SU(2) \approx Sp(1) \approx S^3 \rightarrow S^7 \approx SO(8)/SO(7) \rightarrow S^4 \approx HP(1) \approx Sp(2)/Sp(1)\psi Sp(1)$$
$$\|$$
$$Spin(8)/Spin(7) \approx S^7 \rightarrow S^{15} \approx Spin(9)/Spin(7) \rightarrow S^8 \approx \Omega P(1) \approx Spin(9)/Spin(8).$$

Taking the second alternative, the generic σ-model action with the added Hopf term then reads

$$S_{(n)} = \int_M \partial_\mu \vec{N}\cdot\partial^\mu \vec{N}\, d^{2n-1}x + \frac{\theta}{a}\int_M A_{n-1}\wedge dA_{n-1} + \text{ suitable Skyrme terms;} \qquad (4.10)$$

$$n = 4, 8, \quad M = S^7, S^{15}$$

where the unit vector N with K = H and Ω is given (4.3). The composite U(1) ATGF A_{n-1}, nonlocal in N, is local in the 2-spinor K (4.3). Its expression in terms of K will be given later.

The θ-term can be rewritten as

$$S_I = \frac{1}{(n-1)!}\int d^{2n-1}x\, J^{\mu_1\ldots\mu_{n-1}} A_{\mu_1\ldots\mu_{n-1}} \qquad (4.11)$$

i.e. an interaction of the potential A_{n-1} with the topological current $J_{n-1} = -\frac{(n-1)!\theta}{4\pi^2}{}^*F_n$ (n=4, 8). The latter's conservation and expression in terms of N will be shortly deduced solely from the field topology. Since the sources of J_{n-1} are charged solitons, we shall first determine what types of solitons are allowed in our KP(1) models.

In condensed matter physics our σ-models are the familiar Wegner's n-vector models . As field theories of a 5- and 9- unit-vector order parameter $\vec{N}$, they are the quaternionic and octonionic counterparts of the isotropic Heisenberg ferromagnet, albeit in rather exotic higher dimensions and with an added Chern-Simons terms. In consequence the nature and dimensionality of their allowed topological defects should only depend on the dimensionalities of the order parameters and of the compactified spacetime. They should obey the defect formula of Toulouse and Kleman [95].

Consider a topological defect of spatial dimension d' in D-space or D-Euclidean spacetime. To measure its homotopic charge, we need to completely "surround" this defect by a submanifold of dimension r such that d' + r + 1 = D. The meaning of the contribution 1 on the LHS of this last relation is evident for a vortex line; it corresponds to the distance in 3-space (D=3) between the line defect (d'=1) and its surrounding loop (r=1) . The topological charge labels the equivalence classes of the group $\pi_r(S^n)$ of mappings $S^r \to S^n$, of the spatial submanifold S^r into the space of the (n+1) unit vector order parameter $\vec{N}$. With r < n and $\pi_r(S^m) = 0$ for r < m , $\pi_m(S^m) = Z$, it follows that topologically stable ($\pi_r(S^n) \neq 0$) defects must have spatial dimension d' = D - 1 - r = D - (n+1). So there exist no stable defects for (n+1) > D and (n+1)<0 , but for 0 < (n+1) < D, the so called *triangle of defects* in the ((n+1), d) plane, there are defects of various kinds, points, vortices, membranes etc... Furthermore if D > 4 such as in Kaluza-Klein-typed theories <u>and</u> if r > m $\pi_r(S^m)$ is generally nontrivial, even a richer variety of defects are possible.

Applying the Toulouse-Kleman formula to our cases of (D, r=n) = (3, 2), (7, 4) and (15, 8) we find that the allowed topological defects in the CP(1), HP(1) and ΩP(1) σ-models (3.16) to be 0-, 2- and 6-membrane solitons, their topological charges being the generators of $\pi_n(S^n) \approx Z$, n=2, 4 , 8.

Since our solitons are charged 2- and 6- membranes, we expect the associated σ-models to possess a rank 3 and rank 7 topological conserved current $J^{\mu\rho\sigma}$ and $J^{\mu\rho\sigma\alpha\beta\gamma\lambda}$. Their conservation follows solely from the constraint $\vec{N}^2 = 1$, hence $N\cdot\partial_\mu N = 0$, and the fact that n the dimension of the unit vector $\vec{N}$ is less or equal the dimension D of spacetime. Since here (D, n) = (7, 5), (15, 9), the latter condition is satisfied. Indeed

$$\partial_{\mu_1} J^{\mu_n \mu_{n+1} \ldots \mu_D} = 0 \tag{4.12}$$

with $J^{\mu_n\mu_{n+1}\ldots\mu_D} = \varepsilon^{\mu_1\ldots\mu_D}\varepsilon_{\alpha_1\ldots\alpha_n}\left(\partial_{\mu_1}N^{\alpha_1}\ldots\partial_{\mu_{n-1}}N^{\alpha_{n-1}}\right)N^{\alpha_n}$

As with the CP(1) model, these 3 and 7-form conserved currents, suitably normalized, are just the D=7 and 15 Hodge duals of the respective 4- and 8-forms antisymmetric gauge fields $F_n = dA_{n-1}$ appearing in the Hopf invariant action in (3.17): $J_n = -\frac{n!\theta}{4\pi^2} {}^*F_{n+1}$, with the star operation denoting the Hodge dual ${}^*F_{\mu_1\ldots\mu_{n-p}} = \frac{1}{n!}\varepsilon_{\mu_1\ldots\mu_n} F^{\mu_{n-p+1}\ldots\mu_n}$.

The conserved current (4.12) can be converted into a conservation law. Two equalities are used : a) Stokes' theorem $\int_M d\omega = \int_{\partial M}\omega$, ω is a p-form and M is an oriented compact manifold with boundary ∂M, D = dim M = (p+1) and dim(∂M) = p; b) the relation between the divergence and the exterior derivative: $\partial_\alpha \omega^{\alpha}_{\mu_1\ldots\mu_p} = (-1)^{Dp}[{}^*d^*\,\omega]_{\mu_1\ldots\mu_p}$. With the latter identity 4.12) becomes $d^*J = 0$. Its integration over a (D-p+1)-dimensional manifold M with boundary ∂M gives

$$\oint_{(\partial M)} {}^*J = 0 \tag{4.13}$$

If ∂M consists of two spacelike hypersurfaces Σ (with dim(Σ) = D-p), connected by a remote timelike tube ∂T and since the topological current J in our σ-models is localized in space, the integral over ∂T vanishes and (4.13) gives the Lorentz invariant and conserved charge $Q = \int_\Sigma {}^*J$

, its value being independent of Σ. Applied to our KP(1) σ-models where the equations of motion , e.g.(3.12), forces a θ-dependent linear relation between topological charge and flux , for (D, p) =(3, 1) ( ) reduces to the Skyrmion winding number, the generator $\pi_2(CP(1)) \approx Z$

$$Q = \frac{-1}{4\pi}\int_{S^2} d\Sigma^{ij} F_{ij} = \frac{-1}{2\pi}\int_{S^1} dx^i A_i = C_1 \qquad (\text{for } q = p). \tag{4.14}$$

(Note that we are using Roman indices for the spacelike components.) It is also the first Chern index C_1 of the U(1) bundle, the complex $S^3 \rightarrow S^2$ Hopf fibration. For (D,p) = (7,3) and (15,7), the topological charges of the membrane S^4- and S^8- solitons and the generators of $\pi_4(HP(1)) \approx Z$ and $\pi_8(\Omega P(1)) \approx Z$ are similarly given for $\theta = \pi$ by

$$Q = \int_{S^3} J^{0i_1i_2} d\Sigma_{i_3\ldots i_6} = \frac{-1}{4\pi}\int_{S^4} F_{i_3\ldots i_6} d\Sigma^{i_3\ldots i_6} = \frac{-1}{2\pi}\int_{S^3} A_{i_3i_4i_5} d\Sigma^{i_3i_4i_5} \tag{4.15}$$

and

$$Q = \int_{S^7} J^{0i_1\ldots i_6} d\Sigma_{i_7\ldots i_{15}} = \frac{-1}{4\pi}\int_{S^8} F_{i_7\ldots i_{15}} d\Sigma^{i_7\ldots i_{15}} = \frac{-1}{2\pi}\int_{S^7} A_{i_7\ldots i_{14}} d\Sigma^{i_7\ldots i_{14}} \tag{4.16}$$

respectively.

That Q is equal to n, an integer can be seen through the mentioned gauge field connection between our problem and the D=2 complex, D=4 quaternionic and D=8 octonionic instanton. We take for the KP(1) field coordinate, the mapping $K(x) = x^n$, where x is the space position K-number in Σ, K = C, H and Ω. While these maps are not 0-, 2- or 6-membrane solutions to the systems (4.10), they are the simplest harmonic representatives maps : $S^m \to S^m$ (m =2, 4 and 8) with topological number Q= n . As will be clear the charges (4.15) and (4.16) can be identified with the 2nd and 4th Chern index which reflects the relation between the U(1) ATGF and the hidden non-Abelian gauge structure of the σ-models, namely the Sp(1) quaternionic and associated Spin(8) octonionic Hopf fibrations respectively. Though our subsequent analysis of the thin soliton limit deals primarily with the θ-term in (4.10), the Hopf term, to make clear the hidden gauge connection we now consider the HP(1) $\approx S^4$ model in greater detail. An analogous discussion of the ΩP(1) model can be carried out.

As the coset space Sp(2)/Sp(1)x Sp(1), the quaternionic projective line HP(1) can be parametrized in two ways (see Sect.2f.3 for details) . Either we have by two real quaternions q_1 and q_2 with $|q_1|^2+|q_2|^2 =1$, i.e by a 2-component H-spinor $Q^T = (q_1, q_2)$, $Q^\tau Q = 1$, coordinatizing the sphere S^7 or by one quaternionic inhomogeneous coordinate $h = q_2 q_1^{-1}$. An alternative parametrization is by the unit 5-vector **N** defined by the Hopf projection map (4.3) from S^7 to S^4, $\vec{N} = Sc\left(Q^\dagger \vec{\gamma} Q\right) = \left(N = \frac{2h}{1+\bar{h}h}, N_5 = \frac{1-\bar{h}h}{1+\bar{h}h}\right)$. To make manifest the local Sp(1) ≈ SU(2) gauge invariance

$$q_\alpha' = U(x)\, q_\alpha \qquad \alpha = 1, 2 \ ; \ U(x) \,\varepsilon\, Sp(1) \tag{4.17}$$

of the HP(1) model, we introduce the covariant derivative $D_\mu Q = (\partial_\mu + a_\mu)Q$. The holonomic Sp(1) gauge field is $a_\mu = \mathbf{a}_\mu\cdot\mathbf{e} = Q^\tau \partial_\mu Q = \frac{1}{2}\bar{q}_\alpha \partial_\mu q_\alpha = \frac{1}{2}\frac{\bar{h}\overleftrightarrow{\partial_\mu} h}{1+\bar{h}h}$ is purely vectorial and takes the ADHM form [96] for the 1-SU(2) instanton solution. So the first term in (4.10) reads

$$S_{(4)\,0} = Sc\left\{(D_\mu Q)^\dagger (D^\mu Q)\right\} \tag{4.18}$$

and similarly for the Skyrme terms.

As for the Hopf term, we can check that the 3-form $A_{(3)}$ to be the D=4 Sp(1) Chern-Simons form

$$\begin{aligned} A_{(3)} &= \frac{1}{3!}A_{[\mu\nu\lambda]}dx^\mu dx^\nu dx^\lambda \, , \\ &= Tr(A\wedge dA + \frac{2}{3}A^3) \end{aligned} \tag{4.19}$$

$$F_{(4)} = dA_{(3)} \tag{4.20}$$

where $dx^\mu\, dx^\nu = dx^\mu \wedge dx^\nu$ etc... In terms of the 2-spinor Q, they read

$$A_{(3)} = Sc\left\{Q^\dagger dQ\, dQ\, dQ + \frac{1}{3}(Q^\dagger dQ)^3\right\}, \tag{4.21}$$

$$F_{(4)} = Sc\left\{dQ^\dagger dQ + (Q^\dagger dQ)^2\right\}. \tag{4.22}$$

These forms clearly show the *local* nature of the Hopf term when written in terms of quaternionic -valued field Q of the bundle space S^7. It is thus locally a total divergence as is already clear from (4.6).

A parallel derivation of $A_{(7)}$ and $F_{(8)} = dA_{(7)}$ can be done for the D=15 $\Omega P(1)$ σ-model. In fact the connection to the D=8 octonionic instanton problem identifies the 7-form $A_{(7)}$ as the D=8 Chern-Simons term of a Spin(8) gauge field :

$$A_{(7)} = \mathrm{Tr}\left\{ A(dA)^3 + \frac{4}{3}A^3(dA)^2 + \frac{6}{5}A^5 dA + \frac{4}{7}A^7 \right\}. \tag{4.23}$$

The above specifics of the σ-models are sufficient for our analysis. Being essentially nonlinear, our models are analytically highly intractable in their field theoretic details. Besides, there is much arbitrariness in the choice of Skyrme terms which, being higher order in the field. derivatives, control the shorter distance solitonic structure. As in the D=3 case, the latter structure is not relevant to the problem of phase entanglements of the solitons. Only the existence but not the details of the Skyrme terms matter. It is enough to analyse the effective theories obtained in the geometrical Nambu-Goto limit of widely separated membranes.

Referring to [82] for the relevant details, we can show that our membrane solitons have a thin London limit . Thus Polyakov's approximation for the models (4.10) translates into the Chern-Simons-Kalb-Ramond electrodynamics of Nambu-Goto membranes. To regularate the ultraviolet divergences of the theory we can also add a Maxwellian kinetic terms for the antisymmetric gauge field.

To obtain the statistical phase [97], we consider the propagation of two pairs of membranes-antimembrane and compute the phase resulting from adiabatically exchanging the two membranes. We get

$$\left\langle \exp\left(\frac{i}{3!}\oint_{P_1} A^{[\mu\nu\lambda]} dx^\mu \wedge dx^\nu \wedge dx^\lambda\right) \exp\left(\frac{i}{3!}\oint_{P_2} A^{[\mu\nu\lambda]} dx^\mu \wedge dx^\nu \wedge dx^\lambda\right)\right\rangle. \tag{4.24}$$

P_1 and P_2 are S^3 hyper-curves. The functional average $\langle\ldots\rangle$ is taken over the Hopf action (9). As in D=3 case , the resulting phase here is the sum of three phases. The first contribution gives the phase factor $\exp\{2i\,(\pi^2/\theta)\,L\}$ with L being the generalized Gauss' linking coefficient (4.8) for two S^3-loops. We get π^2/θ for the statistical phase. The other two phase factors $\Phi(P_i)$ are given by the expectation value of one hyperloop:

$$\Phi(P) = \left\langle \exp\left(\frac{i}{3!}\oint_{P} A^{[\mu\nu\lambda]} dx^\mu \wedge dx^\nu \wedge dx^\lambda\right)\right\rangle . \tag{4.25}$$

In the London-Nielsen-Olesen limit the effective action reads [98]

$$S = S_0 + \frac{\theta}{(3!)^2 a}\int_{S^7} d^7x\, \varepsilon_{\mu\nu\lambda\alpha\beta\gamma\delta}\, A^{\mu\nu\lambda}\partial^\alpha A^{\beta\gamma\delta} + \frac{1}{3!}\int_{S^7} d^7x\, J_{\mu\nu\lambda}A^{\mu\nu\lambda} , \tag{4.26}$$

where S_0 is free Nambu-Goto action for a 2-membrane, $0 \leq \theta \leq \pi$ and a is a constant to be freely chosen . Direct integration of the equation of motion

$$J_{\mu\nu\lambda} + 2\frac{\theta}{3!a}\varepsilon_{\mu\nu\lambda\alpha\beta\gamma\delta}\partial^\alpha A^{\beta\gamma\delta} = 0 \tag{4.27}$$

with $J^{\mu\nu\lambda}(y) = \int d^3x\, \delta^7(x-y)\,\dfrac{\partial(x^\mu,x^\nu,x^\lambda)}{\partial(\tau,\sigma_1,\sigma_2)}$. , in the Lorentz gauge $\partial^\alpha A_{\alpha\beta\gamma} = 0$, gives

$$i\frac{1}{2\cdot 3!}\int d^7x\, J_{\beta\gamma\delta}A^{\beta\gamma\delta}(x) = = i\,\frac{a}{4\theta}\frac{1}{\Omega_6}\int_{S^3} d\Sigma_{\beta\gamma\delta}\int_{S^3} d\Sigma_{\mu\nu\lambda}\frac{\varepsilon^{\mu\nu\lambda\alpha\beta\gamma\delta}(x-y)_\alpha}{|x-y|^7} \quad . \tag{4.28}$$

Here the double 3-sphere integration is over one and the same hypercurve S^3 ; the phase (4.25) is *undetermined* unless we regulate[99] the short distance divergence say by including the Maxwellian kinetic term for the gauge field $A_{\mu\nu\rho}$. The regularized phase is then

$$\Phi\,(\,P \approx S^3) \;=\; \exp\left(i\,\frac{a}{4\theta}W(P)\right) \tag{4.29}$$

where $W(P) = \frac{1}{\Omega_6}\int_{S^3\times S^3} d\Omega_6$ is the writhe of the Nambu-Goto S^2-membrane tracing the Feynman path P, a S^3 hyper-ribon in 7-spacetimes. A parallel computation gives the same expression as (4.29) for Φ in the octonionic case of the 6-brane S^6 with $P \approx S^7$ in S^{15} -spacetime .

Setting $a = 4\pi^2$, $\Phi(P)=\exp(\pi^2 iW/\theta)$. Invoking White's formula (4.9) we obtain for $\theta = \pi$ the exact S^3- (S^7-) counterpart of Polyakov's phase factor $\Phi(P) = \exp(-\pi i\ T(P))\exp(\pi i\ n)$, T(P) being the generalized torsion for an S^3-(S^7) ribbon P. This phase factor presumably embodies the thin membrane's spin in a functional integral formalism. If this reasonable expectation is realized by an explicit construction à la Polyakov[80] of the spin factor for membranes, we will then have a 7- (15-) dimensional analog of the D=3 Fermi-Bose transmutation. With the value of θ not being fixed by the gauge invariance of the antisymmetric tensor gauge field, we have in general the possibility of fractional statistics and spin via the relation W = -T (mod Z) for solitonic membranes.

Finally, without knowing the short distance soliton structure or performing a detailed canonical quantization of the above $\mathbf{KP}_1$ σ-models, the case for the θ-spin and statistics among our Hopf- membranes can be made on topological grounds. We focus on the topology of the configuration space of fields Γ of the above KP(1) σ-model . In the Schrödinger picture, the space Γ of finite energy static solutions is the *mapping space* of all based preserving smooth soliton maps $\vec{N}(x)$: $x \in S^n \rightarrow \vec{N}(x) \in S^n$, n = 2, 4, 8 . Γ is an infinite Lie group with the nontrivial connectivity property:

$$\pi_0(\,\Gamma = \{\vec{N}: S^n \rightarrow S^n\}) \approx \pi_n(S^n) \approx Z. \tag{4.30}$$

So Γ is split into an infinite set of pathwise-connected components Γ_α, $\alpha \in Z$, corresponding to the various soliton sectors labelled by the charge Q. For our membranes, as with Skyrmions and Yang-Mills instantons, each sector Γ_α has further topological obstructions. G.W. Whitehead showed that all the Γ_α's in Γ have the *same* homotopy type i.e. $\pi_i(\Gamma_\alpha) \approx \pi_i(\Gamma_\beta)$. The relations

$$\pi_i\,(\Gamma_1) \;\approx\; \pi_i(\Gamma_0) \approx \pi_{i+n}(S^n\,) \approx Z \;\text{ for }\; (i, n) = (1, 2), (3, 4), (7, 8). \tag{4.31}$$

are of particular relevance to the question of exotic spin and statistics, for the 1–soliton sector . They result from the Whitehead and Hurewicz's isomorphisms, the latter stating $\pi_i(\Gamma_\alpha) \approx \pi_{i+n}(S^n)$, and reflect the *multi-valuedness* of Γ_i . (4.31) imply the possibility of adding to the KP(1) σ-model action a Hopf invariant $\gamma(\vec{N})$, the generator of the torsion free part of $\pi_{i+n}(S^n)$ ($\pi_3(S^2)\approx Z$, $\pi_7(S^4) \approx Z\oplus Z_{12}$ and $\pi_{15}(S^7) \approx Z\oplus Z_{120}$). Generalizing the CP(1) model {(i,n) = (1,2)}, the nontriviality of these $\pi_i(\Gamma_1)$ implies the possibilities of Aharonov-Bohm effects of a multiply connected configuration space Γ and signals for the membrane solitons the existence of a higher dimensional analog of a θ spin and statistics connection .

In the CP(1) case,upon a 2π rotation P of the Skyrmion or an interchange of two Skyrmions, the Hopf term induces a spin phase factor $\Phi(P) = \exp\{i\theta\} = \exp(i\, 2\pi s)$, s being the soliton spin .The equality $\theta = 2\pi s$ for this process of rotation is a physical realization of the homomorphism:

$$\pi_1(SO(2)) \approx \pi_3(S^2) \approx \pi_1(\Gamma_1) \approx Z. \tag{4.32}$$

It establishes the equality of the kinematically allowed exotic spin to the dynamically induced θ-spin by way of the Hopf term. Notably (4.32) is but a special case of the Hopf-Whitehead J-homomorphism $\pi_k(SO(n)) \approx \pi_{k+n}(S^n)$. Generally we have the following chain of homomorphisms :

$$\pi_i(\Gamma_1) \approx \pi_i(\Gamma_0) \approx \pi_{i+n}(S^n) \approx \pi_i(SO(n)) \approx Z \tag{4.33}$$

with (i, n) = (1, 2), (3, 4), (7, 8) . $\pi_3(SO(4)) \approx \pi_7(S^4) \approx Z$, $\pi_7(SO(8)) \approx \pi_{15}(S^8) \approx Z$. Clearly the most natural physical interpretation of these topological relations is a dynamically induced exotic spin and statistics connection for the 2-and 6-membranes.

The foregoing analysis represents a first assault on the problem of the spin and statistics connection for higher dimensional topological extended objects. It is a small step both in the bosonic functional integral formulation for spinning extended objects and in the study of the θ-vacuum phenomenon in Kaluza-Klein compactification. Paralleling the study of p-branes it would be of interest to supersymmetrize the above theories, to study the canonical quantization of anyonic membranes, attempt the construction of a thin membrane soliton operator, tackle the representation theory of associated algebras of diffeomorphisms with their specific central extensions. In a broader framework the systems discussed here are parts of higher dimensional topological field theories[100].

5. PARTING REMARKS

We have entered into a new phase of extensive developments and applications of algebraic methods to physics. In this review we try to illustrate in the context of field theoy some deep interconnections between topology, geometry, division algebras and the representation theory of certain infinite algebras. The unifying entity is a geometrical phase carried by solitonic excitations realized as projective representations of certain current algebras. Kac-Moody groups and their higher dimensional counterparts have provided the common thread for seemingly disparate areas of physics and mathematics.Why is there such unreasonable effectiveness of mathematics in accounting for physical phenomena ? Writing in the Notices of the American Mathematical Society, Weinberg[101] advanced a tantalizing explanation : " It is because some mathematicians have sold their soul to the Devil in return for advance information about what sort of mathematics will be of scientific importance". If he is right, some of us should perhaps consider taking this Faustian path in order to make significant headway on the problems outlined here.

ACKNOWLEDGMENTS: I particularly wish to thank all the authors, collaborators and friends whose works form the basis of much of this review. Any errors or misrepresentations are of course my own . This work was partially supported by the U.S. DOE under Grant DE-AS05-80ER10713

REFERENCES

1. H. Weyl, Symmetry ,Princeton University Press, Princeton, New Jersey (1952).
2. R. E. Marshak, in AAAS Symposium on " Symmetries across the Sciences -1990",
3. J. Dieudonné, Gazette des Mathematiques (1974) 73.
4. S. Adler and R. Dashen, Current Algebras and Applications to Particle Physics, Benjamin, N.Y. (1968).
5. Y. Ne'eman, Algebraic Theory of Particle Physics,W.A. Benjamin, Inc., N.Y. Amsterdam (1967).

6. S. Treiman, R. Jackiw, B. Zumino and E. Witten, Current Algebras and Anomalies, Princeton University Press, Princeton, N.J.(1985).
7. I. M. Gelfand, M. I. Graev and A. M. Versik, Compositio Math. 42 (1981) 217 .
8. A. Shapere and F. Wilczek, Geometric Phases in Physics, Advanced Series in Mathematical Physics, Vol.5 ,World Scientific (1989)
9. For a review , see R. Jackiw, MIT preprint CTP#1824 (December 1989)
10. Y. Aharonov and D. Bohm, Phys. rev. 115 (1959) 485 .
11. J. Mickelsson, Comm. Math. Phys. 110 (1987) 173 .
12. E. Witten, Comm. Math. Phys. 92 (1984) 455 .
13. P. Jordan, Z. Phys. 93 (1935) 465 .
14. P. Jordan, Z. Phys. (1937) 229 .
15. T. Goto and T. Imamura, Prog. Theor. Phys. 14 (1955) 396 .
16. J. Schwinger, Phys. Rev. Lett. 3 (1959) 296 .
17. S. Coleman, Phys. Rev. D11 (1975) 2088 .
18. S. Mandelstam, Phys. Rev. D11 (1975) 3026 .
19. A. Y. Morozov, Sov. Phys. Usp. 29 (1986) 993 .
20. M. A. Shifman, Sov. Phys. Usp. 32 (1989) 289 .
21. V. G. Knizhnik and A. B. Zamolodchikov, Nucl. Phys. B247 (1984) 83 .
22. S. Novikov, Russian Math. Surveys 37 (1981) 1 .
23. J. L. Cardy, in Phase Transitions and Critical Phenomena C.Domb, J. L. Lebowitz, Eds.,Academic Press, N.Y.
24. C. Itzykson, H. Saleur and J.-B. Zuber , Conformal Invariance and Applications to Statistical Mechanics ,World Scientific, Singapore (1988)
25. A. A. Belavin, A. M. Polyakov and A. B. Zamolodchikov, Nucl. Phys. 241B (1984) 333 .
26. M. Virasoro, Phys. Rev. D1 (1970) 2933 .
27. R. Bott, Enseign. Math. 23 (1977) 209 .
28. A. A. Kirillov, in Lecture Notes in Math. 1982 (Springer) 101
29. A. M. Polyakov, ZhETF Lett. 12 (1970) 538 .
30. V. Kac, Infinite Dimensional Groups with Applications,Springer-Verlag (1985).
31. P. Goddard and D. Olive, in Advanced Series in Mathematical Physics ,World Scientific, Singapore (1988).
32. V. G. Kac and D. H. Petersen, Adv. Math. 53 (1984) 125 .
33. C. M. Sommerfield, Phys. Rev. 176 (1968) 2019 .
34. H. Sugawara, Phys. rev. 170 (1968) 1659 .
35. I. G. Frenkel and V. G. Kac, Inventiones Math. 62 (1980) 23 .
36. I. Frenkel, in Proceedings of the International Congress of Mathematicians -1986, Eds. 821
37. R. S. Ismagilov, in Proceedings of the International Congress of Mathematicians (PWN , Warsaw, 1983), pp. 861.
38. U.-G. Meissner, Phys. Rep., 161C (1988) 213 .
39. J. Gipson and H. C. Tze, Nucl. Phys. B183 (1981) 524 .
40. I. Bars, in Vertex Operators in Mathematics and Physcs, J. Lepowski, S. Mandelstam and I. M. Singer, Eds.,Springer-Verlag (1985) , p. 373
41. S. K. Bose and S. A. Bruce, Univ. of Notre Dame Preprint 1990.
42. B. L. Feigin, Russian. Math. Surveys 35 (1980) 239 .
43. J. Mickelsson, Comm. Math. Phys. 97 (1985) 361 .
44. L. Faddeev and S. Shatashvili, Theor. Mat. Fiz. 60 (1984) 206 .
45. N. K. Pak and H. C. Tze, Ann. Phys. 117 (1979) 164 .
46. E. Witten, Nucl. Phys. B223 (1983) 433 .
47. G. Adkins, C. R. Nappi and E. Witten, Nucl.Phys. B228 (1983) 552 .
48. A. P. Balachadran, V. P. Nair, S. G. Rajeev and A. Stern, Phys. Rev. Lett. 49 (1982)1124 ; Phys.Rev. D27(1983)1153 .
49. A. P. Balachandran, in Proceedings of the Yale Theoretical Advanced Study Institute, Vol.1, M. Bowick and F.Gürsey , Eds. ,Word Scientific (1985), pp.1-81
50. T. H. R. Skyrme, Proc. Roy. Soc. A260 (1961) 127 .
51. T. H. R. Skyrme, Proc. Roy. Soc. A262 (1961) 237 .
52. T. H. R. Skyrme, J. Math. Phys. 12 (1971) 1735 .
53. S. G. Rajeev, Phys. Rev. D29 (1984) 2944 .
54. V. Bargmann, Ann. Math. 59 (1954) 1 .
55. E. Witten, Nucl. Phys. B223 (1983) 422 .

56. J. Mickelsson and S. G. Rajeev, Com. Math, Phys. 116 (1988) 365 . See in particular J. Mickelsson and S. G. Rajeev, Preprint UR-1178 -ER-13065-636 (July 1990)
57. A. N. Pressley and G. Segal, Loop Groups and Their Representations (Oxford University Press, Oxford, U.K. (1986).
58. M. Lüscher, Nucl. Phys. B236 (1989) 557 .
59. J. Mickelsson, Phys. Rev. D32 (1984) 436 .
60. V. Arnold, Ann. Inst. Fourier 16 (1966) 319 .
61. J. Marsden and A. Weinstein, Physica 7D (1983) 305 .
62. G. A. Goldin, R. Menikoff and D. H. Sharp, Phys. Rev. Lett. 51 (1983) 2246 .
63. I. Bars, USC Preprint USC-89/HEP14 (Oct.1989).
64. I. Bars, C. N. Pope and E. Sezgin, Phys. Lett. B210 (1988) 85 .
65. T. A. Arakelyan and G. K. Savvidy, Phys. Lett. B214 (1988) 350 .
66. D. B. Fuks, in Cohomology of Infinite -Dimensional Lie Algebras, F. I. Kizner, Ed. Nauka, Moscow (1984), pp. 329.
67. V. Arnold, Mathematical Methods of Classical Mechanics ,Springer-Verlag New York, Inc. (1978)
68. I. Bakas, Phys.Lett. B228 (1989) 57 .
69. F. Figueirido and E. Ramos, Phys. Lett. 238B (1990) 247 .
70. B. L. Feigin and D. B. Fuchs, Funct. Anal. Appl. 16 (1982) 114 .
71. F. Gürsey and H. C. Tze, Lett. Math. Phys. 8 (1984) 387 .
72. F. Gürsey, Private Communication and Yale preprint in preparation 1990
73. V. P. Nair and J. Schiff, (Columbia, 1990),
74. Y. Eisenberg, Phys. Lett. B236 (1989) 349 .
75. Y. Eisenberg, Phys. Lett. B235 (1989) 95 .
76. E. Witten, Comm. Math. Phys. 121 (1989) 351 .
77. For the earliest papers on "anyonic" phenomena see G. A. Goldin, R. Menikoff and D. H. Sharp, J.Math.Phys. 22 (1981) 1664 ; Phys.Rev. Lett. 51 (1983) 2246; G. A. Goldin and D. H. Sharp, Phys. Rev. D 28 (1983) 830; F. Wilczek, Phys. Rev. Lett. 48 (1982) 1144 ; Phys. Rev. Lett. 49 (1982) 957.
78. E. Witten, Comm. Math. Phys. 117 (1988) 353 .
79. A. M. Polyakov, Mod. Phys. Lett. A3 (1988) 325 .
80. A. Polyakov, Les Houches Lectures 1988).
81. C. H. Tze, Int. J. Mod. Phys. A3 (1988) 1959 .
82. C. H. Tze and S. Nam, Ann. Phys. 193 (1989) 419 .
83. F. Wilczek and A. Zee, Phys. Rev. Lett. 51 (1983) 2250.
84. M. J. Bowick, D. Karabali and L. C. R. Wijewardhana, Nucl.Phys. B271 (1986) 417 .
85. G. Semenoff and P. Sodano, Nucl. Phys. B328 (1989) 753 .
86. D. Karabali, CCNY Preprint CCNY-HEP-90/51
87. G. Calugareanu, Rev. Math. Pures Appl. 4 (1961) 588 .
88. F. H. C. Crick, Proc. Natl. Acad. Sci. USA 73 (1976) 2639 .
89. F. B. Fuller, Proc. Natl. Acad. Sci.USA 68 (1971) 815 .
90. H. Hopf, Fund. Math. 25 (1935) 427 .
91. H. Hopf, Math. Ann. 104 (1931) 637 .
92. C. Godbillon, Elements de Topologie Algebrique,Hermann, Paris (1971).
93. J. H. White, Am. J. Math. 91 (1969) 693 .
94. J. F. Adams, Ann. of Math. 72 (1960) 20 .
95. G. Toulouse and M. J. Kleman, J. Phys. Lett. 37 (1976) L-149 .
96. M. F. Atiyah, N. J. Hitchin, V. G. Drinfeld and Y. I. Manin, Phys. Lett. A65 (1978) 185 .
97. J. Grundberg, T. H. Hanson, A. Karlhede and U. Lindstrom, Phys. Lett. B218 (1989) 321 .
98. R. I. Nepomechie and A. Zee, in Quantum Field Theory and Quantum Statistics . (Fradkin Festschrift), I. A. Batalin et al., Eds., Adam Hilger, Bristol (1986).
99. T. H. Hansson, A. Karlhede and M. Rocek, Phys. Lett. B225 (1989) 92 .
100.G. T. Horowitz, Commun.Math. Phys. 125 (1989) 417.
101.S. Weinberg, Notices of the AMS (1986) 728 .

ALGEBRA AND GEOMETRY

IN THE THEORY OF MIXED STATES

Armin Uhlmann

University of Leipzig
Department of Physics
Leipzig, Germany

1. INTRODUCTION

It is my intention to show some aspects of the very the rich theory of mixed states. A more or less subjective selection of the topics for this purpose was necessary. A certain complement to this selection is reviewed in[1,2,3].

My very aim is an introduction to the idea of extending the concept of *topological phases* a la Berry[4] and Simon[5] to the spaces of mixed states. However, a considerable fraction of the material is 'common knowledge'. This is in particular true with chapters 2, 3, and 5. Also the content of the other parts (except of chapter 9 and the last topic in chapter 8) are essentially known though, perhaps, not so widely. The introduction to Berry's phase within chapter 3 touches only those properties the generalization of which to mixed states is known to me.

The problem of the topological phase is explained as a natural problem of transporting phases along curves (not necessarily and loops) within state spaces. The *naturality* is expressed by the fact that the result depends only on the curve along which the transport takes place, and on the general setting of Quantum Theory.

In this paper I generally consider finite dimensional objects (Hilbert spaces, algebras) mainly in order to minimize technicallities, and (hopefully) allowing easier reading. As a matter of fact almost all what is to be said allows for several extensions to much more general situations. One then, however, enters large areas of complicated and partly unsolved mathematical questions.

2. ELEMENTARY DEFINITIONS

The (pure) states of a quantum system can be described by the vectors of an Hilbert space $\mathcal{H}$. In accordance of what has been said previously we assume

$$\dim \mathcal{H} = n < \infty$$

where n is the dimension of $\mathcal{H}$ as a complex linear space, while its geometrical dimension is of course $2n$. The scalar product of $\mathcal{H}$ will be denoted by $< .,. >$.

An operator of $\mathcal{H}$ is nothing than a complex linear map from $\mathcal{H}$ into $\mathcal{H}$. In the finite dimensional case there is no need for further requirements.

Maps can be composed and this is equivalent to perform the product, AB , of two operators, A and B. Hence, in addition of being a complex linear space, the operators constitute an associative algebra (or ring). It is an algebra over the complex numbers, simply because one

Symmetries in Science V, Edited by B. Gruber *et al.*
Plenum Press, New York, 1991

can multiply an operator by a complex number to get another operator. If A is an operator on $\mathcal{H}$, one denotes by A^* its hermitian conjugate. Because of the antilinear operation $A \mapsto A^*$ which satisfies $(AB)^* = B^* A^*$ one may call the algebra of all operators a **-algebra*. (Indeed, it is the most elementary noncommutative *-algebra one can imagine.) There is a standard notation: The algebra of all bounded operators acting on an Hilbert space is denoted by $\mathcal{B}(\mathcal{H})$. In our case of finite dimension every operator is bounded trivially.

An operator A is called *positive* (more accurate *positive semidefinite*) if $< \psi, A\psi >$ is non-negative real for all $\psi \in \mathcal{H}$. One then writes $A \geq 0$. Clearly, A is positive iff A is hermitian and all its eigenvalues are non-negative. A is called *strictly positive* (or *positive definite*) if A is positive and one of the following conditions is fulfilled: a) A^{-1} exists, b) $< \psi, A\psi >= 0$ implies $\psi = 0$, c) all eigenvalues are positive.

For every positive real number $s > 0$ there is one and only one positive s–th power A^s of a positive operator A. As an hermitian operator is uniquely given by its eigenvectors and its eigenvalues, its definition may be given by

$$A^s\psi = a^s\psi \quad \text{if} \quad A\psi = a\psi$$

$$A^s\psi = 0 \quad \text{if} \quad A\psi = 0$$

Remembering $a^s = s \ln a$ with real $\ln a$, this definition could be extented to all complex numbers. This is well done for strictly positive A but comes otherwise in conflict with the notation of the inverse A^{-1} if $\Re s < 0$. In the latter case one speaks of the *operational defined* s–th power if one uses the definition above.

Evidently

$$\lim A^s \to P \quad \text{for} \quad s \to +0$$

is a projection operator. P is called *support* of A and is denoted by supp$A$, while I shall call *carrier* of A the subspace $P\mathcal{H}$ whereon A acts non-trivially. (This distinction of support and carrier is not generally in use.)

Now consider an arbitrary operator A . Then AA^* and A^*A are positive. The support of AA^* (of A^*A) is the left support, l–supp, (resp. right support, r–supp,) of A . There is the notation

$$\text{carrier}(A) := \text{carrier}(A^*A) \quad \text{and} \quad \text{range}(A) := \text{carrier}(AA^*)$$

The orthogonal complement of carrier(A) is the *kernel*, ker(A), of A . It will be convenient later on to use the *polar decomposition* of an operator in the (unconventional but equivalent) form

$$A = (AA^*)^{1/2} U$$

where U is a partial isometry (i.e. UU^* and U^*U are projection operators) with

$$UU^* = \text{supp}\, AA^* = \text{l–supp}\, A$$

$$U^*U = \text{supp}\, A^*A = \text{r–supp}\, A$$

3. PURE STATES

As is explained in every textbook, the vectors ψ of $\mathcal{H}$ describe pure states — with the exception of the zero vector. The same state is described by ψ as well by $\lambda\psi$, where λ is any non–zero complex number. This means the pure states are in one to one correspondence to the rays $\{\lambda\psi, \lambda \in \mathcal{C}\}$, or, equivalently, to the 1–dimensionsional subspaces of $\mathcal{H}$. These 1–dimensional subspaces form a smooth manifold called complex projective space, $\mathcal{CP}^{(n-1)}$, or, remembering its inventor, the first Grassmann manifold of $\mathcal{H}$. It is, however, somewhat easier to handle operators than subspaces or rays. Therefore one equally well characterizes a pure state by the projection operator projecting $\mathcal{H}$ onto the 1–dimensional subspace in question. Thus the points of $\mathcal{CP}^{(n-1)}$ may and will be in what follows defined as the operators P of the form

$$P = P_\psi = \frac{|\psi >< \psi|}{< \psi, \psi >} \tag{1}$$

It follows from (1) for every operator A

$$PAP = \frac{<\psi, A\psi>}{<\psi,\psi>}P \tag{2}$$

where the coefficient on the right hand side is the *expectation value* of A if the system is in the state P .

A given 1–dimensional projection (1) does not only characterize a pure state uniquely, it is at the same time an observable asking wether the system is in the state (1) or not. Hence given two pure states, P_1, P_2, one considers the expectation value of P_2 in the state P_1 . This expectation value is the well known *transition probability* :

$$\text{tprob}(P_1, P_2) = \frac{<\psi_1, P_2\psi_1>}{<\psi_1,\psi_1>} = \frac{<\psi_1,\psi_2><\psi_2,\psi_1>}{<\psi_1,\psi_1><\psi_2,\psi_2>} \tag{3}$$

which is a symmetrical quantity in its arguments.

Geometrically, the transition probability governs the relative position of two points of the space of pure states: Two pairs of pure states are unitarily equivalent if and only if they have equal transition probability (see below).

At that point it seems we have lost all relative phases. But this is not so. We lost the *superficial* phases, all others remain. This can be seen by solving the following problem: Given two m-tuples

$$P_1, P_2, \ldots, P_m \quad \text{and} \quad Q_1, Q_2, \ldots, Q_m \tag{4}$$

of pure states. The question is wether there exists a unitary operator U with

$$Q_j = UP_jU^{-1}, \qquad j = 1, 2, \ldots, m \tag{5}$$

One may assume that P_j (resp. Q_j) is given by the normalized vector ψ_j (resp. φ_j) . If

$$<\psi_j, \psi_k> = <\varphi_j, \varphi_k> \quad \text{all } j, k$$

then the map $\psi_j \mapsto \varphi_j$ can be extended isometrically to a map of the linear span of the ψ_j onto the linear span of the φ_j . As both spans are of equal dimensions it is elementary to extend the map to an isometry of $\mathcal{H}$ onto $\mathcal{H}$, i.e. to a unitary map U with $U\psi_k = \varphi_k$. But φ_k and $\epsilon_k\varphi_k$ with $|\epsilon_k| = 1$ define the same pure state. Therefore it is sufficient (and necesary) for the validity of (5) to have

$$<\psi_j, \psi_k> = \epsilon_j\epsilon_k <\varphi_j, \varphi_k> \quad \text{all } j, k$$

with some phase factors ϵ_j . This is true iff for all $2 \leq k \leq m$ the numbers

$$<\psi_{i_1}, \psi_{i_2}><\psi_{i_2}, \psi_{i_3}> \cdots <\psi_{i_k}, \psi_{i_1}> \tag{6}$$

remain unchanged after replacing the ψ–vectors by the φ-vectors. A number (6) is either zero or it is the only non-zero eigenvalue of the partial symmetry

$$P_{i_1}P_{i_2} \cdots P_{i_k} \tag{7}$$

One concludes that (5) is valid if and only if

$$spec\{P_{i_1}P_{i_2} \cdots P_{i_k}\} = spec\{Q_{i_1}Q_{i_2} \cdots Q_{i_k}\} \tag{8}$$

for all subsets $i_1, \ldots, i_k$ of $1, \ldots, m$ with $k \geq 2$. (*spec* denotes "spectrum of".) In particular, an essential relative phase appears in describing the relative positions of at least three pure states.

4. TRANSPORT OF PURE STATES

A partial isometry of rank one, V, can be written as

$$V = |\psi_{\rm out}><\psi_{\rm in}|, \quad VV^* = P_{\rm out} \quad V^*V = P_{\rm in} \tag{1}$$

with two normed vectors, and it can be interpreted as annihilating $P_{\rm in} = |\psi_{\rm in}><\psi_{\rm in}|$ and creating $P_{\rm out} = |\psi_{\rm out}><\psi_{\rm out}|$. Given the in- and out-states this operation is fixed up to a phase factor because there is no other invariant then tprob$(P_{\rm in}, P_{\rm out})$. This slight arbitrariness cannot be removed without introducing a new structural element.

This new structural element is a curve, $\mathbf{c}$, connecting smoothly $P_{\rm in}$ and $P_{\rm out}$:

$$\mathbf{c} : s \mapsto P_s, \quad 0 \leq s \leq 1, \quad P_0 = P_{\rm in}, \quad P_1 = P_{\rm out} \tag{2}$$

By the aid of the construction at the end of chapter 3 it is now possible to fix the phase factor in dependence on $\mathbf{c}$. To this end one takes subdivisions

$$1 > s_1 > s_2 > \ldots > s_m > 0 \tag{2}$$

of the parameter s of the curve and perform

$$V = V(\mathbf{c}) := \lim P_{\rm out} P_{s_1} P_{s_2} \ldots P_{s_m} P_{\rm in} \tag{4}$$

where the limiting procedure is taken over finer and finer subdivisions of (3). To calculate V one uses a *lifted* path

$$\mathbf{c}^{\rm lift} : \quad s \mapsto \psi_s, \quad 0 \leq s \leq 1, \quad \text{with} \quad P_s = |\psi_s><\psi_s| \tag{5}$$

of unit vectors with which (4) is converted into

$$V = |\psi_1><\psi_0| \lim <\psi_1, \psi_{s_1}><\psi_{s_1}, \psi_{s_2}> \ldots <\psi_{s_m}, \psi_0> \tag{6}$$

It is convenient to require

$$<\psi_s, \dot{\psi}_s> = 0 \tag{7}$$

before performing (6). (The dot indicates s-differentiation.) A lift (5) fulfilling (7) is called a *parallel lift*. Already in the early days of the adiabatic theorems of Born, Fock, and Oppenheimer condition (7) has been in use, and presently it is known as the condition for the *natural parallel transport* in the unit ball of $\mathcal{H}$ after the work of Berry and Simon. For the time being, however, (7) appears as a technical device, and the result of (6) does *not* depend on it.

From (7) and two times differentiability of (5) one estimates by the help of Taylor's theorem

$$|1 - <\psi_s \psi_t>| \leq (s-t)^2 \,\text{const.} \tag{8}$$

where the constant is independent of s and t. But this estimate implies in (6) that the limit goes to one. Hence assuming (7) one gets

$$V(\mathbf{c}) = |\psi_{\rm out}><\psi_{\rm in}|, \quad \psi_{\rm in} = \psi_0, \quad \psi_{\rm out} = \psi_1 \tag{9}$$

Now one may relaxe from condition (7) and obtains

$$V(\mathbf{c}) = |\psi_{\rm out}><\psi_{\rm in}| \exp \int <d\psi, \psi> \tag{10}$$

where the integral can be taken over an arbitrary lift (5).

For the proof one shows the indepenmdence of expession (10) with respect of regaugings $\psi_s \mapsto \epsilon(s)\psi_s$ with unimodular $\epsilon(s)$. Then it remains to see that (10) reduces to (9) for parallel lifts (7).

$V(\mathbf{c})$ does not depend on the way $\mathbf{c}$ is parametrized. It is a function of the curve as a 1-dimensional submanifold of $\mathcal{CP}^{n-1}$ and of its orientation. Reversing the orientation, and hence interchanging the in- and out-state gives the curve $\mathbf{c}^{-1}$. Obviously

$$V(\mathbf{c}^{-1}) = V(\mathbf{c})^* \tag{11}$$

Let $\mathbf{c}_1$ and $\mathbf{c}_2$ be two curves for which the out-state of second coincides with the in-state of the first. Then $\mathbf{c}_1\mathbf{c}_2$ is a well defined curve and

$$V(\mathbf{c}_1\mathbf{c}_2) = V(\mathbf{c}_1)V(\mathbf{c}_2) \tag{12}$$

V as defined by (4) satisfies a differential equation where the variable is the out-state. Considering the curve

$$\mathbf{c}_s : t \mapsto P_s, \quad 0 < t \leq 1 \tag{13}$$

and the corresponding

$$V_s := V(\mathbf{c}_s) \tag{14}$$

one uses (12) and (4) or (10) to arrive at

$$\dot{V}_s = \dot{P}_s \dot{V}_s \quad \text{or} \quad \mathrm{d}V = (\mathrm{d}P)V \tag{15}$$

Here the total differential notation in the second equation expresses the irrelevance of the choice of the parameter: $\mathrm{d}V$ restricted to a curve $s \mapsto V_s$ results in $\dot{V}_s \mathrm{d}s$.

It is $P_s V_s = V_s$ according to (13) and (4). Thus, surpressing indices,

$$V^*\dot{V} = V^*\dot{P}V = V^*P\dot{P}PV$$

But differentiating $P = P^2$ results in $\dot{P} = \dot{P}P + P\dot{P}$. Multiplying by P gives $P\dot{P}P = 0$. Finally one gets

$$V_s^*\dot{V}_s = 0 \quad \text{or} \quad V^*\mathrm{d}V = 0 \tag{16}$$

There is an interesting geometrical interpretation of (16). The manifold of rank one isometries can be considered as a fibre bundle with the manifold of rank one projection operators as base manifold. The bundle projection, π, is given by

$$\pi : V \mapsto VV^*$$

A curve $\mathbf{c}$ in the base space can be lifted to a path in the rank one isometries. (16) is the condition for being a parallel lift. (16) implies (15) and vice versa, while (10) gives the explicite construction of parallel lifting.

It should be obvious, last not least, that a closed curve (2), a loop with $P_{\text{in}} = P_{\text{out}}$, gives Berry's construction. It is $V(\mathbf{c}) = \epsilon P_{\text{in}}$ with Berry's phase factor

$$\epsilon = \text{Berry}(\mathbf{c}) = \exp \oint < \mathrm{d}\psi, \psi > \tag{17}$$

while the exponent (divided by i) is the Berry phase.

(17) is the trace of $V(\mathbf{c})$ for a closed loop. Otherwise (10) indicates that this trace depends generally only on the curve, wether this curve is closed or not, and takes values within the unit disk.

5. STATES

Pure states are not the only ones which can be attained by a quantum system with Hilbert space $\mathcal{H}$. On the contrary they form only a thiny (i.e. low dimensional) subset within the state space. The latter will be introduced now.

A *state*, ϱ, is a prescription that associates to every "observable", A, an *expectation value* $\varrho(A)$

$$A \mapsto \varrho(A), \qquad A = A^* \in \mathcal{B}(\mathcal{H}) \tag{1}$$

fulfilling three axioms:

1) Linearity.

$\varrho(A)$ is a real linear functional on the linear space of the hermitian operators.

2) Positivity.

$$\varrho(A) \geq 0 \quad \text{if} \quad A \geq 0 \tag{2}$$

3) Normalization.

$$\varrho(\mathbf{1}) = \mathbf{0} \tag{3}$$

Now every operator A can be uniquely decomposed into an hermitian and an antihermitian part, i.e. $A = A_1 + iA_2$ with hermitian A_j. One therefore extends the definition of ϱ by

$$\varrho(A_1 + iA_2) = \varrho(A_1) + i\varrho(A_2)$$

to get a complex linear functional on $\mathcal{B}(\mathcal{H})$ and not only on its hermitian part. One then calls $\varrho(A)$ an *expectation value* even if A is not hermitian — though, physically, there is no measurement apparatus for that if the hermitian and the antihermitian part of A do not commute. With this abstraction one usually replace the first requirement by

1') Linearity.

$\varrho(A)$ is a complex linear functional on $\mathcal{B}(\mathcal{H})$

and so I will do in this paper always.

The normalization condition is good for two things. At first it prevends ϱ to be identical zero. Namely, like for state vectors in the Hilbert space the zero is never a state. Secondly, every state is counted one time. If the normalization condition is neglected one calls an objekt with axioms 1 and 2 a *positive linear functional.* (By the by, there is no logical distinction between "functional" und "function" if one not bounds the former to linearity, what is not usual in Physics.)

The set of all states is called *state space* and will be denoted by

$$\Omega \quad \text{or} \quad \Omega(\mathcal{H})$$

This is a closed subspace of the linear space $\mathcal{B}(\mathcal{H})^*$ of all linear functionals on $\mathcal{B}(\mathcal{H})$, and it gains very natural its structure from this imbedding. It is also useful to consider Ω as a subset of the real linear space of real linear functionals on the hermitian part of $\mathcal{B}(\mathcal{H})$.

It results from the axioms that Ω is a *compact convex* set: Let $p_1, \ldots, p_k$ be a set of reals with

$$p_i \geq 0 \qquad \sum p_j = 1 \tag{4}$$

Then for every choice $\varrho_1, \ldots, \varrho_k$ of states the linear form

$$\varrho = \sum p_j \, \varrho_j \tag{5}$$

is again a state. This explains the *convex* (or *affine*) structure of Ω. One calls (4), (5) a *convex linear combination* or, in more physical terms, a *Gibbsian mixture* of the states $\varrho_1, \ldots, \varrho_k$ with *weights* or *probabilities* $p_1, \ldots, p_k$. If (5) is valid and $p_j > 0$ then one calls ϱ_j *dominated* by ϱ.

An often more convenient representation of states is given by densitiy operators and this description is isomorphic to that given by linear functionals. Here the starting point is the existence of exactly one linear functional τ satisfying

$$\tau(\mathbf{1}) = 1 \quad \text{and} \quad \tau(AB) = \tau(BA) \tag{6}$$

for all linear operators A, B acting on the Hilbert space, and it is necessarily given by the trace

$$\tau(A) = n^{-1} \operatorname{tr} A, \quad n = \dim \mathcal{H} \tag{7}$$

The uniqueness is a consequence of the statement that the trace of an operator A is zero if and only if A is a linear combination of commutators $BC - CB$. That in turn is equivalent to state that the complexified Lie-algebra of SU(n) is simple and consists of all traceless operators. But also explicit verification is not difficult.

τ as defined by (6) and given by (7) is a state, the unique *tracial state* of our state space: The trace of a positive operator different from the zero operator is positive. While the definition of the trace does not depend on the Hilbert space structur, the property to be a positive functional does. There would be no state at all if the *-operation is given by an indefinite, non-degenerate hermitian scalar product.

Next, every $B \in \mathcal{B}$ defines one and only one linear functional

$$A \mapsto \mathrm{tr}(AB) \tag{8}$$

and because $\mathcal{B}$ and $\mathcal{B}^*$ have equal dimensions, every linear functional can be gained by a suitable B. Moreover, $\mathrm{tr}(AB)$ will be positive for all positive A iff B itself is positive. Hence to every state ϱ there is just one operator, ϱ_{op}, with

$$A \mapsto \varrho(A) = \mathrm{tr}\, \varrho_{\mathrm{op}} A \tag{9}$$

and

$$\varrho_{\mathrm{op}} \geq 0 \qquad \mathrm{tr}\varrho_{\mathrm{op}} = 1 \tag{10}$$

Every such ϱ_{op} is called *density operator*, and there is a complete correspondence between states and density operators. Having this in mind no explicit notational distiction will be made between state and density operator, and the index *op* will be neglected usually. As the eigenvalues of a density operator are not negative and sum up to one they constitute a probability vector.

The most simple density operator is a 1-dimensional projection operator P characterizing a pure state.

6. SOME CONVEX GEOMETRY

Above I considered states as individual objects which form, as a compact comvex set, the state space Ω. One may reverse this setting and look at a state as a point of the state space, and one may ask to what degree its position in Ω determines its properties. In other words, one wonders how much is encoded in the inner geometry of Ω, forgetting for a moment the nature of its points as linear, normed, and positive functionals or density operators. The symmetry group Aut(Ω) for this task consists of the *affine automorphisms* ϕ , i.e. linear mappings from Ω onto Ω fulfilling

$$\phi(\sum p_j \varrho_j) = \sum p_j \phi(\varrho_j) \tag{1}$$

for all convex linear combinations. Every unitary operator, U, gives rise to an element of Aut(Ω) by

$$\varrho \to \varrho^U \quad \text{with} \quad \varrho^{\mathrm{U}}(\mathrm{A}) = \varrho(\mathrm{UAU}^{-1}) = (\mathrm{U}^{-1}\varrho\mathrm{U})(\mathrm{A}) \tag{2}$$

These transformations form a subgroup $\mathrm{Aut}_{\mathrm{U}}(\Omega)$ of Aut(Ω). In fact, $\mathrm{Aut}_{\mathrm{U}}$ is a normal subgroup of Aut of index two as will be seen soon.

According to general terminology a state ϱ is called *extremal* iff for every of its decompositions (5-4), (5-5) one necessarily has $\varrho = \varrho_k$ whenever $p_k > 0$. The set of all extremal elements of a convex set is called its *extremal boundary*.

The pure states are exactly the extremal points of Ω.

Indeed, every density operator allows for a decomposition

$$\varrho = \sum \lambda_j P_j \tag{3}$$

as a convex set of 1-dimensional projection operators. Well known examples of (3) provide spectral decompositions whith mutually orthogonal projectors where the λ_j are the corresponding eigenvalues of ϱ. The decompositions (3) show Ω as the convex span of the pure states. But if P is both, pure and extremal, then so does P^U with unitary U because both properties are stable against the subgroup Aut_U of unitary transformations. But the unitary transformations act transitve on the set of pure states. Hence every pure state has to be extremal.

An extremal point of a convex set is a particular case of what is called a *face*. A convex subset Ω_0 of Ω is called a *face* of Ω if $\varrho \in \Omega$ implies $\tilde{\varrho} \in \Omega$ whenever $\tilde{\varrho}$ is dominated by ϱ.

One knows the following: There is a one-to-one correspondence between faces and subspaces of $\mathcal{H}$. For every face Ω_0 there is a subspace $\mathcal{H} \subset \mathcal{H}$, with

$$\Omega_0 = \{\varrho : \text{carrier}\varrho = \mathcal{H}_0\} \tag{4}$$

The infinite-dimensional case is much mor complicate[1]4

Let now be $P_1, P_2 \in \Omega$ two different pure states and let a denote their transition probability:

$$P_1 P_2 P_1 = a P_1 \quad \text{with} \quad a = \text{tprob}(P1, P_2) \tag{5}$$

Considering a Gibbsian mixture (convex sum)

$$\varrho = \mu_1 P_1 + \mu_2 P_2 \tag{6}$$

one gets by taking its square and then the trace

$$\lambda_1^2 + \lambda_2^2 = \mu_1^2 + \mu_2^2 + 2a\mu_1\mu_2 \tag{7}$$

where λ_1, λ_2 denote the eigenvalues of ϱ. As the latter sum up to one as is the case with the coeficients μ_j equation (7) is equivalent to

$$\lambda_1 \lambda_2 = \mu_1 \mu_2 (1 - a) \tag{8}$$

From these elementary observations follow some important conclusions. At first, if ϱ is given by (6), one may ask for all possible decompoitons of ϱ as a comvex sum of two pure states. Then the product of the coefficients $\mu_1\mu_2$ takes its minimum if and only if $a = 0$, i.e. iff the coefficients of the Gibbsian mixture coincide with the eigenvalues. Hence the eigenvalues of ϱ are determined by its position in Ω.

Because of this, however, (7) or (8) shows that trans(P_1, P_2) is uniquely fixed by the relative position of the two pure states within Ω. Thus what is physically relevant of the Hilbert space structure of $\mathcal{H}$ is encoded in the geometry of Ω.

In particular it should be possible to measure transition probabilities of pure states by performing mixtures. (Let $\mu_1 = \mu_2$ then the transition probability equals $(\lambda_1 - \lambda_2)^2$.)

Furthermore, the transition probability between two pure states remains unchanged under the action of an element of $\mathrm{Aut}(\Omega)$. Applying now a well known theorem of Wigner one concludes that every element of $\mathrm{Aut}(\Omega)$ is induced either by a unitary or an antiunitary transformation of $\mathcal{H}$ and the affine automorphisms of the state space reflect the essentials of the Hilbert space transformations.

It is interesting to note that an explicite characterization of the semigroup of affine endomorphisms is not known. Physically it is meaningful to consider restricted classes of them (in particular the duals of comletely positive unital transformations of $\mathcal{B}$). However, a remarkable exception can be found at the end of chapter 8.

Now one may pose the questions: How to extend the concept of transition probability to general states, and how to extend equations (7, (8) to them. I shall explain the answer to the first question later, and give a partial answer to the second.

If ordered in decreasing order, (7) or (8) gives

$$\lambda_1 \geq \mu_1 \geq \mu_2 \geq \lambda_2$$

and $\lambda_1 = \lambda_2$ iff $a = 0$, i.e. iff $P_1 P_2 = 0$. Considering the general case

$$\varrho = \sum \mu_j \varrho_j, \qquad \mu_1 \geq \mu_2 \geq \dots \geq 0 \tag{9}$$

and denoting by $\lambda_1 \geq \lambda_2 \geq \ldots \geq 0$ the eigenvalues of ϱ in decreasing order, one finds[15,2]

$$\text{for all k}: \quad \sum_{j=1}^{k} \lambda_j \geq \sum_{j=1}^{k} \mu_j \tag{10}$$

and equality holds for all k iff (9) is an orthogonal decomposition with the eigenvalues of ϱ as coefficients. (If the λ- and the μ-vectors are of different length, one adds zeros to compensate this.)

For the proof of (10) one proceeds in the following steps. At first a set of inequalities of the form (10) defines a partial order within the probability vectors appearing in decompositions (9) of a fixed density operator. To every such vector there is at least one maximal element with respect to the partial order. Assumings such a maximal element is not an orthogonal decomposition it follows at least for two pure states, P_i, P_k with $i < k$ a positive transition probability. But this is impossible: Replacing $\mu_i P_i + \mu_k P_k$ by their orthogonal decomposition shows (after some calculations) that a not orthogonal decomposition cannot go with a maximal element of the partial order.

It is more complicated to derive handy relations similar to (7) or (8) for a general decompostion (9). To do so one has to characterize the relative positions of the pure states involved. This demands to know all the numbers

$$\operatorname{tr} P_{i_1} \ldots P_{i_k}$$

constructed from subsets of the $P_1, P_2, \ldots$. This is clearly possible in an implicite form by calculating the trace of the powers of the density operator ϱ as given by (9).

7. Subsystems. Reductions

As before let $\mathcal{B}(\mathcal{H})$ be the algebra of operators acting on $\mathcal{H}$. Within sektion 3 the pure states have been described by normed vectors $\psi \in \mathcal{H}$ or by 1-dimensional projections $P = |\psi><\psi|$. This usage of the word *pure* is bound to an assumption: The algebra $\mathcal{B}(\mathcal{H})$ should be the algebra of observables. To be more accurate, the algebra of observables should contain to every pair of different 1-dimensional projections (or rays) at least one hermitian operator which distinguish them by its expectation value. However, a *-subalgebra of $\mathcal{B}(\mathcal{H})$ with this properties coincides with $\mathcal{B}(\mathcal{H})$.

The conceptual perhaps cleanest way to distinguish a ***subsystem*** is by choosing a *-subalgebra $\mathcal{A}$ of $\mathcal{B}(\mathcal{H})$ i.e. one which is generated by its hermitian operators.

Heuristically this can be understood as following. There may be distinct states (pure or not) which, if restricted to a given subsystem, cannot be distinguished any more. Indeed, if this would not be true, every state of the subsystem could be uniquely extended to the bigger system, every state could already be distinguished from all others by the subsystem. But then there is no possibility to explain in what system and subsystem are different, and both had to be considered as identical.

Hence, looking from a genuine subsystem, some states can be no longer distinguished though they are originally as seen from the system. Therefore, an operator with different expectation values for a pair of such states cannot be an observable of the subsystem. Consequently, a subsystem can be characterized by a subset of observables of the full system. This is the key idea in the concept of subsystems.

It is, however, not completely deducable from this considerations that observables of a subsystem can be given by the hermitian elements of a *-subalgebra $\mathcal{A}$ of the system's observables. But here this assumption will be accepted and, moreover, $\mathcal{A}$ should contain the identity map of $\mathcal{H}$, i.e. it should be a unital subalgebra.

Even then the notation of a subsystem is a rather general one ranging from commuting subalgebras describing compatible measurements to *proper* subsystems for which $\mathcal{A}$ is a *factor* for which the centre contains the multiples of the unit element only. A theorem of Wedderburn allows to ennumerate all unital *-subalgebras $\mathcal{A}$ of $\mathcal{B}(\mathcal{H})$. In such an algebra there exists an orthogonal and central set $Q_1, \ldots Q_m$ of projection operators ($Q_j = Q_j^2 = Q_j^*$) with

$$1 = \sum Q_j, \quad Q_j Q_K = 0 \quad \text{if} \quad j \neq k \tag{11}$$

which define subalgebras

$$\mathcal{A}_j = \{A \in \mathcal{A} : Q_j A = A, \quad Q_k A = 0 \text{ otherwise}\} \tag{2}$$

with

$$\mathcal{A} = \mathcal{A}_1 \oplus \mathcal{A}_2 \oplus \ldots \oplus \mathcal{A}_m \tag{3}$$

and with centres consisting of the multiples of Q_j only: They are factors. Let

$$\mathcal{H}_j = Q_j \mathcal{H} \tag{4}$$

denote the carrier of Q_j. (1) implies

$$\mathcal{H} = \mathcal{H}_1 \oplus \mathcal{H}_2 \oplus \ldots \oplus \mathcal{H}_m \tag{5}$$

and $\mathcal{A}_j$ can be regardes as a *-subalgebra of $\mathcal{B}(\mathcal{H}_j)$ because the carrier of every element of $\mathcal{A}_j$ is contained in $\mathcal{H}_j$ according to (2). Be!ng a factor of $\mathcal{B}(\mathcal{H}_j)$ there exists a direct product decomposition

$$\mathcal{H}_j = \mathcal{H}_j^{(x)} \otimes \mathcal{H}_j^{(y)} \tag{6}$$

such that

$$\mathcal{A}_j = \mathcal{B}(\mathcal{H}_j^{(x)}) \otimes \mathbf{1}_j^{(y)} \tag{7}$$

In this relation $\mathbf{1}_j^{(y)}$ denotes the identity map of $\mathcal{B}(\mathcal{H}_j^{(y)})$, and the algebra (7) consists just of all operators

$$A_j^{(x)} \otimes \mathbf{1}_j^{(y)} \tag{8}$$

where the first factor runs through all elements of $\mathcal{B}(\mathcal{H}_j^{(x)})$.

It is fairly clear, how a *state* of $\mathcal{A}$ is to define. One literally repeats the definition of section 5: A state of $\mathcal{A}$ is a real linear functional on the real linear space of hermitian operators of $\mathcal{A}$ which takes non-negative values for positive operators, and is equal to one for the unity of $\mathcal{A}$. Of course one can and does extend these functionals uniquely to a complex linear one defined on all $\mathcal{A}$.

The states of $\mathcal{A}$ form a convex set (an affine space) $\Omega_{\mathcal{A}}$, the *state space* of $\mathcal{A}$ or, equivalently, of the *subsystem* defined by $\mathcal{A}$. Now one *defines* the pure state of $\mathcal{A}$ to be the extremal states of its state space.

If ϱ is a state of $\mathcal{B}(\mathcal{H})$ its restriction

$$\omega = \varrho_{\mathcal{A}} : \omega(A) = \varrho(A) \quad \text{all} \quad A \in \mathcal{A} \tag{9}$$

is a state of $\mathcal{A}$. This state, as defined by (9), is called *the restriction* of ϱ on $\mathcal{A}$. If ω is the reduction of ϱ, ϱ is called an *extension* of ω.

One knows by general extension theorems (of Hahn Banach type) for positive functionals that every state of $\mathcal{A}$ is a reduced state and can be gained as the reduction of a state of $\mathcal{B}(\mathcal{H})$. Thus, the reduction of states is a map from Ω, the state space of $\mathcal{B}(\mathcal{H})$, onto $\Omega_{\mathcal{A}}$, the state space of $\mathcal{A}$.

Generally, the reducing map does not directly respect the concept of purity. If ω is pure, its extension ϱ may not be pure. (This can occur also classically.) If ϱ is pure its reduction may not be pure. This is an important fact showing the influence of the superpostion principle of Quantum Physics. Even more, every state can be considered as the reduction of a pure state. This is contrary to all what is known from Classical Statistical Physics.

The next issue is to ask for the concept of density operators for staes of $\mathcal{A}$. It should not depend on the way, $\mathcal{A}$ is embedded within $\mathcal{B}(\mathcal{H})$ or within another algebra. To this end one needs a particular trace, i.e. a linear functional of $\mathcal{A}$ satisfying

$$\mathrm{tr}_{\mathcal{A}}(AB) = \mathrm{tr}_{\mathcal{A}}(BA)$$

and taking value one for all minimal projection operators of $\mathcal{A}$. By the last condition this trace is uniquely defined and called *canonical trace* of $\mathcal{A}$. Using the general form of $\mathcal{A}$ explained by (3) and (7) its existence can be shown by construction.

Given now ω, a state of $\mathcal{A}$, there is an operator $\omega = \omega_{op}$ in $\mathcal{A}$, called *density operator*, such that

$$A \mapsto \omega(A) = \mathrm{tr}_{\mathcal{A}}(A) \quad \text{for all} \quad A \in \mathcal{A} \tag{10}$$

In case the state ω appears as the reduction of a state of a larger system, the density operator ω is called the *reduced* density operator.

8. EXTENSIONS AND PURIFICATIONS

It is evident how to reverse the procedure explained in chapter 7, and how to ask for extensions of the given system, i.e. for embeddings of $\mathcal{B}(\mathcal{H})$ in a larger observable algebra, and extending the states of $\Omega(\mathcal{H})$ appropriately. Though elementary, it is a key ingredient that an observable (operator) of a system remains an observable in every larger system.

In extending $\mathcal{H}$, $\mathcal{B}$, Ω I restrict myself to cases where $\mathcal{B}(\mathcal{H})$ as a subalgebra of the larger system becomes a factor. The general case is obtained by performing direct sums of such extensions.

Thus let

$$\mathcal{H}^{\mathrm{ext}} = \mathcal{H} \otimes \mathcal{H}' \tag{1}$$

describe a system composed of our original one, $\mathcal{H}$, and a supplementary one, $\mathcal{H}'$. Then $\mathcal{B}(\mathcal{H})$ is embedded in $\mathcal{B}(\mathcal{H}^{\mathrm{ext}})$ by

$$A \in \mathcal{B}(\mathcal{H}) \mapsto A \otimes \mathbf{1}' \in \mathcal{B}(\mathcal{H}^{\mathrm{ext}}) \tag{2}$$

where $\mathbf{1}'$ is the unit element of $\mathcal{B}(\mathcal{H}')$. In the same spirit the map

$$A' \in \mathcal{B}(\mathcal{H}') \mapsto \mathbf{1} \otimes A' \in \mathcal{B}(\mathcal{H}^{\mathrm{ext}}) \tag{3}$$

is a unital isomorphism of $\mathcal{B}(\mathcal{H}')$ into $\mathcal{B}(\mathcal{H}^{\mathrm{ext}})$. The decomposition (1) induces in this way uniquely these two embeddings (2) and (3), and as long one sticks to (1) it is possible and convenient to consider $\mathcal{B}(\mathcal{H})$ and $\mathcal{B}(\mathcal{H}')$ as subalgebras of $\mathcal{B}(\mathcal{H}^{\mathrm{ext}})$. This convention will be used in the following. Then $\mathcal{B}(\mathcal{H}')$ is the commutant of $\mathcal{B}(\mathcal{H})$, i.e. the set of operators of $\mathcal{B}(\mathcal{H}^{\mathrm{ext}})$ commuting with every element of $\mathcal{B}(\mathcal{H})$, and $\mathcal{B}(\mathcal{H})$ is the commutant of $\mathcal{B}(\mathcal{H}')$.

An element $\tilde{\varrho}$ of $\Omega^{\mathrm{ext}} = \Omega(\mathcal{H}^{\mathrm{ext}})$ is called an *extension* of ϱ iff $\tilde{\varrho}$ gives ϱ if restricted to $\mathcal{B}(\mathcal{H})$.

Given $\varrho \in \Omega(\mathcal{H})$ the set of all extensions of ϱ to Ω^{ext} will be denoted by $\{\varrho\}^{\mathrm{ext}}$.

$\{\varrho\}^{\mathrm{ext}}$ is a convex compact subset of Ω^{ext}. According to Minkowski it is therefore the convex hull of its extremal points. However, no effective procedure is known to ennumerate these points. This is a common difficulty arising mathematically from the positivity condition, and from the point of view of Physics from the superposition principle: In a composed quantum system there are much more states then obviously visible from the subsystems. Here the difficult question is the following. Given density operators ϱ_j and ϱ'_j from $\Omega(\mathcal{H})$ and $\Omega(\mathcal{H}')$. For what coefficients α_{jk} is the operator

$$\sum \alpha_{jk} \varrho_j \otimes \varrho_k$$

a positive one?

Because of these circumstances I will pose a not so ambitious question and ask for the set of pure states within $\{\varrho\}^{\mathrm{ext}}$. Clearly, every such pure state is automatically extremal in $\{\varrho\}^{\mathrm{ext}}$ while the reverse assertion is wrong.

Let $\psi_1, \psi_2, \ldots, \psi_n$ and $\psi'_1, \ldots, \psi'_{n'}$ be a complete orthonormal system of $\mathcal{H}$ and $\mathcal{H}'$ respectively. Every normed vector $\varphi \in \mathcal{H}ext$ can be represented as

$$\varphi = \sum_{j=1}^{n} \sum_{k=1}^{n'} c_{jk} \psi_j \otimes \psi'_k, \quad \sum |c_j k| = 1 \tag{4}$$

To reduce $|\varphi><\varphi|$ to $\mathcal{B}(\mathcal{H})$ one has to perform with $A \in \mathcal{B}(\mathcal{H})$

$$<\varphi, A\varphi>=\sum \bar{c}_{jk}c_{ir} <\psi_j \otimes \psi'_k, (A\psi_i)\otimes\psi'_r>$$

to obtain

$$<\varphi, A\varphi>=\sum_k \bar{c}_{jk}c_{ik} <\psi_j, A\psi_i>, \tag{5}$$

This should be equivalent to

$$\varrho(A) = \operatorname{tr}\varrho A = \sum_{j,i} <\psi_i, \varrho\psi_j><\psi_j, A\psi_i> \tag{6}$$

If the linear operator C from $\mathcal{H}'$ into $\mathcal{H}$ is now defined by

$$<\psi_i, C\psi'_k>= c_{ik} \tag{7}$$

then one gets

$$\varrho = C\,C^* \tag{8}$$

The result is: There are pure states in $\{\varrho\}^{\text{ext}}$ if and only if

$$\operatorname{rank}\varrho \leq n' = \dim\mathcal{H} \tag{9}$$

Therefore *every* state $\varrho \in \Omega(\mathcal{H})$ has an extension in Ω^{ext} which is pure if $n \leq n'$. Every pure extension of $\varrho \in \Omega(\mathcal{H})$ is called a *purification* of ϱ.

Looking at (7) one immediately sees: If there is one purification then there are many. With an operator C of (7) also every other one given by

$$C^{\text{new}} = CU, \quad U \in \mathcal{B}(\mathcal{H}') \tag{10}$$

with a unitary U will give the same ϱ. Thus for the purification ambiguity a *gauging* with the group of all unitarities of $\mathcal{B}(\mathcal{H}')$ is responsible. It is possible (see later on) to construct a gauge field theory to handle this situation. While one has

$$U\{\varrho\}^{\text{ext}}U^* = \{\varrho\}^{\text{ext}} \tag{11}$$

with unitary $U \in \mathcal{H}'$, it remains to remark the following. Let $\tilde{\varrho} \in \{\varrho\}^{\text{ext}}$ and let $\in \mathcal{B}(\mathcal{H}')$ be a partial isometry the right support of which contains the carrier of $\tilde{\varrho}$. Then $U\tilde{\varrho}U^*$ is again in $\{\varrho\}^{\text{ext}}$. However, the partial isometries are not equipped with the structure of a group but only with that of a groupoid.

The next aim is to introduce the notation of transition probability for general states. Let ϱ_1, ϱ_2 denote two states of $\Omega = \Omega(\mathcal{H})$. In a suitable extended system (1) one can consider purifications φ_1, φ_2 of them which give rise to operators C_1, C_2 as indicated by (4) and (7). A similar calculation yields

$$<\varphi_1, \varphi_2>= \operatorname{tr} C_1^* C_2 \tag{12}$$

This number depends on the choice of the purifications. Both operators, C_1, C_2, can be gauged independently according to (9). This changes the number (12). This means that one cannot control completely by a system what is going on in a larger system. In particular, the absolute square of (12) gives the a priori probability that ϱ_1 goes over to ϱ_2 if probed in the extended system by the *obsewrvable* P_2 wether the state which is originally P_1 is equal to the *state* P_2.

Though (12) and $\operatorname{tprob}(P_1, P_2)$ is not determined by ϱ_1 and ϱ_2, one can correctly speak of the supremum

$$\operatorname{tprob}(\varrho_1, \varrho_2) = \sup \operatorname{tprob}(P_1, P_2) \tag{13}$$

where P_1, P_2 run through all possible pairs of simultaneous purifications of ϱ_1 and ϱ_2. This supremum is *by definition* called the *transition probability* of ϱ_1, ϱ_2, see[16].

The calculation of (13) can be performed as following. There are polar decompositions

$$C_j = \varrho_j^{\frac{1}{2}} U_j, \quad j = 1, 2 \tag{14}$$

where uniquness is achieved with maps U_j from $\mathcal{H}'$ onto carrier(ϱ_j) in $\mathcal{H}$.

Now

$$\operatorname{tr} C_1^* C_2 = \operatorname{tr} \varrho_1^{\frac{1}{2}} \varrho_2^{\frac{1}{2}} U_2 U_1^* = \operatorname{tr} |\varrho_1^{\frac{1}{2}} \varrho_2^{\frac{1}{2}}| U U_2 U_1^* \tag{15}$$

In this relation the polar decomposition

$$\varrho_1^{\frac{1}{2}} \varrho_2^{\frac{1}{2}} = |\varrho_1^{\frac{1}{2}} \varrho_2^{\frac{1}{2}}| U, \quad |\varrho_1^{\frac{1}{2}} \varrho_2^{\frac{1}{2}}| = (\varrho_1^{\frac{1}{2}} \varrho_2 \varrho_1^{\frac{1}{2}})^{\frac{1}{2}} \tag{16}$$

has been used. However, U, U_1, U_2 have operator norm one, and so the operator norm of their product cannot exceed one either. This has the consequence that the supremum will be reached if

$$U U_2 U_1^* \geq \operatorname{supp} |\varrho_1^{\frac{1}{2}} \varrho_2^{\frac{1}{2}}| \tag{17}$$

Hence

$$\operatorname{tprob}(\varrho_1 \varrho_2) = (\operatorname{tr}(\varrho_1^{\frac{1}{2}} \varrho_2 \varrho_1^{\frac{1}{2}})^{\frac{1}{2}})^2 \tag{18}$$

and there are indeed always purifications P_1, P_2 of ϱ_1, ϱ_2 with which the supremum of (13) is attained.

I call *parallel* purifications those pairs P_1, P_2 which purify simultaneously ϱ_2, ϱ_2, and for which (13) becomes an equality. The same notation (i.e. parallel purification) applies for pairs of normed vectors φ_1, φ_2, of the extended system if there pure states purify simultaneously, and if in addition there scalar product is real and not negative.

(17) can also expressed by

$$C_1^* C_2 = C_2^* C_1 \geq 0 \tag{19}$$

(18) indicates the possibility to calculate the transition probability in the original system. Without proofs I describe some properties arising in this connection[17–22].

Let ν be a linear function on $\mathcal{B}(\mathcal{H})$ satisfying

$$|\nu(A^* B)|^2 \leq \varrho_1(A^* A)\, \varrho_2(B^* B) \tag{20}$$

The typical example is

$$\nu(A) = \operatorname{tr} A C_2 C_1^* \tag{21}$$

With C_j given by (14). It can be shown that always

$$|\nu(\mathbf{1})|^2 \leq \operatorname{tprob}(\varrho_1, \varrho_2) \tag{22}$$

and, using an ansatz (21), that equality can be obtained in (22). If ν satisfies (20), if the carriers of φ_1 and φ_2 coincides, and

$$\nu(\mathbf{1}) > 0 \quad \text{with} \quad |\nu(\mathbf{1})|^2 = \operatorname{tprob}(\varrho_1, \varrho_2) \tag{23}$$

is valid then ν is uniquely determined. If the carrier condition is not fulfilled the uniqueness question is more complicated.

It is also possible to get the transition probability by an infimum. Surprising enough one gets

$$\operatorname{tprob}(\varrho_1, \varrho_2) = \inf \varrho_1(A)\, \varrho_2(A^{-1}) \tag{24}$$

where A runs through all the positive operators of $\mathcal{B}(\mathcal{H})$.

The pecular role of the transition probability is further underlined by the statement: Let ϕ be an affine endomorphism of Ω. Then

$$\operatorname{tprob}(\phi \circ \varrho_1, \phi \circ \varrho_2) \geq \operatorname{tprob}(\varrho_1, \varrho_2) \tag{25}$$

9. PARALLEL TRANSPORT

To treat the problem of chapter 4 in the case of general states[24] is now more or less straightforward: Given a curve of states (respectively density operators)

$$s \mapsto \varrho_s \qquad \text{with} \quad 0 \leq s \leq 1 \tag{1}$$

defined on $\mathcal{H}$, one looks for a purified lift

$$s \mapsto P_s = \frac{|\varphi_s><\varphi_s|}{<\varphi_s, \varphi_s>}, \quad 0 \leq s \leq 1, \tag{2}$$

in an extended system with Hilbert space $\mathcal{H}^{ext}$ as defined in chapter 8.

To restrict as much as possible the arbitrariness of such a lift the simple idea is to require parallelity for 'infinitely neighboured' purifications. The Hilbert space metric

$$\| \varphi_s - \varphi_t \| = \sqrt{< \varphi_s - \varphi_t, \varphi_s - \varphi_t >} \tag{3}$$

will be used to make this attempt correct.

Assuming at first the normed vectors φ_s and φ_t to be purifications of ϱ_s and ϱ_t then, according to the defining equation (8-13), it is

$$\| \varphi_s - \varphi_t \|^2 \leq 2 - 2\text{tprob}(\varrho_s, \varrho_t) \tag{4}$$

and the equality sign holds for parallel lifts[23,24]. The latter case gives the metric of Bures[25] in the state space one is starting from:

$$\| \varrho_s - \varrho_t \| = \sqrt{2 - 2\text{tprob}(\varrho_s, \varrho_t)} \tag{5}$$

Returning to (4) and concerned with 'infinitely neighboured' states it is tempting to use

$$\lim \epsilon^{-1} \| \varphi_{s+\epsilon} - \varphi_s \| = \sqrt{< \dot\varphi_s, \dot\varphi_s >} \tag{6}$$

and to require a minimal right hand side in order that (2) is a parallel lifting of (1). For normed purifying vectors we then are done. It is, however, convenient to relaxe from norming the purifying vectors. In this slightly more general case the following seems to be very natural:

A lift (2) is called a *parallel* purification of (1) if the expression

$$\frac{< \dot\varphi, \dot\varphi >}{< \varphi, \varphi >} \tag{7}$$

attains for every value of the parameter s its minimum with respect to all purifications of a given curve of states (1).

For the transport of frames of degenerated eigenstates a similar proposel, though not worked out at all, has been made already by Fock in the appendix of[26]. As a first extension of Berry's idea it is discussed, however from quite another side, by Wilczek and Zee[27]. Their theory concerns the case of a path of density operators (1) which is proportional to a path of projection operators of fixed rank.

In (7) a fixed parametrization of the original curve (1) is assumed. The essentials are parameter independent of course. This can be made obvious by considering the line element

$$\sqrt{\frac{< \dot\varphi, \dot\varphi >}{< \varphi, \varphi >}} ds \tag{8}$$

There is, by the by, a slight difference between the Bures metric (5) and the metric given by the line element (8) because of the occurence of the denominator in (7) and (8).

The next aim is to get conditions for the curve

$$s \mapsto \varphi_s \qquad \varphi_s \in \mathcal{H}^{\text{ext}} \tag{9}$$

which are necessary for attaining the minimum of the expression (7) while (2) is a purification of (1).

Using the notations of chapter 8, in particular the decomposition (1) and its consequences, the following is obvious: For every hermitian operator B' of $\mathcal{B}(\mathcal{H}')$ one considers $\mathbf{1} \otimes B' = B$. Then with (9) the curve

$$s \mapsto U(s)\varphi_s \quad \text{with} \quad U(s) = \exp(isB) \tag{10}$$

will not give a value smaller than (9) if inserted into (7). The resulting inequality reads

$$0 \leq < B\varphi, B\varphi > +i[< \dot{\varphi}, B\varphi > - < \varphi, B\dot{\varphi} >] \tag{11}$$

If this ineqality is valid for all allowed B it implies

$$< \dot{\varphi}, B\varphi >=< \varphi, B\dot{\varphi} > \quad \text{for all} \quad B = I \otimes B' \tag{12}$$

Since the condition (12) is linear in B, and hence in B', and since the linear span of the hermitian operators contains all operators, (12) holds for all operators of the form $\mathbf{1} \otimes B' = B$. Remarkable enough every of the relations (12) is a transport condition like (4-7) which is only slightly masked by the arbitrariness of the norms. Namly if for a particular B one has $< \varphi, B\varphi >=$ const. then (12) sharpens to $< \dot{\varphi}, B\varphi >= 0$.

The necessary condition (12) does not reflect what is going on with the norm of the vectors. This is natural since it is the same with (2). If now (9) fulfills (12) and is normed, one can probe (7) with a new curve

$$s \mapsto \lambda_s \varphi_s, \qquad \lambda_s > 0 \tag{13}$$

In the expression (7) this will act as a substitution

$$\frac{< \dot{\varphi}, \dot{\varphi} >}{< \varphi, \varphi >} \quad \to \quad \frac{< \dot{\varphi}, \dot{\varphi} >}{< \varphi, \varphi >} + (\frac{\dot{\lambda}}{\lambda})^2 \tag{14}$$

The net result is: The necessary condition for (9) to minimize (7) such that (2) becomes a parallel lift of (1) is the following: The curve (9) fulfills (12) and the norm of its vectors is constant, i.e. is independent of the parameter.

But is this necessary condition also sufficient? To decide this question one needs more general deformations of a curve (9) which minimizes (7). One needs not only curves (10) of unitaries but smooth curves of arbitrary partial isometrics

$$U_s = \mathbf{1} \otimes U'_s, \quad \text{with} \quad U_s \in \mathcal{B}(\mathcal{H}') \tag{15}$$

They give rise to deformations

$$\varphi'_s = U_s \varphi_s \quad \text{with} \quad U_s^* \varphi'_s = \varphi_s \tag{16}$$

A certain role will be played by the curves of projection operators

$$Q_s = U_s^* U_s \quad \text{and} \quad Q'_s = U_s U_s^* \tag{17}$$

Its first role is in fixing U_s uniquely by one of the relations (16). Within all partial isometries fulfilling one (and therefore the other) relation (16) there is one for which the carriers Q_s are as small as possible. This choice will be made further on. Now

$$< \dot{\varphi}', \dot{\varphi}' >=< \dot{U}\varphi, \dot{U}\varphi > + < U\dot{\varphi}, U\dot{\varphi} > + < \dot{U}\varphi, U\dot{\varphi} > + < U\dot{\varphi}, \dot{U}\varphi > \tag{18}$$

The last two terms of (18) can be rewritten as following. By (12)

$$< U\dot{\varphi}, \dot{U}\varphi > = < \dot{\varphi}, U^*\dot{U}\varphi > = < \varphi, U^*\dot{U}\dot{\varphi} > \tag{19}$$

Hence

$$< \dot{U}\varphi, U\dot{\varphi} > + < U\dot{\varphi}, \dot{U}\varphi > = < \varphi, \dot{U}^*U\dot{\varphi} > + < \varphi, U^*\dot{U}\dot{\varphi} > = < \varphi, \dot{Q}\dot{\varphi} > \tag{20}$$

Inserting into (18) yields

$$< \dot{\varphi}', \dot{\varphi}' > = < \dot{U}\varphi, \dot{U}\varphi > + < U\dot{\varphi}, U\dot{\varphi} > + < \varphi, \dot{Q}\varphi > \tag{21}$$

Examination of the last two terms yields

$$< U\dot{\varphi}, U\dot{\varphi} > + < \varphi, \dot{Q}\varphi > = < \dot{\varphi}, Q\dot{\varphi} > + < \varphi, \dot{Q}\varphi > = < \dot{\varphi}, \dot{\varphi} > \tag{22}$$

where the last equality sign follows from differentiating $\varphi = Q\varphi$. Combining (21) and (22) provides

$$< \dot{\varphi}', \dot{\varphi}' > = < \dot{U}\varphi, \dot{U}\varphi > + < \varphi, \varphi > \tag{23}$$

Thus the condition (12) is not only necessary but also sufficient under the *additional assumption* that the projection operators (17) depend *smoothly* on its parameter s. In the case at hand, where the Hilbert spaces are all finite dimensional, this additional assumption is fulfilled iff (1) is smooth and of constant rank[11].

There is a further consequence if both, φ_s and φ'_s, are minimizing (7) because (23) then supplies $\dot{U}\varphi = 0$. The last is equivalent with $\dot{U}U^*\varphi'_s = 0$ as is shown by (16). Since the right support of U^* is choosen as small as possible it is the closed span of all $A\varphi'_s$ with $A \in \mathcal{B}(\mathcal{H}) \otimes \mathbf{1}'$. As these operators A commute with $\dot{U}U^*$ one concludes $\dot{U}U^* = 0$. Multiplying from the left with U shows $\dot{U}Q = 0$. Now differentiating $UQ = U$ finally yields

$$\dot{U} = 0 \quad \text{or} \quad U_s = \text{const.} \tag{24}$$

This important results reads: If the curves of normed vectors $s \to \varphi_s$ and $s \to \varphi'_s$ both give parallel lifts of (1), i.e. if both curves minimize (7), then there is a s-independent partial isometry U of the form (15) such that

$$\varphi'_s = U\varphi_s \quad \text{with} \quad U^*\varphi'_s = \varphi_s \tag{25}$$

It follows that

$$< \varphi_0, A\varphi_1 > = < \varphi'_0, A\varphi'_1 > \quad \text{for all} \quad A \in \mathcal{B}(\mathcal{H}) \otimes \mathbf{1}' \tag{26}$$

In particular one can define correctly

$$\text{Berry}(s \mapsto \varrho_s) = < \varphi_0, \varphi_1 > \tag{27}$$

for smooth curves (1) of constant rank.

If (2) is a lift of (1) it is possible to switch to another representation. To every normed curve (9) there is according to (8-7) und (8-8) a curve

$$s \mapsto C_s, \quad 0 \le s \le 1 \tag{28}$$

of operators (maps) from $\mathcal{H}$ into $\mathcal{H}'$ satisfying

$$\varrho_s = C_s C_s^* \qquad \text{for all } s \tag{29}$$

It is now simply a matter of translation to rewrite some of the results above in terms of these operastors. (12) occurs to be equivalent to[23]

$$C_s^*\dot{C}_s - \dot{C}_s^*C_s = 0 \tag{30}$$

See also[28,29,30] for further discussions and calculations. As (29) implies norm one for the associated curve of vectors of $\mathcal{H}^{ext}$, (30) is another necessary and sufficient condition for paralleltiy if the supports of (1) change smoothly. If (30) is valid, i.e. for a parallel lift, (26) now reads

$$C_0\, C_1^* \qquad \text{depends only on } s \mapsto \varrho_s \tag{31}$$

and the value (27) is the trace of $C_0\, C_1^* \in \mathcal{B}(\mathcal{H})$.

Now I return to arbitrary lifts. Comparing (29) and (30) it is obvious that a *local gauge transformation*

$$C_s \mapsto C_s U_s, \qquad U_s \in \mathcal{B}(\mathcal{H}') \tag{32}$$

with unitary U_s will not change (29). (This is also true for the larger class of partial isometries if relevant support condition are fullfilled. But this will be ignored for simplicity.) Therefore one may ask wether the parallelity condition (30) can be understood as a parallel transport of a gauge theory living, so to say, somehow between $\mathcal{H}'$ and $\mathcal{H}$ but mostly on $\mathcal{H}'$. To make this vague idea handable one first converts (30) into an operator valued differential form

$$C_s^*\, dC_s - dC_s^*\, C_s \tag{33}$$

the integral curves of which are the parallel ones. This form, however, will not be a connection form with respect to a local gauge transformation (32).

Such a connection form, called $\mathbf{A}$ will be defined implicitely by[31]

$$C_s^*\, dC_s - dC_s^*\, C_s = C^* C\, \mathbf{A} + \mathbf{A}\, C^* C \tag{34}$$

$\mathbf{A}$ is a differential 1-form with values in $\mathcal{B}(\mathcal{H}')$. It depends by its very definition on operators mapping $\mathcal{H}'$ into $\mathcal{H}$ and their adjoints.

The definition is completed by

$$< \psi', \mathbf{A}\psi' >= 0 \quad \text{if} \quad C\psi' = 0 \quad \text{with} \quad \psi' \in \mathcal{H} \tag{35}$$

$\mathbf{A}$ is undefined (singular) at tangential elements $\dot{C}$ in the direction of which the rank of CC^* is changing. These directions must be excluded. With this restriction in mind, (34) and (35) determine $\mathbf{A}$ *uniquely*. This can be seen very easily by sandwiching (34) with a complete orthonormal frame of eigenvectors of CC^*.

The uniqueness yields

$$\mathbf{A}^* + \mathbf{A} = 0 \tag{35}$$

and, the main point of its introduction, $\mathbf{A}$ behaves with respect to *local* gauge transformations $C \to CU$ as

$$\mathbf{A} \mapsto U^* \mathbf{A} U + U^*\, dU \tag{36}$$

which is proved by applying a local gauge to (34) and (35) and using the uniqueness property.

The connection form determines a *covariant differentiation* for (manifolds of) maps from $\mathcal{H}'$ into $\mathcal{H}$ and their adjoints. For instance one has

$$DC := dC - C\mathbf{A} \tag{37}$$

and

$$DC^* := dC^* + \mathbf{A}C^* \tag{38}$$

Of course (37) and (38) are adjoints one from another. The change in sign is due to (35).

It is of some use to introduce a further differential 1-form, $\mathbf{G}$ with values in $\mathcal{B}(\mathcal{H})$ by

$$DC = \mathbf{G}C \tag{39}$$

By this $\mathbf{G}$ is defined for vectors of the form $C\psi'$ only, and the definition will be completed soon. Substituting

$$dC = C\mathbf{A} + \mathbf{G}C \tag{40}$$

which comes from (37) and (40) into (34) results in

$$C^{*}(\mathbf{G}-\mathbf{G}^{*})C=0 \tag{41}$$

Thus $\mathbf{G}$ is is hermitian if resricted to the subspace of the vectors $C\psi'$.

The definition of $\mathbf{G}$ can now be completed without destroying (39) and (41) by

$$\mathbf{G}=\mathbf{G}^{*} \quad \text{and} \quad <\psi,\mathbf{G}\psi>=0 \quad \text{if} \quad C^{*}\psi=0 \tag{42}$$

($C^{*}\psi=0$ is equivalent with the existence of a ψ' that produces $\psi=C\psi'$.)

$\mathbf{A}$ is obviously zero if restricted on a parallel curve. Because of (49) the parallel shifts are integral curves of

$$\mathrm{d}C-\mathbf{G}C \tag{43}$$

as well. It is therefore interesting to determine $\mathbf{G}$ alternatively.

Taking the total differential

$$\mathrm{d}(CC^{*})=\mathrm{d}C\,C^{*}+C\,\mathrm{d}C^{*} \tag{44}$$

and replacing on the right hand side the total differential by the covariant ones, (37) and (38), one gets

$$\mathrm{d}(CC^{*})=\mathrm{D}C\,C^{*}+C\,\mathrm{D}C^{*} \tag{45}$$

where the terms with $\mathbf{A}$ have cancelled. It remains to insert (39) and to replace CC^{*} by ϱ according to (8-8) or (30) to obtain

$$\mathrm{d}\varrho=\mathbf{G}\varrho+\varrho\mathbf{G} \tag{46}$$

REFERENCES

1. Fano, U., Rev. Mod. Phys. 29 (1957) 74
2. Wehrl, A., Rev. Mod. Phys. 50 (1978) 221
3. Bluhm, K.: Density Matrix Theory and Applications., Plenum Press, New York, 1981
4. Berry, M. V. Proc. Royal. Soc. Lond. A 392 (1984) 45
5. Simon, B., Phys. Rev. Lett. 51 (1983) 2167
6. Born, M., Z. Phys. 40 (1926) 167
7. Born, M. and Fock, V., Z. Phys. 51 (1928) 165
8. Born, M. and Oppenheimer, J., Ann. d. Phys. 84 (1927) 457
9. Aharonow, Y., Anandan, J., Phys. Rev. Lett. 58 (1987) 1593
10. Littlejohn, R.G., Phys. Rev. Lett. 61 (1988) 2159
11. Uhlmann, A., Annalen d. Phys. 46 (1989) 63
12. Berry, M., Quantum Adiabatic Holonomy. In: Anomalies, phases, defects (ed. M.Bregola, G.Marmo, G.Morandi), Bibliopolis, Naples, 1990
13. Geometric Phases in Physics. (ed. A.Shapere, F.Wilczek), World Scientific Publishing Co., Singapure, 1990
14. Akemann, C.A., Pedersen, G.K., Facial Structure in Operator Algebra Theory. Preprint, Kopenhavns, June 1990.
15. Uhlmann, A., Rep. Math. Phys. 7 (1975) 449
16. Uhlmann, A., Rep. Math. Phys. 9 (1976) 273
17. Raggio, G. A., Lett. Math. Phys. 6 (1982) 233
18. Araki, H. and Raggio, G. A., Lett. Math. Phys. 6 (1982) 237
19. Alberti, P. M., Lett. Math. Phys. 7 (1983) 25
20. Alberti, P. M. and Uhlmann, A., Lett. Math. Phys. 7 (1983) 107
21. Cantoni, V., Comm. Math. Phys. 44 (1975) 125
22. Uhlmann, A., Ann. d. Physik 42 (1985) 524
23. Uhlmann, A., Rep. Math. Phys. 24 (1986) 229

24. Uhlmann, A., Parallel Transport and Holonomy along Density Operators. In: "Differential Geometric Methods in Theoretical Physics" (ed. H.D.Doebner, J.D.Hennig), World Sci. Publ., Singapore 1987, p. 246 - 254
25. Bures, D. J. C., Trans. Amer. Math. Soc. 135 (1969) 199
26. Fock, V., Z. Phys. 49 (1928) 323
27. Wilczek, F. and Zee, A., Phys. Rev. Lett. 52 (1984) 2111
28. Dabrowski, L., Grosse, H., On Quantum Holonomy for Mixed States. Preprint, Wien, UWThPh-1988-36
29. Dabrowski, L., Jadcyk, A., Quantum Statistical Holonomy. Preprint, Trieste, 155/88/FM
30. Dabrowski, L. A Superposition Principle for Mixed States? Preprint, Trieste, 156/88/FM
31. Uhlmann, A., Gauge Field Governing Parallel Transport along Mixed States. Preprint, Wien, UWThPh-1990-25

GROUP STRUCTURES AND THE INTERACTING BOSON APPROXIMATION FOR NUCLEI

Brian G. Wybourne

Physics Department, University of Canterbury
Christchurch, New Zealand

1. Origins of the exceptional groups in physics

The exceptional groups, or more precisely, the exceptional Lie algebras , were made explicit in Elie Cartan's thesis[1] of 1894 by his classification of the complex semisimple Lie algebras. Cartan noted that in addition to the four classical Lie algebras A_n, B_n, C_n and D_n , respectively associated with the classical Lie groups SU_{n+1}, SO_{2n+1}, Sp_{2n} and SO_{2n}, there were five exceptional Lie algebras, G_2, F_4, E_6, E_7 and E_8 whose adjoint irreducible representations were of degree 14, 52, 78, 133 and 248 respectively.

It is interesting to note that even before Cartan's classification amateur mathematicians in the United Kingdom and Russia[2] had established what is now known as the E_8 root lattice. Such lattices are relevant to various problems in coding theory.

The first attempt to use the exceptional groups in physical problems was in Racah's elegant treatment[3] of the atomic f-shell in 1949. His work has found wide application in the study of the atomic and solid state properties of lanthanides and actinides and in nuclear shell theory. Racah realised that the 14 angular momentum states of an electron, with spin, in a f-orbital could be regarded as spanning the vector irrep of a unitary group in 14 dimensions. Clearly the angular momentum states must be embedded so that under $U_{14} \to SO_3$ we have $\{1\} \to [3]$. For an $n-$electron configuration f^n the complete set of antisymmetric states will span the column irrep $\{1^n\}$ irrep of U_{14}. The question then arises as to what groups can be embedded in U_{14} that contain the physical spin SU_2 and rotation SO_3 groups as a subgroup. The introduction of the symplectic group Sp_{14} gave rise to the well-known seniority group that was to lead to the concept of pairing in nuclei and in superconductivity theory. The spin part of the wave function could be regarded as spanning irreps of SU_2 while the orbital part irreps of SO_7 leading naturally to the group chain

$$U_{14} \to Sp_{14} \to SU_2^S \times [SO_7 \to SO_3^L]$$

The irreps of the various groups in the chain lead to a partial classification of the states of the f^n-configuration. Racah realised that a richer classification scheme would arise if a group could be placed between the SO_7 and SO_3 groups. Such a possibility is severely restricted if we are to retain SO_3 as the physical rotation group. Racah observed that the exceptional group G_2 filled the requirement uniquely. We note that the existence of G_2 in this chain implies that the SO_3 6-j symbol

$$\begin{Bmatrix} 5 & 5 & 3 \\ 3 & 3 & 3 \end{Bmatrix} = 0.$$

During the late 1960's we worked extensively on the properties of the exceptional groups though there seemed little interest among physicists in such esoteric structures. We revived our interest in the late 1970's looking at the very practical properties of calculating the properties of their irreducible representations. In particular, my then young student, Mark Bowick and I spent the summer of 1976 looking at relationships between the exceptional groups and their maximal compact Lie subgroups[4] as there seemed to be a growing interest among physicists in grand unified theories where every possible group structure seemed fair game. In the early 1980's Ron King of Southampton extended the work of Bowick and I in a series of three papers [5-7], two being with his student Al-Qubanchi, which clearly established a systematic labelling of the irreps in terms of those of their maximal classical subgroups. Since that time King and I have maintained a collaborative study of the properties of the exceptional groups among other topics [8-12].

Today most physicists have heard of the exceptional groups largely as a result of the development of string theories where the awesome group E_8 plays a key role. Today I want to discuss another area of physics where one might see a role for , at least some, of the exceptional groups , namely in the interacting boson model of nuclei. I shall commence by first looking at the group structures that arise when the IBA(Interacting Boson Approximation) is extended beyond the simple *sd*-bosons of Arima and Iachello[13].

2. Group structures and the simple IBA

Racah's work on the atomic shell model was rapidly applied to the nuclear shell model by Jahn[14], Flowers[15] and Elliott[16]. In a series of papers J.P.Elliott[16-18] discussed the group structure of the SU_3 nuclear shell model and in particular the $sd-$shell. Elliott noted that the primary group of interest was U_6 and discussed various possible subgroup structures identifying in particular the group chains

$$U_6 \rightarrow SU_3 \rightarrow SO_3 \tag{1}$$

$$U_6 \rightarrow SO_6 \rightarrow SO_5 \rightarrow SO_3 \tag{2}$$

$$U_6 \rightarrow SU_5 \rightarrow SO_5 \rightarrow SO_3 \tag{3}$$

The Elliott model was based on fermionic nucleons and yet he had arrived at a set of group structures identical to those that were to arise in Arima and Iachello's first version of the interacting boson approximation. The simple IBA $sd-$boson approximation was an almost instant success due to its ability to describe significant properties of nuclei in terms of a few parameters. The calculation of these properties were extremely simple especially when the real complexity of the nuclei is considered. Nevertheless, one cannot realistically consider a model limited to just $sd-$bosons to have sufficient degrees of freedom to account for observed high spin states of nuclei. As a consequence many variants of the original IBA have been proposed. These variants are naturally more complicated and can be expected to involve more complex group structures, especially as higher spin bosons are introduced. It is not my intention today to review the experimental evidence for high spin states and properties of nuclei that are discordant with the simplest IBA but rather to discuss the group structures that can arise when the set of bosons is enlarged to include higher spin bosons. I shall barely touch upon the important topic of the microscopic foundations of the IBA.

3. Group structures and high spin bosons

The symmetric irreps of SU_3 may be conveniently labelled $\{n\}$ and are of degree

$$x = (n+1)(n+2)/2 \tag{4}$$

Under restriction to the subgroup SO_3 we associate the *odd* parity bosons with *odd* values of n and *even* parity bosons with *even* values of n. We find that under $SU_3 \to SO_3$ we have

$$\{n\} \to \begin{cases} s+d+\ldots+n & \text{n } even \\ p+f+\ldots+n & \text{n } odd \end{cases} \tag{5}$$

The set of bosons associated with a given value of n transform collectively as the vector irrep $\{1\}$ of U_x. Here we must ask "What subgroups G exist such that

$$U_x \supset G \supset SO_3 \tag{6}$$

for n *even* or n *odd* ?"

A given embedding is specified by giving the decomposition of the vector irrep.

4. Structures for even bosons

In the case of even bosons we may readily establish the following group chains

$$U_x \supset SU_3 \supset SO_3 \tag{7}$$

with the vector irrep of U_x decomposing as

$$\{1\} \to \{n\} \to [0] + [2] + \ldots + [n] \tag{7a}$$

$$U_x \supset SU_{n+1} \supset SO_{n+1} \supset SO_3 \tag{8}$$

with the vector irrep now decomposing as

$$\begin{array}{llll} \{1\} \to & \{2\} \to & [2] \to & [2]+[4]+\ldots+[n] \\ & & [0] \to & [0] \end{array} \tag{8a}$$

$$U_x \supset SU_{n+2} \supset Sp_{n+2} \supset SO_3 \tag{9}$$

with the vector irrep now decomposing as

$$\begin{array}{llll} \{1\} \to & \{1^2\} \to & <1^2> \to & [2]+[4]+\ldots+[n] \\ & & <0> \to & [0] \end{array} \tag{9a}$$

We note that Eq.(8) implies a decomposition for the vector irrep $\{1\}$ under $SU_{n+1} \to SO_3$ of $\{1\} \to [n/2]$ while Eq.(9) implies a decomposition for the vector irrep $\{1\}$ under $SU_{n+2} \to SO_3$ of $\{1\} \to [(n+1)/2]$.

Finally, for n *even* we can also have

$$U_x \supset U_1 \times SU_{n(n+3)/2} \supset G \supset SO_3 \tag{10}$$

with the vector irrep $\{1\}$ decomposing as

$$\begin{array}{llll} \{1\} \to & \{-1\} \times \{1\} \to & \{?\} \to & [2]+[4]+\ldots+[n] \\ & \{n(n+3)/2\} \times \{0\} \to & \{0\} \to & [0] \end{array} \tag{10a}$$

5. Examples for even bosons

Let us, by way of example, list the cases that arise for $n = 2, 4, 6$.

(a). $n = 2$

$$U_6 \supset SU_3 \supset SO(3) \quad (11a)$$

$$U_6 \supset SU_4 \supset Sp_4 \supset SO_3 \quad (11b)$$

$$U_6 \supset SU_5 \supset SO_5 \supset SO_3 \quad (11c)$$

These are of course the familiar Arima-Iachello structures associated with *sd*-boson models.

(b). $n = 4$

$$U_{15} \supset SU_3 \supset SO_3 \quad (12a)$$

$$U_{15} \supset SU_5 \supset SO_5 \supset SO_3 \quad (12b)$$

$$U_{15} \supset SU_6 \supset Sp_6 \supset SO_3 \quad (12c)$$

$$U_{15} \supset U_1 \times SU_{14} \supset SO_5 \supset SO_3 \quad (12d)$$

These group chains are relevant to *sdg*-boson models.

(c). $n = 6$

$$U_{28} \supset SU_3 \supset SO_3 \quad (13a)$$

$$U_{28} \supset SU_7 \supset SO_7 \supset G_2 \supset SO_3 \quad (13b)$$

$$U_{28} \supset SU_8 \supset Sp_8 \supset SO_3 \quad (13c)$$

$$U_{28} \supset U_1 \times SU_{27} \supset E_6 \supset G_2 \supset SO_3 \quad (13d)$$

These chains are relevant to IBM models based on *sdgi*-bosons. We note that the chain given by Eq.(13b) involves the exceptional group G_2 while that of Eq. (13d) is exceptionally exceptional involving not only G_2 but also E_6.

6. Structures for odd bosons

For odd bosons we readily find the following structures

$$U_x \supset SU_3 \supset SO_3 \quad (14)$$

with the vector irrep of U_x decomposing as

$$\{1\} \to \{n\} \to [1] + [3] + \ldots + [n] \quad (14a)$$

$$U_x \supset SU_{n+1} \supset Sp_{n+1} \supset SO_3 \quad (15)$$

with the vector irrep now decomposing as

$$\{1\} \to \{2\} \to <2> \to [1] + [3] + \ldots + [n] \quad (15a)$$

$$U_x \supset SU_{n+2} \supset SO_{n+2} \supset SO_3 \quad (16)$$

with the vector irrep decomposing as

$$\{1\} \to \{1^2\} \to [1^2] \to [1] + [3] + \ldots + [n] \quad (16a)$$

We note that Eq.(15) implies a decomposition for the vector irrep $\{1\}$ under $SU_{n+1} \to SO_3$ of $\{1\} \to [n/2]$ while Eq.(16) implies a decomposition for the vector irrep $\{1\}$ under $SU_{n+2} \to SO_3$ of $\{1\} \to [(n+1)/2]$.

7. Examples for odd bosons

We now list the cases for $n = 3, 5$

(a). $n = 3$

$$U_{10} \supset SU_3 \supset SO_3 \tag{17a}$$

$$U_{10} \supset SU_4 \supset Sp_4 \supset SO_3 \tag{17b}$$

$$U_{10} \supset SU_5 \supset SO_5 \supset SO_3 \tag{17c}$$

These are structures relevant to pf-bosons.

(b). $n = 5$

$$U_{21} \supset SU_3 \supset SO_3 \tag{18a}$$

$$U_{21} \supset SU_6 \supset Sp_6 \supset SO_3 \tag{18b}$$

$$U_{21} \supset SU_7 \supset SO_7 \supset G_2 \supset SO_3 \tag{18c}$$

Again we see the emergence of the exceptional group G_2, this time in the case of pfh-bosons.

8. An SO_{10} odd-even boson system

The set of 16 states of *spdf*-bosons span the vector irrep $\{1\}$ of U_{16}. This irrep remains irreducible under $U_{16} \to SO_{10}$. Under SO_{10} the 16 states span the basic spin representation Δ_+. This observation suggests the possible use of a scheme that resembles the simplest extension of the SU_5 of grand unification schemes in particle physics. The complete group scheme can be written as

$$\begin{array}{lllll}
U_{16} \to & SO_{10} \to & U_1 \times SU_5 \to & SO_5 \to & SO_3 \\
\{1\} & \Delta_+ & \{3\} \times \{1^4\} & [1] & d \\
 & & \{\bar{1}\} \times \{1^2\} & [1^2] & p+f \\
 & & \{\bar{5}\} \times \{0\} & [0] & s
\end{array} \tag{19}$$

We note that there are even and odd parity states but they occur in different $U_1 \times SU_5$ irreps. A set of N bosons will span the symmetric $\{N\}$ irrep of U_{16}. These irreps may be readily decomposed by noting the spin plethysm[19,20]

$$\Delta_+ \otimes \{N\} = \sum_{m=0}^{[N/2]} [\frac{N}{2}, (\frac{N}{2} - m)^4] \tag{20}$$

The remaining branching rules may be found by standard means or in our case rapidly evaluated using SCHUR. The U_1 number determines the parity of the states which may be readily found by the 'Pascal triangles'

$$\begin{array}{ccccccccccccc}
 & & & & \textit{even} & & & & & & \textit{odd} & & \\
N = 0 & & & & 0 & & & & & & & & \\
1 & & & 3 & & \bar{5} & & & & & \bar{1} & & \\
2 & & 6 & & \bar{2} & & \overline{10} & & & 2 & & \bar{6} & \\
3 & 9 & & 1 & & \bar{7} & & \overline{15} & 5 & & \bar{3} & & \overline{11}
\end{array} \tag{21}$$

The SO_{10} eigenstates involve the mixing of configurations of the *same* parity. It would be interesting to see if such mixings are realised in any nuclei.

In the scheme just discussed parity is explicitly conserved. This may be compared with the coulomb like case for *spdf* bosons under

$$U_{16} \to SO_{10} \to SO_4 \to SO_3 \tag{22}$$

where the vector irrep decomposes as

$$\{1\} \to \Delta_+ \to [3] \to s + p + d + f \tag{23}$$

In that scheme states of different parity occur in the same irrep and hence it is necessary to project out eigenstates of parity. This situation is well known in atomic physics[21].
bf 9. U_{28} and *sdgi* bosons

Peter Pieruschka has made a preliminary study of the role of the exceptional E_6 and G_2 groups in the group chains of Eqs. (13b) and (13d) in his M.Sc. thesis[22]. A detailed paper has been recently completed with Morrison and I[23]. I shall only sketch the main details. As with any such group scheme one starts with constructing the group generators in terms of coupled pairs of creation and annihilation operators symmetrised with respect to the smallest groups, $SO_3 \supset SO_2$. The commutators of these operators are formed and isoscalar factors evaluated to give the appropriate linear combinations symmetrised with respect to the higher groups in the chain. This latter step forces relationships among the various $6-j$ symbols of SO_3. Thus Racah's single vanishing $6-j$ symbol is replaced by vanishing linear combinations of $6-j$ symbols such as

$$55 \begin{Bmatrix} 5 & 5 & k \\ 6 & 6 & 2 \end{Bmatrix} + 120 \begin{Bmatrix} 5 & 5 & k \\ 6 & 6 & 4 \end{Bmatrix} - 34 \begin{Bmatrix} 5 & 5 & k \\ 6 & 6 & 6 \end{Bmatrix} = 0 \tag{24}$$

where the above identity holds for $k = 3, 7, 9$ Similar identities were noted many years ago by Wadzinski[24] in his construction of the exceptional group F_4 in a SO_3 basis. The existence of the Regge symmetries[25] for $6-j$ symbols implies many other such relationships.

The various branching rules were readily constructed using SCHUR and the relevant isoscalar factors by similar software. Model Hamiltonians are then constructed in terms of the Casimir invariants whose eigenvalues for the various irreps were also found using SCHUR. The boson contents of the states were determined and the properties of the **M1**, **E0**, and **E2** electromagnetic transition operators studied. Special attention was given to the detailed study of the SU_7 spectral limit with specific application to the states of the nucleus ^{200}Hg. The fit of the energy levels and the **M1** and **E2** transitions suggests that these systems warrant more detailed study and particularly of series of nuclei showing SU_7 symmetry.

Concluding remarks

The interacting boson model offers a rich variety of group structures. Some are well known while others remain to be explored. While we have indicated the principal structures involving simple groups there are of course many possible structures involving direct product groups and for systems involving different partitions of the bosons. Each structure involves a different basis and hence the possibility of generating different spectra in terms of the Casimir invariants. The overlapping of states belonging to these different bases warrants further study.

The exceptional groups have entered in the *sdgi* and *pfh* boson states and provide a useful simplification to otherwise very complex many-boson systems. The exceptional group G_2 plays a key role with the group E_6 a less obvious role. The two sets of *pfh* and *sdgi* bosons may be combined in a U_{49} group using the group chain

$$U_{49} \supset SU_7 \times SU_7 \supset SU_7 \supset SO_7 \supset G_2 \supset SO_3$$

While we have restricted the scope of this talk to the interacting boson model of nuclei many of the ideas are directly transferable to problems in atomic and molecular physics and even to esoteric phenomena such as the Jahn-Teller effect in solid state physics.

Acknowledgements

It has been a pleasure to participate in this symposium in such magnificent surroundings. We are all indebted to Bruno Gruber. The writing of this talk was completed in the Faculty of Mathematical Studies, University of Southampton while holding a Hartley Visiting Award.

REFERENCES

1. E.Cartan,*Sur la structure des groupes de transformations finis et continus*, thesis, Nony, Paris (1894).
2. N.J.A. Sloane, *Sci. Amer.***250(1)**, 116 (1984).
3. G. Racah, *Phys. Rev.***76**, 1352 (1949).
4. B.G. Wybourne and M.J. Bowick, *Austr. J. Phys.***30**, 259 (1977).
5. R.C. King and A.H.A. Al-Qubanchi,*J. Phys. A:Math. Gen.***14**, 15 (1981).
6. R.C. King and A.H.A. Al-Qubanchi,*J. Phys. A:Math. Gen.***14**, 51 (1981).
7. R.C. King, *J. Phys. A:Math. Gen.***14**, 76 (1981).
8. R.C. King, Luan Dehuai and B.G. Wybourne, *J. Phys. A:Math. Gen.* **14**, 2509 (1981).
9. G.R.E. Black, R.C. King and B.G. Wybourne, *J. Phys. A:Math. Gen.* **16**, 1555 (1983).
10. B.G. Wybourne, *J. Phys. A:Math. Gen.***17**, 1397 (1984).
11. R.C. King and B.G. Wybourne, *J Phys. A:Math. Gen.***18**, 3113 (1985).
12. R.J. Farmer, R.C. King and B.G.Wybourne, *J. Phys. A:Math. Gen.* **21**, 3979 (1988).
13. A. Arima and F. Iachello, *Ann. Phys. N.Y.***99**, 253 (1976).
14. H.A. Jahn, *Proc. Roy. Soc. London,***A201**, 516 (1950).
15. B.H. Flowers, *Proc. Roy. Soc. London,***A212**, 248 (1952).
16. J.P. Elliott, *Proc. Roy. Soc. London,***A245**, 128 (1958).
17. J.P. Elliott, *Proc. Roy. Soc. London,***A218**, 345 (1953).
18. J.P. Elliott and B.H. Flowers, *Proc. Roy. Soc. London,* **A229**, 545 (1955).
19. D.E. Littlewood, *The theory of group characters*,(Oxford:Clarendon) (2nd edn 1950).
20. B.G. Wybourne, *Symmetry principles and atomic spectroscopy*, (New York:Wiley) (1970).
21. P.H. Butler and B.G. Wybourne, *J. Math. Phys.***11** 2512 (1976).
22. P.W. Pieruschka, *M.Sc. thesis University of Canterbury* (1990).
23. I. Morrison, P.W. Pieruschka and B.G. Wybourne,*J. Math. Phys.* (submitted).
24. H.T. Wadzinski, *Il Nuovo Cimento*,**X,62B**, 247 (1969).
25. T. Regge, *Il Nuovo Cimento*,**11**, 116 (1959).

PARADIGMS OF QUANTUM ALGEBRAS

Cosmas Zachos*

High Energy Physics Division, Argonne National Laboratory

Argonne, IL 60439-4815, USA (zachos@anlhep)

This is an informal overview of versions of quantum algebras which are currently finding applications in physics. Special attention is given to the quantum deformations of SU(2) and illustrations of general principles.

1. Introduction

Quantum Algebras, or QUE-(quantized universal enveloping)-algebras, are remarkable mathematical structures (noncommutative, noncocommutative Hopf algebras) which have been figuring in

i. 2-d solvable model S-matrices and solutions to their Yang-Baxter factorization equations [55, 90, 26, 45, 46, 23, 42, 33, 8].

ii. Anisotropic spin chain hamiltonians [74, 3].

iii. 3-d Chern-Simons theory Wilson loops [99, 39, 62, 89].

iv. Chiral vertices, fusion rules, and conformal blocks of RCFT [1, 67, 36]; 2-d gravity [35]; related applications of knot theory to physics [49, 88].

v. q-strings and group-theoretic interpretation of q-hypergeometric functions [82].

vi. Nonstandard quantum statistics [37].

vii. Heuristic phenomenology of deformed molecules and nuclei [44, 77].

Quantum algebras become relevant in physics where the limits of applicability of Lie Algebras are stretched: they describe perturbations from some underlying symmetry structure, such as quantum corrections or anisotropies. They are currently being explored with a view

*Work supported by the U.S.Department of Energy, Division of High Energy Physics, Contract W-31-109-ENG-38.

to new applications in a broad range of contexts. There are several outstanding reviews of the subject, which also cover much of its interesting history and illuminate particular aspects of it [25, 46, 26, 63, 61]. Here, I opt instead for a briefer, more eclectic, illustrative, and less historical introduction to these ideas. It is based on explicit prototypes, mostly addressing quantum deformations of SU(2), and techniques that may facilitate and encourage new applications.

2. Deformation of SU(2)

Consider the algebra of SU(2):

$$[j_x, j_y] = ij_z \qquad [j_y, j_z] = ij_x \qquad [j_z, j_x] = ij_y\ , \tag{2.1}$$

or, equivalently, for $j_x = (j_+ + j_-)/\sqrt{2}$, $j_y = -i(j_+ - j_-)/\sqrt{2}$, $j_z = j_0$,

$$[j_0, j_+] = j_+ \qquad [j_+, j_-] = j_0 \qquad [j_-, j_0] = j_-\ . \tag{2.2}$$

The Casimir invariant is

$$C \equiv j_x^2 + j_y^2 + j_z^2 = j_+j_- + j_-j_+ + j_0^2 = 2j_+j_- + j_0(j_0 - 1)\ . \tag{2.3}$$

Now suppose we mar the isotropy of this spherical expression by *deforming* it to:

$$C_q(j) \equiv j_+j_- + j_-j_+ + \frac{q + 1/q}{2}\left(\frac{q^{j_0} - q^{-j_0}}{q - q^{-1}}\right)^2 , \tag{2.4}$$

where the real or complex $q - 1$ parameterizes the amount of anisotropy. q may be thought of as a phase, as in RCFT, or as $e^{\hbar}$, following historical development; in that case, the last term in C_q amounts to

$$\cosh\hbar\ \left(\frac{\sinh(\hbar j_0)}{\sinh(\hbar)}\right)^2,$$

which goes to the classical/isotropic limit as $\hbar \to 0$, i.e. $q \to 1$. Define, in general, the "*q-deformation of x*": $[x]_q \equiv (q^x - q^{-x})/(q - q^{-1})$, so that $[x]_q \to x$ as $q \to 1$. Thus, the last term above amounts to

$$\frac{[2]_q}{2}\ [j_0]_q^2\ .$$

Is most of the symmetry of the operator C_q gone (beyond the residual axial j_0)? It turns out in fact that it may be salvaged, provided the universal enveloping algebra of SU(2) is used in a suitable deformation. Define, with [55, 25, 45] new operators J_a which satisfy

$$[J_0, J_+] = J_+ \qquad [J_+, J_-] = \frac{1}{2}\ [2J_0]_q \qquad [J_-, J_0] = J_-\ , \tag{2.5}$$

which has (2.2) as its classical limit $q \to 1$. All of its generators now commute with C_q, written as

$$C_q(J) = 2J_+J_- + [J_0]_q[J_0 - 1]_q\ . \tag{2.6}$$

(2.5) is not a Lie algebra anymore, which forestalls its Lie-exponentiation to a group. It is a more general algebra: a Hopf algebra, which is to say that it is endowed with the following structures.

I. **Coproduct** Δ. This is an algebra homomorphism that corresponds to the composition of angular momenta, i.e. it specifies tensor (co)multiplication of representations. In the above example, it is [91, 45]:

$$\Delta_q(J_0) = J_0 \otimes \mathbb{1} + \mathbb{1} \otimes J_0 \qquad \Delta_q(J_\pm) = J_\pm \otimes q^{J_0} + q^{-J_0} \otimes J_\pm\ , \tag{2.7}$$

so that the $\Delta(J)$ satisfy the algebra (2.5), like a "total angular momentum". This coproduct is coassociative, but not cocommutative, since, defining the permutation map $\sigma(a\otimes b)\equiv b\otimes a$, you may note that $\sigma(\Delta_q)=\Delta_{1/q}\neq\Delta_q$. (This is an equally good coproduct, and still others are discussed below.) A given coproduct such as Δ_q determines the other two structures which, however, will not be crucial for this discussion:

II. **Counit** ϵ. This algebra homomorphism reverses the effect of the above comultiplication. Given the multiplication map $m(a\otimes b)\equiv ab$ for spaces of matching dimension, the counit satisfies

$$m\Big((\epsilon\otimes 1\!\!1)\Delta(J_a)\Big)=m\Big((1\!\!1\otimes\epsilon)\Delta(J_a)\Big)=J_a.$$

Here, it is $\epsilon(J_a)=0,\quad \epsilon(1\!\!1)=1\!\!1$.

III. **Antipode** S. This is an algebra antihomomorphism, $S(J_aJ_b)=S(J_b)S(J_a)$, s.t.

$$\sigma\Big(\Delta(S(J_a))\Big)=(S\otimes S)(\Delta(J_a));\qquad m\Big((S\otimes 1\!\!1)\Delta(J_a)\Big)=m\Big((1\!\!1\otimes S)\Delta(J_a)\Big)=\epsilon(J_a).$$

Here, it is easy to check $S(J_\pm)=-q^{\pm1}J_\pm,\quad S(J_0)=-J_0$. Note the familiar classical limits of all of the above maps.

For generic q not equal to 1, the representation theory of this deformation, as detailed later, is in one-to-one correspondence with the representation theory of its classical limit, here the theory of angular momentum. Just as composing representations and taking functions of their Casimir invariants for SU(2) yields invariant hamiltonians, parallel comultiplications for $\mathrm{SU}(2)_q$ provide a variety of invariants, out of which, for instance, important spin-chain hamiltonians have been identified to be invariant under $\mathrm{SU}(2)_q$ [74, 3]. In (I) above, the alternative coproduct $\Delta_{1/q}$ was introduced, which is in fact equivalent to Δ_q via a similarity transformation: $\Delta_q=R_q\Delta_{1/q}R_q^{-1}$. This universal R-matrix of Drinfeld, with $R_q^{-1}=R_{1/q}$, leads to solutions of the Yang-Baxter equation, which is not reviewed here, as it is covered in detail in the reviews of [46, 26, 53].

There are several alternate deformations of SU(2) available [90, 100, 99, 27]. Each one has its distinctive invariants and representation theory, and all are related among themselves. To map them onto each other, one may first map them to this prototype deformation discussed, or to their common classical limit SU(2), as described next.

3. Deforming functionals and representation theory

The term "deformation" used above may, in fact, be made explicit [18]. Rewrite the classical invariant operator C, (2.3), as $j(j+1)$, where j is the formal operator $(\sqrt{1+4C}-1)/2$. Then, by dint of the commutation relations of SU(2), the functionals

$$J_0=Q_0(j_0)=j_0\qquad J_+=Q_+(g)=\sqrt{\frac{[j_0+j]_q[j_0-1-j]_q}{(j_0+j)(j_0-1-j)}}\,j_+\qquad J_-=(Q_+(g))^\dagger \tag{3.1}$$

satisfy the commutation relations of $\mathrm{SU}(2)_q$, (2.5). The maps $Q_\pm$ are functionals of all three SU(2) generators $g:\ j_0,\ j_+,\ j_-$, since they depend on the operator j. Moreover, (2.6) then amounts to $[j]_q[j+1]_q$, i.e. a function of the classical invariant C. Conversely, for generic q (not a root of unity), one may further solve for j if only the J_a's are given:

$$2j+1=\operatorname{arccosh}\Big((q+1/q+(q-1/q)^2C_q)/2\Big)\Big/\ln q\,. \tag{3.2}$$

Consequently, the functionals (3.1) are invertible, and their inverses Q^{-1} provide a realization of SU(2) in terms of quantum algebra generators, with the classical Casimir expressible as a function of the quantum one, C_q. These maps then provide realizations of each algebra in terms of the other. Thus, functions of C_q are also invariant under SU(2), while functions of C are also invariant under $SU(2)_q$[2]. As a result, these deforming maps specify the representation theory of each; e.g. when representations of SU(2) are substituted into (3.1), they yield the corresponding representations of $SU(2)_q$ of the same dimension. This underscores the general result that the representation theory of $SU(2)_q$ for generic q reduces to a "distorted echo" of the representation theory of SU(2) [83, 57, 96]. Functionals of broadly analogous type have also appeared in [45, 84, 71, 60, 14, 75, 27].[3]

Having referred the representation theory of the QUE-algebra to the representation theory of SU(2), the above map links the respective composition laws for representations. It thus specifies a coproduct, which appears different from (2.7). The map-induced coproduct simply *classicizes the* $SU(2)_q$ *representations through the inverse maps* Q^{-1}, *it composes them at the classical level, and then it quantizes the answer through* Q. More specifically, in the classical addition of angular momenta, two parallel operators tensor-multiply to an operator satisfying the same SU(2) commutation relations; this operator is a reducible representation of SU(2), the reduction (and diagonalization of the cocasimir) effected by the Clebsch-Gordan operator $\mathcal{C}$:

$$\Delta(g) = \mathbb{1} \otimes g + g \otimes \mathbb{1} = \mathcal{C}(g_1 \oplus g_2 \oplus g_3 \oplus ...)\mathcal{C}^{-1} \,. \tag{3.3}$$

Thus, the invertible map Q from SU(2) generators g to $SU(2)_q$ generators $G = Q(g)$ induces the following tensor coproduct of G's

$$Q(\Delta(g)) = Q\Big(\mathbb{1} \otimes Q^{-1}(G) + Q^{-1}(G) \otimes \mathbb{1}\Big) \,, \tag{3.4}$$

which obeys $SU(2)_q$ quommutations, since its argument obeys SU(2) [18, 75]. Now the same Clebsch operator $\mathcal{C}$ will automatically also reduce the coproduct (3.4): $\mathcal{C}^{-1}Q(\Delta(g))\mathcal{C} = G_1 \oplus G_2 \oplus G_3 \oplus ...$; this reduced coproduct is an equivalent one, since any similarity transformation on a coproduct will isomorphically produce an expression also satisfying the same algebra. The antipodes specified by the map (3.1) evidently amount to mere sign flips, just as in the classical algebra, and thus also appear different from (III); the resulting counit is likewise identical to the classical one.

The map-induced coproduct discussed is quite difficult to handle in some cases, and is not well-defined for q equal to a root of unity, as discussed later. How does it relate to the prototype Δ_q of the previous section? For generic q, that coproduct Δ_q reduces to a direct sum by the unitary q-Clebsch operators $\mathcal{C}_q$. Such coefficients are covered in [95, 50, 73, 71, 5, 85, 52, 80], while the q-Wigner-Eckart theorem is worked out in [5, 71]. Consequently,

$$Q(\Delta(g)) = \mathcal{C}\mathcal{C}_q^{-1}\; \Delta_q(G)\; \mathcal{C}_q\mathcal{C}^{-1} \equiv U_q^{-1}\; \Delta_q(G)\; U_q \,. \tag{3.5}$$

The induced comultiplication is thus related to (2.7) by a similarity transformation introduced in [17], $U_q = \mathcal{C}_q\mathcal{C}^{-1}$, the unitary operator that converts $\mathcal{C}$ to $\mathcal{C}_q$. Similarly, as already mentioned, Δ_q transforms to its double $\Delta_{1/q}$ through the operator $R_q = U_q U_{1/q}^{-1} = \mathcal{C}_q \mathcal{C}_{1/q}^{-1}$, which converts $\mathcal{C}_{1/q}$ to $\mathcal{C}_q$: $\Delta_q = R_q \Delta_{1/q} R_q^{-1}$. Some discussion of the broad equivalence class of coproducts is given in [17]. The inverse functionals, in an unfolding of (3.3), moreover specify a non-cocommutative coproduct for classical SU(2) [15], which reduces by $\mathcal{C}_q$ instead of $\mathcal{C}$ and thus also transforms to the standard one (3.3) by the U matrix. For generic analysis see [34].

It is worth illustrating the above general statements by substitution of unitary irreducible representations of SU(2) into formulas (3.1). The J_-'s follow from hermitean conjugation.

[2]An extension to spin-chain hamiltonians, [9], contingent on their complete decomposition to irreducible blocks, uncovers SU(2) symmetry in anisotropic spin chains.

[3]Beyond the functionals sketched so far, various *non*invertible functionals are available which connect SU(1,1) with the centerless Virasoro algebra [29], or the classical SU(2) current algebra with $SU(2)_q$ [43], and others.

The doublet representation (Pauli matrices):

$$j_0 = \frac{1}{2}\begin{pmatrix} 1 & 0 \\ 0 & -1 \end{pmatrix} \qquad j_+ = \frac{1}{\sqrt{2}}\begin{pmatrix} 0 & 1 \\ 0 & 0 \end{pmatrix} \tag{3.6}$$

maps to itself for this deformation: $J_0 = j_0, \quad J_+ = j_+$. This is a special feature of the defining representation in this particular deformation. Note $C_q = 1 - [1/2]_q^2$. The **3**:

$$j_0 = \begin{pmatrix} 1 & 0 & 0 \\ 0 & 0 & 0 \\ 0 & 0 & -1. \end{pmatrix} \qquad j_+ = \begin{pmatrix} 0 & 1 & 0 \\ 0 & 0 & 1 \\ 0 & 0 & 0 \end{pmatrix} \tag{3.7}$$

maps to $J_0 = j_0, \quad J_+ = \sqrt{(q+1/q)/2}\ j_+ = \sqrt{[2]_q/2}\ j_+$. The **4**:

$$j_0 = \begin{pmatrix} 3/2 & 0 & 0 & 0 \\ 0 & 1/2 & 0 & 0 \\ 0 & 0 & -1/2 & 0 \\ 0 & 0 & 0 & -3/2 \end{pmatrix} \qquad j_+ = \begin{pmatrix} 0 & \sqrt{3/2} & 0 & 0 \\ 0 & 0 & \sqrt{2} & 0 \\ 0 & 0 & 0 & \sqrt{3/2} \\ 0 & 0 & 0 & 0 \end{pmatrix} \tag{3.8}$$

maps to

$$J_0 = j_0 \qquad J_+ = \begin{pmatrix} 0 & \sqrt{[3]_q/2} & 0 & 0 \\ 0 & 0 & [2]_q/\sqrt{2} & 0 \\ 0 & 0 & 0 & \sqrt{[3]_q/2} \\ 0 & 0 & 0 & 0 \end{pmatrix}, \tag{3.9}$$

and so forth.

To illustrate coproducts (2.7,3.4), consider the $\mathbf{2} \otimes \mathbf{3}$ case. Classically, by (3.3) and (3.6-7),

$$\Delta(j_0) = \text{diag}\,(3/2, 1/2, -1/2, 1/2, -1/2, -3/2)\,,$$

$$\Delta(j_+) = \begin{pmatrix} 0 & 1 & 0 & 1/\sqrt{2} & 0 & 0 \\ 0 & 0 & 1 & 0 & 1/\sqrt{2} & 0 \\ 0 & 0 & 0 & 0 & 0 & 1/\sqrt{2} \\ 0 & 0 & 0 & 0 & 1 & 0 \\ 0 & 0 & 0 & 0 & 0 & 1 \\ 0 & 0 & 0 & 0 & 0 & 0 \end{pmatrix} \tag{3.10}$$

reduce by

$$C = \begin{pmatrix} 1 & 0 & 0 & 0 & 0 & 0 \\ 0 & \sqrt{2/3} & 0 & -1/\sqrt{3} & 0 & 0 \\ 0 & 0 & 1/\sqrt{3} & 0 & -\sqrt{2/3} & 0 \\ 0 & 1/\sqrt{3} & 0 & \sqrt{2/3} & 0 & 0 \\ 0 & 0 & \sqrt{2/3} & 0 & 1/\sqrt{3} & 0 \\ 0 & 0 & 0 & 0 & 0 & 1 \end{pmatrix} \tag{3.11}$$

to $\mathbf{4} \oplus \mathbf{2}$ blocks — the classical limit of (3.14) below. The same C also reduces $Q(\Delta(j_+))$.

However,

$$\Delta_q(J_+) = 1/\sqrt{2}\begin{pmatrix} 0 & \sqrt{[2]_q/q} & 0 & q & 0 & 0 \\ 0 & 0 & \sqrt{[2]_q/q} & 0 & 1 & 0 \\ 0 & 0 & 0 & 0 & 0 & 1/q \\ 0 & 0 & 0 & 0 & \sqrt{[2]_q q} & 0 \\ 0 & 0 & 0 & 0 & 0 & \sqrt{[2]_q q} \\ 0 & 0 & 0 & 0 & 0 & 0 \end{pmatrix} \tag{3.12}$$

reduces instead through

$$C_q = \begin{pmatrix} 1 & 0 & 0 & 0 & 0 & 0 \\ 0 & \sqrt{[2]_q/[3]_q q} & 0 & -q/\sqrt{[3]_q} & 0 & 0 \\ 0 & 0 & 1/q\sqrt{[3]_q} & 0 & -\sqrt{[2]_q q/[3]_q} & 0 \\ 0 & q/\sqrt{[3]_q} & 0 & \sqrt{[2]_q/[3]_q q} & 0 & 0 \\ 0 & 0 & \sqrt{[2]_q q/[3]_q} & 0 & 1/q\sqrt{[3]_q} & 0 \\ 0 & 0 & 0 & 0 & 0 & 1 \end{pmatrix} \tag{3.13}$$

to $\mathbf{4} \oplus \mathbf{2}$ blocks,

$$C_q^{-1}\Delta_q(J_+)C_q = 1/\sqrt{2} \begin{pmatrix} 0 & \sqrt{[3]_q} & 0 & 0 & 0 & 0 \\ 0 & 0 & [2]_q & 0 & 0 & 0 \\ 0 & 0 & 0 & 0 & 0 & \sqrt{[3]_q} \\ 0 & 0 & 0 & 0 & 1 & 0 \\ 0 & 0 & 0 & 0 & 0 & 0 \\ 0 & 0 & 0 & 0 & 0 & 0 \end{pmatrix}. \tag{3.14}$$

Naturally, the q-cocasimir diagonalizes to $C_q^{-1}\Big(2\Delta_q(J_+)\Delta_q(J_-) + [\Delta_q(J_0)][\Delta_q(J_0) - 1]\Big)C_q$ $= \mathrm{diag}\,([3/2][5/2], [3/2][5/2], [3/2][5/2], [1/2][3/2], [1/2][3/2], [3/2][5/2])$, which bears the expected functional relationship to its clasical limit. The reader ought to check all corresponding classical limits.

The two quantum coproducts are related by

$$U_q = \begin{pmatrix} 1 & 0 & 0 & 0 & 0 & 0 \\ 0 & c(q) & 0 & s(q) & 0 & 0 \\ 0 & 0 & c(1/q) & 0 & -s(1/q) & 0 \\ 0 & -s(q) & 0 & c(q) & 0 & 0 \\ 0 & 0 & s(1/q) & 0 & c(1/q) & 0 \\ 0 & 0 & 0 & 0 & 0 & 1 \end{pmatrix}$$

$$c(q) = \frac{\sqrt{2[2]_q/q} + q}{\sqrt{3[3]_q}}, \qquad s(q) = \frac{\sqrt{[2]_q/q} - \sqrt{2}q}{\sqrt{3[3]_q}}, \tag{3.15}$$

and therefore

$$R_q = \begin{pmatrix} 1 & 0 & 0 & 0 & 0 & 0 \\ 0 & \frac{[2]+1}{[3]} & 0 & (\frac{1}{q}-q)\frac{\sqrt{[2][3/2]}}{[3]} & 0 & 0 \\ 0 & 0 & \frac{[2]+1}{[3]} & 0 & (\frac{1}{q}-q)\frac{\sqrt{[2][3/2]}}{[3]} & 0 \\ 0 & (q-\frac{1}{q})\frac{\sqrt{[2][3/2]}}{[3]} & 0 & \frac{[2]+1}{[3]} & 0 & 0 \\ 0 & 0 & (q-\frac{1}{q})\frac{\sqrt{[2][3/2]}}{[3]} & 0 & \frac{[2]+1}{[3]} & 0 \\ 0 & 0 & 0 & 0 & 0 & 1 \end{pmatrix}$$

$$= (\frac{1}{q} - q)\frac{2[3/2]_q}{[3]_q}(J_+ \otimes J_- - J_- \otimes J_+) + 2(1 - \frac{[2]_q + 1}{[3]_q})J_0^2 \otimes J_0^2 + (1 - \frac{[2]_q + 1}{[3]_q})J_0 \otimes J_0 + \frac{[2]_q + 1}{[3]_q}\,\mathbb{1} \otimes \mathbb{1}\,. \tag{3.16}$$

Dramatic new features emerge as q becomes an Nth root of unity, hence $[N/2]_q = 0$, however [58, 81, 1, 86, 32], which is of special relevance to RCFT. Inspection of the deforming functionals (3.1) indicates that:

a. *The dimensionality of the irreducible representations is bounded above by N.* (The constraints are actually twice as stringent for even Ns, since the effective period is $N/2$—see the above references). $J_\pm$ become nilpotent, $J_\pm^N = 0$, which may be seen from the vanishing products $[j_0+j][j_0+j-1]....[j_0+j+2-N][j_0+j+1-N]$ resulting inside the square-roots of the Nth power of (3.1) through the identity $j_+ f(j_0) = f(j_0-1) j_+$. Thus there is only a *finite* number of irreducible representations for $SU(2)_q$. Consequently, it is necessary that large irreps of SU(2) map to reducible representations of $SU(2)_q$, as the raising/lowering within a representation is interrupted by the zeros inside the square-roots of (3.1). For example, observe that $[3]_q = 0 = [3/2]_q$ for $q = \exp(2\pi i/3)$. The **4** representation J_+ now has only one nontrivial entry and $J_+^2 = 0$; the middle commutator in (2.5) breaks up, so the representation reduces: $\mathbf{4} = \mathbf{1} \oplus \mathbf{2} \oplus \mathbf{1}$.

b. The invariant operator C_q *does not label representations uniquely anymore.* E.g. for odd N, the invariant for any representation of dimension $2j+1$ coincides with that of dimension $2j'+1 \equiv nN-(2j+1)$, integer n, or dimension $2j+1+nN$. Such representations with identical Casimir operators can mix into *indecomposable* but not irreducible representations, provided the collective *q-dimension*, $\sum q^{2j_0} = [2j+1] + [2j'+1]$, of the composite representation vanishes [74]. Pasquier & Saleur term such representations "type I". Full reduction fails by dint of the divergence of C_q [17, 101][4]. For example, for $q = \exp(2\pi i/3)$ again, the nonunitary **6** of eq. (3.12) is reducible, but not decomposable to a **4** and a **2**, as their collective q-dimension vanishes: $[4]+[2] = 0$. Specifically, since $J_- = J_+^t$, the norm is $v \cdot v = v^t v$. The six states $a^t = (0,0,0,0,0,1)$, $d^t = (-1,0,0,0,0,0)$, $b \equiv J_+ a$, $c \equiv J_- d$, $b'^t = q\sqrt{2}(0,0,1,0,-i,0)$, $c'^t = q^2\sqrt{2}(0,i,0,0,-1,0)$, contain the doublet of zero-norm states b and c, which only transform to each other: $J_+ b = c/\sqrt{2}$, $J_- c = b/\sqrt{2}$, $J_- b = 0$, $J_+ c = 0$. However, as evident above, a and d are not singlets, and may, in turn, be reached from elsewhere: $J_- b' = a$, $J_+ c' = d$; $b \cdot b' = c \cdot c' = 1$, and $J_- c' = (b+b')/\sqrt{2}$, $J_+ b' = (c+c')/\sqrt{2}$. However, the r.h.s of (3.14) decomposes completely, so divergence of C_q is necessary, and likewise of U_q, but not of R_q. The reader would profit from working out more examples so as to develop facility for applications.

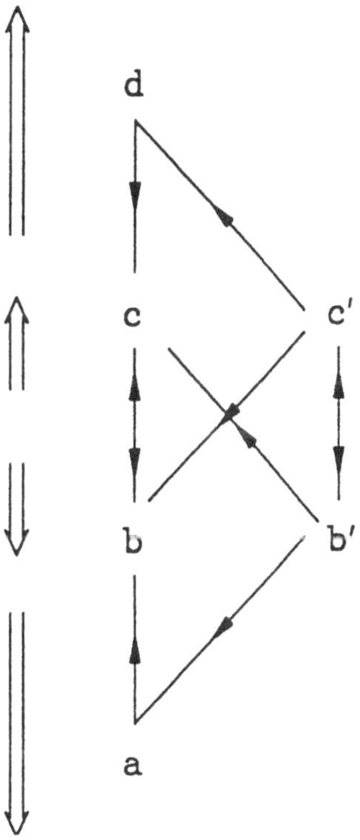

c. If, in addition, unitarity is required, substantially more stringent constraints ensue on the allowed dimensionalities of the irreducible representations [66]. $SU(2)_q$ and $SU(1,1)_q$ are linked, as unitary representations of one are "antiunitary" ones ($J_+^\dagger = -J_-$) of the other and vice-versa. The dimensionalities of these unitary/antiunitary representations are given by Takahashi-Suzuki numbers, while there is also a class of irreducible representations of indefinite hermitean conjugation signature. (E.g. the **4** for $q = \exp(2\pi i/5)$. Again, the reader may wish to practice with (3.9)).

[4]This is also implicit in [80].

4. Other deformations of SU(2), and generalizations to other algebras

The deforming functionals exemplified above are by no means unique. Nonhermitean functionals are also found in [45, 18] and, in general, any nonsingular similarity transform of the functionals discussed will also do.

There is a number of interesting alternative deformations of $SU(2)_q$ which have arisen in several contexts, listed below:

i. Sklyanin's trigonometric deformation [90, 60]:

$$[S_0, S_3] = 0 \; ; \qquad [S_+, S_-] = 4S_0S_3 \; , \qquad [S_3, S_\pm] = \pm(S_0S_\pm + S_\pm S_0) \; ,$$

$$[S_0, S_\pm] = \pm\tanh^2 u \, (S_\pm S_3 + S_3 S_\pm) \; , \quad \text{where} \quad S_0^2 - S_3^2 \tanh^2 u = 4\sinh^2 u \; , \quad (4.1)$$

with classical limit $u \to 0$ (upon rescaling of the generators), and an invariant

$$C_u = S_+S_- + S_-S_+ + S_3^2\Big(\frac{2\cosh 2u}{\cosh^2 u}\Big) \; . \qquad (4.2)$$

ii. Woronowicz's deformation [100] has a linear r.h.s., but "quommutators" in lieu of commutators:

$$[V_0, V_+]_{s^2} \equiv s^2 V_0 V_+ - \frac{1}{s^2} V_+ V_0 = V_+ \qquad [V_-, V_0]_{s^2} = V_-$$

$$[V_+, V_-]_{1/s} \equiv \frac{1}{s} V_+ V_- - s V_- V_+ = V_0 \; . \qquad (4.3)$$

The invariant,

$$C_s = 2\Big(V_-V_+ + \frac{(1 - V_0(1 - 1/s^2))}{s(s - 1/s)^2}\Big) \Big/ \sqrt{1 - (s^2 - 1/s^2)V_0} \; , \qquad (4.4)$$

strictly commutes with the generators, i.e. $[C_s, V] = 0$. In the classical $s \to 1$ limit, this reduces to the operator (2.3) plus the divergent constant $(1 - s)^{-2}/2 - 1/8$.

iii. Witten's first deformation [99]:

$$[E_0, E_+]_p \equiv pE_0E_+ - \frac{1}{p}E_+E_0 = E_+ \qquad [E_+, E_-] = E_0 - (p - \frac{1}{p})E_0^2 \qquad [E_-, E_0]_p = E_- \; . \qquad (4.5)$$

The Casimir operator which commutes with all generators is:

$$C_p = \frac{1}{p}E_+E_- + pE_-E_+ + E_0^2 \qquad [C_p, E] = 0. \qquad (4.6)$$

iv. Witten's second deformation [99]:

$$[W_0, W_+]_r \equiv rW_0W_+ - \frac{1}{r}W_+W_0 = W_+ \qquad [W_+, W_-]_{1/r^2} = W_0 \qquad [W_-, W_0]_r = W_- \; . \qquad (4.7)$$

Observe the symmetry $W_0 \leftrightarrow -W_0$, $W_+ \leftrightarrow W_-$, $r \leftrightarrow 1/r$. For arbitrary functions f, it follows that $W_+f(W_0) = f(r^2W_0 - r)W_+$ and $W_+f(W_-W_+) = f(W_+W_-)W_+ = f(r^4W_-W_+ + r^2W_0)W_+$, and their $+/-$ symmetric analogs. As a result, by virtue of

$$C_1 = 2W_-W_+ + \frac{2}{r^2(r + 1/r)}W_0(W_0 + r), \qquad C_2 = (1 - (r - 1/r)W_0)^2,$$

$$[C_i, W_\pm]_{r^{\pm 2}} = 0 \qquad [C_i, W_0] = 0 \; , \qquad (4.8)$$

a Casimir operator which commutes with all generators is:

$$C_r = C_1/C_2 \qquad\qquad [C_r, W] = 0. \tag{4.9}$$

Observe that (ii, iii, iv) have SU(2) as their $s = 1$, $p = 1$, and $r = 1$ limit, respectively, and SU(1,1) as their $s = -1$, $p = -1$, and $r = -1$ limit.

v. The two deformations (ii), (iv) are special limits of a 2-parameter generalization of [27],

$$[I_0, I_+]_r = I_+ \qquad [I_+, I_-]_{1/s} = I_0 \qquad [I_-, I_0]_r = I_- \ , \tag{4.10}$$

upon $r \to s^2$, or $s \to r^2$, respectively. The corresponding invariant is

$$I_{r,s} = C_1/C_2 \ , \tag{4.11}$$

$$C_1 = 2I_-I_+ + \frac{2}{r-1/r}\Big(\frac{1}{s-1/s} - \frac{1-(r-1/r)I_0}{s-r^2/s}\Big) \ , \qquad C_2 = (1-(r-1/r)I_0)^{\ln s/\ln r}.$$

Linear combinations of C_1 and C_2 are equally acceptable in the numerator of the above invariant, which leads to the limit (4.9).

vi. The cyclically symmetric deformation [72, 27]:

$$qXY - q^{-1}YX = Z \qquad qYZ - q^{-1}ZY = X \qquad qZX - q^{-1}XZ = Y \ , \tag{4.12}$$

with a cubic invariant

$$C_q = (q^3 + 2q^{-1})(XYZ + YZX + ZXY) - (q^{-3} + 2q)(XZY + ZYX + YXZ) =$$

$$= \frac{2q^4 + 5 + 2q^{-4}}{(q-1/q)(q^2 - 1 + q^{-2})}\Big([X,Y]_Q, Z]_Q + [Y,Z]_Q, X]_Q + [Z,X]_Q, Y]_Q\Big), \tag{4.13}$$

where $Q^2 \equiv (q^3 + 2q^{-1})/(q^{-3} + 2q)$. The Casimir invariant goes to the conventional one in the classical limit—the vanishing of the denominator of the coefficient exactly compensates for the collapse of the Q-determinant to the Jacobi identity [101].

Deforming maps which map the representation theories (including the special limits q – roots of unity in the previous section) to that of each other, either directly, or via SU(2) are also available. E.g. consider (iv) above. A map to the prototype deformation (2.5) is [18]

$$W_0 = \frac{r^{-J_0}}{r+1/r}(r^{1+j}[J_0 - j]_r + r^{-1-j}[J_0 + j]_r) = \frac{1}{r-1/r}\Big(1 - \frac{r^{2j+1} + r^{-2j-1}}{r+1/r} r^{-2J_0}\Big)$$

$$W_+ = r^{-J_0}\sqrt{\frac{2r}{r+1/r}}\sqrt{\frac{[J_0+j]_r[J_0-1-j]_r}{[J_0+j]_q[J_0-1-j]_q}}\, J_+ \ , \tag{4.14}$$

for which the Casimir invariant (4.9) reduces to[5]

$$C_r = \frac{2\,[2j]_r[2j+2]_r}{r^2(r+1/r)(r^{2j+1}+r^{-2j-1})^2} \ . \tag{4.15}$$

By virtue of (3.1), it is evident that (4.14) is identical with its $q = 1$ limit and also represents, in fact, a map from (2.2) to (4.7). Moreover, note the substantial simplification when $q = r$:

$$W_0 = \frac{1}{r-1/r}\Big(1 + \frac{[2j]_r - [2j+2]_r}{[2]_r} r^{-2J_0}\Big) \ , \qquad W_+ - r^{1/2-J_0}\sqrt{\frac{2}{[2]_r}}\, J_+ \ , \tag{4.16}$$

[5]This amounts to (5.10) of [99], up to a factor of $2r^{-2}/[4]_r$.

which, e.g., allows a rapid inspection of the limit when r is a root of unity; as in the previous section, the zeros of $W_\pm$ dictate breakup of large representations and impose the same bounds on dimensionalities of irreps.

Analogous functionals exist in the literature for each of the above deformations: (i.) [81, 60]; (ii.)[92, 18, 83]; (iii.) [71, 18]; (iv.) [18]; (v.) [18]; (vi.) [27, 101, 15]. In this sense, these deformations are "equivalent" being all equivalent to SU(2). Normally, invertibility is lost for q = root of unity, but several of the direct maps among deformations survive, and map the respective modular representations discussed in the previous section to each other, as exemplified for (iv). [34] discuss equivalence more generally, as well as obstructions and the cocycle structure of such deformations.

Generalizations to the other Lie Algebras beyond SU(2) are available [78, 20, 24]:

a. $SU(1,1)_q$ mentioned already, [4]: unitary irreps for generic q discussed by [65, 54, 38]; [101] probes modular representations.

b. $SU(N)_q$ and their affine (Kac-Moody) extensions [25, 45, 79]; [94] investigate the representation theory; [19, 2] study the cyclic representations; symmetric representations also discussed in [93, 75] via q-oscillator realizations described below.

   There exists an intriguing deformation of the Moyal algebra [28]:

$$q^{\mathbf{n}\times\mathbf{m}} J_{\mathbf{m}} J_{\mathbf{n}} - q^{\mathbf{m}\times\mathbf{n}} J_{\mathbf{n}} J_{\mathbf{m}} = \left(\omega^{\mathbf{m}\times\mathbf{n}/2} - \omega^{\mathbf{n}\times\mathbf{m}/2}\right) J_{\mathbf{m}+\mathbf{n}} + \mathbf{a}\cdot\mathbf{m}\ \delta_{\mathbf{m}+\mathbf{n},0} \tag{4.17}$$

   which holds promise for practical applications. The indices are 2-vectors with integer entries, $\mathbf{m} = (m_1, m_2)$, $\mathbf{m}\times\mathbf{n} = m_1 n_2 - m_2 n_1$, and $\mathbf{a}$ is an arbitrary 2-vector characterizing the center. The classical limit is the Moyal algebra [30], which identifies with a maximally graded basis of $SU(N)$ for $\omega = e^{2\pi i/N}$ (the natural generalization of the Pauli matrices to $N > 2$): in this cyclotomic case, all indices identify mod N, and consequently there are only N^2 different J's. For nontrivial q, in the limit $N \to \infty$, this provides the generalization of (vi) to $SU(\infty)_q$, upon proper rescaling of the generators. This is the quantum version of the Poisson Bracket.

c. $SO(N)_q$, $Sp(N)_q$, in [78, 47, 68, 48].

d. The exceptional algebras have been approached in [78, 51, 59].

e. The graded algebras: $Osp(2|1)_q$ is detailed in [56, 22, 87, 7, 16]; $Osp(2|2)_q$ in [21]; $Osp(N|2)_q$ in [13]; $Sl(N|M)_q$ in [12, 7].

f. A candidate for the q-deformation of the Virasoro algebra has been proposed [18] and investigated [13, 76, 69], but, in the absence of a coproduct, it is not known to be a Hopf algebra. The operators $Z_m = x^{-m}\,(r^{2x\partial} - 1)/(r - 1/r)$ satisfy the deformation of the centerless Virasoro algebra

$$[Z_m, Z_n]_{r^{n-m}} = [m-n]_r\ Z_{m+n}\ . \tag{4.18}$$

   The operators Z_1, Z_0, Z_{-1} comprise the $SU(1,1)_q$ analog of (iv). In the limit $r \to 1$, Z_m yields the standard Virasoro realization $x^{1-m}\partial$. It does not appear possible to introduce a satisfactory center [76].

g. It is possible to map $SU(2)_q$ to the q-Heisenberg algebra. Consider the following formal contraction of [10] and [70]:

$$b \equiv q^{J_0} J_- \sqrt{2(q-1/q)}\ , \qquad\qquad b^\dagger \equiv J_+ q^{J_0} \sqrt{2(q-1/q)} \tag{4.19}$$

so that

$$[J_0, b^\dagger] = b^\dagger\ , \qquad\qquad [b, J_0] = b \tag{4.20}$$

and hence

$$bb^\dagger - q^2 b^\dagger b = 1 - q^{4J_0}. \tag{4.21}$$

The last term on the r.h.s. vanishes e.g. for $|q| > 1$, $J_0 << 0$ in an infinite-dimensional representation (Schwinger's contraction), to yield the q-oscillator algebra [60, 5, 54]

$$bb^\dagger - q^2 b^\dagger b = 1. \tag{4.22}$$

If the number operator $N \equiv \ln\left(1 + (q^2-1)b^\dagger b\right)/\ln q^2$ is introduced [60], hermitean for real q, s.t.

$$[N, b^\dagger] = b^\dagger\ , \qquad\qquad [N, b] = -b\ , \tag{4.23}$$

this q-oscillator algebra can be mapped to the alternate form

$$\alpha = q^{-N} b, \qquad \alpha^\dagger = b^\dagger q^{-N} \qquad \Longrightarrow \qquad [\alpha, \alpha^\dagger] = q^{-2N} \tag{4.24}$$

with $\alpha^\dagger \alpha = (1 - q^{-2N})/(1 - q^{-2})$; or else the hybrid form

$$a = q^{-N/2} b, \qquad a^\dagger = b^\dagger q^{-N/2} \qquad \Longrightarrow \qquad aa^\dagger - qa^\dagger a = q^{-N}\ , \tag{4.25}$$

with $a^\dagger a = [N]_q$. These q-Heisenberg algebras provide a natural language for nonstandard quantum statistics [37].

In general, maps with exponentials of number operators serve to rewrite quommutator algebras such as those listed at the beginning of this section, (ii-vi), to commutator ones with a more complicated right-hand-side; for instance [15],

$$[H_0, H_\pm] = \pm H_\pm \qquad\qquad [H_+, H_-] = \frac{s^{1+2H_0}(1 - r^{-2H_0})}{r - 1/r} \tag{4.26}$$

deforms to (4.10) of (v) via

$$I_+ = s^{-H_0} H_+\ , \qquad I_- = H_- s^{-H_0}, \qquad I_0 = \frac{1 - r^{-2H_0}}{r - 1/r}\ . \tag{4.27}$$

Deforming functionals for the q-Heisenberg algebra (4.25) are [76]:

$$a^\dagger = \sqrt{\frac{[N]}{N}} A^\dagger, \qquad\qquad a = A\sqrt{\frac{[N]}{N}}\ , \tag{4.28}$$

where the classical oscillator algebra is

$$[A, A^\dagger] = 1, \qquad\qquad N = A^\dagger A\ , \tag{4.29}$$

consistent with the above.

The Jordan-Schwinger realization [6] of, e.g., $SU(N)$ generators via classical oscillators: $J^a = A_i^\dagger T_{ij}^a A_j$, where T_{ij}^a are fundamental representation matrices, serves to produce symmetric representations. Substitution of q-oscillators a_i for the classical A_i's yields a realization of

$SU(N)_q$ [93, 75, 5, 60] which produces "q-symmetric" representations, and affords a practical glimpse into their structure.[6] A more general treatment of oscillator and spinor representations for all unexceptional q-algebras is available in [40].

5. Miscellany and outlook

All the structures discussed in the above survey are, of course, associative (and coassociative). To confirm associativity in quommutator algebras, one must verify the "braid-Jacobi" (Yang-Baxter) relations. This consists [64, 76] of using the quommutator algebra to permute the operators in a trilinear product $J_1 J_2 J_3$ in two alternate ways:

$$\begin{array}{ccccccc} & \nearrow & 132 & \longrightarrow & 312 & \searrow & \\ 123 & & & & & & 321\ . \\ & \searrow & 213 & \longrightarrow & 231 & \nearrow & \end{array}$$

Comparing coefficients of the resulting terms of each order in 321 reached by the two pathways indicates whether associativity is a direct consequence of the algebra, as is the case in all algebras listed above, with the exception of the q-Virasoro candidate in (f) of the previous section, for which extra quadratic constraint relations result (and so are necessary for associativity).

The link of QUE-algebras to q-groups differs from the standard connection between Lie Algebras and Lie Groups. [26, 84][7] provide realizations of q-groups in terms of $SU(2)_q$ generators with the proper q-group relations. Nevertheless, the classical limit of these realizations is virtually trivial, in that it does not determine the conventional "exponential composition" of the Lie algebra — a workable q-deformation of that exponentiation for all representations is still unavailable. Conversely, [65] construct the QUE-algebra generators out of q-group elements. Perhaps illustrative, a particular realization I find in the spirit of that setting is the following.

Faddeev and Takhtajan's $SL_q(2)$ q-group [63] of uni-q-modular 2×2 q-matrices

$$\begin{pmatrix} a & b \\ c & d \end{pmatrix}$$

is specified by the following component relations:

$$qab = ba \qquad qcd = dc \qquad bc = cb$$
$$qac = ca \qquad qbd = db \qquad ad - bc/q = da - qbc = 1\ . \tag{5.1}$$

The following formal functionals of q-group entries:

$$J_0 = \ln\sqrt{bc}/\ln q\ , \qquad J_+ = \frac{i\,2^{-1/2}}{q-1/q}\, a\sqrt{1+\frac{1}{qbc}}\ , \qquad J_- = \frac{i\,2^{-1/2}}{q-1/q}\sqrt{1+\frac{1}{qbc}}\, d \tag{5.2}$$

reproduce the $SU(2)_q$ algebra commutation relations:

$$[J_0, J_+] = J_+ \qquad [J_-, J_0] = J_-\ ,$$

[6] also see the treatment of $SU(1,1)_q$ by [54], $SL(n|m)_q$ by [12], and $Osp(1|2N)_q$ by [31].
[7] Also see [100, 97, 92, 98] for the connection to noncommutative differential geometry.

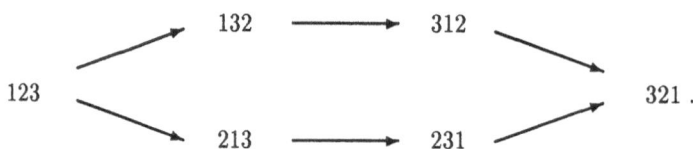

$$[J_+, J_-] = \frac{-2^{-1}}{(q-1/q)^2}\Big((1+q/bc)ad - (1+1/qbc)da\Big) = [2J_0]_q/2\ , \tag{5.3}$$

so this is a nonhermitean[8] realization of the q-algebra in terms of the q-group $SL_q(2)$ elements. Conversely,

$$bc = q^{2J_0}\ , \qquad a = -i(q-1/q)J_+\sqrt{\frac{2}{1+q^{-2J_0-1}}}\ , \qquad d = -i(q-1/q)\sqrt{\frac{2}{1+q^{-2J_0-1}}}J_-\ . \tag{5.4}$$

Extensions beyond $SU(2)_q$. Not much has been carried out so far in the way of systematic generalization of the above observations, techniques, and structures to the q-algebras beyond $SU(2)_q$, such as the ones mentioned in the previous section. Investigation of the various deforming functionals, modular representations, and model-building applications of such algebras emerges as an extraordinarily promising desideratum.

Applications. Selected applications were listed at the beginning of this overview[9]. Given the wealth of deformations, invariants, and ready reference to classical SU(2), further applications with intriguing prospects may include: construction of q-solvable potentials and use of q-algebras for spectrum-generation; construction of spin-chain hamiltonians with the alternate deformations listed as their degeneracy algebras; and several other opportunities to perturb beyond some underlying Lie algebraic structure.

I wish to thank T. Curtright, P. Freund, J. Uretsky, and R. Slansky for conversations; and the Ohio State University and the Aspen Center for Physics for their hospitality as this overview was taking shape.

References

[1] L. Alvarez-Gaumé, C. Gomez, and G. Sierra, Phys.Lett. **220B** (1989) 142; Nucl.Phys. **B319** (1989) 155.

[2] D. Arnaudon and A. Chakrabarti, Ecole Polytechnique preprint, June 1990.

[3] M. Batchelor, L. Mezincescu, R. Nepomechie, and V. Rittenberg, J.Phys. **A23** (1990) L141.

[4] D. Bernard and A. Leclair, Phys.Lett. **227B** (1989) 417.

[5] L. Biedenharn, J.Phys. **A22** (1989) L873; *Proceedings of the 1989 Clausthal Workshop*, Springer-Verlag, Berlin, 1990; L. Biendenharn and M. Tarlini, Lett.Math.Phys. **20** (1990), in press.

[6] L. Biedenharn and J. Louck, *Angular Momentum in Quantum Physics, (Encyclopedia of Mathematics and Its Applications* **8**), Addison Wesley, 1981.

[7] A. Bracken, H. Gould, and R. Zhang, Mod.Phys.Lett. **A5** (1990) 831.

[8] N. Burroughs, Comm.Math.Phys. **127** (1990) 109.

[9] D. Caldi, A. Chodos, Z. Zhu, and A. Barth, Yale preprint YCTP-P13-90, July 1990.

[10] M. Chaichian and D. Ellinas, J.Phys. **A23** (1990) L291.

[11] M. Chaichian, D. Ellinas, and P. Kulish, Helsinki preprint, HU-TFT-90-16, March 1990.

[8] $C_q = -(\sqrt{q} - 1/\sqrt{q})^{-2}$.

[9] For less compelling applications, see [44, 77, 11].

[12] M. Chaichian and P. Kulish, Phys.Lett. **234B** (1990) 72.

[13] M. Chaichian, P. Kulish, and J. Lukierski, Phys.Lett. **237B** (1990) 401.

[14] T. Curtright, Miami preprint TH/1/89, to appear in *Physics and Geometry*, L-L. Chau and W. Nahm, eds., Plenum, 1990.

[15] T. Curtright, Miami preprint TH/3/90, to appear in the *Proceedings of the Argonne Workshop on Quantum Groups*, T. Curtright, D. Fairlie, and C. Zachos (eds.), World Scientific, 1990.

[16] T. Curtright and G. Ghandour, Miami preprint TH/8/89.

[17] T. Curtright, G. Ghandour and C. Zachos, Argonne preprint ANL-HEP-PR-90-08, March 1990, to appear in J.Math.Phys.

[18] T. Curtright and C. Zachos, Phys.Lett. **243B** (1990) 237.

[19] E. Date, M. Jimbo, K. Miki, and T. Miwa, Kyoto preprint RIMS-703, June 1990.

[20] C. De Concini and V. Kac, Pisa SNS-math. preprint 75, May 1990.

[21] T. Deguchi, A. Fuji, and K. Ito, Phys.Lett. **238B** (1990) 242.

[22] C. Devchand, Freiburg Preprint, to appear in *Physics and Geometry*, L-L. Chau and W. Nahm, eds., Plenum, 1990.

[23] H. de Vega, Int.J.Mod.Phys. **A4** (1989) 2371.

[24] V. Dobrev, Sofia preprint, INRNE-TH-90, August 1990, these proceedings.

[25] V. Drinfeld, Sov.Math.Dokl. **32** (1985) 254; *Proc.Int.Cong. Mathematicians*, Berkeley 1986, (1987) 798.

[26] L. Faddeev, N. Reshetikhin, and L. Takhtajan, Alg.Anal. **1** (1988) 129; also in *Braid Group, Knot Theory and Statistical Mechanics*, C. Yang and M. Ge (eds.), World Scientific, 1989.

[27] D. Fairlie, J.Phys. **A23** (1990) L183.

[28] D. Fairlie, to appear in the *Proceedings of the Argonne Workshop on Quantum Groups*, T. Curtright, D. Fairlie, and C. Zachos (eds.), World Scientific, 1990.

[29] D. Fairlie, J. Nuyts, and C. Zachos, Phys.Lett. **202B**, 320 (1988).

[30] D. Fairlie and C. Zachos, Phys.Lett. **224B** (1989) 101; D. Fairlie, P. Fletcher, and C. Zachos, J.Math.Phys. **31** (1990) 1088.

[31] R. Floreanini, V. Spiridonov, and L. Vinet, Phys.Lett. **242B** (1990) 383.

[32] A. Ganchev and V. Petkova, Phys.Lett. **233B** (1989) 374.

[33] M-L. Ge, Y-S. Wu, and K. Xue, Stony Brook preprint, ITP-SB-90-02.

[34] M. Gerstenhaber and S. Schack, Proc.Acad.Sci.USA **87** (1990) 478.

[35] J-L. Gervais, Comm.Math.Phys. **130** (1990) 257; Phys. Lett. **243B** (1990) 85.

[36] C. Gómez and G. Sierra, Phys.Lett. **240B** (1990) 149.

[37] O.W. Greenberg, to appear in the *Proceedings of the Argonne Workshop on Quantum Groups*, T. Curtright, D. Fairlie, and C. Zachos (eds.), World Scientific, 1990.

[38] B. Gruber and A. Klimyk, Kiev preprint, July 1990 and these proceedings; A. Klimyk, Kiev preprint ITP-90-27E.

[39] E. Guadagnini et al., Phys.Lett. **235B** (1990)275.

[40] T. Hayashi, Comm.Math.Phys. **127** (1990) 129.

[41] Bo-Yu Hou, Bo-Yuan Hou, and Z-Q. Ma, Beijing preprints BIHEP-TH-89-7; 8.

[42] H. Itoyama, Phys.Lett. **140A** (1989) 391.

[43] H. Itoyama and A. Sevrin, Stony Brook preprint, ITP-SB-90-12, February 1990.

[44] S. Iwao, Prog.Theo.Phys. **83** (1990) 363.

[45] M. Jimbo, Lett.Math.Phys. **10** (1985) 63; **11** (1986) 247; Commun.Math.Phys. **102** (1986) 537.

[46] M. Jimbo, Int.J.Mod.Phys. **A4** (1989) 3759.

[47] N. Jing, to appear in Inv.Math.

[48] M. Kashiwara, Kyoto preprint RIMS-676, December 1989.

[49] L. Kauffman, Int.J.Mod.Phys. **A5** (1990) 93.

[50] A. Kirillov and N. Reshetikhin, in *Infinite Dimensional Lie Algebras and Groups*, Marseille 1988 Meeting, V. Kac (ed.), World Scientific, 1989.

[51] I. Koh and Z-Q. Ma, Phys.Lett. **234B** (1990) 480.

[52] T. Koornwinder, Proc. Kon. Ned. Akad. Wetensch. **A92** (1989) 97; H. Koelink and T. Koornwinder, ibid. **A92** (1989) 443.

[53] Y. Kosmann-Schwarzbach, Mod.Phys.Lett. **A5** (1990) 981.

[54] P. Kulish and E. Damashinsky, J.Phys. **A23** (1990) L415.

[55] P. Kulish and N. Reshetikhin, J.Sov.Math. **23** (1983) 2435.

[56] P. Kulish and N. Reshetikhin, Lett.Math.Phys. **18** (1989) 143.

[57] G. Lusztig, Adv.Math. **70** (1988) 237.

[58] G. Lusztig, Cont.Math. **82** (1989) 59.

[59] Z-Q. Ma, Beijing Preprint, BIHEP-TH-90-14, May 1990.

[60] A. Macfarlane, J.Phys.**A22** (1989) 4581.

[61] S. Majid, Int.J.Mod.Phys. **A5** (1990) 1.

[62] S. Majid and Y. Soibelman, Cambridge preprint, DAMTP/90-12, July 1990.

[63] Y. Manin, *Quantum Groups and Non-Commutative Geometry*, Cent.R.Math.-**1561**, Univ. Montréal, 1988.

[64] Y. Manin, Comm.Math.Phys. **123** (1989) 163.

[65] T. Masuda et al., Lett.Math.Phys. **19** (1990) 187; 195.

[66] L. Mezincescu and R. Nepomechie, Phys.Lett. **246B** (1990) 412.

[67] G. Moore and N. Reshetikhin, Nucl.Phys **B328** (1989) 557.

[68] T. Nakashima, Kyoto preprint RIMS-682, February 1990.

[69] F. Narganes-Quijano, Brussels preprint, ULB-TH-90/01, Mach 1990.

[70] Y. Ng, J.Phys. **A23** (1990) 1023.

[71] M. Nomura, J.Math.Phys. **30** (1989) 2397; J.Phys.Soc.Jap. **58** (1989) 2694; ibid. **59** (1990) 439; 1954; 2345.

[72] A. Odesskii, Funct.Anal.Appl. **20** (1986) 152.

[73] V. Pasquier, Comm.Math.Phys. **118** (1988) 355.

[74] V. Pasquier and H. Saleur, Nucl.Phys. **B330** (1990) 523.

[75] A. Polychronakos Florida preprint, UFTP-89-23, November 1989.

[76] A. Polychronakos Florida preprint, UFIFT-HEP-90-14, 1990, to appear in the *Proceedings of the Argonne Workshop on Quantum Groups*, T. Curtright, D. Fairlie, and C. Zachos (eds.), World Scientific, 1990.

[77] P. Raychev, R. Roussev, and Yu. Smirnov, J.Phys. **G16** (1990) L137.

[78] N. Reshetikhin, Steklov preprint LOMI-E-4-87, E-17-87 (1988).

[79] N. Reshetikhin and M. Semenov-Tian-Shansky, Lett.Math.Phys. **19** (1990) 133.

[80] N. Reshetikhin and F. Smirnov, Comm.Math.Phys. **131** (1990) 157.

[81] P. Roche and D. Arnaudon, Lett.Math.Phys. **17** (1989) 295.

[82] L. Romans, in *Strings '89*, R. Arnowitt et al. (eds.), World Scientific, 1990, contains a substantial bibliography.

[83] M. Rosso, Comm.Math.Phys. **117** (1988) 581.

[84] M. Rosso, C.R.Acad.Sc.Paris, **304** (1987) 323.

[85] H. Ruegg, J.Math.Phys. **31** (1990) 1085.

[86] H. Saleur, *Number Theory and Physics*, p. 68, Springer Proceedings in Physics **47**, J. Luck et al. (eds.), Springer Verlag, 1990.

[87] H. Saleur, Nucl.Phys. **B336** (1990) 363.

[88] H. Saleur and D. Altschüler, SphT/90-041, June 1990.

[89] G. Siopsis, Texas A&M preprint, CTP-TAMU-33/90, March 1990.

[90] E. Sklyanin, Funct.Anal.Appl. **16** (1982) 263.

[91] E. Sklyanin, Uspekhi.Mat.Nauk. **40** (1985) 214.

[92] A. Sudbery, to appear in the *Proceedings of the Argonne Workshop on Quantum Groups*, T. Curtright, D. Fairlie, and C. Zachos (eds.), World Scientific, 1990.

[93] C-P. Sun and H-C. Fu, J.Phys. **A22** (1989) L983.

[94] K. Ueno, T. Takebayashi, and Y. Shibukawa, Lett.Math.Phys. **18** (1989) 215.

[95] L. Vaksman, Sov.Math.Dokl. **39** (1989) 467.

[96] L. Vaksman and Y. Soibelman, Funct.Anal.Appl. **22** (1988) 170.

[97] S. Vokos, B. Zumino, and J. Wess, Berkeley preprint UCH-PTH-89/25.

[98] Wess and Zumino, preprint CERN-TH-5697/90, April 1990.

[99] E. Witten, Nucl.Phys. **B330** (1990) 285.

[100] S. Woronowicz, Publ. RIMS-Kyoto **23** (1987) 117; Comm.Math.Phys. **111** (1987) 613; **122** (1989) 125; **130** (1990) 381.

[101] C. Zachos, ANL-HEP-CP-90-43, July 1990, to appear in the *Proceedings of the Argonne Workshop on Quantum Groups*, T. Curtright, D. Fairlie, and C. Zachos (eds.), World Scientific, 1990.

INDEX